Mars – A Warmer, Wetter Planet

Springer
London
Berlin
Heidelberg
New York
Hong Kong
Milan
Paris
Tokyo

Jeffrey S. Kargel

Mars – A Warmer, Wetter Planet

Springer

Published in association with
Praxis Publishing
Chichester, UK

Dr Jeffrey S. Kargel
Geologist and Planetary Scientist
Flagstaff
Arizona
USA

SPRINGER–PRAXIS BOOKS IN ASTRONOMY AND SPACE SCIENCES
SUBJECT *ADVISORY EDITOR*: John Mason B.Sc., M.Sc., Ph.D.

ISBN 1-85233-568-8 Springer-Verlag Berlin Heidelberg New York

Springer-Verlag is a part of Springer Science + Business Media (*springeronline.com*)

British Library Cataloguing-in-Publication Data
Kargel, Jeffrey S.
Mars: a warmer, wetter planet. – (Springer-Praxis books in astronomy and space sciences)
1. Mars (Planet) – Climate 2. Mars (Planet) – Water 3. Mars (Planet) – Geology 4. Space colonies
I. Title
523.4′3

ISBN 1-85233-568-8

Library of Congress Cataloging-in-Publication Data
Kargel, J. S. (Jeffrey Stuart), 1958–
Mars: a warmer, wetter planet / Jeffrey S. Kargel.
p. cm. – (Springer-Praxis books in astronomy and space sciences)
ISBN 1-85233-568-8 (acid-free paper)
1. Mars (Planet)–Climate. I. Title. II. Series.

QB641.K23 2004
523.43–dc22 2003065305

Project Copy Editor: Alex Whyte
Cover design: Jim Wilkie
Typesetting: BookEns Ltd, Royston, Herts., UK

Printed in the United States of America on acid-free paper

Contents

List of Illustrations

Color section between pages 304 and 305

Preface

At publication time, a flotilla of spacecraft has just arrived at Mars; the robotic craft join *Mars Global Surveyor* and *Mars Odyssey*, both of which continue to operate, taking new observations and helping to relay data from NASA's rovers that are traversing the surface of the Red Planet. Europe operates *Mars Express*, a large and hugely successful orbiter. The Japanese *Nozumi*, unfortunately unsuccessful, was sent to study the flow of material and energy in the upper atmosphere and ionosphere of Mars. The British *Beagle 2* lander unfortunately was lost, but at least it arrived on the surface of Mars, a technological feat in itself, though its biological experiments must await a future reflight. The most remarkable thing about this flotilla is the dramatic expansion of international participation. We can now say that Mars exploration is a global enterprise. Amid all the divisions between East and West, New Europe and Old Europe and America, and between former Cold War rivals, in science and a spirit of exploration, we have one united planet exploring another, as never before.

Why should this be? Why is there such a focus on space exploration, and with all there is to explore in space, why Mars? The question is often asked of me, "Why is exploration of Mars important?" Is it important at all? Its importance is not in the elucidation of the geologic history of water or climate or any of these scientific things – those are fun and educational, and occasionally Mars exploration may spur a technological innovation or yield an insight into Earth's climate, or do something else that helps Earthlings; and one day, not too far away, water and ice and Martian climate will be pivotal for successful development of a new world. However, for now, the importance of Mars exploration is that it is a stepping stone to understanding ourselves. Particularly significant is the issue of whether we on Earth, in all this Solar System and perhaps far beyond, alone have those qualities of life and consciousness. Could our Earth be the only living world of intelligent beings amid billions harboring creatures in possession of consciousness of self and Universe. Most crucially, space exploration generally spurs an introspective look at ourselves and our stewardship of Earth as nothing else can. That is what is really important. Mars happens to be the planet easiest to explore.

One author's book will not change this world or Mars. Though I have a seriousness scattered through this book, the foundation and the façade of my purpose here is to have some fun! Mars can be a serious academic and philosophical topic, but it's not *that* serious! This book is about the first documented extraterrestrial planetary hydrological cycle; the first documented extraterrestrial glaciers; the first suspected ancient extraterrestrial oceans; the physical influences of water on the rocky surface of the planet; the linkages between Martian hydrosphere, cryosphere, lithosphere, and environment; and the significance of water, ice, and climate history for the future exploration and settlement of Mars. However, greater than the science of Mars is the sheer beauty and fun of it. That is the motivation driving both my Mars research and this book. I wish readers to contemplate Mars with an artful, exploratory, and joyous spirit, with full recognition that what is not yet known or what is interpreted inaccurately here concerns a reality that is more wondrous than all of us together – with or without our facts straight – could ever realize.

The greatest joy, however, is what we have right here on Earth.

NOTE ADDED IN PROOF (29 MARCH 2004)

Two incredible findings by the Mars science community have been reported in recent weeks: (1) NASA's *Opportunity* rover, sent to Mars to "follow the water," has found the water (remnants of a seafull of it) and an abundance of physical evidence favoring conditions that might have allowed surface life to prosper! (2) Researchers working with the European *Mars Express* mission, along with Earthbound telescopic observers, have reported the detection of atmospheric methane, a possible chemical trace of life's existence on Mars!

1. The seashore

In a nutshell (as presented by the Mars Exploration Rover science and engineering teams and developed further in Chapter 11), the rover has discovered layered sulfate salt-rich rocks at Meridiani Planum. These rocks were deposited or chemically altered by a highly acidic, salty sea, which I shall informally call Mare Meridiani for lack of an official term for what is still more a concept than a geographically and geologically characterized place. This finding is consistent with orbital observations of unusual mineralogies and nearby landforms appearing to be wave-cut shores.

Mare Merdiani could induce a fanciful spirit in the best of us. Alas, a warning is in order, as Mare Meridiani was no Earthly sea, according to our latest understanding. Humans: Beware! No sunbathing and no swimming permitted! If a daring foolish human could have dived into Mare Meridiani unprotected (never mind the impossibilities of dealing with low pressure, a low-oxygen atmosphere, and harsh ultraviolet environment), her/his skin would immediately have been subjected to a momentary excruciating pain of shocking-cold salty acid solution; a lemon-sharp sour taste would be the least of the diver's concerns. Though severe and macabre, the effects of the cryogenic acid would be mercifully fast. The diver's initial shock of painful error would be followed immediately by freezing of skin and then of blood and every organ. The frozen diver would not notice his gradual osmotic dehydration, or her subsequent dissolution into this sea of natural abiotic digestive juice. The calcium phosphate and carbonate of the bones and teeth might become poorly fossilized first by reacting to gypsum, and eventually being replaced by ferric sulfates, silica, and ferric oxides.

Okay, enough gruesome imagination. We know of no life on Earth that would tolerate Mare Meridiani's combined conditions of extreme acidity (pH –1 to +3), extreme salinity (possibly over 10 times Earth's seawater), and extreme low temperature (somewhere between 210 K and 260 K; or –81 to +9 °F, depending on composition and climate). But then Earth has never presented life with such a combination of conditions. It is known that some terrestrial microorganisms thrive under conditions as cold or nearly as cold in subglacial lakes, polar deserts, permafrost, and arctic sea ice; others in conditions as acid or nearly as acid, in Rio Tinto (Spain), acid lakes in Australia, and sulfurous volcanic crater lakes; and some as salty or nearly as salty, such as the Dead Sea. The question is whether all of these extreme conditions can be satisfied by life simultaneously. This brings us to the methane find.

2. Martian gas-passers?

NewScientist.com ("Methane on Mars could signal life," 29 March 2004) and BBC are reporting that three teams of scientists, working independently, have discovered a trace of methane in the Martian atmosphere. The European Space Agency has announced that Vittorio Formisano (Institute of Physics and Interplanetary Science, Rome) and his team of scientists have detected 10.5 parts per billion of methane in the Martian atmosphere based on measurements made by the Planetary Fourier Spectrometer (PFS) on board the *Mars Express* spacecraft. This result confirms a detection made last year by Michael Mumma (Goddard Spaceflight Center) using Earth-based telescopes and another similar detection recently made by Vladimir Krasnopolsky (Catholic University of America, Washington, DC) and colleagues. These independent results are unequivocal, according to the scientists involved. NewScientist.com quoted Formisano, "*The fact that methane is present on Mars means that there must be a source*," because methane will be photolytically destroyed in the Martian atmosphere in just a hundred years or so.

The observed quantity of methane – producing a CH_4 partial pressure in the atmosphere of just 7×10^{-6} Pa, representing about 300,000 metric tonnes throughout the atmosphere – would have to be replaced at a rate of a few thousand metric tonnes per year. This compares to Earth's release of about 600 million metric tonnes of methane per year. Scaled to the different planetary surface areas, the disparity between Earth and Mars, on a per-area basis, is reduced to about 5 orders of magnitude – still no small difference!

Martian methane production is small, but may be highly significant. Methane on Earth is primarily attributed to life. We now have the first biogeochemical clue to suspect that Mars may actually harbor life, thus adding to abundant geological evidence that the conditions necessary for life may have existed. It is possible that thermodynamic equilibrium or kinetic processes in magmatic gases can explain a minuscule amount of Martian methane, and that these gases are being released either actively or in delayed fashion due to crustal storage of the gas, without involvement of life. However, it is possible that active methanogenic microbial processes are occurring at a rate very low by comparison to Earth. An intermediate possibility is that the methane may have ancient biogenic origins and is stored in and slowly released from methane clathrate hydrate. The partial pressure of methane may be controlled by the continued existence and slow decomposition of methane hydrate coupled with photolytic destruction that never allows the gas pressure to build up in the atmosphere. Both current and past biological activity on Earth contributes methane to the atmosphere at rates that, due to constant photolytic destruction, cannot allow methane pressures to build in the atmosphere anywhere close to equilibrium with methane hydrate. Methane hydrate currently accounts for a few percent of Earth's methane releases; the same storehouses and processes could explain the Martian methane release rate when Earth's production is scaled to lower crustal temperatures and lower vapor pressures of methane hydrate on Mars.

Methanogens (methane-producing microorganisms) on Earth are primarily microscopic, single-cell, hyperthermophilic Archaea. Their life is hot and steamy in hot springs and submarine hydrothermal "black-smoker" vents. Methanogens might exist deep in the Martian crust and in volcanic hydrothermal systems, where conditions are hot and wet. Mars might have evolved psychrophilic (cold-loving) methanogenic or sulphate-reducing microbes dwelling near the surface. However, methanogenic systems (both biogenic and abiogenic) are anoxic: oxygen is poison. So are other oxidants, such as sulfate salts and ferric iron. Concentrated sulfuric acid, a strong oxidizer, probably would be incompatible with methanogenesis. Thus, the source of Martian methane is unlikely to be related to the oxidized evaporitic mineral assemblage and lakewater of Mare Meridiani.

These ideas are speculative, but we might now be a large stride closer to finding extraterrestrial life. My extraterrestrial bio-agnosticism, mentioned later in the book, is shifting a bit toward optimism. A strong observational foundation has been established upon which we can build specific biogeochemical investigations for future space missions. However, it reinforces a need for caution with regard to possible interplanetary cross-contamination. Caution, unfortunately, carries a price tag. However, the chance that Earth life could adapt (without deliberate genetic engineering) to Martian conditions is very low, and the chance that it could undergo transgenic interactions is even lower; anyway, several spacecraft already are on Mars and have carried potentially viable terrestrial microbiota there. So far as future Mars sample returns are concerned, there are some obvious issues; again, successful transfer of endemic Martian life to Earth, if it can occur at all (not likely), probably would have occurred already due to natural Martian meteorite delivery to Earth over geologic time. It would be prudent, however, to revisit issues of planetary protection considering the fact that biological transfer probably already has taken place, most likely with no effect whatsoever. There remain some small risks related to a more in-depth future exploration strategy that meteorite and simple lander-based deliveries of random microbes would not entail. The cost of too-hasty exploration of Mars, without suitable sterilization or some other precautionary measures, could be regrettably high. Nevertheless, these are statements based on current understanding – a highly incomplete knowledge. Caution does not mean a red light, but, in my opinion, it should mean a review of procedures and adoption, as needed, of modified approaches to exploration. An adaptive Mars exploration strategy will therefore require increased attention to biological and biogeochemical investigations – all constructed and executed with proper consideration of planetary protection. NASA, ESA, and other space agencies need to be proactive in revising their exploration strategy to take into account recent discoveries their space missions have made and the logical possibilities for extant life on Mars.

Foreword

Debate over the origin and evolution of the Moon has remained lively and productive since the last human exploration 32 years ago. More recently, issues related to the evolution of Venus, Mercury, and especially Mars get a larger share of lunar and planetary scientists' and the public's attention. How much do the known, probable and possible events in lunar history constrain what may have occurred on and in Mars? Or, indeed, on and in the Earth?

Controversy still surrounds a number of significant issues in lunar history, particularly the origin of the Moon some 30 million years after the Solar System's origin 4.567 billion years ago, the timing of the periods of violent mega-impacts that ended about 3.8 billion years ago, and the composition and evolution of the deep interior below about 550 km. On the other hand, logical presumptions on these lunar issues and others based on the results of the six *Apollo* exploration missions to the Moon and remote-sensing data collected before, during, and after *Apollo* offer potential insights into several aspects of the history of Mars and other terrestrial planets. At the very least, the impact history for the inner solar system recorded on the Moon can be used to constrain some of the history of these planets.

For example, the terrestrial planets may have had chondritic bulk compositions before they developed metallic cores. Seismic, isotopic, and elemental evidence indicates that the lower mantle of the 1,735-km-radius Moon, i.e., material below about 550 km, never melted significantly and is relatively undifferentiated as compared to the upper mantle. Most researchers agree that the upper mantle is highly differentiated or modified from its primordial characteristics by having crystallized from a magma ocean. Logical extension of these data and the initially slow process of accretion both would suggest that the Moon and other terrestrial planets initially had early chondritic cores, or protocores, about 1,200 km in radius. The assembly of these protocores and their assembly into planets virtually the size of the present terrestrial planets probably took place around 10 million years after the Sun itself formed. The initial assembly of the planets would have involved a comparatively mild series of impact events, with low impact speeds due to the small gravitational attraction of the incomplete planets and due to the slight gravitational "stirring" in the nebula and accretion disk. To this point, the origin of Mars would have been very similar to that of Earth; both would have formed relatively cool protocores. Although details of the accretion processes of planets in helicentric orbit and of the Moon in geocentric orbit are different, the same basic sequence applies, where accretion of a cool protocore was followed by much hotter accretion of the outer layer.

The existence of relatively cool protocores beneath the magma oceans of the fully accreted terrestrial planets might delay, to varying degrees, the migration of iron–sulfur liquid to form metallic cores in these planets. In the case of the Moon, remnant magnetic fields at the antipodes of the four youngest large lunar basins suggests that metallic core formation and an active dynamo-driven magnetic field was delayed by its protocore until about 3.9 billion years ago, the estimated age of the oldest of these four basins. Because of greater gravitational potential to drive core formation, and increased heating because of it, one would expect that there would have been progressively less delay in core formation in the sequence from Moon, Mercury, Mars, Venus, and Earth. For Mars, we can document that there was significantly less delay of the onset of a core dynamo, and presumably core formation, than for the Moon. Remnant magnetic striping in the Southern Uplands of Mars occurred and ended before the end of heavy bombardment 3.9 billion years ago and probably before the last of the extremely large basins (such as South Pole Aiken) formed about at 4.2 billion years. These relationships indicate that dynamo activation and perhaps displacement of the chondritic protocore of Mars by a metallic core occurred significantly earlier than in the Moon. In Mars, unlike in the Earth, continuous but less vigorous convective mixing of the mantle may not have broadly dispersed the displaced protocore. The remnants of that protocore may now reside primarily beneath the Southern Hemisphere, accounting perhaps for a 3-km offset in the center of mass, relative to the center of figure, toward the south pole of Mars and hemispheric differences in geological provenance also may be the result of this selective displacement of the protocore. Increasing isotopic and chemical evidence also exists on broad heterogeneities in the composition of the Earth's mantle, some of which may record incomplete mixing of the Earth's protocore after displacement during metallic core formation.

It is clear that the inner Solar System was a very violent place between 4.6 and 3.8 billion years ago. Whether or not one concludes that most of this violence occurred over less than 100 million years, the so-called "cataclysm," or in stages over the entire 800-million-year period, as I have maintained, the surfaces of the terrestrial planets were very different from today. Intense impact cratering would have produced abundant glassy and pulverized silicate material in the upper several kilometers of such planets. On the water-rich terrestrial planets, such as Mars and the Earth, this material would have altered rapidly to clay minerals, probably with great variations in mineral composition, structural dimensions, and environmental niches. Crystal structural patterns of broad variability on the surfaces of the clay mineral grains, possibly in association with sulfide minerals may have assisted in the aggregation of complex organic molecules, possible precursors to the first replicating forms of such molecules. Indeed, replication may have first been symbiotic with the forms, growth, and/or expansion of clay mineral structures. Tubular forms of clay minerals may also have been initially incorporated in the earliest single cell organisms to assist in the movement and retention of fluid, only to be replaced later by organic cell walls.

The evidence of isotopic fractionation by organic processes associated with 3.8 billion year old terrestrial rocks may not be a timing coincidence. The end of the large basin forming events in the inner Solar System also appears to be at about 3.8 billion years. Although simple organic replication, and possibly single-cell organisms, may have existed on Earth prior to ~3.8 billion years ago, the catastrophic effects of large impacts appear to have prevented significant biological activity until after that time. The cratering history of the Moon has alerted us to the potential pervasiveness of clays on the early hydrous crusts of Earth, Mars and Venus. Mars, therefore, may be the arrested crucible of early organic processes now lost to us on Earth. Just as we look to the Moon for insights derived from solidified remnants of an early, arrested magma ocean phase, we may look to Mars for solidified remnants of an early water ocean phase.

The Moon's evolution, on the other hand, does not appear to constrain the history of volcanic activity on the terrestrial planets. Nor would this be expected due to the profound differences between them relative to pressure and temperatures over time. Volcanic constructional features and remnant magnetic striping identified in the Martian Southern Uplands, and partial erasure of the striping by large impact crater formation, such as Hellas, indicates the early existence of surface or near-surface magmatic activity. Such activity may have been triggered by the upward displacement of Mars' chondritic protocore discussed above, and associated pressure release melting and devolatilization of that material and partial melting of overlying Martian mantle due to volatile fluxing.

Up to 3.8 billion years ago, intense impact cratering and solar wind erosion probably depleted primordial water initially retained at the Martian surface and in the atmosphere. Unlike the Moon, where water apparently did not recondense in the Moon-forming impact-ejected material, significant reservoirs of water and other volatiles in the Martian mantle would have remained to be mobilized later by magmatic activity. First, Mars' protocore would have retained some of its primordial water. Second, high-pressure crystallization of hydrous silicates, particularly sodium-rich amphibole and mica, during the Martian magma ocean stage, and their retention in the mantle would represent an additional water reservoir. Thus, whereas the Moon's early differentiation offered little, if any, major role for water, on Mars water may have been introduced into segregated mantle and crustal reservoirs at high rates during the early magmatic episode.

Two possible strand lines for ancient oceans on Mars have been identified, including one from a possible early northern ocean. This line has highly variable elevations, creating doubt as to its reality as an actual strand line. On the other hand, the variable elevations appear to be where a level line would have been most affected by uplift of the Tharsis Bulge and its antipodal response in Arabia Terra. As effects of Tharsis Bulge post-date the large basin stage of impact cratering as well as the early volcanic period, this possible strand line may well be evidence of an early northern ocean. If the correlation of an early northern ocean with an early volcanic period is correct, it would indicate that this volcanism was after the formation of most of the extremely large basins. Some of such basins, in aggregate, probably form the Northern Lowlands of Mars. More clearly evident than the strand line for an early northern ocean is a later one that largely but not totally post-dates the crustal deformation associated with the Tharsis–Arabia bulges. This second northern ocean has a well-defined and largely level strand line. Its water would have been evolved during Tharsis-Elysium and related volcanic activity and persisted after the major deformations associated with that activity. This ocean may have been in part remobilized water from the earlier ocean stored as ice and water in the Martian subsurface. The hematite and sulfates present in Meridiani Planum suggests that epithermal sulfide deposits may have formed during this period, only to have been extensively oxidized as climate changed.

With respect to ice and water, it has long been evident from various geological forms that water ice was present in much of the Martian subsurface particularly at high latitudes. Liquid water (the hydrosphere) likely existed below the ice (the cryosphere) due to heat-flow-induced temperature increases with depth. The recent epithermal neutron data from *Mars Odyssey* more precisely defines the distribution of water ice. The boundary region between the hydrosphere and the cryosphere of Mars is a stable, long-term ecological niche that could harbor simple life forms, including their evolutionary derivatives, since a more Earth-like environment disappeared from the Martian surface a few billion years ago. Certainly, this niche would be no more hostile to life than similar subsurface environments that exist on Earth today.

Clearly, Mars has much to teach us about the Solar System and our own Earth – not just the geologically ancient past, but perhaps the near future of our struggle to survive on this planet and to extend our presence beyond Earth, and the distant future of the terrestrial planets. The lunar volcanic chronology, including cessation of volcanic activity and slow cooling of the interior, may offer a glimpse into the future activity of these planets, first Mars and then Earth and Venus, as their internal heating slowly decays and stored heat dissipates. More immediately, an exciting area of debate today concerns the best strategy to begin the human exploration and potential settlement of Mars in order to take advantage of its lessons. The exploration of the Moon by *Apollo* astronauts formed the foundations for the future growth of an Earth–Moon economy into which Mars can be integrated in the longer term. Most importantly, for future inhabitants of the Earth, we know that non-radioactive fusion energy resources (solar wind derived Helium-3) exist in the soils of the Moon. These potentially commercial resources provide both a long-term and environmentally benign alternative to the terrestrial use of fossil fuels and the economic foundations for lunar and Martian settlement. Further, by-products of the extraction of Helium-3 from lunar soil can sustain the future travelers and settlers of deep space.

In spite of the record of the *Apollo* astronauts, the questions probably will always be asked: "Why humans in space? Wouldn't robots be better and safer?" History and personal experience on the Moon have convinced many that during exploration, humans will provide instantaneous observation, interpretation, and assimilation of the environment in which they work and a creative reaction to that environment unavailable from any other source. Human eyes, experience, judgment, ingenuity, and manipulative capabilities are unique in and of themselves and are highly additive in synergistic and spontaneous interaction with instruments and robotic systems. Most of the questions raised by the *Spirit* and *Opportunity* rovers would have been resolved quickly, and more lines of investigation suggested, if a human had been on site and doing what good field geologists do with their eyes, feet, and hammer. Due to inherent communication delays and the cost of returning samples to Earth, human exploration of Mars will depend on people and facilities on or near Mars for mission support and sample analysis. On Mars, as for the development of the Moon's resources, cost considerations favor the rapid development of Martian self-sufficiency and early settlement.

These comments and speculations are not in the realm of science fiction. Because of what Americans accomplished in the last half of the twentieth century, *Apollo* bent the curve of human evolution into the future. The psychological, technological and survival bonds holding humans to the Earth have been broken. This new evolutionary status in the universe now permits us to live on the Moon and Mars. Generations now alive can determine if humankind will take advantage of this new status – that is, they will determine whether or not we will begin the settlement of the solar system and provide for a new birth of freedom beyond the Earth.

Harrison H. Schmitt

Background

Harrison H. Schmitt is a field geologist and flew as the Lunar Module Pilot on *Apollo 17* in December 1972. He later served as a United States Senator from New Mexico. Since 1983, Dr Schmitt has been active in business, lunar research, and education and has served as an adviser to various governmental entities. He also is currently Chairman and co-founder of Interlune-Intermars Initiative, Inc. Interlune is a commercial enterprise with the goal to develop Helium-3 fusion power, fueled by lunar Helium-3. This has been proposed as an early twenty-first century alternative to fossil fuel and fission power plants. Interlune also believes that the commercialization of lunar resources, not only Helium-3 but hydrogen, oxygen, water and food, will be the most cost-effective means of returning science to the Moon and of initiating the human exploration and settlement of Mars.

Although supplies of both fossil and fission fuels would be adequate to support a projected factor of 8 growth in terrestrial energy demand by 2050, the indefinite use of such fuels does not appear to be a good environmental or economic choice. Within the proposed lunar Helium-3/fusion power initiative, major objectives are to provide investors with a competitive rate of return; protect the Earth's environment by using clean energy from space; develop other resources from space that will support future near-Earth and deep-space activities and human settlement; and develop reliable and robust capabilities to launch payloads from Earth to deep space at a cost of $1,000 or less per kilogram.

Past technical activities related to access to, and operations in, deep space and to terrestrial mining and processing, provide a strong base for initiating this enterprise. Recent progress in the development of inertial electrostatic confinement (IEC) fusion at the University of Wisconsin indicates that this approach to fusion has significantly more near-term applications and commercial viability than other technologies pursued by the fusion community in the past.

Interlune believes that international law relative to outer space is permissive relative to properly licensed and regulated commercial endeavors, i.e., lunar resources can be extracted and, once extracted, the resources can be owned by a commercial entity but national sovereignty cannot be asserted over the mining area. Attaining a level of sustaining operations for a core business in fusion power and lunar resources will require about 10–15 years and $10–15 billion of private investment capital as well as the successful marketing and profitable sales of a variety of applied fusion technologies. Success also depends to some degree on the US Government being actively supportive in matters involving research and technology development and tax, regulatory, and international law but no more so than is expected for other commercial endeavors with compelling national interests at stake.

Former astronaut and former Senator Schmitt now leads Interlune, Inc., seeking to shape the next giant leap of Mankind into energy security here on Earth and outward through the Solar System.

Acknowledgments

I offer my deepest gratitude to those around the world for the compelling vision to explore Mars, and to NASA and other space agencies for transforming that vision into reality. Gratitude is also directed to Malin Space Science Systems for *MGS* MOC imagery, and the *MGS* MOLA and *Mars Odyssey* Gamma Ray Spectrometer and THEMIS teams, the Mars Exploration rover team, to NASA, ESA, JPL, ASU, Cornell, DLR, and USGS for providing timely access to data. MOC and THEMIS images are identified throughout the book either by their formal image numbers or by their press release numbers.

This book would not have been possible without the work of the whole Mars science community, including many students who have labored through long days and nights to bring our understanding of Mars to where it now is. Special thanks to the late Eugene M. Shoemaker for impelling planetary science into a viable research field, and to the late Carl Sagan for bringing the wonders of outer space into our living rooms. I am indebted to my mentors and academic advisers: Kenneth Foland (Ohio State University) and Robert Strom, Vic Baker, and John Lewis (University of Arizona). I also thank John Rummel, Andrew Schuerger, and Carlton Allen for enlightened discussions on planetary protection. My gratitude also goes to every scientist, philosopher, artist, and theologian who has wondered about the Universe around them, to every teacher of every inquisitive child, to my Grandmother for engendering in me an appreciation of science, my parents for teaching me about our planet of many cultures, and especially to my lovely wife and children for their willing sacrifices during years of writing this book and many family explorations of Mars analogs.

Finally, I am thankful to the Praxis–Springer team, including Clive Horwood (Publisher of Praxis), Alex Whyte (project copy editor), Jim Wilkie (cover designer), and John Mason (Praxis science consultant) for the expert assistance given to this book.

This book is dedicated to
ISAIAH ALEXANDER DEL GIORGIO

Introduction

"We need to think like Martians."
(Hugh H. Kieffer)

SIMPLY WARPED

Some of my most joyous moments of science outreach have been in Fifth Grade classrooms, the age when I became very much defined by nature love and gripped by the *Apollo* lunar program. Every fifth grader knows that outer space is a seemingly limitless domain of every extreme and unfamiliar attribute. It is exceedingly inhospitable or mind-numbingly bleak, intangibly bizarre in strange beauty, or terrifying in its pulverizing, vaporizing, ionizing, sterilizing energy or its vast and cold emptiness. Outer Space is far different from everyday life and thus is intellectually challenging for its abstract qualities; certainly that is part of the awesome wonder that space holds for all of us. Grade schoolers have a fascination in outer space partly for the same reasons that dinosaurs, volcanoes, and warfare may fascinate. Each of these is deadly dangerous and, to some, compelling in its power, but certainly these are not phenomena that many people would care to live with. The Lowellian–Wellsian popular view of Mars fit this concept of outer space in multiple ways, in that it once was the abode of an ingenious and aggressive, evil civilization vastly beyond our own in its power. This concept of the Red Planet, of course, no longer holds sway.

Mars now fascinates the way that the distant shores of the New World once beckoned the restive imagination of Europe – the New World was the ultimate challenge, but a potentially survivable one, and it offered hope to prosper along new, undreamt horizons that could be found upon arrival. The new fascination of Mars to grade schoolers and scientists alike arises because there are attributes of and places on Mars that are so very much like parts of Earth. Mars indeed has a beautiful alien quality, but in many ways it is so relatively Earth-like and comprehensible, and so comparatively near Earth, that it can be visited and understood by Earthlings with the aid of our machines. One can even imagine living there.

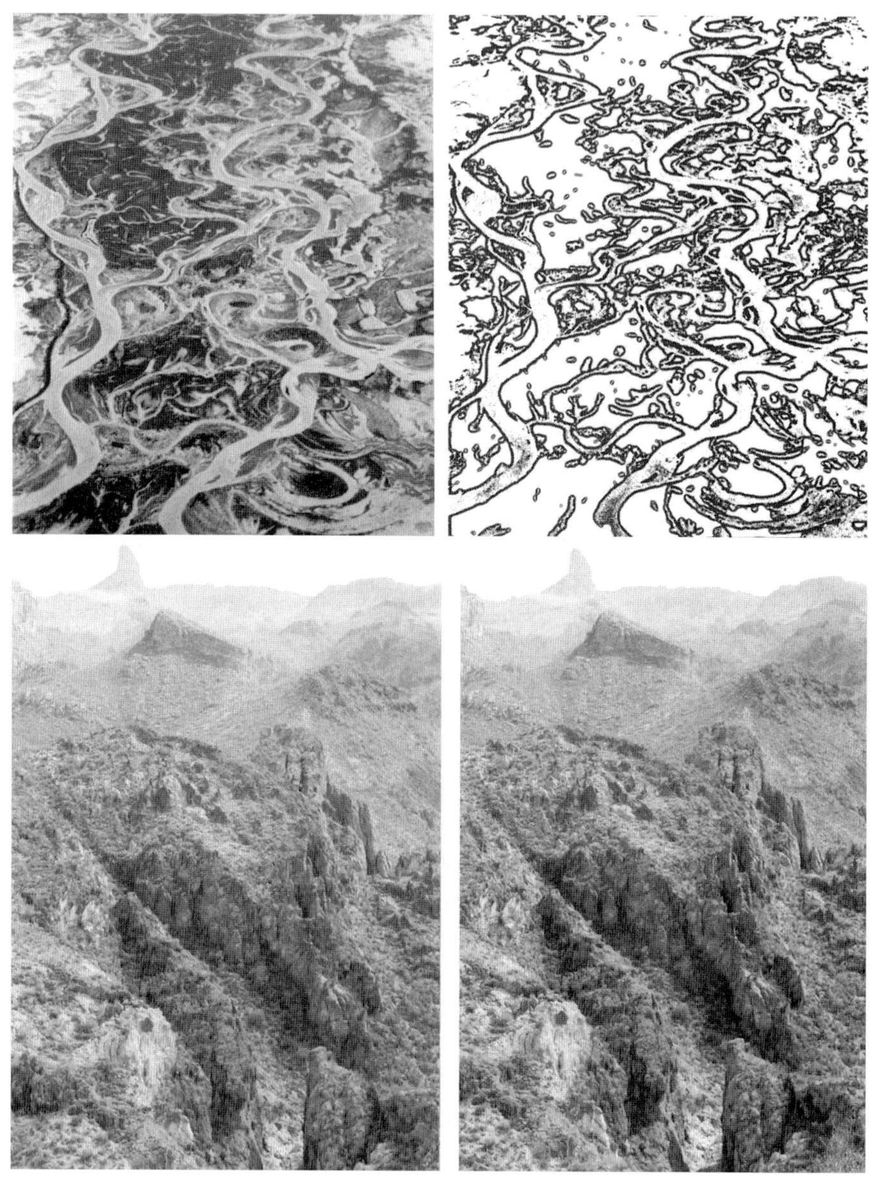

Earth has a bewildering variety of water-formed or water-modified landscapes which are rare on Mars. Two examples: (*Top pair*) Wild river meanders and abandoned meanders (oxbows) on the permafrost plain of western Alaska. The left scene is an oblique airphoto; the right image is a specially filtered version which brings out the meander structure. (*Bottom*) Faulted and throughly water-dissected landscape of Arizona's Superstition Mountains. The two images form a stereo pair; fuse the two scenes with your eyes into one image, and it will pop out in three dimensions. Mars has many water-formed or water-modified terrains which Earth also has, and some which we might not have. These are the subjects of this book.

The tangibility of something so distant is what motivates many in the planetary science community and separates us from the astronomers, who of course study objects even more distant. The peculiar draw of planetary science either creates or attracts a rather peculiar crowd. John Nye, perhaps the most famous living glaciologist, recently contributed to his third Mars Polar conference. At the conference dinner, Nye, by then inducted into planetology, took to the floor and offered some notes about the Mars community specifically. With his characteristic dry, witty humor, and to a roar of appreciative laughter, Nye noted of the collective group, "*You have your own characteristics. You may not know it, but you do.*" Indeed, I think we are all aware of our group's oddities, which are borne of the scientist's love of nature combined with a draw to a planet that we are certainly free to think we understand. In fact, we are like Fifth Graders in our adoration of Mars and other worlds. And so I initiate my story about Mars. Hopefully as much as half of what I offer actually is partly true, but in any event Mars assures a fun story.

During some of Earth's more anomalous geologic ages, such as the present Quaternary/Recent, a few percent of H_2O has become solid across much of the surface, producing some stunning Mars-like glacial and permafrost landscapes. On Mars H_2O is normally mostly solid; the anomalous times have been when there was some liquid present on and near the surface. The area of overlap of this complementarity is where we can say Mars is Earth-like, or Earth is Mars-like.

Much of the Martian story is written in familiar water- and ice-worn landscapes and water-formed minerals and rock strata, signaling to Earthlings that, despite non-ideal environmental conditions today, at times Mars has been more Earth-like than anywhere else known in outer space; in fact, it remains so. The sum of landscapes and their origins is, to leave no doubt, uniquely Martian. Nothing there is really just exactly like anything here, and that some of what exists on this planet has no rough analog on Mars. However, on the scale of the many extreme and hostile conditions that this Solar System offers, Mars is almost made to be comprehended and then beaten, cooked, and manipulated into a brand new home. Thus, whether or not Mars is today a living world, the chances are quite strong that it will be. Whether Humanity will or should choose to so manipulate another world in this way is uncertain, but if this is possible anywhere else in the Solar System, Mars is surely the place. No arguments of lunar proximity or low asteroid gravity can contend with the fact that Mars is a water world, slightly colder than Earth.

Half the diameter of Earth, with about the same area as Earth's land mass, an atmosphere 1% as dense as Earth's and a gravity 40% of Earth's, Mars is the fourth planet from the Sun, and second closest in space to Earth. Mars, however, is closest in my heart and soul to our own blue–green world. This book does not cover Mars from front to back and pole to pole or cross every canyon and crater and concept. It is about water, ice, and climate, as written in the landscapes of Mars, and about revolutions in Mars science that started with the *Mariner 9* mission. With each new successful mission, this revolution has been building steam (but even more cold water and ice).

Albert Einstein's and Stephen Hawking's gravitational warp in space–time underlies the phenomenological richness of our Universe, and that includes ordinary geology as well as cosmology, but Sir Isaac Newton's continuum physics and his version of gravity does well enough for geology; there are no black holes here, but without gravity there would be no geology. Solar radiation-driven evaporation and global atmospheric circulation, rain-caused erosion gullies, glacier-carved mountain valleys, wind-blown sand dunes, water-lain sediment strata, salt-domed anticlines, mantle-folded mountains, lava-constructed volcanoes, and impact-formed craters all have gravitation as a chief root cause if traced back far enough. Even the solar and radiogenic energy that fights Earth's and Mars' gravity is gravity-related. The cosmic quest for ultimate nuclear stability (fusion of hydrogen in the Sun and fission of radioisotopes in the planets) and the drive toward global planetary differentiation are ultimately gravitationally driven or caused by conditions that gravitation sets up. These processes establish gravitationally unstable conditions at planetary surfaces. The relaxation of these conditions drives geology and sculpts planetary surfaces: water flows, glaciers creep, rocks tumble, mountains collapse, continents crush, and depressurized mantle minerals weather.

Geomorphology is largely about the competition between the forces that build up and wear down land surfaces and particularly the three-dimensional planetary surfaces that result. This competition is completely different depending on the activity of water. Disequilibia generated by mantle/crustal processes and atmospheric dynamics, though arising from different immediate sources of energy, are closely coupled at the surfaces of Earth and Mars, though this is not the case for all planets. On Earth solar heating drives oceanic moisture and energy from tectonic/topographic basins to the poles and to tectonically/volcanically elevated terrain, where precipitation accumulates on the land surface and then flows back to the sea and major inland basins. Thus, tectonic and volcanic processes can organize the deposition of solar energy by way of the gravitational potential energy carried by precipitation. Once deposited, some fraction of the precipitation runs off at speeds related to surface slope, concentrating the release of that potential energy on the solid surface layer. Shearing flows of water and ice pluck rock fragments, abrade grains, lift dust, and move sand, and then deposit these grains. The amazing diversity of landscapes on Earth and Mars attests to the sensitive material responses to the global energy flows mediated by water and ice.

Cosmologists have long marveled at the generation of heavy baryons and galactic structure and galactic clusters and superclusters from a Universe that began as a nearly homogeneous energy field. Fluctuations of millikelvin in the background 3 K microwave radiation field betray primordial Big Bang heterogeneities. Humans, and a bazillion other details, emerged from those slight differences, which may as well be called God's whispers given the ratio of what resulted to what is understood. Geologists and planetologists work in a parallel Universe to that of the cosmologists, humbled by something of much less immensity: we marvel that the planets' highly structured surface morphologies and crustal structures arose spontaneously from initial homogeneity or randomized heterogeneity of the solar nebula and accretion disk. A long and tortuous but logical path extends from there to the erosion of glacial cirques and stream valleys.

Grappling with time – from the nanoseconds of chemical reaction relaxation times to four gigannums of planetary convulsions – is something that actually unites geologists, planetary scientists, and astronomers (including cosmologists). Our work not only demands attention to all time, but we revel in it. Although most other people rarely consider such miniscule and immense expanses of time, pondering the roles and ravages of time is, however, an intellectual pursuit in which almost everyone engages now and then. A stroll through ancient Rome or an ancient redwood forest offers every visitor a window into happenings transcending the individual human experience by a factor of 50 or more. And yet the expanse of time afforded by those experiences is less than one millionth of all geologic time; human written history and the life and times of the oldest living forests is to geologic time as a millimeter is to a kilometer of length.

Deciphering histories of worlds may seem impossibly more complex than deciphering the recent history of a redwood forest, but in fact one of the most important logical techniques is identical – the use of superposition ("which tree fell first?"). Planetary science, as with forestry, involves the use of other tools, including, in both cases, studies of the flow of energy and matter; and that is the investigative pursuit in which we find some of the most striking comparisons and contrasts between Earth and Mars.

IMPOSSIBLE FEAT

The efficiency of coupling solar energy to H_2O evaporation and precipitation currently is orders of magnitude less on Mars (where CO_2 is the bigger story climatically) than on Earth. It's a matter of vapor pressures. We can also consider the efficiency of coupling internal energy to the surface via geologic processes. Both are inefficient systems on Mars, by comparison to Earth, explaining why Mars still retains tens of thousands of impact craters dating from as long ago as 4 billion years. A lot of solar and internal energy produces very little work on Mars. However, the relative efficiency of solar/atmospherically driven processes versus internally driven processes is perhaps not very different from Earth; we see ample signs of tectonic/volcanic and water/ice modifications on both worlds, and there is excellent evidence of coupling of these internally and atmospherically driven processes.

The particular importance of H_2O for the geology of Earth and Mars cannot be doubted, but it is not universally the most geologically active fluid. On some worlds water is essentially absent and utterly unimportant (e.g., the Moon and Mercury); on other worlds H_2O is confined to one major phase that is mostly isolated from the surface (e.g., aqueous sulfuric acid clouds of Venus); and, on some, H_2O is sequestered in an inactive icy phase (e.g., many icy satellites). If we lived on a planet of abundant solid methane and other non-polar ices and a methane—nitrogen atmosphere and ocean, liquid ethane–methane in the interior might be as effective a solvent as water is on Earth; such a world might be Saturn's moon Titan, where water might be as solid, insoluble, and inert as quartz is on Earth, but where the rain is methane. Neon Moons may literally occur somewhere far outside Country and Western bars (in some distant Solar System), but if this the case, then water, if present at all, may have little importance there except as bedrock. But Earth is no Titan, and neon has almost no role here beyond Las Vegas; water is the condensible atmospheric constituent here, and its stability and geologic effects make our world different from Venus.

The myriad geomorphic influences of water are seen almost everywhere on Earth and across most of Mars. Water is the foremost volatile driving activity on Mars and Earth because of its electrically polar and chemically amphoteric nature, hydrogen bonding, and unique negative volume change on melting. These properties control water's reactivity with silicate igneous minerals, the physical weathering of rocks, the vapor equilibrium and pressure–temperature conditions of melting, the viscosity of solid and liquid phases, and the gravitational stability of the liquid phase beneath floating ice. H_2O's high latent heats of fusion and sublimation mean that it is an efficient sink and carrier of energy, and so it has a major influence on climate and geothermal heat flow. The erosion potential allowed by water is further enhanced by the just-right abundance of H_2O on Earth and Mars, which allows water and ice to cover and fill part but not all of the crust, thus promoting erosive surface flows of rivers and glaciers. Most important is (1) water's exchangeability at Martian and terrestrial crustal and surface temperatures between three major pure phases, and (2) its long-term stability in solid hydroxides and hydrates within the crust. On Mars and Earth, water vapor is globally transportable through the atmosphere; in liquid and solid states H_2O can flow over and erode the surface and deposit sediment; and in liquid form it can flow through the shallow subsurface and pool in basins. In each form it reacts and bonds with silicate minerals or forms trapped fluid inclusions; water is then released from crustal hydrated and hydroxylated minerals as temperature rises or pressure decreases. In stabilizing phyllosilicates, water helps to make rocks that can fold under compressive stress; it makes rocks that can melt to form andesitic and rhyolitic magma and continental granitoid crust. Under chemical attack at the surface, rocks variously dissolve, recrystallize, or crumble under the influence of water. Water also stabilizes life, which fundamentally alters chemical equilibria and kinetics of weathering and solution processes.

Contrast that with the stability of water in basically just three minor forms on Venus: as sulfuric acid droplets in the clouds, as a bare trace vapor constituent of atmospheric gas, and as hydroxyl of tremolite in the crust. H_2O is a trace material there. Of course the clouds and vapor are important in the radiative balance, but water is not otherwise a geologically significant substance on Venus. It may not always have been this way. When the Sun was a cooler star and Venus a younger world, Venus was a damp planet. However, as the Sun grew hotter over time, water vapor and clouds helped Venus to create the conditions for its own demise: surface liquid boiled away long ago, and from the atmosphere H_2O was lost. Now we see on Venus the products of volcanism and tectonism, though 400 million years old, preserved as though they were made yesterday, all because liquid water and ice are absent.

The volatility of water and the conditions on the surfaces of Earth and Mars and in their crusts means that this substance is prone to repeated cycling among its three pure phases in the atmosphere, surface, and subsurface. Such a difference from

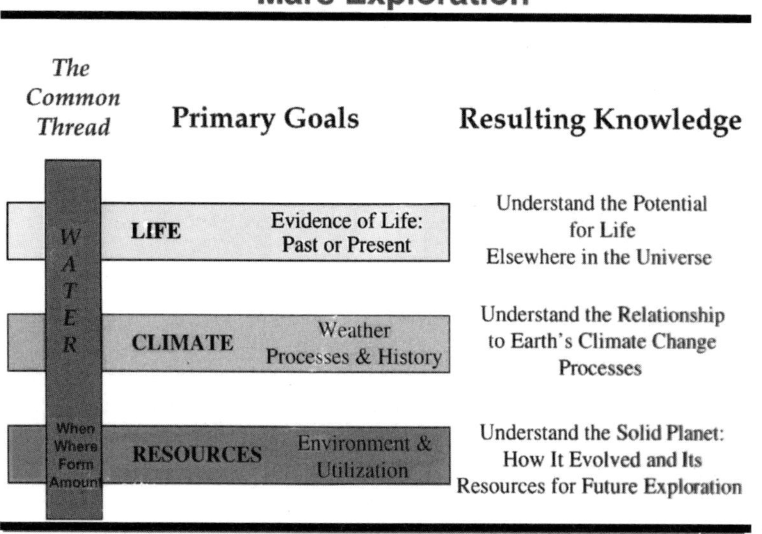

"Follow the water" is NASA's official Mars exploration theme. Courtesy NASA/JPL.

Venus and the Moon! The Martian *hydrologic cycle* is key to the geologic past and human future of Mars. The surface transfer of huge volumes of water or ice, if often repeated or long sustained, requires replenishment of the source. My five-year-old daughter, not having yet grasped the concept of a closed water cycle, once looked up at the mighty Yellowstone Falls and observed water pouring over the cliff, source unseen; the incessant flow seemed a puzzling, impossible feat of nature; finally convinced that some unseen process added water back to its source, the mystery was half solved. The Mars community today, while applying far more sophisticated observations and analytical tools, is not much more advanced in understanding Martian hydrology. We see signs of the one-way flow of hundreds of thousands, even millions of cubic kilometers of water and ice occurring multiple times in the distant past, and we see vast icy deposits, but no one has witnessed the replenishment leg of the Martian water cycle for anytime but the present epoch of feeble recycling.

We can observe primarily some things about the Martian hydrologic cycle: the transit of water molecules through the atmosphere and that occurring between surface and atmosphere (but only during the present era); the present repositories and distribution of ice at the visible surface (through imaging and reflectance spectroscopy); and the upper meter (by neutron spectroscopy), and the geological manifestations of the past movement of condensed water molecules over the surface. We lack observations of what is happening in the subsurface (with exceptions that can be inferred from surface morphology), and we lack any direct observations of what happened with atmospheric transport before the age of instrumented astronomy and space exploration. Until we can go to Mars with the necessary observational capability, the full extent and detailed behavior of the Martian water cycle is left to every scientist's imagination based on slender observations, studied

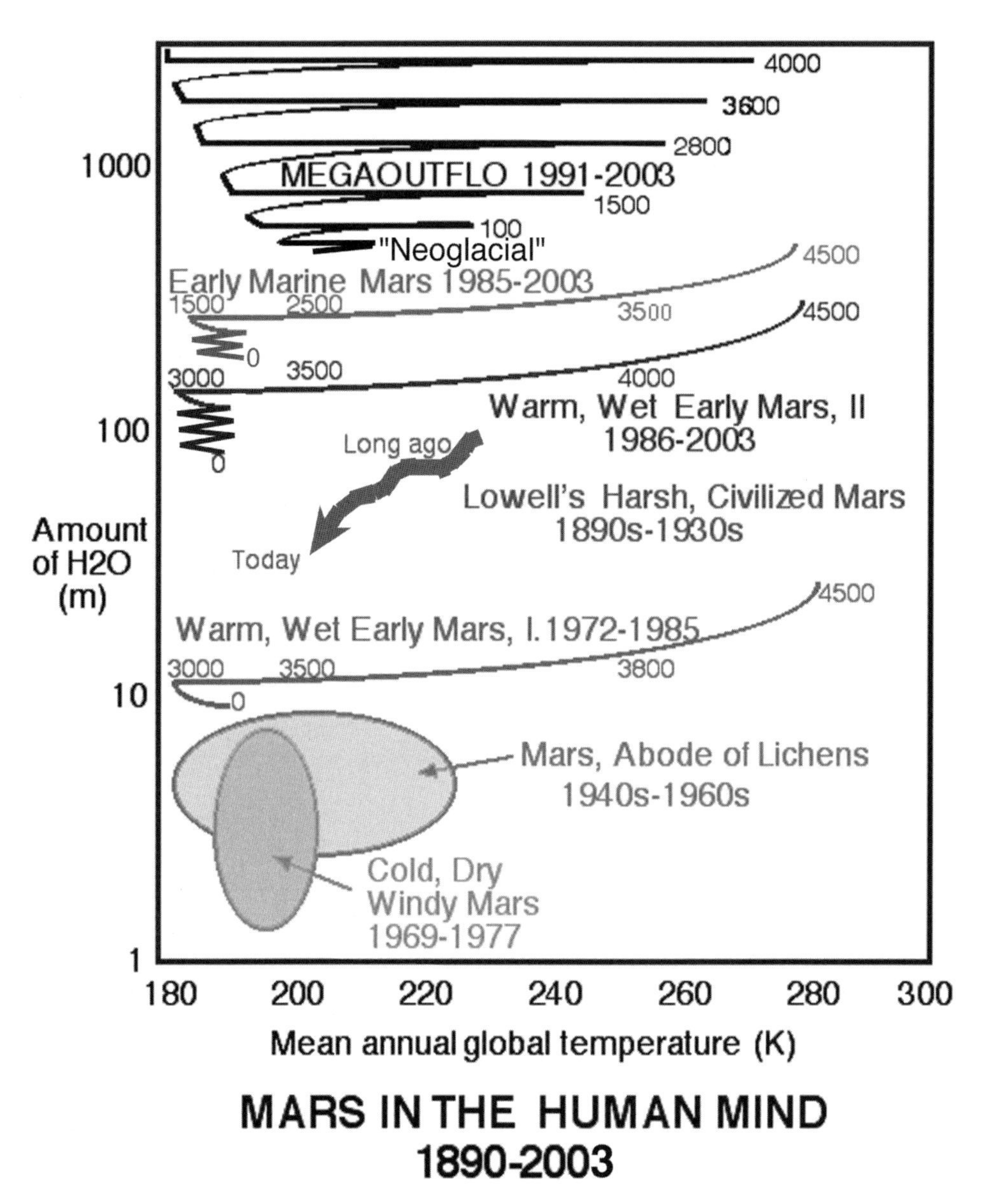

Scientific interpretations of Martian climatic history and hydrologic environment have shifted wildly as we have learned more about Mars and as certain paradigms have risen and fallen or have been forced to morph. The representations here are purely illustrative and should not be taken quantitatively or too literally. The dates of scientific dominance of each hypothesis are approximate, and the numbers labeling each "curve" (in hundreds of millions of years ago) and the graph axes are illustrative.

inferences, and analytical and numerical models. Some of the broad outlines of Martian hydrology, however, are well revealed to us.

The fundamental importance of water to Mars is recognized by NASA, which has adopted the over-arching theme of "Follow the Water" in designing Mars exploration goals and means. This theme has been accepted also by other national and international space programs. To perform some of the needed investigations, the world now has three functioning orbiting Mars spacecraft (*Mars Global Surveyor*, *Mars Odyssey*, and *Mars Express*) and two operating rovers (*Spirit* and *Opportunity*). The simultaneous operation of five Mars spacecraft is unprecedented and the variety of scientific observations being made is far beyond anything else in the history of extraterrestrial planet exploration. Mars is being probed from orbit at radio, thermal infrared, near infrared, visible, X-Ray, and gamma ray wavelengths. It has been zapped billions of times by lasers, its neutron emissions are being measured, its landforms are being imaged in three dimensions, its gravitational and magnetic fields are being mapped in detail. Martian rocks are being abraded by a small drill, its soil rooted up by rover wheels, its surface analyzed up close in visible, infrared, X-Ray, and gamma-ray wavelengths, and microscopes are scrutinizing its finest sand particles. Never before has an extraterrestrial object been studied in such detail, so globally comprehensively, and with such a variety of quality tools. The European *Mars Express* radar-mapping mission has begun to penetrate a million times deeper than the visible and near infrared can and a thousand times deeper than the thermal sounding being done from orbit. We are at the threshold of a new understanding that will look increasingly at the subsurface.

Notwithstanding our successes, the almost demonic ability of Mars to gobble spacecraft continues. The British *Beagle 2* lander and Japan's *Nozumi* orbiter have fallen prey to the whims of Mars and malfunction. *Beagle 2* was designed to provide detailed assessments of surface chemistry and mineralogy, including characterization of possible organic materials up to two meters beneath the surface. *Beagle 2* apparently landed, but disappeared without a trace, not unlike the fates of several other American and Russian probes. *Nozumi* was equipped to explore the loss of volatiles from the Martian atmosphere but failed shortly before orbit entry. These lost objectives remain as worthy as ever and are for the future to try again.

NASA and ESA offer and aggressive program of future Mars space missions, including an orbiter in 2005 that will image the surface in 10-km-wide strips with a spatial resolution reaching 25 cm, and the *Phoenix* mission in 2007 that will dig a meter-deep trench and explore for organic and volatile materials in a particularly icy part of the far northern plains deposits near the north polar cap. Meanwhile, NASA and ESA are testing a wide range of mission platforms and instruments, including balloon-borne and airplane-borne experiments, advanced rover designs, and many designs for drilling into the ice and rocks of Mars.

MANIC-DEPRESSIVE MARS

As we continue our exploration of Mars, a few questions will dominate our thoughts. Was Mars ever warm and wet (and by what standards of comparison?), or just plain cold and icy for eon upon eon? Was Mars at times "wet," but only by an astrogeologist's bone-dry super-Saharan standard, its ice and water a minor accessory to its constitution and recent history, where the same teaspoon of water has been recycled again and again with profound geomorphic efficiency? Is Mars a sand-blasted, frost-touched world forever at the edge of death? Or is Mars Earth's sister Water World, but so distant from the Sun that its ocean has always been frozen, with slight melting of subsurface ice happening grudgingly, sparsely, rarely, at the whim of episodic volcanic activity and changing interior heat flow? Or is it a planet where a discrete warm and massively wet climatic epoch occurred early in its history under a dense primordial greenhouse atmosphere, and where H_2O has been freeze-dried ever since the collapse of that primordial atmosphere? Have vast quantities of water and ice given a spectacular performance in the distant past, but has the hydrologic system run down as water and ice moved unidirectionally to lower gravitational potential energy? Could Mars have had repeated, geologically short-lived warm, wet episodes recurring through its history? Is Mars a world with a true water cycle, with surface, atmospheric, and subsurface components of flow completing a closed cycle? Of course, there is the fundamental question also of life. The search for life and for habitable environments, including the roles of ice and water in those environments, constitutes an overarching unifying theme of NASA's Mars exploration program.

My own view, right or wrong, is that the geologic evidence is clear – Mars is a manic-depressive world currently in a long depressed phase, but it is not inactive. It is substantially the intense oscillations of hydrologic behavior that have caused the wild and crazy controversy in the Mars community. Mars can be all things or nothing of the sort to each person. Just how Mars behaves depends on when and where in the planet's history (and where on the surface) one looks.

The concept that Mars has a hydrological cycle extending past the barest façade has only recently been accepted, though details remain obscure and controversial in most respects. Prior to the introduction and gradual acceptance of this concept over the past two decades, there was a widespread assumption that water is relatively immobile and unsubstantial on Mars. According to this old view, Mars has a mere trace of exchangeable H_2O cycled through the atmosphere as vapor, and any excess ice is stored like so much inert crustal rock and so is almost irrelevant to Martian geologic evolution. Many Mars scientists would consider nothing so icy or wet as anything on Earth until findings due to the *Mars Global Surveyor* mission forced changes in their outlook. Evidence for the existence and cycling of huge amounts of water and ice is compelling but mostly indirect, as the effects of these movements are seen, but the water itself is now frozen and primarily hidden from direct view by rock and dust.

The story of water and ice on Mars is written in its rocks and landscapes. It is not, however, a comprehensive encyclopedia of information. It is a highly imperfect record. In *On the Origin of Species by Natural Selection*, Charles Darwin wrote of the equivalent incompleteness of Earth's geologic record:

> *The crust of the Earth with its embedded remains must not be looked at as a well-filled museum, but as a poor collection made at hazard and at rare intervals.*

The reading of the rock record of worlds is a science of translation and an art of interpretation, interpolation, and extrapolation to fill in gaps and to draw plausible links in the fragmentary record. The more active a world, the more information exists in its rock record, but more is also lost, and the net result is a less complete record. Mars is no doubt somewhere between Earth and the Moon in terms of the volume and completeness of the rock records. While Mars is not Earth and the landscapes there are in many respects distinctly Martian, there is much about the geomorphology of Mars that is familiar to us from terrestrial experience, especially to those experienced in the distinctive morphologies and water- and ice-related processes of Earth's cold and arid regions. This is largely because a familiar polyphase substance – H_2O – has been acted upon by familiar forces under conditions overlapping the cold, icy regions and cold, arid zones on Earth, and it has acted on similar rocky materials without dense vegetation, extensive rainfall, or the sustained surface flow of liquid water. While hydrogeologic activity operates much more slowly on Mars compared to Earth – or less frequently – there can now be no doubting that ice and water have made their mark.

The roles of volcanism, tectonics, impacts, and wind-driven processes and landforms on Mars have already been very well documented and are only touched on briefly in this work. This lack of close attention to these important geologic agencies is not to say that these processes are somehow less important than previously thought or subservient to ice. It is just that the geomorphic roles of water – especially of ice – have not been adequately addressed in book literature. In fact, these processes all work together.

Before the first spaceborne observations of Mars in 1965, there was nothing in Earth-based observations of the Red Planet that would require it to resemble in any way what we experience and observe anywhere on Earth, though nothing we knew then eliminated this possibility. We find ourselves calling on Earth to explain Mars because this approach seems to explain a lot. Below I offer an explanation for this approach.

There is, of course, always some – often much – uncertainty when we are prevented by logistics from direct observation of our Martian subjects. We must always bear in mind that similar appearances do not prove similar origins. This is true when comparing two terrestrial landforms, and the caution is even more required when comparing Martian landforms with terrestrial ones. Hugh Kieffer advises, "*We need to think like Martians. . . . We need to consider all the physics of the Martian environment, especially the differences from Earth.*" That is a necessary order, but at the same time a tall order, as we each have only Earthly experiences. Imagine, for instance, if Earth did not now have any glaciers at all, but only a rock record of former glaciers; imagine how difficult it would be to conceive of glaciation through studies of these strange landforms and sediment deposits, despite all the analytical power of the computer. I rather doubt we would ever have conceived of glaciation, and if anyone did, and stated so, then that person would soon be out of a job. What Martian processes (or even ancient, extinct terrestrial processes) are there that we cannot conceive of, try as we may to think like Martians? Nevertheless, Kieffer's admonition is one that if we fail to meet it, we will fail to understand Mars for sure. Thinking like Martians with a background of Earthlings' experiences means, in fact, thinking like Earthlings, but applying suitable modifications to account for Martian characteristics. There are many commonalities between Earth and Mars, and we should not lose sight of this fact.

Many scientists, especially those lacking much geomorphological field experience, have a deep distrust of the analog approach to planetary science. They believe instead that the differences among the planets are so substantial that similar appearances of landforms on different planets are apt to be coincidental occurrences of similar forms produced by dissimilar processes. Their distrust leads to a valuable different approach, one that is not only highly theoretical in nature, but also tends to introduce a barrier to deeper understanding unless it is combined with analog thinking.

Our present understanding of Mars is not entirely one of high confidence, but neither are we clueless to the history of the Red Planet, the processes responsible for the origin of surface features and geologic rock units, and the conditions under which they originated. Caution is certainly needed when interpreting the Martian surface based on Earth-trained eyes. Though boots-on-the-ground Martian field experience would be an invaluable aid to reaching more accurate interpretations of the Red Planet more quickly and with more confidence, our lack of this experience does not mean that we are helpless to comprehend something of the Martian reality. The same scientific principles and approaches apply to studies of Mars as to any other field of science. While vast new sets of observations come in spurts and deluges, reaching a true understanding is a goal, never totally achieved, and comes by way of an evolution of knowledge.

Carl Sagan, in the *Demon Haunted World* (Ballantine Books, New York, 1996), wrote:

> *Science is different from many another human enterprise – not, of course, in its practitioners being influenced by the culture they grew up in, nor in sometimes being wrong... but in its passion for framing testable hypotheses, in its search for definitive experiments that confirm or deny ideas, in the vigor of its substantive debate, and in its willingness to abandon ideas that are found wanting.*

Of course, scientific thinking carries enormous inertia, as indeed it should. Widely held ideas normally have a basis in observation and/or theory. This basis is normally significant, even if often incomplete and sometimes erroneous or misleading. The overturning of entrenched, errant scientific thought is an evolution that occurs with the development of new ideas. Any new concept carries with it, if not a burden of proof, a burden of plausibility. While a demonstration of the incorrectness of old and passing ideas should not be a prerequisite of a new science concept or hypothesis, in practice it usually is required to make any headway in an inherently conservative establishment.

While I have expressed my views frankly in this book and at science meetings, and I regret some occasions that were subject to more heat than light, I aspire to Benjamin Franklin's guiding principle. Franklin wrote to a friend on October 14, 1777:

> *I have never entered into any controversy in defence of my philosophical opinion; I leave them to take their chance in the world. If they are right, truth and experience will support them; if wrong, they ought to be refuted and rejected.*

In Franklin's spirit, but providing no revelation to the Mars science community, I see that I might be more wrong than right about many issues; all I or any scientist should do is make the best case that observations merit in our most level opinion, and then let the ideas be tested with time. It is sobering to recognize that geomorphic features of similar appearance on this planet can be formed by different processes. Flowing water and blowing wind can make dunes, yet the implications for environmental conditions of origin are totally different. Wind erosion and glacial erosion can produce streamlined or linear erosional ridges. Glaciers and impact ejecta can produce similar unsorted debris sheets. Volcanic cinder eruptions and the home construction of ant species can make cratered cones of particulates at angle or repose. Sometimes observations are sufficient to discriminate origins, but at other times the needed observations are lacking. How much more so this multiplicity of possible origins and difficulty of discrimination must be when comparing Martian and terrestrial features. We do not even know the limits of possible processes on Mars; we might not even have imagined some processes.

The remoteness and distinctiveness of Mars, however, is not a good reason to distrust the geo-analog, which would be to distrust our ability to make sense of nature and to take into consideration environmental differences that would affect processes and products. Trusting in our logical abilities is what science is about. Along with laboratory experimental analog systems and analytical and numerical computer models, analog studies have a valid place. Earth is our own distinctive world, but its rocks have been subjected to natural experiments covering spatial and temporal scales and processes that cannot be well simulated in the laboratory or by computer or studied real-time in human experience. Some environments on Earth are about as close as we can get to Mars (past and present) without being on Mars and traveling back in time, and so Earth yields needed insights. This practice of using analogs to explain a new phenomenon or to offer other insights is not something that geologists invented. We all do it in everyday life from childhood until death; many elegant math- and physics-based theories are rooted in simple analogs.

The real strength of interplanetary analog-based interpretations is in identification of multiple types of *landforms* arranged nearby one another in logically ordered spatial associations that are similar to well-documented *landscapes* on Earth. Such spatial associations – especially if backed up by synchronicity or logical, rapidly sequenced generation of related landforms – favor explanations involving generation of individual landscape components (landforms) by the operation of a shared process system throughout the landscape. It is the geomorphologist's job to understand what that process system is. Close analogies at only the landform level, without finding analogies in the broader spatial context, carry a significant uncertainty that the correct process has been inferred. Close analogies between Earth and Mars at both the landform and landscape levels carry a much higher likelihood of successful identification of similar origins involving a particular process regime involving environmental and geophysical consistency. If a landscape on Earth is an environmental indicator, then any strong and close analog would imply a comparable environment on Mars within the range of conditions under which the process can operate. When the global occurrences of particular environmentally sensitive landscapes obeys a simple relationship to some understood climate proxy (for example, elevation, latitude, or slope aspect) or geologic terrain, or geologic age, then confidence is enhanced in a parallel interpretation of geomorphology and process.

Most lacking now for Mars studies is knowledge of the temporal associations or sequences of the individual landforms constituting a landscape. This is a major and humbling limitation. In fact, I place very little emphasis in this book on establishment of an accurate stratigraphy and relative geochronology, only because new data show that we are very far from understanding a Martian sequence of events. We usually have only loose control on the average ages of landscapes from impact crater superposition, and occasionally other types of stratigraphic superposition relations offer further constraints on relative ages of individual landforms. In most cases that lack better knowledge, our default assumption is that if multiple types of spatially associated landforms appear consistent with a similar process system and have an appearance of similar age (occur together and have similar extent of degradation or freshness, and no known stratigraphic relations rule out a close temporal relationship), then they are assumed to have been formed synchronously or in a tight temporal sequence by the same process system. This necessary assumption would not go far in geomorphological analysis of terrestrial landscapes, but it is a handicap Mars scientists have learned to accept. It is probably a fair assumption for young terrains on Mars more often than for Earth simply because most areas of Mars have less going on than Earth has.

So caution in interpretation of Mars and in drawing logical inferences is warranted. However, when we observe consistent similarities among Martian and terrestrial features within entire landscapes, and the terrestrial analogs are accepted as environmental indicators, then at some point we must consider whether environments and processes on Earth and Mars might have been similar in key deterministic respects. If we see that Mars is different now from the conditions needed to form particular features as we understand their environmental limits of formation on Earth, then we must ask ourselves whether global or regional planetary environmental change may have elapsed on Mars. Is a comparable change indicated by or consistent with other features and other landscapes? Is there some physical mechanism that might account for the implied changes? Could Mars have been truly Earth-like? When, try as we may, similar terrestrial and Martian landscapes nevertheless exhibit significant differences in morphology and ordering, or the Martian landscape lacks one key landform element contained in an otherwise similar terrestrial analog, then we must not be blind to the differences. We can consider how these differences might relate to differing processes or contrasting planetary conditions without necessarily giving up all insights from the analog.

FIFTH-GRADE ROCKET SCIENCE

Despite many similarities of Earth and Mars at landform and landscape levels, the global geomorphic sum of landscapes on Mars is not equal to the sum of Earth. This is evident from casual inspection and becomes clearer upon detailed scrutiny. Ultimately, we must seek an explanation for these differences as well as for the similarities. The differences require significant differences in processes, in the relative imprint or intensity of various processes, or differences in the way similar processes might manifest themselves on planets of differing characteristics. As Hugh Kieffer (US Geological Survey) emphasizes, the physics of Mars is identical to the physics of Earth, but we have no direct familiarity with certain conditions, processes, and materials on Mars. While we cannot help but bring our terrestrial experiences to bear on key Mars issues, we must attempt "to think like Martians," as Kieffer says. To meet this challenge we must be both analytical in our thinking and artistically, imaginatively creative.

In popular conception, any planetary science is "rocket science" – a matter of rockets, math, computing, and logical derivation of answers by really smart people. In fact, the way studies of Mars are conducted today, qualitative elements enter every aspect, starting with the formulation of outstanding questions by scientists, the public's and government's interest in pursuing answers to those questions, the boldness needed to design exploration goals, the creativity required to develop functional technological approaches to achieve those goals, the subjectivity in deciding what features and landscapes to observe and how to observe them, and the artful interpretation of planetary data upon which logical hypotheses are built and tested. The extraction of scientific meaning from planetary data is in fact a creative community endeavor as well as one based on objective deductions drawn from premises.

The analog-based approach to the geomorphology of Mars is a matter of developing artful scientific comparisons of landforms and landscapes, inferences regarding the mechanisms of origin of individual landforms and landscapes, synthesis with other observations, and logical deductions of conditions of origin. This approach is steeped in familiarity with planet Earth. There are enough independent and consistent lines of evidence that we must finally acknowledge that the Red Planet is an Ice Age analog of the icier realms of mostly Water World Earth. Polar deserts, icy permafrost, and debris-covered glaciers are key points of comparison throughout this book. While this deduction is secure, this interpretive–deductive approach allows a great deal of uncertainty and freedom in acceptable interpretations. There is a lot we simply cannot yet know. The conditions on present-day Mars do not in every case allow the deduced processes; the inference may be drawn that the climate has somehow changed. In other cases, a broad range of possible environmental conditions may satisfy the observations. It is important to consider this whole range rather than focus strictly on the one condition that fits a favorite model.

The interpretations of ice- and water-formed landscapes presented here motivates a profound question raised by every Fifth Grade class I have visited to discuss Mars. *Where did all that water go*? Good science is largely the process of asking the right questions, and this is the right one. While the polar caps have been known since the days of classical astronomy, the landforms of Mars seen ever since the *Mariner 7* flyby have given, in retrospect, good clues that H_2O is still there primarily as ground ice; from the *Mariners* and *Viking Orbiters* to *Mars Global Surveyor* and *Mars Odyssey*, these clues have accumulated and now are overwhelming in what they tell us about Mars. Geomorphology tells us that the ice is still there, mostly at middle and high latitudes, redistributed somewhat from where it was at times in the ancient past, but it is still on Mars, inside Mars, filling craters, smoothing over craters and mountains, in the regolith, and in great ice sheets at both poles. The landforms of Martian middle and high latitudes were formed largely by ice; many apparently are constituted of ice. However, the ice is impure, buried, or sublimated from the near surface. Mars has shown us a great variety of ice-formed features, but the particular suite of features formed through the ages has changed with time. No doubt ice was once distributed differently than it is today; ice once behaved more like the ice of the great ice sheets, glaciers, and permafrost on present-day and Ice Age Earth, whereas today it has a more restricted range of behaviors consistent with a seriously cold and mostly dry climate. The landscapes of Mars tell us that there has been climate evolution over the course of 4 billion years of Mars history, and some significant change has occurred quite recently in geologic terms.

Climatic shifts on Mars are both oscillatory on a small scale, and some rare changes are epochal and shrouded in mystery. With each climatic shift, there were changes in geomorphic processes and altered styles of expression in the landforms produced during each age. We are only beginning to perceive many of these changes, and we have yet to come to a real understanding of their causes.

Except where ice is directly exposed on the surface at the poles, it takes a different type of observing system to discern the buried ice directly. The neutron spectrometer aboard NASA's most recent space mission, *Mars Odyssey*, has at last provided direct detection of abundant ground ice (more correctly, hydrogen) in exactly the places where theory and imaging have indicated the ice should exist. Thus, one of the most important questions about Mars – that raised by the Fifth Graders – has been substantially answered. The ice and water that has over time sculpted and eroded, buried and graded the Martian surface is still there, exactly where many of us in the Mars science community have said it remains based on the indirect imaging evidence and theoretical calculations of ground-ice stability. Landforms that have been identified as ice-related based on morphology are concentrated in regions of the planet – at middle and high latitudes – where ground ice has long been predicted theoretically and where now *Mars Odyssey* has seen the upper meter of huge deposits of ground ice.

To our eyes, Mars appears to be nearly a homogeneous red. Indeed, Ashley

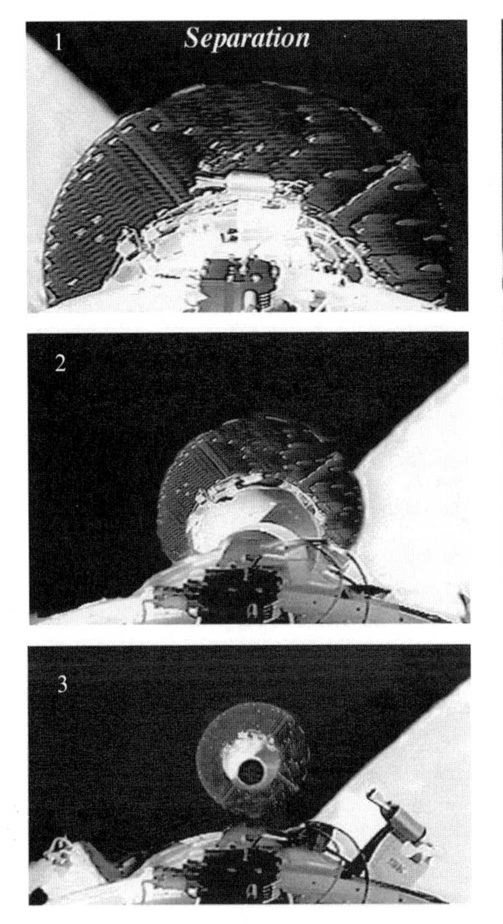

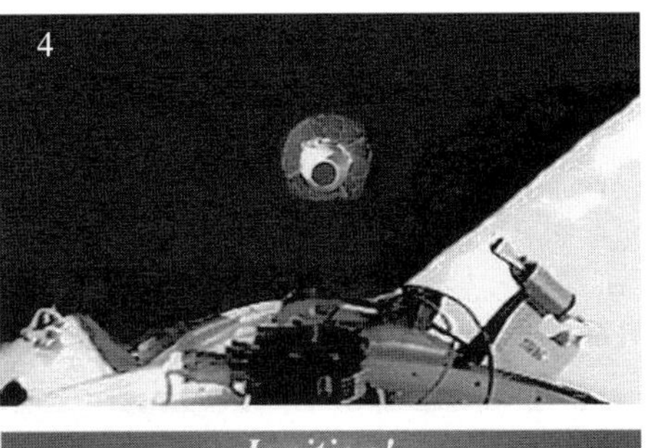

Warp drive engines engaged? Not quite. Rocketcam (TM), based on the Boeing Delta II booster, captured *Opportunity* and its upper stage separating and blasting off toward its surface-roving mission on Mars.
Credit: Dan Maas (Maas Digital)/ Ecliptic Enterprises Corp./Boeing/ NASA.

Davies (JPL) recently quipped, "*Trying to see Mars in multicolor is as tough as seeing Io in black and white.*" So true! However, Mars is far from homogeneous in much of anything. The redness is simply due to the pervasive and overpowering spectral reflectance and absorption properties of iron oxides disseminated in nearly all Martian materials outside the polar ices. Indeed, close scrutiny of THEMIS images by Jim Bell (Cornell University) and others has revealed small blue deposits on Mars. If our eyes could see in thermal wavelengths, or if they could resolve narrow absorption and emission features of other minerals, if we could resolve the discrete gamma emissions from silicon and iron and potassium or the neutrons moderated by ground ice, or if we could discern the long-wave topography of Mars or the heterogeneities in its gravitational and magnetic fields, then we would see a glorious "Jacobs' robe" of heterogeneous structures of Mars, which we can represent in false colors. In fact, our instruments orbiting Mars are allowing us to see in false colors all of these variations of real properties. It is an exciting time to be a Mars explorer, when our senses are brought not only nearer Mars but expanded in ways that biology has never allowed to any human being until now.

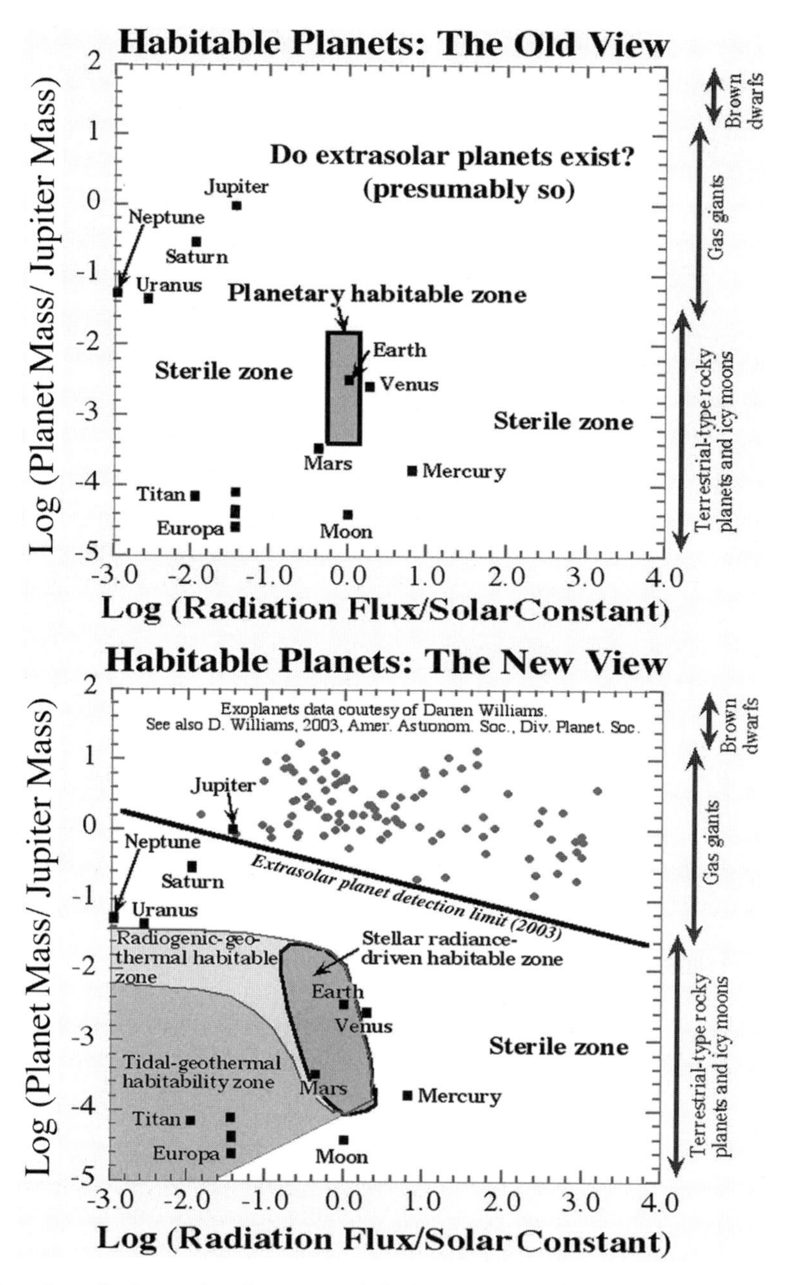

Our view of what makes for a potentially habitable world has expanded with gains in knowledge of Earth's extremophiles, greenhouse atmospheres, global planetary climatic excursions, geothermal environments (including tidally heating situations), and extrasolar planets. The solar constant refers to current solar radiation at Earth's distance. Planet mass on the y axis refers to mass $\times$ sin (inclination) in the cases of exoplanets.

The achievements made by Mars exploration and data analysis have been tremendous in recent years. Mars debates have been a moving target as I have written this book. Some of the last holdouts of ideas born in the 1960s – that Mars geology is written in little more than wind-blown dust and sand – have changed their views and now acknowledge the very major roles of water and ice, in addition to wind, volcanism, impacts, and internally driven tectonics. I have been forced to change my views on some matters, such as that the so-called "thumbprint terrains" of the northern plains are glacial moraines; in fact, they could be dust-mantled moraines but are more likely related to emplacement of mud oceans or massive debris flows. That pullback from the glacial hypothesis notwithstanding, most researchers now accept that glaciation and permafrost processes are major aspects of Martian geology. The concept that even oceans and continental ice sheets may once have existed are now widely discussed, amid deep controversy, by a community that a few years ago would not consider such concepts. We now agree that Mars was and remains ice rich; we know that the surface at times and in places has run deep with liquid water; that climate has made dramatic shifts; and that Mars has been and remains substantially a glacial and periglacial planet. The field no doubt will remain dynamic as new data sets are added by new missions, as old data are reanalyzed with the new, and as old ideas are finally set to rest. More than overturning or proving old ideas, I expect that over the next decade everyone's concept of Mars will continue to be moulded by new revelations of 4 billion years of geology and a few decades of close-up observation.

The subject of a formerly warmer, wetter Mars is the latest and still unfolding revolution in Marsthink. Each new revolution hopefully takes us closer to reality, but it is worth recognizing that science culture, like pop culture, religion, and politics, often takes a wild ride on the fashion bandwagon. Science, of course, exhibits a key difference from these other human institutions; it is nominally an objective endeavor, built on facts, logic, and relentless testing, but in practice science is also highly subjective. Since not all science questions can be addressed due to human limits, the relative priority or importance of particular science questions must be established. This is a subjective process, and assignment of highest priority in planetary science really boils down to what the public and political leaders are interested in. For Mars, the answer is overwhelmingly "life" and the water to which life is tied. For Mars, the story of water is inextricably tied to the history and roles of ice. Herein is part of the Mars story, told with significant interpretive license. It is not my purpose to validate each claim; to present an irrefutable case for formerly warmer, wetter conditions; to lay out quantitatively the links between the Martian cryo-hydrosphere and the rest of the Mars global system; or to provide a comprehensive review of the vast literature on these topics. This book is simply one observer's continuing contribution to a fascinating community-wide discussion about a new paradigm in Mars geology. It has been decades in the making but just recently has taken new robustness, with many details now clarified that were uncertain just four years ago. There have been surprising twists – such as the discovery of youthful gullies – that no one body expected a few years ago.

It remains for future astronauts and Earthbound Mars explorers to prove, refute, or modify the specific model of an ice-driven global process system developed here, and similarly to evaluate the rich variety of alternative ideas advocated by other researchers. Discussions and debates in the conference halls and beer lounges and scientific print media will increasingly shift to quantitative and chronologic aspects of these issues and the development of global process systems linking myriad geophysical, geologic, and geochemical phenomena of an integrated planetary machine. Behind all the discussion will be this intense human curiosity about the magical substance of H_2O, past climate history, and the possibility that life at one time existed on Mars or even still thrives in special subsurface oases. The Mars science community is as creative as ever in addressing this curiosity.

Foremost in this researcher's way of viewing the current state of Mars science, *Mars Global Surveyor* and *Mars Odyssey* observations have added consistent and persuasive support to the idea that Mars is a planet of living, active ice. The revolutionary advances made since the *Viking* era have not yet overturned the fact that we still almost lack solid ground-based observations and long baseline studies of active ice processes and landform development. Until this is corrected, the field will be embued with the most fundamental disagreements, though the nature of the controversies have shifted radically since the arrivals at Mars of *Mars Global Surveyor* and *Mars Odyssey*. It can be argued now that the evidence verges toward proof that Mars is a world touched if not wholly remade over half of its surface by glacial and periglacial processes, but we have far to go in understanding the details of these processes, their chronology, and climatic implications. Not to become too confident that I understand Mars, reality always is quite a bit different from Science, and the reality of Mars is no doubt something different from that presented here or anywhere else. We have good reason to believe that we are closer to the truth of Mars than ever before. Today's Mars is not much like the Mars perceived by any preceding generation. It is the Ice-Age Red Planet, I am sure of that, but that is a detail compared to what is displayed before us in the Heavens. The field will remain dynamic, but no matter what new dramatic discoveries and revolutionary changes in paradigm lie ahead, at least we can say that Science, like Art, Religion, and Humor, provides a further means to appreciate and interpret Reality. If that is all that is achieved by this presentation, I am satisfied.

1

Imagining

INFINITE WORLDS

Modern concepts of Mars as a world and possible abode of life are rooted in ancient times. The relationship is as definite but as abstract as the relationships of present life forms on Earth with successively more ancient life forms. This is inherent in systems where heritage and innovation work together to make the next generation better suited to a changing environment – or at least makes something new that is able to hang on to existence. In evolving Marsthink, the "environment" consists of understood facts and raw observations; the latter has expanded exponentially since the age of modern astronomy began, and present understanding somewhat lags the deluge of data we now have. The inheritance factor in our evolving understanding is the education received from parents, teachers, peers, and the rest of society. Learning with and competition from our peers is a part of the system. Books and journal papers, conference proceedings and seminars, media reports and bar-room and lunch-time discussions have painted for planetologists an evolving picture – accurate or not – of how Mars works as a system. An evolving, innovative element is our continual scientific reinterpretation of observations, modification of old hypotheses, and development of fresh ideas that link new and old observations together much better than any previous ideas. The competition – which kills our ideas or makes them more resilient – is revealed in intellectual discourse with our peers. Contrary to what many impassioned researchers may momentarily believe, there are no absolute right or wrong ideas; there are simply ideas that are better suited to improving or changing observations, or are perceived and judged by our peers as being better (whether they actually are or not). There are also ideas that are seen as not making the grade, and are tossed out or shelved until resurrected.

As with evolving biological systems, where some semblance of the biomolecular processes, morphology, and genetic roots are still evident in the descendants, our modern ideas of Mars (or almost any other topic) are traceable in some way to the ideas of the Ancients. The further we trace, the more distant seems the relationship in detail, but we eventually reach certain fundamentals that tend to be conserved through time. A compulsion to identify with or to extract other matters of relevance from the Great Beyond has driven astronomy and religion for millennia. In the Heavens we seek ourselves, our souls, things familiar from our world, the cause of blessings and travails in our world, and the source and fate of Humanity and our world. For example, the mythological roles of Mars – the celestial orb and God in various forms – in causing or sponsoring war, disease, or bad luck is documented from ancient cultures around the world. Maybe there was a prehistoric antecedent oral tradition linking Old and New World mythologies. Maybe the classical intercultural ideas are independent and rooted in Mars' bloody-red appearance to all color-seeing eyes and its presence in the sky when many ills have befallen humans. Such specific links of Mars in classical cultures and human psychology reach further than I find possible. More generally, this tendency to view the Heavens in supernatural terms is understandable, as there was nothing in the Heavens prior to *Sputnik* and our propagating telecommunications signals that humans can claim they have affected in any way – the Heavens are beyond human influence. *Psalm 19* declares it so simply, "*The Heavens declare the glory of God.*" And in the *Qur'an* it is written, "*Whatever is in the heavens and the earth declares the glory of Allah, and He is the Mighty, the Wise*" (Chapter 57, Al-Hadid: IRON, revealed at Madina).

Though both the heavens and the Earth exhibit spirit-awakening phenomena, there is nothing on Earth that clearly resembles anything in the Heavens in the spaces beyond the clouds when viewed with the unaided eye, except for the shapes and figures we may trace out by connecting the starry dots, and the fire in the Sun. However, for millennia astronomical mythologies have been balanced against another perspective that describes the celestial orbs in worldly terms: the Heavens are explicable as distant and glorified versions of what is familiar to us in this world, though details are bound to differ.

Some ancients of Greece and Rome recognized the stars and planets as being worlds, some of which were even Earth-like. I find this an astounding conclusion, considering that, except for the Sun and Moon, they appear as mere points of light in the night sky. How could the Ancients have conceived of worlds beyond our own or even have defined our own planet with the observations then in hand? Armed with pure philosophical intellect (we call it speculation), modest means of measuring star and planet positions, and having no observations besides those made by otherwise unaided sight, the Greek philosopher Epicurus (341–270 BC) wrote to his contemporary, Pythocles:

> *A world is a circumscribed portion of the universe, which contains stars and earth and all other visible things, cut off from the infinite, and terminating in an exterior which may either revolve or be at rest, and be round or triangular or of any other shape whatever. ...*
>
> *That there is an infinite number of such worlds can be perceived, and that such a world may arise in a world or in one of the intermundia (by which term we mean the spaces between worlds) in a tolerably empty space and not, as some maintain, in a vast space perfectly clear and void. ...*

> *The sun and moon and stars generally were not of independent origin . . . but they once began to take form and grow . . . by the accretions and whirling motions of certain substances of the finest texture, of the nature either of wind or fire, or of both. . . .*

Perhaps most remarkable in this philosophical pronouncement is the vivid description of a cosmic phenomena similar to the modern nebula theory on the origin of the Solar System. The modern versions of nebula theories of star and planet formation have drawn fresh support from *Hubble Space Telescope* images of nebulae caught in the act of forming condensate rings and, perhaps, planets. Rudiments of this model seem to go back at least several centuries earlier than Epicurus. We cannot forget the *Genesis 1* creation story,

> *In the Beginning, when God created the heavens and Earth, the Earth was without form and void, and darkness was upon the face of the deep. The spirit of God moved upon the face of the waters. . . .*

Like the common role of Mars in death and war, creation stories from around the world contain similar language and events: the heavens and Earth emerged from a process of differentiation out of an initially amorphous, dark, and turbulent source, such as a giant flood or vaporous mists, and then evolved under the further influence of megafloods. This commonality could be due to a common source for oral and written traditions, perhaps the ruinous consequences of floods that inevitably affect civilizations often situated on river flood plains and deltas, or it may be due to the formidable presence of the oceans for sailing colonists.

Adherents of myriad religious sects find differences in their beliefs in who or what gets credit and honor for creation and destruction. These differences are commonly used by human manipulators to motivate the killing of others who believe differently. However, if there is one thing besides our biological needs that nearly all humans share, it is a reverent honoring of the physical origins of the Earth and Universe. Although believed facts about origins differ, there is no discordance in that core reverence. The Iraqi Sabaean Mandeans, who believe that John the Baptist is Savior, celebrate New Year's Day in the summer; on that greatest of all days, physical origins began with the formation and placement of the Earth and stars. Scientists don't have a holiday for it, but the discovery of the Beginning by Penzias and Wilson earned them the Nobel Prize. Thousands of scientists work all their lives, many overtime to an extreme, to elucidate far lesser aspects of those times of origin. Whatever our science field or religion, we each have our teachers and hallowed institutions dedicated to pay respect and to learn about the Beginning.

Though the very concept of a discrete genesis is very common, it is by no means ubiquitous in intellectual thought. Epicurus, who so well conceived of planetary origins, was nevertheless convinced of the continuity of existence, even though the form of matter, including whole worlds, is transmutable:

> *. . . nothing comes into being out of what is nonexistent. For in that case anything would have arisen out of anything, standing as it would in no need of its proper germs. And if that which disappears had been destroyed and become nonexistent,*

everything would have perished, that into which the things were dissolved being nonexistent. Moreover, the sum total of things was always such as it is now, and such it will ever remain.

The concept of differentiation of homogeneous media into diverse substances, and the capacity for multiple heavenly bodies to emerge from the chaos, was already well established in myth when Epicurus wrote his decidedly secular philosophy. He further stated,

There are an infinite number of worlds both like and unlike this world of ours.

This infinity of worlds is itself an amazing concept, since even the visible stars can be counted. Perhaps it is easier to deal with the concept of infinite existence extending like unbroken terrain or seas beyond the visible horizon than to consider a limit placed on all existence. The transfer of ideas from teacher to student, and the possible extension of Epicurus' thinking on physical existence of many worlds to animate existence beyond Earth can be seen in the statement of Metrodorus of Chios, a disciple of Epicurus:

To consider the Earth as the only populated world in infinite space is as absurd as to assert that in an entire field sown with millet only one grain will grow.

While there were definite trends of thought right up to the age of instrumental astronomy that tended to remove the Earth and Humanity from the center of everything, there were contrary theories, and still another that rediscovered in the Sun a center of existence that for millennia had been worshiped as a supreme god. One of the most important astronomical advances was the heliocentric theory of Nicolaus Copernicus who, in 1543, not only argued a sound basis for having the Earth removed from the center of the Universe, but practically formed a scientific basis for worship of the Sun as a second-tier God:

And behold, in the midst of all resides the sun. For who, in this most beautiful temple, would set this lamp in another or a better place, whence to illuminate all things at once? For apt indeed do some call him the lantern – and others the mind or the ruler – of the universe. .. Truly indeed does the sun, as if seated upon a royal throne, govern his family of planets as they circle about him. . . . The earth is impregnated by the sun, by whom is begotten her annual offspring.

Until modern scientific instruments were devised in the sixteenth century to extend human abilities to observe the heavens, ancient cosmology and belief systems could not advance solidly much further than the point to which Epicurus and Copernicus pushed the theory of the Cosmos, which had been steadfast in offering inspiration for ages. Giordano Bruno, still lacking telescopic instruments, echoed the thoughts of the Ancient Greeks, offering ideas that still resonate among modern scientists:

There is not merely one world, one Earth, one sun, but as many worlds as we see bright lights around us.

(*On the Infinite Universe*, 1584)

Such ideas could not then be proven, but in Europe they threatened the power of the Catholic Church, which envisioned different, more special roles for the Cosmos. For issuing such heresies, especially for espousing the possibility of extraterrestrial mortal life, Giordano Bruno was burned alive in 1600. Within decades, overwhelming evidence brought to the fore by the telescope would force the Catholic Church to change its stance to one of dismissive acceptance.

Now, 23 centuries after the advent of Epicurean philosophy, Humankind's search for origins of our physical and animate existence – and a parallel quest for an understanding of the Heavens – remain the chief psychological charges that impel modern space exploration and astronomy. Life of some form in outer space is a logical extension of, or perhaps a precursor to, a philosophy that recognizes the plurality of worlds and seeks familiarity amid strangeness and even friends among distant strangers.

WATER: THE RIGHT STUFF FOR LIFE

Philosophical speculations and religious beliefs on extraterrestrial life do not constitute science in the modern sense until the physical conditions on other worlds are understood and consideration is given to the reasonable physical limits and microbiological basis of life. Once that step is taken, the original philosophical or religious basis for driving the investigation becomes irrelevant to the progress of the quest, although answers to key science questions may very well drive further development of the underlying philosophical doctrine. A main underpinning of planetary science and astronomy is that the Cosmos is everywhere made of the same "star stuff," as Carl Sagan called it – the same naturally-occurring elements. René Decartes, inheriting the knowledge of Galileo, wrote

"*... that the earth and the heavens are all made of one same matter, and that though there were an infinity of worlds, they would be made only of this matter.*"

Without this universal point of similarity, there would be no understanding of anything in the Cosmos, much less a basis on which to consider extraterrestrial life. While Descartes had conviction, he lacked proof, but that proof was to come.

Modern scientific recognition of the hostile surface environment of Mars precludes present-day surface life there, but this has not hindered studies of possible ancient habitats and possible current subsurface ecosystems. Mere microbes on Mars – past or present – would be of immense human and scientific interest, most importantly because they would support the long speculation that life is common in the Universe. It would be confirmation that our philosophical faith, going back to Epicurus and the authors of *Genesis*, was well placed.

The special role of water as necessary for all life and as a marker of physical conditions on planetary surfaces was not explicitly investigated until the nineteenth century. The Italian astronomer Cassini discovered in 1666 that Mars has bright white features at its poles, but their composition was not clearly expressed until 1719, when Giacomo Maraldi said that they are probably made of ice. British astronomer

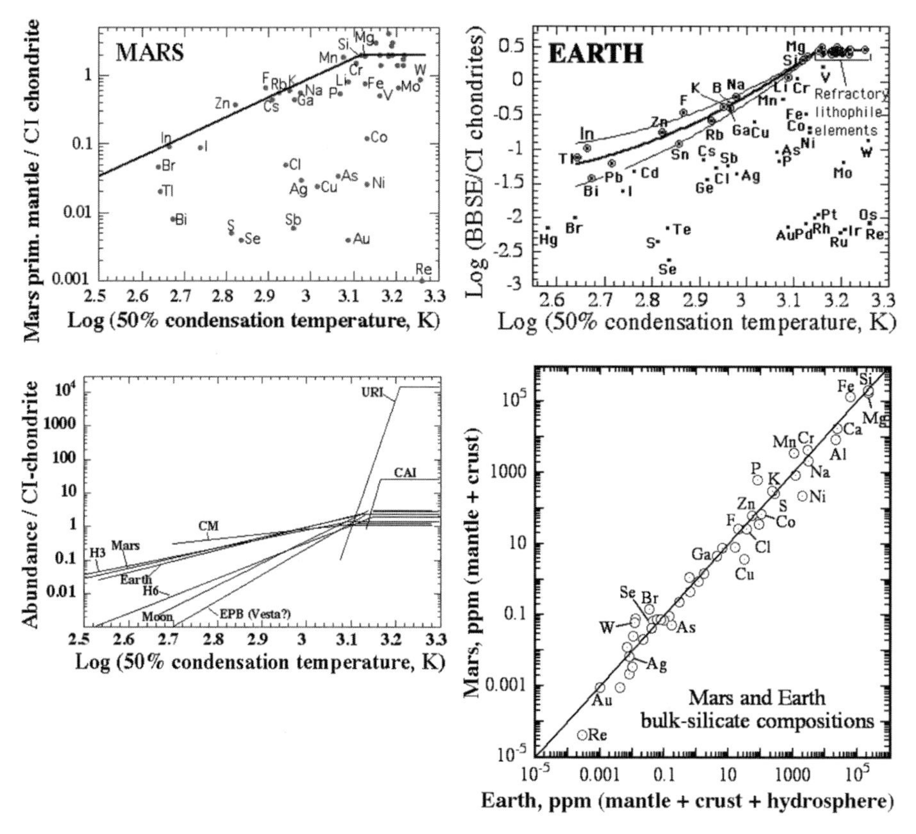

Top two diagrams show similar abundances of elements in Earth and Mars. The CI chondrite reference composition represents nearly complete condensation of solar-composition material minus hydrogen and helium. The sloping lines or curves (including three alternative fits to Earth's data) represent volatile depletions ("volatility trends") in Mars and Earth; they have been drawn through points for elements that have a chemical aversion to metallic cores. These elements provide an indication of the bulk planetary compositions (the summations of oceans, crusts, mantles, and cores), since siderophile core-forming elements presumably have bulk planetary abundances also near the sloping trends. Elements well below these trends may have been sequestered by the planets' cores. Alternatively, some common salt-forming elements may have been lost by large impacts into primordial oceans. *Lower left:* The volatility trends of Mars and Earth are very similar. The volatility trends on this diagram pertain to the following: EPB Eucrite Parent Body (asteroid Vesta); CM, H3, and H6 types of carbonaceous chondrite and ordinary H-chondrite meteorites; CAI and URI calcium-aluminum-rich inclusions and ultrarefractory inclusions in carbonaceous chondrites. *Lower right:* Bulk-silicate Earth and Mars fall almost on the line indicating identical compositions; the combined effects of volatile fractionations in the solar nebula and core formation were similar for most elements in both planets. The first-order similarities in bulk-silicate compositions would produce roughly similar crustal compositions, a point that the Mars Odyssey gamma-ray spectrometer team recently has been making from the perspective of their surface composition analyses.

William Herschel went further: he discovered that Mars has an atmosphere and agreed that the polar caps are probably made of snow and ice, noting also that they wax and wane with the seasons. Thus, evidence of phase changes of the water molecule, with all that means for physical conditions, was provided. Herschel studied the spin-axis obliquity of Mars and considered climatic implications, finally stating in 1784 that "*the analogy between Mars and the Earth is, perhaps, by far the greatest in the whole Solar System.*"

Thus was born the modern scientific concepts of comparative planetology and planetary analogs, and specifically the recognition of Mars as an Earth-like world. This approach was taken to an extreme by Percival Lowell, whose observations from an astronomical observatory on the other side of my town stimulated a world of dreamers. The golden age of classical Mars astronomy had begun, and it did so with a study of the Martian cryosphere and that substance all important to life, water. Refinements in the knowledge given to us by Herschel remain one of my life's chief pursuits and is the central theme to NASA's Mars exploration strategy.

OUR FAVORITE MARTIANS

I shall return in Chapter 8 to serious scientific issues of life on Mars, but shall devote all the chapters before then to the issues of water and climate on Mars. Before I do so, it is worth considering Mars in pop culture, since it is not only a direct reflection (often distorted) of the scientific knowledge of Mars, but is also a chief driver of the politics that commands funding of Mars exploration and research. This is a point that scientists would do well to take seriously, however frivolous and unfounded some popular concepts of Mars may seem and however inaccurate mass media's portrayals of Mars commonly are. Life on Mars – whether or not there is a single microbe of it – has driven billions of dollars into the space program and hundreds of millions of people into the theaters time and time again. Mars is fun. It's serious ... and it's big business.

Starting with Herschel's writings, popular and scientific concepts of extraterrestrial life took on new fervor. As astronomical technologies to map surface features on Mars steadily improved throughout the nineteenth century, the need for explanations – scientific or fanciful – of those features also increased. The Italian astronomer Schiaparelli drew maps of Mars showing many dark splotches and lineaments, which he named *canali* ("channels"), a descriptive and broad interpretive term that inevitably spurred a dramatic inference. By 1894 the American astronomer Percival Lowell and a few others had begun a decades-long campaign of Mars mapping. Lowell and some other astronomers saw in the Martian markings the handiwork of intelligent beings who were desperate for survival on a world undergoing environmental change. As he thought the markings showed a network of verdant growth adjacent to canals dug to transport water from the melting polar caps to arid but liveably warm low latitudes, Lowell then drew some logical conclusions about the canal engineers. This was logical, of course, but only if the interpretive premise was correct.

Lowell was a dedicated professional astronomer, but he was also effective in relaying his findings and his interpretations to the public. He not only discussed the intelligent inhabitants of Mars, but he deduced from the global structure of the canals (which most astronomers could not see) a global guiding politic. Many of Lowell's contemporaries were unimpressed. James Keeler, later the director of Lick Observatory, remarked critically in 1897 that, for Lowell, "... *the step from habitability to inhabitants is a very short one*."

Not to be dissuaded, Mars was portrayed by Lowell – and then by science fiction writers who based their work on Lowell's writings – as a world so tangible that it and the civilizations on it could be painted on canvas and narrated in stories, though Lowell cautioned that evolution on two worlds would undoubtedly not make inhabitants of similar form. While Jules Verne and others had taken an engineering-based sci-fi approach to writing about human space travel, a new generation of novelists looked at human adventures on Mars and at the aggressions of alien beings visiting humans on Earth. Mars ever since has been a consistent favorite due to its relatively Earthlike conditions and a presumption of life.

The War of the Worlds, by H.G. Wells, hit the streets in 1898, just as Lowell was amassing more observations and developing his fanciful interpretations that served as a basis for the fiction; in a little more than one decade these two instigated a mass culture around the world that believed solidly in Martians. It was arguably the first time in history that planetary science had become directly relevant to the person in the street. Edgar Rice Burroughs started writing about Mars and Martians just four years after Lowell's third book, *Mars as the Abode of Life* (1908, Macmillan). In 1917, the year after Lowell's death, Burroughs brought Lowell's world alive, writing:

> *The water which supplies the farms of Mars is collected in immense underground reservoirs at either pole from the melting ice caps, and pumped through long conduits to the various populated centers.*

Very few astronomers accepted the Lowellian paradigm explicitly, but there were few who doubted that life of some form existed on Mars. Life on Mars was a recurrent theme in science and in the mass media throughout the twentieth century, and continued through the space age. The unwitting collaboration among scientists and science popularizers and fiction artists has only accelerated since Lowell's time.

As our knowledge of Mars has increased, our view of "Martians" has evolved. Long before the *Mariner* Mars flybys, we had begun to understand the Red Planet as a bitterly cold, dry world, more so than Lowell and his contemporaries had understood it to be. As recognition of conditions on Mars hardened by the mid-twentieth century, astronomers pointed to lichens, which can survive in arid and polar climates, as a possible biological analog of Martian life. With the hardening of our understanding of the severity of Martian climate, pop culture's perception of the threat posed by "Martians" softened, unless one is a rock fearful of attack by lichens. Martians became something we could laugh about in, for example, the 1963 television sitcom series, *My Favorite Martian*, produced by the American television network, CBS. During the era of early spaceborne Mars exploration, the difficult environment on Mars became widely known, and any heart-stopping (or plain silly) drama built around threatening extraterrestrials had to be visiting from places much more distant than Mars, or caused by nearby but primordial extraterrestrial life attacking us. The 1960s series *Lost in Space* and *Star Trek* took the first approach, while the 1971 film (based on a Michael Crichton novel) *The Andromeda Strain* took the latter approach. *Star Trek*'s warp drive engines, allowing travel faster than light-speed, was a complication needed to allow one-hour dramas to be built around extraterrestrial life that was far-flung among the stars and nowhere in this Solar System.

The path taken by popular sci-fi space culture since the mid-nineteenth century has been adapted directly from evolving scientific understanding of water on Mars. When Mars was seen as flowing with water, Mars teemed with alien creatures; when Mars dried up and cooled down in prevalent scientific thought, the alien creatures were from somewhere else in outer space. The past three decades of seeing an increasingly wet Mars in science has posed an interesting new twist in Mars pop culture. Mars is now regarded as very dry and cold at the surface, but with abundant water and ice beneath the surface. With that water and ice, novelists, Hollywood, and the tabloid media have seized upon a latent capacity of Mars to harbor life, including intelligent life – either our future descendants or civilizations of the distant past, or both, on a collision course.

Recognizing that Mars has had surface water in its past, a cult phenomenon has appeared where it is believed that Mars was the home of ancient civilizations. With all apparent seriousness, adherents believe that NASA has photographic evidence of these civilizations, and now "they" (NASA, the Pentagon, or some ominous black agency) are engaging in a vast conspiratorial cover-up. According to this hypothesis, some evidence has evaded government censors, such as the famous "Face on Mars" and other features, which are especially abundant in the Cydonia region. Features in these images can be interpreted as monuments of an ancient civilization. Mars geologists are bemused by the insistence of the "Cydonia Clan" (as I call them) that certain peculiar rock formations are due to intelligent design and fabrication. As a group, geologists do their best to ignore the Cydonia Clan, who in turn seem to ignore the fact that natural geologic processes on Mars are fully capable of producing landforms that can leave anthropomorphic impressions on human viewers, especially when the images are of low resolution or taken with deep shadows and processed just right. The Cydonia Clan is simply struggling, as we all are, to make sense of the wonders of the Universe. Personally, I find great positive amusement from the geo-anthropomorphisms in the world and Universe around me, whether it is the Face of Jesus in the Eagle Nebula, or the Face of a human on Mars. In payback for the amusement I have been offered, I point out a few of my own observations below that should be as amusing.

The possibility that intelligent life may once, long ago, have existed on Mars is difficult to argue against, since the current lack of plausible evidence for that concept is different from proven refutation. The Cydonia Clan argue to no end that the evidence could be there; that geologists just turn a blind eye. Since there are more

Cydonians than scientists, I think it is time that we, the scientific community, offer a patient and serious explanation of our view. In so doing, I shall do my best to step into the Cydonians' frame of mind.

Consider this thought experiment: Think of the hundreds of millions of artifacts of human civilization that presently exist and can be observed from low Earth orbit with a small telescope or from an airliner with the unaided eye: this is approximately the resolution of the best imagery of Mars from *Mars Global Surveyor*. Think of our homes, factories, roadways, and tracts of clear-cut forest, all visible at that resolution. Now, mentally, magically, wipe out all living humans, as though from a plague. Within 500 years, no patterns of clear-cut forest would be recognizable to a visiting alien orbital probe as having ever been cut artificially; the Netherlands would be under the sea; and most homes would have collapsed into dust, their foundations filled and overgrown with vegetation. In 500 years, our interstate highways will have been pulverized by weathering and overgrown by forests, buried by sand dunes, or washed into the sea. Within 50 millennia few skyscrapers will still be standing in a form recognizable as artificial. No hydroelectric dam will be intact; only the erosion due to the sudden collapse of some dams will be visible, and in other cases their reservoirs will have long since filled with sediment and then the dams incised from waterfalls that have eroded back to base level. Drastic human alterations to natural ecosystems, including replacement of natural forest assemblages with others, would still be pervasive, but it would not be possible, from Earth orbit, to know what alterations had taken place or to link those altered ecosystems to human activities Within one million years, I doubt that any but the most patient archaeological studies would uncover the least shred of evidence of our existence; a few rover traverses would provide no real chance of uncovering any evidence of past civilization. There might be an out-of-place yardang where the Great Pyramids now stand, with treasures still inside, but they will more likely be deeply buried under the Nile Delta or long since eroded completely by the avulsion of one of the Nile's meanders. Forget the Great Wall; it doesn't have a snowball's chance to last a million years.

Mars has seen the equivalent of several million years' worth of weathering and erosion at average terrestrial rates. Thus, from a purely scientific viewpoint, we cannot exclude even the possibility that intelligent life once thrived on Mars. We simply have no way, presently, to see through the dust and ravages of time to a level of detail where civilized alterations of landforms remain significant on long geologic timescales.

If the Cydonia Clan should find in this statement some semblance of support for their off-the-edge interpretations of geologic landforms, they should now be disappointed. True, I must say that nothing learned about Mars so far can rigorously exclude the possibility of ancient civilizations on Mars hundreds of millions or billions of years ago. To most scientists, this possibility seems an absurd concept lacking the least real evidence or plausible hint in evidence. This concept is not in the realm of science, but it made for excellent philosophy two millennia ago and could still make for a good Hollywood series. I take a slightly more moderated view that rests again in the fact that we still know little about Mars, that what we do know points to drastic environmental change, that environmental and geological conditions are such that we can realistically contemplate and even plan future civilizations there, and that nothing we know now can rule out the possibility that ancient civilizations could have developed and thrived on Mars. However, this remote possibility is not one that interests me or takes my time, simply because the ideas so far presented lack any shred of support and are immersed in a quagmire of selective use of observations and illogical connections and extrapolations of some observations. I shall not go into the details of why the Cydonian theory falls on its Face, but rather I shall emphasize things geological that relate to The Face and its anthropomorphic friends. Before doing so, let's look again at the lighter side.

Hollywood does not take their domain quite as seriously as either the Cydonia Clan or geologists, but they make more money in their pursuits, which lately has merged the richest potential of science and sci-fi. Recurrent themes of recent popular novels and films have included cataclysmic transitions to warm, wet conditions and formation of a respirable atmosphere on Mars, either by natural events or by engineered terraforming. With a new world underfoot, these epochal advances in human adaptability and engineering have huge consequences for human settlement and revival of long-dormant life (e.g., Kim Stanley Robinson's *Red Mars*, *Green Mars*, and *Blue Mars* novel trilogy; and the films *Total Recall*, *Red Planet*, and *Mission to Mars*). These are all ideas that have a basis in current geologic and astrobiologic hypotheses. While it is plain to see that science impacts science fiction, anyone who reads science fiction and ponders technological innovation is struck by the fact that science fiction often precedes and motivates new technologies and discoveries. Surely we are exploring Mars today, rather than Venus, through the courtesy of Percival Lowell, H.G. Wells, Orson Welles, Edgar Rice Burroughs, and Arnold Schwarzenneger. In fact, California voters, as of October 2003, felt that the oversize star of *Total Recall* was the last best hope for California during its recall election. The theme of "total recall," unbelievable as it sounds, was very likely a reason for Schwarzenneger's election in that electoral race-from-Mars.

Whether science or sci-fi, it all amounts to unleashing of the human imagination, with one person's creativity feeding another's. Popular music culture, theater, television, and street language all reflect general public knowledge and misinformation about Mars. They are not all serious, and most in fact are part of modern humor. With more scientific knowledge in hand, we can now laugh at early imaginitive human re-creations of Mars that are traceable to Lowell. The integration of Mars into popular culture is so deep that casual references to planet Mars occurs in English-language sources more than for any other planet. Popular uses of Mars pervade cultures around the world, including candy culture (*Mars* candybar) and the twisted science/technology-driven music culture of Thomas Dolby, English techno-rocker, who offered a rich perspective on modern Western "Pulp Culture" (1988):

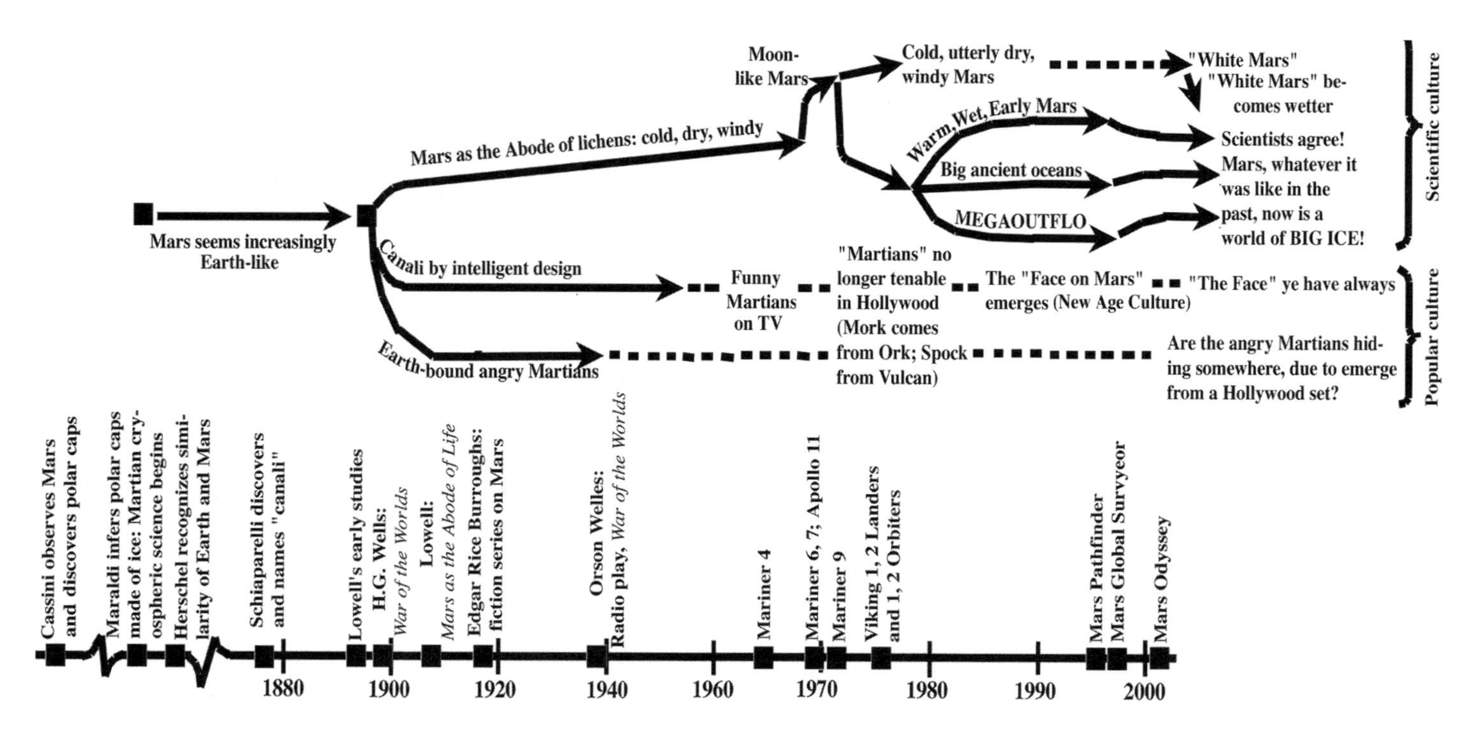

Not even revolutions come from thin air. Revolutions in Mars science and popular Mars culture have a heritage. The lineage is not as simple as depicted here. Ideas borrow from one another. Popular Mars culture has mirrored (using a rather distorted mirror) scientific Mars culture. When science overturned the plausibility of Martians landing in New Jersey, *The War of the Worlds* could not be taken seriously and it could no longer be scary or entertaining. By the 1960s, we had spacecraft telling us for sure that there are no angry Martians. Besides, the Soviets seemed far scarier to Americans, and vice versa. Hollywood had to put Mars on hold until they had a new story to tell. By the 1980s we had learned enough to make up a scary story that was believable. Arnold Schwarzenneger landing on a partly terraformed Mars is believable when scientists say it is. And Arnold is always scary when Hollywood says so.

I drove all over Hollywood,
Looking at the stars.
First I ate my Milky Way,
And then I ate my Mars.
But sucking on a Galaxy,
I noticed something pretty bizarre.
There's not a lot of people there,
Just an awful lot of cars.

The point of my book exactly, but Dolby beat me to it.

Lenny Kravitz in 1999 went to No. 3 with his song, "Fly Away," where he borrowed a rhyme and Dolby's knowledge of modern astronomy, and exclaimed,

I want to fly away!
I want to see the stars,
the Milky Way,
or even Mars!

They say that a little knowledge goes a long way. (Toward making a man rich.)

That same year, Matthew Sweet put out his CD, "Blue Sky on Mars," which featured a jacket dominated by a *Viking 2* lander panorama, but no evident use of Mars in his lyrics; I guess he just thought that picture was cool, and I guess he was right.

Although we Mars specialists and Hollywood may find Mars to be cool, alas, Mars does not figure prominently with lovers. Mars and passion is not an *impossible* link, but it yields the lines most unlikely to succeed. If ever there was a man to translate Mars to make it sound like a romantic walk in a rose garden with a warm evening breeze, it would be Frank Sinatra:

Fly me to the Moon.
Let me play among the stars.
Let me see what spring is like
On Jupiter and Mars.
In other words, hold my hand.
In other words, baby, kiss me.

Few other popular references to Mars are so down to Earth. There are references in modern music to the Girl from Mars, the Beauty Queen from Mars, Teenagers from Mars, the Laziest Men on Mars, and Radioactive Hamsters from a Planet Near Mars. Given the popular use of Mars as the home of the bizarre, I was not so surprised recently at the appreciative response of a new friend, a terrestrial glaciologist, after I had explained what I do for a living. It had not quite yet sunk in that I *really* study Mars, so the simple summary seemed appropriate, "*I am a Martian*," as distinct from my friend, a Terran who studies Chilean glaciers. With a cultural and language gap intervening, this glaciologist instantly deduced that what I meant was that I am a very strange being. Not my meaning, but correct nonetheless. A third effort finally conveyed my actual meaning to my friend's repeated amusement.

One of the more serious philosophical uses of Mars in modern music lyrics includes Sammy Haggar's "Marching to Mars," which caught both the deeper significance and the bandwagon "pulp culture" of Mars:

> *Here we go marching to Mars*
> *On a rainbow bridge, it doesn't seem so far.*
> *Steppin into our Universe, moving towards Life,*
> *To solve the problems on Earth. ...*
>
> *Everybody's marching to Mars ...*
> *... Hollywood's marching to Mars for a grand new movie*
> *With some brand new stars plastered on the silver screen.*
> *Gonna bring it on home, so you won't have to leave ...*
>
> *... Is there life in the Universe?*
> *Yeah, there's life in the Universe.*
> *We'll find God in the Universe.*
> *We'll find God, but we'll find life first.*

The march to Mars is taking place on the internet, too. More common than the casual internet uses of *Mars* are serious scientific and science-fiction uses of the word. Google on "Mars" and (as of August 2002) there are over 8 million hits, by far more than for any planet but Earth (Table 1.1). Mars gets more hits than all other planets, excluding Earth, if one deletes hits for major nonplanetary uses of terms, as in "mercury waste." "Life on Mars" is of particular interest on the internet, somewhere just ahead of "Life in America," whereas "Mars life" strikes up an interest somewhere between "Hollywood film" and "car stereo." The internet agrees with former US President Bill Clinton that "*It's the Economy, Stupid!*", but within the range of Mars topics, public interest says "*It's Life, Stupid!*" It's life – far more than water or geology – that drives the public interest.

The lines between science and sci-fi are seldom entirely clear. Percival Lowell's science is a good example. Another came in 1996 when a scientific discovery was announced by President Clinton: a Martian meteorite may hold fossil evidence of

Table 1.1 Internet hits returned by google searches, August 2002

Search term	No. of hits	Search term	No. of hits
Earth	19,100,000	US *and* economy	4,040,000
Mars	8,270,000	Hollywood *and* film	1,860,000
Mercury	4,210,000	Mars *and* life	1,140,000
Venus	2,970,000	Car *and* stereo	831,000
Saturn	2,440,000	Mars *and* water	612,000
Jupiter	1,890,000	Mars *and* lyrics	124,000
Pluto	1,120,000	Life on Mars (exact phrase)	71,500
Neptune	819,000	Life in America (exact phrase)	66,800
Uranus	456,000	Life in Britain (exact phrase)	30,800

ancient Martian microbes. The President was summarizing the chief findings and interpretations of David McKay and nine other scientists from the Johnson Space Center and four other institutions. While the science basis of this assertion has been roundly challenged and the debate continues, the *very public* announcement has left its mark in science and pop culture. In the scientific realm, the 1996 "bugs-on-Mars" revelation has also had a lasting influence. There has been the formalization of a whole new discipline of planetary science, astrobiology, replete with a new scientific journal and conferences by that title, NASA-sponsored university institutes around the world, and new space missions to go where no one has gone before. The new mood in NASA funding circles since 1996 has loosened the grip on research dollars, which previously had been reserved mainly for those interested in blowing dust and flowing lava. "*Follow the water*" is NASA's new Mars theme – water, the giver of life, the shaper of worlds.

As a result of two recent and continuing missions – *Mars Global Surveyor* and *Mars Odyssey* – in the past four years we have discovered new evidence of recent climatic fluctuations and a veritable ocean of water locked up in subsurface ice. It will be interesting to see where the next thrust in Mars pop sci takes us. Considering the advances in knowledge partly reflected in the following chapters, it's certain to be a wilder place than Hollywood Pulp Culture has yet taken us. More interesting, the next few decades may well confirm Sammy Haggar's thesis and give us proof of the actual existence of life beyond Earth.

PLACE WITH A FACE

There is one matter I must dispense with before becoming too serious. Not being an avid reader of the tabloids, I somehow missed the early years of The Face on Mars. I have a vivid recollection of my introduction to The Face (and The Teeth). I was at a Mars meeting in the late 1980s in Tucson, when a man approached me, presented his engineering credentials, and asked my opinion of his latest work. He had enhanced pictures of The Face in the Cydonia region. "*The what?*" was my response. He proceeded to show me how The Face actually exhibits details, never seen before, right down to teeth. (At least *he* could see the teeth.) It was incontrovertible proof, the engineer informed me. I soon had the inside, fascinating scoop on something NASA had sought to prevent the public from seeing and hearing about, or who were simply still just ignorant. The Cydonia Clan of devotees have also expanded their geographic scope all the way from the Northern Plains, where The Face occurs, to the south polar region, where we can find Inca City. Some wonderful pyramids add to the classical wonders of Mars.

Just for fun, I give a little "face time" to this phenomenon. May I present... a brainy mutant version of *Homo robustus* on Mars. (Though I concede that it might simply be a lower life form, The Mother of All Monstrous Blood-sucking Horrors sure to attack Hollywood.) Inevitably, now that the word is out, the Cydonia Clan will develop amazing new evidence for the origin of hominids on Earth and document the advanced Space Age exploits of our prehistoric ancestors. The

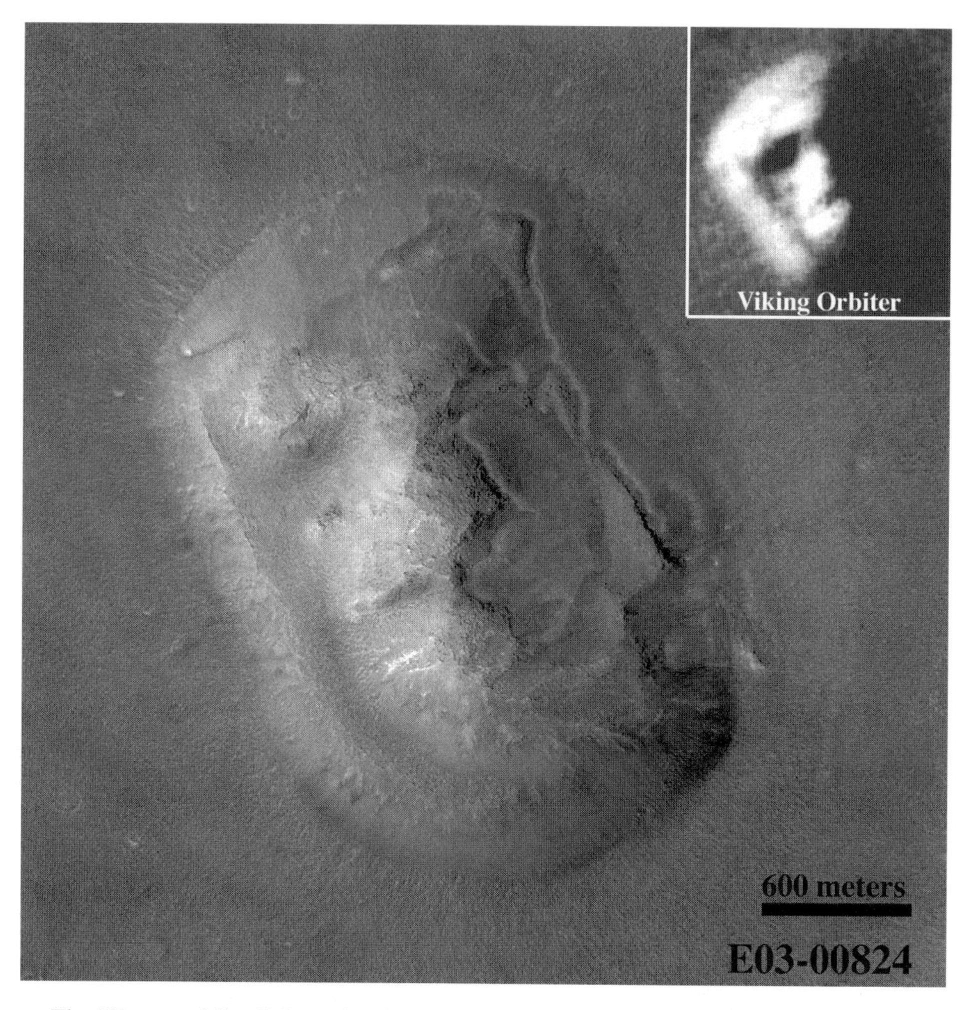

The "Face on Mars," from the Cydonia Mensae region, was imaged in detail by the Mars Global Surveyor spacecraft's MOS instrument. The less detailed view in the inset, upper right, was obtained by the *Viking Orbiter* spacecraft and has stirred decades of rampant, unscientific speculations regarding its supposed origin as a sculpted work of art or industry by a Martian civilization. "The Face", however, is more likely a marvel of geology. Along with hundreds of other similar hills and mesas in this region, it has probably been sculpted naturally with the influence of ground ice; one hypothesis is that the terraces of these features were eroded by an ancient ocean. When the MOC scene is rendered with excessive contrast and then projected into an oblique view, it looks like the *Viking Orbiter* "face".

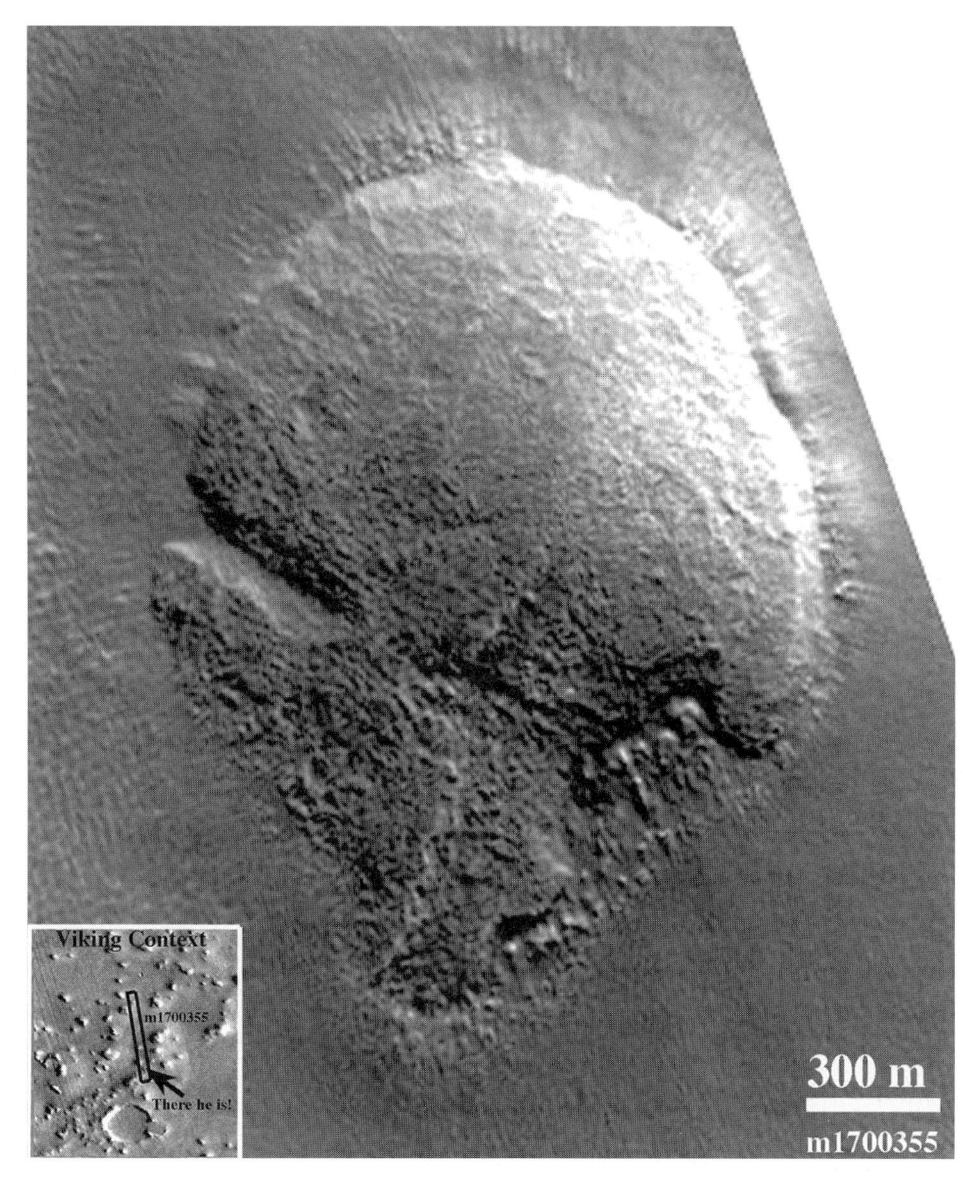

Here's one for the Cydonia Clanists: Erebus Montes, Mars. Alien Roswell Bobblehead? *Homo robustus*? Insect head? Mother of all blood-sucking monstrosities? I could favor any Erebus Monster, but it seems more likely to be a diapir, akin to salt domes and mud diapirs on Earth; its emplacement may be due to thermal instabilities developed in ancient oceanic or mud ocean sediments. Alternatively, it could be an erosional remnant of a deep, ancient crustal structure protruding through the northern plains. "Hair" consists of structures related to the flow of ice-rich debris.

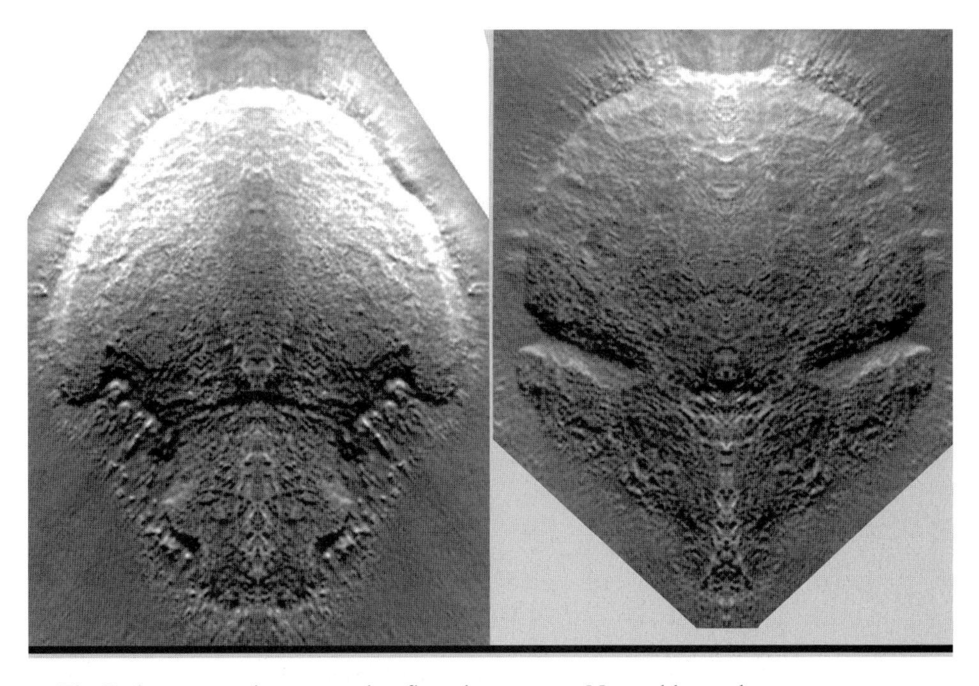

The Erebus mountain-monster has flawed symmetry. No problem: a better monster can be made by taking half of the mountain and copying its mirror image onto itself; by definition, perfect bilateral symmetry emerges. Bilateral and radial symmetry is common in nature; if symmetry is imperfect, humans mentally improve it, as I have done here graphically. A key attribute of all chordates, including vertebrates, is bilateral symmetry. Scary monsters, whether a great white shark or imaginary werewolves, generally are predatory vertebrates or are modeled on them, and therefore are bilaterally symmetric. Monsters' heads – with their mouthparts and the seats of their desires and cunning – are the scariest. (*Left*) "It looks evil!" responded Michelle Minitti, a Mars scientist, when showed this version. Perhaps the evil look is due to the orientations and relative sizes of features, which resemble those of a carnivorous beast's head or skull: large upper mandible and sharp incisors, large nostrils for enhanced aerobic function, eyes for binocular vision, and a robust cranium. (*Right*) This version looks less threatening, a key feature being smaller mouthparts; indeed, it somewhat resembles the Roswell bobblehead alien. Impressions of face and animated being emerge in our minds when presented with such data. Once determined to be an animal, we then assess possible character. When encountering a new face, we immediately evaluate that entity from mere appearance as probable friend or foe or food; if apt to be a foe, we assess its possible physical powers, reasons for being in our presence, and its potential attack strategy, thus enabling defensive tactics to be designed. Wild animal prey and predators must make similar assessments on the basis of sparse data when encountering other animals or humans; mating opportunities, food acquisitions, or survival may hinge on an accurate and timely assessment.

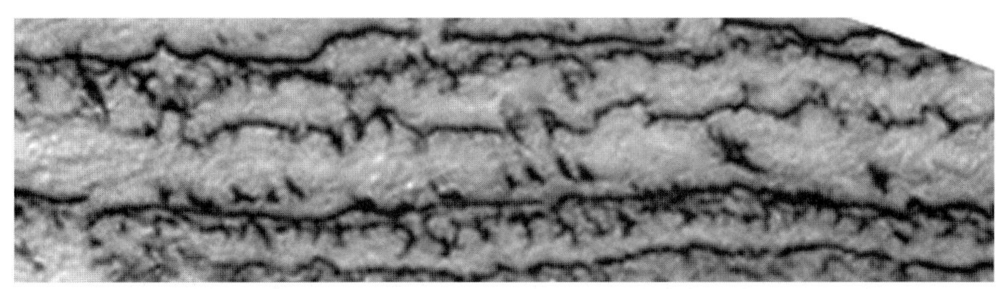

Hieroglyphic writing on Mars? Can you pick out Ra, the hawk-like Sun god, with wings spread wide? These "glyphs" are rewritten each southern spring in the dry ice snows of Mars. MOC M1001487.

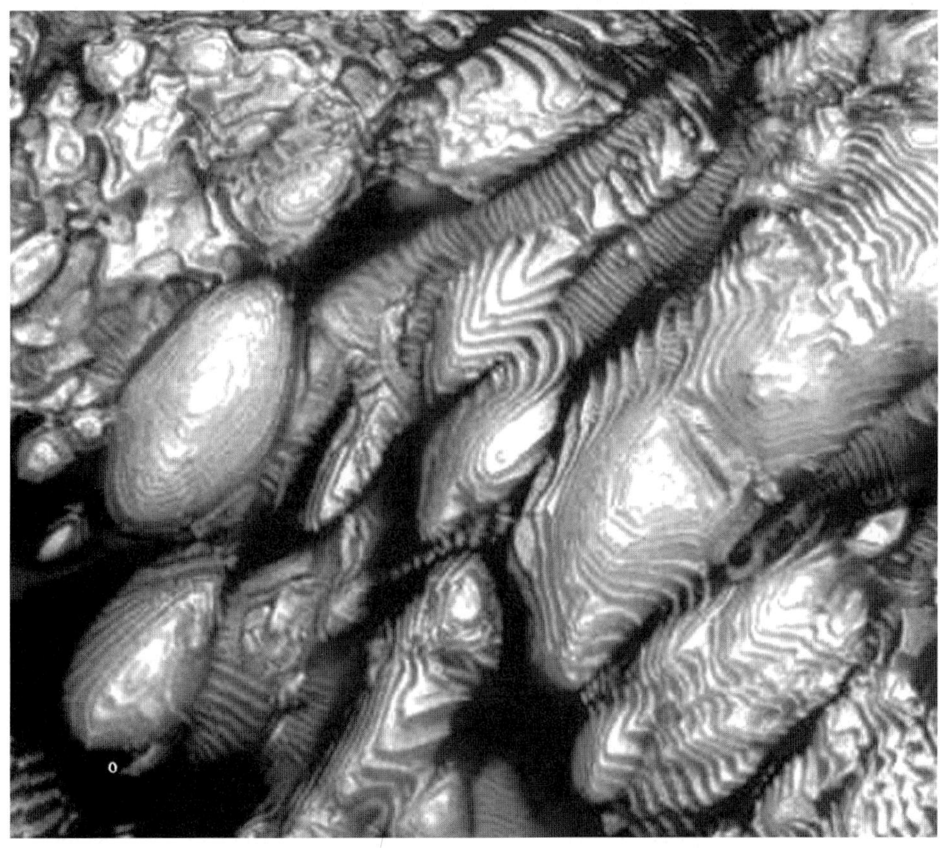

MOC images a herd of zebra on Mars? Not quite! MOC image M1401647 captured a heavily eroded set of uniformly layered rocks, which once formed a continuous sedimentary sheet across the floor of a large, degraded highlands crater near 8°N lat., 7°W long. The scene here is 2,600 m across. The Sun is nearly overhead, so albedo differences are accentuated and shadows muted. The erosion of these layered rocks was due to wind, but their deposition may have been either due to wind or water. Whatever the depositional process and underlying cause, the uniform layering requires climatic modulation.

Raelians, a North-American-based religious group that supposes not only that humans are descended from alien invaders of Earth, but have also recently claimed to have birthed the first human clones, will take heart with this new discovery. Not just that, but did you know that the largest hieroglyphic writing in the Solar System is on Mars? Glyphs are scrawled all around the south polar cap each southern spring. I wonder what mysteries are explained in those glyphs, and I look to the Raelians for the Word. I am sorry to disappoint, but I think the mysteries have more to do with wind speeds and sublimation of volatiles, but never mind me.

If I may wax cyclically serious now, why do the Cydonians see in Martian terrain features evidence of ancient civilizations? Why have I seen some type of face or skull in another mountain? Or a herd of zebra? One reason involves the same psychology that allows us to see sheep and other familiar objects in clouds. The second part of the answer is in Lowell's writings, which the Cydonia Clan takes seriously. We scientists criticize, pointing out that the Cydonians make very selective use of available data, ignoring all the rest. Although the Cydonians' conclusions depart from scientific methods involving rigorous formulation and validation of hypotheses, fundamentally they are engaged in exactly the same search for origins as any scientists. It does not take a psychologist to see that the anthropomorphisms of the Cydonians is a different outlet of the same need to ponder extraterrestrial life and thus to ponder ourselves, as developed in the journal *Astrobiology* and as exhorted by Epicurus 23 centuries ago.

When I drive through Arizona's Sonoran Desert, I see baseball pitchers and families and rootin'-tootin' cowboys in saguaro cacti; in the geomorphic formations of the red rock country near Sedona and Monument Valley I see Pentagons, bells, skyscrapers, and wise monks and nuns, who contend in the desert with fully exposed breasts and erected maleness. One of my personal favorites – I really love her – is Big-Breasted Woman with Long Flowing Hair, a gorgeous sandstone formation in Monument Valley. Upon inquiring with the local Hopi, hoping to discover some mysterious age-old oral tradition of epic wonders – even, perhaps, the root of creation – I learned that at least one Hopi calls her Dolly Parton. My wife, ever the lover of the innocence of babes, sees in Dolly a young woman nurturing an infant. Details aside, she is a remarkable, immense pillar of shapely stone, an actual goddess to behold from the proper frontal vantage. Not to get too excited, from the opposite side, she morphs into the head and trunk of a mighty elephant, a creature otherwise not seen in North America outside zoos and circuses since *Homo sapiens* arrived and did away with the last surviving Ice Age American elephants.

The red rocks near Sedona, Arizona, include a group of wise sages, perhaps monks, engaged in a close personal conference. Perhaps they are discussing my worthiness to enter Heaven or somewhere else in the Great Beyond; or perhaps one of them has a noble plan to save the lone, young woman-with-child in Monument Valley from a life of isolation. We can create stories about these living rocks. New Age culture indeed recognizes a metaphysical power of these rocks, which anyway has a magnetism that draws me there to hike again and again.

One of my favorite hiking destinations just north of Sedona, at the head of Wilson Canyon, offers a 360° spectacle of red sandstone spires and yellow sandstone walls, capped with gray hexagonally columnated lava, like barricades to the mystic lands

We observe and then we create mental images to try to understand. An overzealous forestry industry, though not quite meticulously thorough with tree scraps, laid bare a glacial moraine in northern Ontario. The cleared view allowed for a little science and a little fun. (A) Patrick and Ryan McHenry for scale amid some glacier-faceted boulders; the one they stand on is 7.5 m across. (B) An Iroquois scout. (C) Head of a happy caterpillar. (D) A raging man. The Iroquois scout and the caterpillar survived mere seconds. The raging man is geologically ephemeral; anyway, he exists transiently only when the lighting is right and one is neither too near nor too far. If his fleeting fury is not enough to amaze, you don't need to try too hard and he'll morph instantly into the head and front leg of a stalking mountain lion, with ears cocked forward. The features of the raging man are post-glacial effects of weathering of what had been a smooth, polished surface. We apply rational thought and try to sort through the various mental images we have created, and then decide what is plausible versus what is apt to be a fabrication. Each interpretation, both the plausible and improbable, are products of our creative imaginations rooted in personal experiences.

"Big-Breasted Woman" (a.k.a. "Dolly Parton"): a natural sandstone monument in Monument Valley, Arizona. (*Left*) The perspective of the camera offers a fine front-oblique view. (*Middle*) Profile view shows her long hair and her Mexican dress for a gala event. (*Right*) Back view becomes the head-on view of an elephant!

beyond. Just as you reach a point where surely, you think, no one has gone before, there is a quaint glass jewelry box filled with incense, fused into the rocks to please all who may travel that holy ground. Also available is a very modest engraving in the sandstone, a New Age tip of the hat and offering of the heart to the powers that created these wonders. Certainly, my wife and I both agree that this is a place to exercise one's body and mind and to learn, with or without smouldering aids to greater consciousness of self and Universe; personally, the multihued rocks and blue sky and green cloak of life are enough for me.

Religion or spirituality is as fine a way to respect the wonders of Arizona – or of any place – as any scientific approach. It is better in fact, as it leaves a fundamental, infallible Truth exposed in all its glory, without getting caught up in minutia: the Cosmos is just splendid! That is the real essence of it. But when contemplating technical details of origins, science offers a better approach, as it is based thoroughly in observation, systematic analysis, and logical deduction.

Geologic culture recognizes the rock "formations" of northern Arizona as a part of the region's "layer cake geology," as early explorer/geologist Powell described it. These rocks tell a fascinating story of the advance and retreat of seas and muddy river deltas and sandy deserts, and finally the influences of uplift and erosion to create myriad bizarre rock forms somewhat similar to what we observe on Mars in many places, including the neighborhood of the Face. The amazing natural

Spires, mesas, and buttes of Monument Valley, Arizona, owe their existence to a combination of the recent climate and weathering regime and to a succession of ancient climates during the late Permian and Triassic periods (roughly 250–200 Ma). First, a thinly bedded sequence of soft, shaly mudstones was deposited in shallow-water deltaic conditions during the erosion of the Ancestral Rocky Mountains; this is the Organ Rock Shale, visible as the sloping ground at the base of the steep cliff. Its recent erosion has undermined more resilient cliff-forming de Chelly Sandstone, which is primarily made of cemented eolian dunes. After the cliff-forming sands were deposited, delta-like conditions returned, resulting in the thinly-bedded sedimentary Moenkopi Formation (visible near the top of the mesa in the background), deposited in streams and shallow lakes. Finally, an erosion-resistant cemented gravel bed (Shinarump Conglomerate) was deposited. A replication of this sequence is not necessary to form mesas and buttes on Earth or Mars, but it is necessary to have shifting conditions such that rocks of differing resistance to erosion are deposited and then subjected to differential erosion. A resilient capstone is necessary, but in place of a well-cemented conglomerate, lava beds or ice-cemented permafrost could work. Erosionally weak beds could consist of thawed permafrost. Deposition in oceans or lakes is not necessary to set up these conditions, but it is an ideal situation to do so.

The four main rock formations in Monument Valley are shown well here. Of course, these sedimentary layers once extended unbroken across this whole area. First, a layered plateau was dissected by canyons; canyons widened and coalesced, thus isolating mesas; then mesas narrowed into buttes and spires. A spire's lifetime is short, but lengthened somewhat by the small area that collects rainfall and snow.

A lesson in arid lands erosion, Cedar Breaks National Monument and Bryce Canyon National Park, Utah. (A) Erosional gullies in Cedar Breaks are best developed in one lithologic unit in preference to the units above and below. The gullies form in an incompletely consolidated, fine-grained, easily eroded unit "too soft to form uniform cliffs and too hard to form uniform slopes," according to W.L. Stokes (1986, *Geology of Utah*, Utah Museum of Natural History and Utah Geological and Mineral Survey). A slight cementation promotes channelized water runoff, but the distance scales do not allow development of tributary structure. Short, equally spaced gullies and fin-like interfluves form instead, with by far most of the geomorphic work probably done during especially intense storms once every few years. While preferential seepage at the top of the gullied unit cannot be ruled out, that appears to have no role here. It is simply a difference in lithology and runoff characteristics. This is also likely true on Mars where we see similar forms. (B, C) Erosional remnants can form spectacular "hoodoos," goblin-like towers and pinnacles, such as these formed in the pink Wasatch Formation of Bryce Canyon and Cedar Breaks. In B note that the hoodoos form above and below a rock unit that is gullied, but not so cliff-like as in A. The Wasatch Formation is a sedimentologically varied unit mainly of interbedded sandstone, shale, limy siltstone, and limestone, all deposited in a vast lake of Eocene age.

The progressive development of hoodoos is seen very well in this cliff near Canyonlands National Park, Utah. This is how hoodoo grads are made. It is a process not too unlike that which forms human academic graduates: Start with unrefined masses. Through turmoil and grief, and through exploitation of both strengths and weaknesses (such as their desire to eat), break them down until they stand on their own, facing the world long enough to turn old and gray and crumbly.

sculptures are made of rocks that once stretched in continuous layers, but were dissected grain by grain. Indeed, the monuments and formations of the sandstone terrains of the "Four Corners" (the corners of Arizona, New Mexico, Colorado, and Utah) were made to please by the spatially selective removal of a number of sand grains roughly equal to Avogadro's Number (6×10^{23}, the number of atoms in 12 grams of carbon or of molecules in 22.4 liters of gas at standard temperature and pressure). It would probably take smouldering aids and the music of Thomas Dolby (of "She Blinded Me with Science" fame) to discern the significance of that. Sometimes it's better to marvel in ignorance and not ask any questions.

To make a face

The marvelous rock formations of Arizona's and Utah's Colorado Plateau have many names of humans and of our monumental artifacts. Buddha Temple, Zoroaster Temple, the Three Monks, Dolly Parton, and so on. The complex beauty and form of these structures is derived by erosion from initially quasi-horizontal and laterally continuous beds of sandstone, siltstone, mudstone, limestone, and other sedimentary rocks. A chief attribute of the rocks that makes them susceptible to this type of erosion is a composition of beds of alternating resistance and susceptibility to erosion; absolutely required is a "caprock" that is resistant to erosion, but which can be undermined by erosion of soft or otherwise vulnerable rocks beneath. A chief contributing attribute of the environment is that the climate is arid to semi-arid – rainfall and snowmelt percolates into the rocks, rather than immediately running off the surfaces and gouging the landscape with many river valleys. Freeze-thaw cycles involving expansion of water upon freezing is a primary means by which rocks in temperate and polar regions are progressively wedged apart along fractures, dissected, and sculpted. The aquifer being suitably charged, spring water then variously dribbles or gushes out where highly permeable conduits, such as fractures and lithologic continuities, exist. Typically, this spring discharge occurs beneath massive, slightly permeable rock formations. The spring-fed streams meander downstream, cutting into the rocks. The resulting erosion removes easily weathered rock beneath the massive layers, and so there is undermining. The resistant strata form steep cliffs of mesas and buttes. Eroded sand and other debris are washed downstream or are cleared out by wind. The headward growth of amphitheater-shaped spring-fed sources of these rivers is a process called "sapping" – described in further detail in Chapter 3. Eventually, as progressive sapping and spring-fed streams and wind widens eroded niches and narrows the remnant mesas, the mesas can be wasted into slender spires and bridges and arches and other surreal forms. Our brains want to see these forms as artificial edifices and human figures.

Mesas on Mars have apparently been formed by similar processes, though ice-assisted erosional processes appear to be more prevalent in areas such as Cydonia than water-runoff processes. The result: mountains and hills that can resemble familiar objects. *What are those hieroglyphics on Mars?* They are wind-streaked dust, sand, and frost deposits; the "script" is rewritten with the seasons and shifting dunes. *What is the famous Face?* It has no such resemblance when the Sun angle changes and resolution increases, but it presents an impressive display of structural control of erosion.

MOONSTRUCK BY THE *MARINER* MARS FLYBYS

The astrobiological and water focus of Mars studies since the late 1990s is the new fashion. It represents a complete shift away from the prevalent scientific thought early in spaceborne Mars exploration. There are of course scientific, objective reasons for such shifts. It may help to step back nearly four decades. The first spaceborne views of Mars, from *Mariner 4* in 1965, were transmitted at an agonizing 8 bytes per second. The first images were reconstructed initially on paper using pencil shadings, according to my former PhD academic adviser, Robert Strom (University of Arizona). Those original historic penciled images exist on the top floor of the JPL Administration building, according to Strom. The initial news accounts were "Mariner Transmits First Mars Picture" (*Minneapolis Star-Tribune*, July 16, 1965)

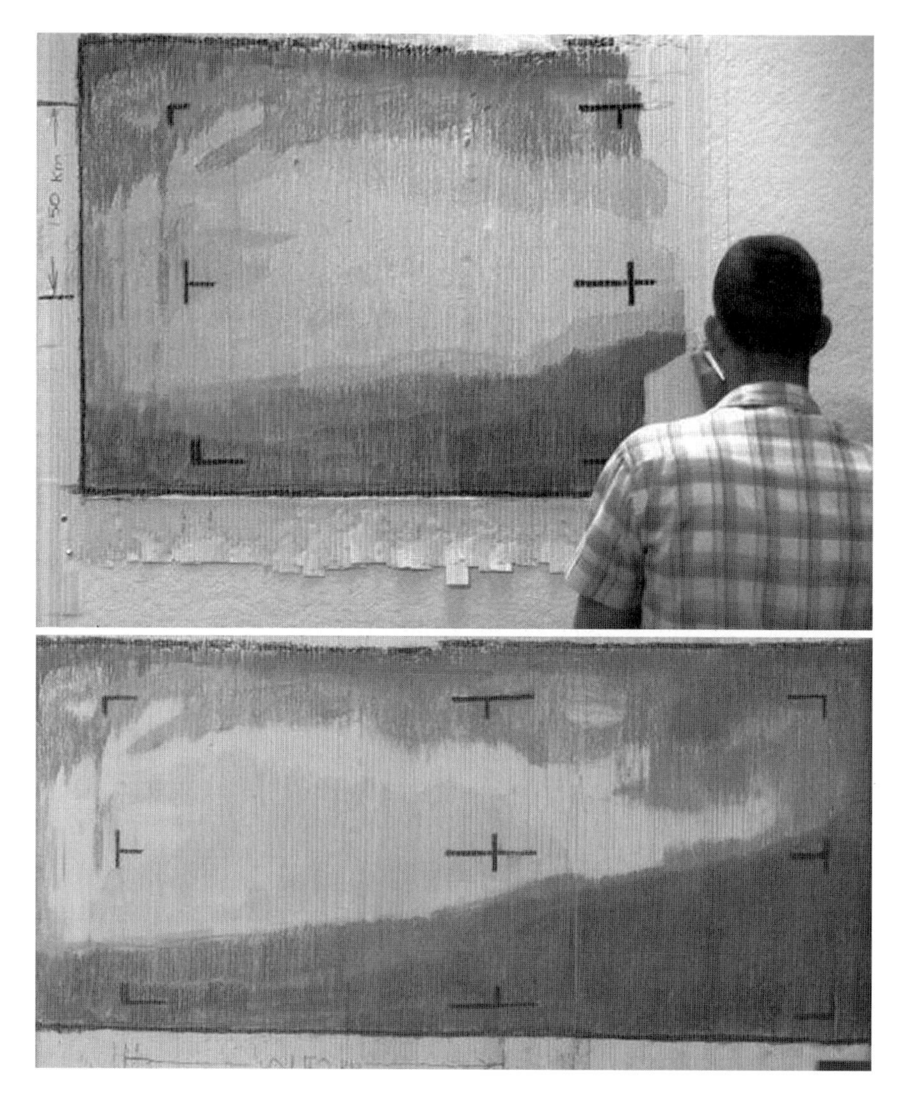

"Rocket science" provides Earthlings' first spacecraft glimpse of Mars. The first *Mariner 4* image of Mars was produced by hand coloring, pixel at a time, in shades of yellow, brown, and red (good, respectable Mars colors; here reproduced in gray scale). Although the spacecraft telemetry came to Earth at the usual speed of light (and 8 bytes per second), the ability to generate hard copy at that time lagged the computers' and printers' ability to spit out 20-foot rolls of digital printout. NASA, still fairly new to the planetary probe business, also had a challenge finding the right balance between a principal investigator's proprietary rights and the public's and researchers' needs for data access. In resolving this issue, Planetary Geology Program chief Steven Dwornik established the national (now international) system of distributed data centers (the Planetary Data Facilities). Photos provided courtesy of David Senske and Torrence Johnson (JPL).

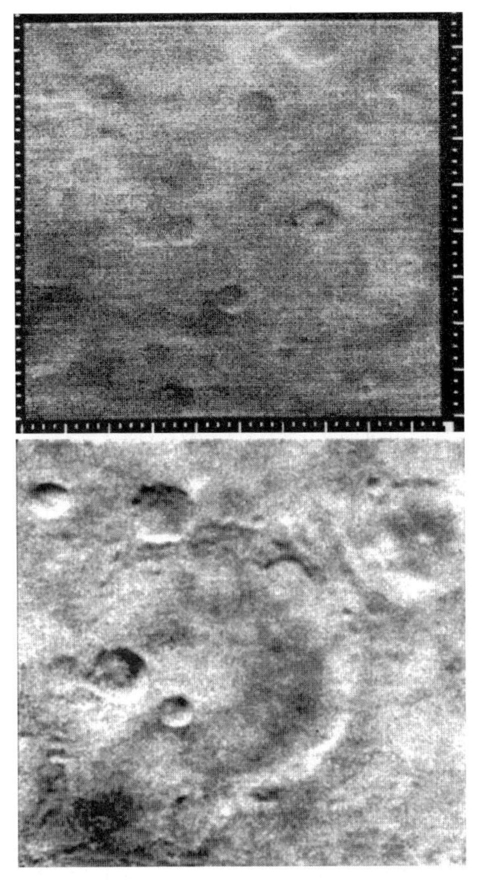

Mariner 4 produced 22 images showing craters and little else that was interpretable. The large, degraded crater dominating the lower image is now named Mariner Crater, which spans about 155 km in Terra Sirenum.

without the picture, because it took so long to downlink, process, and render the data into hard copy images. An AP photo that accompanied my grandmother's news clipping of that article shows two engineers in a staged visual scrutiny of digits on a 20-foot ream of computer paper, with computers in the background. They didn't show the engineers with pencil and paper in hand, creating the first up-close image of Mars. It took 4 days for *Mariner 4* to transmit 22 images of 200 × 200 pixels each. The Mars spacecraft now in orbit are sending as much data (compressed) every 10 seconds, on average, as the entire *Mariner 4* encounter with Mars produced. The *Mariner* pictures, once produced properly, showed fuzzy views of nothing but an impact cratered terrain. In fact, the craters seen in *Mariner 4* images were quickly recognized as having modified (weather beaten) forms relative to what was seen on the Moon, but this important fact was almost lost amid the more basic recognition of a Moonlike cratered landscape. A radio-occultation experiment developed after launch successfully determined that atmospheric pressure was in the range of 4 to 7 mbar. Telescopic

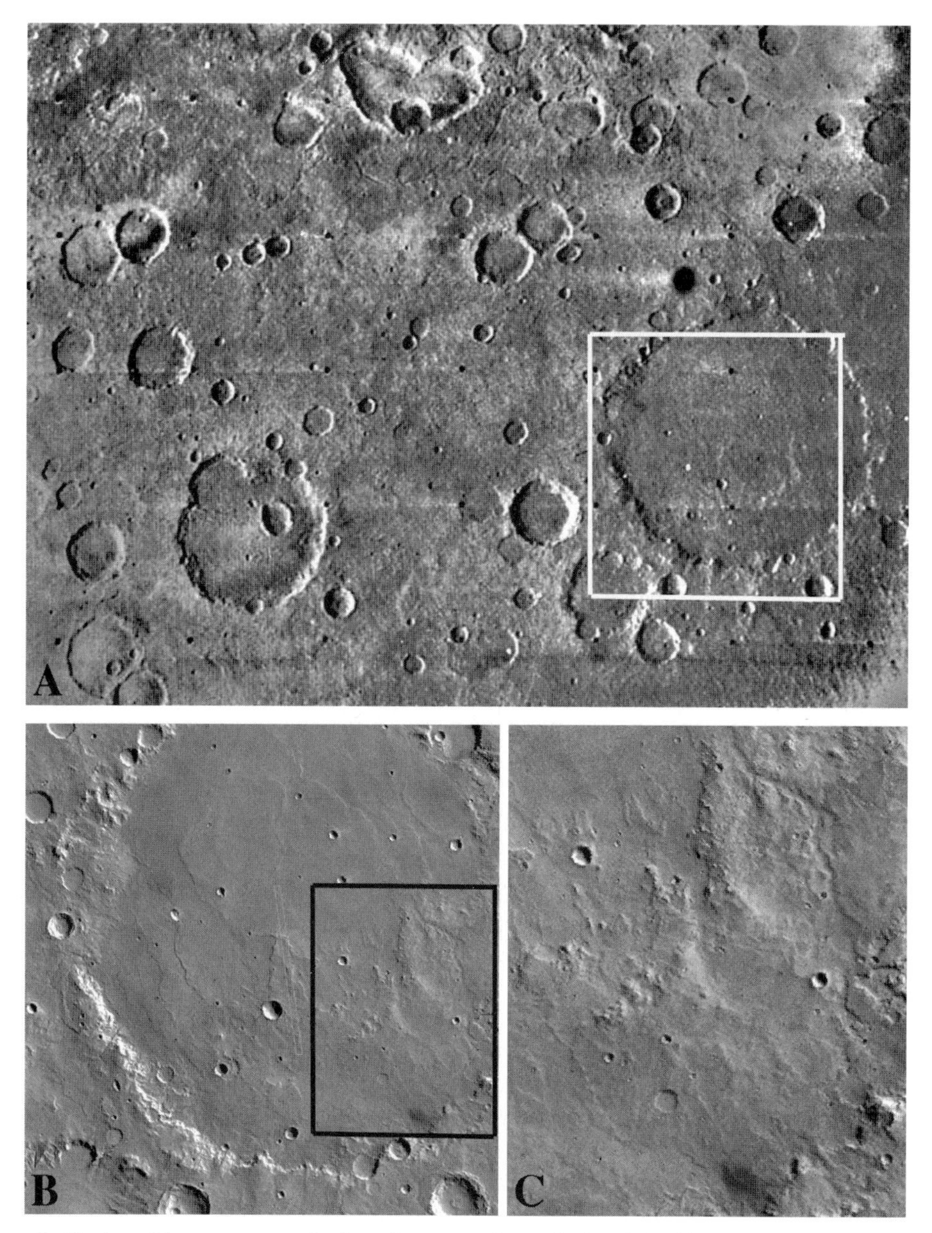

A *Mariner 6* image captured a heavily cratered terrain in 1969 (A). The large crater at right, Flaugergues, is ~220 km across. It lacks an appreciable raised rim and has a flat floor, as is true of many other craters in this scene. We now know that water and/or ice is responsible for much of the terrain degradation, but at the time of the *Mariner* flybys, the Moon-like appearance of Mars was emphasized. Volcanism also was probably an important process in this region according to interpretations of this much better *Viking Orbiter* scene (B and C). Panel C reveals badlands-like erosion.

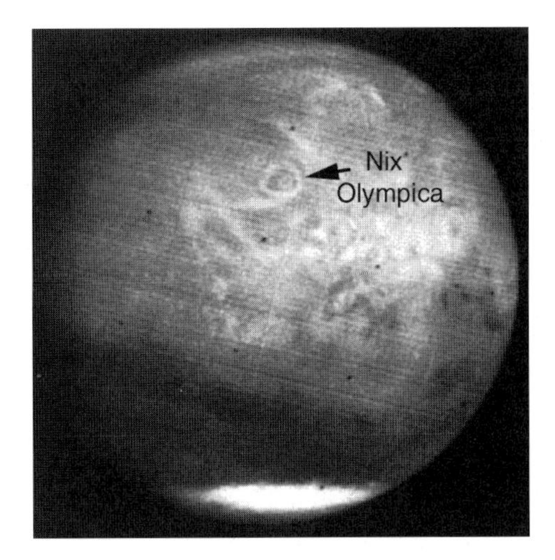

Mariner 7 global approach image. "Nix Olympica" is now known as Olympus Mons. At that time, scientists knew that Mars has craters; so Nix Olympica was confidently interpreted as a giant crater. Note the south polar cap.

observations of Mars had, for the most part, been superseded. Our vivid imaginations based on almost no data were now being replaced by vivid imaginations based on hard data.

Spaceborne exploration of Mars advanced considerably with the *Mariner 6* and *7* missions in 1969, but the discoveries mainly reinforced the slanted perspective of Mars gained from *Mariner 4* owing to coverage of similar terrain types. To understand the magnitude of the shifting science fashion since that era, let's consider just where we were 35 years ago. Everyone on Earth knew about "Mankind's Giant Leap" with Neil Armstrong's small step. Nothing else compares to that, but other things happened that summer. There was the discordance of Woodstock's rock, beating tune after tune against the chop-chop of Hueys over Vietnam's green jungles. Not far from Woodstock and just two months before, the Cuyahoga River was so polluted that it caught fire. Our's was a world in need of serious change.

After the global acclaim for *Apollo 11* had faded somewhat, and while a Massachusetts Senator found himself in trouble as deep as the water in which his secret girlfriend's body was discovered, other momentous events truly "out of sight" went almost unnoticed amid the dogdays of August. The first of Neil Armstrong's and Buzz Aldrin's lunar rock boxes had just been opened, soon to tell an otherworldly story of an orb ripped from mother Earth and subjected to unimaginable impact bombardment and the floatings and sinkings of mineral phases in a global magma ocean. That same week, *Mariners 6* and *7* approached and then whizzed past Mars, snapping fuzzy pictures like tourists on a bus. Over 200 grainy images and other observations shattered once and for all Lowell's creative but ill-founded hypothesis that the Red Planet is the home of hardy plants and industrious beings coping with deteriorating conditions. *Mariner 7*'s trickle of bits and bytes beat senseless any possible similarity of conditions and critters on Mars with those on Earth. Our planet now seemed to be the only, lonely land of the living as far as we could see. On Planet TV, Captain Kirk and *Enterprise* Science Officer Spock explored strange, new worlds and confronted colorful alien civilizations light years and centuries away. The reality in our Solar System was that Outer Space always seemed to come in various shades of bleak. In those black and white images, any remaining lure of the Red Planet was simply that it seemed so God forsaken and un-Earthly. Mars, along with new knowledge of the Moon, highlighted our planet's irreplaceable qualities. Earth became what Carl Sagan would later call the Pale Blue Dot, a small gleaming jewel amid the cruel, cold darkness of outer space.

The new *Mariner* images showed no canals on Mars, and infrared sensors gave added proof that it is a cryogenic world utterly lacking liquid surface water and possessing a tenuous atmosphere only 1% as dense as Earth's. Our modern understanding of severe current-era Martian conditions was established, but the hasty observations of the *Mariner* flyby probes gave a misleading snapshot of the geologic and climatic record of the past, since all of them had passed over cratered terrains of the southern hemisphere. The cratered Martian surface appeared almost as dry and as geologically lifeless as the Moon seen through *Apollo*'s stark images of bone-dry, dusty-gray lunar rocks – "Little Hope for Life," the headline in the *Columbus Evening Dispatch* declared following the *Mariner 7* flyby. Here on Earth, it was a time of broadly based perceptions of hopelessness, where even the most stunning achievements in human history were as though sketched in charcoal on a parchment of cultural and environmental gloom. The fact that we were seeing for the first time – and seeing as a global civilization – vast landscapes on a planet millions of miles away had little impact.

I have a confession from that time in my childhood, my summer between Fifth and Sixth Grades. *Mariner* data in 1969 were personally disappointing. Though I never actually anticipated finding "My favorite Martian," I had expected something green or some other hopeful clue attesting to a quasi-Earthly past and something possibly wondrous for our future. Instead, the curtains seemed to close even on lowly lichens, which had been the astronomers' preferred biological analog according to the books of my childhood. As events unfolded, that was all the better to set the stage for an exhilarating revolution in planetary science brought to us by the 1970s *Mariner 9* orbital mission and the twin sets of *Viking* landers and orbiters. That exciting era of discovery also had a disappointing start with the plunge of *Mariner 8* into the Atlantic Ocean, and then a global dust storm that enshrouded the Red Planet as her sister probe, *Mariner 9*, finally arrived on target. Disappointment was brief, but it reinforced the 1960s view of a cold, dry Mars and added "windy" to its harsh accolades.

THE FOUR WINDS BLOW

On November 13, 1971, after nearly a 6-month interplanetary cruise, *Mariner 9*'s rockets fired at just the right moment, with just the right thrust, for exactly the required period, and with the correct spacecraft orientation, only minutes before she

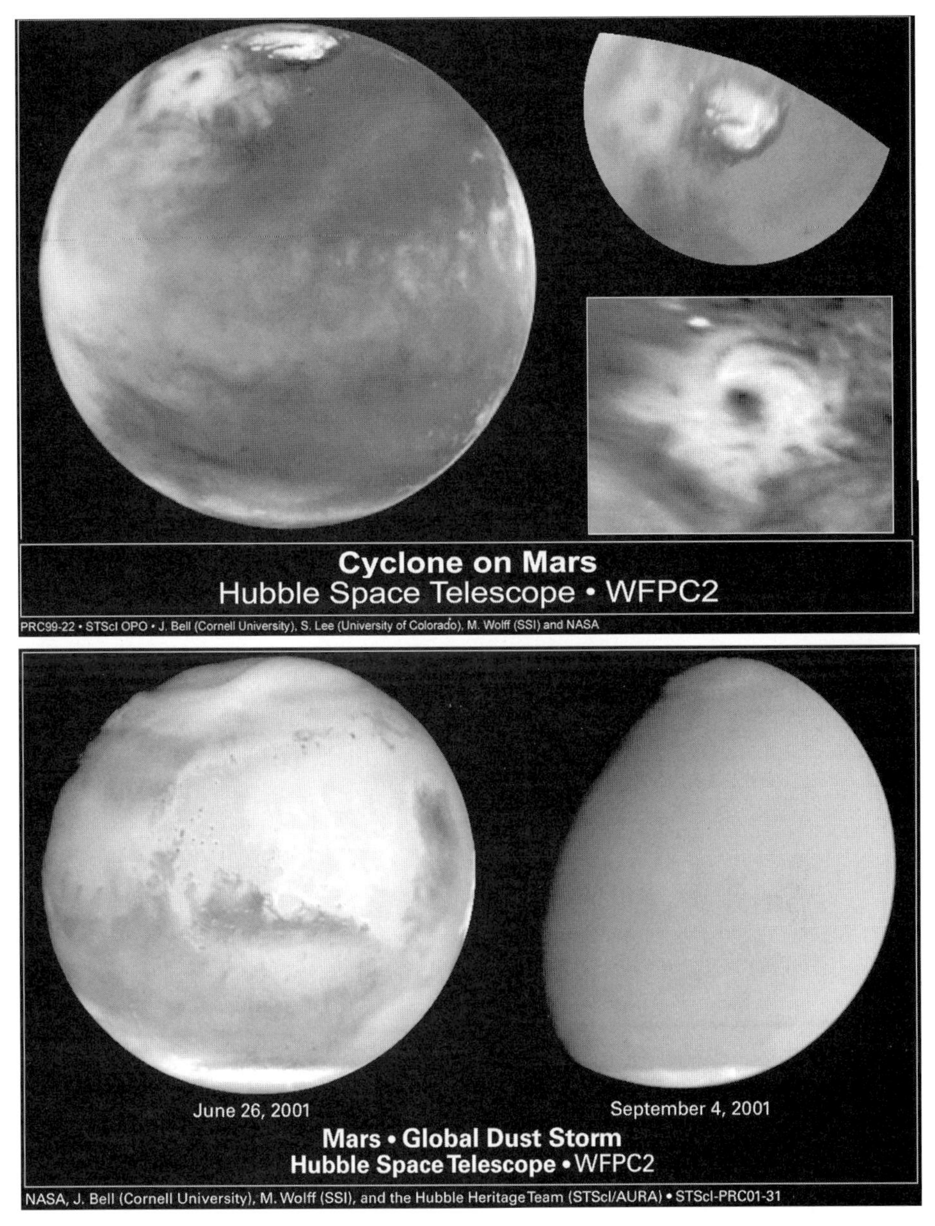

Basically the equivalent of a spy satellite pointed toward the heavens, the *Hubble*'s optics and sensors can monitor the weather on Mars. In the top image (together with its reprojections to different views) a cyclone, 1700 km across, spins water-ice clouds near the north polar cap. In the bottom pair of images, a huge seasonal weather disturbance builds in the June 26, 2001 scene but the rest of the planet is clear; by September 4 a global dust storm had enshrouded the planet.

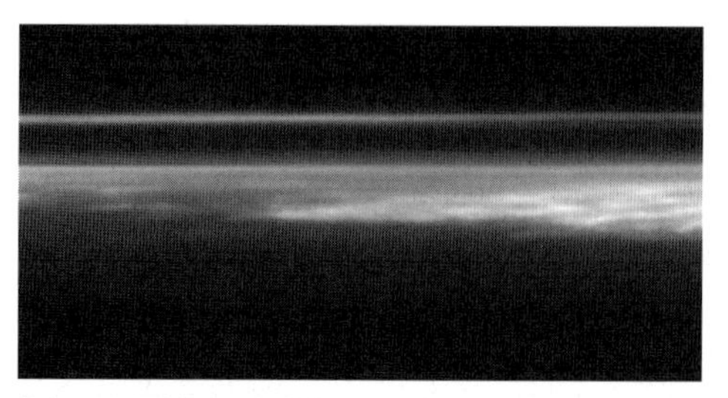

Mars at the crack of dawn. This MOC limb shot of the morning sky catches two cloud decks near 55°N lat. NASA/JPL/MSSS press release MOC2-380.

would have otherwise passed Mars on a hyperbolic trajectory. To have reached the correct orbit around another planet was a remarkable first in the annals of planetary exploration. Just 14 years earlier, struggling to catch up with the Soviet Union's achievements, the US had failed to loft one-thousandth as much mass into an Earth orbit 1/500,000 as far from home as *Mariner 9* was. Now, three decades after *Mariner 9*, attaining an extraterrestrial planetary orbit still remains one of the more difficult challenges in planetary missions.

Mariner 9 had been built precisely to the correct specifications (unlike *Hubble*'s primary mirror, which was fabricated with exacting precision to the wrong specifications). The spacecraft did not explode upon orbit insertion (unlike *Mars Observer* in 1992). There was no human error in converting English to metric engineering units (as doomed *Mars Climatology Orbiter*). Solar panels and communications hardware deployed exactly as they were designed (unlike the handicapping effect of *Mars Global Surveyor*'s broken solar panel, which forced an unplanned two-year long struggle to achieve the correct orbit). Communications gear worked as planned, unlike *Galileo*'s incompletely deployed high-gain antenna (which rendered that spacecraft almost mute). And there had been no penny-pinching engineering effort (as may have contributed to the mysterious loss of *Mars Polar Lander* and her *Deep Space* impact probes). Just in case Murphy's Law demanded redundancy, two spacecraft were built for a mission that one could perform – thankfully so, considering the fate of *Mariner 8*.

Although much was not right with the world, those were the days when scientists and engineers dreamed without limits, and budgeteers and taxpayers gave what was needed to make those dreams a reality. *Mariner 9*'s orbit insertion was a dramatic technological feat for rocket science, but planetary science, which had waited during the long interplanetary voyage, had to wait a bit longer. Two months before reaching orbit, astronomers had reported the spread of a globe-girdling dust storm at Mars. By the time *Mariner 9* began to snap pictures there was nothing to snap but the bland tops of yellow dust clouds. Surface winds up to at least 50 m/s (about 112 mph) – probably including a vast swarm of dust devils created by a huge cyclonic vortex – had lofted more than a billion tons of fine dust, which then circulated globally at altitudes up to 50 km (31 miles). Any wild landscapes and geologic processes were hidden beneath dust clouds. Global dust storms occur every few years on Mars, but this one was especially severe. Engineers had left little to chance with *Mariner 9*

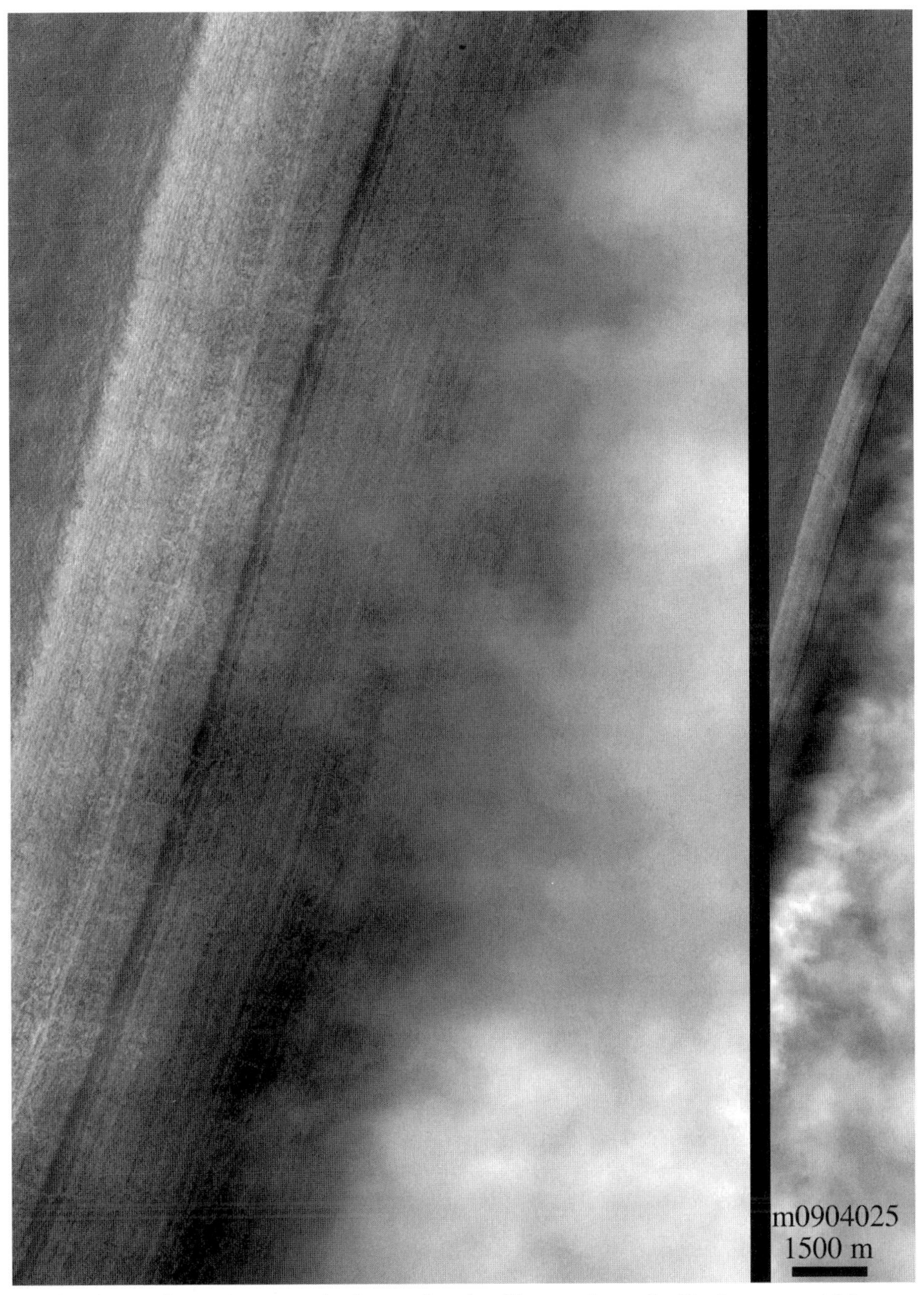

Clouds near layered outcrop in the south polar Chasma Australe. During a recent Mars polar conference there was open debate about whether Martian clouds are convective. They are turbulent at times, and probably convective. This scene is very late southern spring. These clouds may be late-season dry ice clouds.

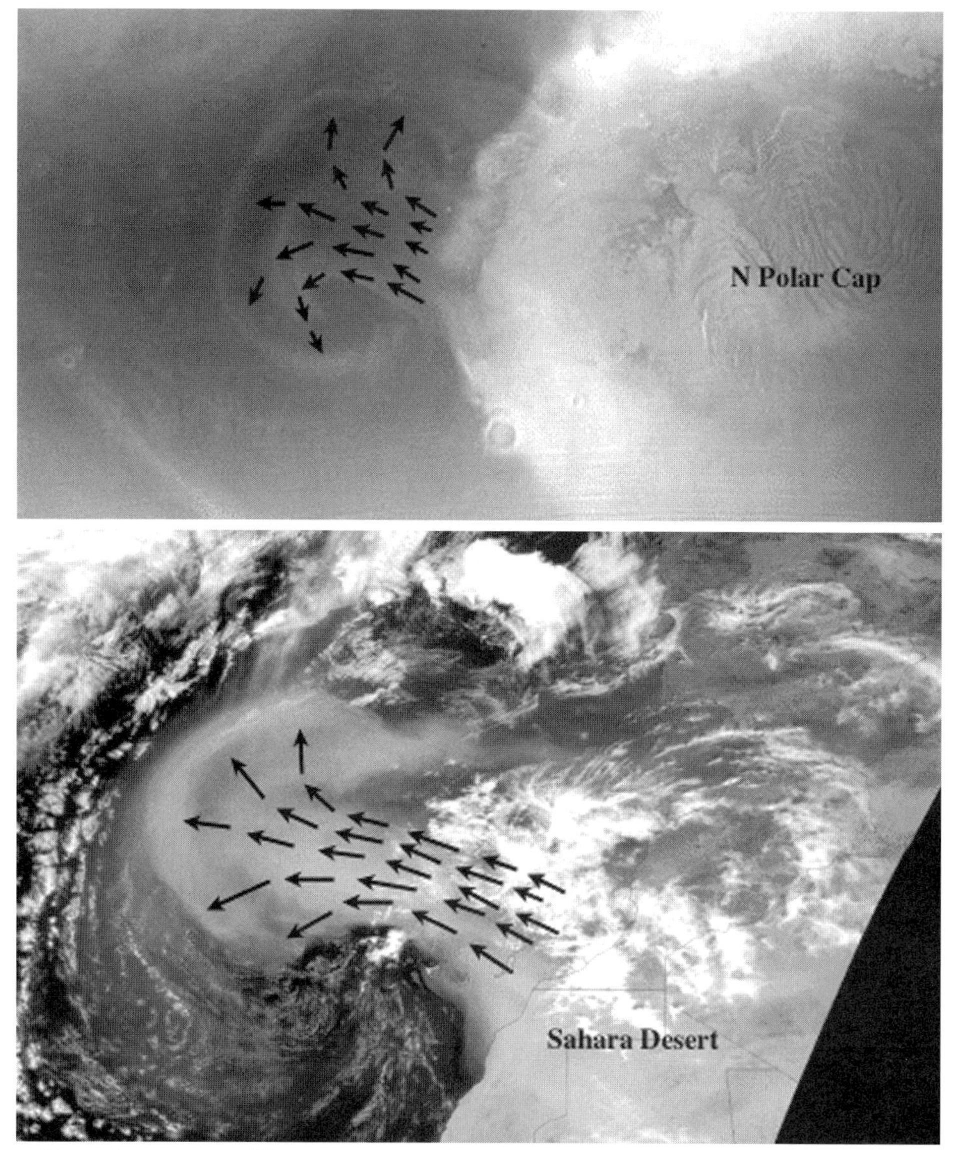

Large dust storms afflict Mars (*top*) and Earth (*bottom*). (A) This spring MOC image captures a dust plume being shed from the sublimating seasonal carbon dioxide cap. Sublimation produces higher pressure, generates sublimation winds, and produces large thermal contrasts between frosted and defrosted ground, and these thermal contrasts produce pressure gradients and drive convective winds. Also, as those experienced in large terrestrial glaciers and ice sheets know, katabatic slope winds can be ferocious. The Martian storm extends 900 km from the edge of the frosted terrain. A similar dust storm, 1,800 km long, over the Atlantic Ocean was shed from the Sahara Desert, as captured in a Sea WiFS satellite image. Credits: NASA/GSFC/ORBIMAGE/SeaWiFS project; and NASA/JPL/MSSS. MSSS release 2-249.

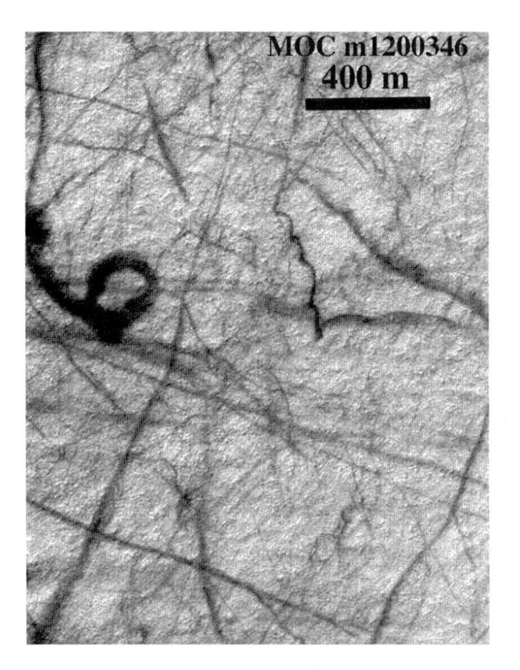

Dust-devil streaks play out over southern Promethei Terra at the height of the southern summer. –69.4°S lat., 250.7°W long.

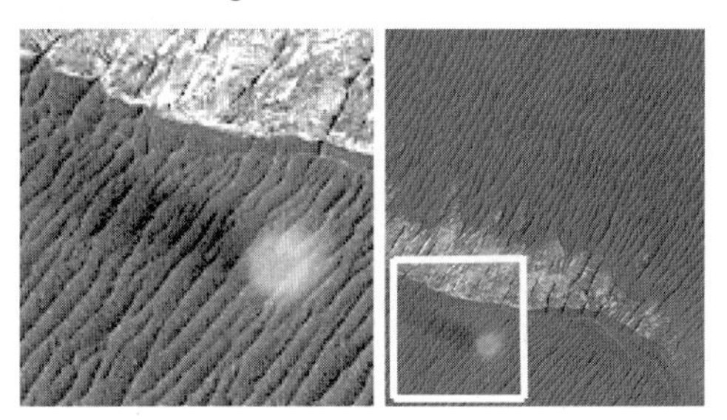

A dust devil, the fuzzy white blob, was caught passing across sand dunes on the floor of Melas Chasma. The dust devil is about 140 m across; from the length of its faint shadow and the solar incidence, its height is determined to be about 600 m. It is not quite optically thick, since the underlying dunes are faintly visible through the dust and its shadow is not as stark as the shadows of nearby cliffs. The total mass of lofted dust is around 100 kg. MOC image M0301869.

except for what could not be controlled – Mother Nature. Fortunately, *Mariner 9* outlived the dust storm, which eventually cleared bit by hazy bit over the ensuing several months.

Most types of geologic activity occurs either extremely slowly or with such patchy irregularity and infrequency that cumulative effects of that activity must be observed in the geologic record and in the landscape, with the activity itself commonly not discerned in a typical human lifetime. Phenomena witnessed in action and with time

Dust devil races across the Navajo Reservation, USA. This dust devil is only 10 m or so across and around 30 m high, a fairly typical size. Though puny in these measures compared to much bigger Martian dust devils, this small one may contain as much dust mass, because the lofted particles are much larger than what the Martian winds can raise.

constants for significant dynamic change on the order of a human work shift to a human lifetime typically attract our attention. Earthbound observers witnessed the development of the 1971 Martian dust storm and, with *Mariner 9* in orbit, monitored its global-scale dynamics. Hence, with the 1960s *Mariner* flybys, and again with *Mariner 9*, we measured low water vapor partial pressures attending low temperatures, and in 1971 we were shown a close-up view of a windy, dust-shrouded Mars to complement the craters seen on earlier missions. It was like a dusty, frozen moonscape with added wind. When the dust storm cleared, much of what we observed were dust streaks and sand dunes. Wind seemed to be the story of Mars. Mars was cold, dry, and windy almost beyond human imagination. Geologists then applied the classical geological principle of "Uniformitarianism" – that the present is the key to the past – to conclude that Mars has always been more or less as it is today.

As we have learned more about Mars in three decades since *Mariner 9*, we have found ample support for the old idea that wind is, and for eons has been, an extremely important geologic agent on Mars. Fresh dust mantles, eroded dust sheets, sand dunes, and wind-scoured rocks abound on the surface of the Red Planet. Its red color is partly due to red dust that coats almost everything. Red dusts even paints the sky. Whenever a spacecraft visits Mars, it always finds ample new evidence that dust and sand is an environmental reality on that world. The *Mars Pathfinder* spacecraft observed the local passage of dust devils, and it saw the dust pulled from thin air by its magnets. *Mars Global Surveyor*, from orbit, has imaged dust devils in action, their myriad swirling traces in some regions, and vast sand seas elsewhere. Mars is indeed a world where everyday processes have wind at the top of the list of what's happening now.

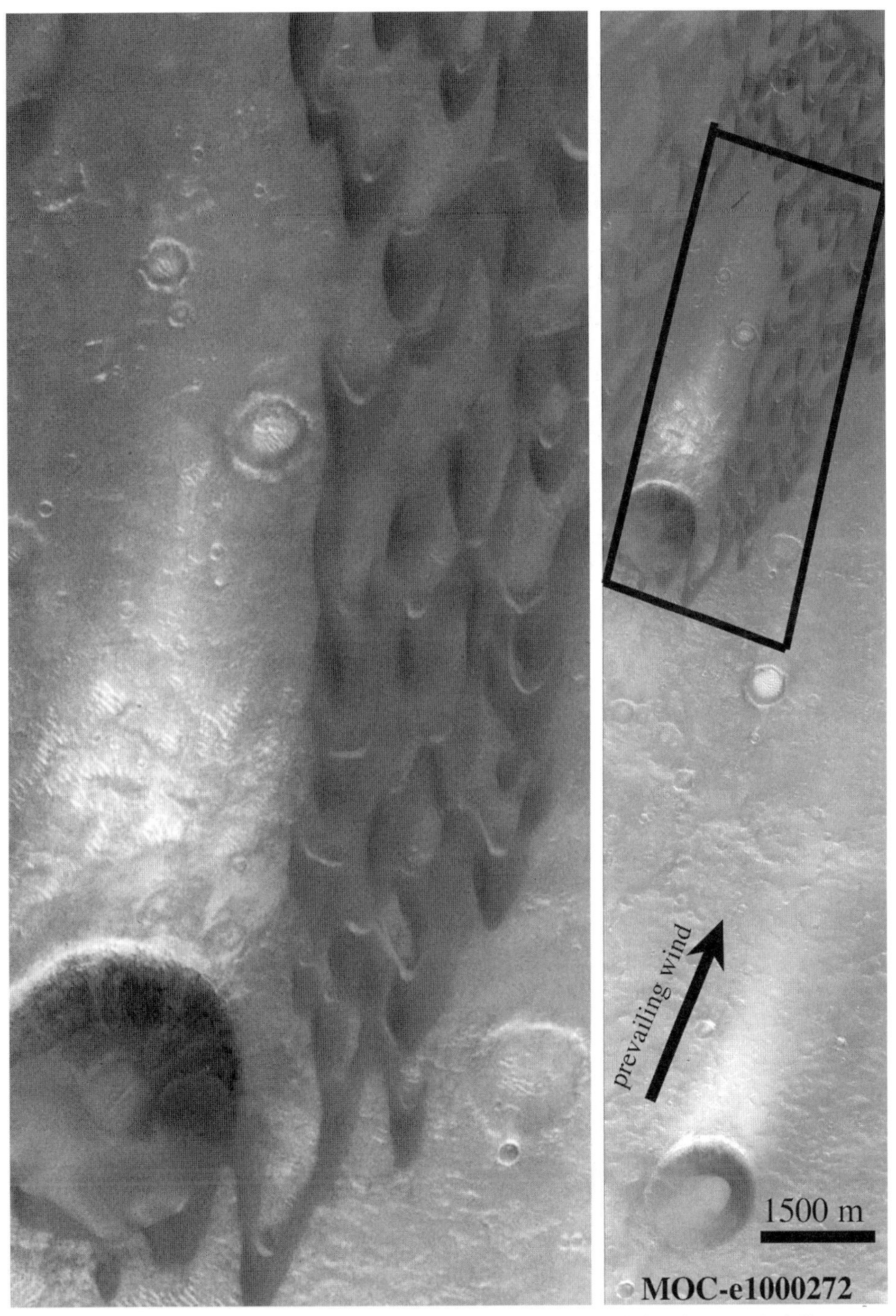

A small field of dark barchan dunes and bright wind-streaked dust in Herschel Crater, Mars. –15.3°S lat., 228.9°W long.

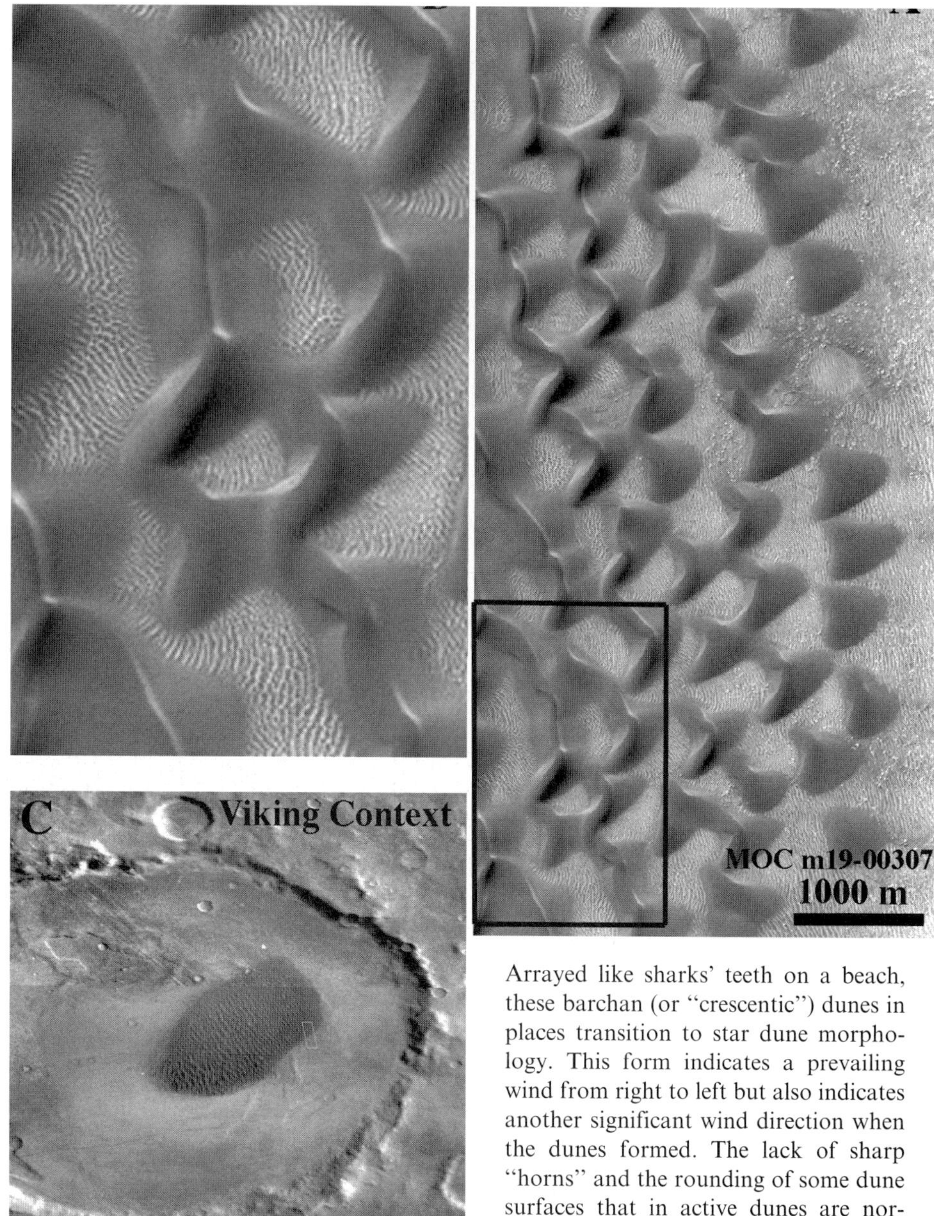

Arrayed like sharks' teeth on a beach, these barchan (or "crescentic") dunes in places transition to star dune morphology. This form indicates a prevailing wind from right to left but also indicates another significant wind direction when the dunes formed. The lack of sharp "horns" and the rounding of some dune surfaces that in active dunes are normally sharp suggests that possibly these dunes are not active. Smaller dunes with a wavelength of about 35 m occur in interdune areas of the big dunes. Location: Proctor Crater, –47.1°S lat., 329.3°W long., Noachis Terra near Hellespontus region.

Dune field at Great Sand Dunes National Park and Preserve, Colorado. (A) Dunes are up to 230 m high and cover 78 km^2 with 4.8 km^3 of quartz sand. Dunes rest on a larger sand sheet covering 466 km^2 plus a salt-cemented sand deposit of 311 km^2. (B) Dunes usually have small sand ripples over their surfaces, here made more visible by dark minerals. (C) Active wind-blown saltating sand blurs dune outlines. Mars has thousands of dune fields as large as this, and many are vastly larger.

2

Watershed

PERIGLACIAL PARADIGM WAITS FOR ITS TIME

In 1968, as the scientific image of an Earth-like Mars was crumbling, Princeton researcher R. Smolukowski predicted that ground ice might exist close to stable equilibrium with Martian atmospheric humidity, or that it could survive for billions of years beneath a protective layer of soil. It made perfect sense that Mars should have recognizable permafrost terrain, but what *Mariner* flyby images showed, instead, were craters. And more craters. However, most large Martian craters, though clearly produced by impacts, did not bear a close resemblance either to fresh or impact-battered lunar craters. Many had flattened floors and low, rounded rims or an absence of discernible rims. Something had acted to round off or completely erode the rims and fill in the bowls. Despite a theoretical basis that existed in the late 1960s that could have been used to support an interpretation involving periglacial modifications of craters, the *Mariner* flyby images were instead interpreted within the context of the impact-battered Cold, Dry, Windy Mars paradigm, which then commanded the Mars scientific community. Degradation of craters was explained by wind erosion and dust blankets. Until recent years, many researchers continued to favor this interpretation, thinking it to be conservative and inherently most probable, and allowing minor roles, if any, to ice- and water-based degradation.

Early water vapor measurements made by the *Mariner* flyby probes should have raised some serious questions on how dry Mars actually was. By one way of looking at it, Mars is exceedingly dry. The amount of water vapor in the atmosphere is by any Earthly standard very small – typically a global average of only tens of precipitable microns (roughly one-thousandth of an inch) – representing the average amount of ice that would be deposited globally if all the water vapor in the atmosphere was condensed; the warmest and most humid summer days in the most humid areas of the polar caps have 10 times that much atmospheric moisture. Okay that's *really* dry! However, in another sense it is not so much that Mars is dry, but that it is cold. For the temperatures measured, the amount of water vapor is reasonably close to saturation. The relative humidity varies typically from about 5% to 100%; it's a very Earth-like range of humidities, skewed just a bit to the drier side. A water-snow blizzard cannot occur on present-day Mars not for lack of humidity but for lack of the relatively warm temperatures needed to hold and release a lot of moisture. The strong temperature dependence of the partial pressure of water vapor over cold ice assures this. With occasional water-saturated conditions, Mars has occasional tenuous ice clouds and fogs. Traces of water-ice frost condense on winter nights down to the tropics. In fact, Maria Zuber recently reported that the MOLA instrument on *Mars Global Surveyor*, operating as a radiometer since it lost its ability to serve as an altimeter, has found indications of fall-season water-frost brightenings of the high-latitude subpolar northern hemisphere surface starting as early as mid-summer! This situation is not unlike some areas of Earth, especially middle- and high-latitude deserts, where late-summer frost is common. There is no water ocean on Mars supplying abundant moisture, but conditions reaching saturation means that there are icy sources of water vapor on Mars exposed on or very near the surface in many areas.

SURPRISE!

The *Mariner* flyby missions provided no evidence of the canals mapped by Lowell, and evidence that could have indicated ice or water processes had instead been interpreted by most scientists in terms of wind. However, the *Mariner 9* orbital mission presented, no pun intended, a watershed in our understanding of Mars. It was with considerable surprise that *Mariner 9*, and then the *Viking Orbiters*, showed us freeze-dried river beds of all types and sizes almost as soon as the global dust storm dissipated. Impact craters were discovered that looked as though their ejecta were emplaced by watery debris flows due to impact melting. Striated landscapes and lobate forms suggested the creep of ground ice. The signs of water and ice were there to see, not everywhere, but in many places. In the decades since *Mariner 9*, the more we see, the more water and ice features we find. By far the most compelling features found by *Mariner 9*'s survey were the outflow channels, features that continue to intrigue as increasingly fascinating views have been obtained.

ANYTHING BUT WATER

Many researchers saw familiar landforms in *Mariner 9* images and recognized evidence for familiar ice- and water-driven processes. Though specific process models along these lines vary widely, the general implication of this class of "Blue Mars" models is for large amounts of ice and the existence (at least transiently) of liquid water on the surface in the past. Other researchers, recognizing that the Solar System has an extreme set of environments, took this same evidence and applied

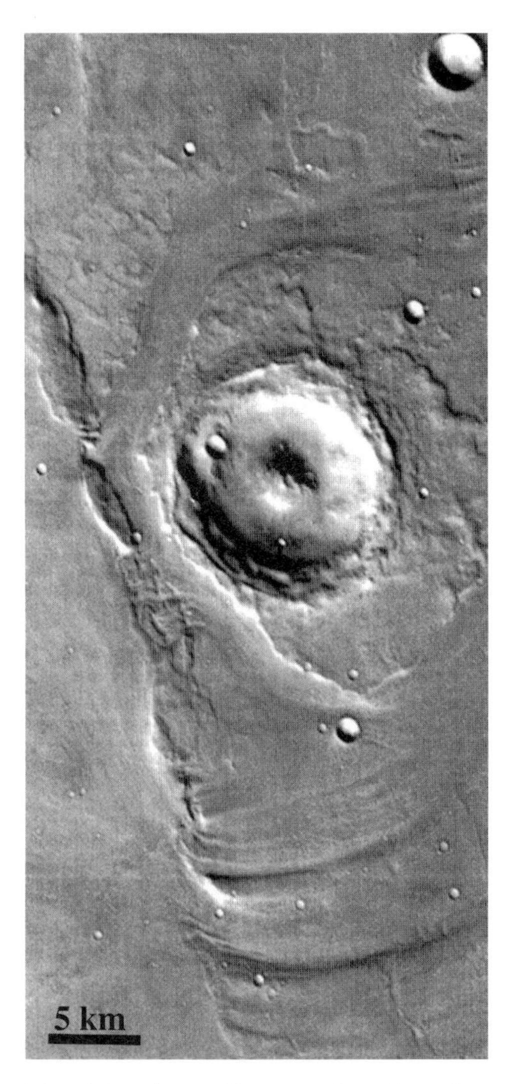

Maja Vallis, one of the mighty circum-Chryse outflow channels, ravaged Mars along a 1,700-km floodway (or glacial ice stream) before reaching this point. The erosive flows continued onward, but their erosional potential changed to deposition soon after this point in western Chryse Planitia. Perhaps floodwaters dissipated and dumped their sediment loads. The *Viking 1 Lander* settled in 1976 onto Chryse Planitia about 170 km downstream in a bouldery area, presumably deposits emplaced by the Maja flows. The flow here obviously was lower left to upper right, with damming by the tectonic ridge, overflow and erosion at nick points, and diversion around the crater. THEMIS image 100863002. 19.8°N lat., 49.6°W long.

exotic mechanisms in keeping with the un-Earthly conditions known to prevail on Mars. River-like features were perhaps carved by wind; if not wind, then a liquid other than water. Erosion of channels by a condensed surface-flowing fluid medium was evident to most researchers, and to the majority of them the most plausible substance was always liquid water. However, some others sought an explanation with exotic substances. Mars could be comet-like in its volatile constituents, having ice frozen like a rock but possessing other abundant substances even more volatile than ice. If so, Mars might have had rivers of liquid CO_2 or alkanes – anything but liquid water! Fluidization of impact ejecta blankets was perhaps by entrainment of traces of atmospheric gas; features seeming to suggest the flow of icy permafrost actually may be sand dunes or dust blankets; the polar caps could be 99% dust with a thin veneer of frost; and so on. Proponents of such ideas tended to think of liquid water as having such narrow stability limits, and Mars to be so utterly cold and dry, that something other than liquid water must have been responsible for the Martian landscapes. Mars, they thought, was very much like the Moon with just a whisp of an atmosphere added.

Mechanisms involving alternatives to ice and water could not explain certain features and soon fell by the wayside, but so did some of the water- and ice-related hypotheses. For instance, the wind hypothesis for the origin of outflow channels failed to provide a satisfactory explanation for erosional terraces and "high-water marks" in Martian channels, the transport and deposition of large boulders, and the chaotic nature and origin of collapse of the sources of the outflow channels. Thus the proposed eolian origin of Martian channels had no sustaining power and it withered and blew away. Without a sound foundation, most of these non-water ideas were soon forgotten.

After several years of debate, most scientists had become convinced that water and ice really are required to explain the observation of *Mariner 9* and the *Viking Orbiters*. The great Mars debates did not end then, but the discussions shifted to consider (a) whether the erosional medium responsible for channels and valleys was ice or liquid water, (b) how much water was involved, (c) whether the climate was formerly warmer, and (d) if it ever rained on Mars. One thing was clear: Cold, Dry, Windy Mars provided no explanation for many of the most interesting features.

SIMPLY, UNIFORMLY, MANIFESTLY WRONG

The Cold, Dry, Windy Mars model, though having logical gaps, is not without a basis in physics and observations of Mars. It explains some aspects very well. Adherents to this model maintained that Mars, being such a small world and so far from the Sun, cooled inside and outside rapidly after formation. Since then, by this model, the Red Planet has been nearly as geologically dead as *Apollo* astronauts found Earth's Moon to be. Mars retained little interior heat to cause intense volcanism, or to force volatile outgassing and formation of a dense atmosphere or massive hydrosphere, or to drive other types of internal geologic activity. Mars lacked the high gravity needed for full retention of atmospheric volatiles, and so it was far from having either the substance or the energy to make a Water World comparable to Earth. According to this perspective, Mars was a place where tenuous traces of an atmosphere blew hard and dust storms raged, where a minuscule trace of H_2O was locked as a thin polar frost, and where sporadic impacts scarred the surface. Dust locally accumulated in thick blankets, but little else of interest ever happened – *ever*, eon upon eon. One key assumption was that planets acquire their basic characteristics, determine their geologic processes, and build their atmospheres early in their histories, and do not change appreciably after that. If Mars now is a cold, dry, windy world, then by the Uniformitarian Principle it always has been so, according to this interpretation. Implicit assumptions are that Space Age humanity has witnessed more or less a complete spectrum of geologic processes representative of 4 billion years of Martian geology, and that Mars has not evolved as a geologic system over this time.

As mortal observers with limited capacities to measure and analyze, and with only a few centuries of scientific observations of Mars and a few decades of close-up observations, we must process our empirical data with as little extrapolation as possible. The Cold, Dry, Windy Mars paradigm was a plausible extrapolation, given the available observations up to January 1972. In hindsight, this conceptual model was an incomplete rendering of too-few data using a uniformitarian "principle" laden with assumptions. This paradigm was a 4-billion-year extrapolation based on grossly incomplete knowledge of the present. Given our theoretical understanding of Mars, a permafrost-based paradigm might as easily have satisfied the observations through the early 1970s, but as fate ruled, it was not to be developed that way.

It has taken a human generation to make the needed observations indicating beyond plausible doubt that water and ice are major active components of the Martian crust, that global climate change occurs on Mars, and that the uniformitarian principle is no more applicable to Mars than it is to Earth. To indict the Cold, Dry, Windy Mars paradigm is not to say that Mars fails to live up to all those descriptors. Mars is just so much more, and as a global system it has evolved through the eons. The present is clearly not the key to *all* of the Martian past. The recognition of this fact has taken place along a long winding road of fresh discoveries and changing retrospectives, and *Mariner 9* was just the first turn in the road away from cold, dry, and windy.

WHOLE WIDE WORLD IN MY HANDS

By Christmas Day, 1971, with the first orbital mission circling Mars, our knowledge of that world was still mainly from three flyby missions and distant telescopic observations. Mars seemed moonlike, but with dismal red craterscapes in lieu of dismal gray ones and cryogenic dust-laden winds replacing the hardest vacuum. Later that winter, after a billion tons of dust had settled and the Martian atmosphere had finally cleared from its raging, globe-girdling storm, *Mariner 9* turned planetary science upside down. Gigantic terrain features, such as the Olympus Mons volcano and the canyons of Valles Marineris, boasted of a planet alive (now or long ago) with lava outpourings and Marsquakings and bold geologic features to tell the story.

For me, those days of discovery were – with *Apollo* at the Moon and *Mariner 9* at Mars – as though I was a child expeditionary with Meriwether Lewis and William Clark, but I was trekking the Solar System instead. Cosmic revelations and the wildest new lands came to me in the newspapers and on television. Not all of the geologic underpinning of that stark beauty was then within my grasp, but something about that planet was comprehensible even to a child. During my stint as a deliverer of the *Columbus Evening Dispatch*, I waited day by day on Merwin Road for grainy black-and-white photos of the Red Planet's remarkable surface, which appeared on the front page at least once a month. On those days, each time I pulled a rubber band around a copy, landscapes millions of miles away were in my hands. Indeed, I had a job of great privilege, learning of faraway lands even before my father or anyone in my neighborhood arrived home from work.

Martian geologic features attest to an awesome story, which in certain details sometime resemble parts of our own planet's history but on the whole Mars is dramatically different stretching back through the eons. Written in each geologic landform is a line or a chapter of a bigger story. The science community continues to read Mars, each investigator adding his own interpretation and writing a different beginning, middle, or end of the story, imagining a different cause, or a different unseen inferential effect. The most remarkable trend of the past decade is that we have begun to understand that Mars is a global system trying to establish an equilibrium of matter and energy in the face of processes continually fighting to push the system to disequilibrium. The condensed forms of matter that are most heaved and blown and otherwise affected by this disequilibration are the volatiles: H_2O and CO_2 and whatever loose or soluble material those volatiles can take along and redeposit. But the global system includes much more than the volatiles – volcanic outpourings, impacts, and crustal fracturing are all a part of the story, as is the Sun and the cosmic dance of the planets.

NIX OLYMPICA

Mariner 9's first faintly visible glimpse of the surface seen through the dusty atmosphere was a discovery in itself. Nix Olympica and sister blemishes in the Tharsis region appeared first, while the rest of the planet remained shrouded in dust and mystery for a little longer. Nix Olympica had been known to Earth-bound astronomers since Schiaparelli in 1879. Writing in 1954, de Vaucouleurs noted intermittent white clouds around Nix Olympica and the Tharsis region; he believed that this transient feature was orographically induced atmospheric condensation on high mountains – an observation and interpretation that proved to be correct. The improved coverage of Nix Olympica by *Mariner 7* showed a concentric ring-like surface structure. By the time *Mariner 9* approached Mars, Nix Olympica had already been reinterpreted as a gigantic impact crater – an idea consistent with the

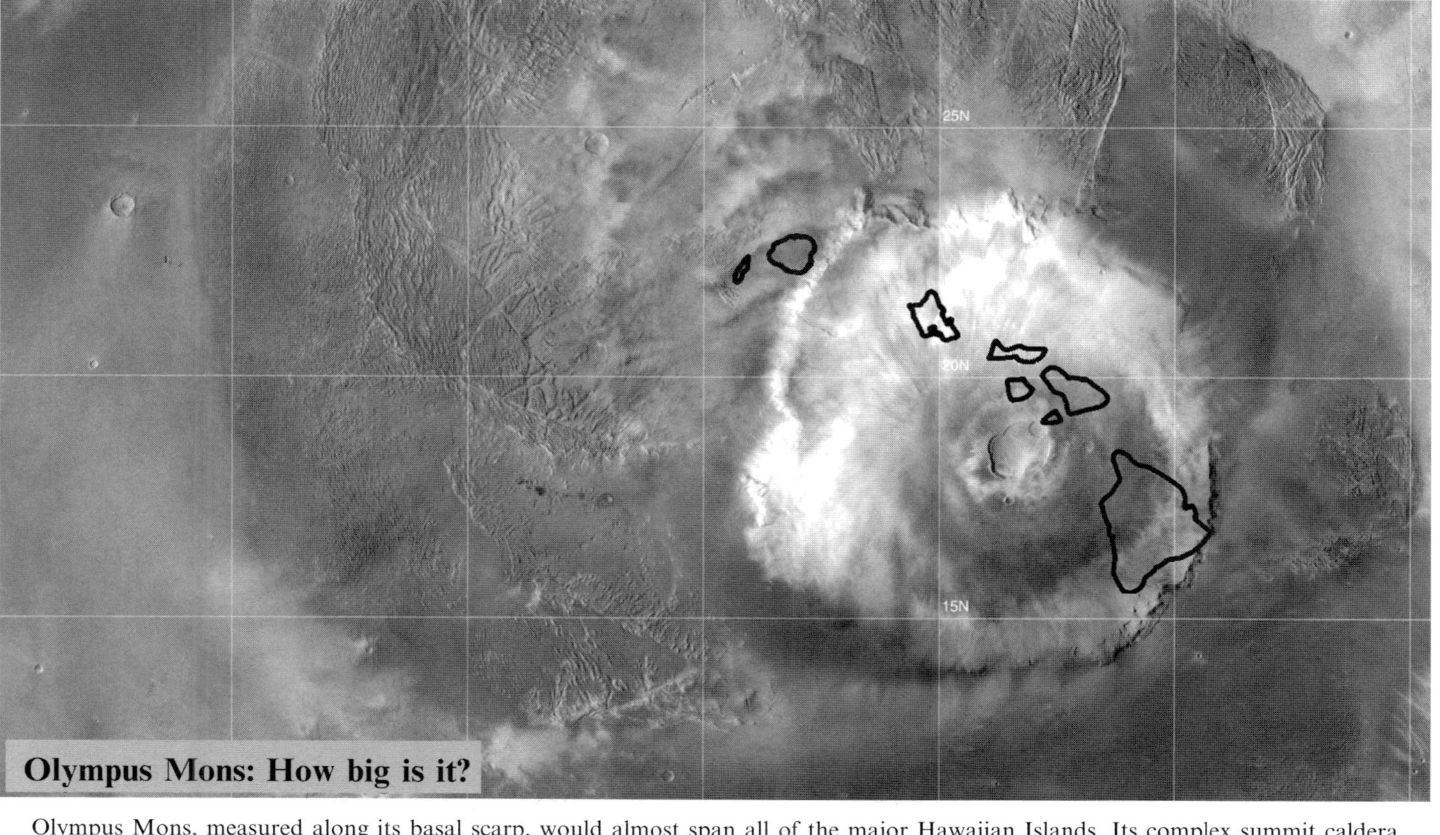

Olympus Mons, measured along its basal scarp, would almost span all of the major Hawaiian Islands. Its complex summit caldera alone would easily hold the Big Island's largest volcano, Mauna Loa. Measured across the 1500-km diameter of the aureole deposits, this structure exceeds the area of Alaska by about 10%; alternatively, it exceeds the combined areas of Germany, France, Spain, Portugal, and the UK.

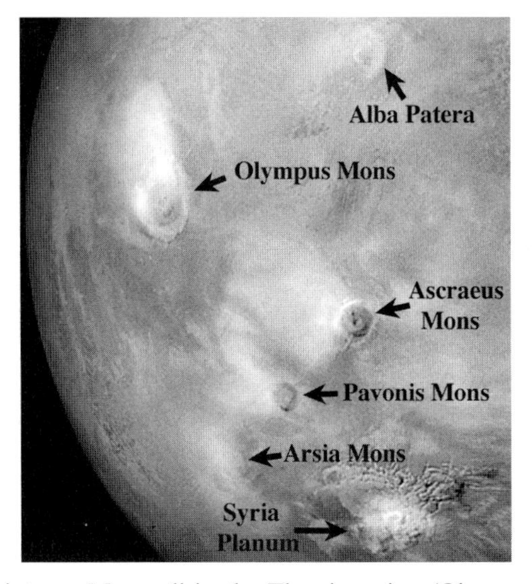

The five highest points on Mars, all in the Tharsis region (Olympus Mons, Ascraeus Mons, Pavonis Mons, Arsia Mons, and Syria Planum), commonly develop orographic clouds due to air being forced to higher altitude, where it cools and becomes saturated. These are ice clouds. Alba Patera is not as high as the others, but it, too, may have some slight cloud development. MOC image from press release 2-366, NASA/JPL/MSSS.

fresh illusion of Mars as a moon-like world. The huge "crater" of Olympus morphed with the gradual passing of the dust storm. The fact that Nix Olympica appeared to *Mariner 9* before the curtain of dust lifted from the rest of the planet – and then seemed to grow as dust progressively settled – demonstrated positive relief, as first suspected by de Vaucouleurs.

By January 1972 the main physical characteristics of Nix Olympica were revealed: flanks covered by lava flows, a basal scarp, and a summit marked by a complex caldera. This was no crater; it was the tallest, most massive volcano in the Solar System. Soon renamed Olympus Mons, this volcano towers 27 km (17 miles, over 88,000 feet) above adjoining plains and would span Arizona's girth. Several sister volcanoes in the same region of Mars – a massive volcanic plateau named Tharsis – and elsewhere on the planet are nearly as high. One, Alba Patera, is even wider, though not nearly as tall.

Olympus Mons – Throne of Mars

Olympus Mons, the Solar System's largest known volcano, is almost a hundred times more massive than Mauna Loa (Hawaii), Earth's largest. Both are known as shields for their resemblance to the profiles of Roman warriors' shields. Shield volcanoes, common on Venus, Earth, and Mars, are due to mantle "hotspots." Where anomalous heat or volatiles exist in the mantle, ascending plumes or buoyant blobs of partly molten rock feed the crust and surface with magma. Relatively fluid basaltic lava flows accumulate one on top of another, building a broad-based mega-volcano.

Why are shield volcanoes on diminutive Mars so gigantic? Three factors work together. First there is a lack of plate tectonics on Mars. Mantle plumes commonly persist for a hundred million years on Earth and over a billion years on Mars. On Earth, crustal/lithospheric plates shift thousands of kilometers on such long time scales. Thus, a long volcanic ridge and chain of volcanoes, such as the Hawaiian archipelago and its submarine extension known as the Emperor Seamounts, is built on the moving lithosphere above each of Earth's hotspots. On Mars, the plate tectonic conveyor belt does not function well, so lava flows tend to pile up in one place. In fact, the mass of Olympus Mons is about the same as that of the entire Hawaiian–Emperor volcanic chain. The second reason is that thermally driven solid-state mantle convection cells may be larger and fewer on Mars than on Earth, potentially yielding a larger supply of lava at each Martian hotspot. Finally, lower gravity of Mars (38% of Earth's) allows mountains to be two and half times higher on Mars than on Earth without collapsing under their own weight, if other things are equal.

As we shall see later, the immensity of Martian magma delivery processes turns out to be a key to the behavior of Martian water and other volatiles usually locked in cold storage in the crust. By one hypothesis, advocated by Vic Baker, myself, and others, Mars may sit volcanically fairly idle for hundreds of millions of years, and then erupt furiously. In so doing, these volcanic-geothermal episodes disturb the hydrologic balance and cold-storage of Martian volatiles, and upset the climatic regimen, thereby unleashing a cascading sequence of events.

Due to its great elevation, immense area, and crisply bold landforms, the Tharsis Plateau was the first of three major physiographic regions of Mars to emerge from the dust and be mapped by *Mariner 9*. Five times the area of Tibet, this immense, uplifted structure is a platform upon which the Tharsis volcanoes were built. Olympus Mons, huge as it is, has just a fraction of the total igneous mass in the Tharsis region.

The discovery of Olympus Mons and other huge Martian volcanoes was the first among many discoveries by *Mariner 9* that showed the Red Planet to be a half-size world with outlandishly outsize features. In 1972, it was also recognized that Tharsis sits astride vast Martian northern lowlands, covering about one-third of the globe, and heavily cratered uplands covering most of the remainder and dominating the southern hemisphere. The highlands–lowlands physiographic dichotomy is the most important geophysical crustal feature and is among the most enduring mysteries of Mars; it is also the story of the next chapter.

500 GRAND CANYONS

Another early superstar of the *Mariner 9* mission was Valles Marineris – a canyon system spanning 4,300 km (2,700 miles), greater than the distance from New York to Los Angeles or Glasgow to Damascus. Ten kilometers (6 miles) deep in places, the

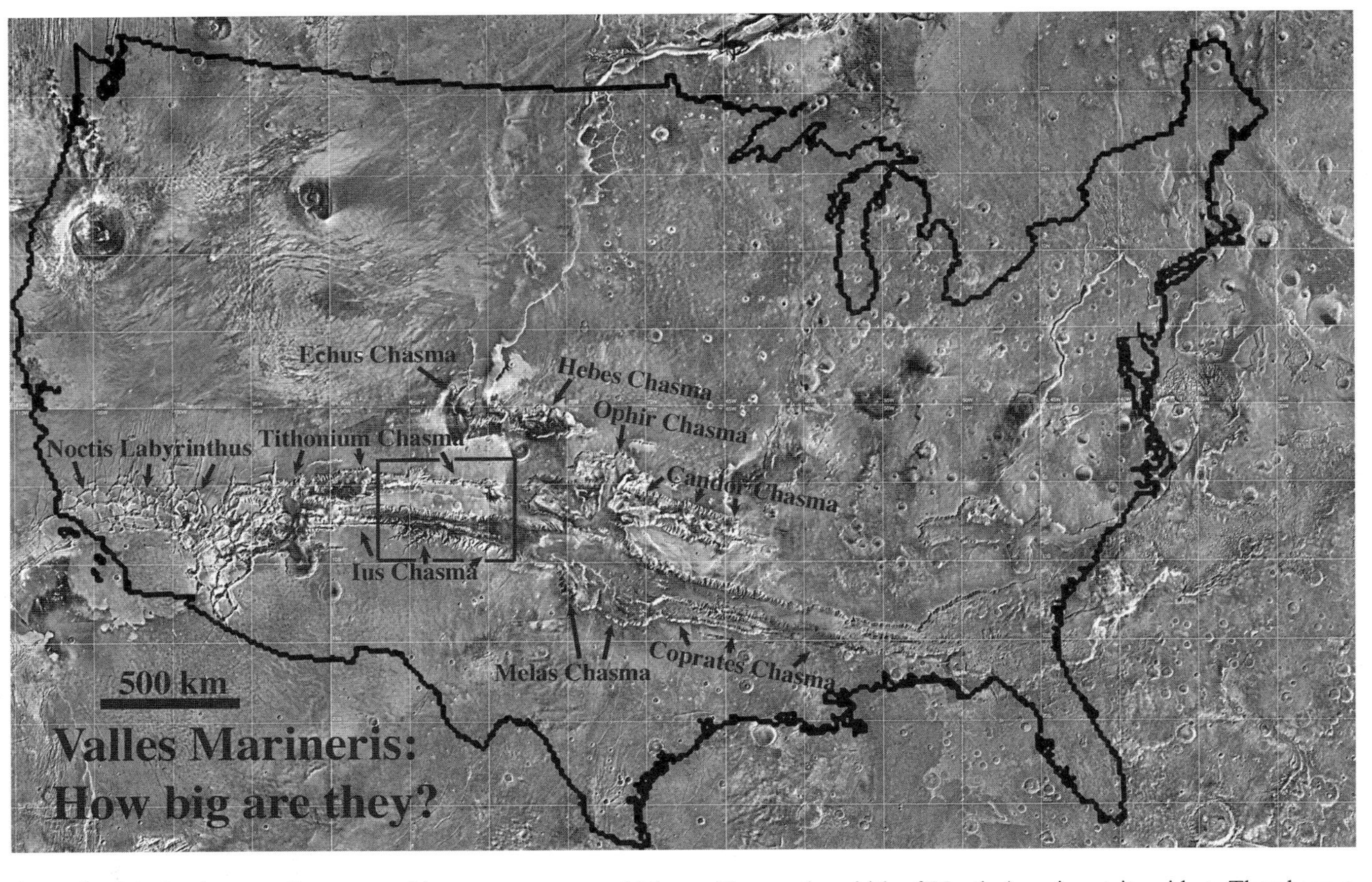

The Valles Marineris are a diverse set of immense canyons which would span the width of North America at its widest. They have a volume equivalent to more than 500 Grand Canyons.

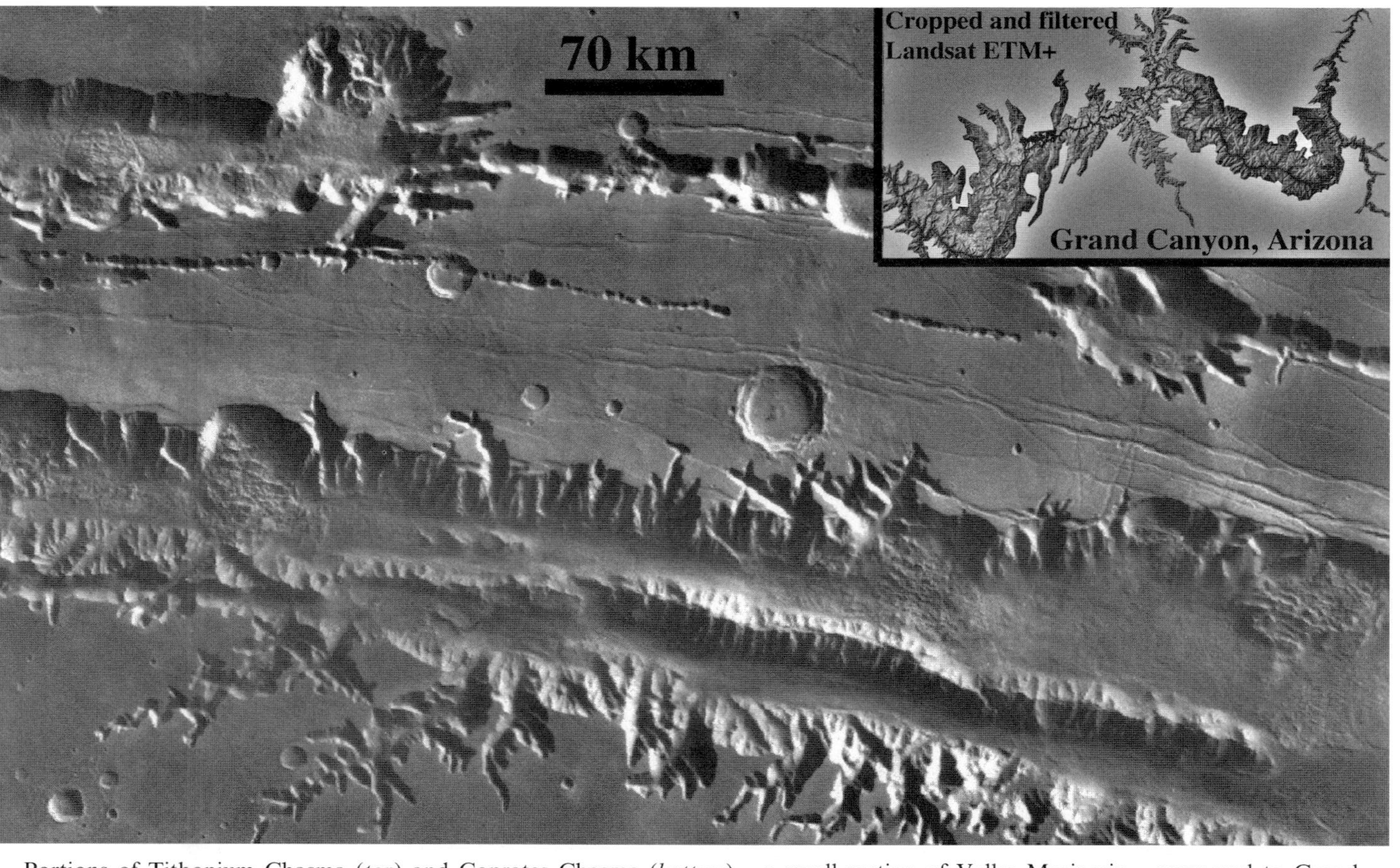

Portions of Tithonium Chasma (*top*) and Coprates Chasma (*bottom*) – a small section of Valles Marineris – compared to Grand Canyon, small scale.

Mars Odyssey THEMIS image of west Coprates Chasma.

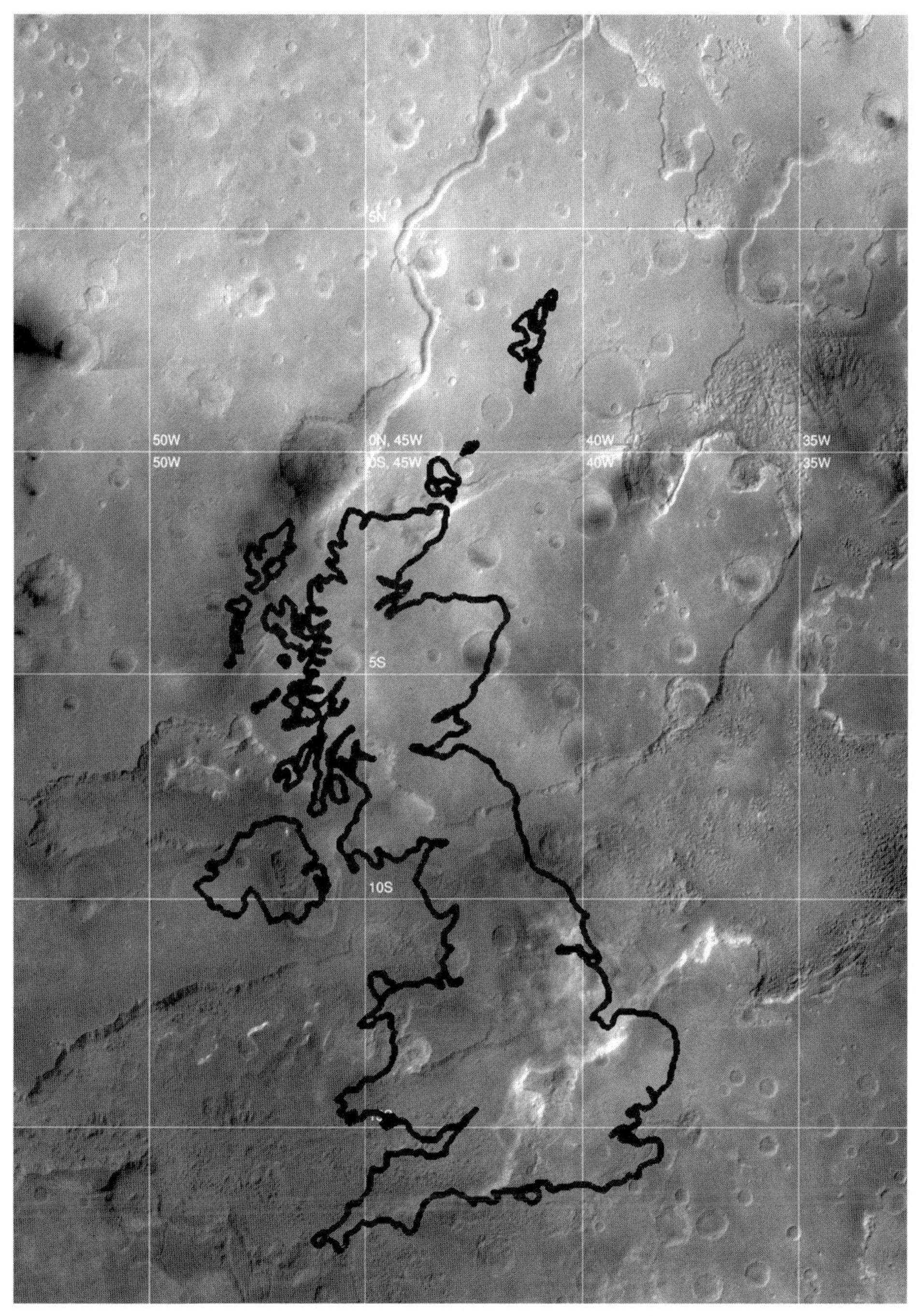

The eastern Valles Marineris degenerate eastward into chaotic terrain, which feed into outflow channels that drained into the Chryse Basin. This mosaic shows a portion of this region's chaotic terrain in relation to the size of Great Britain.

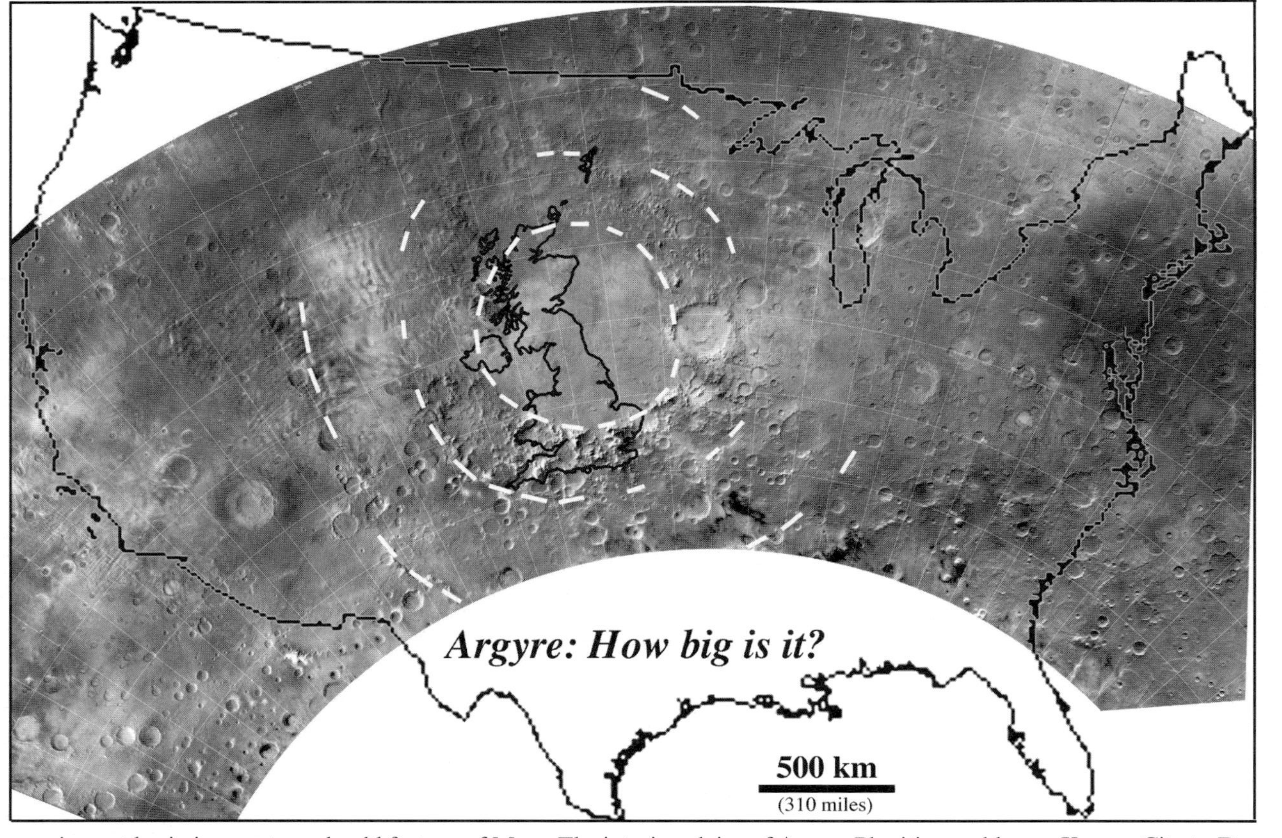

The Argyre impact basin is an extremely old feature of Mars. The interior plains of Argyre Planitia would span Kansas City to Denver, its middle mountainous ring would span from the south of Cornwall, England, to the north of the Shetlands, and its incomplete outer scarp would span from Dallas to the Canadian border. Mosaic of MC-25, 26, and 27 MOC wide angle mosaics.

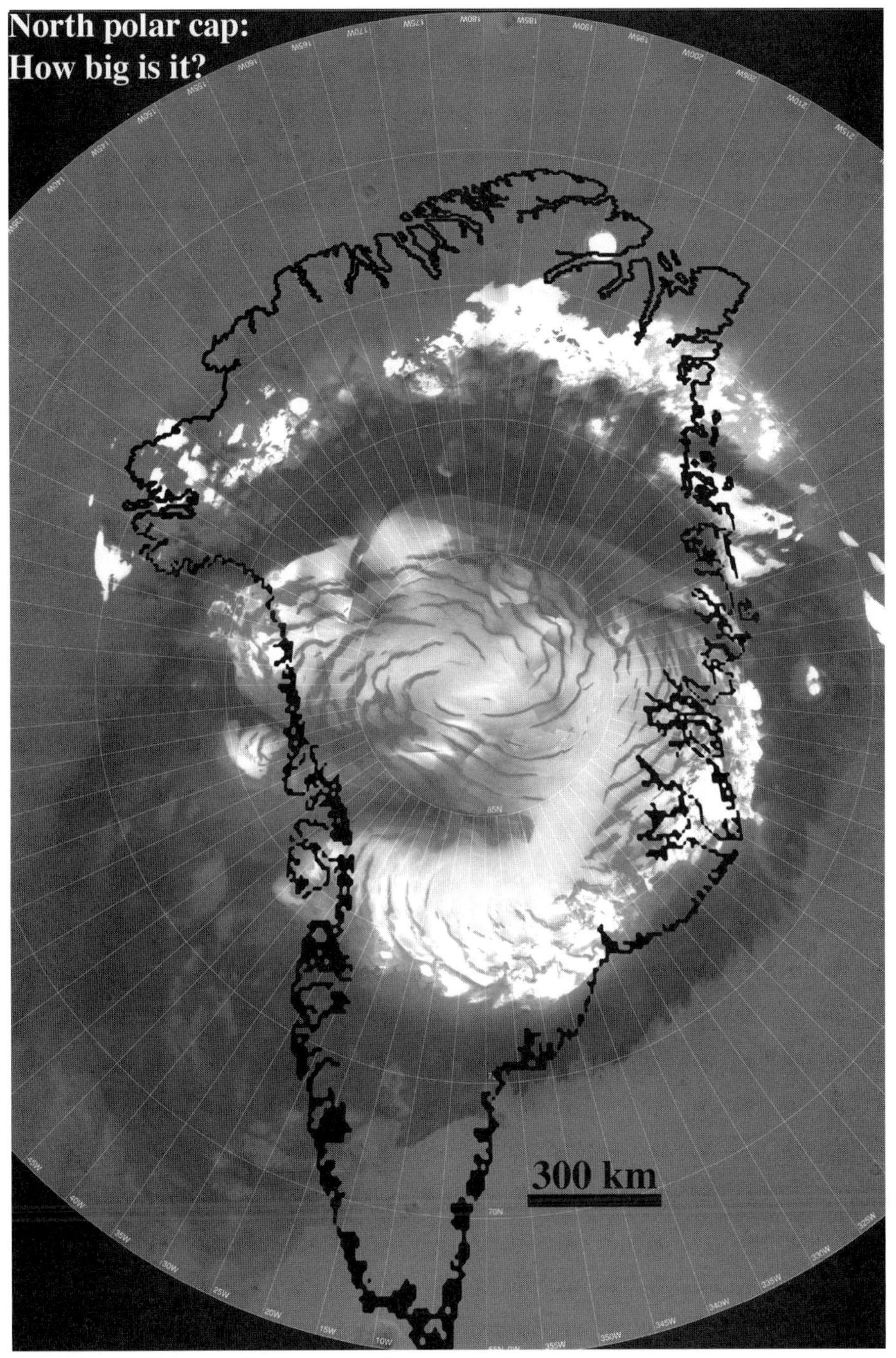

The north polar cap of Mars would span Greenland from east to west, or Paris to Rome. MOC MC-1.

canyons of Valles Marineris could swallow the volume of approximately 500 Grand Canyons along with its sister and tributary canyons of the Colorado River. The Valles Marineris expose immense layered sequences of rocks and with them billions of years of planetary history, though we await surface exploration of these sequences to understand this history. Along with Tharsis, the Valles Marineris have played key roles in hydrologic events that have touched almost the whole surface of the planet, a significance that the feeble imitation we know as the Grand Canyon cannot begin to approach.

In 1972 and 1973, Earthlings rode an exhiliarating rollercoaster of intriguing scenes of Martian landscapes and hypotheses for their origin. The astounding roles of internal geologic forces and water emerged early in the *Mariner 9* mission. Unimaginable outpourings of water produced by far the largest flood systems known in the Solar System; amazingly, these start from underground sources known as the chaotic terrain. Not quite so unexpected, in view of the craters already observed on flyby missions, *Mariner 9* found immense impact basins comparable to the very largest on the Moon. And known since Cassini's telescopic observations, *Mariner 9* produced the first detailed knowledge of the north and south polar caps, including their unique large-scale quasi-spiral structures. In addition to these mega features, a variety of smaller Martian landforms – such as the comparatively puny small-valley networks – delivered a humbling shock to planetary scientists whose Mars-think had emerged during the 1960s.

Valles Marineris: Canyon King of the Solar System

Inevitable analogies are made between the Valles Marineris and the Grand Canyon, where a ragged incision exposes a rocky "layer cake." However, the origins and development of the two canyon systems are more different than similar. The Grand Canyon developed principally by erosion due to a river cutting through a gradually uplifted layered plateau. The Valles Marineris formed by crustal extension and collapse of a layered plateau due to stresses imposed by the juvenile Tharsis igneous complex.

The Valles Marineris are part of a modified system of canyons and faults radiating from Tharsis. Planetary geologists James Dohm and Victor Baker (University of Arizona) have mapped thousands of Tharsis-related faults scarring nearly one-third of the planet. By careful analysis of cross-cutting structural relations, they have deduced a pulsating history of uplift shifting from one locus to another. This process has spanned most of Martian history and probably continues. These canyons comprise a giant system of heavily modified rift valleys, like a distant geological relative of the Earth's intracontinental rifts, such as the East African and Rio Grande Rift valleys. Lacking Earth-like subduction zones and fold belts, and the absence of other plate tectonic machinery by which Mars could create and destroy mobile plates, there was no means of achieving full-scale separation of old tectonic plates or of generating new ones. Rifting aborted, but geologic modification continued.

Volcanism, mechanical erosion, landslides, downhill creep of debris, stream seepage, gigantic floods and lake sedimentation, settling of dust in thick layers, and occasional impacts have all greatly modified the Valles Marineris but have not obscured their underlying tectonic fabric. Fracture-aligned systems of enormous sink holes in this region could indicate sublimation of subsurface ice or dissolution of limestone by ground water or melting of deeply buried massive ice deposits in the crust. There is recurrent thinking in the Mars science community, based on careful observations and modeling, that water and ice are in some way related to the destabilization of the rocky walls of Valles Marineris and to the long runout of the landslides.

The canyons have several distinct expressions. In the west they form a polygonal labyrinth of interconnected troughs. In the central region, the canyons attain their most dramatic and complex form; interior layered mesas mark the canyon bottoms; and spiny outcrops of hard layered rocks form the main canyon walls, from which alluvial fans and giant landslides were shed. Branching theater-headed tributaries splay out along the sides of the canyons. The eastern canyons – the sources for the Solar System's largest known floods – are composed of chaotically collapsed materials with scattered knob-like hills marking the remnants of what once stood as a high plateau.

OCEAN-SIZE BASIN, NO WATER

As scientists analyzed thousands of *Mariner 9* images, one thing became clear very rapidly – by far most of the large impact craters were concentrated on about half the planet, especially the southern hemisphere. Mars possessed a heavily cratered terrain and a vast, sparsely cratered plain dominating the northern hemisphere. What these images did not clearly reveal was the global-scale topography, which was soon shown by radio tracking data, Earth-based radar studies, and atmospheric pressure measurements: Mars has an ocean-size depression in the northern hemisphere primarily in the area lacking abundant big craters, and a heavily cratered highlands region dominating the southern hemisphere.

The discovery of this topographic and geologic dichotomy on Mars was one of the most important findings in 40 years of spaceborne planetary exploration. These distinctive geophysical provinces are for Mars what the continents and ocean basins are for Earth. The mean elevation difference between the two main geophysical units on the two planets actually is about the same, about 6 km. However, the more we have learned, the more we recognize that the underlying geophysical structure and geodynamic causes of the dichotomies on Mars and Earth are not very similar. For billions of years Mars has lacked a plate tectonic cycle such as that responsible for the origin of Earth's ocean basins. If Mars ever had plate tectonics, it operated only briefly early in Martian history and was a very feeble mechanism quite dissimilar to Earth's plate tectonics. For the topic of this book, the key fact is that Mars lacks an ocean of water in its ocean-size basin. Or does it?

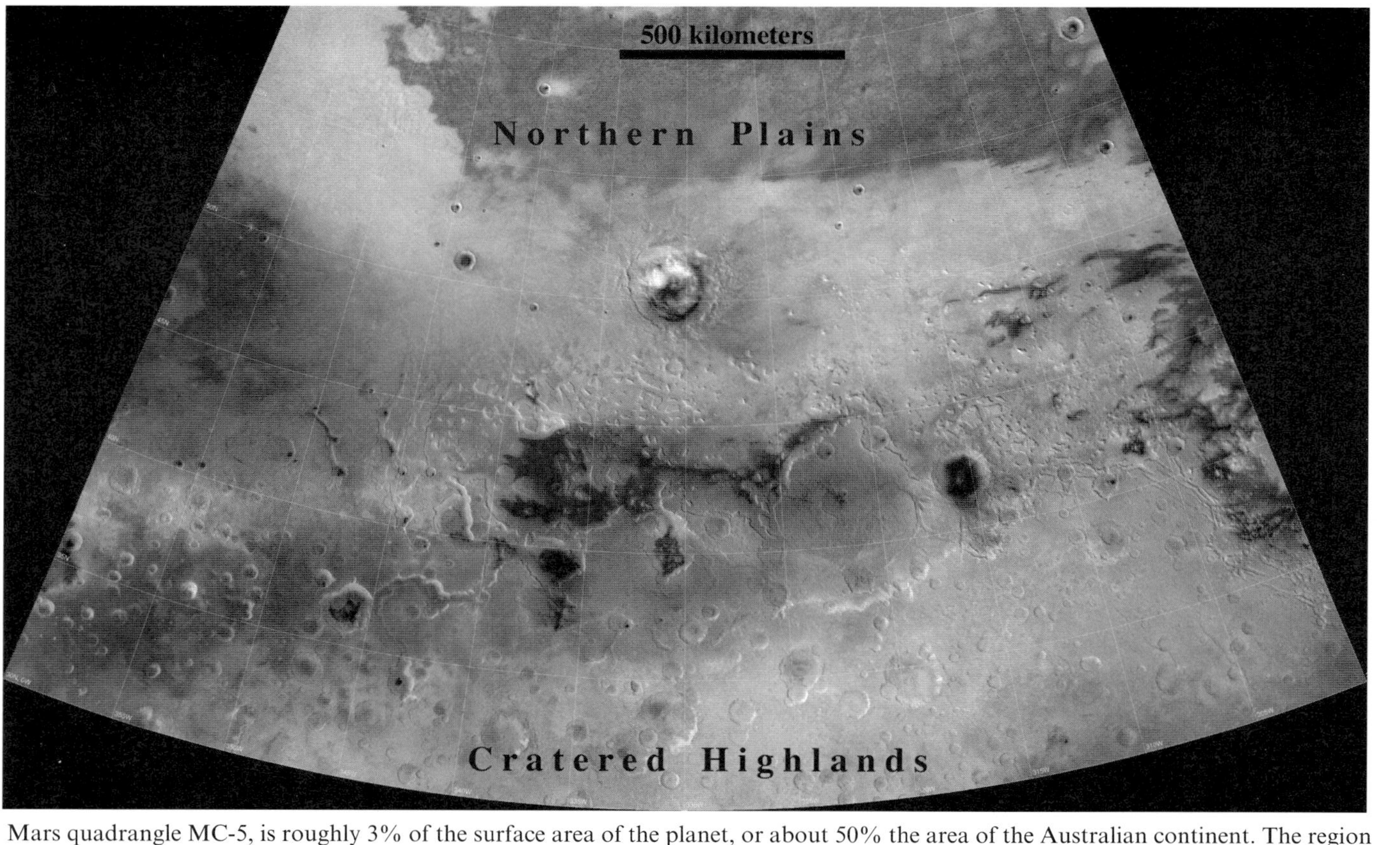

Mars quadrangle MC-5, is roughly 3% of the surface area of the planet, or about 50% the area of the Australian continent. The region illustrates well the age dichotomy between the older, heavily cratered highlands and the younger, more sparsely cratered northern plains. The dichotomy boundary is of the "fretted" type over the eastern five-sixths of this quad, whereas it is what is defined by Tim Parker (JPL) as the "gradational" type in the western one-sixth.

ERASING THE BLITZ

Mars presents us with much more than a cratered landscape, but craters do abound on some terrains. Some craters rival the Solar System's largest. Hundreds of thousands of craters from less than 1 km to more than 2,000 km across are scattered widely over many parts of the planet, especially the cratered uplands. The freshest craters resemble the multi-ringed basins, peak-ring craters, central-peak craters, and simple bowl-shaped craters on the Moon. *Apollo* lunar exploration and dynamical models of asteroid orbit evolution have taught us that Earth, Moon, Mars, and other rocky objects of the inner Solar System were all scarred by a horrific blitz of impacts up to about 3.8 billion years ago. (It may have lasted a bit longer on Mars.) A slower bombardment has continued at each object since then. Though superficially resembling the lunar surface in places, the Martian cratering record is distinct. Crater statistics and the degraded forms of most craters – barely discernible battered, buried, and flattened remnants – gave the first geologic indication that the Red Planet's history is not like the Moon's.

In the spaces between planets, innumerable asteroid-on-asteroid impacts have produced a nearly fractal distribution of potential impactors. This is manifested in the size–frequency distribution of craters produced by asteroids on solid bodies in the Solar System – small craters are produced in far greater numbers than large ones. Although there are good reasons to expect some differences in the details of lunar and Martian crater production size–distribution functions, to a first order they are similar. Screening of meteors by the tenuous Martian atmosphere has a negligible effect on preventing craters larger than 100 m, so this factor can be ignored. Observed crater populations across most of Mars do not nearly match the expected crater production function. Small craters are generally grossly deficient, though the degree of deficiency varies according to each region's geologic history. This effect is revealed also in Viking Orbiter and previous images. Although this aspect has been recognized since the 1960s *Mariner* flybys, many Mars geologists have, for some reason, ignored this important piece of evidence. Consequently, the youthful ages of some climate-related resurfacing events was not widely acknowledged until the past few years.

The underabundance of small craters is easily explained. Small craters have been erased preferentially. On the Moon, the main geologic processes that modify and eventually erase craters is subsequent impacts and volcanic burial. The same degradational processes also occur on Mars. However, in addition, on Mars more efficient degradational processes chiefly involve a combination of water-driven chemical weathering and erosion and burial due to water, ice, and wind, which are all absent on the Moon. These Martian processes are of interest chiefly because they involve the atmosphere and specifically H_2O in its various forms. The oldest craters in the Noachian highlands of Mars are especially degraded: they have erosionally flattened rims and partly infilled bowls; they also have sand dunes, gullies, signs of regolith creep, and debris-covered glaciers. In some cases, these craters are barely recognizable as due to impact, except for the circularity of structure. We see this on Earth; but rarely do we see impact craters so pristine as the famous Meteor Crater, where 50,000 years of ephemeral rainfall runoff and snowmelt under semi-arid conditions have already removed part of the rim and incised the wall with small gullies. Where we see craters on Earth, they are most often vaguely discernible as circular structures. The condition of Haughton Crater on Devon Island is typical: 22.4 million years of erosion by glaciers, snowmelt runoff, and cold-spring sapping under severe permafrost conditions – plus an episode of crater-lake activity – have reduced the rims and infilled the crater and produced a valley system similar to the form and scale of small-valley networks in the Martian cratered highlands.

While the Moon certainly possesses some attributes that are of help in understanding Mars – particularly its impact cratering and single-plate thick-shell tectonics – the scientific community has only recently fully overcome the many misleading aspects of the lunar analogy. Jean Tricart was among the first to directly criticize the lunar analogy. In 1988, he wrote:

> *Investigations on Mars Landforms carried out by geomorphologists are few, contrasting with the numerous papers written by geophysicists, who have been too much influenced by the "Moon Model,"* ...

Tricart also summarized numerous landforms that appear to require an active role of liquid water and soft, viscous, warm ice. He recognized the role of volcanism in interactions with the Martian icy permafrost. He also proposed the first model – the antecedent of the MEGAOUTFLO hypothesis – whereby volcanic emissions of radiatively active gases produced a transient modified climate and accelerated geomorphic processing of the surface.

In the remainder of this book we shall discuss Martian ice and water processes, their effects and climatic and biological implications, and how, in the future, humans may take advantage of these processes and rock deposits. As water is so essential for our own existence, it is these aqueous and icy aspects of Mars that so tantalize scientific and popular thinking of that planet. The role (or absence) of volatiles on planetary surfaces is all important to their geology; and the fact that H_2O is both abundant and close to multiple key phase transitions on Mars is the fundamental underlying basis for what we observe on the planet.

3

Warm, Wet, Early Mars

GULLY WASHERS

Martian valley networks, in retrospect discernible in *Mariner* flyby images, were first discovered in *Mariner 9* orbital images, clarified in *Viking Orbiter* scenes, and are now being shown in fine detail in *Mars Global Surveyor* MOC images. These landforms ("small-valley networks") and smaller, newly recognized gullies are among the most compelling indications of ancient Martian hydrologic activity and climate-change history, though every specific explanation has been controversial from the start. Small-valley networks and eroded gullies on mountain peaks, scarps, and crater rims, and sediment infill of associated basins, are widespread across certain areas of Mars, though absent elsewhere. I shall say more on the detailed form and global distribution of these features, especially of the small gullies, in Chapter 7, but here I offer just a few observations and interpretations of the old small-valley networks. One thing, however, needs to be said now – though termed "small," the ancient valley networks are by no means insignificant erosional features. Some drainage basins cover tens of thousands of square kilometers, and individual valleys are typically a few kilometers across and a few hundred meters deep. For readers who have visited famous Oak Creek Canyon, near Flagstaff and Sedona, Arizona, imagine that, with sparse tributaries (such as my favorite West Fork), but erosionally widened, somewhat filled in, rim-worn, and much longer. Arizona's red rocks can certainly get one in the Mars mood.

An early idea proposed by the late Carl Sagan (Cornell University) and others, and still supported by some researchers, was that small valley networks formed in the Martian highlands during ancient periods of rainfall runoff. However, it was soon recognized, as first shown by David Pieri, that most Martian valleys are different from the fluvial valleys found in most places on Earth. That was the swing of the pendulum back to something a little less wet; today the scientific pendulum is easing back to precipitation runoff, but that shall be discussed further in Chapter 7. As Pieri and others have observed, the geometrical integration, mathematical hierarchy, and

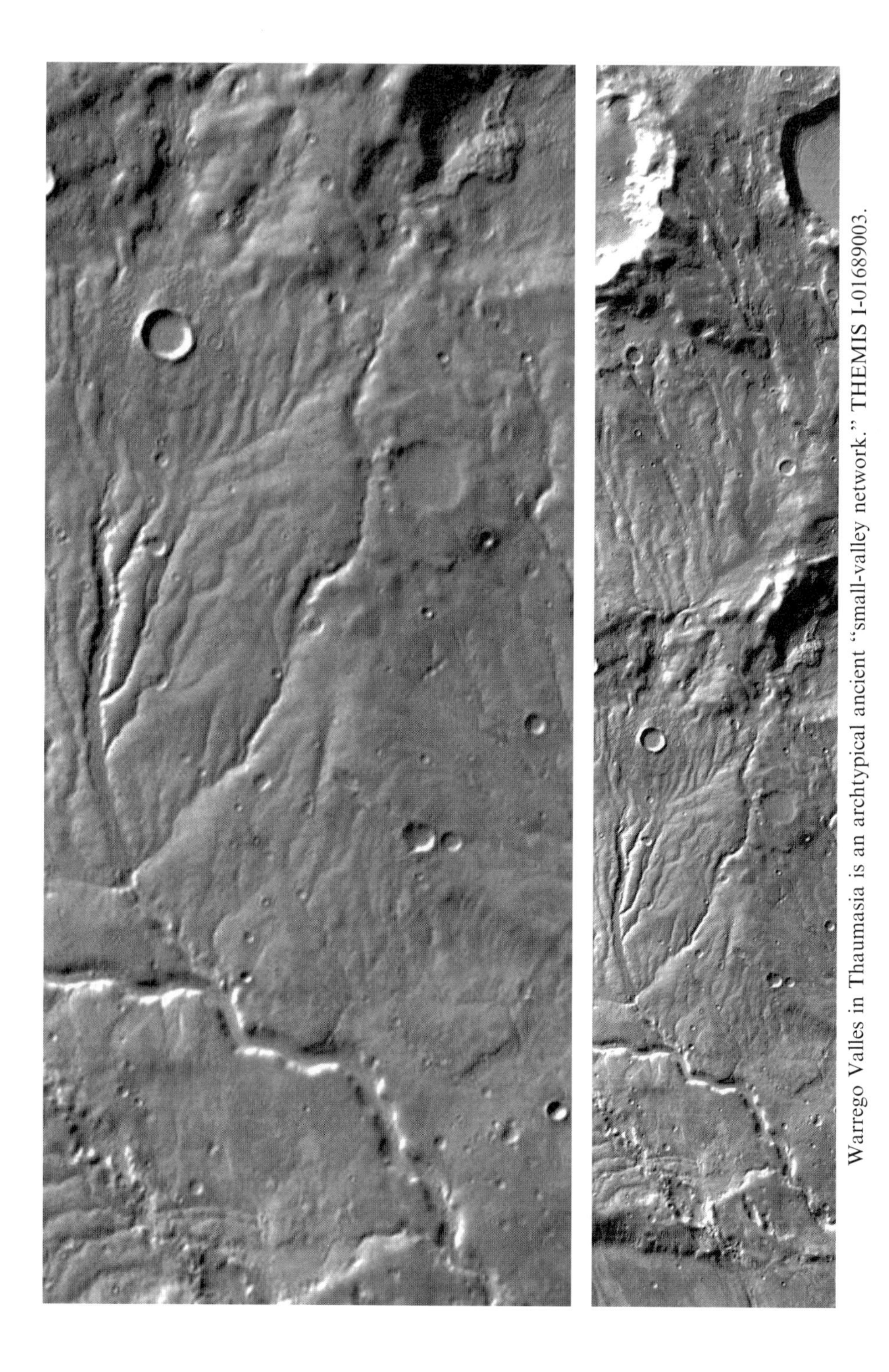

Warrego Valles in Thaumasia is an archtypical ancient "small-valley network." THEMIS I-01689003.

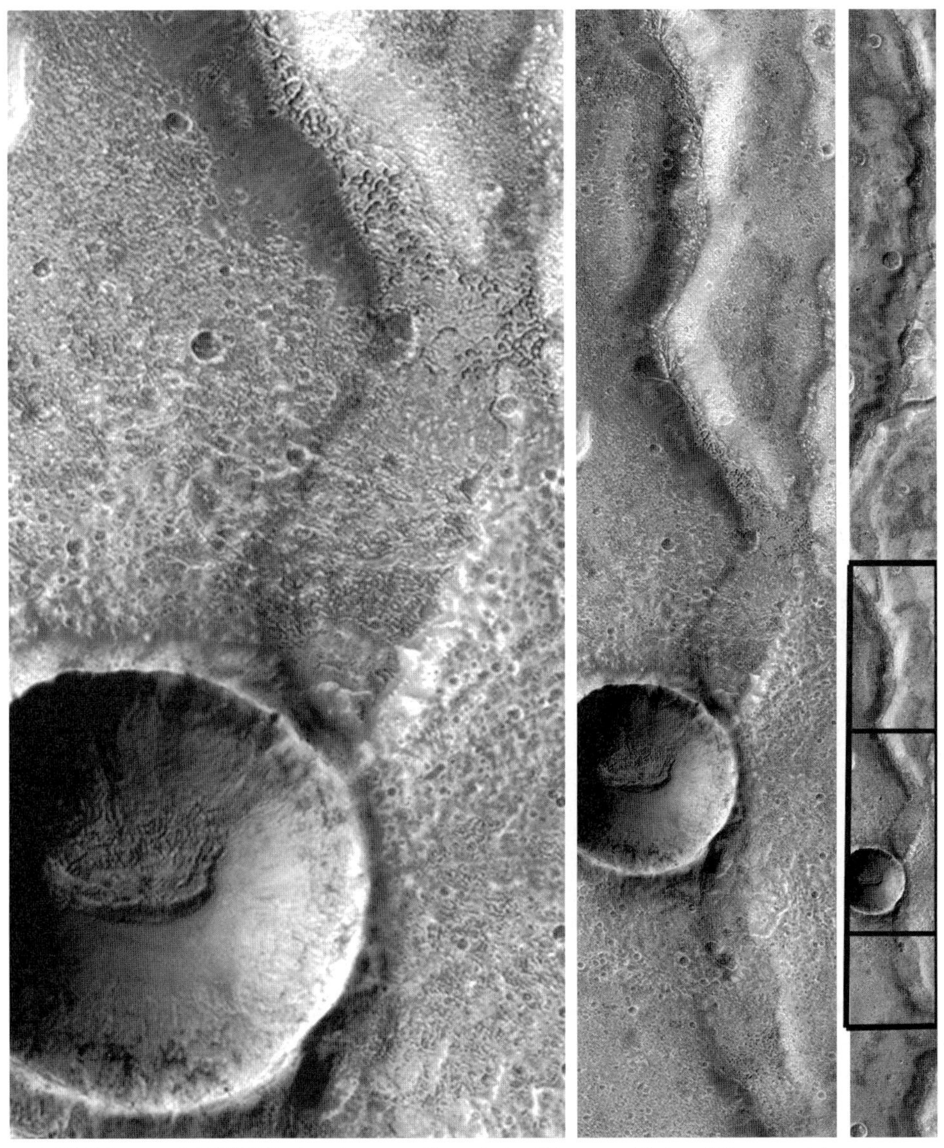

MOC image M0803363 showing a valley of the Warrego Valley (Thaumasia) region. This terrain has the typical fine-scale pitted texture of the Martian middle latitudes, a texture believed to be due to recent sublimation of a surficial ice deposit. The deposition and then sublimation of ice represents two relatively recent climatic events. This valley network, however, is ancient, since many impact craters are superposed over the valley. The valley system contains a thicker ice deposit, which also is pitted and apparently has partly sublimated. In places, the valley-fill deposit looks as though it could be the remnants of glacier-like flows. The biggest crater shown here contains a glacier-like deposit, possibly an overflow of the glacier-like mass filling the valley. This mass might have been a debris flow rather than a glacier. Scene width 2.9 km.

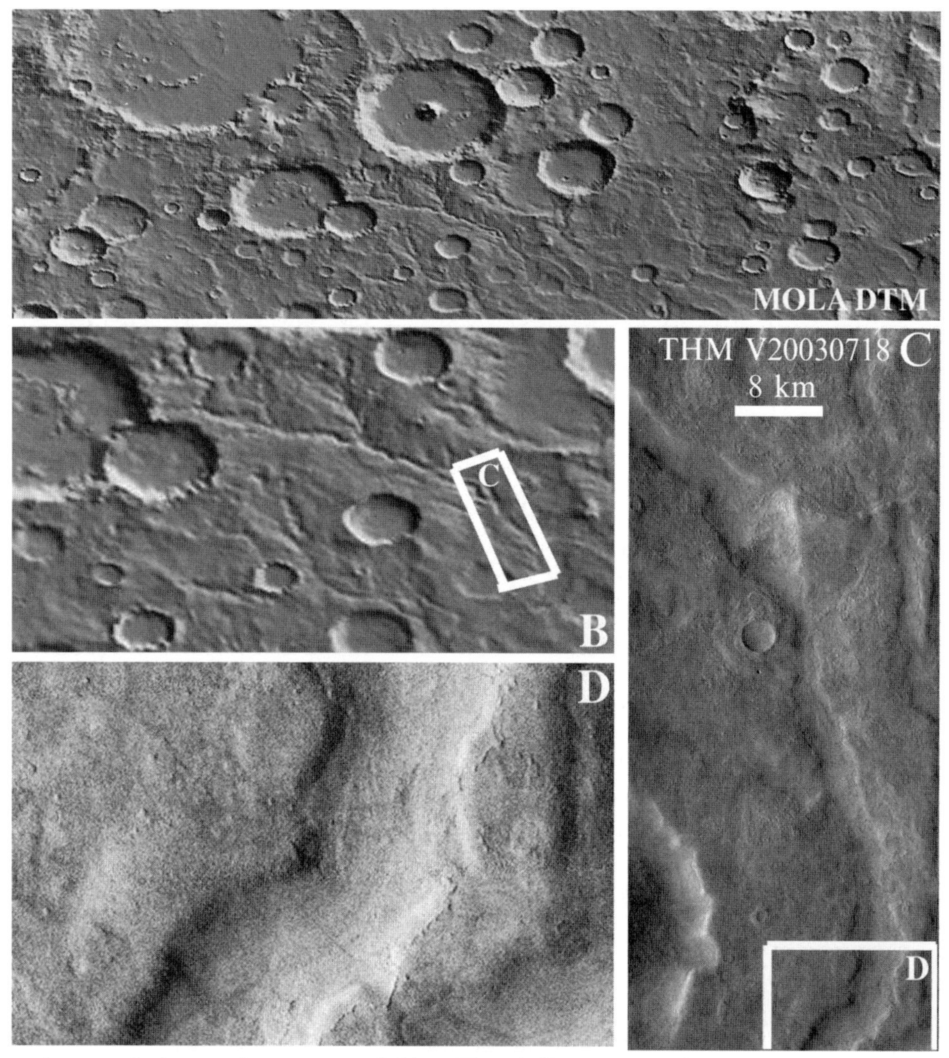

Low resolution can be a very good thing. There is little one can do to make this heavily cratered terrain in Terra Cimmeria look nice except to look at it from a long distance. If there was one place on Mars that looks like the lunar highlands, this is it. It is old and ravaged by time. It has taken a beating by Martian weather, but herein lies the difference from the lunar highlands, whose weather consists only of solar storms and micrometeoroids. (A) Shaded-relief image produced from a MOLA digital terrain model. Gullied crater rims, ancient valley networks, and breached craters are evident. (B) A valley network, Tader Valles, enlarged to limit of resolution (940 m per pixel). (C) THEMIS image V20030718 showing the weather-worn details. (D) Enlarging it and enhancing contrast only makes it look worse. While this poor geomorphic expression could be due to its extreme age, there are very few small impact craters; more likely, a lack of crisp details is due to a recent ~100 m mantle of ice/dust. The valley's age then is unknown. –49.4°S lat., 151.4°W long.

form of valley tributaries as preserved and observable today are generally more consistent with formation by groundwater spring discharge or snowmelt from local sources than by areally extensive snowmelt or rainfall runoff. Among the most marked distinctions is that the preserved record of dissection of plateau surfaces on Mars is usually not as substantial as on Earth in areas of high rainfall runoff; not seen (until recent data) are narrow and ever more finely dissected interfluvial drainage divides ("interfluves"). Mars most commonly has broad undissected interfluves. Furthermore, tributaries on Mars start nearly full size at amphitheater-like sources. This is in sharp contrast to most places on Earth, where stream valleys can normally be traced up one tributary after another; they become narrower and shallower all the way until they show as small rivulets eroded by flowing fingers of ephemeral water. Unless the smaller tributaries have been obliterated, the same is not the case on Mars.

I have long wondered whether blanketing by dust and slow churning of the regolith (and add to that small impacts) might not explain the lack of a dense network of very small-scale tributaries. Temperature oscillations can cause soils to heave and settle and gradually creep; these oscillations are a factor of several more extreme than on Earth, and creeping Mars soils have a luxury of time that Earth's hyperactivity does not lend. While this is perhaps part of the explanation, it is very clear that vast, high-order dendritic stream systems, such as the Mississippi and Amazon basins, are absent on Mars and probably have never existed there. This remains true even with more recent observations (discussed in Chapter 6) that tend to show a more highly dissected highlands surface. Rainfall, if it ever occurred on Mars, has likely not been a long-term part of the Mars story any time in the last 4 billion years. (There are some, such as Bob Craddock and Ted Maxwell, Smithsonian Institution, who will heartily disagree.) One other thing is apparent about most, though not all, of the small-valley networks. They are extremely old and degraded by impacts and weathering. Whatever processes formed these valleys, it operated primarily a very long time ago.

SAPPING AND RECHARGE

The attributes of Martian valley networks and some other valley systems on Mars, such as the tributary canyons of Valles Marineris, suggest a very different type of hydrologic system where the surface hydrologic component was dominated by spring discharge or localized snowmelt. Rainfall appears to have played little if any role on Mars, unless it was very early in Martian history. This model, with several variants, has gained general acceptance by the Mars science community as the explanation for most, but not all, fluvial valleys on Mars.

The limited range of morphologies of Martian valleys and channels has produced still ongoing controversy; Steven Squyres (Cornell University) and James Kasting (then at Pennsylvania State University, but now at NASA Ames) summarized much of the observational and theoretical evidence in a critique of the Warm, Wet, Early Mars concept in an article in *Science*, using the inquisitive title, "Early Mars: How

Warm and How Wet?" Those researchers suggested that greenhouse warming was not sufficiently robust to drive vigorous above-ground cycling of water and proposed instead that impact-driven heating provided the energy for local Martian hydrothermal systems; spring runoff driven by this energy source can explain the water-driven erosion, and the underground circulation can also replenish the aquifer. The problem with this model, as we have seen increasingly well with *Mars Global Surveyor* images, is that some water-carved features were formed long after – in many cases billions of years after – the latest large impacts in many regions

Teresa Segura (NASA Ames) and colleagues have circumvented this problem by proposing that large impacts somewhere on the planet put red-hot impact heated rock and dust into the atmosphere, thus producing a globally modified transient climate, in which it rained widely for geologically brief periods of time. Their model works for the largest basin-forming impacts, but those are too ancient to explain the younger valley networks. Smaller impacts would generate similar transient climates, but the rainfall would be too brief and too little to erode valley networks. Furthermore, both of these impact models do not account for the latitudinal distribution of several classes of water-formed features.

Squyres and Kasting were correct about one important point. Martian valley networks have no good analogs in humid temperate and humid tropical regions of Earth. However, a type of erosional process known as sapping can explain the Martian valleys, as David Pieri and others have shown. According to a widely accepted interpretation of the Martian valleys, the arid, semi-arid, and polar desert regions of Earth provide many fine analogs. Fluvial valleys on the Colorado Plateau, for instance near the Grand Canyon, have spring discharge as their dominant source of surface flow. The springs drive the sapping process. The close resemblance of these valleys to Martian valleys and gullies, especially where layered rock sequences are eroded, has evoked spring-head erosion and sapping as the chief cause of Martian valleys and gullies. In these valley systems, ground water seeps or gushes from the rocks along stratigraphic contacts, especially along cliffs or permeable rocks at points just beneath massive, impermeable rock formations. Erosion of water-weakened rocks at the point of discharge causes undermining of overlying rocky cliffs, which collapse along vertical extension fractures, causing the headward growth of the valley. The eroded rock would plug the valley except that continued runoff then transports sediment downstream, clearing the valley for further erosion of rocky debris along the valley walls and at the valley source.

One terrestrial region rich in sapping analogs is the author's area of northern Arizona. Many of the major canyons of this area – such as Oak Creek Canyon and the tributaries of the Grand Canyon – originate mainly by spring sapping. Overland runoff in these areas is suppressed by the fractured characteristics of surface rocks and their otherwise impermeable character. Abundant groundwater or snowmelt and a suppression of overland runoff are thought to be crucial in the development of this type of valley, which has few tributaries, almost undissected interfluves, and amphitheater valley heads. In some permafrost terrains, similar sapping valleys are formed; surface water cannot readily percolate or erode through the frozen permafrost, but it can transfer to subpermafrost aquifer through thaw lakes and

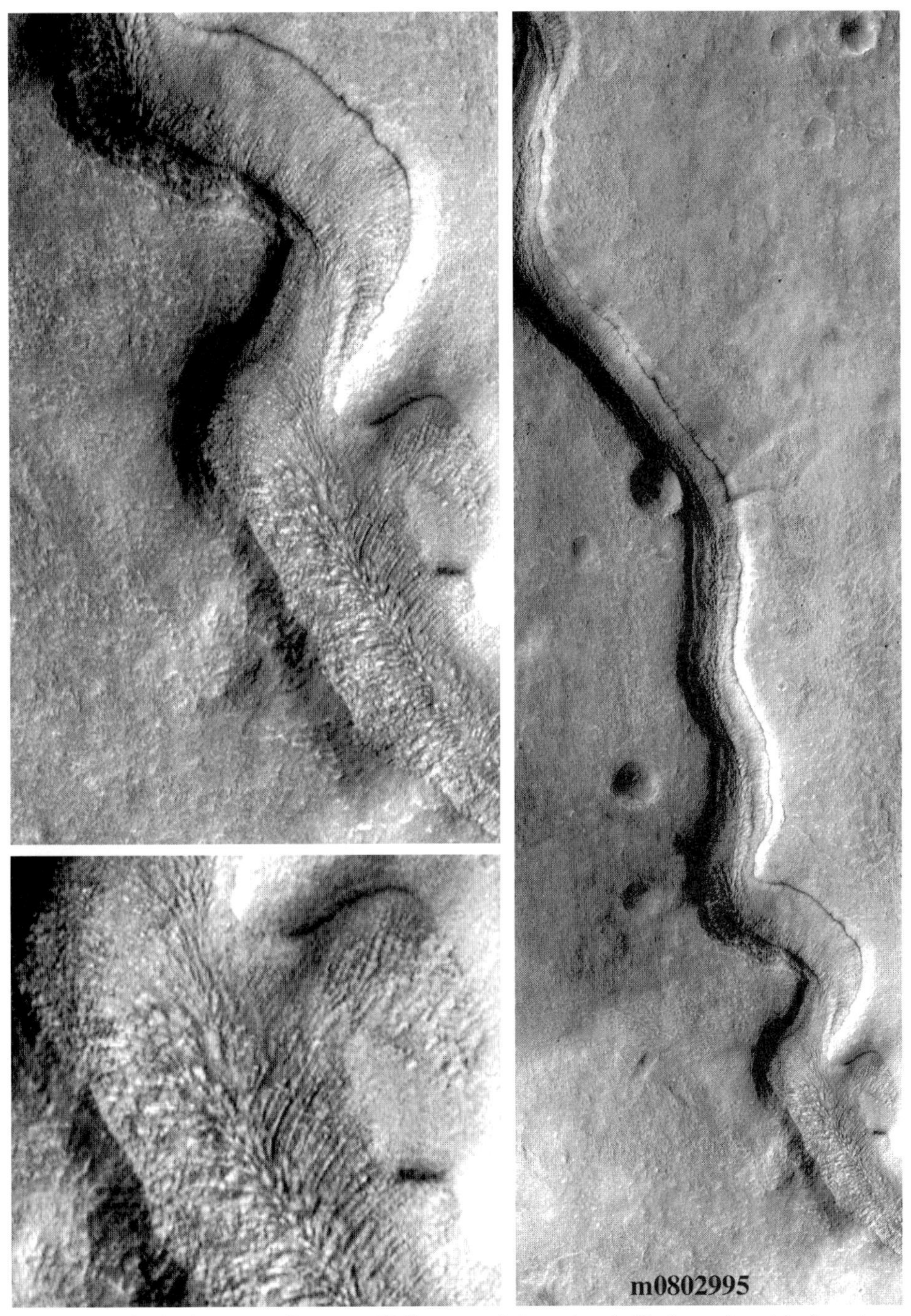

A fairly fresh sinuous valley near Hellas containing what is likely an ice-rich deposit. This texture is common for glacier-like flows around 30–38° latitude (north and south) that have undergone deep sublimational etching.

Sapping by a tributary to the Green River, Utah, is promoted by the contrasting mechanical strengths and permeabilities of the White Rim Sandstone (strong but permeable) and the underlying Organ Rock Shale (mechanically weak and easily eroded, but relatively impermeable to water). Rain and snowmelt percolate through the sandstone, and then run off underground at the bedding contact with the Organ Rock Shale. The flowing water erodes sediment and flushes it out at springs; this subsurface erosion is known as piping. Such springs can be amazingly reliable, flowing at the same rate during drought and wet years, but on decadal timescales they do respond to climatic shifts.

other truncations of a permafrost structure. From there it can flow underground and emerge at springs, where the sapping process complements erosion caused by snowmelt runoff. The permafrost zone, mechanically reinforced by ice cement, then forms cliffs at the head of sapping valleys, acting somewhat like the resistant cliff-forming formations in northern Arizona.

On Mars a similar process is likely, although in some cases the transport process may have also included solid-state glacier-like transport and muddy debris flows – a process further described in Chapter 7. Sapping valleys on Earth in fact derive their water from a variety of sources – spring discharge tends to dominate in many cases, but usually there is also some contribution from rainfall runoff and, in some climates, from snowmelt. Not surprisingly, proposed models of valley-sapping processes on Mars differ also in the degree of liquid versus solid water transport, and the roles of spring discharge versus surface snow melt; even rainfall runoff has been suggested, although this seems unlikely.

There is no reason to require any one source to have dominated everywhere on Mars where these valleys occur. Furthermore, on Mars we have the uncertain roles

These spectacular stereo pairs give a sense of the great power of lithology to control erosion. Both scenes look onto a resistant caprock known as the White Rim Sandstone, a late Permian formation of desert eolian sand dunes in southern Utah. The bottom pair look out over the Green River, which has cut through the Organ Rock Shale, a mud-, silt-, and sand-rich red-bed unit of highly oxidized shallow-water sediments. Erosion of the Organ Rock Shale is undermining the cliff-forming White Rim Sandstone. The top panels show some classic sapping valleys in the near field.

This sapping valley at Mesa Verde National Park and World Heritage Site, Colorado, has an amphitheatre source (A) related to spring sapping; sloughing of boulders due to freeze-thaw (B); and a typical fluvial V-profile (C). The cliff-forming Cliffhouse Sandstone is a Late Cretaceous (~87.5 Ma) near-shore and beach sandstone. The underlying slope-forming Menefee Formation is shale, coal, and sandstone rich in plant fossils indicative of a swamp, marsh, or delta. The Menefee Formation is impervious to water, which percolates through the permeable Cliffhouse Sandstone, then runs underground along the formational contact, producing springs and sapping. The pre-Columbian Anasazi culture built villages of stone and mud (A).

of a modified ancient climate or dissolved solid or gaseous impurities in the water in promoting spring activity and runoff. Perhaps most significant for the Mars debate – though not debated as much as it ought to be – is the cause of groundwater recharge, especially atmospheric versus internally driven geothermal mechanisms. The issue of recharge is as important as the immediate source of runoff, because a prodigious long-term source of water is needed to carve such valleys. It is in fact difficult to achieve the needed supply of water without some form of atmospheric precipitation – either rainfall or melting snow, of which the latter seems more probable.

Small-valley networks on Mars were once thought to be exclusively among the oldest Martian features, since most occur in heavily cratered Noachian terrains, and many appear to be themselves heavily degraded. This inference has been challenged by many researchers, but it held sway long enough to become a central tenet of the rallying Mars paradigm of the 1980s, "Warm, Wet, Early Mars." By the end of the *Viking Orbiter* missions, this revolutionary new concept of Mars had nearly supplanted the 1960s notion of a Mars that has been cold, dry, and windy throughout its history.

RAINS ON THE PLAINS OF COLD, SUNNY MARS?

Martian valleys give no indication of any close climatic and geomorphologic analog of humid temperate or tropical conditions and heavy rainfall runoff as we know on Earth. Most Mars experts agree that the preserved record of fluvial valley development on Mars generally does not indicate an advanced or mature stage, although subsequent erosion over billions of years possibly has obscured the full extent of fluvial erosion in some very ancient and weather-beaten highland terrains. These facts together suggest that it may never have rained on Mars. Rainfall would not be expected if the main Martian hydrologic equivalent of Earth's oceans is a huge body of ice. Sublimation and melting of ice would produce water vapor at partial pressures up to the 6-mbar vapor pressure at the triple point of water. If humid air masses so formed became warm enough to rain, relative humidity would drop below the dew point. It could rain if the temperature of water sources (such as a northern ocean) became greater than the triple point of water; or if water sources at the triple point generated humidity, which then flowed to lower elevations, where it warmed and compressed without much mixing with drier air. These are unlikely scenarios.

GOING, GOING, GONE

While this book is concerned most with the geologically recent climatic oscillations of Mars, a more intense, very early period of erosion is indicated by the subdued relief of many large craters dating from the Noachian, also known as the period of heavy impact bombardment (about 3.8–4.2 billion years ago if synchronous with heavy bombardment of the Moon). Early in this period is when the largest well preserved impact basins on Mars, such as Hellas and Argyre, were formed. Many craters tens of kilometers across, mostly dating from the middle to end of the Noachian, have been nearly obliterated by erosion. The cause of this erosion is not entirely clear, because most discrete erosional landforms are themselves obscured by processes occurring over 4 billion years of existence. The erasure of hundreds of thousands of craters over several kilometers across is indicated by statistical analysis of crater populations, which show the smaller craters in much smaller numbers relative to those expected as scaled from the populations of large craters. The missing craters originally were hundreds of meters deep (or deeper). Typical erosion and burial across vast expanses of Mars amounting to 500–1,000 meters is needed to erase these missing craters. Flattening of the profiles of most remaining large craters (reduction of rims and partial infilling of bowls) is an intriguing and consistent hint supporting a much greater cumulative extent of ancient erosion than that betrayed by the obvious remnants of gullies and valleys carved across the crater rims. The proof will lie in the sediments that we infer to exist within the crater bowls. In some instances partial erosion has revealed their layering, and relics of cross-cutting valleys are visible in certain areas. Some craters are mere ghostly, fragmented relics or vague circular, heavily gullied scarps. Others, however, retain some semblance of their original form, and erosional features are well preserved.

Many surviving Martian crater remnants from that ancient erosional period resemble the water- and ice-battered relic of a large impact structure known as Haughton Crater on Devon Island. This crater, 20 km across and originally about 2,500 m from floor to rim, now has a surface expression just about 200 m deep. This crater, though large, is too small for major isostatic relaxation, so it would appear that over 2 km of erosion has occurred. Channels and valleys and small gullies incise the rim, entering one side of the crater and exiting the other. Such dramatic erosional leveling of crater relief has occurred in 23.4 million years of glacial and periglacial conditions since the impact, roughly 1 m of differential erosion (rim erosion rate minus floor erosion rate) per 10,000 years. This erosion rate is the same order of magnitude as, or maybe a factor of 2 or 3 slower than, erosional degradation, under temperate semi-arid conditions, of Meteor Crater, Arizona, according to the work of the late David Roddy (US Geological Survey).

These terrestrial examples give some idea of the cumulative degree of erosion that the most ancient terrains on Mars have undergone. Since most of that erosion probably occurred during an interval of a few hundred million years early in Martian history, erosion rates were perhaps an order of magnitude slower than that on Devon Island if sustained for, say, a quarter billion years; or faster than that if erosion was intensely active for only a fraction of this time. In any event, such deep erosion and high erosion rates (compared to Martian rates averaged over geologic time) appear to be unique to that early period of time (early to middle Noachian). Subsequent to the period of heavy impact bombardment, the cumulative extent of erosion over a period of about 4 billion years was less than during that early period. It is those scenes of heavily degraded cratered highlands and their remnants of ancient valleys that have motivated the concept of Warm, Wet, Early Mars. As the Haughton Crater analog and the back-of-the-envelope estimates above would indicate, that early period of Martian erosion does not require conditions any warmer or wetter than Devon Island for a period of tens to hundreds of millions of years. However, as indicated by the drastic slowdown in erosion since the Noachian, there must have been a sharp climatic shift toward drier and/or colder conditions around the end of that distant era. This "Warm, Wet, Early Mars," like "Cold, Dry, Windy Mars," reflects more than a little reality. But these together are not the sum of Martian climatic evolution, as there have been many more recent periods of intense hydrologically driven erosion, though these episodes are restricted to certain regions (described later).

The rainfall runoff interpretation of small Martian valleys and gullies has been variously criticized and supported by researchers over the last quarter century. Ancient conditions allowing rainfall, or at least stable perennial surface water some 4 billion years ago, has become popular NASA doctrine – the Warm, Wet, Early Mars paradigm. The hypothesized clement period is thought to have been during the ancient Noachian era, because valleys seem to be developed most abundantly in the heavily cratered highlands. It was, until recently, usually stated that annual average temperatures above 273 K (32° F) were implied, but this is clearly not a requirement, as illustrated by the Haughton Crater and Devon Island landscape, which has snow-fed fluvial and proglacial valley networks but has primarily existed in hard permafrost or glacial conditions since its origin 23.4 million years ago. Controversy surrounds each specific interpretation of each important Martian feature; any climatic inference drawn from any model invariably draws criticism and counterarguments. Multiple alternative processes thus reduce or obscure the climatic significance of small-valley networks and gullies. For now I draw a more general conclusion, to be developed in later chapters – stabilization and flow of surface and subsurface water (in any form and in combination with any impurities) effectively explains small valleys and gullies; it is possible, but very difficult, to explain their existence without liquid water, but a seriously cold climate (albeit warmer than Mars is today) is admissable. Liquid carbon dioxide, for instance, has been proposed, but this explanation raises a host of severe difficulties more problematic than those implied by liquid water.

Water erosion of valleys and gullies comes with a special bonus – this surface flow process also effectively explains the obliteration of most small craters and the eroded crater rims and flattened floors of most remaining craters. This explanation allows a wide range of sedimentologic and climatic regimes, though in succeeding chapters I shall be more specific. But think "Devon Island."

Conditions that were conducive to fluvial erosion were not restricted to early periods of Martian history. The discovery during the *Mars Global Surveyor* mission of very youthful small-scale gullies and debris flows on Mars (described in Chapter 7) has turned the Mars community on its head. No one expected this, because these gullies are demonstrably among the youngest geologic features on Mars, perhaps having formed within the past million years.

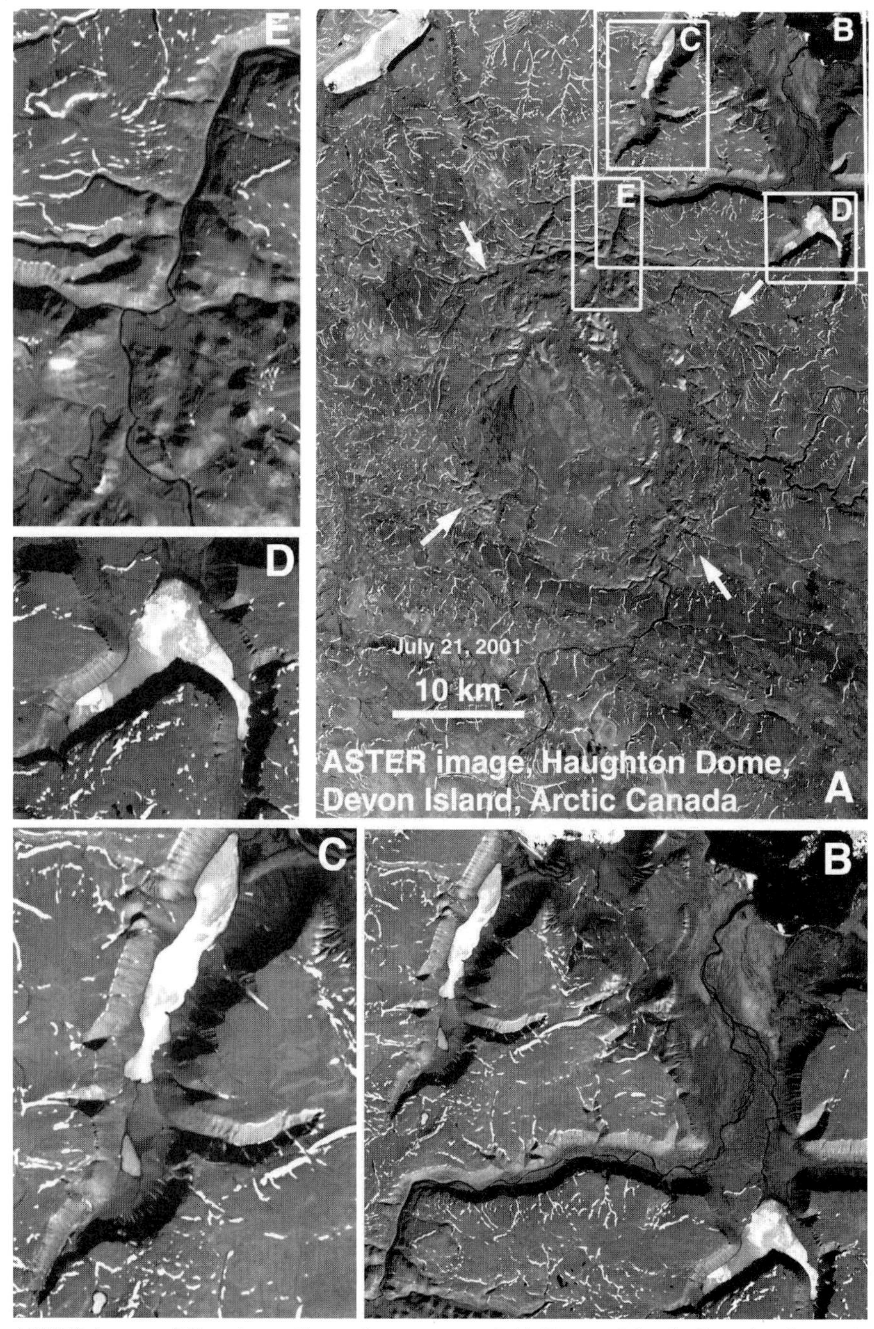

ASTER image of Houghton Crater (arrows, panel A), Devon Island, Canada, shows the effects of 23 million years of erosion. Sapping valleys also have localized snow accumulations, which may sustain summer-season melting and runoff when the permafrost active layer is thawed and can be more easily eroded.

This crater is almost rimless. Gullies and glacier-like deposits occur on the inside crater wall. Noachis Terra.

OUTFLOW FROM SHEER CHAOS

In one of the greatest finds of the Space Age, *Mariner 9* and the *Viking Orbiters* showed us immense freeze-dried river beds carved by the Solar System's largest-ever flash floods. Unlike dendritic highland small-valley networks, these magnificent "outflow channels" start as extra-super-size features, usually at sites of massive ground collapse, known as chaos or chaotic terrain. Outflow channels generally lack tributaries, but instead consist of multiple-braided channels. The most spectacular and well-documented outflow channels originated in chaos near the Tharsis and Valles Marineris region and emptied into the northern plains, especially into the Chryse Basin. Floods that were a hundred miles across in places, hundreds of feet deep, and running for up to 2,000 miles carved canyons half a mile deep, gouged potholes and scours a hundred feet deep, and etched tear-drop shaped islands the size of Long Island. Martian chaotic terrain at the sources of outflow channels apparently formed by rapid, massive fluid withdrawal (usually assumed to be water) from the subsurface. Estimation of the discharge rates and cumulative discharge from outflow channels has been one of the consuming passions of many Mars researchers. The actual answers are uncertain, as they depend on assumptions such as the sediment:water ratio or the vertical incision rate of the erosive flows. But all the answers are enormous numbers, as can be appreciated from the scale of these channels. The other consuming passion of these researchers is then the logical implications for water ponding and sediment deposition in the northern plains and other major basins where outflow channels emptied. Was there an actual ocean? A mud ocean? A series of isolated or interconnected lakes? These are key questions and the main topic of Chapter 4.

Outflow channels have spurred many ideas regarding their origin, but none has dominated discussion as much as a model involving catastrophic outburst floods, first developed by Vic Baker based on Mariner 9 observations, and developed further in various forms by Michael Carr, Ken Tanaka, and many others. Carr's estimates of discharged water volume was a leading basis for the ocean hypothesis of Baker et al. In fact, all those proposing oceans in the northern plains since Baerbel Lucchitta, H.P. Jöns, and Tim Parker in the mid-1980s have linked the sedimentary deposits and hypothesized ocean of the northern plains to outflow channel discharges.

In 1982, Baerbel Lucchitta hypothesized that plastically deforming glaciers may have carved the outflow channels, citing as evidence the erosional effects of ice streams on Earth. She has recently added particularly strong arguments based on ice-stream erosion of the shallow seafloor where Antarctic glaciers have sculpted streamlined forms, grooves, and ridges practically identical in form and scale to what is seen on Mars. Her proposed similarities between the outflow channels and Antarctic ice streams, though compelling in many respects, did not fare so well against the prevalent view that the outflow channels are fundamentally flood features. One difficulty in distinguishing these different origins is that floods and glaciers can produce streamlined forms and scours that in many respects are very similar in appearance and scale. The underlying reason for this similarity, according to John Shaw, a glacial geomorphologist at the University of Alberta, is that

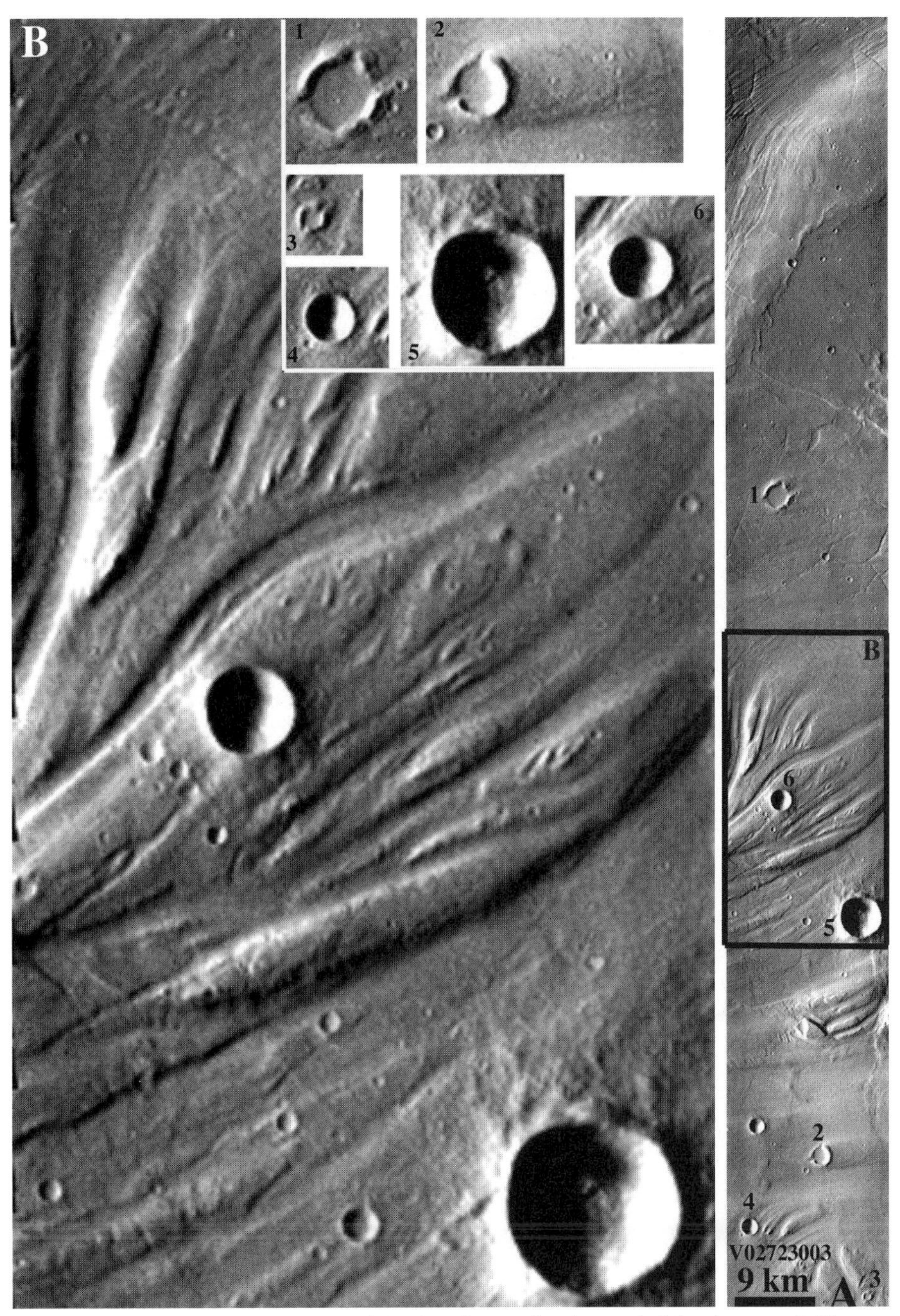

THEMIS image V02723003, Band 3, showing a deeply eroded landscape of Chryse Planitia beyond the mouth region of Maja Vallis. Affected is a swath over 140 km across. Insets show six impact craters; craters 1–3 pre-date the erosional flow, and craters 4–6 post-date it.

THEMIS image V00963003, Band 3, of the mouth of Maja Vallis. As the flood was in progress, it was almost like a jet nozzle. Blocks up to 800 m across were moved. As the flow waned, a sedimentary deposit, perhaps a debris flow, covered the channel floor. Maja's great age (Hesperian) is discernible from the sprinkling of small impact craters. 17.8°N lat., 306.2° long.

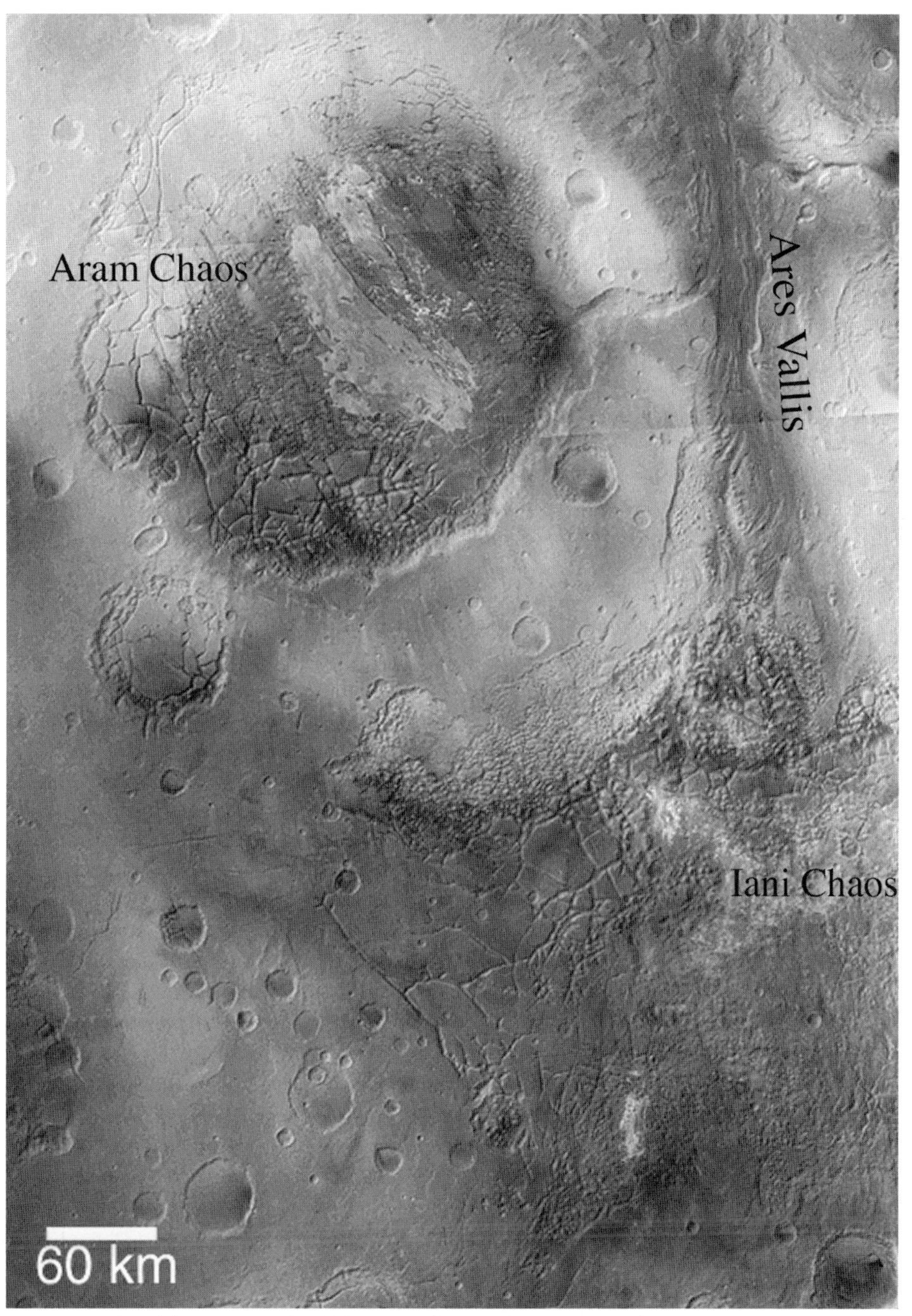

THEMIS red image mosaic of Aram and Iani Chaos, the main sources for Ares Vallis. Such chaos formed by massive groundwater withdrawal and terrain disintegration, possibly initiated by permafrost melting and CO_2 release. 0.5°N lat., 20° long. NASA/JPL/MSSS, press release MOC2-344.

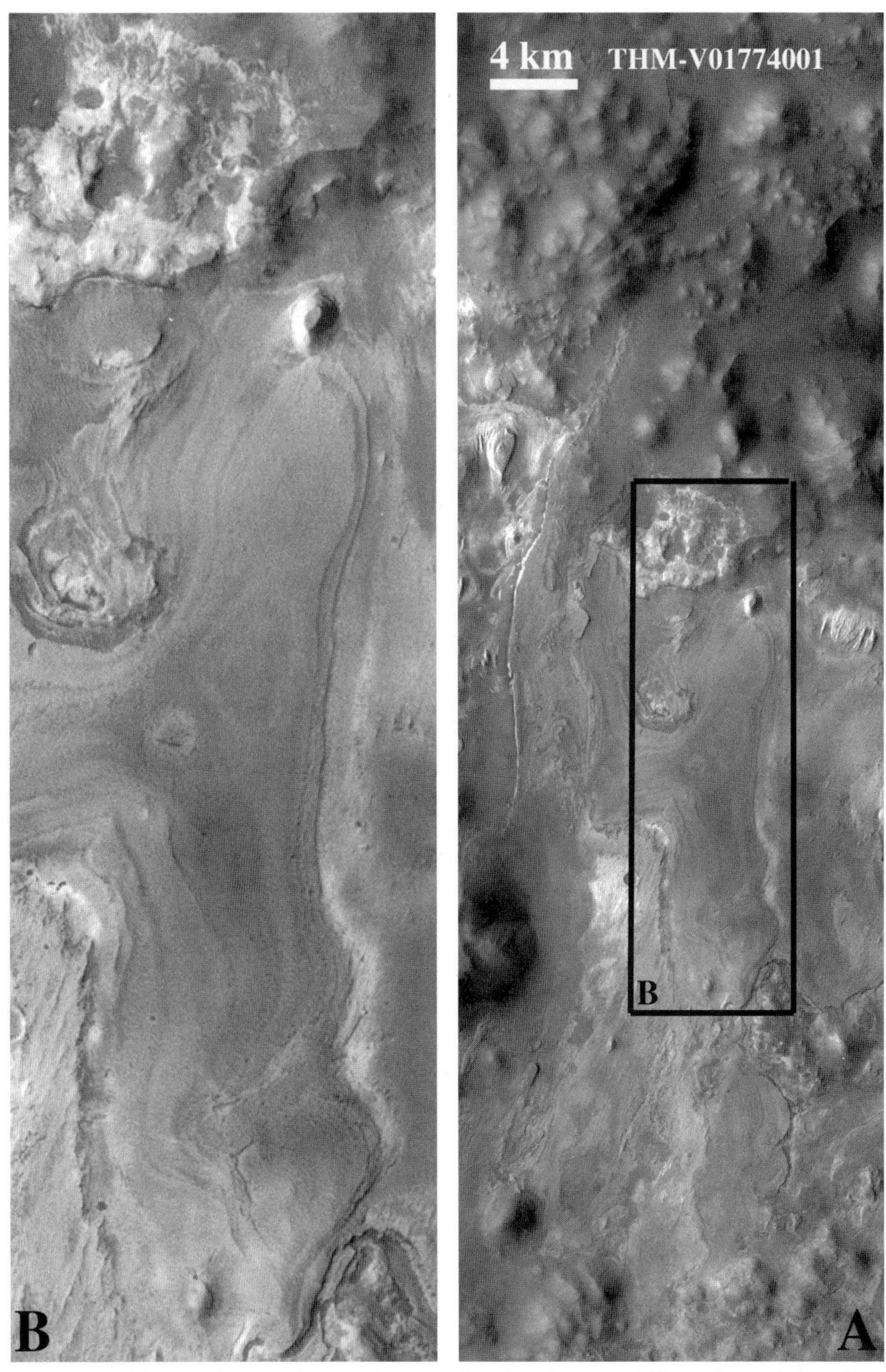

THEMIS image of part of Iani Chaos, showing knobby hills and mud-flow-like deposits. This is a 400-km-diameter region of collapse centered near 2.8°S lat., 17.5 long.

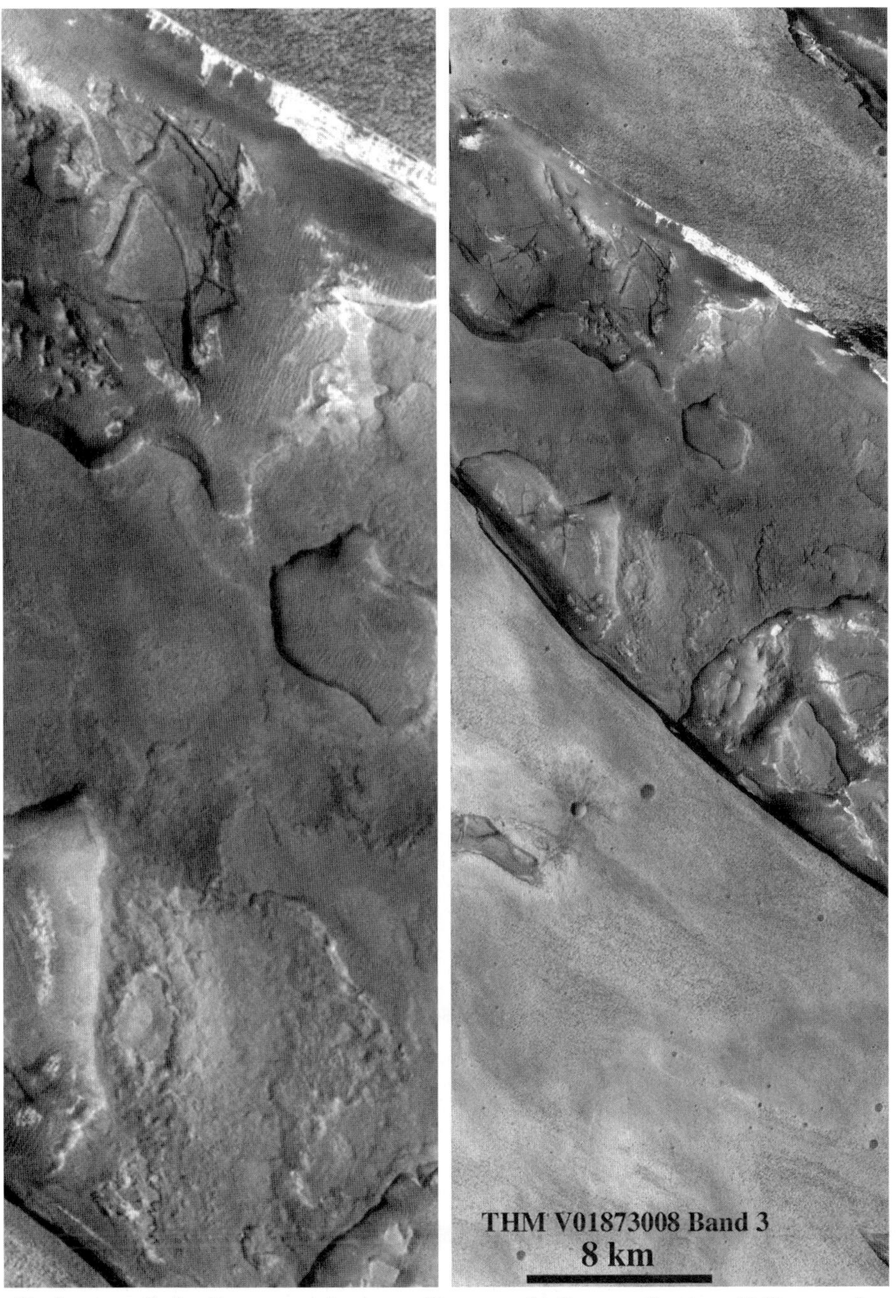

Fault-controlled collapse zone in Aram Chaos, a chief source for Ares Vallis, a major outflow channel on Mars. The outflow channel formed substantially by fluid withdrawal and terrain disintegration in such chaotic terrains. Bright deposits in some depressions and on some scarps could be salts deposited by water or steam venting, or it could be ice due to recent water venting. THEMIS V01873008B3.

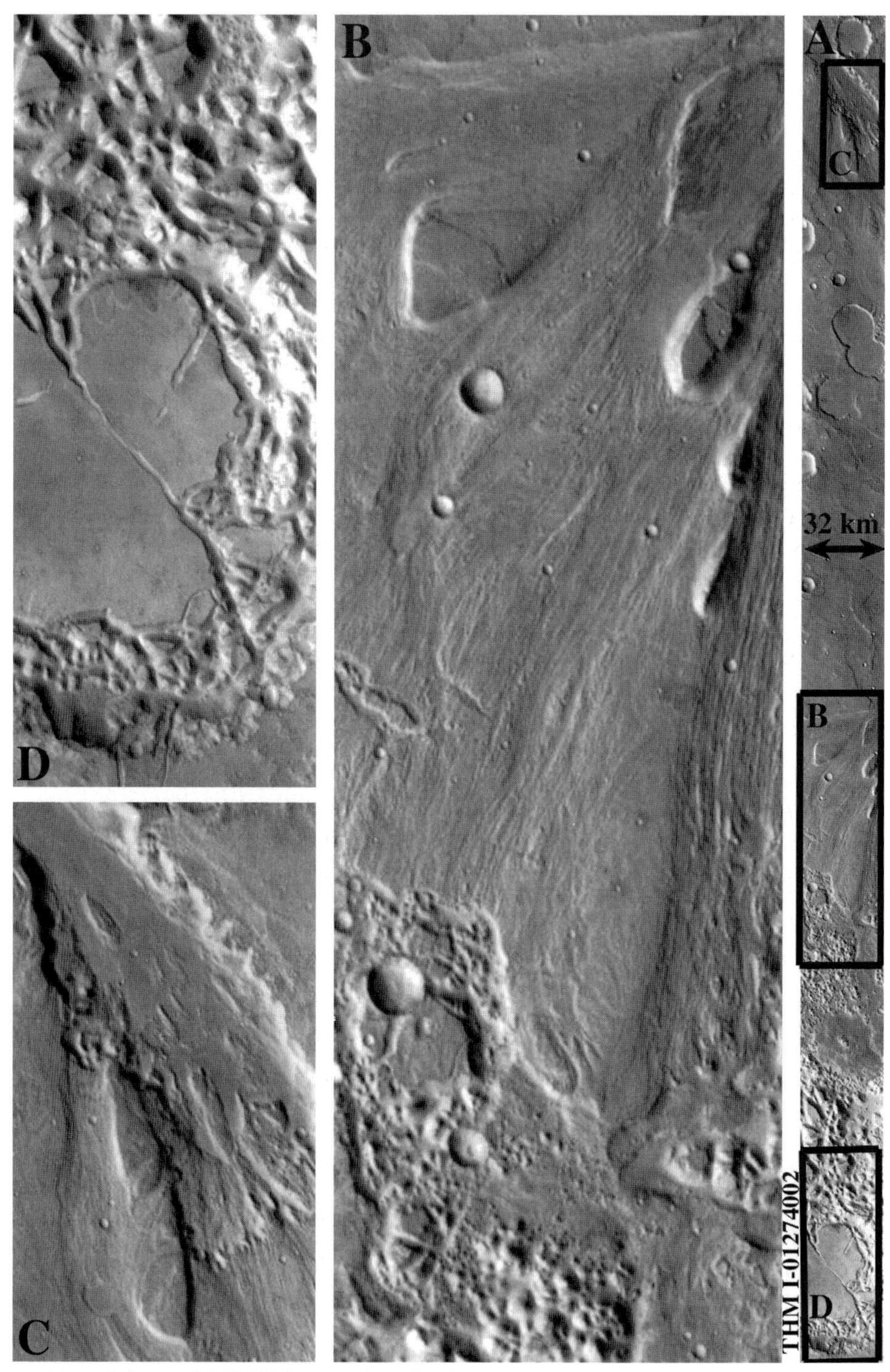

Mars Odyssey THEMIS image of Hydaspis Chaos and Tiu Vallis (outflow channel)

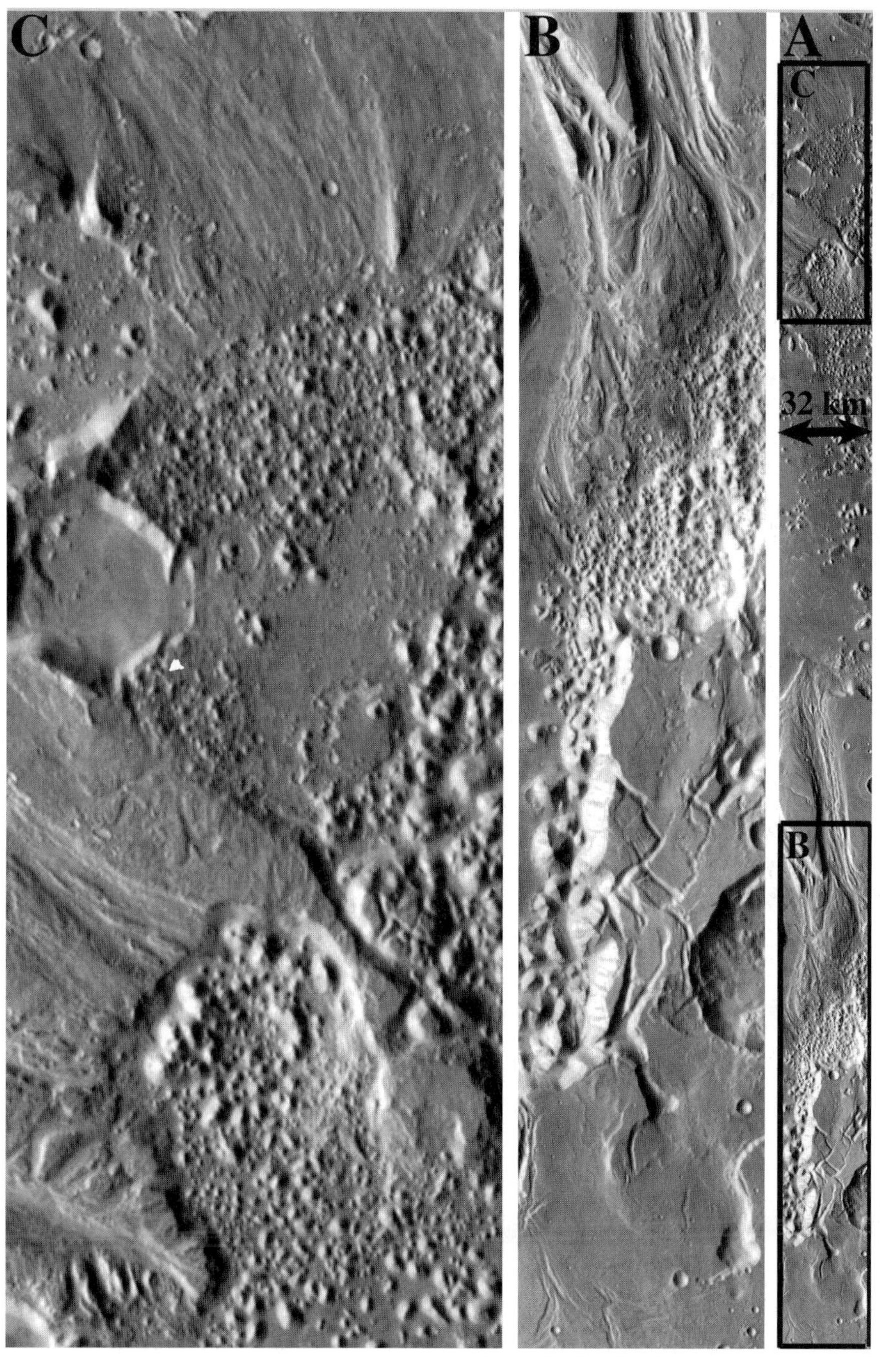

Mars Odyssey THEMIS image of Hadraotes Chaos and Tiu Vallis (outflow channel). Chaos regions lend a definite impression that a liquid upwelled from below. THEMIS I-02123002.

terrestrial ice sheets produce their streamlined bed forms, grooves, and ridges by subglacial outburst flood processes. Shaw's ideas remain controversial in the Earth science community; this controversy is sufficient to give pause to any rush to judgment on the formation of Martian outflow channels. However, still a problem for the glacial ice stream model of the outflow channels, there is the vexing observation that no normal terrestrial-style accumulation zone for the supposed glaciers was present. Instead, they were sourced at subterranean levels, as indicated by fissures and the collapse features evident in chaotic terrains at the sources of outflow channels. Thus floods seems a more likely fundamental cause, though rapid freezing and ice processes would be inevitable. Furthermore, observations of the deposits produced by these channel-forming flows indicate a material more like debris flows (Chapters 4 and 6). The rapid extraction of the erosive material from large volumes of crust requires a fluid form of the material. However, wholesale ingestion of Martian upper crustal material, fluidized by liquid water and perhaps liquid or vaporous carbon dioxide, would no doubt have included ground ice. The likelihood is that the flowing mass would have been partly frozen, and so a composite origin seems likely.

Other early hypotheses explained the outflow channels with floods of liquid carbon dioxide or hydrocarbons. Even wind was suggested at one time. As these models were considered, it became only more evident to most Mars researchers that floods of water or massive water-bearing debris flows were the basic driving force behind these huge channels. The water-flood (or the related debris-flow) hypothesis has become a cornerstone of nearly all current Mars models.

The sources of the circum-Chryse outflow channels in the Valles Marineris area has spurred an interesting hypothesis. According to one view initially championed by Baerbel Lucchitta (US Geological Survey, Flagstaff) and further developed by others, Valles Marineris at one point may have filled brim-full with a lake, which probably was ice covered. If so, this lake may have had a volume more than 1 million cubic kilometers, about 15 times more voluminous than Hudson Bay. Layered rock sequences observed within the canyons may be partly sediments (plus landslides, lavas, and volcanic ash) deposited in the lake. Eventually, a breach at the eastern end of Valles Marineris contributed to massive sediment-rich floods that carved enormous winding outflow channels. Recent reports from THEMIS observations that a layer or dike of olivine – a mineral quickly altered to "iddingsite" under aqueous conditions – occurs near the bottom of Valles Marineris has reduced the likelihood that a vast lake existed in this region for any great amount of time.

HOT MUD BATHS OF ELYSIUM

My Norwegian ancestors came from a town built around therapeutic springs and mud baths. On Mars, the best analogs might be found at Elysium. This volcanic plateau has on its flanks many large channels of apparent water origin, and presumably there have been many smaller, still undetected points of geothermal interactions of heat with ground ice. In many cases in Elysium, Tharsis, and east of Hellas, large channels emerge full size from regions of fractures and chaotic collapse

THEMIS image of a small portion of Kasei Vallis, a huge outflow channel system that debouched into the Chryse Basin. It is generally believed to be a flood channel, but ice sculpture may have contributed. If floodwaters formed the channel, they were probably hyperconcentrated slurries or debris flows. The system is 3300 km long and about 400 km wide. The portion shown in the THEMIS scene above, right (33 × 192 km) covers just 0.5% of the channel system. In many places the channels are over 1 km deep.

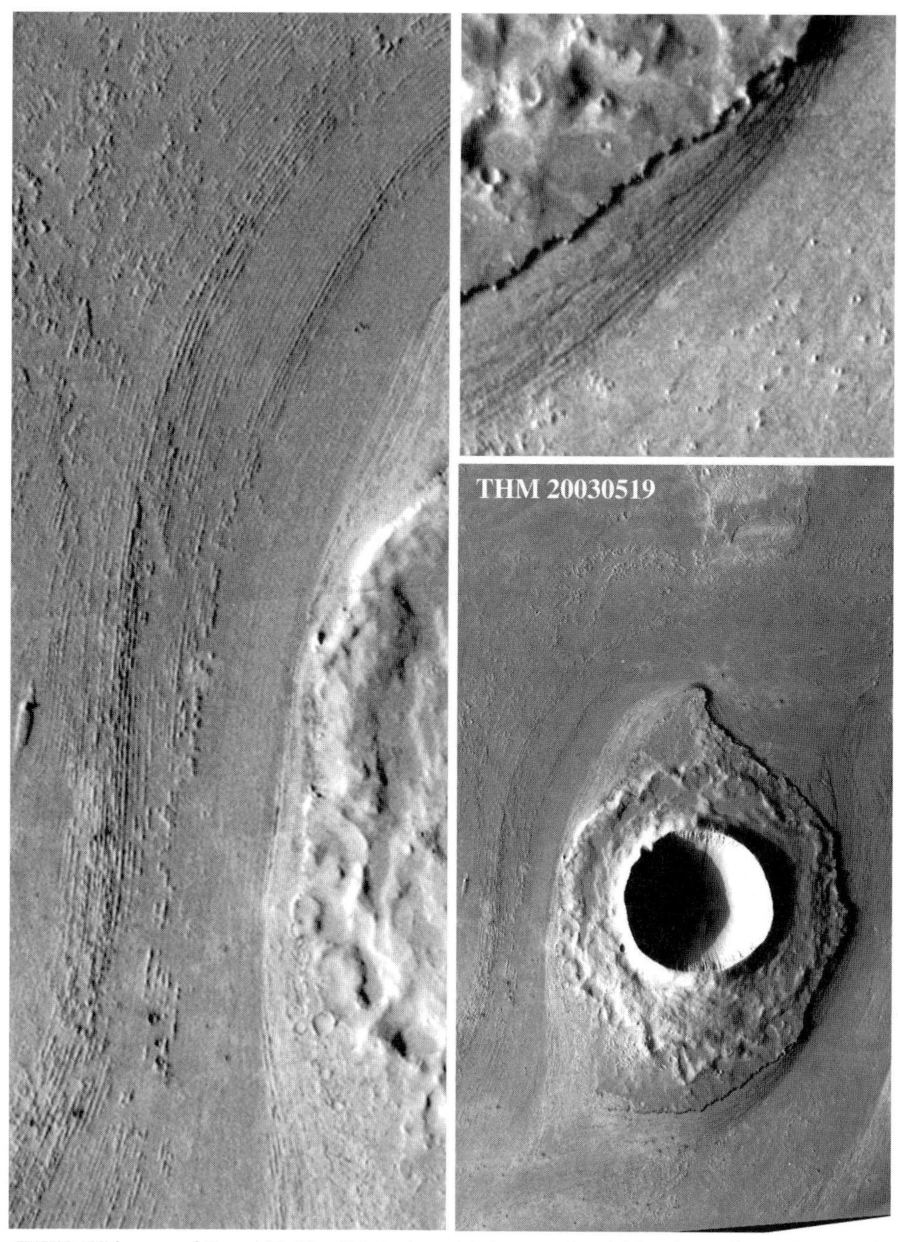

THEMIS image of Kasei Vallis. (A) A sigmoidal streamlined island was formed around an impact crater. (B) The island is clearly eroded up to a high-elevation limit (the prominent cliff). The finer cliffs or lineations appear at a glance to be eroded layers, but more likely they are nonstructural erosion features. (C) Similar fine erosional grooves, flutes, and drumlinoid features trace out the flow lines of the erosive medium. These are some of the morphological indicators that Baerbel Lucchitta has used, at similar scales, to support a dominant ice stream erosional process. See Lucchitta (2001).

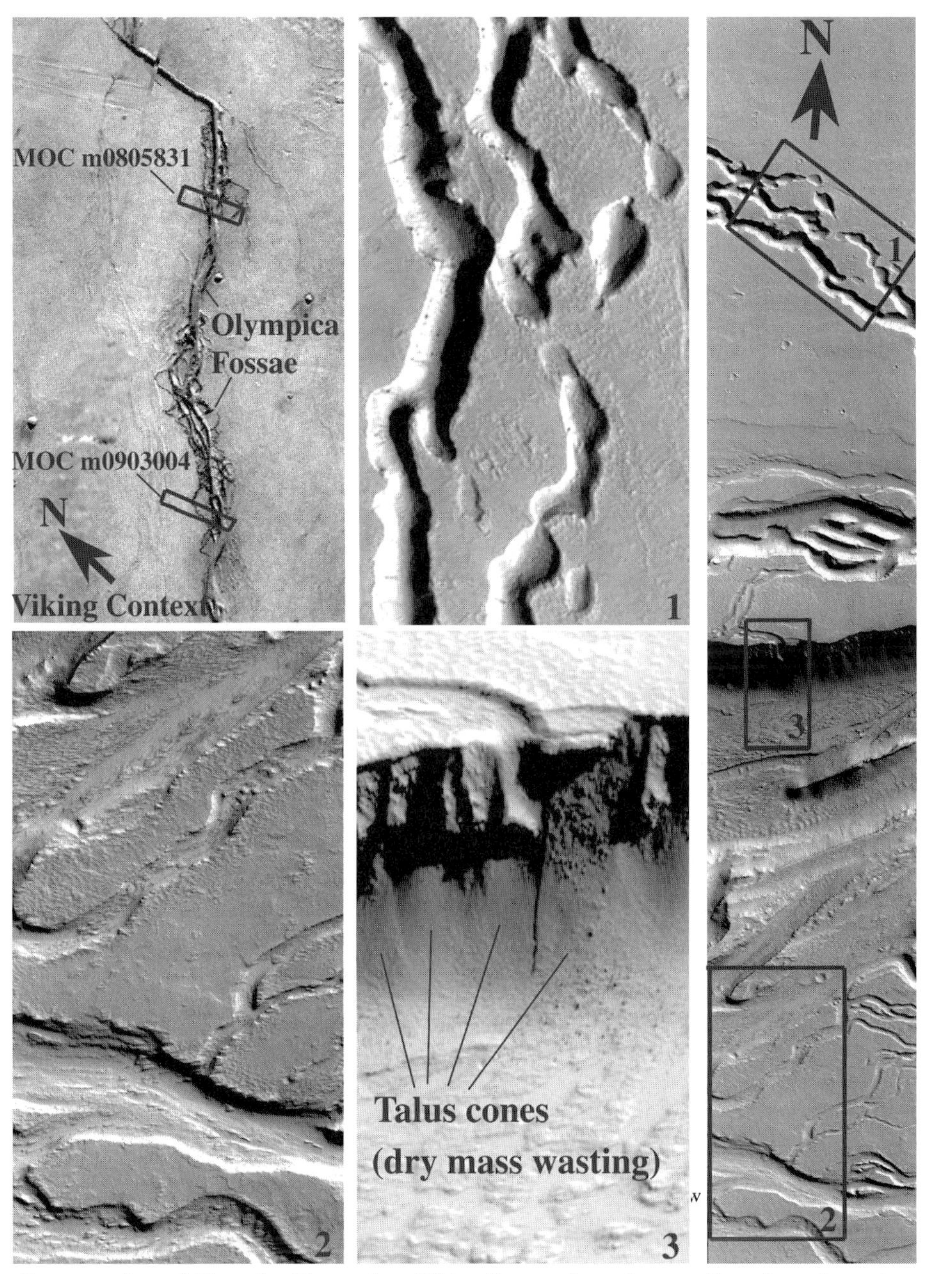

Source region of Olympica Fossae (see *Viking* Context image), a small outflow channel on the volcanic Tharsis Plateau. The MOC image (m0805831) and enlargements reveal the youthful age (Amazonian) and excellent preservation of this channel system. Even so, it has undergone some degradation, such as mass wasting along major cliffs. Like other outflow channels, this one emerged full size from underground sources. Unlike most others, which generally start from chaotic terrain, this one emerged from fissures.

Braided channel deposit, ~ 1 m across, in sand; flow to right. Martian outflow channels are up to 10^5 times longer, wider, and deeper, and although scaling is different, some fluid dynamical processes are the same; this is shown in similar morphologies. However, there are also fundamental differences. Here, well-sorted fine sand dominates the sediment load, whereas on Mars the catastrophic flows caused everything from micronic dust-size particles to kilometer-size crustal blocks to be transported and deposited. Flow cavitation probably plays little or no role here, but it was a key process for the Martian channels. Probably the biggest difference is that sediment transport here involves mainly rolling, sliding, and bouncing of sand grains over the bed, whereas in the Martian outflow channels, flows involved mainly dense suspensions. There is a difference also in source: here the flows had sources in surface water, but the Martian outflow channels originated from underground sources at chaotic terrain.

terrains. It is very clear that the volcanic structures have played a critical role in the origin of the water-carved features. Geothermal heating in these volcanic regions, including heat from shallow magmatic intrusions, has no doubt played a role in producing the water or at least thinning the permafrost that had confined any groundwater. Also possibly significant is the tectonic uplift of permafrost-confined aquifers. Such an uplift would generate a high hydraulic head that, combined with thinning and weakening of the permafrost, could cause sudden outbursts.

On Earth, volcanoes and glacier ice historically have been a volatile and dangerous mix. Abrupt increases in geothermal flow due to volcanic eruptions or shallow magmatism can cause partial melting of the glaciers. Subglacial lakes may form and suddenly drain by melting a conduit or by hydraulically lifting the ice. The resulting flood is called a jökulhlaup (an Icelandic term). In other common types of interactions, geothermal heating can cause the glaciers to break off in an ice

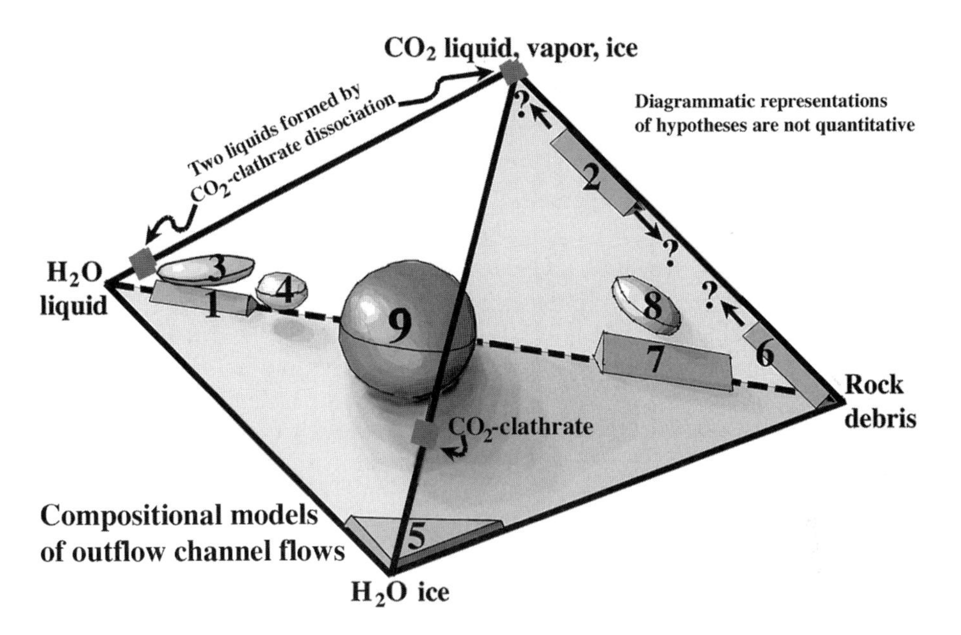

Many hypotheses have been presented in attempts to explain the Martian outflow channels. While there is no disputing the necessity of having vast quantities of a fluid or plastic material move erosively over the Martian surface so as to erode the immense outflow channels and deposit the eroded rock in the northern plains, that is where the agreement ends. Disagreement begins, first, with the issue of the erosional agent's composition; second, over the time interval over which the flows were active and the flows' precise volumes, and third, over the causes of the outbursts. There is no settling the second and third matters until the first – composition – is resolved. This diagram illustrates the wide range of ideas about the composition of the erosional agent. It is minimally made of two physical components – the fluid or plastic material(s) and the eroded rock. Thus, every model involves rock debris as a component of the flows, although the rock could have been carried in suspension or as bedload or in heterogeneous mixture with the fluid or plastic substance. This diagram, like the Mars community as a whole, discounts or neglects the possibility, as suggested soon after the channels' discovery, that wind or condensed hydrocarbons may have eroded them. The remaining fluid or plastic substances include ice, liquid water, and carbon dioxide. Within this "tetrahedron of possibilities," the numbered solid objects represent hypotheses that are still viable or discussed in the Mars science community: (1) dominantly water; (2) liquid carbon dioxide; (3) CO_2-saturated liquid water; (4) CO_2- and ice-rich water; (5) glacier ice; (6) CO_2-vapor assisted dry flows; (7) water-based debris flows; (8) H_2O- and CO_2-based debris flows; (9) generous mixtures of all these materials. This concept has been pushed most notably by myself, Vic Baker, and Ken Tanaka. See also these papers (listed in the Bibliography at the end of the book): Baker (1973), Milton (1974); Dobrovolskis (1975); Clark and Mullin (1976); Lucchitta (1981); Lucchitta et al. (1985); Jöns (1986); Parker et al. (1989, 1993); Baker et al. (1991, 2000); Head et al. (1999); Kargel (2000); Hoffman et al. (2001); Tanaka et al. (2001, 2002).

avalanche or to flow rapidly and thin, thus reducing pressure on the volcano's magma chamber. Volatiles dissoved in the magma or acquired by ingestion of glacial meltwater then can exsolve suddenly and cause the magma to explode. When the water and ice from the glacier mixes with rock debris and volcanic ash, a huge and devastating flood or debris flow, or a steam- and air-lubricated flow, called a lahar, may be unleashed, sometimes killing thousands of people. While glaciers are usually involved on Earth, the physical processes suggest that permafrost ice might achieve the same result when affected by volcanism.

On Mars, similar processes – minus the interactions with human beings – appear to have occurred on a scale orders of magnitude greater than is common on Earth. The further likely addition in Martian permafrost of large amounts of CO_2 clathrate or CO_2 in pure liquid form or dissolved in groundwater (Chapter 6) – all far more volatile than pure ice or water – can create even more highly unstable and powerfully explosive conditions when it interacts with volcanos.

Jökulhlaups and lahars are caused by catastrophic processes not directly tied to climate. Thus, the water phase represented by these events does not carry implications for the ancient Martian climate, but the huge scale of the volcano-ice features so produced are a testimony to the vast reserves and wide distribution of ground ice on Mars – a fact that certainly is closely tied to climate.

RIVER-FED LAKES OF A BYGONE ERA

Among the most exciting and formative papers in my experience was one presented by Steven Jankowski and Steven Squyres (Cornell University) in 1993 in *Icarus*. On the basis of *Viking Orbiter* photos, they mapped the distribution and organization of small-valley segments. Rather than forming an arbitary distribution, they were shown to concentrate within certain highland intercrater basins, forming in some cases fairly complete systems of closed (inward) radial drainage or systems that were not closed but had basins with connected inlet and outlet valleys. Some of these basins also had peculiar highland chaotic terrains. Associated with erosional channels and former crater lakes are layered and sometimes terraced basin deposits that appear to represent sedimentary rocks – probably lake deposits. With just a bit of interpretive license, these landscapes appear to form complete erosional–depositional systems associated with drainage into closed or overflowing basins, though certainly not closed hydrologic systems. These basins represent high-priority targets for future spacecraft exploration, including a NASA rover due to land shortly before publication of this book.

While, in 1993, we did not have the accurate topographic information we now possess, regional slopes could be inferred from the nature of the drainage patterns. The inferred hydrologic systems were analogous to those of lake basins and their drainage systems here on Earth. While most of these supposed lake and river systems were demonstrably very ancient, they imply an era when water flowed and pooled freely on the surface and probably for a very long period of time. While the climatic implications of these hydrologic systems are subject to interpretation, the

Eroded outcrops of a series of uniform sedimentary layers in a crater, 2.3 km across, on the floor of the giant crater Schiaparelli. Schiaperelli, an equatorial impact basin, may once have been filled by a lake or an eolian dust sheet. Schiaparelli is on the boundary of Arabia and Sabaea Terrae, near Meridiani Planum, where vast layered hematite-rich plains occur. NASA/JPL/MSSS Press Release MOC2-403.

implications for sustained water activity are profound. The driving force behind the mobilization of liquid water seems not to have been volcanic or geothermal but an energy source more broadly distributed. A warm climate driven by solar insolation and, perhaps, heat-trapping greenhouse gases in the atmosphere was instantly one plausible interpretation. Abundant new knowledge of Mars gained in the last decade has bolstered the idea that river-fed lakes once occurred widely in several regions of Mars (Chapters 6 and 7).

CREEPY

On the basis of *Viking Orbiter* images, Gerry Schaber, Michael Carr, and Baerbel Lucchitta (all at USGS), in several separate works with other colleagues, identified a number of landforms on Mars consistent with the appearance and scale of ice-caused landforms on Earth. Lucchitta observed fine-scale polygonal patterns that resemble the polygons common in many terrestrial permafrost landscapes, and each of these workers described glacier-like lineated deposits in some valleys of Mars, lobate

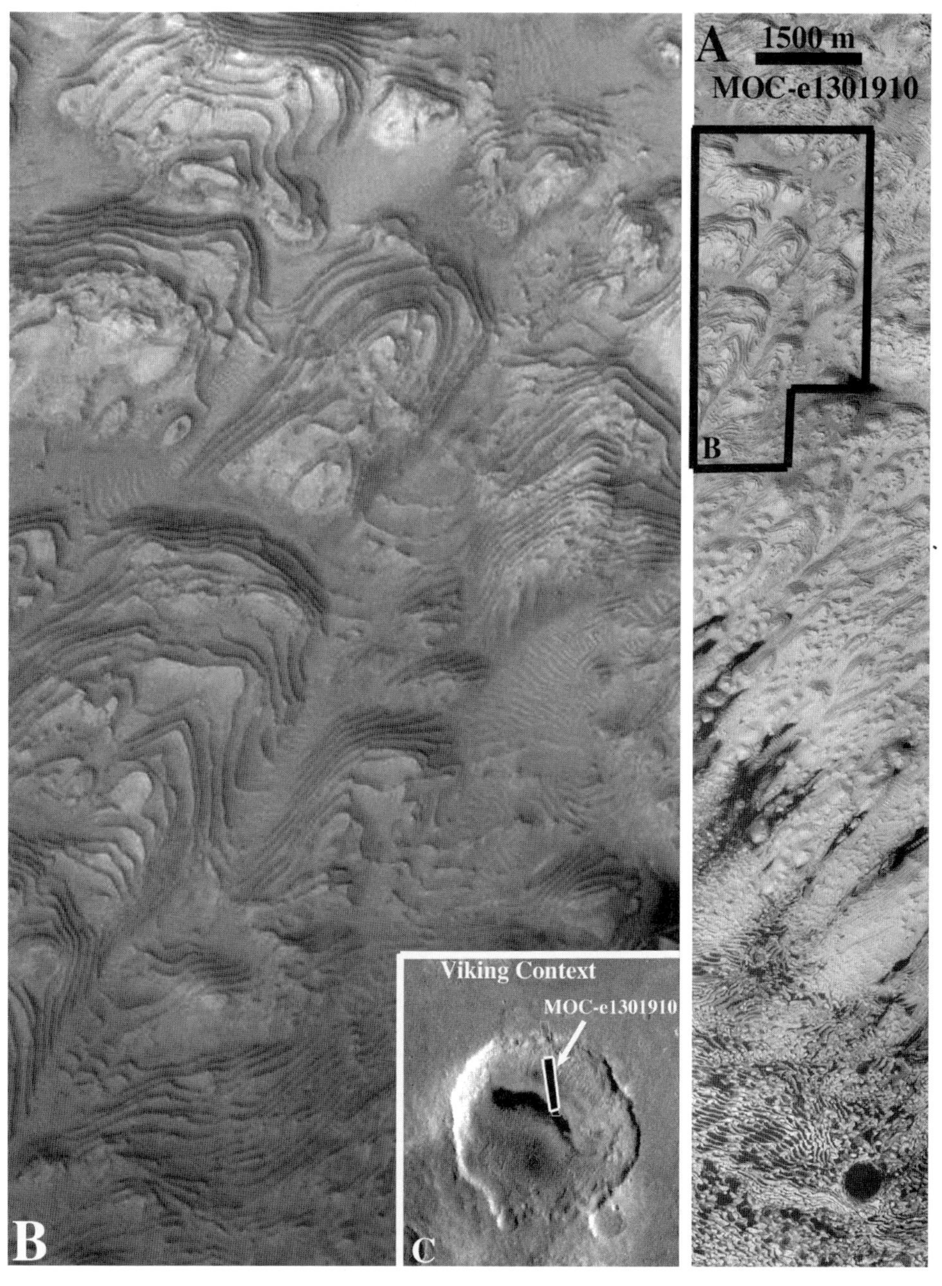

Rock layers are widespread in western Arabia Terra. Suggestively, this region lies within Tim Parker's largest proposed ocean shoreline. The uniformity of layers indicates a cyclic process. Volcanism does not usually exhibit such remarkable uniform cyclicity, but climatic modulation of deposition often does. Wind-storm lofting and deposition of dust or aqueous deposition of clastic or chemical sediments could explain these rocks.

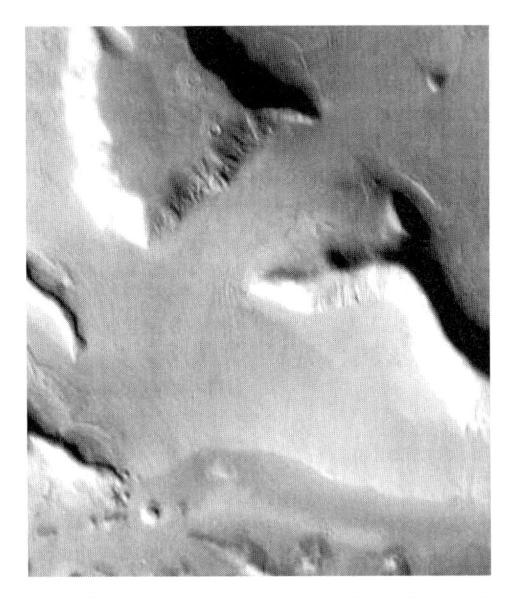

"Lobrate debris aprons," such as these in the fretted terrain, inspired the first interpretations of relic iceflow features on Mars. Michael Carr, Gerald Schaber, and Baerbel Lucchitta (all at USGS), Steve Squyres (Cornell, then at NASA Ames) and others considered ice to be a key ingredient of these landforms. There was early consideration given to the possibility of glacier-type flow, but generally these were regarded at first to be the products of periglacial permafrost creep. The early interpretations gave little emphasis to details such as creep mechanisms. When mechanisms were suggested, there was little consideration of the specific climatic and hydrologic implications. It was recognized by Squyres and Carr these are exclusively a phenomenon of the middle latitudes, a point that has been amply supported by more recent studies based on improved imaging observations of recent missions. From THEMIS I-02221006.

glacier-like aprons around many mountains, and concentrically structured fillings in many craters. Most of the well-preserved Martian craters between 4 and 40 km across exhibit ejecta resembling mud flows, which are interpreted by these researchers as due to entrained water produced by impact melting of ground ice. From *Viking Orbiter* images, Steven Squyres (Cornell University) first documented that much of the middle and high latitudes of Mars has a softened appearance, including the lineated valleys, lobate debris aprons, and concentric crater fill. While major ancient terrain features, such as large craters and mountains, are not obliterated by the softening, they are heavily degraded, especially in the middle and high latitudes. Squyres and Carr (USGS), interpreted the terrain softening, concentric crater fill, lobate debris aprons, and lineated valley fill as due to slow, ice-driven creep of rubbly regolith and sediment common in permafrost areas of Earth. These modifications (collectively sometimes grouped as "terrain softening") are absent at low latitudes but are particularly predominant at middle latitudes. This latitude control is a clear indication that a climate-related phenomenon is the root cause.

A number of researchers initially dismissed some of these mid-latitude modifications as due to blanketing of the surface by dust and sand. By itself, the latitude control would not exclude climate-related dust deposition as the cause of terrain softening. Lucchitta and Squyres and Carr identified a host of specific types of textures that pointed more specifically to ice-driven creep and could not have been caused fundamentally by wind-transported/deposited sediment. Wind-deposited dust or sand could have contributed to the observed record, but lacking a major ice component such materials would not produce the observed features, such as lobate, convex-up profiles and flow lineations.

Although it seems that no ideas regarding landform interpretation of Mars ever gains complete acceptance, evidence from *Mariner 9* and *Viking Orbiters* interpreted as due to creep and freeze–thaw of permafrost became fairly well accepted during the 1970s and 1980s, becoming another key component of the Warm, Wet, Early Mars concept. Analogies to Earth's permafrost have proven persuasive in general, though the finer details and specific models are always subjective and controversial. Among the most compelling observations is the geographic distribution of hypothesized creep-related features. Poleward of about 35° latitude (the exact latitude varying widely according to slope, slope aspect, and geologic controls), the Martian crust appears to be enriched in ice to within a few feet of the surface, but at lower latitudes, ground ice is absent or restricted to great depths, as though a relic of the past. This relationship between the location of icy permafrost features and latitude is almost as predicted from theoretical studies of ice stability in Martian permafrost.

MARTIAN INK BLOTS

The erosional states of impact craters depend on crater size, age, and the specific processes that have affected that region. Thus, some Martian craters – either very young ones or craters in regions that have not seen much geologic activity – are morphologically fresh. Uneroded craters on the Moon and Mars differ in one main respect. The ejecta blanket of nearly every fresh Martian crater between 4 and 40 km in diameter appears like a giant mud splatter or has an ink-blot pattern; they apparently were emplaced in surface-hugging fluid-like flows, rather than emplaced as a dense spray of particles like the lunar crater ejecta. Craters exhibiting such "fluidized ejecta blankets" are known as rampart craters.

Fluidized ejecta blankets are thought by most Mars researchers to have been emplaced as mud flows or water-laden debris flows. Groundwater or impact-melted or vaporized permafrost ice apparently becomes entrained in the rocky debris thrown out by impact; as it hits the surface, it continues to flow until finally coming to a halt and forming a mud-flow-like pattern. The global occurrence of fluidized ejecta on Mars is thought to be a proxy for the occurrence of condensed H_2O – ice or liquid water – in impact-excavated material. This type of crater ejecta occurs almost everywhere on Mars for fresh craters of a particular size range, suggesting that ground ice (or groundwater) is universally present at some depth interval. However, the size of craters exhibiting such ejecta differs from one area to another and is

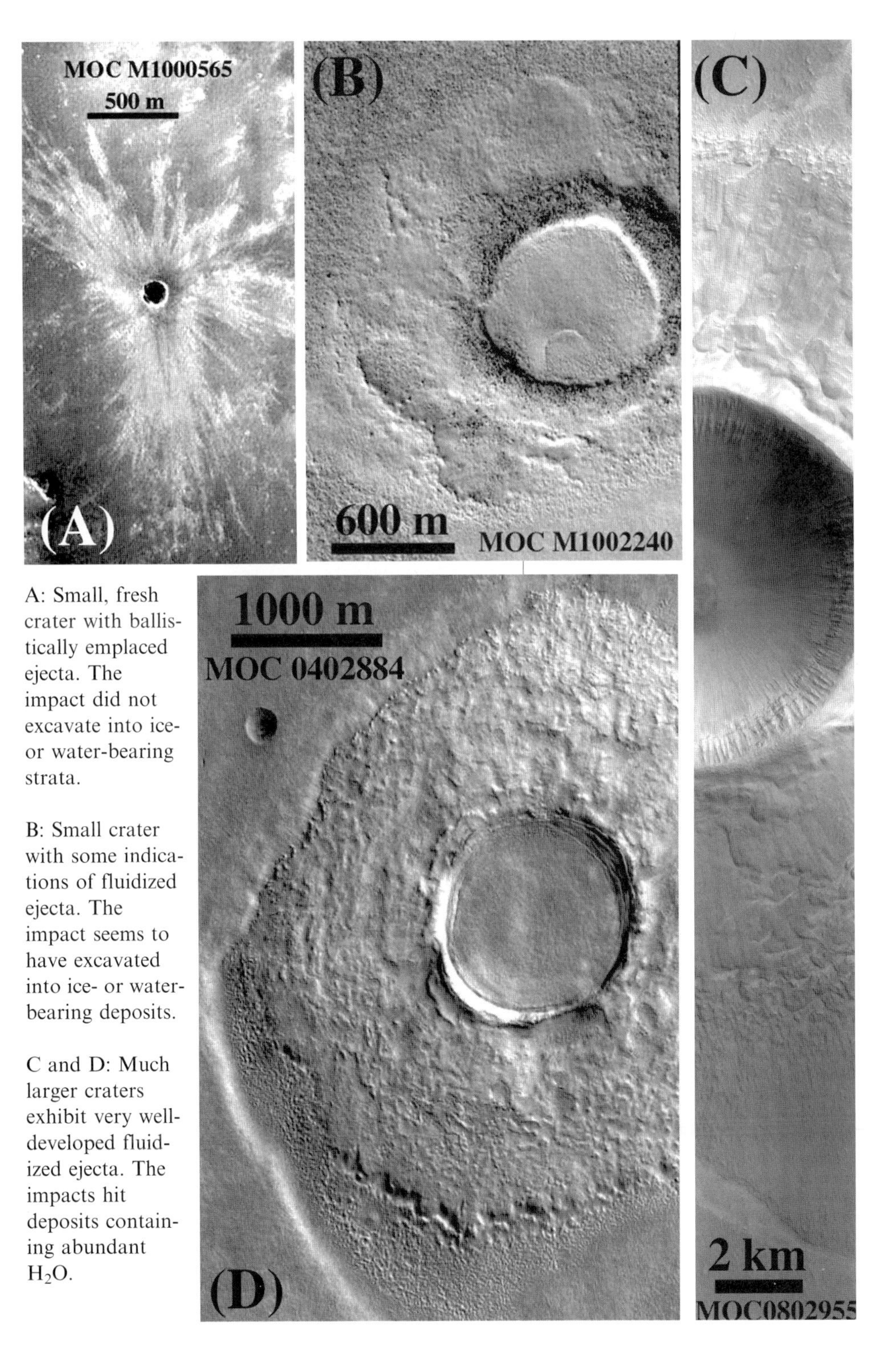

A: Small, fresh crater with ballistically emplaced ejecta. The impact did not excavate into ice- or water-bearing strata.

B: Small crater with some indications of fluidized ejecta. The impact seems to have excavated into ice- or water-bearing deposits.

C and D: Much larger craters exhibit very well-developed fluidized ejecta. The impacts hit deposits containing abundant H_2O.

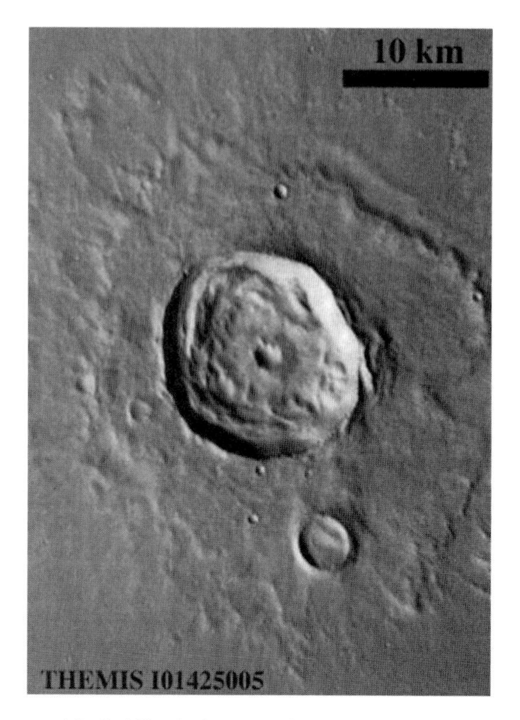

A typical large crater with fluidized ejecta. This crater has incipient development of a double ejecta blanket, with one lobate blanket overlying another. The interior also exhibits some interesting features: a central pit in place of a central peak normally expected for a crater of this size, and an interior lobate deposit of what looks much like a fluidized ejecta blanket. None of these features is found on the Moon. Work by Nadine Barlow, Ruslan Kuzmin, myself, François Costard, and others has shown that the crater size limits, occurrence, and geographic distribution of various fluidized ejecta morphologies correlate well with latitude. The exact meaning of the different morphologies is unclear, but a relationship to near-surface condensed volatiles – especially ice and/or groundwater – seems assured. A model that had maintained that atmospheric gas entrainment was responsible just does not seem adequate to the observations. In the case of the central pit, it is possible that the dynamic response of the impacted target initiated a central peak, as usual, but that the central peak itself was fluid-like and was unable to maintain itself; it promptly and violently collapsed under its own weight, sending fluidized fall-back ejecta across the crater floor. Nadine Barlow and others have speculated that double-lobed ejecta might relate to penetration through a ground-ice permafrost zone into a groundwater aquifer. It will take considerable geophysical and surface geological exploration to ferret out and confirm any such details.

generally latitude dependent. Crater depth, which scales to crater diameter by a simple relation, is an approximation to the depth of excavation by the impact. The fact that the smallest craters in many areas lack fluidized ejecta indicates that in these areas the upper layer is devoid of, or deficient in, ice. The largest craters also lack well-developed fluidized ejecta, probably indicating that at great depths rocks also lack ice or water. There appears to be a depth interval at every latitude where ice or ground water is prevalent. The minimum depth to the icy layer decreases as latitude increases, with a sharp transition from deep to shallow at around 35° both north and south of the equator.

Using dimensional scaling relationships known for typical Martian craters, and considering uncertainties in how much water is needed to fluidize ejecta, it appears that an icy permafrost exists at typical depths starting at 500–1,000 m at low latitudes and extends to depths of about 2,000–4,000 m. At latitudes greater than ±35° from each pole, fluidized ejecta occurs for smaller craters, indicating ice at far shallower depths, locally just tens of meters or less. A surface layer appears desiccated, especially at low latitudes, as theory very well predicts for the present climate. The deeper limit of icy permafrost may be set by the gravitational compaction of regolith and elimination of pore space in the Martian crust. A possible role of subsurface liquid versus solid H_2O – including a melting isotherm – in controlling the morphology of fluidized ejecta has been examined but is really very difficult to determine with confidence. While fluidized ejecta do not indicate anything about ancient temperature conditions on Mars, they do seem to suggest a very widespread distribution of subsurface volatiles, notably water (ice or liquid). This implies that volatile abundances globally are quite high and that, at least, at great depths water and ice are distributed widely – more widely than present climate models would allow if equilibrium was found to hold. Possibly ground ice remains extant (not necessarily stable) at depth even at the equator from a very ancient period when tropical humidities were higher; or possibly groundwater has flowed in the subsurface from the high latitudes to the lower latitudes. Either way, crater ejecta on Mars indicate that it really is a "water planet."

LONG, LONG AGO, A WATER WORLD

One chemist/astronomer– Carl Sagan – decided in favor of a liquid-water origin of small valleys on Mars very early in the debate, and many other prominent Mars scientists followed. Geologists came into the fray, and carried with them another bias borne of a different perspective and some common sense derived from looking at the world (our world) around them. It is a bias that profoundly shapes my own perspective of Mars. By analogy to water-carved features on Earth, liquid water was implicated in the Martian channels and valleys.

The story soon emerged that Mars – though now a cold, dry, windy planet – once had great volumes of water and a climate such that liquid water was stable. The new paradigm held that Mars at one time was relatively Earth-like. ***B***illions and ***B***illions of years ago (as Carl Sagan would say) Mars was a more Earth-like world where ice melted and water flooded and seeped and dribbled over the surface.

THE FAINT-YOUNG-SUN PARADOX

The emerging vision of a once-watery Mars posed a serious dilemma. Mars is now so remote from the Sun that water is frozen solid (in equilibrium with the barest trace of water vapor), and the radiation environment billions of years ago was much worse. The Sun has steadily brightened with time, and running the clock backward made the Sun an even fainter object, delivering only 70% as much heat and light 4 billion years ago as it does today. Yet, Martian geology indicates that liquid water was present.

The dilemma of the so-labeled Faint-Young-Sun Paradox also applies in modified form to Earth. On Mars, limits to the existence of liquid surface water are from the atmospheric temperature and pressure and relative humidity, and total pressure averages 6 mbar, near the triple-point vapor-pressure of H_2O. Thus, liquid water could exist at the lower elevations on Mars, if only a warm spot could be found where ice exists. At higher elevations, under current conditions, pure ice would sublimate before it could melt. However, a doubling of total atmospheric mass – for instance, from partial sublimation of the South Polar Cap – would raise the atmospheric surface pressure above the triple-point pressure of water, and this would make it possible to have a puddle of ice water, magically emplaced across broad areas of the surface at noon at the right time of year to prevent it from boiling.

Temperature and ice availability is a bigger issue than pressure for liquid-water stability on Mars. The surface temperature exceeds the melting point of pure ice around the summer solstice near the equator and at middle latitudes on equator-facing slopes – precisely the areas that lack ice at shallow levels. Furthermore, the melting point is only attained for a few hours of the day, and night-time temperatures are still so cold that melting temperatures never penetrate more deeply than a few centimeters, where there is no ice to melt. Thus, we can be certain that in the present era pure ice never melts at the surface, except for a few drops at special places. Those special places might include, for instance, mid-latitude cliffs at low elevations where recent slumps and landslides may have exposed buried ice along slopes favorable to a brief noon-time summer thaw. Such liquid water would be a mere trace technicality with no geomorphological or biological significance. With a faint early Sun, the temperature problem is much more severe, as James Kasting (NASA Ames) and many others have found. The occurrence of ice-saturated atmospheric conditions (high humidities) is also a major issue. As ice needs to exist in the right places in order to melt, and as the air needs to be humid to prevent a rapid evaporation of the water, this is an issue that will be addressed further in the following chapters.

EARLY MARTIAN GREENHOUSE

There are only three plausible solutions to the creation of temperature conditions that are more amenable to the melting of ice – reduce the melting point of water by dissolution of natural antifreezes (such as salts present in the soil), localize the flow of internal heat (such as at volcanoes and hot springs), or somehow make solar heating more effective. Although we shall return later to the other mechanisms, a natural planetary greenhouse seemed the answer to Carl Sagan and others in the 1970s. A solar greenhouse in our back yard reduces advection of heat by wind and convection currents, allows sunlight to enter, and slightly reduces outgoing infared radiation. If we add a little chemical heat from decaying biomass, a lengthy growing season results.

The greenhouse namesake of the planetary atmospheric blanketing mechanism is not strictly physically accurate. In agriculture and gardening, the greenhouse achieves its purpose by restricting advective dissipation of solar heat by wind and thermal convection. In planetary atmospheres the prime function of the atmospheric greenhouse blanket is due to a molecular mechanism causing increased infrared scattering and opacity relative to that of incoming visible radiation. The term "greenhouse" conveys the basic idea even if not the science. The radiative activity of Earth's trace atmospheric constituents – carbon dioxide, water vapor, methane, and ozone, plus the radiation scattering caused by clouds – make our planet's surface an average of 30 K (54 °F) warmer than it would be if our atmosphere was absent. Mars also has a CO_2 greenhouse atmosphere, but as it lacks Earth's abundance of greenhouse gases other than CO_2. Today's frigid Martian greenhouse produces just a few degrees of warming, with many spectral windows wide open due to its inability to hold much H_2O, another greenhouse gas. However, Carl Sagan and others knew that planetary greenhouses can be far more robust – in the extreme, Venus is crushed under a pressure of 95 atmospheres of mostly carbon dioxide, roasting the surface at 760 K (844 °F), which is some 450 K (over 800 °F) warmer than it would be if it had no greenhouse atmosphere.

One well-known obstacle to a watery Mars is the current low atmospheric pressure. With a mere doubling of atmospheric pressure, Mars would exceed the triple-point pressure of ice water over most of the surface, but the additional gas would be almost inconsequential to mean surface temperatures. Several bars of CO_2 pressure would be needed for a robust Martian greenhouse and a wet surface 4 billion years ago. Placing and maintaining this amount of CO_2 into the atmosphere is not a trivial matter. Calculations by Jim Kasting (then at NASA Ames Research Center) showed that with so much carbon dioxide, cloud condensation in the upper atmosphere and at the poles may have caused both an increase in planetary albedo (allowing less heating of the surface) and a collapse of both the atmospheric density and the greenhouse effect. Other scientists have determined that the greenhouse can be made even more robust by traces of other radiatively active gases or by the infrared scattering of fine-grained CO_2 clouds in the upper atmosphere.

In a posthumous 1997 publication in *Science*, Carl Sagan, working with Christopher Chyba (then at the University of Arizona), found that an early reducing methane–ammonia-rich atmosphere on Earth and Mars would, through ultraviolet-driven photochemical processes, produce a high-altitude aerosol haze that would tend to protect the lower atmosphere from solar ultraviolet rays. With the bulk of the atmosphere's methane and ammonia so protected from destruction, it would linger long enough and build up to sufficiently high levels that it could provide the necessary greenhouse heating to make a warm early Mars and Earth. No final resolution of these matters has been reached, but some variant of the greenhouse possibility remains one of the best solutions going for the existence of a warm, wet Mars (recently or long ago).

SEEING MARS ACROSS THE GEO-SPECTRUM

The outsize features of Mars may suggest a world that has somehow outdone our own, but one whose activity is all or nearly nothing. While the drama of Martian geologic history far exceeds all natural events on Earth in recorded human history, the modern student of Earth knows that our world has also its bag of unbelievable tricks and employs them as frequently as Mars.

James Hutton, a founding father of modern geology, believed that "*the present is the key to the past*." This "Uniformitarian Principle" was established on the basis of features in the rock record that have direct parallels in modern, active processes that any patient geologist can observe. Many observed processes have left their marks in rocks for anyone to see and to construct into hypothetical rock stories, which can be used to draw logical inferences on important issues, whether exploring for gold or looking for clues to the beginnings of life.

Hutton's point was a major achievement in making Earth's rocks comprehensible. However, uniformitarianism misses the equally important points that (1) many of the most enduring features produced by erosional, sedimentary, tectonic, impact, and volcanic processes, and the geomorphic work so caused, is accomplished in short periods of time by mega events; and (2) planets are generally in a state of long-term gradual change and, in some cases, sudden epochal change. These changes can alter the rates and ratios of key processes and bring planets into entirely new process realms. On Earth, we have had changes, for instance, from a reducing to an oxidative atmosphere, sudden excursions to snowball and hothouse climatic conditions, and the drying and then abrupt refilling of the Mediterranean Sea (the Mediterranean Messinian Crisis).

Uniformitarianism makes no accounting for epochal change, yet evidence has steadily accumulated that it is a potent reality of planetary evolution. Furthermore, uniformitarianism acknowledges processes that geologists personally observe and document, but it fails to account for important processes that may not occur frequently enough or quickly enough to be observed in the lifetimes of geologists. Uniformitarianism is a fine start toward understanding the rock record, but it acknowledges only a small part of the intensity/scale–frequency spectrum of observed processes; as any basis of concept modeling, it misses entirely a range of processes that may occur episodically and so rarely that geologists do not live long enough to witness them.

There is a modern public recognition, lacking in Hutton's view, of the great and frightening breadth of Earth's phenomenological geospectrum. The ancients recognized this spectrum and wrote geological and meteorological mega-calamities into religious texts. The power of Earth (or of God working through Earth) to create and sustain the diversity of life is widely praised, but equal time is given to the Earth's fearful power to ruin lives *en masse*. The public knows very well that uniformitarianism and this year's news go only so far to explain Earth. We fear the "perfect storm," the seismic "Big One," and careening, crashing asteroids as never before experienced by civilization. Many of us pray for divine deliverance from processes that are coded into the physical natural system and are in fact completely unavoidable. In a carry-over from the animalistic human need to prepare for the very worst that Humanity, Nature, or the Evil One can throw at our tribe, we pay talented actors and actresses tens of millions of dollars to show us how terrifying it would be, and how Nature can be defeated, if that perfect danger ever crosses our village path. The public craves an understanding of the physical limits – if there are any – of natural calamities.

Amazingly, the Cold, Dry, Windy Mars adherents took decades to recognize that the present Mars is the key to only a small part of the phenomenological spectrum of the Martian past. Evidence of much grander events occurring in eons past, transient though they may have been, are written in the rocks and landscapes. In many ways the Martian record is thinly coded and simpler to read than Earth's. Mars has its own unique and unimaginably broad geospectrum; the present is a small slice of this spectrum, even though it may represent almost 95% of geologic time.

Hutton's Earth, while more 'scientific' than the biblical Noachian view of Earth, and certainly demystifying important aspects of Earth's rock record, was in other regards a step backward from that offered by the authors of *Genesis*. There is no hint that geologists and planetary scientists have at last seen our own world and Mars through the full spectrum of the process rainbow, but surely we are no longer seeing these worlds in the narrow-band monochrome of our own life experiences or snapshots by the *Viking Landers*; nor are we seeing these worlds purely in terms of catastrophic God- and Satan-driven acts of incomprehensible devastation. Certainly geology has evolved beyond Hutton. This evolution in thought is forced upon us by the rock records of both Earth and Mars.

While uniformitarianism falls far short of explaining real worlds, pure catastrophism, epitomized by the biblical view of the Noachian flood, neglects the substantial changes to Earth that are occurring day by day. These sure-but-steady processes include, for instance, the incessant creep of mantle rocks (occurring at about the same rate that fingernails grow) and glaciers (one to four orders of magnitude faster than the growth of fingernails) and the incessant pick-pick-pick due to freeze–thaw weathering (causing mountains to wear down at a rate about one to two orders of magnitude more slowly than the growth of fingernails).

Earth is neither a Huttonian uniformitarian planet, nor a Noachian world; Earth has a complete geoprocess spectrum, and so does Mars. However, if we consider Earth's geoprocess spectrum to be relatively "blue," implying a high activity of the high-frequency everyday processes, the Martian geoprocess spectrum is relatively "red," implying a greater relative domination of the rock record and landscapes by infrequent high-magnitude events. Mars is more the catastrophic world, and less the uniformitarian world, by comparison to Earth. Martian geomorphology and sedimentology is almost surely even more prone than Earth to control by rare, high-magnitude excursions from eons of boring "normalcy." The reason is simple in theory and well-supported in observation. First, being a one-plate planet, Mars has difficulties releasing internal heat in a steady way, so it is prone to sudden surges of igneous and tectonic activity. Second, when global surface temperatures are freezing cold, liquid water is absent at the surface and atmospheric transport of water vapor is reduced; very little hydrogeologic action happens (compared to Earth's prevalent watery surface), even over tens of millions of years. Once in a blue Martian moon, conditions are such that global temperatures and the distribution of ice permit liquid water to do its work. Suddenly, a lot of geologic work overprints eons of comparative nothingness. Sometimes surface-driven conditions conspire with (or are driven by) internal processes to trigger an autocatalytic chain reaction. The sum result is a Mars overwhelmingly dominated by extreme events. Comparable extreme events occur on Earth, but their rough edges are muted and destroyed by the everyday slumping, creeping, quaking, crumpling, wearing, and burial we all know from our lives and newspapers.

Much geological action on Mars appears to be spasmodic or episodic; a lot of geomorphic work may be done in short epochs interspersed with far less active eons. Even if Mars was once much warmer than today and had transient liquid oceans or mud seas and vast ice sheets, the erosion record of Mars tells us that the cumulative duration of any possible marine-dominated global circulation/precipitation regime is at most a few million years (a short time, geologically speaking).

Some chief underlying bases for the efficacy of mega events on Earth is the low viscosity of the atmosphere (10^{-5} Pa-s on Earth at sea level) and liquid water (10^{-3} Pa-s), the vertically stratified and laterally heterogeneous input of solar energy, and the heterogeneous distribution of sources of water vapor (such as the oceans). Consequently, Earth's atmosphere has large thermal contrasts, a high Reynolds Number, and a heterogeneous water vapor saturation content. These turbulent and locally soaking-wet conditions alternate around the globe with unsaturated weather conditions. The atmosphere is unsettled by high winds and large fluxes of water vapor. The irregular topography of Earth then interacts with the atmosphere to produce large variations in precipitation rates and larger variations in erosional potential.

Geologically and hydrologically active worlds such as Earth and Mars thus have a heterogeneous distribution of landscapes formed through the combination of interior and environmental processes. Euclidean geometry, trigonometry, and celestial mechanics describe the incident solar radiation flux in smoothly varying functions of latitude and time. Geology and climate are not as tidy. Mathematical chaos theory describes the day-to-day transport and surface deposition of water and thermal and mechanical energy, though predictable order emerges at seasonal and longer timescales. It is no surprise that the erosion of Earth's land surface and sedimentary deposition in basins is spatially heterogeneous and complexly linked to latitude and Earth's geology and tectonics. Mars, in each of these regards, does not fundamentally differ from Earth. Hence, we can qualitatively understand how the Martian surface exhibits variability in the occurrence and depth of fluvial erosional dissection, glacial erosion, periglacial processes, and basin sedimentation. Planetary geologists are still putting together this complex picture. There are correlations of geoprocesses with latitude, elevation, and geologic terrain. Mars climatologists, including Robert Haberle (NASA Ames) and Michael Mellon, are now developing global climate models that explicitly seek to understand the distribution of ice- and water-related landscapes.

CONSENSUS

Mars today has only about 6 mbar of mostly CO_2. By any measure it is a cold, dry, windy planet – a place that many researchers regard as having almost nothing happening today geologically, except that "almost nothing" blows ferociously hard. This is *not* the climate generally regarded as characterizing ancient times. A consensus emerged toward the end of the 1970s that Mars had an early warm epoch, characterized by a dense CO_2 atmosphere, and then billions of years ago the thick atmosphere was somehow lost and the climate irreversibly changed to the current state. Thus, the Warm, Wet, Early Mars concept does not replace our present Cold, Dry, Windy Mars concept, but adds to it. Similarly, newer concepts of Martian climate and hydrologic history, discussed in subsequent chapters, do not replace these older concepts, but add to them.

4

Oceanus Borealis and the Austral Ice Sheet

THE FAUL GUY

Possibly the most prescient big idea of the early *Mariner* 9 period - albeit poorly substantiated at that time – never saw publication. Early post-dust storm Mariner 9 images of Olympus Mons revealed a huge basal scarp and giant landslides around the volcano. In 1972 Henry Faul (deceased 1981), a professor and expert in geochronology at the University of Pennsylvania, was the first to determine, using stereophotogrammetry applied to Mariner 9 images, the true height of the basal scarp of this mammoth volcano-about 5–7 km. (The volcanic construct itself towers an additional 20 km.) Faul did not hesitate sharing his educated opinion, which is outlined in an unpublished manuscript ("The cliff of Nix Olympica") in my possession:

> "*Morphologically it has the appearance of a very large sea cliff.*"

He further noted that Olympus Mons with its cliff also resembles shield volcanoes formed within thick glacial ice sheets, so-called table mountains. Faul considered but rejected wind erosion and faulting as the principal agencies of the scarp's formation. He allowed possible processes known from Earth's watery and icy realms: landslide collapse into a sea, erosion by waves, or volcanic growth of Olympus Mons in an ice sheet or frozen ocean. Henry Faul closed his manuscript with a speculation that if there was a sea, then fossilized remains of life may exist in marine sediments deposited therein. Faul was the first space-age Mars oceanographer, glaciologist, and astrobiologist, all in one.

Faul's ideas were big pants to wear on a body of little actual data, yet that cliff – and similar ones at other Martian shield volcanoes – is a major geological feature that was in need of an explanation. The manuscript lacked many details that one would ordinarily want to see to support such hypotheses. The manuscript lacked a full basis of observation, but most critically, Faul's ideas did not conform to the Cold, Dry, Windy Mars paradigm that then commanded the Mars science community. The paper was rejected for publication. I obtained the manuscript through a long, indirect route. Ken Foland, who later became my M.S. thesis adviser at Ohio State was a 1970s Penn colleague of Henry Faul with whom he many times discussed Mars issues. Faul gave a copy of his manuscript to Foland who relayed a copy to me in the mid-1990s. Try as he did, Faul failed to gain further NASA funding to pursue his ideas on Olympus Mons and other Mars science matters. He went on to other professional endeavors, including an authoritative book on the history of the geological sciences; his Mars career was over.

Faul's ideas had to await more data and further years of pondering before others would continue where he left off. Richie Williams, a glaciologist at the US Geological Survey, was the next in line to be battered for having dared to suggest that the Tharsis volcanoes were affected by glaciation – an idea that has recurred again and again as data have accumulated. Williams' ideas were suppressed at USGS, because they were too far off the edge of the times, too unsubstantiated. The ideas, however, did not die. Over the decades, researchers, including Henry Moore (USGS), Vic Baker (University of Arizona), and others have pursued similar ideas regarding Martian shield volcano morphology.

The original working hypotheses regarding Olympus Mons (marine or glacial formation of the basal cliff and aureole deposits) remain under discussion by the Mars science community, which has further applied similar models to the aureoles and basal cliffs of other Tharsis volcanoes. René Battistini (Université d'Orléans, France) presented the first thorough glacial interpretation of the Tharsis volcano auroles in a work that also emphasized the role of chaotic terrain at the sources of outflow channels. That hypothesis recently has been re-investigated and advocated by Jim Head (Brown University) and associates. Head has presented a compelling case that each of the Tharsis volcanoes exhibits moraines produced by cold-based glaciers. Taking a different approach, and assuming a landslide origin, Keith Harrison and Robert Grimm (University of Colorado) have modeled the flow of the slide material. However, to match the morphologic observations of the Olympus Mons aureole and landslides in Valles Marineris, a fluid-supported rheology is required. Harrison and Grimm do not advocate a specific mechanism, but point out that entrained pore water or flash-vaporized CO_2 could supply the needed pore fluid pressure. The implication is that an ample supply of condensed volatiles existed at these volcanoes. Faul's and Williams' ideas have thus been reincarnated in new, more robust current forms based now on a richness of data that would have been difficult to imagine 30 years ago. These concepts constitute an important part of ongoing scientific discourse regarding a significant Martian enigma. In due time the hypotheses will be found to be correct or incorrect.

Had I reviewed Faul's manuscript on Mars with the knowledge available in 1972, I may have been disappointed with the tenuous observations used to support a seemingly incredulous couple of interpretations. In fact, Faul employed the useful approach of *multiple working hypotheses* – a conservative methodology designed to cover all bases, though Faul hit on only two alternative explanations and in those ideas he failed to mesh with the dictates of conventional wisdom or even to make adequate acknowledgments of conventional wisdom. As Carl Sagan would say, extraordinary claims require extraordinary proof, something Henry Faul did not present to the satisfaction of reviewers. What Faul possessed were data showing the

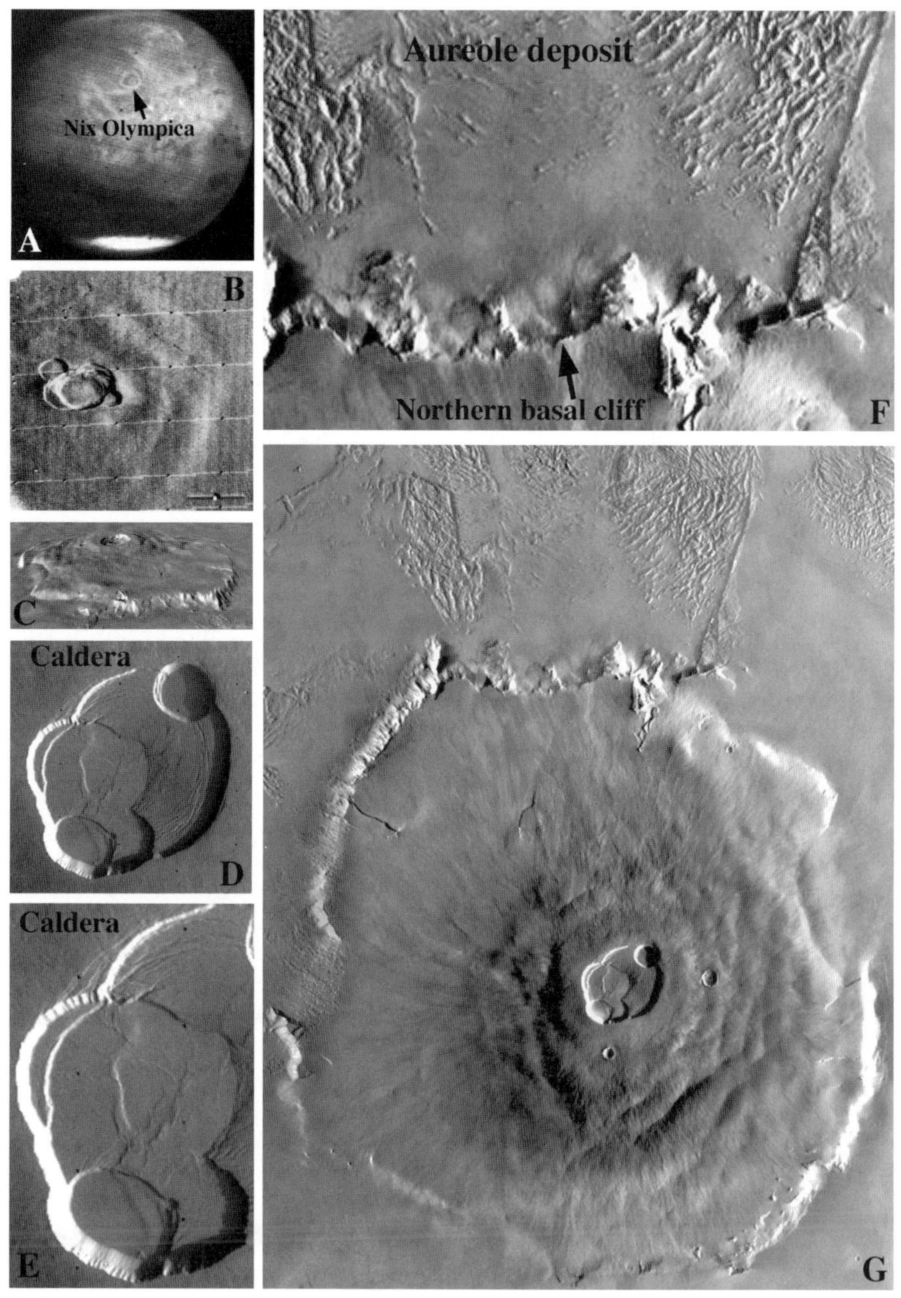

Olympus Mons through the Space Age. (A) *Mariner 7* saw Nix Olympica as a ring-like feature, thought then to be an impact crater. (B) *Mariner 9* discovered it to be the Solar System's largest shield volcano. (C, D, E and F) *Viking Orbiter*, and now *Mars Global Surveyor* and *Mars Odyssey* have seen it in amazing detail; these views hardly touch on the detail.

highest volcanic cliffs in the Solar System. Armed with this result and an unabashed willingness to step up and take a bold swing at a big problem, Faul fouled out in the referees' opinions. Unlike sports, however, where evolving truth is recognized second by second, truth in science is often less tangible, is always slower to be recognized, and is more lengthily debated. As we shall see, there remain valid reasons to reconsider Henry Faul's hypothesis. Even Venus provides back-handed support of sorts. We now have a global radar image of Venus, where huge shield volcanoes exist; oceans have never lapped against their bases, ice sheets have never touched their summits, and they lack basal scarps similar to those common on Mars and Earth. In my own assessment, there are adequate explanations for the cliff of Olympus Mons that do not require an ocean or ice sheet, but a key role for water in their development is central to most of the viable explanations of the scarp's origin. In the final analysis – yet to be written – Henry Faul may prove to have been right, or wrong, but he was certainly honest and bold in his expert judgment, and, in retrospect, unduly punished for his creativity.

FAUL'S LEADS POINT THE RIGHT WAY

Being an established and accepted member of a club aids science publication as much as it helps in any endeavor where community acceptance is needed. Carl Sagan – already a prominent planetary scientist in 1972 – and his coworkers interpreted sinuous light and dark markings in some *Mariner* Mars flyby images of the south polar region in terms of glacial moraines (D. Belcher, J. Veverka, and C. Sagan, *Icarus*, **15**, 241–252). Their paper had even less quantitative analysis than Faul's manuscript, yet it was published in a prestigious journal, *Science*. Although Sagan's many other contributions have become classics, that particular one has been forgotten. As it turns out, this paper was a false lead, though it led in the right direction. A few years before Sagan's untimely death, he sent me a letter in which he graciously pointed out that he and some coworkers had proposed the existence of glacial moraines 20 years prior to my first thoughts on Mars glaciers. I had missed his paper. Indeed, Belcher, Veverka, and Sagan had a published journal article pointing out south polar moraines in *Mariner 7* images. On checking the features against subsequent, more detailed, *Viking Orbiter* and *Mariner 9* scenes, however, it was clear that what appeared as possible lateral and end moraines of some bygone glacier in fuzzy *Mariner 7* images were in fact troughs and scarps in the present-day south polar layered deposits. I was obliged to apologize for my lapse in recognizing his claims, but I also pointed out the nature of the features now that they were better resolved. Carl just as graciously conceded his error.

This little story does not end there. Following Sagan's death, and with benefit of improved imaging by *Mars Global Surveyor*, I reinterpreted the troughs and valleys and other features of the south polar cap in terms of extant glacial deformation, as described later. The south polar cap is unquestionably unlike Earth's glaciers and ice sheets in several regards, especially the types of alpine valley glaciers suggested by Sagan and others. However, sand and dust accumulations in the spiral polar valleys appear to be emplaced partly through processes of ice flow and ablation; and this process – though only partly similar to what we commonly see on Earth – appears to fit the definition of moraines. Thus, in my new opinion, I owe Sagan a letter conceding that he had indeed been partly right all along. Sagan always had a vision of the Universe and our place in it that transcended space-time. I feel certain that he would be satisfied that this small part of the ongoing discussion of Mars is taking place after he has passed on. The important point is that there is a vibrant discussion. People are right, people are wrong, but nature always remains on top and wondrous. Sagan's passing is a mere technicality requiring others to take up where he left off – a concept that fits his perspective of a unitary and time-transcendant human existence.

Henry Faul's failure to get published and Carl Sagan's success is not to say that fame and eloquence or a track record is the sole formula for publication, though none of these can hurt. For one thing, Sagan's popularity had not yet crested. Distinctly not famous, Richard Kane (a school teacher) and some of his students at Garden City High School presented evidence in 1973 for glacial erosion on Mars; their paper, "Alpine Glacial Features on Mars," was published as the cover-page feature article in *Science*. I keep a copy of that cover page and key excerpts highlighted on my office door to remind visitors that no student of science necessarily lacks the qualifications and expertise to make useful contributions to professional science. Those authors interpreted the light and dark markings in the south polar area as the result of alpine glacial erosion and flow of moraine-striped glaciers. Those features were later resolved in better images as gigantic pits, which probably formed by sublimation of ice (not necessarily – but possibly – dead-ice glaciers). Again, this may have been a case of premature interpretations of form and process based on grainy images. However, the latest explanations of these features are still related to massive polar ice bodies – not at all the way conceived by the authors, but possible glacial ice masses nonetheless.

Was Garden City High School offering another false lead, and was it pointing in the right direction? How could this be? Very simply, it is interpretation being driven by common sense, which is not a bad thing. Observations tell us that Mars is a cold planet and that it has polar caps. Common sense – born of terrestrial experience – trains Earth-thinking people to consider polar glacial phenomena as an analog of what is seen near the Martian poles. Common sense exhibited by Kane and Sagan and their respective coworkers was the right sense, but the data then available were sparse, of poor quality, and their specific interpretations were incorrect.

These papers, for all their faults, presaged a broad range of new models of the Red Planet, which lately have taken the general label, "Blue Mars," after Kim Stanley Robinson's *Red Mars – Blue Mars – Green Mars* fictional trilogy. Unfortunately, these theoretical and observational studies had little impact on prevalent thinking in the early *Mariner 9* era, which focused on volcanism, wind-blown dust and sand, and fearsomely cold conditions. In part, evidence for ice and particularly for warm, flowing and melting ice, was not widely accepted, and such ideas were soon forgotten by the Mars Establishment.

While the vigorous eolian environment of Mars is unquestioned and ubiquitous in its touch, the current Mars exploration program is, in fact, being driven mainly by ice and water, past and present. In this regard, those ideas of early Blue Mars researchers – though probably not completely accurate technically – were on target with NASA's "follow the water" theme. They were, however, three decades too early. Science has caught up with common sense, and space-program politics is seeking legitimate advantage in it.

MARS: PLANET ICE

The recognition of water and ice as prime geomorphic agents on Mars came early in the *Mariner 9* mission. However, other possibilities, such as wind, liquid carbon dioxide, or liquid alkanes, were considered but generally discounted by the Mars community. So why would Mars be a world of water and ice? Why were there not rivers or glaciers of liquefied/solidified dry ice or liquefied natural gas? After all, if liquid surface water requires only a modest change in surface temperatures to make it stable, liquid carbon dioxide would require only an increase in atmospheric pressure to just over 5 bars. Liquid CO_2 may exist as liquid even today at most latitudes beneath the overburden of just a hundred meters or less if sealed from the atmosphere.

The foremost reason for implicating H_2O is that it is roughly five times as abundant in our Solar System, on a molecular basis, than all the carbon compounds together, and on Mars, as on Earth, this relative abundance ratio is apt to be even greater. The ubiquity and abundance of water was not a necessity. Water is abundant because of the abundance of oxygen, which greatly exceeded the amount that could be taken up by oxidation of metals in the solar nebula; reaction of excess oxygen with hydrogen, the major gas, caused water to be produced in greater abundance than oxides and other compounds of carbon. Oxygen's high abundance, and with it a great quantity of water, was forged by the advanced state of stellar and supernova nucleosynthesis in our part of the galaxy. So fortunate for life, we live in an oxygen-rich, hence, water-rich, Solar System. Other condensed volatiles, though not as abundant as water, are also widespread and important components of our Solar System. Except for sun-baked Mercury (where at least a little ice even exists at the sun-shaded poles) and Venus (where even a trace of water exists in the clouds and perhaps some minerals) and some small, lonely, heat-processed asteroidal metallic cores and lavas, this is an aqueous and icy Solar System. Mars is not an exception, so it should not be difficult to understand a high geomorphic activity of ice and water.

We have no direct determinations of the total abundance of ice and water at Mars. We do see ice widely distributed, at least near the surface (but an unknown abundance at depth), from the poles to the middle latitudes. Being midway between water world Earth and the icy satellites, Mars almost surely must be, gram for gram, as H_2O-rich as Earth. We can look to the icy satellites and comets of the outer Solar System, where 40–70% of their mass is ice, and conceive of a Mars that was endowed with even more water, gram for gram, than Earth. An Earth ocean amount of water, scaled to the smaller Martian mass and surface area, would suggest a global layer of ice about 1300 m deep. The polar caps account for only a few per cent of this amount, and the atmosphere less than one-ten-millionth of the total. If Mars is anywhere near as water-rich as Earth, gram for gram, and if most of the water is not contained in hydrated rocky minerals, then the Martian crust must contain ice as a major mineral phase; and what is not solid ice (or clathrate) must be mostly groundwater.

A RIVER RUNS THROUGH IT

The many water-carved features discovered by the *Mariner 9* project, further clarified by the *Viking Orbiters*, and now by *Mars Global Surveyor* and *Mars Odyssey*, indicate a world that at one time ran deep. In just about every Mars system model – even those that implausibly deny water a major role in Martian geologic development – gullies, valleys, and channels are at the crux of key issues.

As soon as *Mariner 9* began to show us freeze-dried river beds and valleys, this favored substance immediately spurred concepts of a formerly Earth-like Mars. Disagreement has raged ever since as to exactly how Earthlike Mars was. The role of impact heating or internal geothermal heating in producing liquid surface conditions have been suggested repeatedly as an alternative to some super-efficient mixture of greenhouse gases or special conditions of orbital obliquity and eccentricity. Recently, Nick Hoffman (University of Melbourne, Australia) and some researchers have reconsidered the possibility of a radical alternative to H_2O itself; the water-like liquid may have been some other substance, such as liquid carbon dioxide. In fact, I have been out front along with Hoffman and others to support the idea that at least the water is apt to have been highly carbonated – a feature that has important dynamical and geomorphological implications.

Accepting as a premise the overarching interpretation that these features were formed by a flowing condensed fluid, there is one necessary implication that sets aside the deep disagreements as to specific processes and histories: The flow of huge volumes of condensed fluid over the Martian surface at high flow rates requires the buildup and storage of gravitational potential energy in the fluid; some features require the fluid's sudden release (gradual and sustained in other cases) so that it could flow turbulently over the surface to a state of lower potential energy. This is a basic requirement, and it holds regardless of the volumes, rates, and durations of fluid flow – and regardless of what type of fluid was involved.

This requirement does not limit the source of the high potential energy state, and it does not implicate any particular history, which are the aspects where most of the disagreements about Mars geology currently exist. It is, however, a very broad and critically important constraint on any successful model of global Mars hydrologic history. Much of the Mars science community's effort has been dedicated to developing an understanding of the outflow channel, small-valley network, and gully erosion processes; much less attention has been devoted to development of explanations for the high potential energy of the erosional fluid before the channels were formed. The eroded volumes of outflow channels and the erosional degradation of the highlands implicate a huge volume of erosional liquid. Michael Carr estimated that 10 times as much water volume as eroded rock is needed for the outflow channels. For small-valley networks, where flows were less intense but more sustained than those of outflow channels, much higher water:rock ratios are needed. The typical one kilometer cumulative average erosional depth in the highlands implies an amount of erosional fluid equivalent to hundreds of kilometers over that whole eroded region. This water was not present all at once, but was cycled again and again. The large liquid:rock ratios necessitate recharge of the liquid within a surface or subsurface reservoir, and unless some limitless deep mantle reservoir of a mystery liquid (fountains from hell) is involved, recycling of crustal liquids seems to be required if these valleys and gullies are anything like those known on Earth. Recycling means that there has been a reverse flow either through the atmosphere or through the subsurface.

Martian hydrology, and that of Earth, is like a gravitational potential energy battery. Release of this energy does work, forming valleys and channels, eroding peaks and high plains, and infilling the low spots with sediment and water/ice, but what energy source recharges this battery? A simple one-shot recharging system could be the tectonic subsidence of the northern plains or uplift of the highlands. This process took ground water and ground ice that was at first near gravitational equilibrium and lifted it several kilometers relative to the sagging basin, thus forming an unstable, perched, fluid-charged aquifer (or "liquifer") in the highlands, which eventually drained into the northern plains. It probably was not a one-shot process, however, as water flows occurred episodically for billions of years. Furthermore, this mechanism does not explain the small-scale systems of gullies and other features in the cratered highlands. True recycling, presumably through the atmosphere, seems to be required. We shall consider other possible fluids later, but for the present we shall accept what ultimately seems most logical: that is, the fluid is/was primarily water with various additions of salts and gases.

What caused such huge volumes of water to be released in such a short time, and what caused this release event(s) to be delayed hundreds of millions or billions of years following the formation of the perched aquifers and then, for similar long expanses of time, to intercede between subsequent drainage events? There are not yet any definitive answers. Martian hydrology – though it has produced landscapes similar to those on Earth – does not, in its totality, portray much similarity to the behavior of Earth's hydrologic system. However, some fundamentals are identical, because they are based on the laws of physics and the behavior of similar geological materials.

There are two major sources of potential energy and pathways for water recharge. This can be external or internal to Mars: through the atmosphere, driven by solar heating; or through the subsurface, driven by geothermal heat. There is, in addition, solar heat-driven evaporative pumping, which can draw water upward to replace water lost at the surface by evaporation, or which can induce thermohaline convection of ground and surface waters. Solar energy can evaporate water and sublimate ice, drive atmospheric currents, and transport water vapor from low to high elevations, where precipitation may then redeposit the water or ice. The effectiveness of solar energy in warming the surface and driving hydrologic processes on the surface on a planet that is as distant as Mars from the Sun is largely dependent on the effectiveness of the atmosphere in retaining heat. The main internal source of energy is due, primarily, to the decay of natural radioactive isotopes of potassium, uranium, and thorium and, secondly, to residual heat from planetary accretion and differentiation. Internal heat of any origin can be transferred by either conduction or advection in magma or warm fluids. In the context of the cryosphere, general background levels of geothermal heat flux can cause water vapor to sublimate from ice or evaporate from deep groundwater and then be cold-trapped in the colder regolith layers near the surface. Steve Clifford (Lunar and Planetary Institute, Houston) has considered these systematics in detail. Higher levels of geothermal heat flow, for instance near a volcano, can produce large-scale groundwater convection, including the potential for cold and hot springs. A combination of these energy sources can also be effective in groundwater recharge: first, solar energy and atmospheric transport may produce a slow accumulation of glaciers; second, when a critical thickness is reached, geothermal heating may cause basal melting; and, third, the pressurized liquid water is then injected into the ground below.

These fundamentals are shared by Earth. Models of Mars geology, hydrology, and climatology are largely related to this range of energy sources and how they work together in the Mars system. The energy sources are similar; but the relative roles of external and internal water transport processes and the way these processes link with the global system are different. Many of the great ideas and great Mars controversies of recent decades have pertained to these mechanisms of water transport and the global linkages. A primary point of contention has been the key matter of the abundance of water, in any form, on Mars. For some researchers, water/ice is present in oceanic abundances; for others, water is a sparse commodity, so Mars makes do with wind.

Victor Baker (University of Arizona), using the catastrophic floods unleashed by Glacial Lake Missoula 13,000 years ago as a partial analog, estimated that the floods that carved the Martian outflow channels were active at peak discharges of several cubic kilometers per second, rates sustained perhaps for a few weeks or months. The Missoula floods, by comparison, unleashed 2,000 cubic kilometers at a peak rate of 0.017 cubic kilometer per second for a couple days, eroding canyons and removing 200 cubic kilometers of rock. Even the catastrophic filling of the Black Sea 8,000 years ago through the Straits of Bosporus – a possible source of the Noah flood story – falls short by orders of magnitude in total discharge compared to the Martian floods. In Earth's known geologic history, only the catastrophic filling of the Mediterranean Sea 5.5 million years ago can compare. The magnitude of these floods cannot be imagined except by comparison to far smaller floods of historical interest, e.g., the 1997 flood of Minnesota's Red River Valley (up to 4,000 cubic meters/second at Grand Forks, ND) due to unexpected rapid snow melt, which caused billions of dollars in damages; the 1993 flood of the Mississippi basin (up to 30,000 cubic meters per second at St. Louis, 5 times normal), which caused $12 billion damage and took 47 lives; or a 1981 glacial lake outburst flood in Nepal (up to 3,000 cubic meters/second), which took out small villages, a key bridge on the road to Tibet, and a small hydroelectric power plant. These disastrous flood discharges amounted to 5–6 orders of magnitude smaller than that of a typical Martian outflow channel event.

As people recognized the immense scale of the Martian outflow channels and the floods that created them, it was also recognized that Mars is not quite the extreme desert world it was previously thought to be. Michael Carr (USGS) produced a widely cited estimate that the Martian global inventory of H_2O is equivalent to a global ocean as much as 400 m deep. This estimate is based on the minimum volume of water needed to carve the channels that debouched into the Chryse Basin, and then extrapolation to the global scale. Although the assumptions inherent in any such calculation lead to large uncertainties, Mars does indeed appear to be a "water-rich planet," in Carr's words.

As amazing as the crisply preserved circum-Chryse outflow channels are, the MOLA has mapped the topographic signature of a more ancient, much larger outflow channel that debouched to the west of Tharsis. Details were presented by James Dohm and Vic Baker (University of Arizona); the scale of this feature, though very real, is almost beyond belief. It completely dwarfs Kasei Vallis and other outflow channels. While ice may have contributed to its formation, the total volume of water or ice required to form it could have filled the entire northern plains to the brim.

Many outflow channels are Hesperian age features. Although the largest channel is Noachian, many are Hesperian and several small ones – several orders of magnitude smaller than the largest but still dwarfing the Lake Missoula floods – are Amazonian. The peak in outflow channel activity during the Noachian and Hesperian eras suggests that Mars has since lost much of its water or that most of it has been sequestered in some inactive form or has pooled at low elevations for the latest eon of time. Although scientists have produced steadily increasing estimates of the rate at which hydrogen – a photolysis product of water – is being lost from Mars, escape does not seem a possible way to reconcile the requirements for a very wet Mars with the fact that it is not wet today. Rather, water must be mainly frozen and stored in the subsurface or covered over with dust, lava rocks, and sand.

NEPTUNE ON MARS

The huge quantities of water required to form the Martian outflow channels raise a logical question. What was the immediate aftermath of the floods? The water presumably would have pooled in low-lying regions, forming oceans or large lakes whose area, depth, and volume depended on the scale of each flood episode and the topographic characteristics of the basins into which the water drained. The northern plains, with several sub-basins, the Valles Marineris, the Argyre and Hellas impact structures in the southern hemisphere, and many smaller craters became filled with sediment-rich water. We can see the sedimentary remains of these former lakes, and in some cases the water remains in frozen form. Many crater–lake basins have inlet and outlet channels, and with some we can trace a sequence of flood filling and overflowing from one basin to another. The sedimentary and erosional records of these systems is complex, and it is likely that the layered rock sequences contained by the basins include airfall components of eolian dust and volcanic ash as well as water-lain clay, silt, sand, and boulders.

No matter what one may think about specific hydrogeologic and sedimentary processes on Mars, the northern plains, Argyre, and Hellas, by every account, represent vast sinks of materials that were transported from the highlands. Similarly, Earth's ocean basins represent sinks for sediment transferred from the continents. One main difference between Earth and Mars is that Earth, through the plate tectonic mechanism, continually destroys and recreates the ocean basins, and the sediment stored on oceanic crust is either subducted into the mantle, where some of its mass is recycled as andesitic and granitoid magma into volcanic island arcs, or scraped up and composited into accretionary wedges of metamorphic and sedimentary rocks at convergent plate margins. By contrast, on Mars, the basins just sit there passively (to first order) and fill up, as their rims and extended drainage basins wear down. Since surface transport – by rivers, glaciers, or other condensed fluids or plastically deformable masses – always moves from points of high to low gravitational potential energy, the main mechanism for removal of Martian sediment out of the basins is erosion and transport by wind; small amounts of rock material theoretically could be removed from basins in dissolved form by groundwater. Ultimately, although it may take circuitous routes, wind-borne sediment is mostly eventually redeposited in basins; thus, wind is not a process by which there can be net large-scale sediment removal from basins. Thus, Martian basins are sedimentary repositories of the Martian geologic chronicles.

Mars preserves a sedimentary record as old as the basins – a record over 4 billion years old, going back through 90% or more of Solar System history. Earth's intracontinental basins and sedimentary platforms locally contain sediments going back one-eighth to one-half the age of the Solar System, while metamorphosed sediments preserve a chemically and physically distorted record of sedimentation going back 3800 million years (not quite as far as the Martian record). On Mars, wind, rather than being an agent of net removal of sediment, probably has contributed to the mass transfer and infilling of the northern plains and other major basins. However, judging from the geomorphic record of Mars, ancient rivers (at least rivers of supersize debris flows) – especially the outflow channels – appear to have dominated the transport of sediment to the major basins. We see the results in the deeply eroded outflow channels, and we see it in the sedimentary-type geologic formations in the northern plains, Hellas, Argyre, Valles Marineris, and smaller basins.

The first serious, published suggestion that the Martian northern plains might present sediments deposited in an ancient ocean was offerred by Baerbel Lucchitta and her colleagues at the US Geological Survey. She published a fascinating but little-noticed paper in 1985 with the provocative title, "*An ancient ocean on Mars?*" Her ideas were formalized the following year in a lengthier paper written with coworkers: "*Sedimentary Deposits in the Northern Lowland Plains, Mars.*" Her ideas were based on several observations, notably the apparent mantling of the northern plains and their relationship to the termini of major outflow channels that trend

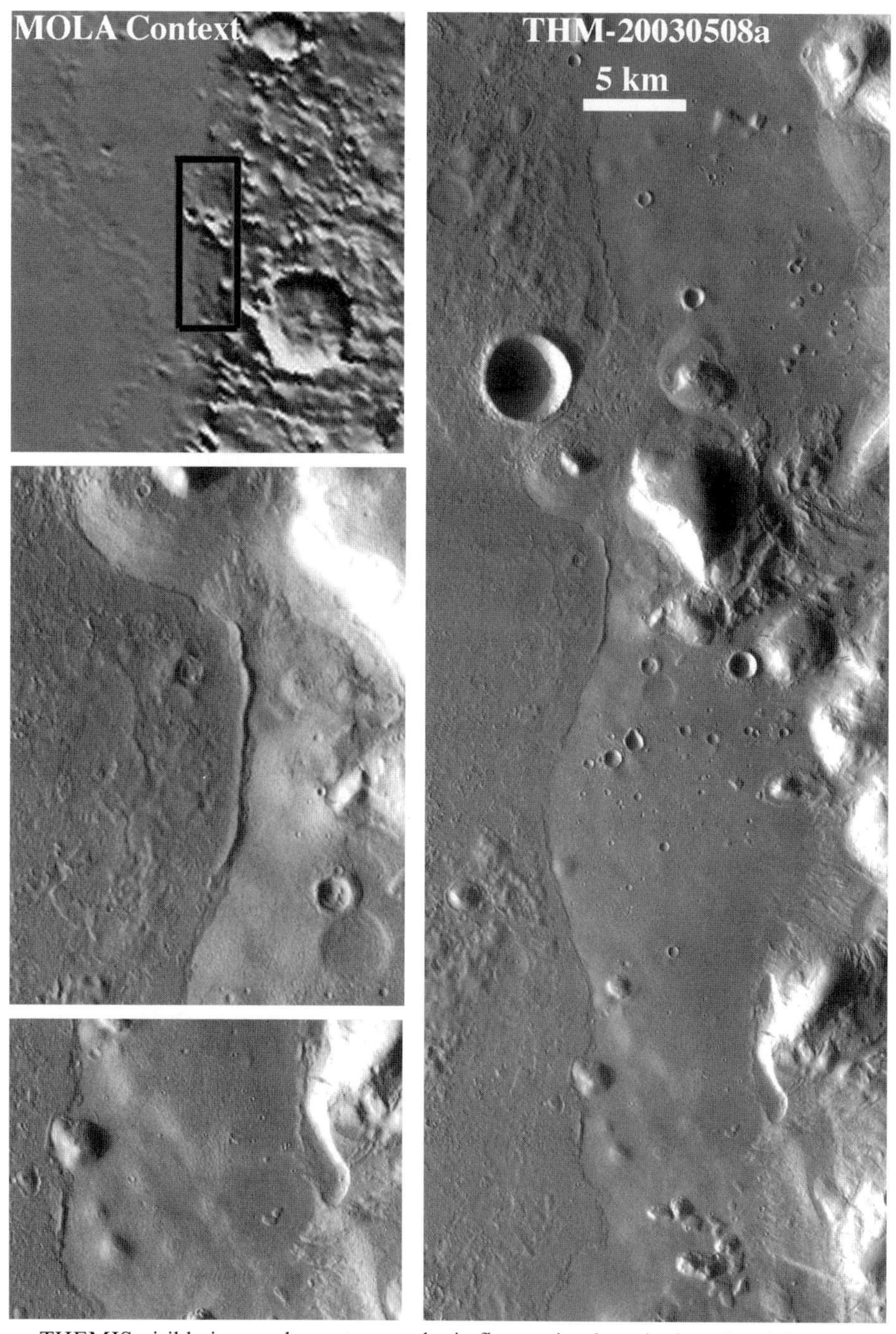

THEMIS visible image shows two geologic flow units deposited on low-lying ground. The older unit embays the uplands at the right, and the younger unit forms the pronounced contact in the bottom left panel. Lava flows are likely for both units, but debris flows are also possible. 36.5°N lat., 217.6° long.

The dark unit may well be lava, but given this deposit's location near the terminus of one of the many channels leading from the Elysium volcanic plateau, and considering the many signs of volcano-ice interactions at Elysium, this is probably a lahar – a sedimentary deposit of mixed volcanic ash and previously deposited lava and various sedimentary materials, all aqueously altered to some degree at temperatures ranging very very hot to icy cold.

Highlands/lowlands boundary zone near Nilosyrtis Mensa includes a "thumbprint terrain" (hummocky plains) form of the interior plains unit, which was attributed by Tim Parker to the youngest, smallest ocean. This terrain unit has a lobate edge, consistent with a debris-flow mode of emplacement. This is an infrared image, so what appears dark is cold, and bright is warm. The local temperature pertains to slope and slope aspect and to thermal inertia (related to rock abundance, soil grain size, and soil cementation). THEMIS image I03268002, 43.0°N lat., 285.7°W long.

from the highlands into the northern plains. Other phenomena, observed and developed by other researchers but linked by Lucchitta into a coherent framework of marine sedimentation, included fluidized ejecta blankets and giant polygons. Both phenomena were believed to reflect the existence of ice-rich deposits. In addition, a variety of features attributed by Lucchitta to glaciation supported a role of ice in the northern plains. Her paper was not very specific about the types of sediment transport and deposition processes or how (or if) these processes linked with each other, and she did not explore any of the possible global relationships and implications. Nevertheless, Lucchitta offered a bold suggestion that has proven fruitful for further development. Although her papers did not receive much immediate attention, they were a spark that ignited a variety of other bold ideas challenging, in the most direct and severe way, the applicability of the Cold, Dry, Windy Mars paradigm.

In the same period when Lucchitta was suggesting major roles for glacial and marine processes in the northern plains and the adjoining transition zone, a young researcher from Lucchitta's German homeland, Heinrich-Peter Jöns, made an amazing and insightful observation which, at first, was not embraced by the Mars science establishment. Through the mid- and late-1980s Jöns presented several posters and abstracts at the annual Lunar and Planetary Science Conference. He highlighted a narrow ridge-like feature or major stratigraphic contact mapped around almost the entire perimeter of the northern plains, and his interpretation was that this feature was the edge of a vast northern mud ocean. Due possibly to the novel interpretation or to his insistent manner of presentation and poor command of English, he was unsuccessful in convincing many people that he had really discovered something important. Personally, I did not then know what to make of the features, but they were clearly real features of enormous extent, though not quite continuous around the whole perimeter of the northern plains. A mud ocean of such vast extent seemed to be a bit ludicrous in my opinion at the time, but something dramatic had to be the implication. Jöns' observations were indelibly etched in my mind.

It was later found that Lucchitta and Jöns were considering different parts of the same story. But they were not alone. Timothy J. Parker, in 1986 a graduate student at the University of Southern California, was also starting to think oceanically. Although he was not the first, he has gone furthest in developing the idea of a northern ocean – some would say he went too far into the deep end. Initially, his concept made no progress with other scientists. In a submitted manuscript, like Lucchitta's 1985 paper, he had the boldness to suggest an ocean directly in the title, but that was simply too much for reviewers. It took another three years of developing his ideas, and a skillful strategy that included teaming up with an eminent Mars researcher and burying the true intent deep within a paper with an abysmally mundane title: "Transitional Morphology in West Deuteronilus Mensae, Mars: Implications for Modification of the Lowland/Upland Boundary." After digging through a huge number of Parker's careful observations, the reader soon discovers that he was really talking about ancient erosional shorelines of an ocean and near-shore marine deposition. Landscapes in the transitional boundary between the highlands and lowlands of Mars are explained by Parker according to a marine model adapted with little modification from terrestrial coastal geomorphology and oceanography. After an extensive study of geological and geomorphological relationships along a large sector of the highlands/lowlands boundary, Parker concluded:

> *... sediment deposition in either a liquid or ice-covered sea provides the best explanation for the observed boundary relationships. Many of the observed morphologies resemble those found along terrestrial lake shores in cold environments. These include shore features such as wave terraces, beach ridges, and arcuate or cuspate patterns and coastal plains features such as desiccation or ice-wedge polygons and pingos.*

Parker's landmark 1989 paper was represented with a style that gave him more than his 15 minutes of fame, and even more criticism. In that and subsequent papers on this topic, Parker has combined excruciatingly careful and abundant observations with bold interpretations. The typical high quality of Parker's basic observations has stood the test of time and has never really been contested. However, in the opinions of many researchers, his interpretations have never been very well substantiated despite the quality of his observations.

During casual discussions with Mars researchers one curious thing emerges. It is their great haste to make known where they stand on "the ocean" and any particular researcher's specific hypothesis: It's a ludicrous idea; so-and-so's hypothesis is an outstanding idea (and was my idea first); it is not as big as my ocean and is far too small to fit the big evidence; it's bigger than my ocean and far too big; so-and-so's ocean is too watery; it's too rocky; it's too icy; the climate is too rainy and warm; it's too old; too young; etc. There is a lot of observational real estate to be claimed on Mars and a lot of intellectual property, and people are claiming it where they fit in. Everyone sees the evidence differently. People are making their views known and exercising their proprietary rights to full advantage; they are staking their claims (or seizing others' in some cases) in new research papers and proposals and new Mars mission proposals. It is a dynamic field, with all the jockeying of a horse race. To be sure, collegial and professional relations prevail, but there is a palpable sense in the Mars community that major ideas about Mars are being sorted out. There is a common denominator: everyone sees the northern plains as a vast repository of Martian climate and process records, whether there was an ocean or not.

Tim Parker's large personality no doubt is a key ingredient that enabled his ocean hypothesis. Parker's planet-love and creative but logical approach to data analysis – the main common denominators of planetary scientists – extends into his home life and hobbies, also a common character trait. Parker makes his own telescopes to support his amateur astronomy habit. He has built a 12.5-in Cassegrain reflector with a teak plywood tube (mirrors figured by Ed Beck of Enterprise Optics) and a 6-in f/10 Jaegers refractor, again with the optics commercially obtained but the tube made from birdseye maple and mahogany. Asked about why he still plays with his elaborate toys when all manner of Mars spacecraft and Mars mission data are available to him, Parker responded, *"When all the synoptic cameras in orbit around*

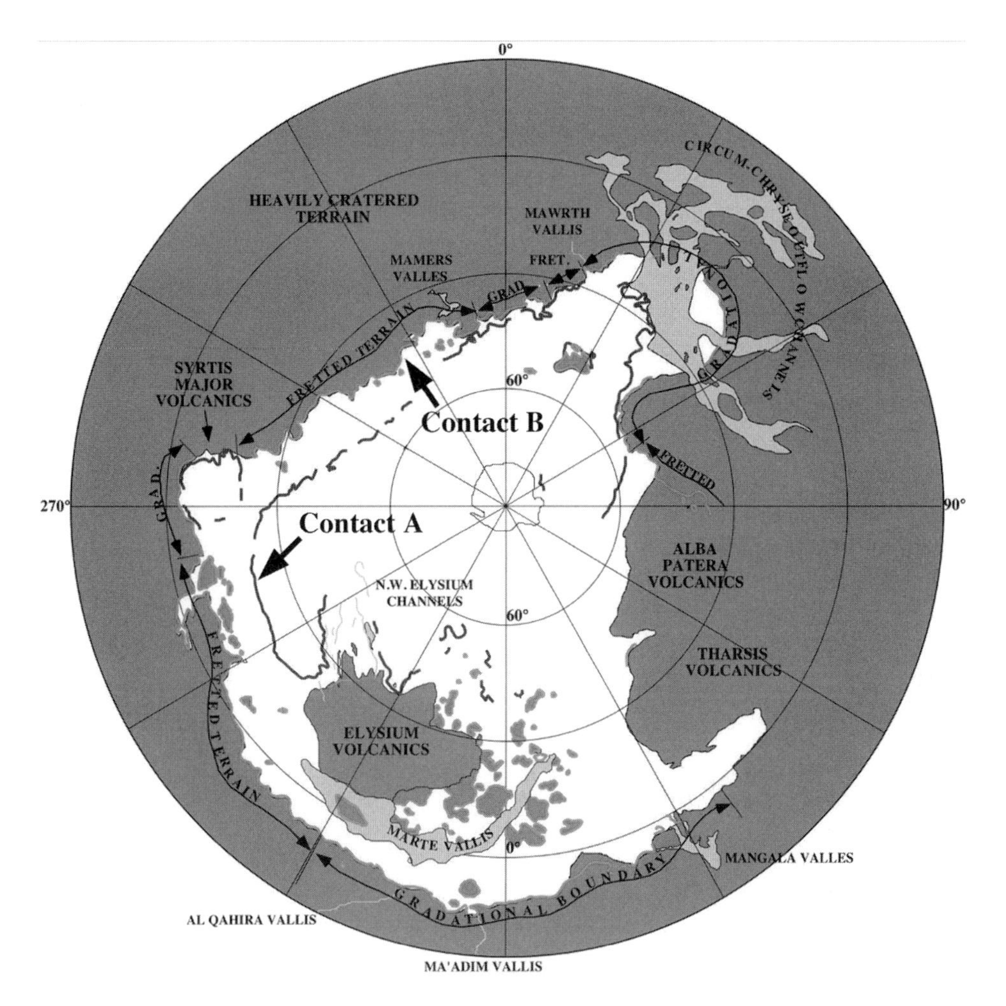

Were Tim Parker and colleagues off the deep end or onto the biggest story Mars has ever told? The Mars ocean controversy may be resolvable within a decade, since many new missions are flying and more are due to fly. The key will be analysis of the sedimentary deposits and elucidation of the processes responsible for sedimentary transport and erosion of the fretted and gradational boundaries. The original hypothesis of Parker et al. was that there are two distinct oceans separated in time, the younger one (represented by Contact A) smaller than the older one (represented by Contact B). Modified after Parker et al. (1989).

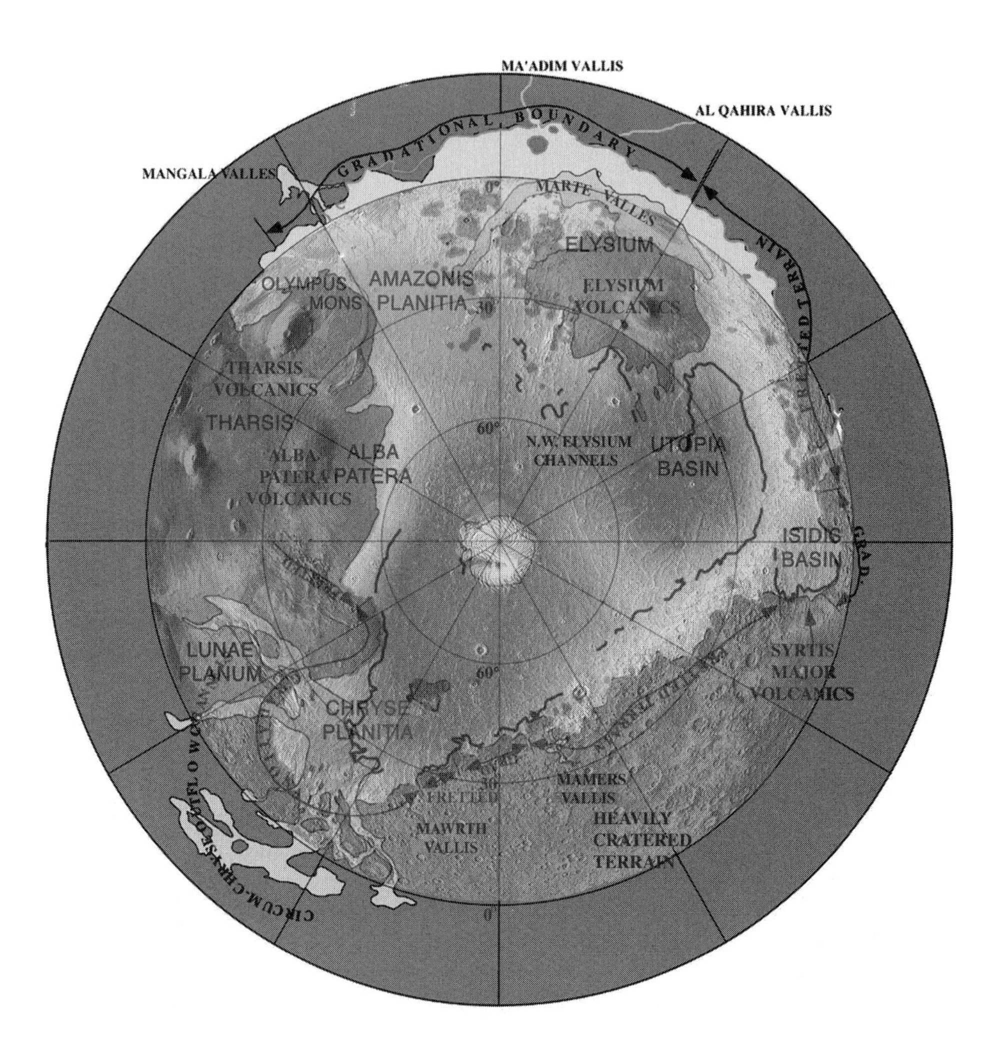

Tim Parker's original "shoreline" map (previous page) superposed on Jim Head's northern hemisphere topographic map and converted to gray scale. Regardless of the origin of northern plains basins and the causes of their infilling and smoothing, the outflow channel systems and fretted terrain boundaries identified by Parker et al. constitute some of the most important types of evidence indicating a water-rich (and icy) Mars. The relationship of these terrains with respect to the northern plains cannot be a mere coincidence. Any successful model must relate the deposition and smoothing of the northern plains to the chief boundary terrains and the processes that are evident there, including ice flow and probably catastrophic flooding. More generally, the hypotheses that can relate these features include mass flows of any combination of the following, but certainly not rock in isolation: rock, ice, liquid water, and carbon dioxide. Data provided courtesy of Tim Parker and Jim Head. See also Parker et al. (1989) and Head et al. (1999).

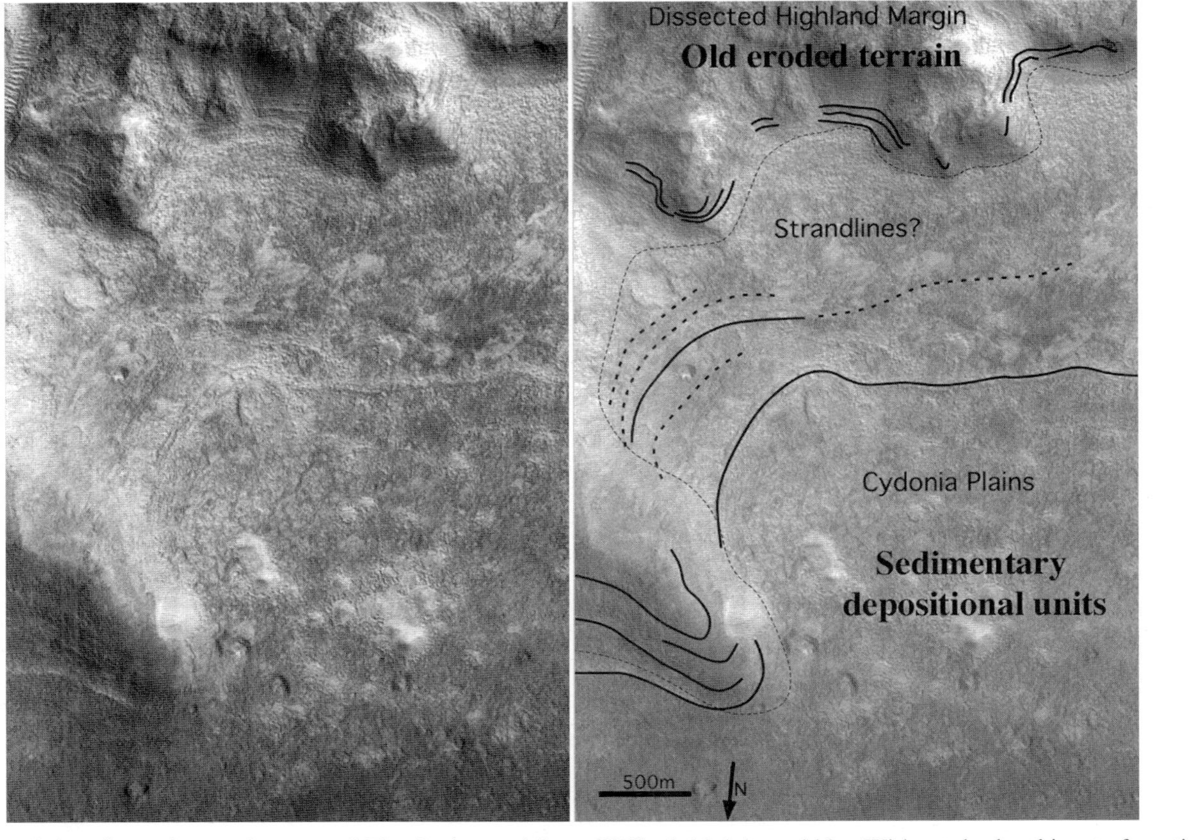

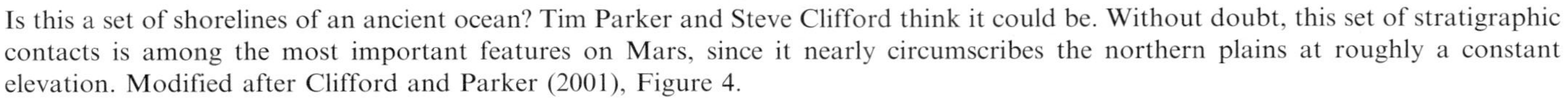
Is this a set of shorelines of an ancient ocean? Tim Parker and Steve Clifford think it could be. Without doubt, this set of stratigraphic contacts is among the most important features on Mars, since it nearly circumscribes the northern plains at roughly a constant elevation. Modified after Clifford and Parker (2001), Figure 4.

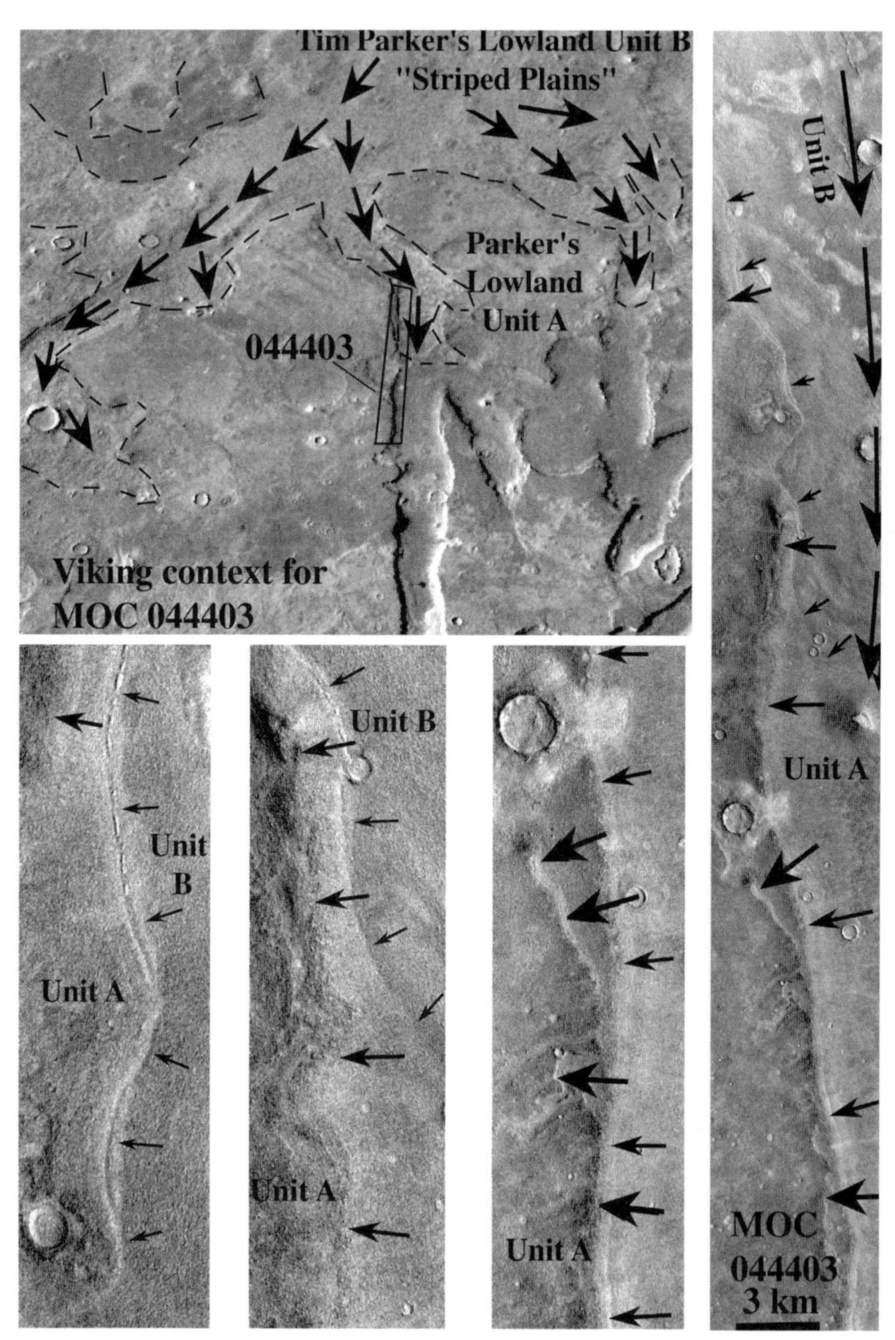

Edge of an ocean or a mud ocean? The sedimentary formational contacts and morphologies here – originally mapped by Tim Parker and colleagues (approximated by dashed lines) – seem more consistent with a mud ocean, as first suggested by H.-P. Jöns. However, we are not looking at the pristine morphologies.

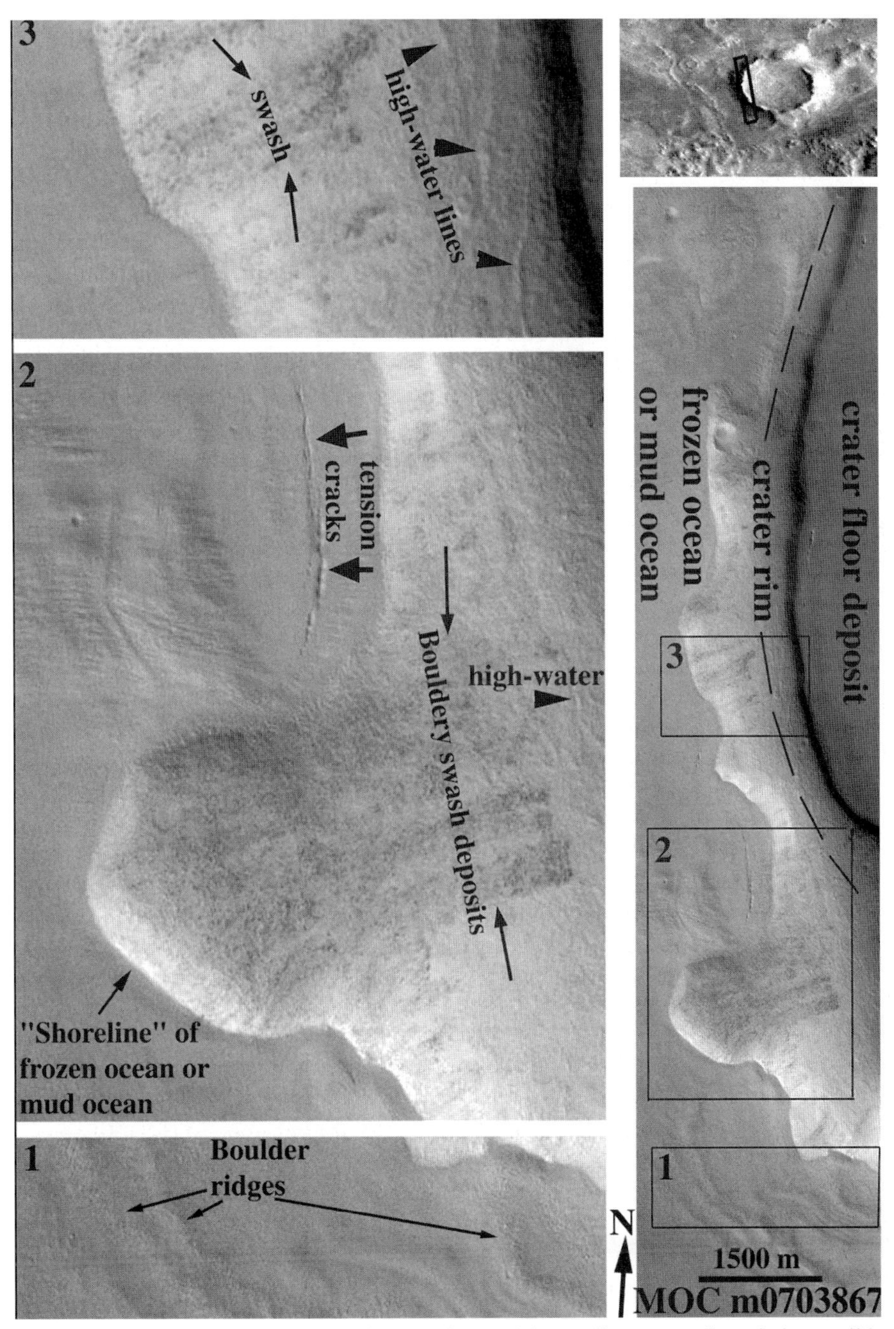

Evidence of rapid emplacement of a northern plains sedimentary deposit is possibly evident here, where features can be interpreted as due to bouldery sediment washed 2 km (horizontal) up steep terrain. The rapid surge of sediment must have involved speeds over 60 m/s (135 mph).

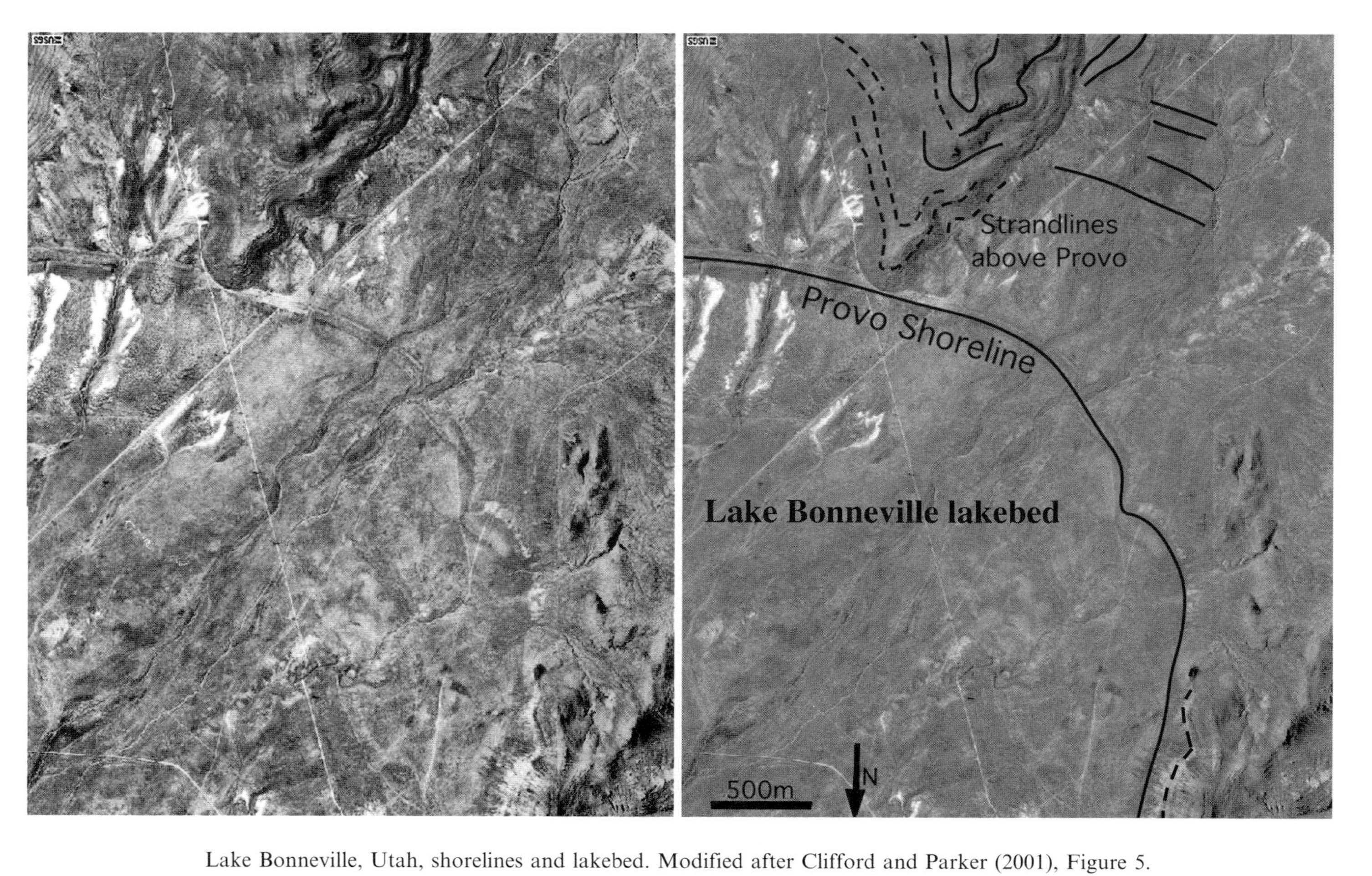

Lake Bonneville, Utah, shorelines and lakebed. Modified after Clifford and Parker (2001), Figure 5.

Late Glacial period Lake Bonneville shore terraces in the vicinity of Provo, Utah. In both scenes lake deposits blanket the valley floor. For scale in B, note houses on the terrace. Similar features are common on Mars, but care always has to be taken to separate layering outcrops from erosional or depositional terraces. Lake Bonneville was fed by both glacial and nonglacial streams. The lake's salinity fluctuated along with lake level. After Pleistocene glaciations ended, climate dried, and so did the lake.

Mars die, I'll still be able to roll my 'scopes out of the garage and fire up the webcam [and view Mars]."

When science finally overcomes psychological barriers to revolutionary ideas, it commonly does so with as much chaos as gusto and bravado. The chaos is because the new ideas tend to leave old ones in limbo. Rarely are leap-frogging advances in science so thoroughly presented that all potential objections are satisfactorily addressed, and Parker's 1989 paper was no exception. Leaping advances in science are hardly ever welcomed. For one thing, big new thoughts usually introduce big new errors that are almost as large as the advances, making them vulnerable to criticism. Parker's zealous pursuit of the marine hypothesis faced the usual problem, that of multiple plausible interpretations of morphology. Furthermore, Parker's hypothesis initially lacked a globally coherent model that could allow for an ocean to form and then be disposed of. Many of the initial criticisms have been addressed by Parker and

The Alaskan arctic shore of the Beaufort Sea, shown here in early August near Barrow, could be what Martian shorelines looked like before finally freezing over completely. Wind-driven ice jams and wave actions during the summer thaw season are responsible for eroding the coastal bluff. On Mars, this type of strandline could be possible only if the climate permitted open water for decades to millennia. Lacking such a warm climate, a rapidly emplaced ocean would start freezing completely at its surface in a matter of weeks, and erosional beaches as seen here would not have had time to develop.

colleagues over the past decade, but few researchers have ever fully accepted the oceanic interpretation, while not necessarily dismissing it entirely. Nevertheless, we saw the initiation of Martian marine geology in the 1980s. It has taken new space missions to uncover some major faults of the early ideas and to highlight some of the truths in these landmark papers. A sure indication of a worthy hypothesis is when increasing numbers of people apply increasing effort to test and develop the hypothesis, and when increased scrutiny with more types of data fail to kill the concept or lead it into new incarnations. This certainly characterizes the Mars ocean hypothesis. We still cannot lean on that idea as though it is a proven episode of Martian geologic evolution, but the more we learn, the more it seems to be plausible. Of course the hypothesis has morphed; and if you ask Ken Tanaka, or Vic Baker, or Tim Parker, or Jim Head to characterize the ocean and its implications, each will describe a different type of ocean and a different set of climatic and process implications.

Specific interpretations by these people differed, but each cited associations between outflow channels and lobate-edged lowland deposits, polygonally fractured plains, and other peculiar features occurring in the lowest parts of the northern plains. The idea was that the floods fed water and debris into the northern plains,

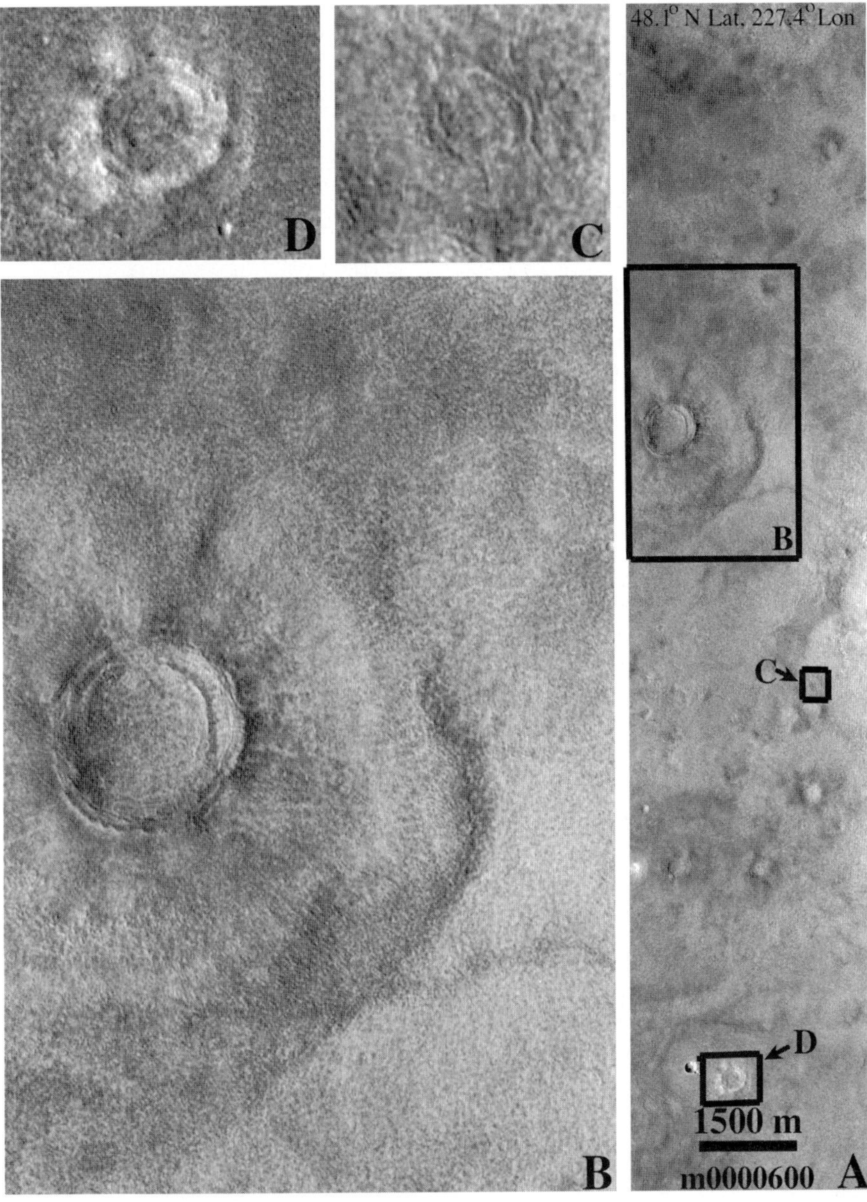

This widespread northern plains deposit, the late Hesperian age knobby member of the Vastitas Borealis Formation, actually is extremely smooth. Craters are flattened and in many cases almost wiped out by what is probably ice-driven relaxation. Neutron and gamma-ray data from *Mars Odyssey* have recently been modeled as indicating over 70% (by volume) ice beneath just 5 cm of rock debris. *Viking Lander 2* found dust, salt-cemented soil clods, small-scale polygons, and rocks. The ice may have formed by freeze–thaw driven ice segregation during high obliquity; or it may be a remnant of an ocean or debris-covered glacier.

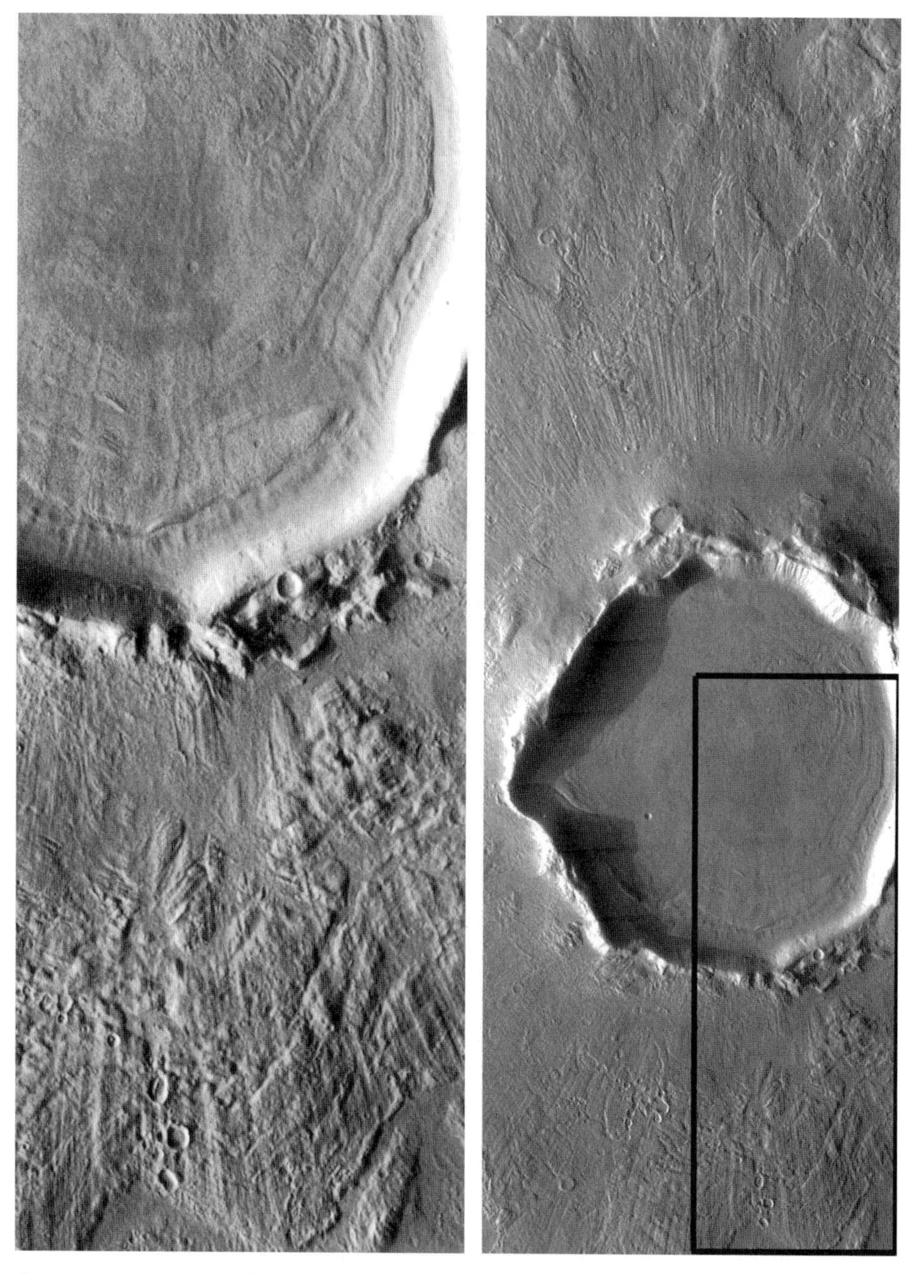

Impact craters are not always circular. Meteor Crater is almost a square. Craters formed by grazing incidence can be elliptical. This crater, however, has been distorted after impact, perhaps by closure of surrounding icy materials. Closure can explain the crater's shape and compressional ridges developed in interior deposits ("concentric crater fill"). This crater morphology occurs in hypothesized oceanic or mud-ocean areas. THEMIS image 20030520.

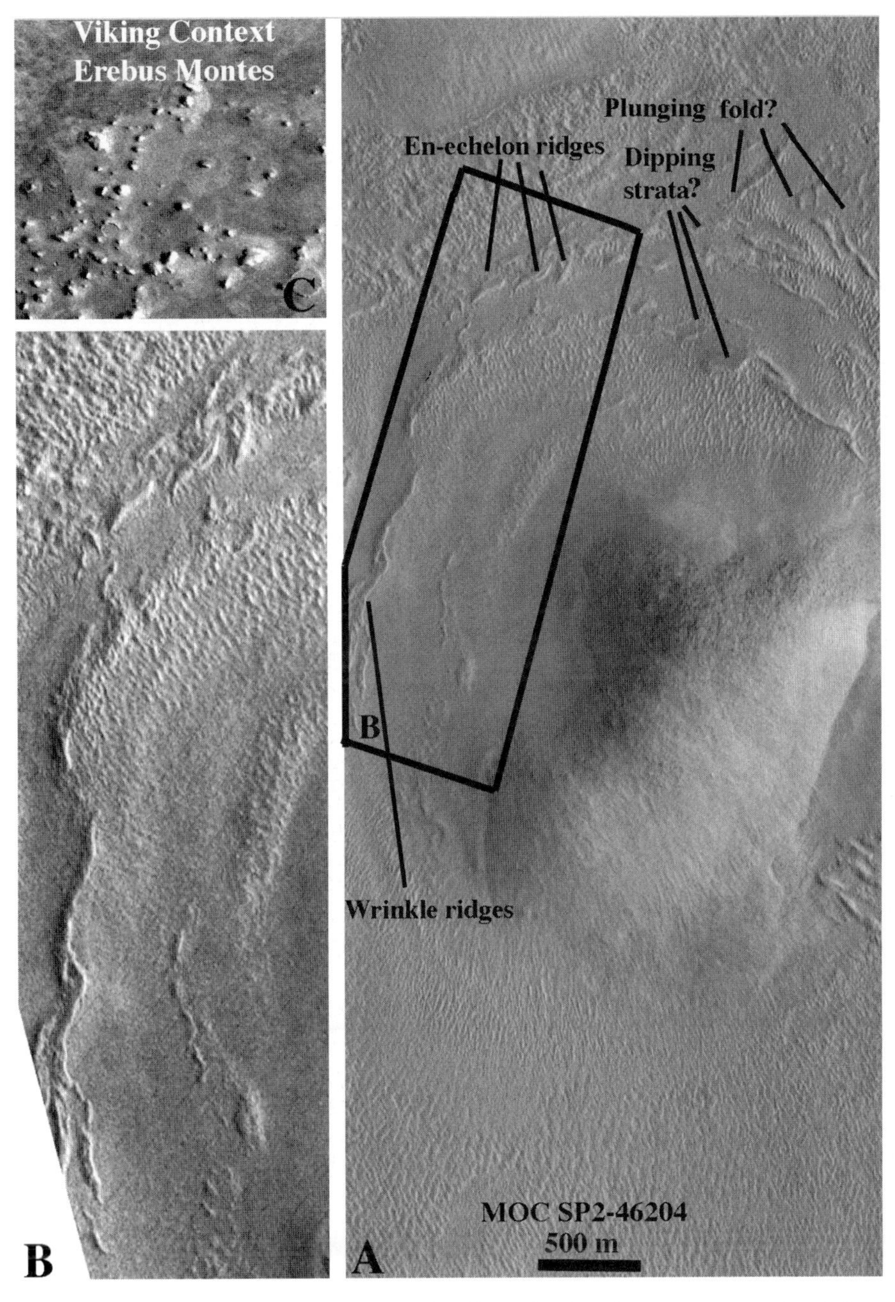

Near the "Erebus Monster" (Chapter 1), this and other domes were emplaced from below by intrusion or diapirism. The giveaway clues are the concentric ridges. These domes occur in sedimentary basins and may consist mostly of ice, mud, or salts related to deposits laid down by former lakes, oceans, or glacial ice sheets. 41.6°N lat., 170.0° long.

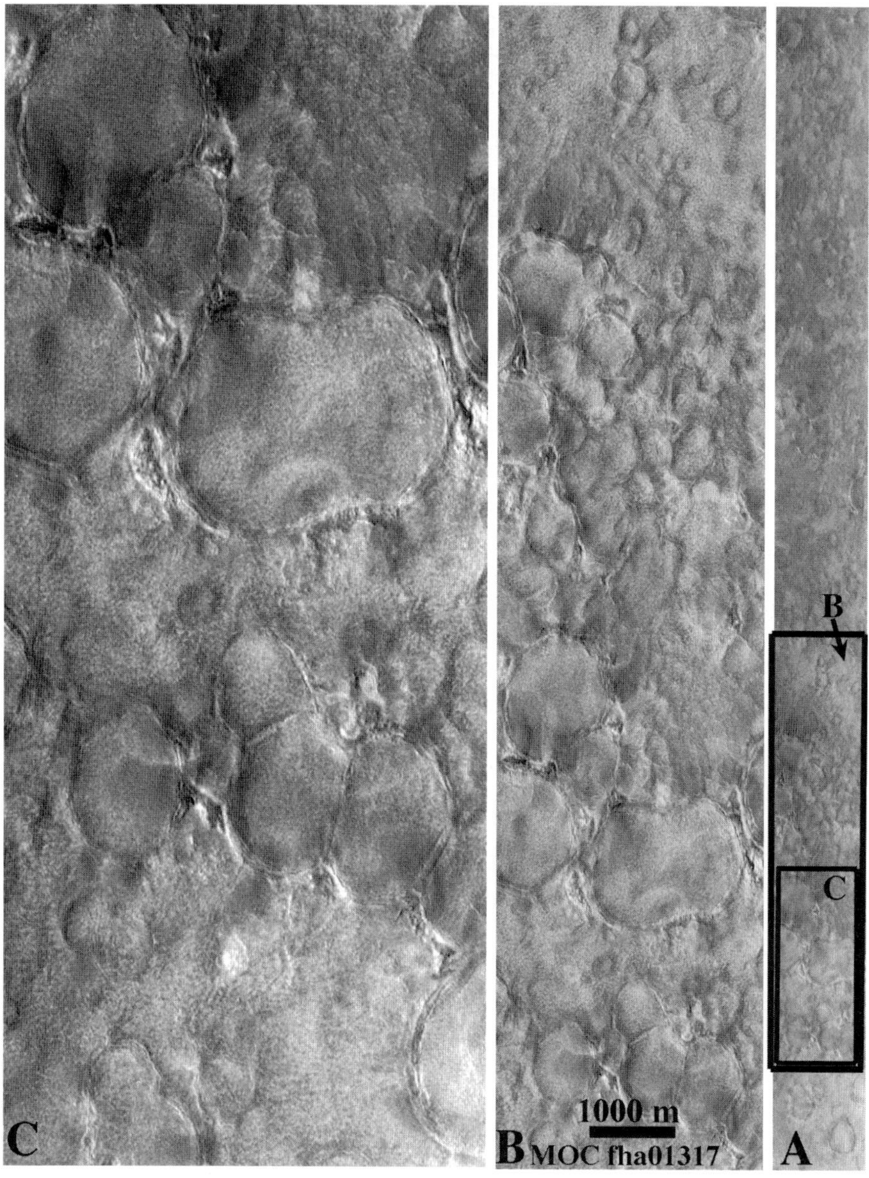

Northern plains deformation 100 km west of the *Viking Lander 2* site may have been caused by either (1) vertical ascent and intrusion of diapirs, or (2) vertical compression and lateral flattening and ductile stretching of layered deposits to form boudins. In case 1, buried mud, salt, or ice may have become bouyant after being loaded with sediment, with the mud, salt, or ice then intruding into deformable sedimentary rocks; the ridges between diapirs represent this sediment. Boudinage (case 2) is discussed in Chapter 7; applied to the northern plains, this process would involve burial of a sequence of less and more viscous ductile material; the ridges represent the less viscous material. 48.2°N lat., 228.4° long.

and the sediment and outflow channels still exist as testimony to those dramatic events billions of years in the past. The ocean would have been similar to an ice-covered Antarctic ice shelf in Lucchitta's view, or a warmer ocean that might have tempted Martian sunbathers in Parker's initial concept; or it may have been a rapidly freezing sea in my models and in Jim Head's latest version of the northern plains ocean, or was a peculiarly Martian type of mud ocean according to Jöns or Tanaka. The concepts differ in water quality and quantity, marine processes, and environmental conditions, but all pertain to a northern body of water and sedimentary deposits dubbed *Oceanus Borealis* by Dr Victor Baker (University of Arizona).

TAKE IT WITH A GRAIN OF SALT

There is much interest in the fate of the boulders, cobbles, pebbles, sand, silt, and clay which outflow channels would have dumped into the northern plains, Hellas, and Argyre. There is much debate about whether these deposits will form sorted deposits (coarse material dropping out first and closest to the mouths of outflow channels; finer material traveling further); or whether hyperconcentrated flows would have lacked time to segregate by grain size, resulting in massive debris flows. Surely some segregation of water and reworking and sorting of sediment would occur, but these two end-members highlight the extremes that geologists are discussing. However, that concerns only the suspended and bedload of clastic material. The debate has hardly considered the fate of the dissolved load – the salts. Considering how salty the Martian soil seems to be, and that any long-trapped groundwater would have weathered the subsurface rocks considerably, we must expect a high salinity in any outflow channel effluent. This will not be freshwater. It would probably make Earth's seawater seem refreshingly pure.

Geologic deposition of salt minerals due to freezing is not as common on Earth as due to evaporation, but it does occur in places such as Antarctica and Tibet. Since any Martian ocean or big lake would probably have frozen in the first instance, I present here the results just for one model based on freezing of a seawater brine. I assume that the brine and clastic sediment is able to segregate, so the bottom layer includes the boulders, sand, etc. Although the surface pattern predicted by the model looks a lot like a classical marine depositional basin structure on Earth, such as the bull's-eye pattern of the Michigan Basin, there are key differences. The model presented here is much simpler than what elapsed in most terrestrial basins.

First, Earth's basin water levels rose and fell hundreds of times over the life of each typical basin, each time with a change in composition and a concordant shift in the input of clastic sediment from nearby land masses; hence, the deposits are finely layered. The Mars ocean deposits might not be so finely layered. The model shown here is a "one-shot" injection of brine; it forms one sequence of rocks with just a few massive layers. The evidence from outflow channels would indicate several injections, therefore there would be multiple sequences, but not hundreds as we see on Earth. In addition, there probably were volcanic eruptions between flooding

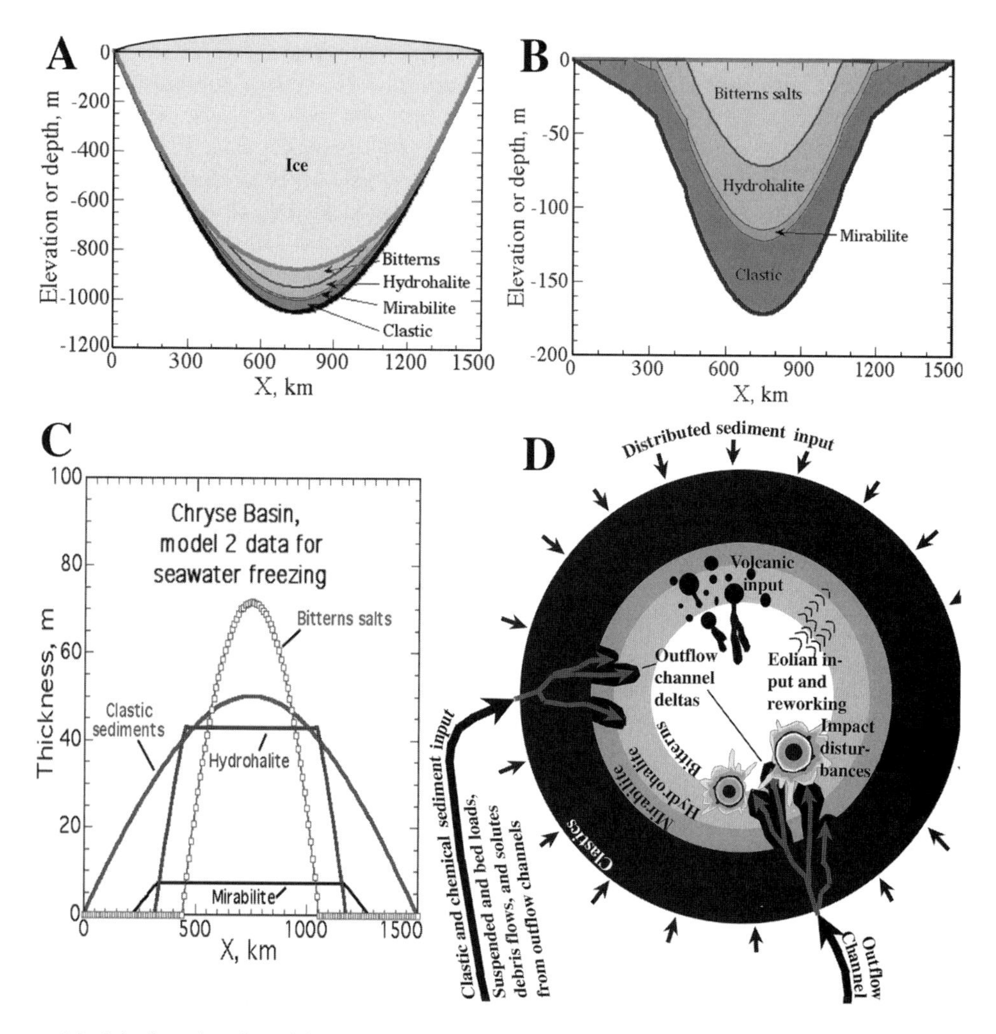

Model of marine deposition in an idealized Martian basin. The basin form was taken as a sinusoid of revolution about 1 km deep at the center. Earth's seawater was chosen as the initial brine, which is not very likely for Mars but is conservative in initial solute content. The clastic load is arbitrarily set at 10% by mass. The basin is filled with brine, clastic sediment settles out, and the brine starts freezing by fractional crystallization. Ice floats, and solutes are concentrated in the residual brine, which eventually saturates. Salts sediment onto the basin floor in a predictable chemical sequence. (A) Initial frozen stratigraphy in cross section. (B) Ice may be lost by sublimation; evaporite stratigraphy shown with basin slope removed. (C) Thicknesses of salt beds. (D) Possible bull's-eye map-view pattern of salts, clastic sediments, and incidental deposits. An actual Martian basin is expected to have a higher initial solute concentration, making thicker salt beds, and a higher abundance of sulfate salts relative to chlorides.

episodes, unsorted debris flows, dust blankets, and impact crater ejecta that would make reality more complex than in the simple model.

Another significant difference is in the composition of the dominant layers. Most terrestrial marine basin sequences include thick carbonate and gypsum deposits, but these are low-solubility salts that could not make up much of the mass of the Martian basin sequences. Why should there be this difference? One difference is between the effects of freezing versus evaporation: freezing produces greater masses of highly hydrated salts, whereas anhydrous or less hydrated salts are produced by evaporation at higher temperatures. Also, on Earth, marine deposition takes place with a continuous or often-repeated influx of seawater, plus a dilute freshwater input of more solutes. For the typical lifetime of over 50 million years for a marine basin, evaporation may total 50,000 km of water! During this time up to 5 km of low-solubility salts may also be deposited. The low-solubility salts precipitate in large amounts in terrestrial basins, while the high-solubility salts hardly ever have a chance to precipitate at all, though when they do, they form thick beds very rapidly. On Mars, rare episodes of massive flooding produces negligible quantities of the low-solubility salts and large amounts of ice and high-solubility phases. But some things are strongly in common with basins on Earth: there must be a tendency for coarse materials to be deposited near the edge of the basin and mouths of major channel inlets, for finer clastics to be deposited further into the basin, and for a crude sequence of various salts to be deposited. Thus, one landing site per basin is not enough. We need to think big, or we shall miss the key points of Martian basin sedimentation, miss the logical places to look for life, and miss the serious mineral resource exploration needed in advance of sending humans.

MOUNTAIN MEN AND WOMEN OF MARS

It was in the mid-1980s, just as these ideas of Martian oceans were being broached and other ideas of widespread icy permafrost were being developed into solid concepts, that I started attending planetary science meetings. That period struck me as a time of still-stodgy resistance by the established elite to bold ideas, but one where bold thinking was being presented and heard, though widely ridiculed in the conference hallways and after-hours lounges. Despite the lack of acceptance of bold new ideas, planetary science still seemed to be the ultimate in freedom, like that which an American "Mountain Man" of the 1800's would have felt – freedom to go it alone against all odds and through all the insults delivered by Man and beast and storm. A hardy man or woman could live the intellectual wild life and thrive in isolated wonderment of Mars; or lacking a thick skin and perseverance could die a premature professional death, as some did, soon forgotten by the world somewhere beyond the frontier. One could dream large thoughts alone or in association with a spartan few like-minded fellows, fully in the contemplation of the vastness and beauty of Nature. Although I didn't meet them until 1989 and 1990, two such hardy fellows (Vic Baker and Tim Parker) were the very images of rough-hewn, wild and woolly Mountain Men of the American Old West. Baerbel Lucchitta, though in the same scientific mould of independence, projected quite a smoother, more lovely physical image, but she, too, lived this difficult professional life. Not only were their personalities large, but their ideas were vast, and none of these would simply go quietly to their professional graves because the elite liked other ideas.

In the early 1980s, Lucchitta's comparative planetology approach, using Earth–Mars cold-climate analogs, first drew me into Mars science research. I recall her presentation of Martian and terrestrial features side by side and at the same scale. I have been heavily influenced by Lucchitta's concepts of permafrost and glaciation. By 1989, I had already been instructed by Vic Baker in a planetary geology course and was quite familiar with his interpretations of unimaginably huge floods debouched through (and forming) the Martian outflow channels. During the autumn of that year, he gave a peculiarly philosophical (very-Vic) presentation at a Mars conference. He was clearly grappling with some strange aspects of Martian geology which indicated vigorous ancient hydrology but seemed to involve mainly subsurface transport of water and occasional surface outbreaks of the fluid, unlike Earth, where the greatest fluxes of water occur through the atmosphere and over the surface. He seemed poised to make either sense or nonsense of that planet, and then stopped just short of conclusions that would have done one or the other. There was something still missing in his presentation, as indeed there is in just about every presentation about Mars; solid conclusions were as elusive in Baker's talk as they were in his summary philosophy. Baker's was an amazing performance to witness, but not so easy to follow, except that his overarching message was plain: Mars was weird, Mars was wet, and something very bizarre had happened there from the ground up. That is still the million-dollar thought today.

5

Mammoth Ice Age

EUREKA!

It was, as close as I can recall, 48 years to the day after the attack on Pearl Harbor. I was taking a break from lab research for my dissertation on the topic of cryovolcanism on icy satellites. I have always been a restless researcher, drawn by each amazing world, but never to the exclusion of the others. At that time the main publicly accessible planetary image archives were in paper print form, and one archive was in my department in my adviser's lab. Not liking my broom-closet office, I made the Planetary Data Center my office. A few times each week I spent some hours foraging through file cabinets of *Viking Orbiter* images, one imaging sequence after another, in a systematic scrutiny of more than 40,000 images. These were humanity's pictures – a showcase for a distant and otherwise inaccessible piece of real No-Man's Land. This digital photographic archive is our generation's equivalent of Lewis and Clark's accounts of the great and natural North American West. Every trip to those filing cabinets was like a day experienced with the great explorers of the classical age of discovery. The Mars image archives are pure observation stripped of every interpretive touch, leaving all the scientific and artistic interpretation to each image analyst.

About every other sequence of images presented something stunning to ponder, but on that December day I came across one particularly gripping image. That singular scene of 800 pixels square has captivated me for 14 years.

One particular *Viking Orbiter* image, what I call the Discovery Image, showed a portion of southern Argyre, a 1,000-km-wide impact basin. This single image spanned over 200 km of some of the most stunning scenery and dramatic geology that Mars offers. At the bottom of the image were long, stark shadows and sunlit northern faces and peaks of the Charitum Montes, a mountainous ring of the Argyre impact basin, towering 5,000 to 7,000 m (16,500 to 23,000 ft) in local relief. In the upper part was a vast smooth plain marked by sinuous, braided features. Clearly, I thought, this was the most perfect system of dried river channels I had seen in weeks

Portion of MC-26 MOC wide-angle mosaic with major place names in and near the Argyre Basin.

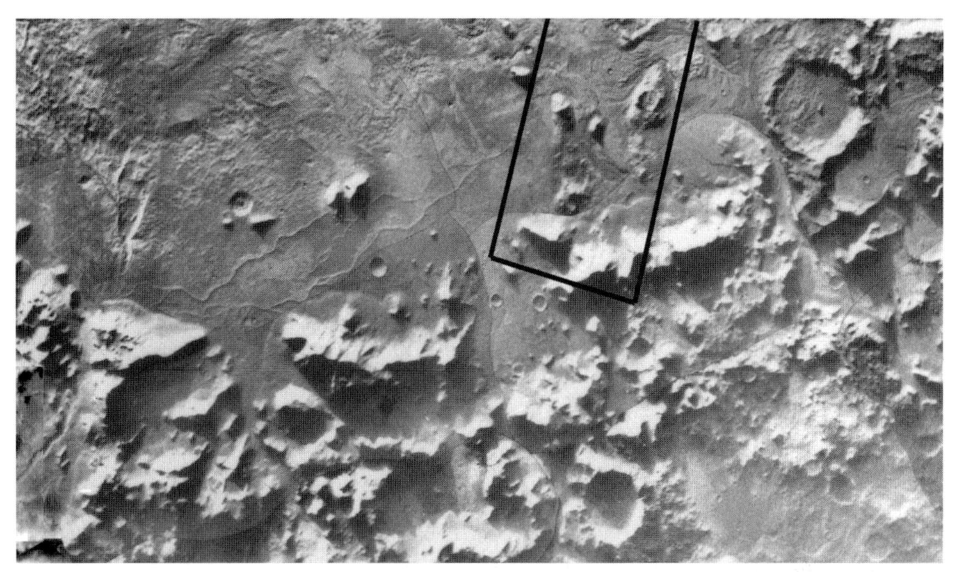

Mosaic of *Viking Orbiter* images of southern Argyre Planitia (*top of mosaic*) and the Charitum Montes (*bottom part*) with low-resolution gaps filled with MOC wide-angle images. The sinuous, braided ridge system of Argyre is plainly visible left of the center of the image in the plains. Also apparent are lobate debris aprons surrounding the massifs and through-cutting valleys that connect the highlands to the basin floor.

of image browsing. The positive and negative relief then clarified, based on which way the sun was shining. To my amazement, these were not channels at all – they were *sinuous, braided ridges*!

Measurements from VO and later MOLA data have shown these ridges to be about 100 to 200 m high and 2 to 3 km wide. The ridge system splays across an expanse of plains 300 km broad, seeming to be sourced at the mouth of a gigantic canyon that slices clear across the Charitum Montes. The overall pattern of braids was exactly what flowing water characteristically produces, but in positive relief. This did not make sense if these were ordinary river beds. Flows of water and all condensed channel-forming fluids are enslaved to the laws of gravity, and running water erodes channels, not ridges. Normally.

As the ridge morphology of these braids became evident, I recalled a field trip taken as an undergraduate student to study the leftovers of the Pleistocene Laurentide Ice Sheet in Ohio. Amid forests and corn fields was a sinuous gravelly ridge – an *esker* (from Old Irish *escir*, meaning "ridge"), a type of subglacial stream deposit of sand and gravel. *Are these Martian ridges also eskers?* This was a short question with big meaning: was there once a great ice sheet where the ridges now stand? Had a mighty subglacial river melted an inverted channel (or tube) in the ice, depositing a bed of sediment in its ice-walled, ice-roofed channel? Had the ice sheet then melted or sublimated, so as to reveal these river beds as sedimentary ridges? Did Martian climate shift one way and then another in grand fashion, much as Earth's climate had done through the Coal Ages and Ice Ages and interglacials?

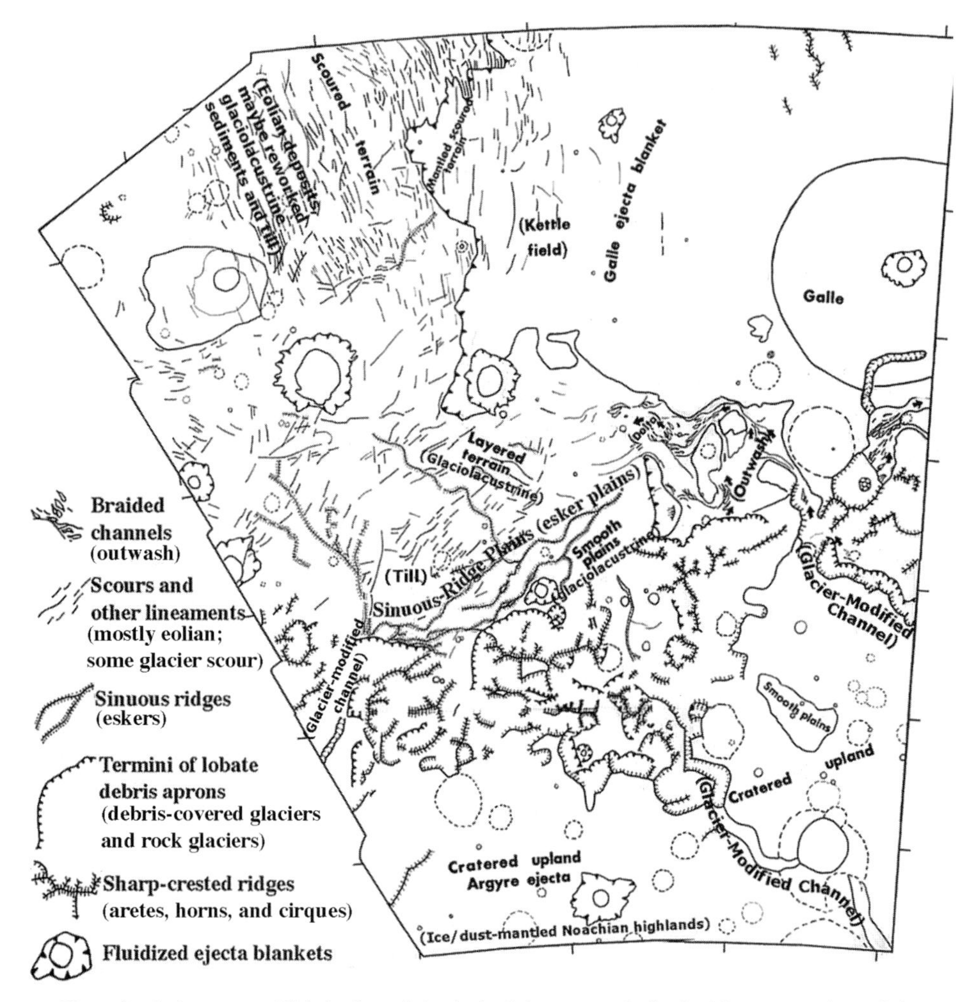

Hypothesis-in-a-map. This is the original glacial geomorphological interpretation of the Argyre basin given by Kargel and Strom (1990), with modifications to represent updated interpretations. A direct glacial linkage of some features has been weakened (what had been seen as scours are now seen to be eolian features, though, speculatively, they may be related to reworking of glacial deposits). In general, the glacial hypothesis has been greatly strengthened by new observations, and glaciation in Argyre is now more strongly linked to ice processes in the highlands south and east of Argyre (as first suggested by Kargel and Strom, 1992b, and strongly supported by recent work by Jim Head and colleagues). Conversely, the former case for glaciation is Hellas is now more ambiguous than seemed previously.

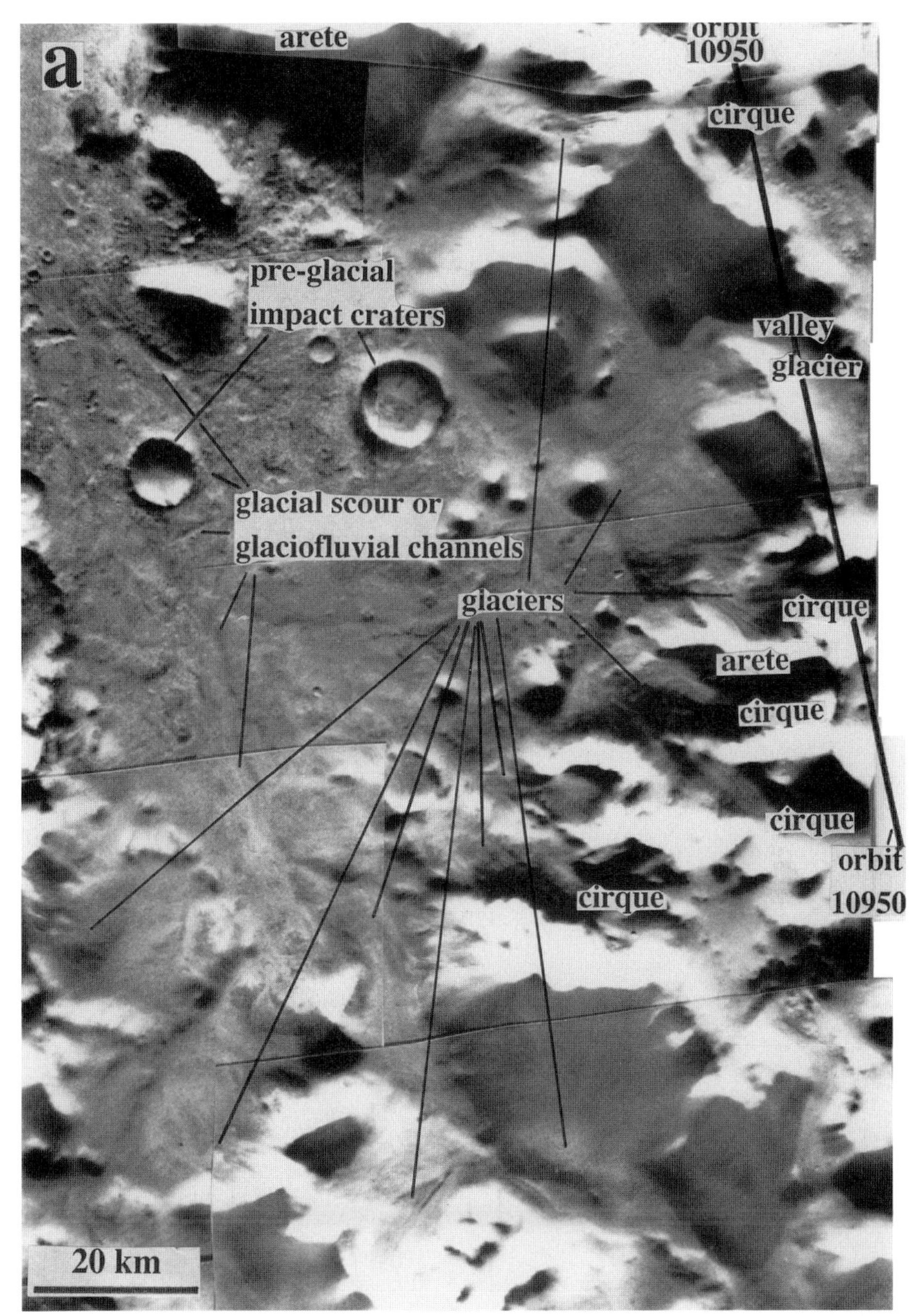

Dramatic relief in the Charitum Montes is interpreted in this *Viking Orbiter* image mosaic in terms of glaciation. Some details of the MOLA altimetry track are shown, described, and explained in Chapter 6.

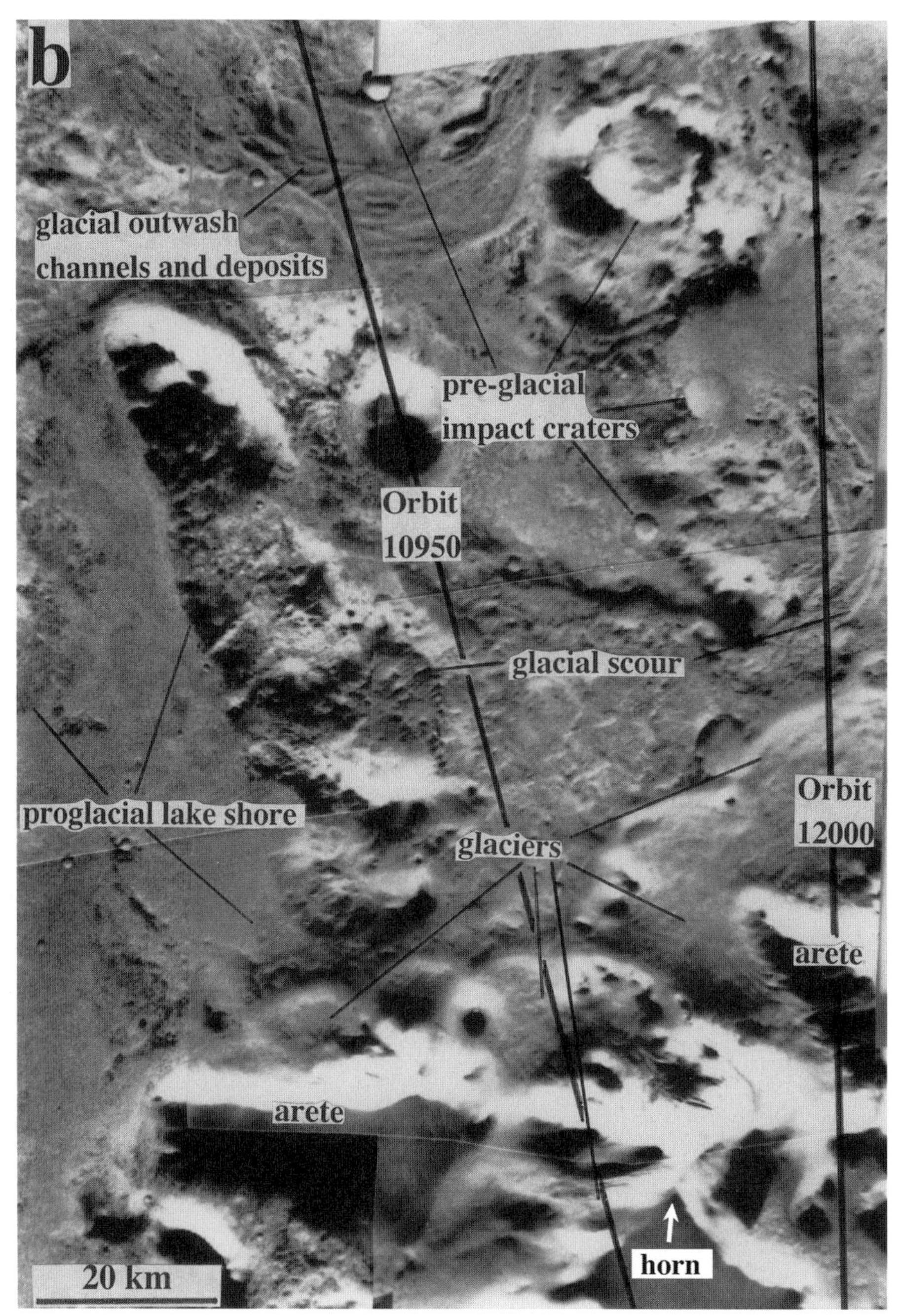

Rugged topography along the contact zone between the Charitum Montes and southern Argyre Planitia abounds with features interpretable as having glacial origins, as indicated here. MOLA topographic transects are shown, described, and explained in Chapter 6.

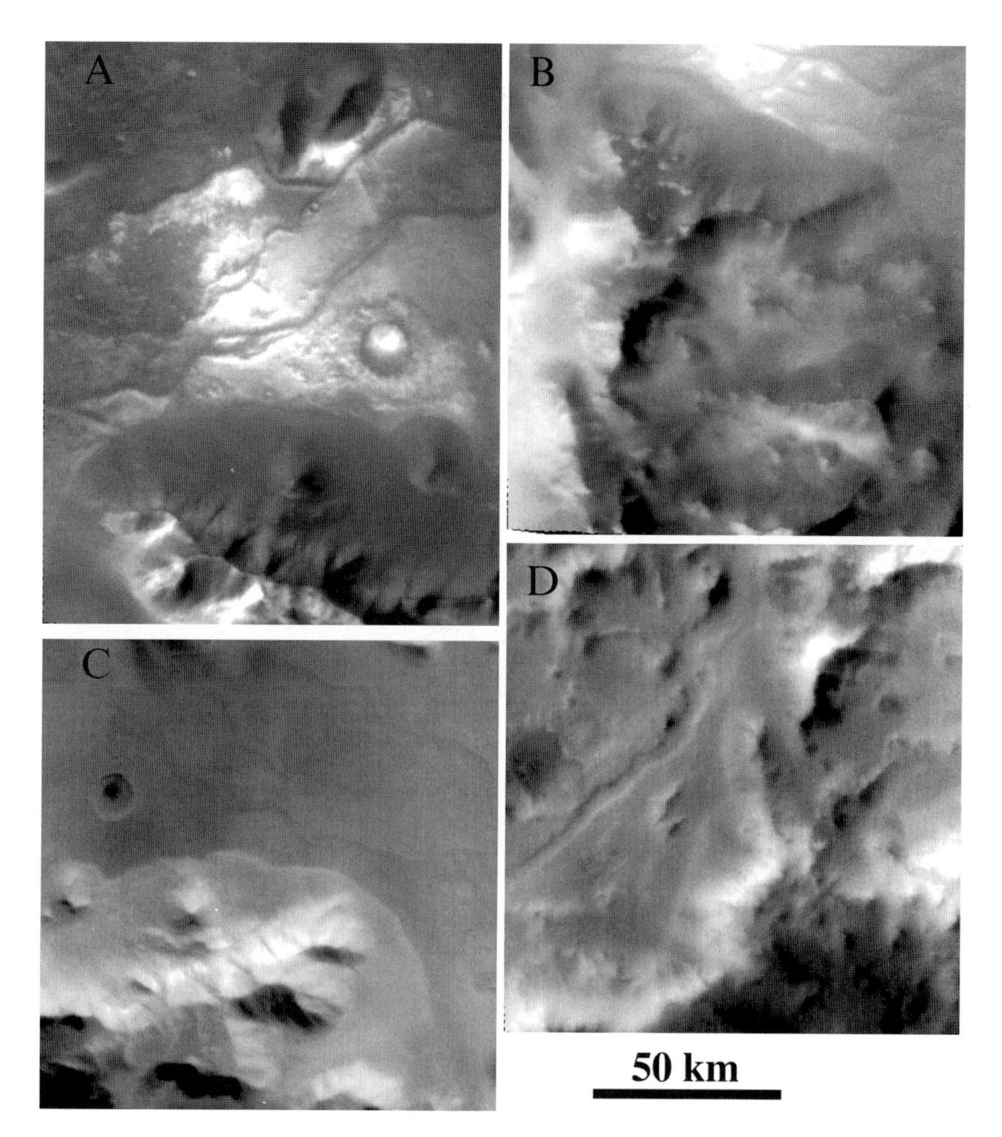

The MOC wide-angle camera has provided synoptic repeat coverage showing seasonal changes in frost cover and morphologic features that in Viking Orbiter images had been obscured in shadow or were on saturated sunlit slopes. (*Upper left*) MOC WA0400319 highlights cirque- and arete-like mountains, esker-like sinuous ridges, and glacier-like lobate debris aprons. (*Lower left*) MOC m1401577 shows lobate debris aprons and cirque-like mountain sides; the top of the massif appears not to have been glaciated. (*Upper right*) MOC 0307673 shows amphitheater-like mountains with compound cirque-like structure. (*Lower right*) MOC WA10-00100 shows a channel eroded into a wide valley that cuts across the Charitum Montes. The mouth of this channel is buried by a lobate debris apron, but this channel probably terminates in the sinuous ridge system of southern Argyre Planitia.

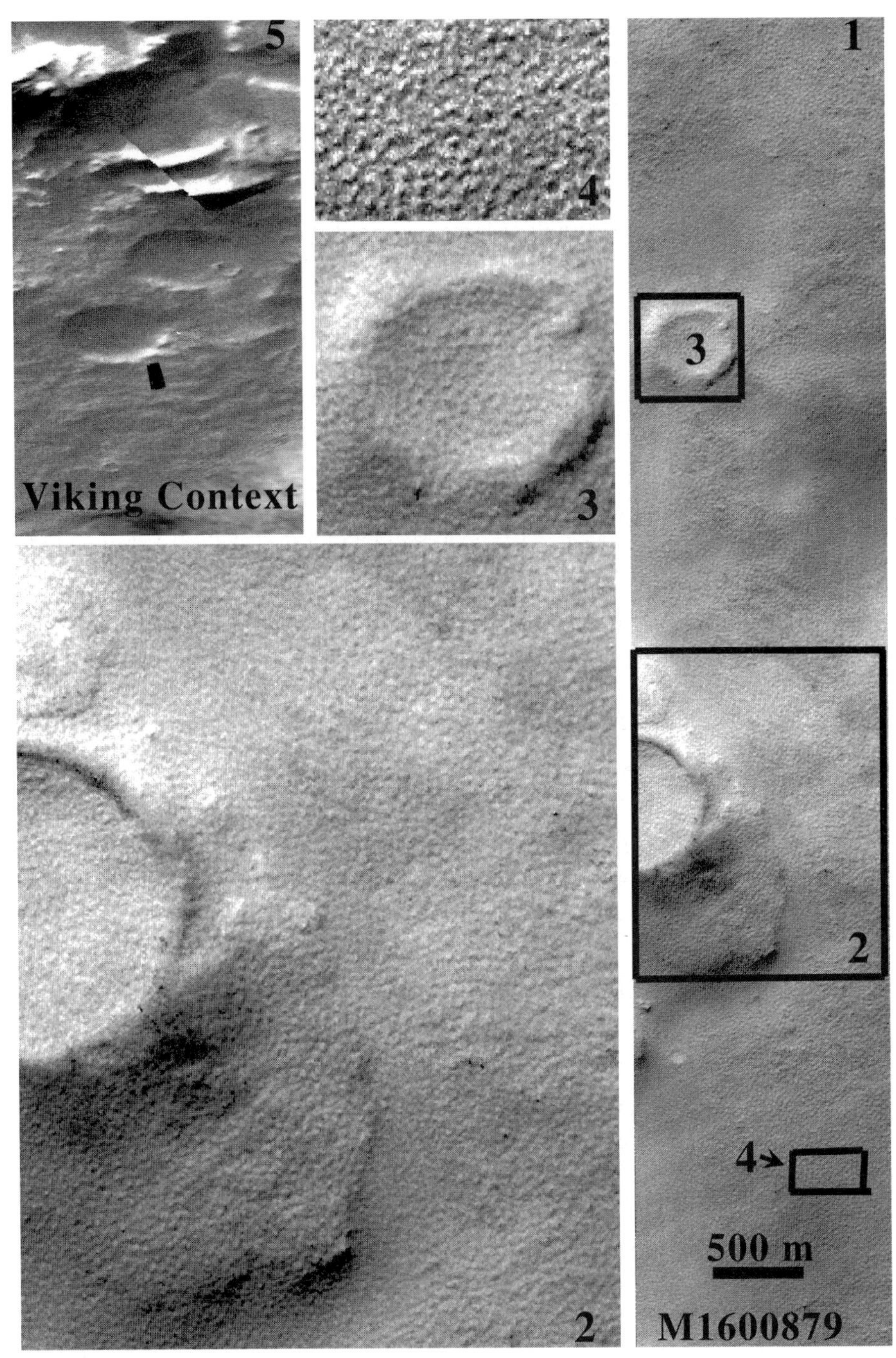

The cratered uplands south of Argyre are heavily mantled. The mantle material appears to be an icy, periglacially modified deposit roughly 80 m thick – nearly enough to bury craters 400–600 m across.

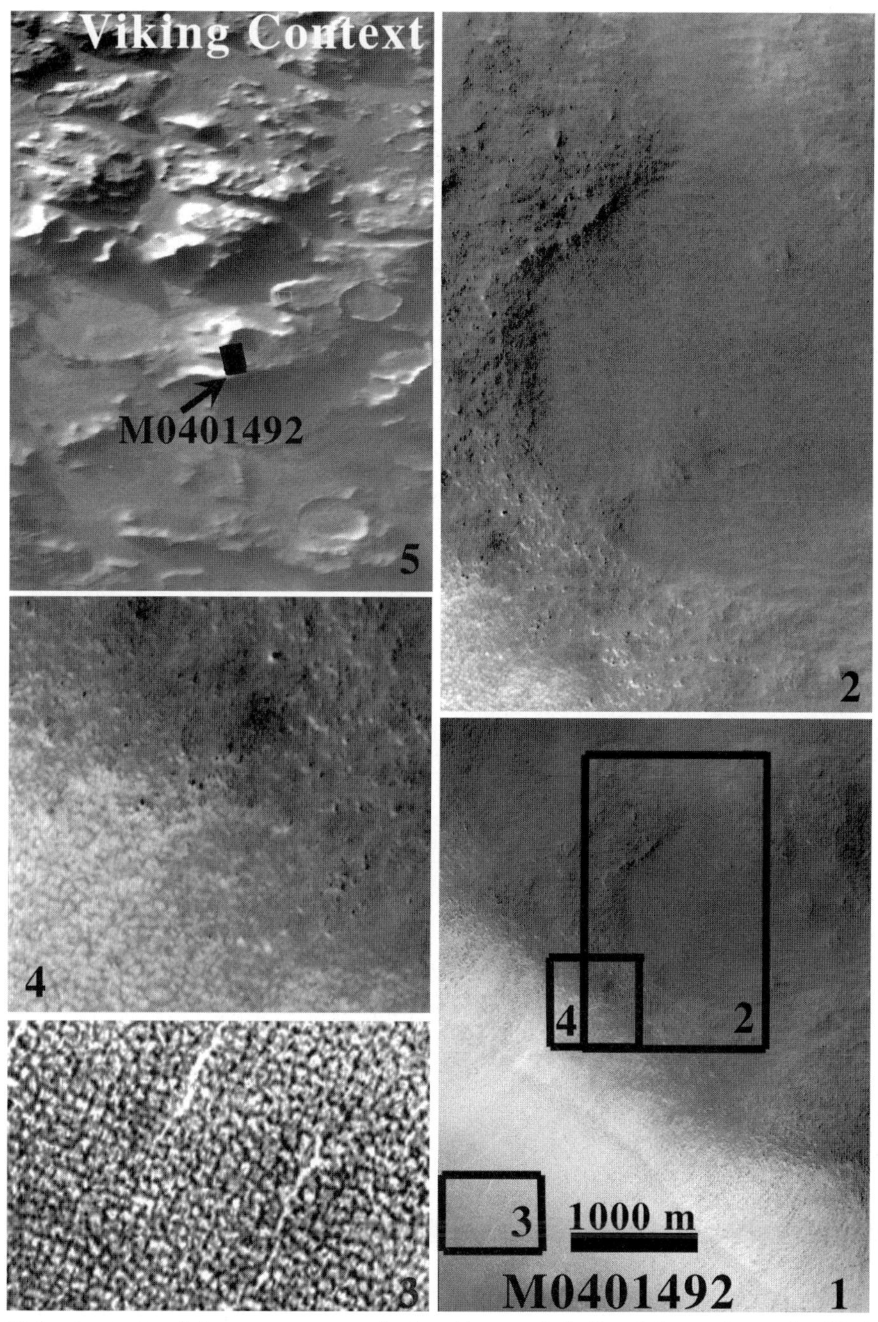

Uplands south of Argyre appear to be heavily mantled. Panel 2 may be an ice-filled cirque or breached impact crater. Panels 3 and 4 appear to be permafrost patterned ground. Except for the filled crater or cirque, the north side of the mountain ridge has thin mantling; on the south side it buries boulders.

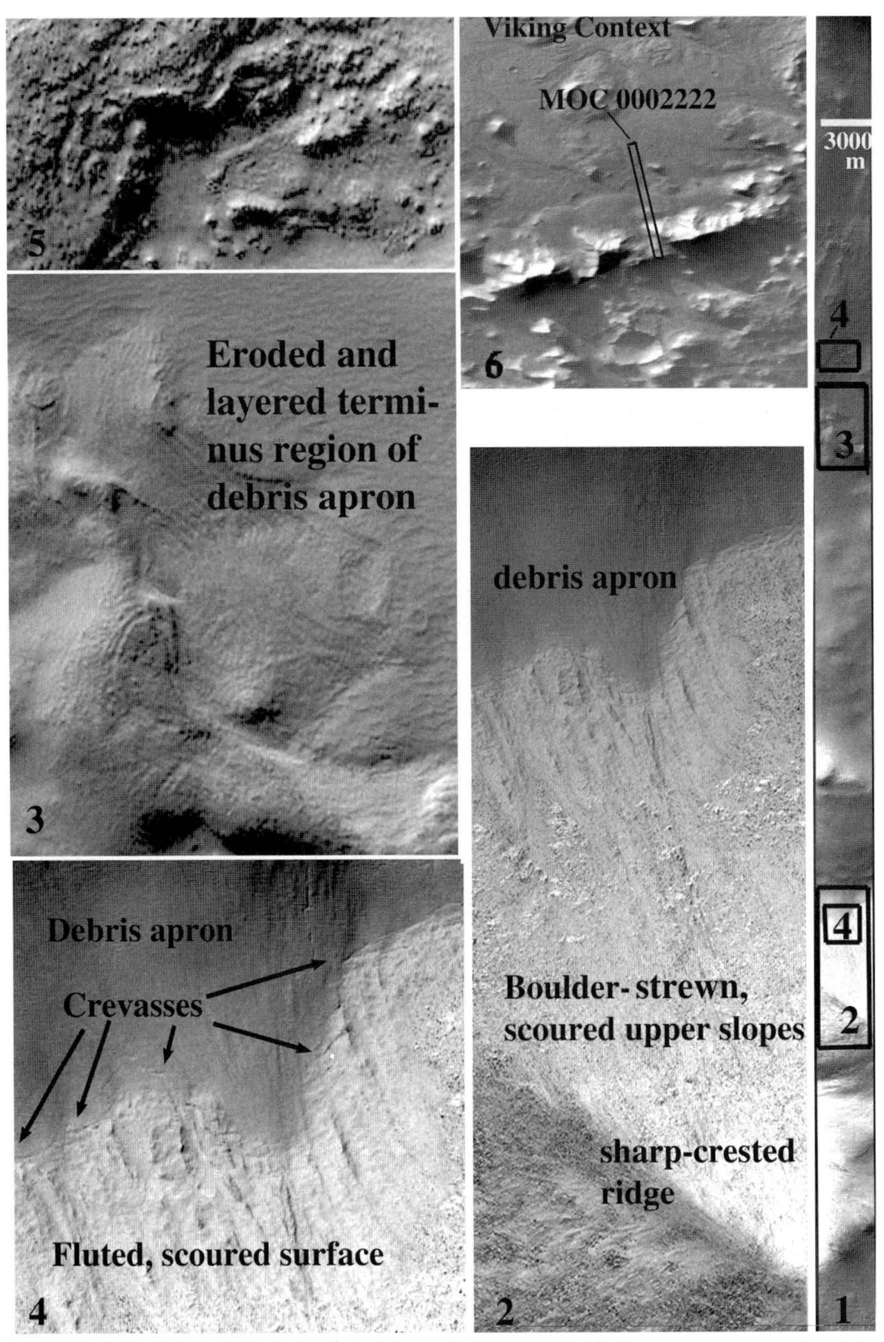

A lobate debris apron in the Charitum Montes is associated with fluted and boulder-strewn areas of the upper mountain slopes, is crevassed, is layered in its terminal region, has a hummocky texture like that of glaciers, and is marked at its terminus by a transverse ridge of boulders.

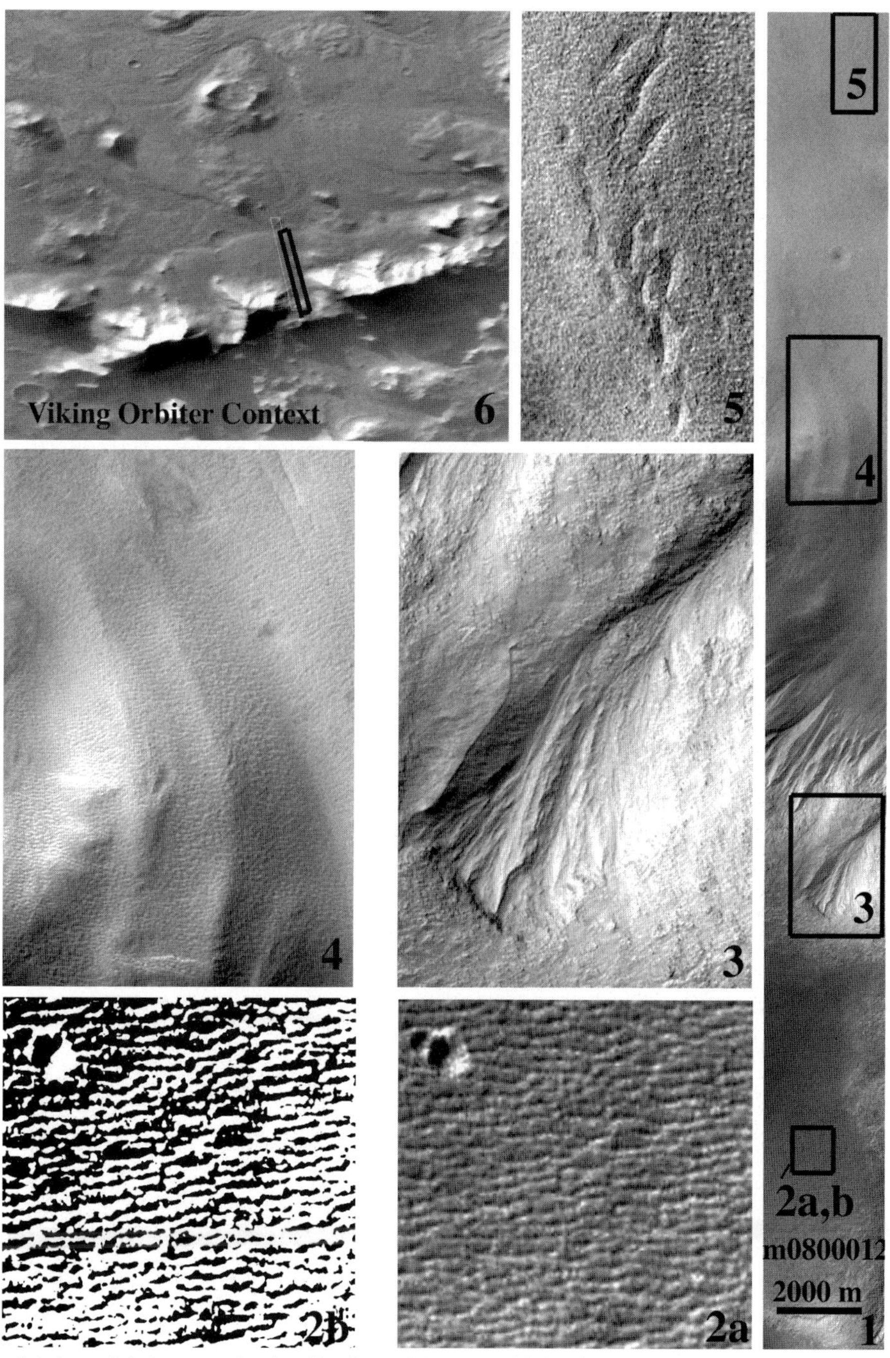

A lobate debris apron in the Charitum Montes exhibits signs of sublimation (panels 2a and 2b), melting (panel 3), flow (panel 4), and brittle deformation (panel 5) – all behaviors of glaciers.

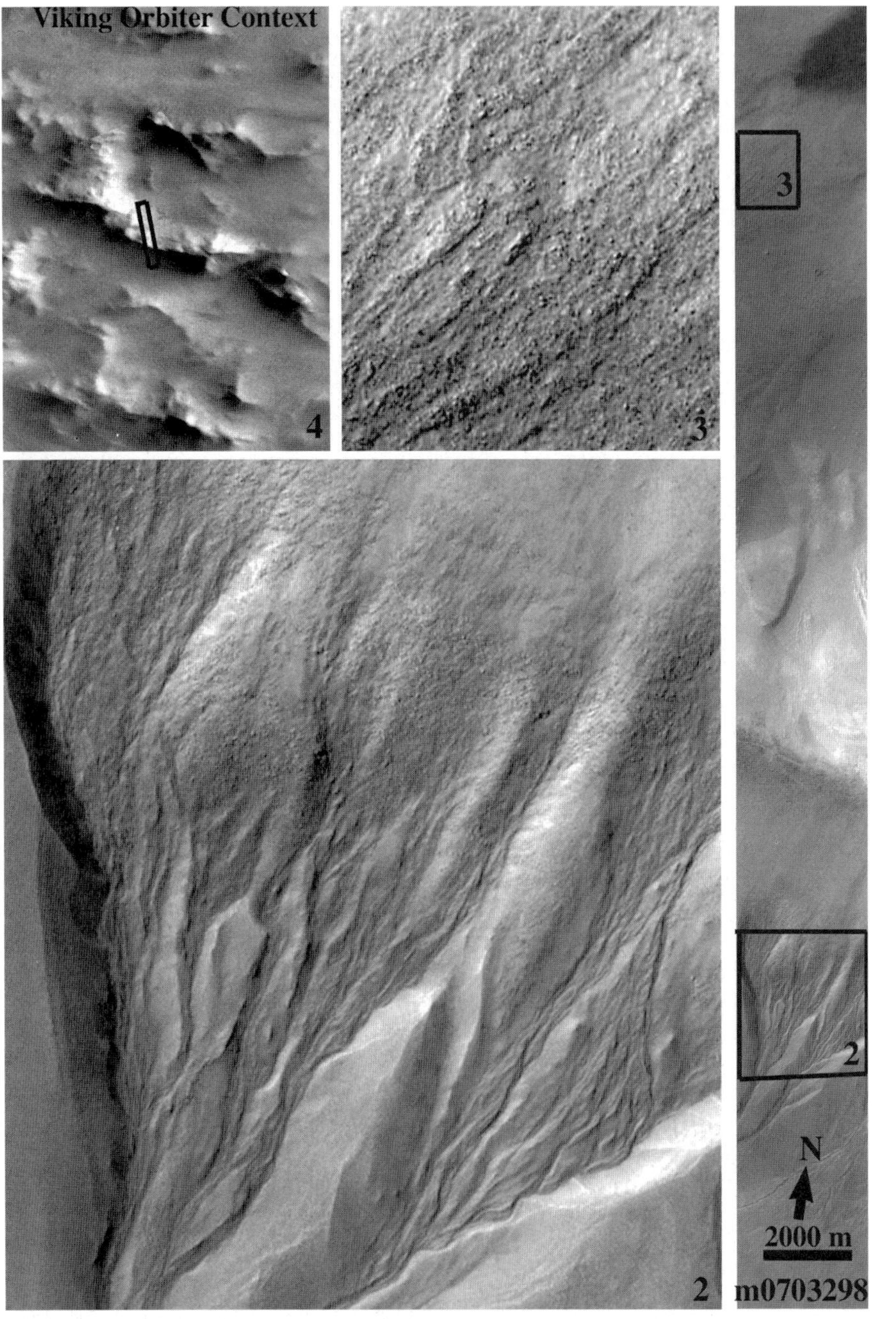

Massif in western Argyre shows features interpreted as glacial alpine scultpure (*Viking* Context), moulded bouldery glacial till (panel 3), and meltwater runoff (panel 2). The mountains appear still to be ice mantled. The lack of a strong hydrogen signature in GRS/HEND data require more than 1 m of debris cover.

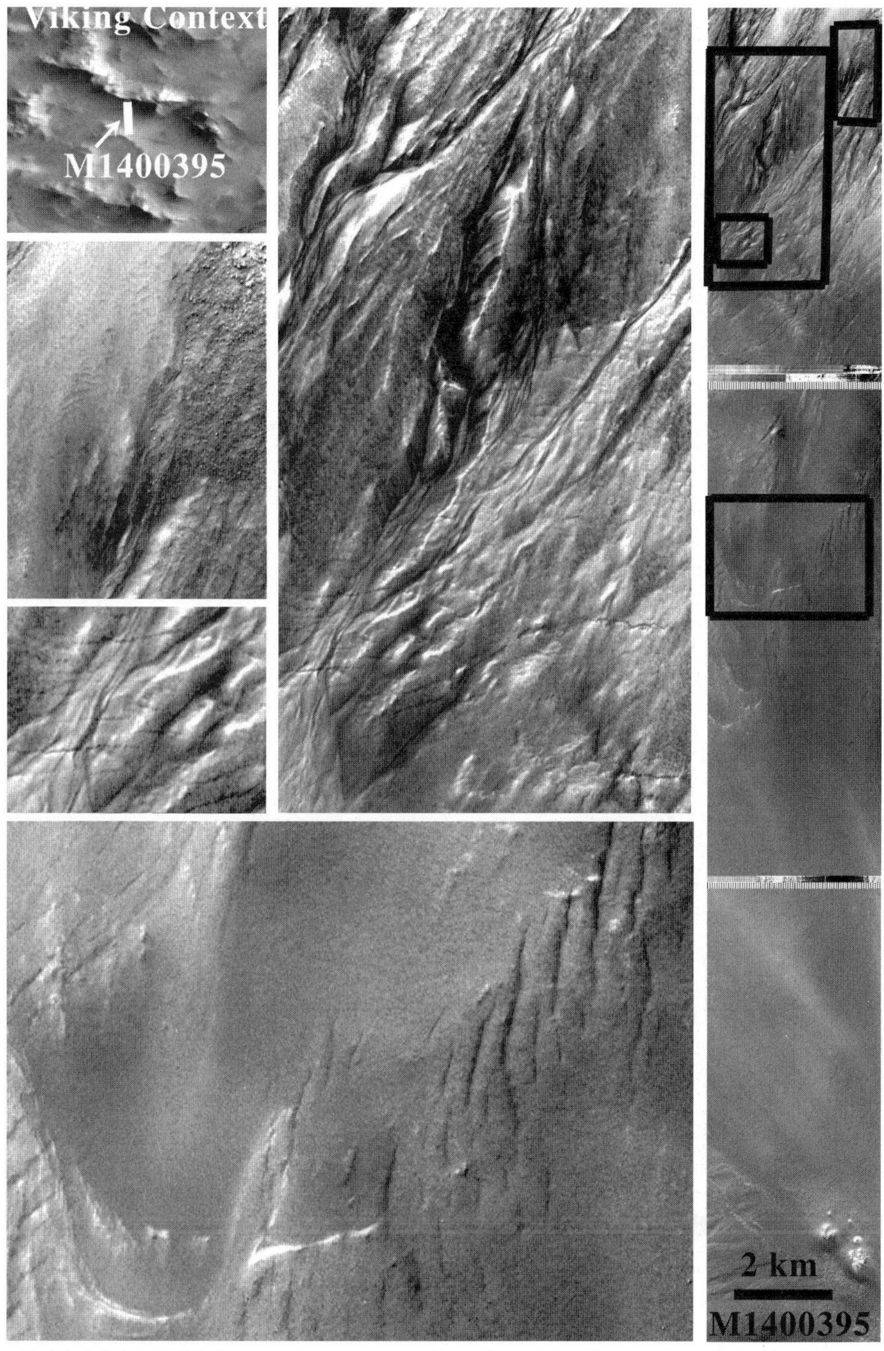

Gullied lobate debris aprons on mountains of western Argyre are themselves cut by crevasses. This, like other lobate debris aprons, has the hallmarks of a debris-covered glacier.

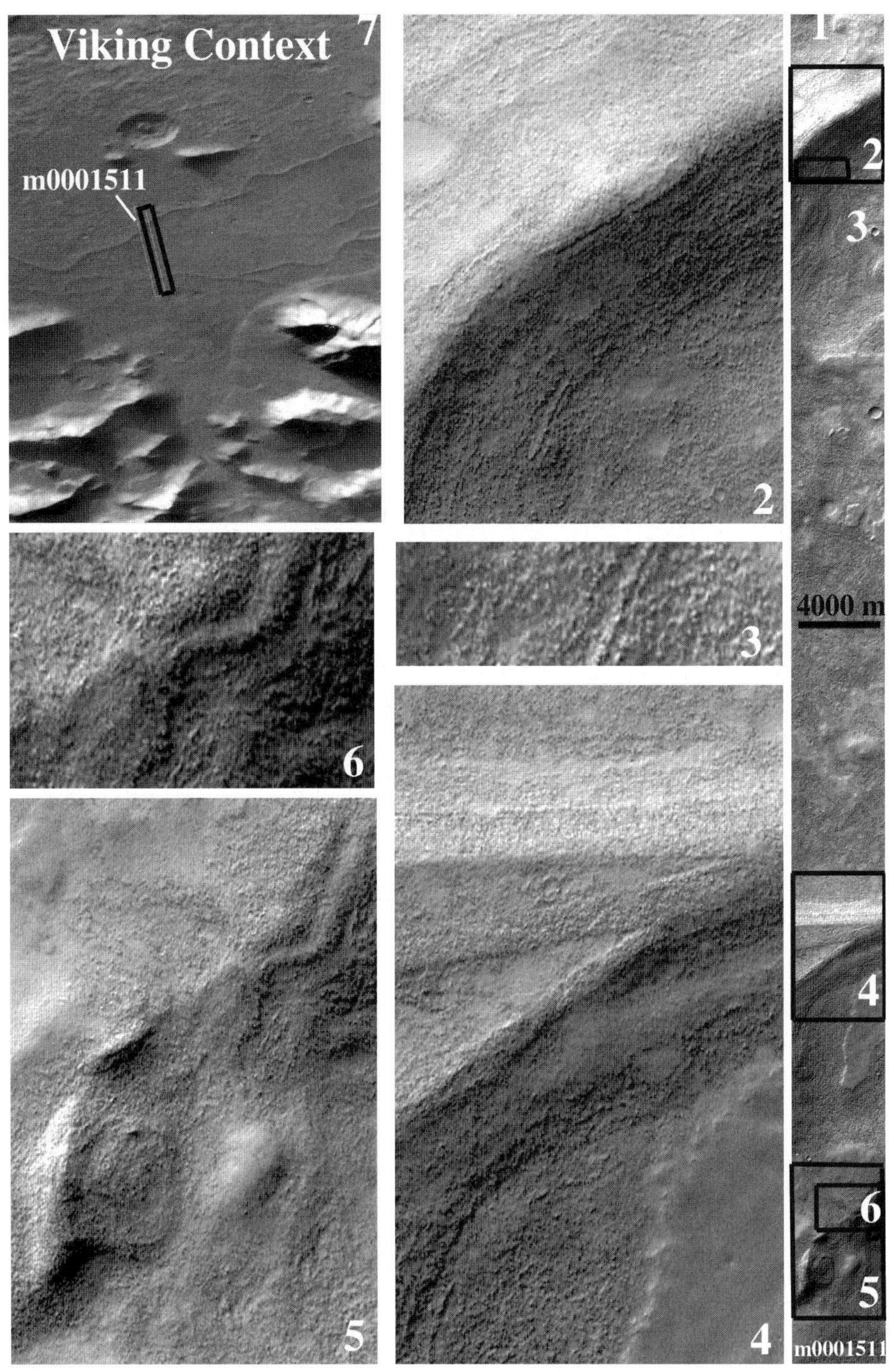

Bouldery sinuous ridges, like eskers, come in many shapes: flat-topped, sharp-crested, and discontinuous.

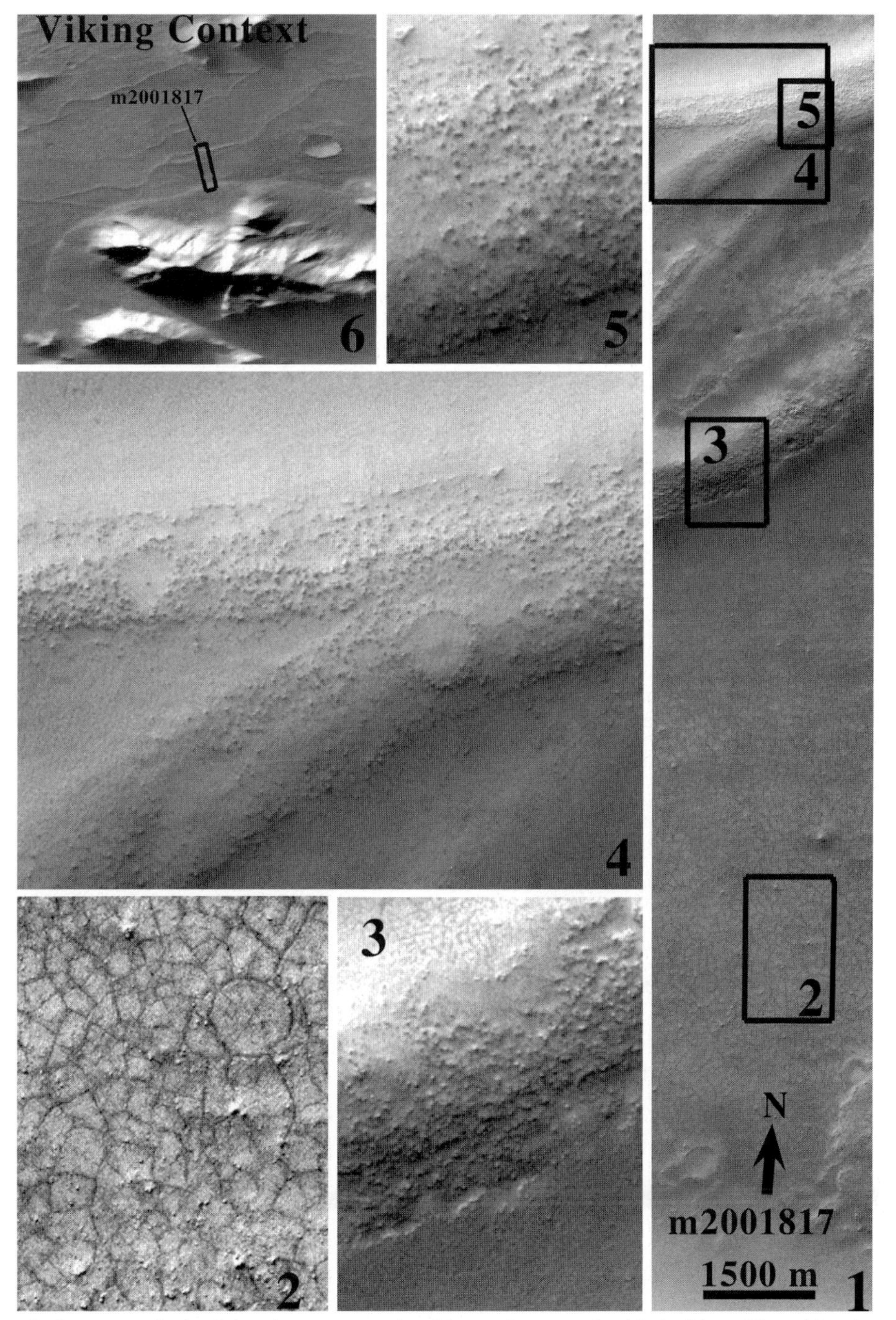

Polygon-cracked plains lap onto a bouldery, sinuous, braided ridge. The ridge is probably an esker or esker-like glacial deposit; the polygon plains may be periglacially modified glaciolacustrine plains.

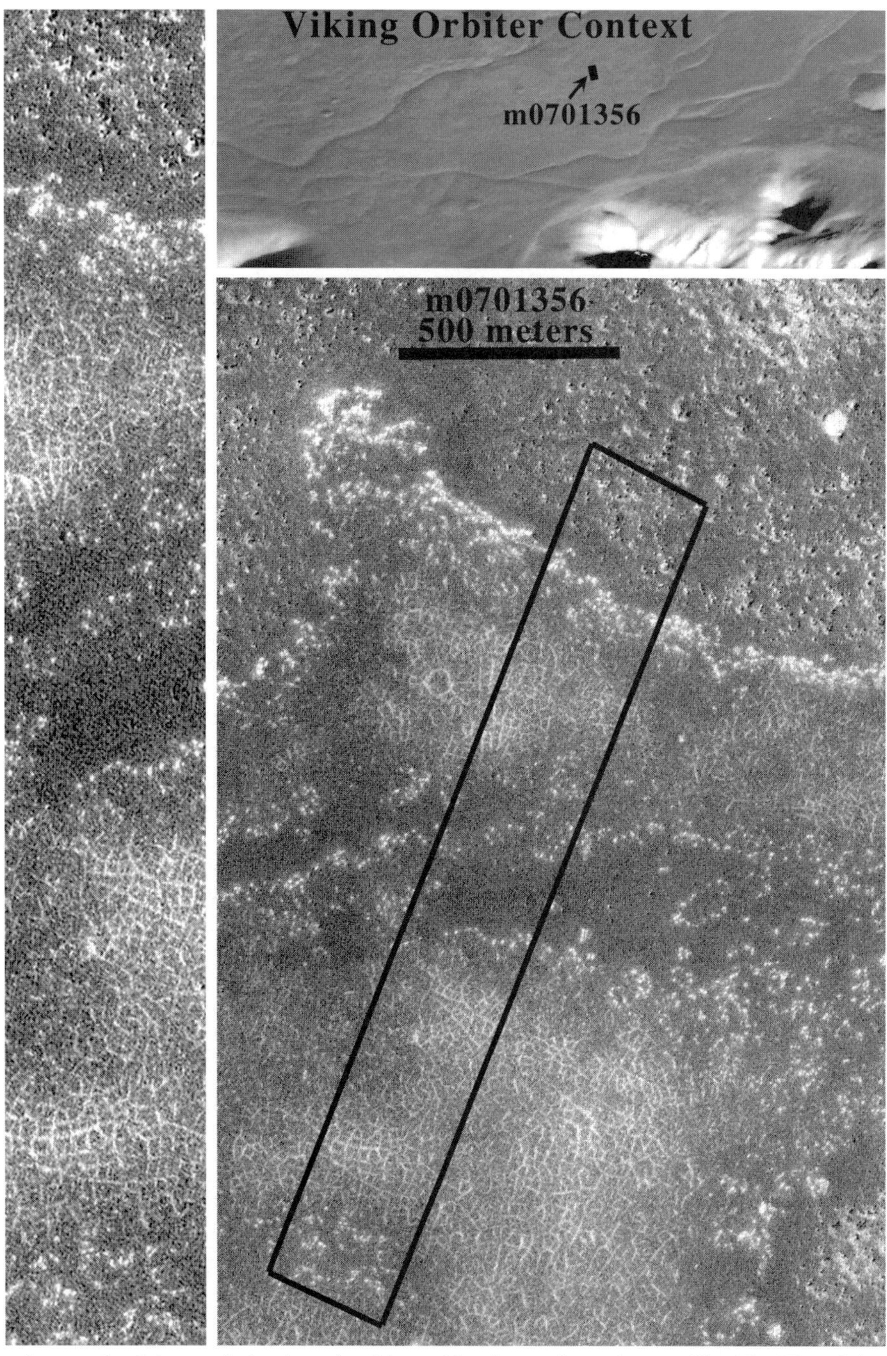

A smooth plain nearly buries a bouldery ridge in southern Argyre Planitia. The plain, probably a late-stage glacial lake deposit, is polygonally cracked due perhaps to desiccation or thermal contraction ice-wedging.

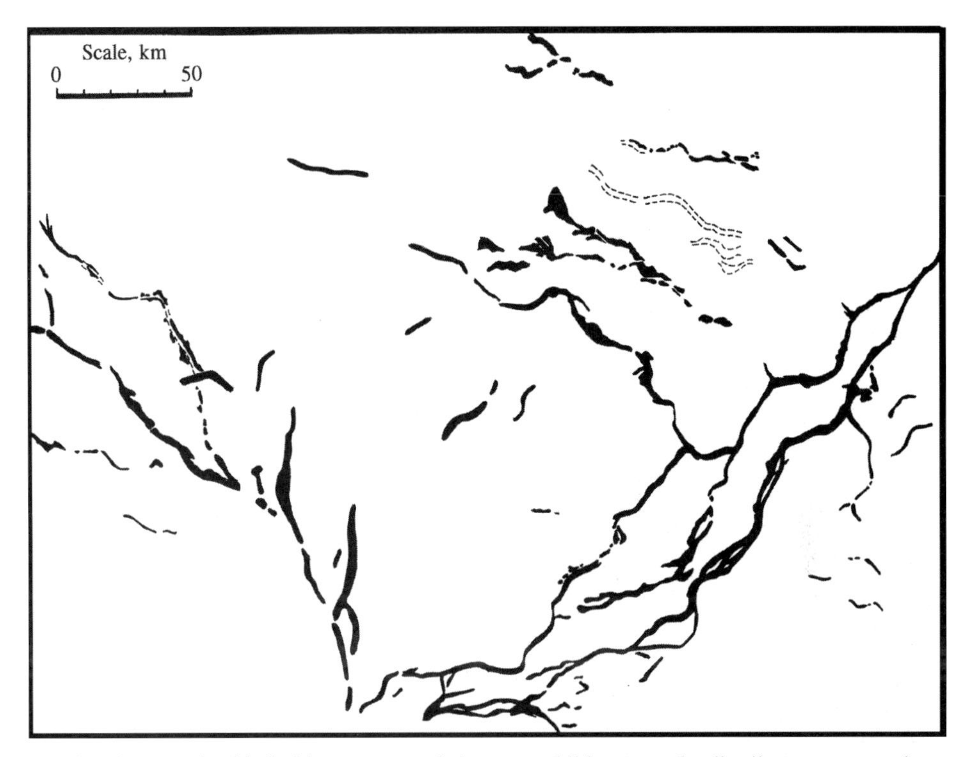

The sinuous, braided ridge system of Argyre exhibits a crude distributory network radiating from a major inlet channel. However, some of the ridges could be moraines instead of eskers. Tim Parker has favored a lacustrine strandline interpretation. However, their extreme bouldery composition (up to 20-m boulders) and height (up to 80–100 m) may pose a dilemma for that model; however, the interpretations and arguments go round and round. Older interpretations that the ridges might be lava flows, wrinkle ridges, or sand dunes are clearly eliminated by the new data. There now is agreement that the ridges are made of bouldery sedimentary rocks and their deposition is related in some way to channels that cut across the highlands through the Charitum Montes and onto the basin floor. Deposition as hyperconcentrated slurries, as suggested by Ken Tanaka and others, could explain the huge boulders. Some of these features are troughs (where dashed) instead of ridges. This map is based on *Viking Orbiter* image analysis; additional minor ridges and a need for other revisions is evident in improved imagery obtained from *Mars Global Surveyor* and *Mars Odyssey*.

The interpretation and vast implications came across in seconds, with a sudden rushing. The addiction was instant, and the research became a career focal point.

These braided Martian ridges needed some discussion, and so I talked to a fellow graduate student, Virginia Gulick (now at NASA Ames), who urged that I should discuss the issue with geomorphology Professor Victor Baker about it. And so I organized a meeting of Baker, my adviser Robert Strom, and several graduate students. Baker agreed that the sinuous ridges may be eskers. We discussed the natural variability in form, scale, and geometric integration of terrestrial eskers.

Stewart River eskers in/near Eskers Provincial Park, British Columbia, Canada. Panels A, B, and C show a sandy, esker-like delta deposited in a proglacial lake ahead of the retreating Cordilleran ice sheet. D and E show the eskers proper, which occur as a braided/branching system of ridges set within a subglacially eroded tunnel channel (F, a cross-channel view). The eskers' crests expose gravel and cobbles, although the bases of the eskers are mantled with late-glacial lake clay.

Cross-bedded sand and gravel in the Darling Esker, Central Minnesota.

Whatever the actual nature of these ridges, they should not have been produced in isolation. Eskers are always associated with other glacial landforms, and that was the first clue about what to do next.

The hypothesized eskerine nature of these ridges was important not so much for the origin of the ridges, but for the vast ice sheet whose existence and disappearance is implied, and for the climatic implications that the comings and goings of ice ages imply. Whether Earth or Mars, eskers are products of something much larger. If these Martian ridges were eskers, it meant that Mars was once unlike anything conceived by those who purport to understand it. It meant that outsiders such as Parker and Lucchitta and Jöns may have been on the right track to consider the big waters, big mud, and big ice, but lots of other people had to be very wrong. It implied not only ice, but living and melting ice, and that was a real problem for some people. Applied publicly to Mars in 1990, the word "esker" had a broader meaning: "instant huge controversy."

Talk about eskers

An *esker* is a bed of coarse sediment laid down by a meltwater stream that flows in an ice tunnel generally located at the bed of a glacier or ice sheet. It is commonly energetically favorable for turbulent, pressurized basal meltwater streams to erode thermally/mechanically upward and sideways into the ice than to erode mechanically into the rocky substrate, and thus a sediment-floored, ice-roofed tube is formed at the glacier's base. Sediment typically accumulates in the tube at the bed, as it does in any stream in accordance with hydraulic conditions, but in a subglacial environment the tube may be situated above the rocky substrate. As the icy confines of the perched river bed ultimately melt away, an esker may be left behind as a ridge. Eskers may form by englacial (within-glacier) or supraglacial (on top of glacier) flows, but the resulting sediment ridges left by a wasted glacier are then chaotic and often difficult to recognize. More often, significant englacial or supraglacial sediment is washed down to the sides or front of the glacier (contributing to moraines or washed further downstream) or to the bed (forming usual subglacial eskers).

In other cases, eskers are formed right at the snout of a retreating glacier where a subglacial meltwater stream discharges and dumps sediment. As the high-pressure stream of water or hyperconcentrated slurry suddenly exits the tube and the flow spreads out, a cone or fan of sediment is dumped. As the glacier retreats, a chain of sediment fans or hillocks with the appearance of a broken, sinuous beaded necklace may be produced along the course of the former subglacial waterway.

Eskers may form by continuous flows or may form during transient subglacial water movements, for example, when a supraglacial lake suddenly drains into crevasses and then across the glacier bed. Such pulsed deposition may give rise to large-scale layering or contribute to a bead-and-fan aspect of eskers.

The sediment of eskers may comprise sand, gravel, or boulders, and often the esker contains size-graded beds, cross beds, and other indications of deposition from water under conditions of variable flow. The tube shape may be affected by fluctuations in hydraulic pressure and meltwater supply and may change longitudinally downstream. The tube may expand vertically or laterally as the turbulently flowing, pressurized water generates heat and thermally erodes the ice. The subglacial stream may erode mechanically into the substrate, or the tube may contract as the weight of ice plastically collapses the tube when hydraulic pressure declines. Like thermomechanically eroded lava tubes, ice-walled esker-forming meltwater tubes in glaciers commonly have the idealized shapes of subway tunnels, but they can also have more complex shapes, often being broad and low or irregular.

The cross-sections of eskers commonly mimic that of their tubes, though collapse following melting of ice may modify their shapes. Eskers thus formed may be flat-topped, sharply crested, or rounded in cross-section, and either continuous or discontinuous longitudinally. They may also change shape downstream. They may form sinuous ridges on an otherwise flat plain: a ridge within a trough (tunnel channel) eroded into a plain; a ridge connecting trough segments; a sharp ridge on top of a broad linear plateau of sediment; or a sinuous arrangement of vague hummocks. Eskers may occur as isolated single ridges, multiple quasi-parallel ridges, or complex braided, distributary, or tributary complexes. Eskers may be made of fairly homogeneous and poorly sorted sediment or may be well graded or consist of well-formed beds of differing grain size. Eskers are always oriented approximately in the direction of slope of the surface of the glacier, as that is the direction of maximum hydraulic gradient, but eskers may cross low topographic divides. Unlike normal stream beds, the water flow direction may go uphill for short distances.

The sizes of eskers vary as widely as the sizes of glacial drainageways. Eskers just decimeters high and a few meters long have been described, as have eskers hundreds of kilometers long, over a kilometer wide, and tens of meters high. In glacial landscapes pocked with lakes, eskers commonly form the most continuous patch of high ground on which to build roads and railways. They also serve as valuable sources of sand and gravel for construction, as so many are mined. Eskers are commonly associated with other landforms produced during the last dying days of melting glaciers. Chaotic hilly moraines and sublimation pits (kettles) are common, as are flat expanses of lake sediment (kames) deposited in hollows in the ablating ice.

WHITE CHARITUM CHRISTMAS

Any interpretation of geomorphic process carries with it the burden of finding consistent observations, such as other nearby landforms that allow consistent interpretations; the cost of not presenting a coherent, broad perspective may well be the credibility of the proposed formation process. This is especially true for any hypothesis that would break an existing paradigm in thought. If these ridges were eskers, there must be other glacial features close by, and my task now was to identify them. If they were not eskers, this, too, should be written in the landscape.

The success or failure of the glacial hypothesis – or any other explanation for these features – would be in the identification of multiple landforms that can be interpreted as the product of a whole process system. This higher-level interpretation is still photogeology, and it does not offer a means to prove the interpretation; however, it helps build the case, heightens the level of discussion *vis-à-vis* other possible mechanisms, and tightens the implications of the interpretation.

The *Viking Orbiter* Discovery Image was one among several dozen scenes showing about half of the Argyre impact basin in detail down to anywhere from 50 to 250 m per pixel. That December, numerous other types of features were identified which plausibly could represent a variety of related glacial processes. My task, spurred on by scientific responsibility and a January deadline for submission of abstracts to a major planetary science conference, was to document the observations and clarify the interpretation. It soon became apparent that the whole landscape at Argyre could be interpreted as a modification of an ancient impact basin by processes of ice accumulation, glacial flow and erosion, glacial sediment deposition, and ice ablation.

The Argyre story, while only partly told here and no doubt incorrect in some respects, is a chief underpinning of the Mars glaciation story and seems to me a key to understanding broader issues of oceans and global climate evolution. Because of the importance of the Argyre region, and the need to keep images together instead of distributing them throughout the book by feature category, I offer a special section of MOC and other images in preceding pages. Some of the feature types shown in these new images, such as polygons and details of the sinuous ridges, are discussed in Chapter 7.

If the sinuous ridges on the floor of Argyre were indeed eskers – meaning that an ice sheet had once stood there – the logical place to look for the source of ice was in the nearby Charitum Montes. Indeed, the sinuous ridges seemed to splay out from the mouth of one major valley that cut clean across the Charitum Montes, as though a huge piedmont glacier once spread from this valley in a gigantic lobe. Once the step had been taken to consider glacial processes, evidence for glacial erosional dissection of the Charitum Montes was apparent, even at a glance. The mountain tops are, in many places, deeply indented by amphitheater-like alpine basins, similar to glacial cirques. Mountain tops mostly have sharp crests, like those typical of glacial aretes. Some cirque-like basins hold lobate-fronted lineated deposits (lobate debris aprons) that resemble debris-covered glaciers on Earth, suggesting that remnants of glaciers are still present. Some cirque-like basins have, in place of lobate debris aprons, complexly braided plains that resemble glacial outwash plains.

Cross-cutting valleys have, in places, what appear to be U-shaped profiles – a hallmark of alpine glaciation (more about this in Chapter 6). One channel system, whose source is in the cratered highlands far to the east of Argyre, cuts cleanly across large craters and across the Charitum Montes. This channel terminates on the floor of the basin in a spectacular braided delta or outwash plain. The highlands source of this channel network actually begins around crater Green as a system of dendritic ridge-in-trough features, interpreted here as eskers and tunnel channels. Remnants of abundant large impact craters are in nearly every case breached, as though a glacier had once incised the rim and spilled out of – or into – the crater.

The morphology of the Charitum Montes is unique, and they certainly offer some of the most imposing vistas on Mars, with some peaks rising 7000 m from the basin floor, higher than the Alps and nearly as high as the Himalaya. Heavy glaciation of the Argyre impact basin provides an explanation for the morphology. While there is only a partial morphologic analog to the Himalaya, since the structure of valleys is very different and root-mean-square slopes far lower in the Charitum Montes, the Himalaya do suggest a possible mechanism for generation of anomalous relief. The Argyre impact, which initially almost penetrated the entire crust and excavated a basin about 50 km deep, made a 5-km-deep hole after later processes, including both isostatic adjustment and sedimentary infilling, had their effects. However, isostacy and spatially heterogeneous erosion can work together to produce extreme local relief.

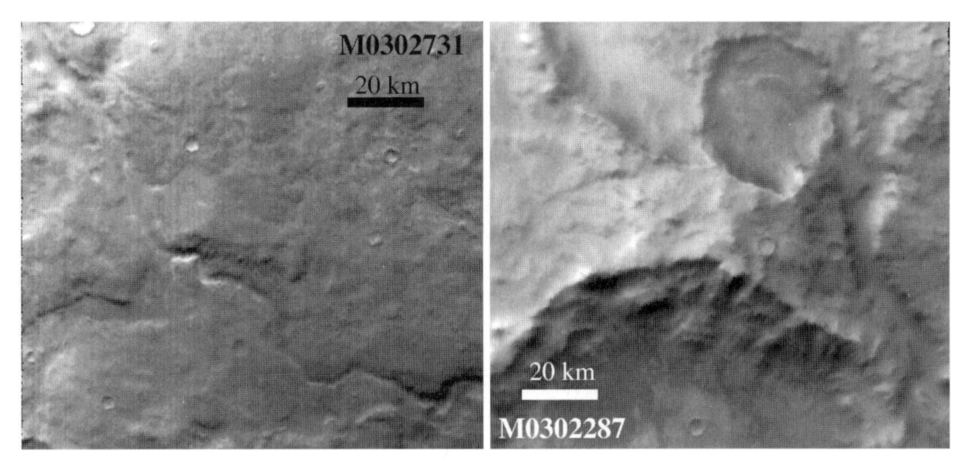

Oceanidum Fossa/Doanus Vallis, along with other channel/valley systems, drained from Noachian highlands and polar cavi into Argyre Planitia. This valley has a U-shaped cross-section, according to MOLA data, and it terminates in a braided pattern of channels or deposits in the basin floor. It remains an open question whether these channels/valleys transported and were formed by ice (favored by Kargel), water (favored by Tim Parker and Jim Head), debris flows, or perhaps CO_2-charged debris flows (favored by Eric Kolb and Ken Tanaka). It is also uncertain whether the valleys formed by fluids in equilibrium with the climate, or whether purely catastrophic processes were involved that may have had only an indirect relation to the climate or no relation at all. The U-shaped channel cross-section and breaching of craters favors a glacial interpretation, or at least late-stage glacial modification. MOC wide-angle images.

Cirques, aretes, and horns actively being formed by glaciers in the Alaskan panhandle.

Cirque amphitheater eroded by a Pleistocene glacier, Rocky Mountain National Park, Colorado. The alpine relief at this locality was caused first by ice-cap glaciation (where the whole mountain was covered by ice and eroded to a gently undulating surface), and then by smaller cirque glaciation.

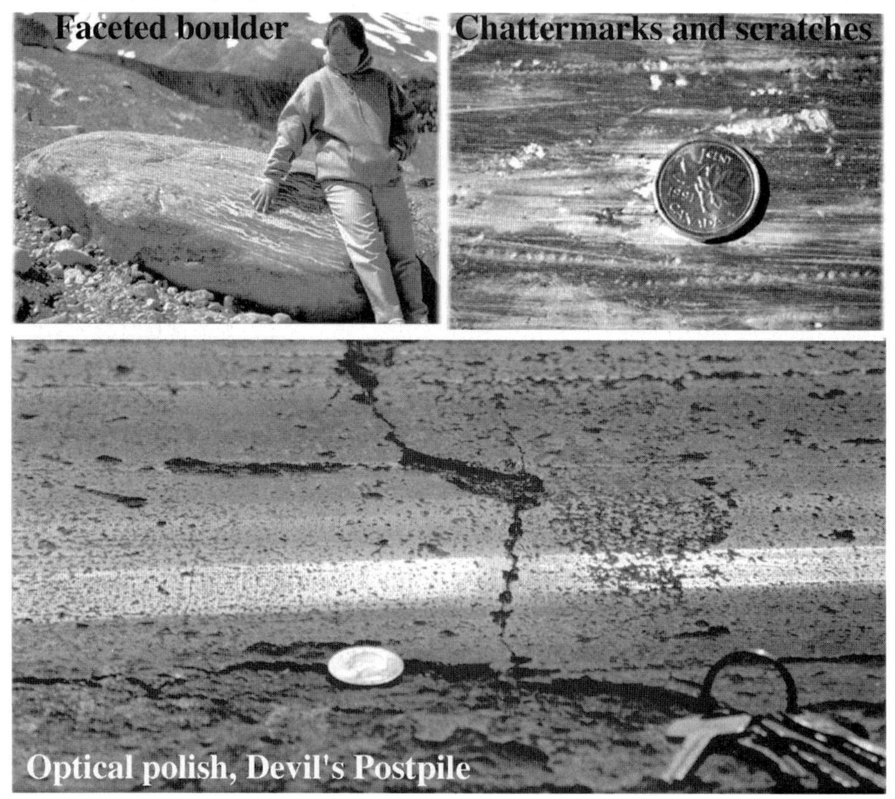

(A) Smoothed boulder, (B) chattermarks and scratches, and (C) optical polish produced by glacial abrasion. This boulder had been exposed by retreat of the Athabasca Glacier just a few years before.

Glacial abrasion produces grooves and ridges spanning 10 orders of magnitude in size: from the microscopic scratches that produce an optical polish to large streamlined ridges and troughs.

Typical small-scale glacial sculptural and abrasional forms near Athabasca Glacier, Jasper National Park. (A) A small roche moutonnee in dolomite with optically polished surface (glove for scale). As typical, this structure has a smoothly eliptical domical or bulbous, convex-up form interrupted by a ragged crag at the down-flow side of the structure. (B) A close-up of scratches, chattermarks, and other imperfections on this polished surface (fingertips of the glove for scale). (C) A large crescentic gouge on a boulder ($2 coin for scale). (D) The crescentic gouge isolated from the image and with a parabola fit to it.

A current model, developed by Michael Bishop and Jack Shroder (University of Nebraska at Omaha) and others, for the Nanga Parbat massif (Pakistan) and the K2, Karakoram and Hunza region of the Himalaya, seeks to explain the anomalous high relief as due to tectonic uplift in response to reduction of crustal and lithospheric mass by glaciers, rivers, and mass movement. A key to their mechanism is that these massifs are of a scale where they can respond to isostatic forces as a block, while the scale of individual peaks and ridges is too small to adjust isostatically. The rate of uplift is linked to the rate of erosional denudation and unloading of the lithosphere. A cycle of uplift and erosion can be sustained because of the thick buoyant crust existing along the Indian–Asian plate suture zone where the massifs occur. Furthermore, the landscape exhibits highly differential erosion where cold-based ice and permafrost protects high altitude ridges and summits, but where wet-based ice in adjacent cirque basins and valleys erodes rapidly. In areas with well-established fluvial systems, rapid incision and transport of sediment facilitates erosion and then uplift. As uplift proceeds, precipitation increases, and so the process self accelerates and local relief grows up to a limit imposed by the strength of local rocks. This is called a "tectonic aneurysm."

On Mars there is no alpine relief anywhere that is comparable in mean and maximum slopes to what we see in the Karakoram, although the vertical dimension of relief is similar and glacier-type erosional forms are similar (but have broader horizontal dimensions). On Mars, with a much thicker lithosphere (100-300 km, compared to 30–100 km for Earth), any possible tectonic aneurysm would operate on a broader spatial scale, and glaciers would erode on broader scales. For equivalent degrees of vertical glacial incision, the result would be alpine topography with broader gross features (glacial valleys, ridges, and summits) but lower mean and maximum slopes. The glacially incised southern rim mountains of the Argyre Basin – the Charitum Montes – may well have been subjected to a tectonic aneurysm comparable to that of the Karakoram, thus explaining the inordinately deep incision of those mountains, the great alpine heights (5,000–7,000 m local relief), and the unique morphology. The thin lobate debris aprons on the flanks of the massifs could not have been very effective in eroding the massifs, but basal ice of thicker glaciers in the valleys were able to melt and erode the valleys, emptying the sediment into the basin. The Argyre basin itself is isostatically compensated, and the rim as a whole would have slowly achieved compensation, but the massifs and valleys are too small to be individually compensated. Thus, according to this interpretation, selective erosional unloading of the southern rim of Argyre, with its over-thickened ejecta-laden crust, caused a tectonic aneurysm. If the climate had been warmer with melting more widespread, and if antecedant stream erosion had provided topographic control of glacial erosion (instead of impact craters), the relief would have evolved over time to something resembling the Himalaya. The story in Argyre remains enigmatic in many respects, and much hinges on the detailed stratigraphy – what happened when.

GLACIAL POSTMORTEM: DEATH BY MELTING

Having discerned plausible evidence for a major ice accumulation region in the mountains of southern Argyre, the next step was then to look at the terminations of the sinuous ridges, for there might be the terminus of the ice sheet and additional evidence for deposition. One set of sinuous ridges, set amid a smooth plain, seems to fade out to the northeast, as though buried by the smooth plain. This plain has a sharp contact with the mountainous uplands. The contact, in different places, is of two types: in some places lobate debris aprons – rock-strewn glaciers perhaps – appear to flow over the edge of the smooth plain, meaning that relative age sequence, from oldest to youngest, is: sinuous ridges ... smooth plain ... lobate debris apron. Elsewhere, the smooth plain appears to onlap (overlap) the upland of the Charitum Montes wherever lobate debris aprons are absent. To the east, the smooth plain appears to give way to a delta-like system of braided channels and former islands, perhaps braided outwash, mentioned previously. To the north, the smooth plains and sinuous ridges transition to a chaotically pitted and hummocky terrain in which layering is visible on a scale of several tens of meters.

There are several ways to make smooth, layered plains deposits. On Earth's continents, the chief processes are volcanism and sediment deposition by standing water – lake plains, oceanic sediment, or river flood plains. With *Viking Orbiter* imagery alone, there was no way to make a direct and certain determination of which, if either, mode of origin of the smooth plains applied to Argyre Planitia. Dust deposition, another plains-formation process, does not explain Argyre's plains, because they exhibit a sharp depositional contact with upland units; such a contact is not ordinarily produced by dust fallout. Rather, a gravitationally-bound fluid was involved. Lava plains are usually associated with volcanoes, which are not wholly evident in Argyre and certainly not in the area of the smooth plain. Lake plains are usually associated with river channels and other sources of water, including, sometimes, glaciers, for which a case can be made. A glacier-fed lake is simply one interpretation consistent with the other evidence.

A consistent glacial scenario is that the smooth plain was constructed by sediment emptied from the glacier drainage systems represented now by braided eskers and braided outwash plains. Ultimately, the proglacial lake buried the terminal reaches of the eskers and outwash in lake sediment. Here was evidence for a former lake plain produced in the final throes of ice-sheet glaciation in Argyre. Additional evidence is in the rich variety of pits of all sizes in central Argyre Planitia. These pits indicate the removal of material; they do not indicate the process of removal, which could be the result of wind erosion of dust – perhaps loess (dust stripped from glacial outwash and redeposited by wind) or silt and clay deposited in the lake – or sublimation of stagnant, buried glacier ice. In any case the pits, while not definitely indicating any link to glaciation – are consistent with an emerging picture of a regional glacial cycle. An ice sheet appears to have accumulated on the cratered highlands and rim mountains of Argyre. It emptied into Argyre across the Charitum Montes and, in the process, carved the mountains in classic alpine fashion, scoured the landscape, eroded many craters and breached their rims, and formed several

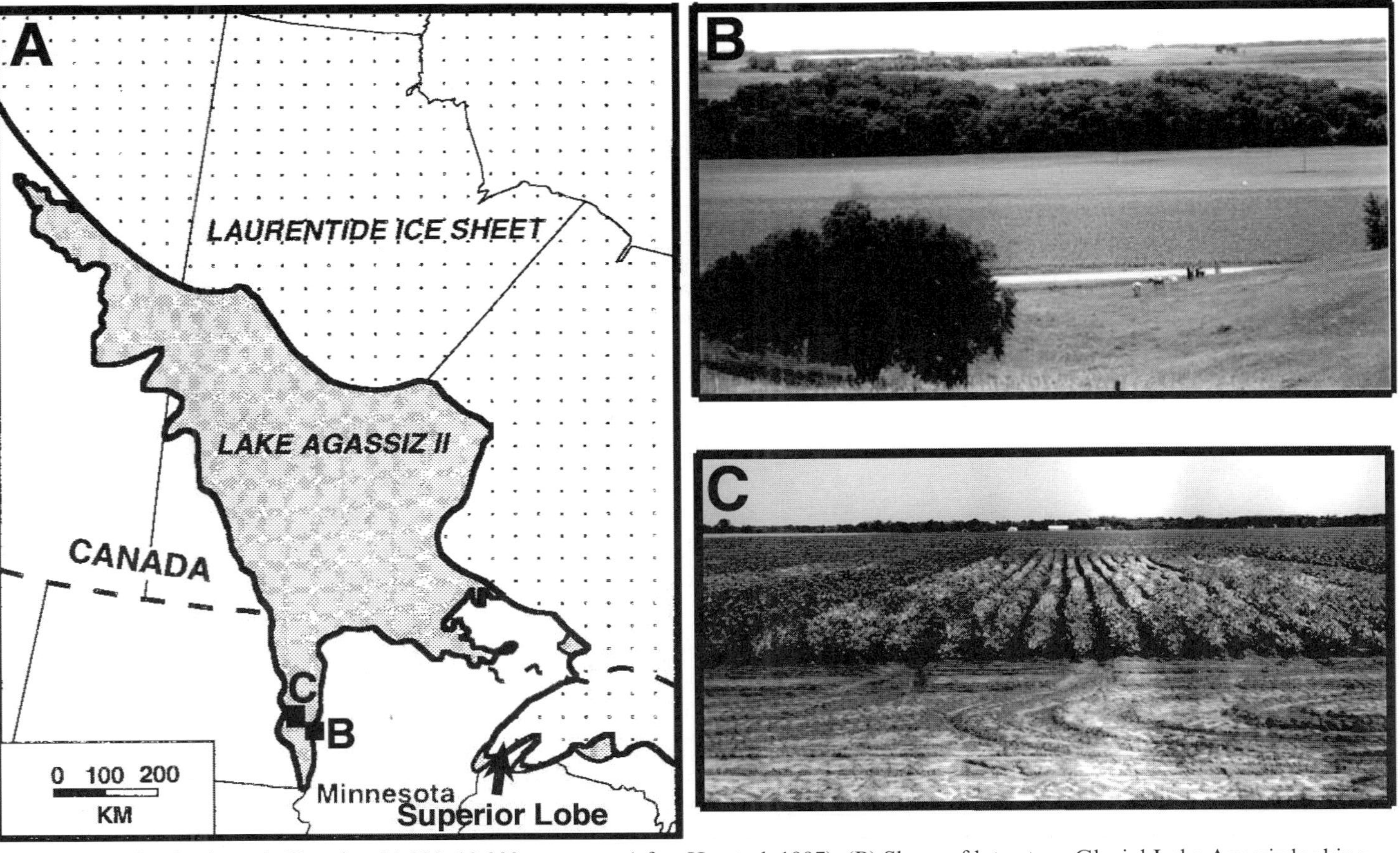

(A) Map of Lake Agassiz II region 11,000–10,000 years ago (after Hu et al. 1997). (B) Shore of late-stage Glacial Lake Agassiz looking onto silty plains of the lake. (C) Floor of the former lake is among the richest agricultural areas in North America. Photos by J. Kargel.

A drainage channel tens of meters deep winds its way across an iceberg-littered floor of Van Cleve Lake. The lake has started to refill, floating apartment high-rise-size bergs.

channels that cut across the southern and southeastern sectors of the mountainous rim. Within the basin, the ice sheet underwent massive melting, forming eskers, ice ablation pits, outwash plains, and glaciolacustrine plains.

Similar landscapes of smooth bouldery plains, esker ridges, tunnel channels, and outwash plains exist across much of the northern half of North America. According to John Shaw (University of Alberta), immense floods coursed beneath and beyond the fronts of the melting Laurentide and Cordilleran ice sheets. Exactly what happened in Argyre is not clear, just as the demise of Earth's Pleistocene ice sheets remains controversial, but it seems rather difficult to explain Argyre's landscape satisfactorily without the involvement of an ice sheet.

The scenario outlined above forms the basis of the painting by Michael Carroll reproduced in Plate 13. This painting was originally done for a popular-science article in 1992 by Bob Strom and myself in *Astronomy* magazine, titled "Ice Ages of Mars."

While photogeology does not prove this scenario, recent imaging and other data are consistent with glaciation. Other than glaciation, few processes can achieve the type of dissection of crater rims observed. Few processes can generate U-shaped valleys; few can generate amphitheater-like embayments in the sides of mountains, or the sharp ridge crests; and few can generate sinuous, braided ridges, or ridges within troughs. It is not a case that nothing else can explain these features, but one process system – glaciation – can do all these things, and also make pits, scours, channels, smooth plains, and other observed landforms. It seems highly doubtful to this observer that any other process system can explain so much so simply. Anyway, the story told by my sketch map interpretation of the basin, prepared over the Christmas holiday, 1989, was of an ice sheet that had accumulated in the Charitum Montes and the highlands south and east of Argyre; the ice sheet had flowed into the Argyre basin, scouring the peaks and valleys along the way, and ultimately melting and depositing huge masses of sediment on the basin's floor.

A DIFFERENT TOUCH

The general story told by the Argyre landscape in many ways is a classic picture of uplands glacial erosion and lowlands glacial deposition. The trouble is that this is Mars, and all we have learned of Mars tells us that it has never really been totally Earth-like. Furthermore, the preserved glacial record has aspects that are different than those seen in terrestrial glacial landscapes. The Martian sinuous ridges are huge and perfect – not longer than many esker systems on Earth, not more complex, and not differently shaped, but they tend to be more continuous and uniform, wider than most, and definitely, on the whole, are gigantic and highly complete. They are absolutely beautiful. Perhaps they are too marvelous for glaciation as we know and love it, according to Steve Metzger (Desert Research Institute, Reno), who is simultaneously a champion and critic of the glaciation hypothesis. Surely something formed them, and a theoretical Mars without Big Ice is severely handicapped in explaining these ridges. The cirque basins, too, are many times larger than most alpine cirques on Earth. These quantitative and qualitative details are significant, and the critics are helpful to point out the differences, because they tell us something about the process: if the process was glaciation, it was different in some way to the process experienced on Earth.

Extra-large cirque sizes, for instance, might reflect a generally deeper level of glacial erosion on Mars over longer periods of time. Or they may indicate that Martian ice has tended to be more viscous – colder – and has deformed on longer length scales than ice typically does on earth. Perhaps lower gravity tends to make ice deform on longer length scales and thus form broader cirques. For an ice that would be Newtonian in its behavior, there would be no reason to expect this type of behavior, but for a non-Newtonian power-law rheology (which ice actually has), this is the type of behavior that is predicted. Cold ice would make fewer, larger cirques (if it makes cirques at all), whereas warm ice would make larger numbers of smaller cirques.

Something else is totally un-Earthly about Argyre's alpine glacial erosion. The pattern of glacial erosion is clearly controlled by impact craters. On Earth, preglacial fluvial valleys do this job. This is perhaps as indicative as anything else that the preglacial history of Argyre was not one dominated by eon upon eon of erosion by rainfall, but by eon upon eon of meteoritic infall and, mostly, slow erosion by ice. This is also indicated by other observations, but it all forms a consistent story that Mars is not and never has been a rainy planet for any length of time. Keep this thought right beside thoughts of oceans and ice sheets and a Mars modified from the ground up. . . .

GLACIATION GOES GLOBAL

Once the glacial interpretation of Argyre was developed, a next logical step was to look elsewhere on the planet. Baerbel Lucchitta had previously proposed glaciation as a cause of outflow channels and various features in the northern plains. Several researchers, going back to Henry Faul, had proposed that the major shield volcanoes were influenced by glacial ice sheets. The ideas were out there, so the task at hand was to reconsider the evidence proposed by these researchers, and to consider possible additional evidence. Increasing new evidence, as presented lately by Jim Head and colleagues, supports a glacial process occurring on the Tharsis volcanoes. Especially compelling is the nature of ridges on the aureoles, interpreted as moraines. I have found important elements of Lucchitta's glacial interpretations – especially regarding the northern plains and fretted terrain – to be compelling possibilities. More generally, the low latitudes have a paucity of features that, by any stretch, can be attributed to glaciation. Plausible glacial terrains are widespread but occur mainly poleward of $\pm 30^\circ$ and primarily poleward of $\pm 40^\circ$. Though the landscape is unique in each candidate glacial region, most regions tell a story that is more or less consistent: a period of ice accumulation and flow, then massive melting and stagnation of the ice sheets, followed by wholesale or partial disappearance of the ice. The story of Martian glaciation, however, remains chronologically poorly documented, and is still unconvincing to some critics.

Strong evidence for the geologically recent flow of rock glaciers or debris-covered glaciers abounds in the mountains east of Hellas. These mountains represent the highly degraded remnants of a mountainous ring of Hellas, analogous to the Charitum Montes at Argyre but much more fragmented by erosion and plains filling between massifs. The lobate debris aprons are even better exhibited there than in the Charitum Montes. Many lobate debris aprons possess longitudinal or transverse corrugations, steep or convex-up rounded flow termini (hence, their lobateness) and other textures that indicate flow. A near absence of superposed impact craters indicate that these are quite young flows. Some of the mountains indicate glacial sculpture, including cirques and aretes. Although the flow of lobate debris aprons is now unquestioned, and there are some alpine glacial erosional forms, the overall glacial assemblage in this region is much less complete than at Argyre. Eskers, moraines, and scour features are absent. Pit analogs of kettles are present on some flow deposits, supporting a basic icy constitution. Very similar features abound in the fretted terrain in the broad contact zone between the northern plains and cratered highlands (shown and discussed in Chapter 7).

In Hellas proper, as at Argyre, evidence suggests that ice accumulated on the highlands south of the basin, then flowed northward into the basin. The ice sheet, according to this interpretation, scoured the rocks along the way, and finally underwent massive stagnation and lake formation deep in the basin. The glacial interpretation of the Hellas landscape, viewed at *Viking Orbiter's* 200 m per pixel resolution, is about as perfectly consistent with idealized patterns of erosion and deposition that one may encounter in nature. The trouble has been that the *Viking Orbiter* coverage of Hellas was never very good, being mostly a factor of 3 lower in resolution than the coverage of Argyre. Thus, considerable interpretive license was required for this or any other interpretation. Not surprisingly, the glacial interpretation first offered for Hellas drew more fire than that proposed for Argyre. Other researchers, notably Ken Edgett (Malin Space Science) and Jeff Moore (NASA Ames), preferred a wind-based paradigm of eolian erosion and deposition, or preferred to see evidence for a huge, deep lake but not for ice (Jeff Moore and Tim Parker). As more evidence has been accumulated, I have to concede to being confused by what has been found. There is, indeed, strong evidence for a vast lake – possibly more voluminous than that ever present in the northern plains – having occurred in Hellas. In MOC images, there is plenty of evidence of the type of chaotic kettles in the basin deposits that one expects for glacial deposits, but the scours so evident in *Viking* images are heavily mantled or are simply not visible at higher resolutions now available. In fact, high on the volcanic shield summits south of Hellas, which Kargel and Strom (1992b) inferred was the accumulation center, there are still thick, presumably icy thermokarstic deposits – possibly the remnants of the ice sheet. Thus, I am still reserving my final assessment for the future, but I suspect that glaciation and lake processes will ultimately be seen as dominant processes in the evolution of Hellas.

In the northern plains, gigantic fingerprint-like whorled ridges ("thumbprint terrain") and sinuous ridges were mapped in patterns resembling glacial moraines and eskers of the North American mid-continent. In several places scattered across the northern plains, shallow troughs, many containing narrow medial ridges, are associated with these whorled ridges, in close analogy to the tunnel channels and eskers associated with whorled recessional moraines of the Laurentide glacial landscape. Shallow, scalloped pits occur in Utopia Planitia that resemble the thermokarstic kettle lake basins of Minnesota. On the whole, the landscape in parts of the northern plains seemed to attest to a style of glaciation less like that in high-relief areas of the southern hemisphere of Mars but distinctly like that of the terminal reaches of the Laurentide Ice Sheet during its retreat phase and in its aftermath. Lacking on Mars are the vast fields of drumlins that mark large portions of the terrain affected by Laurentide glaciation. In the northern plains, possible glacial landforms are primarily associated with sites where ancient oceans or lakes may have accumulated. Thumbprint terrain has the widest distribution of all the possible glacial landforms. This terrain type is predominantly located within a few hundred kilometers of the edge of the northern plains. A possibility is that oceans/lakes evolved to a glacial condition; another possibility is that ice sheets spread across the northern plains from high latitudes toward the edge of the basins.

Another possibility, which is as likely as some variant of glaciation, is that the glacial hypothesis developed for the northern plains is simply wrong. On close inspection of the MOC images, few have the bouldery appearance expected for moraines; many are very smooth, and some show plausible indications of a diapiric type uplift. Perhaps thumbprint terrain represents tectonic squeeze-ups and extrusions of mud, ice, and other unlithified material originally deposited as a mud ocean or series of gigantic debris flows. Or perhaps they are compressional ridges developed where debris flows encountered the edge of the basin. These explanations are consistent with the interpretations of Ken Tanaka and colleagues (including me) in a series of recent papers in *Geology*, and are also consistent with the earlier mud-ocean hypothesis of H.-P. Jöns. Tanaka lately has suggested that groundwater was forced from the highlands and inflated the northern plains, causing hydraulic overpressurization, ridge formation, and mud extrusion. As yet another alternative, perhaps thumbprint terrain indeed consists of moraines, but they have been draped by ocean-related mud flows or blankets of sediment carried in suspension by a water ocean, consistent with both the glacial hypothesis and the ocean hypothesis (but with the ocean coming after the ice). With little doubt, thumbprint terrain carries the identity of whatever process dominated deposition in the northern plains.

OUTRAGE

The source of water needed to feed the ice sheets and the cause of climatic upheavals that could have caused the ice sheets to come and go were new, controversial, problems raised by the glacial interpretation. Water in the amount required – a veritable ocean's worth of the stuff – would surely have left its mark elsewhere. Motivated by the novel glacial interpretation of diverse landforms, Bob Strom and Vic Baker read the 1989 paper in *Icarus* by Tim Parker and his colleagues – the first "shorelines" paper. In true confession, I at first had skipped this paper for its 'boring' title, "Transitional something, something, something." Baker and Strom surmised that Parker's work contained evidence for the source of the water needed for the glacial ice sheet that I inferred had once existed. According to ideas brewing in our research group, underlying volcanic-igneous activity at Tharsis caused immense volumes of water to be emptied from the crust of Mars through chaotic terrain and outflow channels to a northern ocean, and then evaporated and recondensed as snow in the highlands. And so an outrageous hypothesis of a Martian hydrologic cycle was born.

One process on complex planets is always linked to another, and formation of one type of landform is accompanied by other types that may differ in morphology but which carry a common genetic thread and logical association with other related landforms. Geomorphologists use the term "landscape" to encompass any ensemble of landforms related by a single dominant process or process system. Gradually it has been recognized that common threads linking broad suites of Martian geomorphic features suggest that ice- and water-related process systems, together with links to the silicate crust and mantle, have produced widely varying features in logically ordered landform associations and sequences. This is a higher-level application of Ockham's Razor. This new process paradigm in thought represents a single common explanation for a multitude of features that otherwise would require, if we lacked a plausible connective thread, multiple independent explanations. Plate tectonics represents a comparable common thread in Earth science.

Our attempt to provide a comprehensive explanation for Mars involves episodes of internal geologic activity, especially pulses of volcanism and attendant geothermal activity, which have repeatedly mobilized millions of cubic kilometers of condensed volatiles, triggering periods of glaciation and intense permafrost activity. There are several variants of this basic concept. Our model was dubbed MEGAOUTFLO by Dr Victor R. Baker (University of Arizona), an acronym standing for Mars Environmental Glacial Atmospheric OUTburst FLood Oscillations (Baker et al., 2000, *Lunar and Planetary Science*, XXI). During each of several such MEGA-OUTFLO events, the global climate was severely modified and underwent geologically rapid climatic oscillations due to planetary gravitational and solar influences, such as astronomical cycles of changing obliquity and orbital eccentricity, and the solar Hale and lower-frequency cycles of activity.

There is some evidence that Mars might now be in a climatic regime that includes lingering oscillating effects of a recent major MEGAOUTFLO event. For one thing, the considerable indications of present-era ice- and water-related activity (considered in Chapter 7) seem, in some vague and unquantified sense, to be occurring at rates that could not be sustained over all geologic time and still preserve a record extending billions of years in the past.

A key underpinning of the MEGAOUTFLO hypothesis is that glacial and permafrost processes were in the past, and remain, major influences on Martian landscapes. Most of my efforts in Mars studies, including my contributions to the MEGAOUTFLO hypothesis, have been in the recognition of cold-climate ice-related processes and landforms on Mars.

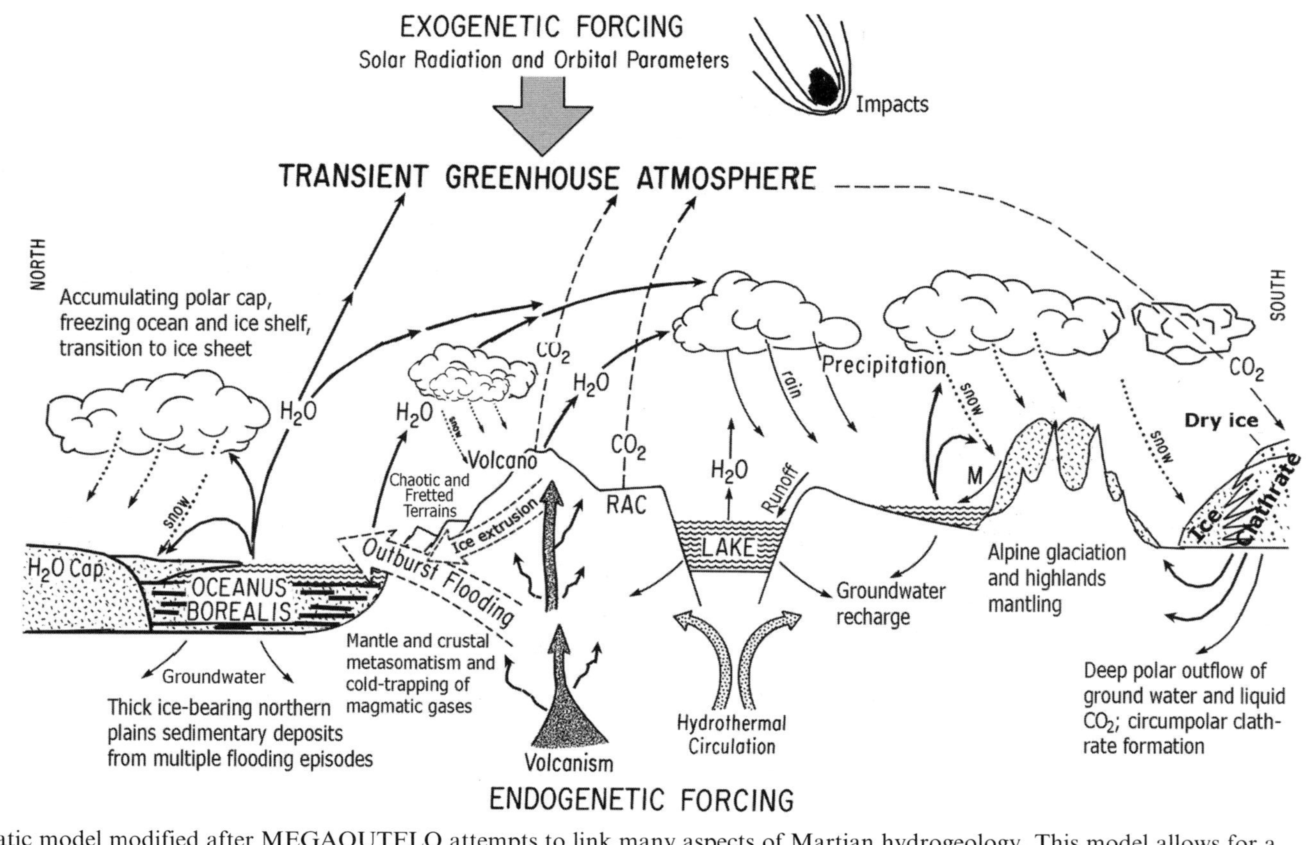

A schematic model modified after MEGAOUTFLO attempts to link many aspects of Martian hydrogeology. This model allows for a little less climatic drama, less liquid water at a given time, and more widespread glacier and permafrost ice than originally considered, but the basic elements of MEGAOUTFLO still pertain, with some added twists as working hypotheses. See also: Baker et al. (1991, 2000); Kargel and Strom (1992b); Kargel et al. (1995). Figure modified after Baker et al. (1991).

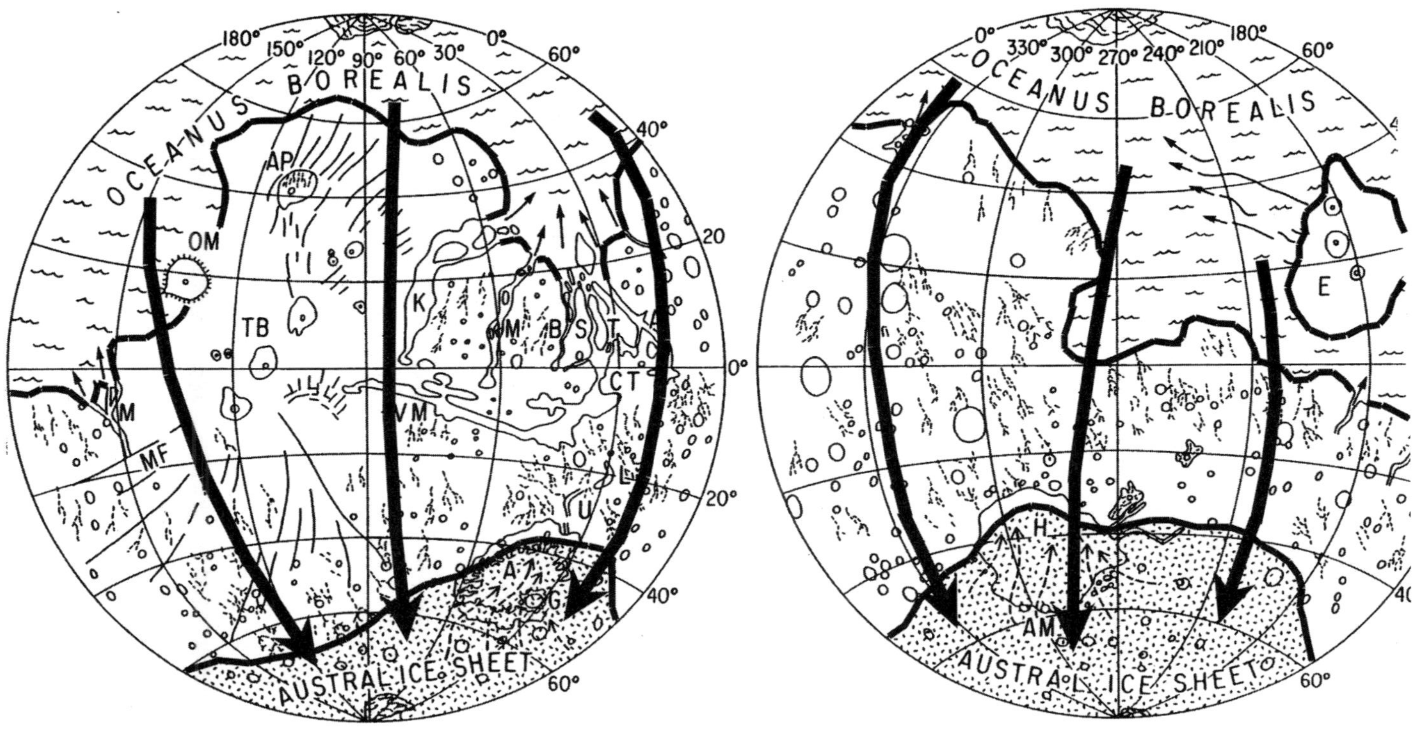

Martian global hydrology, according to Baker et al. (1991) and Kargel and Strom (1992b), involved transient formation of a northern ocean ("Oceanus Borealis") and a southern ice sheet ("Austral Ice Sheet") in response to vigorous volcanic-geothermal destabilization of the permafrost-confined aquifer system and due to magmatic releases of greenhouse gases. The Austral Ice Sheet was, according to this model, a delayed response to formation of the ocean – a large water reservoir – and modification of the climate. Figure, modified to show transfer of water from the northern plains to high and middle southern latitudes, adapted from Baker et al. (1991); see also Kargel and Strom (1992b). For an updated version, see Baker et al. (2000). This model remains a controversial focal point of hypothesis testing.

Following the publication of Parker's paper and my find of landforms appearing as though ice sheets had formed them, our research group at the University of Arizona began to meet on a weekly basis. We discussed Martian glaciation, oceans, and global hydrology. It was the most thrilling period of my academic career. Within weeks, and just in time for the Lunar and Planetary Science Conference (LPSC) in March, 1990, we had quite a story to tell in four sequential presentations. It involved the development and decline of an ancient hydrologic regime, triggered by volcanism and geothermal heating of the crust and volatile outgassing. The cycle culminated in the formation of a densified greenhouse atmosphere, a northern plains ocean (dubbed by Vic Baker "Oceanus Borealis"), a transient modified climate, and a southern hemisphere ice sheet (the "Austral Ice Sheet," as I called it).

Our model of ancient oceans and ice sheets started with an immense episode of volcanism and associated geothermal warming, which led to thinned permafrost and catastrophic rupture of confined aquifers, causing outburst flooding accelerated by exsolution of carbon dioxide. Outflow channels and chaotic terrain formed during these events, and the discharged water and eroded sediment emptied into the northern plains, forming a transient ocean. The volcanic episode and flooding also caused release of CO_2 and other greenhouse gases into the atmosphere, which underwent a sudden warming. Water was transferred globally both through the atmosphere and in the subsurface. Oceanic and glacial responses modified the surface in distinctive ways, until finally the episode subsided as CO_2 condensed out as carbonate rocks and as polar dry ice. This process is episodic and was repeated many times over Martian history, but episodes gradually decreased in vigor as the Martian geothermal heat engine declined over time and as volatiles were eventually lost or sequestered in permanent solid reservoirs. The strength of this hypothesis is that it coherently explains many hitherto enigmatic aspects of Martian geology and makes specific and testable predictions, for instance, that the causes and multiple effects of these processes should be stratigraphically related in a sequence of coordinated and closely timed events. Its vulnerability is that it is based on two controversial interpretations of Martian geology – the glacial and marine hypotheses.

SHEER MADNESS

Positive proof of the glacial hypothesis could not be obtained so readily by photogeology. Developing an outrageous planetological story and convincing one's like-minded coworkers is one thing. Convincing the global planetary science community is a monumentally tougher task fraught with peril to anyone who should try. It requires making a case so compelling that others would accept the implications and turn away from or greatly modify concepts they had been taught or developed themselves.

If Argyre's sinuous ridges are eskers, then that immediately indicates the particular nature of the glaciers that formed them. Eskers are liquid-water-laid features, and so the glacier must have been wet-based. This implication, more than anything else, is what gnawed at those who preferred alternative explanations. Glaciation was wild enough by itself, and the way we had presented it, we had outrageously coupled glaciation to ocean formation. The implication for liquid water was the biggest problem to most critics. Mars, according to them, simply could not be about liquid water and flowing ice or any type of global hydrology that had any resemblance to those processes on Earth.

Other people would soon apply a host of alternative interpretations to the sinuous ridges of Argyre and other similar ridge systems on Mars. They were, perhaps, not eskers, but oceanic coastal landforms, or lava flows, or sand dunes, or dikes of once-wet sediment squeezed up along fractures. Some of these alternative ideas were every bit the equal of eskers as an outrageous hypothesis, but clearly the fight was on from all sides. The fight was waged in manuscript reviews, in conference rooms, in restaurants, and in elevators. The contention over these features was palpable from the moment I said "esker" in a public scientific forum. This word made allies (very few), it made skeptics (very many), and it made for beery discussions (very animated). Of course, I was not alone. I had my group of like-minded colleagues, and I shared the underdog status with others, such as Tim Parker, who were fighting their own battles against the establishment.

Our concept was heard – and immediately assailed. A comment from one friend in the conference room, overhead by another, was simply, "*Madness, sheer madness!*" And so, for the next few years, perhaps it was madness to go up against "Those Who Knew" in the company of a few equally mad colleagues. However, my research group knew that we were onto something very real, even if we grasped the mere bare outlines of the problem and no doubt misunderstood much more than we understood.

The professional toll of the external peer response was career-threatening, but it was managed by the excitement of what Mars presented. Papers were difficult to get through review, and one proposal after another – about six in all from me, and others from Vic Baker and Bob Strom – garnered not a nickel of support. Even people who should have been natural allies – those who had their own radical concepts of big water and big ice – ignored our integrated hypothesis and the glacial hypothesis that seemed to link a warm, wet Mars with its present cold existence. Somehow, through it all, or perhaps because of it, at least I landed a good job in the business. The concept of a global hydrological cycle was offputting to some, while others, who felt they knew Mars inside and out, did not accept that Mars had ever seen a mass of flowing ice. One scientist, who more recently has become a leader of a new wet-Mars movement, stated his unmovable perspective just a couple years ago: "*Consider it a matter of faith: I don't believe there ever was a glacier anywhere on Mars.*" It was a matter of his faith, and that of his coworkers, in a Mars they knew and their limited experience of our own planet, against my faith and that of my coworkers, in the Mars we knew and our limited experience of Martian glacial systems. Both sides considered that the power of the Earth analog and common sense was not misleading; it is just that we chose different analogs, and applied different experiences. It was a conflict of assumptions and a battle of paradigms. He knew blowing dust and sand dunes, and I knew glaciers and eskers.

Outside the small global village of Mars scientists, my small, fractious band of rebel Mars scientists rallied a great deal of support in the separate community of experts in terrestrial cold-climate and marine/lacustrine processes – a community that, despite their relevant experience, for many years carried no weight with NASA funding program managers and planetary science peer reviewers. Some antagonists, particularly the last surviving adherents to the Cold, Dry, Windy Mars camp, did their best to reject publication of this new concept through the peer-review process. That was no conspiracy or misdeed; it was their honest scientific judgment at work. For my original glaciation paper with Bob Strom, and for the ocean paper by Vic Baker and our gang, there were many revisions and resubmissions and responses to reviewers. The literal mass of editorial communications pertaining to my glaciation paper far outweighed, in grams and word length, the final accepted manuscript. Our initial work finally was published in two papers, one in *Nature* in 1991 ("Ancient oceans and ice sheets and the hydrological cycle of Mars" by Baker et al.) and another in *Geology* in 1992 ("Ancient glaciation on Mars," by Kargel and Strom).

The final assessment of this story must await ground validation or refutation. My strong hunch is that Argyre will prove the classic Martian case of glaciation (both ancient and ongoing), but that the stories in the northern plains and Hellas will be much more complex and definitely muddier (pun intended) than my colleagues and I initially considered.

Kim Stanley Robinson drove home the doubts about glaciation in the fictional *Red Mars* (part of the *Red Mars–Green Mars–Blue Mars* trilogy), in an account of future field studies, set in the Charitum Montes, Argyre:

> *But everyone in the trailer seemed to agree that they were not finding any evidence for glaciation ... There were high basins that resembled cirques, and high valleys of the classic U-shape of glacial valleys ... all these features had been seen in satellite photos. ... But on the ground none of it was holding up. They had found no sign of glacial polish ... no moraines... no signs of glacial plucking. Nothing. It was another case of what they called sky areology, which had a history going back to the early satellite photos, and even to the telescopes. The canals had been sky areology, and many more bad hypotheses had been formulated in the same way, hypotheses that only now were being tested with the rigor of ground areology. ...*
>
> *The glacial theory, however, and the ocean theory of which it was part, had always been more persistent than most. First, because almost every model of the planet's formation indicated that there should have been a lot of water outgassing, and the water had to have gone somewhere. And second ... there would be a lot of people who would be comforted if the ocean model were true. ...*
>
> *"The problem is that the oceanic model is not very falsifiable," Simon said quietly. "You can keep failing to find good evidence for it ... but that doesn't prove that it's not true."*

Indeed, Robinson's fictional account makes some good points, though the excerpt above is not the end of that story. However, in the ten years since *Red Mars* appeared in print, considerable new evidence has accumulated, U channels found, moraines identified, and remnant icy flows observed. Glacial processes in Argyre and elsewhere, and other hydrospheric processes, including flooding of the northern plains, are fairly compelling in some areas. The limitation remains, however, that it is primarily still all "sky areology."

TUCSON MAFIA

By the early 1990s, a full Mars rebellion against the Mars Establishment – begun a decade earlier by Baerbel Lucchitta and continued by Tim Parker, the Tucson group, and others – was underway. Certain members of the Mars Establishment labeled our group at the University of Arizona, with some humor, the Tucson Mafia. In fact, we were just a small group of faculty and students who, embracing a different scientific judgment, did not fall into the party line. From our side of the debate, while we had our small-town gang, it was a few members of the Establishment (not to mention names) who behaved like *La Cosa Nostra* of planetology, complete with their Godfathers, money politics, and occasional underhanded manner. Certain criticisms of the ideas of the Mars rebels were unfounded; however, a general science-based critique was warranted, and some elements of the critiques remain valid. Mostly, the criticism was responsible scientific discourse. Key elements of the contending Warm, Wet, Early Mars and Marine-Glacial hypotheses will remain controversial until sky areology can come down to Earth, or down to Mars.

Critics contended that the individual observations can each be explained by nonmarine and nonglacial processes, and that the modified climate and hydrologic cycle so implied is thus unnecessary. Other critics came from within the ranks of the Blue Mars rebellion. Tim Parker objected to the youthful age that our group favored for the most recent major volcanic-hydrologic event. In a 1995 paper in the *Journal of Geophysical Research*, my colleagues and I showed that the age of modification, as indicated by the density of kilometer-size impact craters, of four hypothesized glacial terrains is Early to Middle Amazonian – much younger than the Hesperian to Noachian depositional ages of their rocks (determined from the densities of craters larger than 4 km). This result was based on more than one person-year of intensive labor by students (notably Natasha Johnson) and me. The implication that major modifications occurred to these rock units long after they first formed could not really be doubted; nevertheless, this result was roundly ignored until *Mars Global Surveyor* emphasized even more youthful ages of terrain modification. Indeed, the features that I was interpreting as glacial were interpreted by Tim Parker as marine, and the implication is that whatever actually caused the modifications and formation of these fascinating landscapes, it occurred more recently than Parker and many others would recognize. A problem for some people was that the age of Big Ice was being pushed so late that it conflicted with almost everyone's pet view of when big things happened to Mars.

Tim Parker had a bias, privately expressed, toward truly warm conditions that would not have supported glaciers but could have supported human beachgoers, had other conditions been right; that was another problem on top of my proposed chronology. H.-P. Jöns felt that all the features could be explained by his mud ocean. Ken Tanaka doubted the ocean's existence and the glacial interpretation, and thus saw no need for the global hydrologic cycle. Fully to Mike Carr's credit, I have never been able to pigeon-hole him as a Blue Martian; a Warm, Wet, Early Martian; or a Cold, Not-Completely-Dry eolianist. Carr saw merit in some individual aspects of the Tucson Mafia's ideas, and said so publicly, but he saw no support for the collective set of processes. Those were the perspectives of water- and ice-oriented researchers. For those whose expertise concerned dust and sand, our ideas fared considerably less well. And so it went. (Not well.) But it was a mutual sort of thing. I, for one, have always had some reluctance to embrace the rainfall paradigm advocated for two decades by Ted Maxwell and Bob Craddock (Smithsonian Institution, Washington, DC), and terrestrial-style oceans have never resonated, either. And yet, neither I nor they can get away from all of these ideas. Mars is just too varied, and the evidence too complex, to dismiss these competing (or complementary) ideas easily.

To the graduate students in the Tucson Mafia – at least to me – that period of rapid development of thought regarding the global evolution of a planet was an incredible period of intellectual freedom, despite the heavy handed response from much of the Mars community beyond the University of Arizona. Ideas developed in fits and starts from the initial suggestion by Lucchitta that Mars may once have had oceans. By the early 1990s there were few inhibitions to emboldened scientists, at least those willing to work without funding. The collegial teamwork and our consideration of vast questions and entertainment of vast solutions is the type of activity for which universities were organized. Scientific boldness is something that can go wrong, and often does – it is risky, but bold creativity can also lead to leap-frogging advances in scientific understanding. The type of boldness that was encouraged by Baker and Strom is something only great scientists allow their students.

This community response midway between bitterness, incredulousness, and reticence was not, however, to outlast the *Mars Global Surveyor* mission, and even before the MGS the controversial Blue Mars ideas had their impact. While they were at least very controversial, these new ideas had finally broken the logjam of creative thought that had been brought on by scientific conservatives whose thoughts were lodged firmly in the *Mariner* flyby era. Mars process models then developed rapidly throughout the 1990s claiming evidence of a wetter, icier, and perhaps a warmer past. There was a fresh wind blowing in Mars science. New ideas rejected the established rigid paradigms as inconsistent with Martian geologic features and landscapes that begged to differ. The credit for this fresh environment spreads widely, but I must say – and I speak for the Tucson Mafia – we have a solid sense of accomplishment in having been part of what spurred this new *Glasnost* for planetology.

It remains to be seen how prescient or off target we and others were. The end of this story has not yet been written. New chapters have been written. In 1991 and 1992 David Scott and colleagues at USGS presented results of careful geological mapping that supported large seas within sub-basins within the northern plains (papers with titles such as "Martian Paleolakes and Waterways: Exobiological

Implications"). In 1993, Parker and his colleagues produced a more detailed and globe-girdling analysis of "Coastal Geomorphology of the Martian Northern Plains." In 1995, Kargel and a large international group of researchers added further details in the *Journal of Geophysical Research* to the glaciological interpretation of the Martian northern plains. Ken Tanaka (USGS, Flagstaff) and his coworkers produced a distinct slant on the concept of Martian oceans – much more in line with Jöns' mud ocean – by interpreting the most detailed ever mapping of the northern plains as indicating vast sedimentary deposits produced by mass flows or mud flows. Personally, I think Tanaka is more on target than off, and his ideas will probably be seen as such throughout this decade of discovery.

The new observations concerning Argyre strongly support the glaciation hypothesis. However, the new observations bearing on the possibility of widespread continental-style glaciation in the northern plains and Hellas are much less clear. Hellas exhibits some of the most peculiar terrains on Mars, including one that the MOC team has nicknamed the "taffee-pull" terrain. It seems that there is great evidence for vertical diapiric type of movements and lateral flow, but the nature of the flowing material is not yet clear. It could be ice, mud, or debris. The flows may be related to volcanic destabilization of the eastern margin of Hellas, or they may include debris-covered glacial ice sheets or, as Jeff Moore (NASA Ames) might prefer, perhaps unconsolidated lake sediments. The taffee-pull terrains are in or near the area where I had mapped a proglacial lake, but I am certainly not ready to take a strong position on the nature of Hellas' deposits.

Similarly, the northern plains exhibit much evidence for horizontal flow of material and vertical diapiric type tectonics, but it is not clearly established as related to glaciation. The burden of proof there, unlike Argyre, still rests with the glacial hypothesis, about which I am much less enthusiastic (for the northern plains) than I was a decade ago; rather, Jöns' and Tanaka's mud ocean seems best able to explain what is seen pervasively across the northern plains.

To be fair to the suffering of others and to the critics of all the Blue Mars ideas, no one's ideas of Mars really rallied much support, and everyone who has dared to come in from the cold inevitably has taken a lot of heat for their bold ideas. When balanced with logic and respect, this is as it should be. The last few decades have seen a flourishing of multiple distinct concepts of Martian geology, hydrology, and climate history. There is but one Martian reality, but it remains very much a grand enigma warranting a cacophony of intelligent wonderings. With little doubt, everyone is mostly wrong, while many of us are just a little right, but we have yet to find out who those people are. However, the glory is not in who has been the most right; the glory is in Mars itself.

I also do not wish to give an impression that there had been a complete years-long stagnation of thought until 1989 ever since the first spacecraft reconnaissance of Mars, or that those late *Viking*-era ideas finally satiated the appetite of Blue Martians for new and better ideas. It is apparent to all who work in the field that the fresh injection of new observations finally provided by the *Mars Global Surveyor* and *Mars Odyssey* missions is what was really needed; those data have come just in time to liven up a field that was bound to stagnate again sooner or later if new data had been lacking. With the new data came the first substantial fresh support for the new Blue Mars ideas subsequent to their advent. This is the power of science – it is not merely one religious fervor against another, but it is a dynamic evolution of testable ideas. The energy for this process is scientific discovery, but to accomplish that we need observing platforms near Mars.

I also don't wish to leave an impression that the Mars science community is somehow especially obstinate, stuck in old ways, or unimaginative. Indeed, if there was ever a research field where creativity oozes at the seams, it is planetary science. It is general human nature – a trait that has been genetically instilled in humans for good reason – that holds onto ideas that work, even if they work only partly. Others have thought to remake our knowledge of how planets evolve, such as Alfred Wegner with his continental drift idea, and W.B. Harland's 1964 global Earth glaciation hypothesis (termed "Snowball Earth" by P.F. Hoffman and D.P. Schrag, 2000, *Scientific American*, **68**, 68–75). Such researchers often suffer the consequences of their creativity, and then are lauded (if they are lucky enough to be right) or forgotten (if they are, more often than not, wrong).

ICY WARM

The popular conception of Mars as a cold and dry world is accurate by any ordinary human standard. The association in the public's mind of glaciers and permafrost with cold climates therefore seems, on the surface, to minimize the significance of ice for people who are unfamiliar with ice-related processes or do not recognize fully how utterly cold Mars is at present. It is *liquid water*, the giver of life, that the public cares about. What seems to have gone largely unappreciated – not only by the public but also by some scientific circles – is that ice flows and melts when it is comparatively warm. Ice is the giver of water. Also, in sufficient time, rocky surfaces are sculpted in ways that are totally unique to ice.

Liquid water has left a geomorphic record on Mars that has told us an amazing story, but it has left paragraphs and whole chapters untold or subject to widely divergent interpretations. Glaciers and permafrost fill in many gaps in the story of Mars that water-carved features have resisted telling us clearly. Glacier-like flow of ice on Mars, as on Earth, requires large accumulations of surface ice, and, depending on the circumstances, normally necessitates direct condensation from the atmosphere. Furthermore, ice can melt, and in so doing may leave tell-tale signs that pinpoint temperature conditions precisely. Ice also can undergo gravity-driven brittle fracture under the right conditions. While ice can be an inert rock-like material at low enough temperatures and under low shear stresses – and most of the mass of Martian ice may have behaved as such over most of Martian history – it is evidence of substantial ice activity that is most significant. The propensity of ice to fracture, flow, sublimate, and melt in the right circumstances, and sometimes to do all of these in the same place, but only under a very narrow set of conditions, is what makes ice such an important environmental indicator. As important, glaciers imply sustained conditions that provide for net annual positive mass balance of ice, allowing large

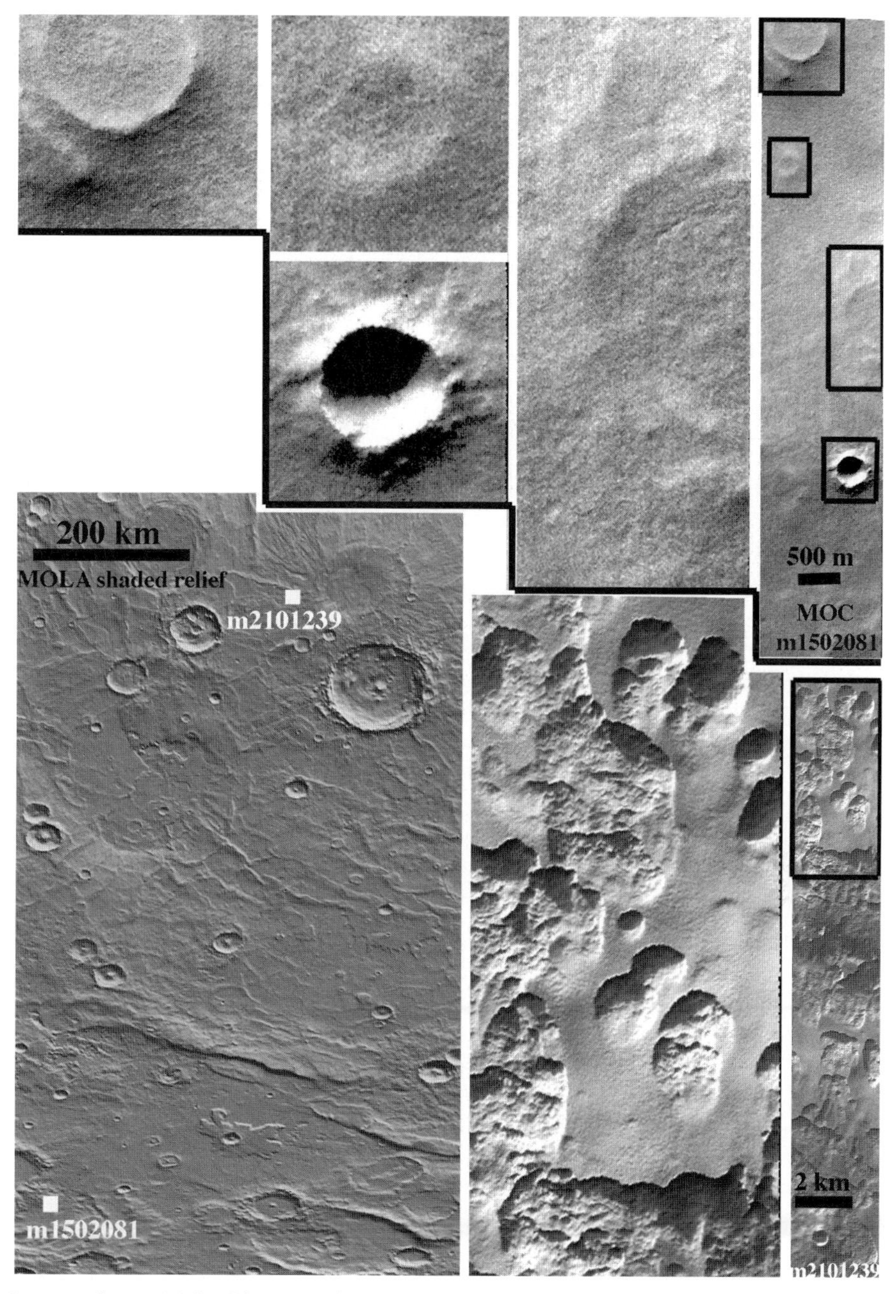

Icy mantles on Malea Planum. The more southerly latitudes exhibit a nearly continuous mantle that almost buries craters between 300 and 1000 m diameter, suggesting a mantle thickness of ~100 m (MOC m1502081, −73.4°S lat., 312.1°W long.). More northerly latitudes have a discontinuous, probably sublimating mantle resembling south polar "swiss cheese" and Utopia thermokarst (MOC m2101239, −58.9°S lat., 304.8°W long.).

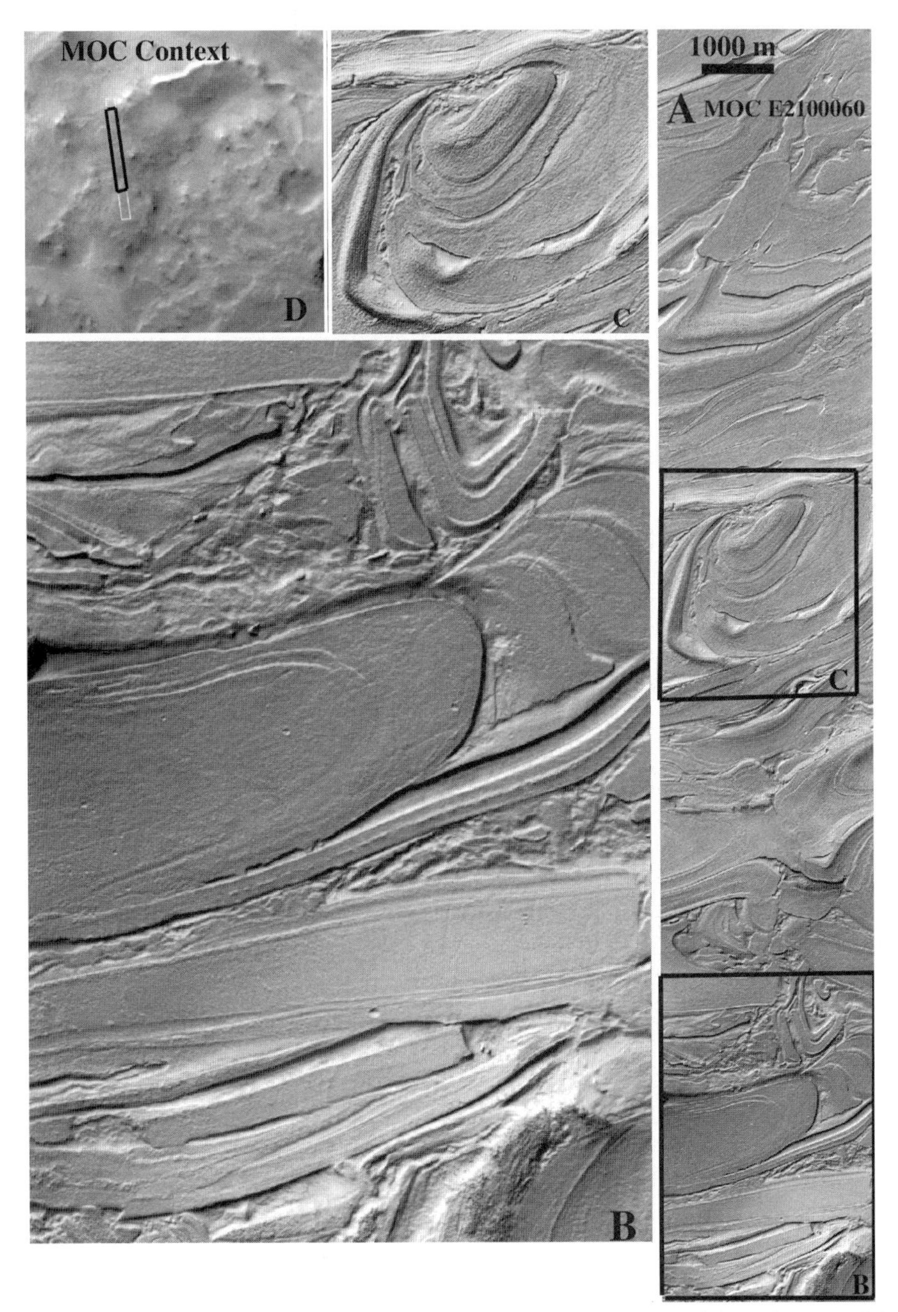

An amazing terrain in northwestern Hellas Planitia involved deposition of a material that was both semi-ductile and semi-brittle. This region is where Kargel and Strom (1992b) favored stagnation of ice sheets, where Jeff Moore and others claim may be a lake floor, and where Ken Tanaka, I, and others have suggested may be a repository of wet volcanic debris. Each model involves volatiles.

masses of ice to build up. The environmental significance of glaciers and permafrost thus adds substantially to that provided by water-carved features, even if this evidence pertains to limited periods of Martian history. Liquid water alone is highly ambiguous regarding timescales of activity, in part because of its low viscosity and consequent ability to do a lot of geomorphic work in a short period of time.

OUTER EDGE OF THE INNER REALM

Glaciation to the extent suggested above – with ice at one time covering 20% or more of the planet – requires an amount of water roughly equivalent to the amount needed to fill the northern plains. That's the better part of a kilometer-deep global layer. How could Mars have become so water-rich in the first place?

Water is among the most abundant substances in the Solar System. There are desiccated worlds in the realm of space near and just beyond Earth's orbit, but there are well-understood explanations for their dryness that would not so readily work for Mars as a whole. The Moon, save for its icy poles, is certainly much drier than any desert bone – because the Moon is the recondensed product of a mega planet-scale impact, and in any case it is too small to hold onto its hydrogen and oxygen products of photolysis of water vapor. Volatiles ejected from the proto-Earth were not able to recondense immediately in the Moon-forming circum-Earth nebula following this giant impact, and so were soon dispersed into space, leaving the Moon to form from refractory material. In the main asteroid belt, and within the Earth-crossing asteroid populations, are metallic M- and S-type asteroids virtually devoid of water because they are the shattered cores, mantles, and lava crusts of mini-planets subjected to igneous temperatures and extreme thermal metamorphism; they are cooked throughout, and they are far too small and too close to the Sun to hang on to any outgassed water. At the inner fringes of the outer solar system, Io is ripped by tidal friction and heated to temperatures more extreme than Earth has encountered at any time in its geologic history (J. Kargel and 23 others, *Eos*, 2003). Mars would not have suffered such tidal torment, but it could have suffered collisional violence that might have stripped it of water. The evidence does not point this way. Mars is no Io, it is no mere asteroidal fragment, and it is nothing like the Moon. It seems in every respect to be an icy water world the equal of Earth, but only half the diameter and colder.

H_2O and CO_2 in various forms have been observed together from Venus to Pluto and beyond. It would be a bizarre set of circumstances that could somehow make Mars – located between the water-world Earth and the icy satellites and comets – a desiccated planet, and still somehow also make it appear to be icy and, at times in its past, wet. A playful God, or One contemptuous of human contemplations, might play this game. Doubting this to be the case, I find that the appearance of Mars is explicable: Mars is very icy and at times is wet.

It is just as clear that Mars is no icy satellite in a thin disguise of dust and lava. All indications from geology and geochemistry and gravity are that Mars is fairly Earth-like in its volatile content, no doubt the most Earth-like world known anywhere in that regard. Mars unquestionably belongs to the inner Solar System, located well inward of the Solar System's nebula condensation "snowline." However, Mars is the outermost of the inner planets, and it is entirely likely that Mars is the most volatile-rich of the five major rocky worlds of the inner Solar System. The direct detection of polar surface ice, shallow buried permafrost ice, and water vapor and ice clouds in the atmosphere are the merest hints that Mars is the other water world among the rocky planets. The geological/geomorphological evidence indicates something of the huge quantity of volatiles, considered on a gram-for-gram basis. We can no longer doubt a prodigious inventory and at least occasional cycling of water or ice on Mars, as deduced from geological evidence.

TWO-VOLATILE PLANETS

Aside from the geological evidence, we have much new cosmochemical evidence and theory tipping us off that Mars is enriched both in both H_2O and CO_2. We see important amounts of chemically bound and mineralogically trapped water and carbon dioxide in Martian meteorites. The isotopic composition of these remnants and the amount of oxygen seen escaping Mars indicates that the Martian crust once had more water than at present. Varied data sources and models point to huge water reservoirs still hidden on the planet, and even greater amounts in the distant past. With that water we also expect some CO_2.

CO_2 is an abundant and widespread substance observed from Venus to the Neptune system. My 'father' in academia, John S. Lewis (University of Arizona), and academic brother, Bruce Fegley, Jr (Washington University), have dedicated much of their careers to understanding how the condensates and volatiles of the Solar System were formed and incorporated into the solid bodies of the Solar System. The planets and moons started with microscopic dust and ice condensate grains, which then accumulated into cosmic hailstones and eventually into comet- and asteroid-size planetesimals. Eventually these planetesimals were accreted by the growing planetary embryos along with their water ice, hydrated and anhydrous silicates, hydroxyl minerals, carbonates, sulfides, and iron–nickel.

The nebula models are successful in explaining to first-order most aspects of the chemistry of planets, comets, and meteorites. However, nebula condensation models do not presently indicate CO_2 forming directly from nebula gas, though its sister oxide, CO, is a major condensate (as CO clathrate) in accepted models of the Uranian and Neptunian systems. There are two possibilities for the origin of the Solar System's carbon oxides outside the special Uranian and Neptunian chemical factories. For terrestrial planets and large moons, CO_2 can be synthesized by high-temperature processes occurring inside the planets, especially under hydrothermal conditions. Secondly, vast quantities of CO_2 and CO are made in the interstellar environment by ultraviolet irradiation of interstellar dust and gas by stars; the energetic radiation blasts apart water molecules and any molecules carbon belongs to; the excited ionized fragments then recombine in haste to CO and CO_2. This is certainly the main source of these volatiles and some other abundant volatiles, such as alcohols and formaldehyde, in comets. Most of the CO_2 in the Solar System may have the same ultimate origin during the starry eons before the Sun formed. This stardust was then swept up by the solar nebula and subjected to incomplete chemical reprocessing.

Whatever actually gave rise to abundant CO_2 on Venus, Earth, and carbonaceous chondrite asteroids, the same thing must have happened to Mars. CO_2 is the major Martian atmospheric gas. It is a major condensed polar volatile, which occurs in Martian meteorites bound into carbonate minerals and as trapped fluid inclusions. Just as the abundance of Earth's atmospheric trace of water vapor is controlled by vapor equilibrium with condensed reservoirs – especially the oceans – Martian water vapor and CO_2 abundances are controlled by vapor equilibrium with condensate reservoirs. With low prevailing temperatures on Mars, it is misleading to think of Mars' "tens of precipitable microns" of atmospheric water vapor as indicating an H_2O-poor planet.

The NASA establishment's perspective now is that Mars is an ice-rich world. That war has been won. A parallel revolution is now underway: recognition that Mars is a two-volatile planet, having abundant CO_2 as well as H_2O. It seems inconceivable that Mars does not have vast reserves of condensed CO_2 (dry ice, liquid CO_2, in aqueous solution, clathrate, and carbonates). Each of us in the Mars community is groping to understand the myriad implications of this. Talk of strange glaciers, strange permafrost, and strange gullies on the strange world of Mars is bound to raise the question of whether dry ice and perhaps liquid CO_2 or salts and brines may be involved in the formation of these features.

The abundances of Martian H_2O and CO_2 are probably somewhat Earth-like. This is such a simple and powerful statement that it needs justification, and for that I go back to the birth of the Solar System, and before, to see that this is a compelling (if unproven) model.

The inner regions of the Solar System are more volatile depleted than the outer realms, as is reasonably understood from (1) solar influences on condensation and selective evaporation of volatiles within the solar nebula, (2) the impact-speed distribution and kinetic energy spectrum of accreting objects (the violence having been greater with the higher orbital speeds at smaller semi-major axes), and (3) solar influences on volatile loss following accretion (due to direct insolation as well as the early T-Tauri phase of elevated magnetic field and mass ejection and consequent electromagnetic induction heating). A crude radial trend in volatile abundances is partly related to a compositional step function, where, at any given point in nebula evolution, a "snowline" existed in the solar nebula around the orbit of Jupiter. As the nebula evolved in pressure and temperature and water vapor abundance, the migration of the snowline made it a fuzzy boundary, more of a sigmoid. To this was added the significant role of hydrated and hydroxylated phases, each with its own evolving condensation limits, and then also the effects of considerable radial mixing of planetesimals stirred up by Jupiter and the other planets during accretion. Because of this mixing, especially where it involved large accreting objects, a significant stochastic variation developed in planetary volatile abundances, but nevertheless the radial trend is evident and strong. In effect, Mars is probably more volatile enriched than Earth.

There are caveats to this tidy view of the Solar System. Earth suffered at least one unimaginable body blow during its early formative years. A Mars-size body collided with Earth in a glancing blow, resulting in the formation of the Moon, yet Earth retained some volatiles. Based on depletions of certain salt-forming elements compared to expectations based purely on nebula condensation models, my own research has suggested that Earth may have had a primordial ocean, much deeper than the present ocean, that was blasted to the cosmic smithereens, possibly by that mega impact. What, then, did Mars experience during its early life? We know that Hellas- and Argyre-size impacts – of which there must have been innumerable comparable hits during the accretion of Mars – must have had important effects on volatile inventory, but the chaotic nature of accretion renders this period very much a mystery; were there far greater impacts earlier in its history, events that blew off most Martian volatiles but left no other remaining trace? We know that Mars, compared to Earth, almost surely experienced a greater proportion of volatile additions during accretion. However, predicting volatile inventories on Earth and Mars also requires knowledge of volatile losses, and we are weak in this subject.

Some researchers consider the violence of Early Earth, and particularly of the Moon's origin, to be fortuitous circumstances that have enabled Earth to have life. Lucky impact parameters left Earth with some water, yet the consequent formation of the Moon constrains rotational obliquity to within a narrow range – a fact that has a stabilizing influence on climate and favoring effect for life. In thinking of such perfect violence, we can imagine that Mars may have suffered something even more horrific, a more direct blow, or multiple blows, that dried it out more thoroughly but without the rare combination of impact parameters that would form a large moon. It's not implausible. Mars did not have to be water-rich, though its location favors water-richness. Geology says that volatile-richness was, indeed, in the cards Mars was dealt. It is also plausible that Mars experienced much less violence than Earth, boding well for volatile abundances but leaving Mars almost moonless and ravaged by chaotic obliquity variations. However, in the case of Mars, its near moonlessness offers the prospect of high-obliquity periods of watery seasonal warmth. Thus, for Earth, a moon is needed for life; for Mars, paradoxically, moonlessness may be needed for surface life.

The story with carbon dioxide is interesting, because, as pure ices, CO_2 is more volatile than H_2O, and in straightforward condensation and accretion of the pure molecular ices, one would expect objects closer to the Sun to have increased ratios of $H_2O{:}CO_2$. Earth's other near neighbor, Venus, once had much more water than it has now. However, models of deuterium isotopic enrichment of its scant remaining water indicates that it never was anywhere near as water-rich as Earth; possibly it had as much as 5% of an Earth ocean, but more likely closer to 0.5%. As Natasha Johnson (NASA Goddard) and Bruce Fegley (Washington University) have found, water may still be tied up in dehydration-resistant metamorphic minerals such as tremolite, but there cannot be much. Yet Venus has over 90 bars of CO_2 in its atmosphere – an amount very close to that tied up mostly in Earth's carbonate rocks, plus that converted to organics, methane, graphite, and our atmospheric trace of CO_2. Venus defies the logic of simplicity, which only says that we must look deeper for answers. A good explanation could be that the carbon now tied up in Venusian CO_2 was, at the time volatiles were accreted, primarily in the form of carbonate, graphite, or other forms less volatile than H_2O. Carbon dioxide could have been released later by metamorphism of carbonates, or graphite may have reacted with oxygen (formed from water by dissociation at magmatic temperatures). Perhaps the death of water on Venus was the birth of carbon dioxide.

Comets and carbonaceous chondrites (types of chemically primitive meteorites) probably made the original major deliveries of highly volatile materials to Earth and Mars and other objects in the Solar System, with a fraction of the incoming volatiles preserved during accretion, somewhat as Jack Schmitt described in the Foreword. Comets are close to 50% H_2O, and CI and CM carbonaceous chondrites contain from 6% to 18% H_2O by mass (including hydroxyl). Earth's hydrosphere amounts to just 0.10% of our planet's mass, and there may be that much water (more or less) still in the mantle. Just a few tenths of a percent comets or a few percent carbonaceous chondrites would explain Earth's quota of volatiles. Mars could not reasonably have escaped accreting comparable or larger amounts of these volatile-rich materials.

Rather than starting with an assumed meteoritic or cometary precursor, an assumption of a systematic order in the Solar System to the abundances and ratios of key elements and volatile compounds retained by the terrestrial planets is as plausible a starting point as any other approach, since these objects have been subjected to much the same global geochemical processes. As work by many geochemists, including John Lewis (University of Arizona), Sasha Basilevsky (Russian Academy of Sciences), me, and many others has shown, Venus is remarkably Earth-like in many compositional parameters of its rocks. This similarity may also extend to volatiles. It has been noted many times that the amount of CO_2 in the Venusian atmosphere is almost identical to the estimated inventory of CO_2 in Earth (mostly in carbonates); roughly comparable amounts of nitrogen and argon also occur in the two atmospheres. If this similarity also extends to water, it would imply that Venus once had about an Earth ocean's worth of water, but lost almost all of it. Either that, or some process prevented water from condensing and accreting while carbon dioxide was admitted; but that would be difficult to understand. David Grinspoon (University of Colorado) and colleagues have reported recently that their studies indicate that a Venusian ocean could have been lost just a few billion years ago.

On the grand scheme of planetary compositional variations, Martian rocks and its mantle and crust are really not too different from Earth. Some details of planetary evolution differ. For example, Mars has a more iron-rich mantle than that on Earth; this can be traced to small differences in initial composition and processes. The more iron-rich Martian mantle, in fact, can be attributed to more extensive oxidation driven by thermal dissociation of a proportionately larger amount of water in the mantle than happened in Earth. The remarkable thing is that the two planets are nearly the same in most measures of composition. The amount of moderately volatile and lithophile elements (i.e., those that neither partition into the metallic core nor constitute atmospheric components), such as sodium and potassium, are nearly the same as in Earth relative to involatile lithophile elements such as magnesium and silicon. There are no widely accepted reasons to suppose that highly volatile materials, such as water and carbon dioxide, should have differed in abundances by large factors in the average original materials accreted by each planet. If there are small differences, the location of Mars closer to the outer Solar System would suggest that the difference should be in favor of larger amounts of volatiles at Mars.

A cosmochemically reasonable postulate, not proven by the arguments above but certainly compelling, would be that Mars contains much more than Earth's share of volatiles (scaled to planetary mass). Geology and the Martian gravity field argue that it really cannot be too much more than Earth's quota (or Mars would begin to look like Europa).

Ice, chemically bound water, and hydroxyl no doubt played a major role in determining the redox state of the Martian mantle and crust, as it has in Earth and probably did in Venus. Thermal dissociation of water, which occurs readily at igneous temperatures (evident, for instance, in the red oxidized cinders of many volcanic cinder cones), produced abundant oxygen and hydrogen; this process probably explains the fact that the terrestrial and Martian mantles contain much oxidized FeO and maybe the Fe_2O_3 of the Martian surface. Indications suggest that Earth formed largely from a mixture of precursors similar to enstatite and ordinary H chondrites; and because that mixture contains less FeO than Earth's mantle, a hydro-oxidation event may be needed. The same process works for Mars. Hydrogen released by dissociated water will also act as a reductant, but much of it is blown out of volcanoes into the atmosphere, and is then rapidly lost to space due to its low molecular mass. Hence, the effect of mantle water dissociation and hydrogen loss is net oxidation of the mantle and crust. This powerful process appears to have left its signature on the mantles of both Earth and Mars, possibly a bit more so on Mars than on Earth. This process also implies that the original amount of H_2O initially retained by Earth was equivalent to tens of oceans' worth; and possibly more than a proportional amount was initially retained by Mars. How much water escaped dissociation and was then released on to the surface and into the upper crust is thus uncertain, but the answer is probably "Quite a lot."

The discussion so far has made the implicit assumption that the major condensible volatiles on Mars are H_2O and CO_2. The compositions of carbonaceous chondrites and comets should entertain a broader consideration of other volatiles. The discussion buttresses the case that H_2O and CO_2 (especially the former) are responsible for the origins of Martian valleys, channels, and permafrost and glacial landscapes. Unlikely are more exotic geomorphic agents, as have been suggested, such as chondrite-derived high-octane petrol (sorry George W. Bush) or a cometary alcoholic brew (sorry Anheuser-Busch). These exotic candidates cannot be dismissed out of hand, but there is no good evidence to support them. It is worth considering the logic chain that points to water acting with carbon dioxide as the main volatile agents active on Mars. It is a long chain, and there is plenty of room for logical or interpretive error, but this, nevertheless, is the case:

1. The atmosphere of Mars is currently often saturated in ice at the poles and locally elsewhere. There exist net annual sources and sinks of water vapor, and over geologic time accumulation, sublimation, and transport rates are significant.
2. The Martian atmosphere also exhibits seasonal polar and mid-latitude saturation in CO_2, and there exist large cyclic interhemispheric fluxes of carbon dioxide, where the evident phases include vapor and solid.
3. With the exception of H_2O and CO_2, no other Martian atmospheric volatiles are known to be near saturation of a condensed phase.
4. Mars undergoes large changes in the global pattern and seasonality of solar insolation related to astronomical cycles (10^5–10^6 year oscillations) of orbital obliquity and eccentricity, and these oscillations drive periodic reversals in the polarity of net global transfers of water vapor and carbon dioxide, where sinks become sources and vice versa. Along with cyclic changes in dust transport, and in conjunction with other processes, the origin of rhythmic sedimentary layers on Mars is thereby explicable.
5. Mars has ample evidence for the existence and flow (current, recent, and long ago) of surface and shallow subsurface solid volatiles at middle and polar latitudes, and for the accumulation and sublimation of significant ice deposits at middle and high latitudes.
6. Mars exhibits ample evidence for having had flowing liquid on its surface and shallow subsurface both recently (in small amounts) at middle and high latitudes, and long ago (in large amounts) at all latitudes.
7. Liquid surface flows represent movement from a high to a lower state of gravitational potential energy.
8. Some Martian features formed by liquid discharge from shallow "liquifers"; sustenance of the flows that formed these features require ground fluid recharge.

Inference: *Martian geomorphology is controlled or affected by volatiles in solid, liquid, and vapor states. CO_2 and H_2O must be considered candidates. Transport processes involve solid, vapor, and liquid states occurring in the atmosphere, on the surface, and in the subsurface.*

9. Considering all cosmochemically abundant molecules in the Solar System, the only major pure substances with melting points in the range of observed Martian surface and shallow subsurface temperatures are H_2O, CO_2, and SO_2; minor substances include propane, butane, ethanol, and methanol, which could exist on Mars, if present, in liquid but not solid form.
10. Of the substances listed above, the only pure substances with triple-point vapor pressures close to Martian ambient surface pressure are H_2O, CH_3OH, and C_2H_5OH. SO_2 and CO_2 currently can exist only in the vapor state on the surface of Mars at regions and times when the temperature attains the substance's triple point, but they could exist as liquids in the subsurface (in pressurized spaces) or on the surface with large increases in atmospheric pressure. H_2O can exist stably in solid form in the polar regions and at shallow depths at middle latitudes, and it can exist transiently as a liquid on the surface. Required for the long-term,

widespread stability of liquid water on the surface is a modest (few millibars) increase in H_2O vapor pressure; required for geomorphic efficacy is a modest increase in summer temperatures to enable prodigious melting of surface ice. Stable, sustained surface flows of liquid carbon dioxide are also possible and is not to be excluded, though it requires three orders of magnitude increase in surface pressure to remain stable at the triple-point temperature. CO_2 vapor-driven density flows are possible on the surface.

11. The alcohols and SO_2 have not been observed on Mars, while H_2O and CO_2 have been seen both in the atmosphere and condensed at the poles and in transient frosts.
12. Hydrogen (in any condensed form) has been observed in close association with the youngest flow-caused features, such as small gullies and lobate debris aprons.
13. H_2O, with or without dissolved impurities, is the most widely stable liquid and condensed solid volatile on Mars and in its shallow subsurface.

Inference: *Mars exhibits global and regional volatile cycles involving solid, liquid, and vaporous H_2O, which are cycled globally and regionally both thermally and gravitationally. Volatile cycling also includes CO_2 solid and vapor.*

14. CO_2 exhibits weak but important interactions with water, for instance dissolving in liquid water and forming solid clathrate hydrates, whose stability is pressure and temperature dependent. Other water-soluble constituents, such as salts, are also abundant on Mars.

Inference: *H_2O, CO_2, and salts would not always occur in isolation of one another. Liquid water in contact with rocks and soils is apt to be impure, with dissolved CO_2 and dissolved salts being abundant. The planet's missing sinks of CO_2 are probably associated with the missing sinks of H_2O.*

15. Impurities in liquid water reduce its melting point and vapor pressure, thus expanding the stability of the liquid phase to broader regions of the surface, greater volumes of the subsurface, and over greater periods of time during the annual and obliquity cycles.
16. Atmospheric H_2O condensates in vapor equilibrium with equilibrated impure surface water and ice would be a solid condensate.
17. While evidence for surface runoff from discrete sources is widespread at the middle and high latitudes on Mars, evidence for distributed runoff from rainfall is either very degraded (extremely ancient) or absent.
18. Certain icy deposits and ice-rich rocks on Mars appear to retain ice, including at fairly shallow levels, dating from deposition in the Hesperian and Noachian. These deposits have existed continuously as permafrost for billions of years.

Inference: *It does not rain on Mars now, and probably never has in 4 billion years, or did so only under special conditions (such as rainfall from transient volcanic eruption plumes of global, water-saturated impact ejecta). Snowfall (traces of H_2O and copious quantities of CO_2) are the dominant form of precipitation on*

Mars today. Any accumulation rates of H_2O in favorable sites are extremely slow by terrestrial standards. Atmospheric temperatures either (1) have never risen above 273 K on a widespread and long continued basis at any time in the last 4 billion years, or (2) the cumulative duration of these periods was less than a few million years, or (3) the Martian atmosphere was extremely dry when and where atmospheric temperatures exceeded 273 K.

19. Vapor distillation of CO_2 and H_2O condensates leads to separate locations of accumulation, except that traces of H_2O can co-condense with CO_2. Thus, relatively pure H_2O and CO_2 deposits may build up over geologic time. They are not always mixed.

    Inference: *Subsurface melting of both ice and dry ice (more or less isolated from each other) is likely. Liquid CO_2 as well as liquid water is probably important in the crust of Mars. Ground-fluid transport and local inter-fluid interactions are probable.*

20. Under conditions likely for condensed-phase CO_2–H_2O interactions on Mars, conditions are such that mutual solubility relations will produce H_2O-saturated liquid CO_2 and CO_2-saturated liquid H_2O, or CO_2 clathrate hydrate. Conditions under which these condensed phases could be present are widespread in the Martian subsurface. Clathrates are stable on the surface at the south pole and in mid-latitude seasonal frosts. However, conditions are expected to oscillate widely around important phase boundaries in the CO_2–H_2O system.
21. Phase changes of condensed H_2O and CO_2 phases involve large volume changes.
22. Flowing solids and liquids of these materials involve a potential for bed erosion, and sediment transport and deposition.
23. The likely Martian inventories of both CO_2 and H_2O far exceed the amounts that are stored in the polar caps. The expected quantities would have to be stored in a volume of crust extending well beyond the polar regions.
24. We see both large- and small-scale youthful landforms plausibly attributed to ice in regions where ice should be in stable equilibrium with the atmosphere and where soil hydrogen is abundant. At low latitudes – where ice could exist metastably at great depths but shallow hydrogen is not abundant – we observe ancient landforms plausibly attributed to melting of deeply buried ice, but no such landforms that are young.

    Inference: *Mars is a two-volatile planet, and the phase relations and transport of vapor, liquid, and solid forms of CO_2 and H_2O (including mixed phases) should have significant geomorphic and chemical effects on the surface and in the silicate crust. Shallow and deep reservoirs of ice exist at middle and high latitudes, and deep reservoirs at low latitudes. The climate would appear to have dried or has become warmer over geologic time, such that relative humidity has decreased on average, and latitude limits of condensate stability have shifted poleward.*

25. Many landforms, including the chaotic terrains and outflow channels, are consistent with violent formation of large quantities of gas and liquid from

crustal volatile reservoirs. Some of these landscapes are specifically associated with volcanic terrains.

Inference: *Mars has a geologic record indicating an important role of degassing and water outburst floods and other processes related to destabilization of crustal reservoirs containing both* H_2O *and* CO_2*. In many cases, geothermal influences are evident.*

26. Mars has a long record of episodic surface modification driven by ice and liquid water. Modifications range in age from the oldest preserved parts of the record to the youngest. Remnants of heavily cratered terrain have survived across half of Mars.

Inference: *Although* H_2O*- and* CO_2*-driven processes have considerably affected the Martian surface, it has not had anywhere near the cumulative extent of* H_2O*-driven geomorphic processing that Earth's surface has.*

Summation (current consensus of most of the Mars community): *Mars, having copious* CO_2 *and* H_2O*, is fundamentally a two-volatile planet, and both volatiles are keys to understanding the geologic and climatic history of Mars. The retention of ancient, albeit degraded, features can be reconciled with evidence for intense geomorphic processing only if there have been extremely lengthy episodes where very little has happened. Thus, it would appear that Mars has very long periods characterized perhaps by the Cold, Dry, Windy paradigm and briefer periods that resemble the Warm, Wet paradigm. Mars, however, remains a hydrologically active world.*

Although this is a long chain of observations and inferences, with ample room to interpret variously in this same general framework, most researchers currently concur on these basic points or arrive slightly differently at the same conclusion. A more controversial extension of the logical chain continues as such:

27. A broad suite of landforms, including some requiring prodigious surface melting, is associated with and related to major volcanically-driven volatile outbursts.
28. CO_2 and H_2O are important greenhouse gases.

Inference: *The sudden emission of large amounts of* H_2O *and* CO_2 *into the atmosphere would prompt a series of climatic and geologic responses. This is the essence of the MEGAOUTFLO model (see Baker et al., 1991 and 2000). Martian geologic history, by this account, is not unlike the lives of warring soldiers – 90% unbearable boredom and 10% pure, unadulterated action and moments of unbridled terror.*

RED, WHITE, AND BLUE MARS

A relative newcomer to the Mars debates, Nick Hoffman, an Australian mining engineer, has taken a novel approach in explaining how undersize and energy-starved Mars could have oversize geologic activity. Hoffman suggests that carbon dioxide is substantially responsible for formation of flood features, fluid-deposited sedimentary blankets, gullies, and other landforms that most geologists attribute to liquid water. He emphasizes the low current temperatures prevalent on the surface of Mars, assumes that the climate was never very different, and has tended to minimize possible heat flow and interior temperatures. Hoffman thus assumes a cryogenic condition of the crust. He suggests that carbon-dioxide fluid-driven processes on Mars form features similar to water-produced landforms on Earth. Hoffman has described his model as "White Mars" to distinguish it from the more watery and icy models collectively termed "Blue Mars."

Hoffman's initial presentations practically excluded any role for water ice, although more recent renditions highlight co-equal roles of CO_2 and H_2O. I have argued above for the CO_2-rich composition of Mars that is a chief basis for White Mars, but I also must point out some limits. It is further necessary to point out that almost everything contained in the White Mars model is rooted in ideas developed over the last several decades by other people. However, like the other models, there is new evidence that can be brought to bear pro and con. One thing is sure: "White Mars" has stirred the pot.

The avid reader of Mars science literature may recognize that I have authored or co-authored a variety of perspectives that include both White Mars and Blue Mars models. Some of my colleagues have asked me, in effect, "Just where do you stand?" In fact, it should be seen that I stand behind both classes of model, but more as a combined model than as extreme alternatives. In fact, Mars is a planet that has two major condensable volatiles, and we should expect to see geologic evidence of the action of both CO_2 and H_2O. The CO_2 is far more important than is generally recognized, but less than Hoffman advocates.

I first met Nick Hoffman at a meeting in 2000 when I presented theoretical model calculations of Martian heat flow and the global subsurface stability of condensed CO_2 in the forms of dry ice, CO_2 clathrate hydrate, liquid CO_2, and CO_2-saturated liquid water. My point that year, my own variant of White Mars, was that condensed CO_2 in each form is theoretically stable at a relatively shallow depth outside the polar regions and throughout the year as long as two conditions are met: (1) there must be a subsurface supply of abundant CO_2 and H_2O, and (2) there must be some way to seal the subsurface reservoirs of condensed CO_2 from the atmosphere. The most likely, abundant, widespread, and effective sealant allowing the pressure needed to maintain buried condensed forms of CO_2 would be water ice (stable at middle and polar latitudes), but salt-cemented regolith or fine-grained clay minerals (especially if water-saturated and then frozen, or at least fully hydrated) would also work. Ice would be especially effective because it is a plastically deformable and impermeable substance, and its propensity to sublimate and recondense aids pore- and crack-filling; due to the volatility of water ice it generally should be absent at low latitudes, according to climate models, and condensed nonpolar CO_2 should follow a similar distribution.

The White Mars model has taken a beating at some science meetings, and in many regards certain criticisms are justified, especially in the extreme form first presented

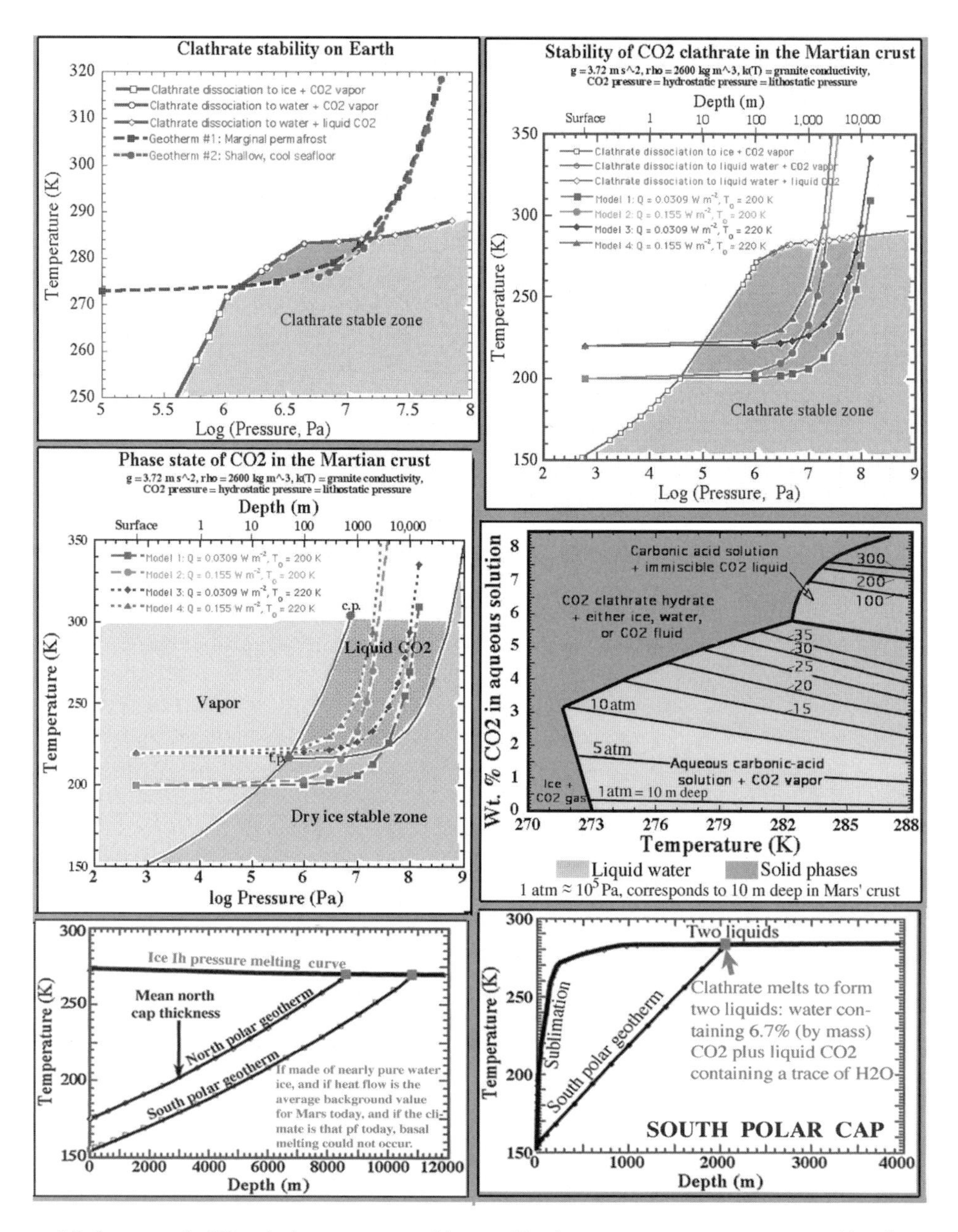

Methane and CO_2 clathrates are stable on Earth over narrow temperature/depth intervals, but on Mars CO_2 clathrate and even dry ice has far broader stable ranges. Clathrate stability in the crust of Mars, however, requires a pressure sealant, such as water ice or salt-cemented regolith. Dissociation of CO_2 clathrate can form liquid water containing 6–7% CO_2 plus a fluid phase of CO_2. Clathrates would be stable in both polar caps starting at shallow depths.

by Hoffman. White Mars has, however, a strong basis, and the concept has evolved to a more sustainable stance from its initial untenable hypothesis. I have little doubt that huge masses of subsurface condensed forms of CO_2 will be found in the coming years and decades, and it will be seen as a major agent of Martian geology alongside, and in combination with, H_2O. Compared to any Mars model where CO_2 is a mere whisp of an atmosphere and thin veneer over the polar caps, Nick Hofmann is more right than he is wrong.

MARS: A GASSY ICY PLANET

So far, I have mentioned large reservoirs of volatiles but few numbers. How well-endowed with volatiles is Mars? While geology provides one means of answering the question (as Michael Carr attempted, coming up with his estimated water inventory amounting to a 400-m-deep global ocean based on eroded volumes of outflow channels), analysis of Martian meteorites offers another direction, and cosmochemistry spanning the breadth of the Solar System provides still another perspective, all yielding in the same answer to order of magnitude. Lately, it would seem that Michael Carr's estimate, once seen as a daring, radical challenge to the Cold, Dry, Windy Mars establishment, is now the most conservative of plausible estimates. Here, I continue a simple Solar-System-wide approach to the issue.

No major object from the Earth to the outer Solar System is known to have more CO_2 than H_2O. It could happen in other planetary systems, around stars formed where debris of carbon stars was abundant, for instance, but in this solar system water is the king of volatiles. Venus, almost alone in the Solar System, is CO_2 dominated, but this is explicable foremost by its unique evolution close to the Sun, where nearly all hydrous and hydroxylated minerals are unstable at prevailing nebula temperatures so close to the Sun, and where most water vapor was swept away with the last remnants of the solar nebula. Secondly, much of the Venusian inventory of water was probably dissociated under magmatic conditions, with hydrogen then outgassed and lost. Thirdly and most important, Venusian water depletion is explained by the extreme CO_2-greenhouse atmosphere, which cooked off the water that Venus accreted, and most of that water was then photolyzed high in the atmosphere, and the hydrogen was mostly lost to space over geologic time.

If Mars completed its accretionary period only with as much CO_2 as Venus and Earth and only with as much H_2O as Earth has, then Mars would have enough CO_2 and H_2O to make a global ocean (if it was all spread uniformly over the surface in liquid form) a few kilometers deep and a layer of dry ice almost 300 m thick. These volatiles would not exist in these forms as uniform oceans and dry-ice layers. In fact, if these two volatiles were mixed together they would form mainly CO_2 clathrate hydrate ($CO_2 \cdot 5\frac{3}{4}\ H_2O$), the CO_2 analog of natural gas (CH_4) hydrate. A clathrate hydrate layer, if CO_2 was fully sequestered in this form, would be 1,300 m thick; there would be excess water ice. CO_2 could as well form a subsurface pressurized zone of liquid carbon dioxide or liquid solution in water (making carbonic acid). These numbers give some idea of the likely volatile richness of Mars and the large amounts of one or more condensed forms of CO_2 in the crust of Mars. Mars may well be even more volatile enriched, but it could not be vastly more enriched, or we would see an object with perhaps a few volcanoes and impact basin rims protruding through a global blanket of volatiles (like Io) or an ice-covered ocean covering all the rock (like Europa). We don't see that, but the more we learn of Mars the more we see something icier than has ever before been recognized, a world where volatiles appear to be major crust-forming minerals.

On Earth, the rarity of CO_2 as a free gas (amounting to just 0.035% of the atmosphere) is due to solubility in water, chemical reactivity with rocks and dissolved cations to form carbonate rocks. There are, to be sure, also small reservoirs of CO_2 in mantle rocks, and more that has been converted to organic carbon, methane, and graphite. Despite these conversions and hidden reservoirs, Earth's crust is no poorer in CO_2 than Venus's atmosphere, and it may have slightly more. Earth contains roughly 100 Earth atmospheres' worth of CO_2 in these various forms. Earth's oceans (plus other minor reservoirs of water) contain about five times that mass of H_2O. While CO_2 and its sister oxide CO are widespread and fairly abundant materials in the Solar System, where abundances are known H_2O is always more abundant by comparable or larger factors. Comet Halley and several other comets, for instance, exhibit about the same H_2O/CO_2 ratio as Earth, and in carbonaceous chondrites (where these volatiles are bound up in hydrate, hydroxide, and carbonate minerals) the ratio is just a little lower. This is understood theoretically as being due to the elemental abundance ratios of the metals, oxygen, and carbon in the Sun and the nearby galactic environment (and the solar nebula). After formation of metal oxides (mostly in silicate forms – the main components of rocks) and oxidized carbon (CO and CO_2), there remained abundant excess oxygen, which reacted with predominant H_2 gas in the nebula and pre-nebula environments to form H_2O, exceeding the amount of carbon oxides by a factor of 3 to 10 (on a mass basis) in our part of the galaxy.

There is no good basis for supposing that Mars might somehow have acquired or developed a volatile assemblage with anywhere near as much CO_2 as H_2O. It is too far from the Sun to have ever gone "Venusian," with its cycle of intense water loss. All this does not imply that CO_2 is inconsequential on Mars. Far from it. In fact, it is inconceivable that the amount of CO_2 presently existing in the polar caps and atmosphere is the whole primordial inventory of CO_2, and it seems improbable that global volatile loss processes have been effective in removing the 13 bars or so that would be implied by a Venusian/Terrestrial quota of CO_2. There is far more CO_2 lurking about in the subsurface than what we see in the poles, unless much of it has been converted to other condensed forms of carbon, such as CH_4 or graphite. Indeed, there may be enough CO_2 on Mars that, if in gaseous form, would be sufficient to make either a truly warm, wet early Mars or a Mars with liquid carbon dioxide rain, depending on climate details.

While climate modelers have long considered a possible dense ancient CO_2 atmosphere, geologists and geomorphologists have barely begun to consider the possible ramifications of that amount of CO_2. Indeed, the big Fifth Grade question, *where did all that water go?* has a Sixth Grade sister question: *where did all that CO_2 go?* The answer in both cases is that it is probably still there in the crust.

WHEN OPPOSITES ATTRACT

CO_2 is not strongly interactive with H_2O because CO_2 is an electrically nonpolar molecule while H_2O is polar, so they mix only to a limited extent and under certain conditions. They associate not by direct electronic covalent bonding, but rather by the much weaker Van der Waals' attraction. These molecules are electronic opposites of one another, almost like oil and water. Weak interactions allow up to a few per cent CO_2 to dissolve in liquid water under CO_2 gas pressure, as is familiar in carbonated beverages. When the two volatiles occur together under cold conditions but low gas pressure, the familiar ice Ih structure may be favored, with CO_2 forming a separate phase. When a critical threshold of gas activity or gas pressure of CO_2 (or of other small, nonpolar molecules) is exceeded, the binary solid called a clathrate hydrate (or clathrate for short) becomes thermodynamically favored more than ice Ih. If the two volatiles can physically get together, the clathrate forms readily. The ordinary ice Ih lattice becomes distorted under the influence of the nonpolar molecule, and the two join to form the clathrate crystalline structure, with each volatile having its specific place in the lattice. The mixed substance resembles water ice to the eye.

Although several structural types of clathrates are known, the ideal stoichiometric formula of the most common Type I clathrate is $X \cdot 5\frac{3}{4}\ H_2O$, where X can be almost any small nonpolar molecule, such as CO_2 (CH_4 is another common clathrate-forming gas, important especially on Earth). The clathrate molecular structures resemble soccer balls (a football in Britain), with the water forming the structure and a single nonpolar molecule rattling around – trapped – inside the structure. Under the right conditions, which today are prevalent at the south polar surface of Mars and probably throughout much of the crust, a mixture of CO_2 and H_2O will form the condensed clathrate structure until one of the volatiles is almost used up and brought to low concentrations in the medium surrounding the clathrate. Clathrate formation can occur between any phase states of CO_2 and any states of H_2O, given the right conditions.

The necessary conditions for clathrate stability are quite broad, but the equilibria between clathrate and free forms of the two volatiles are temperature- and pressure-dependent. At Martian mean annual south polar temperature, 153 K, the equilibrium pressure happens to be the atmospheric surface pressure at the south pole's elevation, which is the best evidence that clathrate may exist there (Kargel and Lunine, 1998, chapter in *Ices in the Solar System*, Kluwer Academic Press), with its abundance and rate of accumulation determined by the very low rate of supply of H_2O to the south polar cap.

At lower latitudes, the temperatures are higher, and so is the equilibrium pressure, such that any clathrate would have to exist in the subsurface at a pressure greater than atmospheric. Water ice- or salt-cemented regolith can serve as a pressure seal, beneath which sufficient CO_2 gas pressures can develop, and then form clathrate-rich permafrost or clathrate glaciers. Permafrost ice similarly provides the pressure seal on Earth allowing CH_4 (rarely, CO_2) clathrate to form. At a sufficient depth (dependent on the geotherm), temperatures exceed the stability limits of clathrate, and so the typical situation is for CO_2- or CH_4-saturated groundwater (with CO_2 reacted to form carbonic acid) to exist beneath the clathrate, with ice developed above the clathrate. Additions of a clathrate-forming gas into the saturated groundwater system – for example, metamorphic or metasomatic fluids emitted from heated rocks or plutonic magmas – would then form more clathrate. The reaction causes a volume expansion of the ice phase in going from ice Ih to clathrate (though the clathrate itself is denser than ice and liquid water, because the clathrate includes caged gas molecules; expansion is caused by formation of a new structure that includes added gas molecules). The net volume change of the reaction, if treated as a closed system, is, however, negative. By Le Chatelier's Principle, this means that, if sitting on the equilibrium curve, any increase in pressure drives further formation of clathrate, and a decrease in pressure destabilizes it. Any increase in temperature while sitting on the equilibrium curve dissociates clathrate, while a decrease in temperature stabilizes it.

As a clathrate-forming fluid phase builds its pressure beneath an imperfect icy permafrost seal, the expansion of ice due to the clathrate reaction may cause the clathrate to fill in any fractures or holes, thus making it a more effective seal. Eventually, the process of conversion of water ice (or water) to clathrate may shut down because the system runs out of CO_2 (and other nonpolar gases) in excess of saturation or because the clathrate itself becomes an impermeable barrier to the diffusion of the gas to unreacted H_2O molecules. Thus, on Mars a common permafrost structure may exist with a surface layer of desiccated regolith, beneath which is a layer of ice-cemented regolith, underneath which clathrate occurs, and a still deeper layer of dry ice, or liquid CO_2, or CO_2-saturated liquid water. Terrestrial clathrate-bearing permafrost characteristically is somewhat like this, but more complex, with ice and clathrate lenses and dikes containing nodules of unreacted ice, and gas-saturated groundwater zones with intermittent lenses and nodules of clathrate.

The quantity of CO_2 that Mars may hide in these condensed forms is potentially enormous. The CO_2 source may be degassing of the mantle or intruded magmas or thermal metamorphism of deeply buried carbonates. Since clathrate is probably distributed principally where water-ice-rich permafrost provides a seal, i.e., potentially over the half of the planet at middle and high latitudes, there could well be permafrost layers many kilometers thick dominated by clathrate or interlayered clathrate and water ice. Any possible excess CO_2 beyond the amount that can bond with H_2O will then be stored as dry ice or liquid CO_2.

GASSY GEYSERS AND CRUMBLING CHAOS

Any sudden rupture of the water-ice seal over thick areas of clathrate-rich permafrost could initiate a catastrophic sequence, especially if CO_2-saturated groundwater is present beneath the clathrate. An impact, seismic activity, or large landslide could be the event that breaks the seal. A rupture may be conditioned by global warming or local geothermal warming and thinning of the stable permafrost zone. Highly pressurized liquid CO_2 or CO_2-saturated water may spurt out of fractures like uncorked champaign, forming mega-geysers. The events could be very similar to the lateral blasts of stratovolcanoes, such as Mount Saint Helens in 1980. The available pent-up energy per unit mass of gas-saturated liquid is smaller than in large gassy magmatic eruptions (mainly because of lower temperature of the gas-saturated groundwater compared to magma, offset only partly by the higher solubility of CO_2 in H_2O than of both those substances in most silicate magmas under relevant pressure–temperature conditions). However, the total available mass of saturated liquid is much larger than in magmatic systems. Thus, the champaigne effect may involve more fluid jetting and slumping of overburden and less turbulent ash than in magmatic eruptions.

Depressurization may cause dry ice to melt or vaporize, and clathrate may dissociate, thus freeing up additional CO_2 vapor or liquid, causing a more protracted and less dramatic but highly effective response. Violent eruptions of fluid forms of CO_2 may mechanically erode the fractures, thus widening them and allowing faster mass transfers through these conduits, hence bigger geyers. Further mechanical erosion and expulsion of condensed volumes could cause massive collapse and disintegration of the terrain, particularly if CO_2-saturated groundwater was stored in large cavernous chambers created by volcanic processes or chemical weathering and dissolution of rocks. The process may be self-accelerating until the system runs out of pressurized fluid CO_2, or when the system becomes choked with debris. If there exists a large-scale slope – for example, near the highlands/lowlands boundary or on the sloping terrain of impact basin rims – the fluid-rich debris may flow away, preventing the source area from becoming choked, and so the process may involve a very large spatial extent and a very thick layer of material.

The result of these processes could be the formation of chaotic terrain, outflow channels, vast debris-flow deposits, transient climatic responses, and eventually retransfer of volatiles from basins to cold-trap locations at high elevations and high latitudes. A similar process may have affected methane clathrate-rich seafloor sediments on the Drake Ridge (Atlantic Ocean), where large chaotic slumped blocks have been associated with clathrate decomposition and massive venting of methane 55 million years ago.

WHERE ARE THE CARBONATES?

The Mars science community, with such a compulsive interest in water and climate on Mars, has been justifiably interested in the inventory and fate of carbon dioxide on Mars, including a long search for carbonates. The MGS TES instrument has, in fact, found only 1% or 2% of carbonates on Mars, according to results recently presented by Phil Christiansen (Arizona State University). This quantity of carbonate could be a minor component of global dust or traces of pore-filling carbonates in rocks. No massive carbonate layers have been found anywhere. This lack of confirmation of what was a widely anticipated material has been hailed by some as an indication that Mars has not had much aqueous activity. However, what it indicates strongly is simply that Mars has not had long periods of time of slow runoff of surface water into long-lasting evaporative or biologically active lakes or seas. In fact, the mineralogy and geomorphology on Mars says the same thing: Mars has not had a hydrologic cycle closely similar to Earth's.

What the TES mineralogy data allow, and the morphology indicates, is that Mars has had important periods of major and minor transient outbursts of groundwater onto the surface. The solubility of Ca–Mg–Fe carbonate minerals under a broad range of expected pH conditions is such that only about one hundredth of a percent of carbonate would be dissolved in the water. Highly water-soluble salts, such as sulfates, could be brought to the surface in much larger quantities, but the carbonates would be mainly left deep underground where they were formed and remained stable. In fact, this model then requires some work to get even 1% or 2% carbonate onto the surface; this, in fact, is an elevated abundance. Meteorite impacts, erosion of deeper in-situ subsurface-formed carbonates, and airborne mobilization of eroded subsurface carbonates, can explain the elevated abundances, as can some degree of evaporative or freeze-driven concentration of solutes. Then the answer to the question *Where are the carbonates?* is that they are there – just underground where groundwater has been most active.

CRYOGENIC MARTIAN ROCK OF AGES

H_2O has an overwhelming influence on the geology of any planet where the trinity of pure phases of this substance can coexist, with each form morphing into the others and cycling again and again across the surface and through the crust. The transmutability of H_2O from one phase to another on Earth and Mars has played a crucial role in the geologic evolution of both worlds. While Mars draws much attention for the roles of H_2O on the surface itself, the subsurface is the primary dwelling place of most Martian H_2O. The very crust of Mars appears to be icy, going back perhaps to the beginnings of Martian time.

Water's abundance in our Solar System is due to the oxygen produced by pre-solar stellar nucleosynthesis. The reaction of nucleosynthetic star-formed oxygen with Big Bang-generated hydrogen occurred in the shock-compressed gas phase during the stellar explosions, on dust grains in the interstellar medium, and then in the gas phase of the solar nebula and by vapor–solid reaction in the nebula. Ice further reacted with silicates in planetesimals, forming carbonaceous chondrite-like rock. Comets probably accreted from icy condensates in the extended atmospheric envelopes of the gas-giant nebulae during the solar nebula phase. Thus, H_2O was incorporated into Mars and Earth as ice, as water of hydration, and as hydroxyl of silicate minerals.

The first major release of these volatiles from the Martian interior (also Earth's) to the crust probably began while rapid accretion was still in progress. Impact shock heating partly volatilized the crashing planetesimals, with refractory elements promptly recondensing and falling out as ash and then becoming reburied. Then heating and igneous differentiation of the interior drove buried volatiles toward the surface along with the low-melting-point eutectic component of the silicate mantle, which formed the crust. On Earth and Mars, volcanic and fumarolic outgassing released vapor, liquid, and supercritical fluid forms of water into the forming crust, onto the surface, and into the atmosphere. Boiling lavas, simmering hot springs, spouting geysers, and slowly diffusing fluid contributed juvenile volatiles to the atmospheres and hydrospheres of our worlds. Post-accretionary comet and asteroid impacts added additional water at later times, but most of the terrestrial planets' water was accreted and outgassed in the first hundred million years of planetary existence. However, as the Foreword by Jack Schmitt suggests, this release of volatiles may have occurred on a more relaxed agenda at Mars compared to what happened at Earth.

Conditions on Earth almost immediately became warm enough, under a greenhouse atmosphere, for most (or all) outgassed water to exist as liquid water and other non-ice forms. Certainly by 4.2 billion years ago, granitoid (probably continental) crust existed (clued in by the presence of zircon sand grains contained in some of Earth's oldest metamorphic rocks). Granites then, as now, were probably produced through hydrous melting of mantle in subduction zones.

On Mars, unless more than 5 bars of CO_2 was pumped into the atmosphere and maintained there, conditions were freezing cold from the beginning while the Martian crust was constructed and as outgassing proceeded. Hydrous fluids emanating from the depths tended to refreeze within the crust instead of being erupted in a flash of steam or warm spring water, as occurs on Earth. Martian ice comprised just one more rock type in the heterogeneous crust.

Early differentiation of Mars would have proceeded with the formation of ice veins and dikes crystallized in association with (and as physical–chemical extensions of) vein-forming carbonates, halides, sulfates, and pegmatite – the aqueously deposited extensions of usual magmatic systems. However, H_2O was an order of magnitude more abundant than those materials. These fluids would have initially segregated from crystallizing magmas deep in the crust, or they may have been produced by metamorphic reactions in buried sedimentary rocks. A fractional crystallization sequence would be common, with a single hydrous fluid precipitating, in succession (for example, along a vertical crack in a shallowing sequence) feldspar-quartz pegmatite, carbonates and anhydrous sulfates, hydrated sulfates and halides, and ice mixed with hydrated salts. Ice and salts also must have formed massive laccoliths, where briny water intruded the crust beneath unyielding permafrost and froze in place.

Episodically through the Noachian, periods of comparative warmth and unleashing of floods by geothermal heat drove the formation of surface lakes in craters and other basins. Lakes, seas, and rivers existed briefly, then were frozen and buried. Crustal ice was uplifted and subsided, was melted and recrystallized, impacted, blown out by volcanic interactions, remobilized, and redistributed a hundred times in a dozen ways in the battered and bruised cratered highlands. Frozen icy lenses and interbeds added to the sediment fill of every major basin. Over time, internal heat and atmospheric transport tended to drive the ice ever higher in the crust and higher in elevation and latitude, even while gravity and surface geology sought to drive it to lower elevations and bury it. Ice accumulated preferentially in the colder and more porous upper crustal levels, at the very low and the very high elevations, at high and middle latitudes, and at cold traps at every favorable site within the crust and on the surface. Thus, an icy crust was formed, ice being a mineral and rock type comprising perhaps well over 10% vol. of the upper 10 km of crust.

A large magmatic province such as Tharsis or Elysium, with perhaps 30 km of lava accumulated over time, may have released 1% of its mass as water. This quantity is common in hydrous lavas on Earth and is an amount supported by hydrogen isotopic analysis of Martian igneous meteorites. That may include both "juvenile" water (newly released from the mantle) and recycled crustal water. If the water was principally cold-trapped in the upper crust covering an area about the size of the volcanic province, an ice layer about 1 km thick could be formed; and it is highly probable that that kilometer of ice would be disseminated through a much larger volume of crust, perhaps about 5 km, filling every pore and fracture with ice and its salty precipitates. Much of this ice was eventually driven off the volcano by geothermal heating, thus explaining volcano-associated outflow channels and chaos.

Since H_2O is abundant in Mars, one can envision ancient crust of the cratered highlands that is absolutely riddled with ice veins. Not only that, but as the scenario offered by Jack Schmitt suggests, prodigious outgassing of juvenile volatiles may continue at Mars, whereas on Earth this period is long past. (Juvenile volatiles are being released from Earth, but in miniscule quantities.) On Mars, ice is a major crust-forming rock type, like basalt, andesite, granite, limestone, and sandstone on Earth. In this context we can understand "melting mountains," expressed glaciers of Arabia's fretted canyons, mega-landslides, chaotic terrain and outflow channels, products of violent super-size volcano–ice interactions, and massive debris flows and climatic upheavals caused when mega-volcanism destabilizes this icy, gassy crust. Solid-state instabilities in large, buried, geothermally warmed and softened ice masses may have produced ice diapirs, where ice-rich sedimentary rock or ice segregations ascended, like salt domes, by buoyant forces (Rayleigh–Taylor instabilities) or as boudins under disruptive conditions of differential principal stress components. We see intrusive domes that may be evidence of this process in the northern plains and elsewhere (the Martian "monster," introduced in Chapter 1). Ice formed thick lenses where ice caps and glaciers grew, subsided, and were buried by accumulations of dust or lava.

Evidence of primordial degassing, such as envisioned in Jack Schmitt's Foreword, may still be extant near the surface of Mars. The glacier-like lobate debris aprons and lineated valley fill at the edge of the uplands and in scattered southern hemisphere regions are young flow features. However, as suggested by Baerbel Lucchitta (USGS) in 1993 at the Workshop on the Martian Northern Plains in Fairbanks, they may represent remobilized ancient icy plateau materials originally emplaced during a primordial period of volcanic outgassing and upward migration of crustal volatiles. Now interbedded with volcanic rocks, the primordial ice may constitute a major rock-forming mineral and form part of the layered plateau sequence. If conditions were right, some of this volatile mass may have melted and formed oceans, while frozen remnants may be slowly extruding from beneath the plateau mass.

Thus, Earth and Mars share important aspects of global differentiation, including formation of a core, mantle, and crust, but different surface conditions caused by slightly different proximity to the Sun caused their hydrospheres to evolve in two directions – one dominated by the liquid phase (Earth) and the other dominated by the solid phase (Mars). This concept of an ancient icy constitution of the Martian crust is a fundamental component of Martian upper crustal geology that has emerged from discoveries made just over the last few years. While Earth also has icy permafrost, the old ice of Mars marks a fundamental geologic distinction from Earth, where volatiles exist mainly in the oceans and atmosphere, or have been sequestered in more familiar types of sedimentary rocks, and only occasionally some fraction of H_2O enters the solid phases of ice. Earth's water (except that still residing stably in the lower mantle) has been cycled again and again, a million times or more through Earth history. On Mars, the component of ice that is actively cycling may have been cycled globally just 1,000 times, while much of it may have been locked in place for 4 billion years.

In some places ice has been much more of an active participant in a hydrological cycle. Most of this book has been about recycled volatiles. Volatiles today may occur in several major Martian reservoirs: primordial volatiles still trapped in the mantle; possible relics of the primordial outgassed hydrosphere and atmosphere, which might still exist in their original condensed form in very old and never-heated parts of the crust; volatiles that have been recycled through the atmosphere and groundwater system many times and redeposited in various forms; and buried crustal volatiles chemically bound within rocks. Deep volatiles generally will be unresponsive to climate due to their isolation from the atmosphere and surface. The near-surface, climate-responsive reservoir of volatiles includes the polar ice caps, shallow permafrost ice, groundwater, and various condensed forms of CO_2. The present-day hydrosphere appears, from isotopic and elemental evidence in Martian meteorites, to contain an exchangeable, dynamic reservoir of near-surface ice, and a less readily exchangeable reservoir of ice and groundwater that is relatively isolated from the atmosphere. The more efficiently exchangeable reservoir presumably includes shallow permafrost ice, especially permafrost near the limits of ground-ice stability, soil-adsorbed volatiles, and the polar caps. The more stable component would include deeply sequestered permafrost ice that may respond to climate on very long timescales or not at all.

Some fraction of the crustal volatiles would, over time, become buried again and convert back into carbonates, hydrates, and hydroxyl phases by aqueous chemistry under what geologists term diagenetic or low-grade metamorphic conditions. As the crust built up, these minerals were more deeply buried and heated to much greater temperatures. Eventual metamorphism of buried carbonate rocks or carbonate-bearing soils, or possibly oxidation of biogenic or abiogenic organic substances could have re-released the buried CO_2 and H_2O, plus other gases, such as SO_2, thus providing a long-sustained and possibly continuing recycled flux of volatiles back into the upper crust and hydrosphere. These volatiles would freeze out very widely in

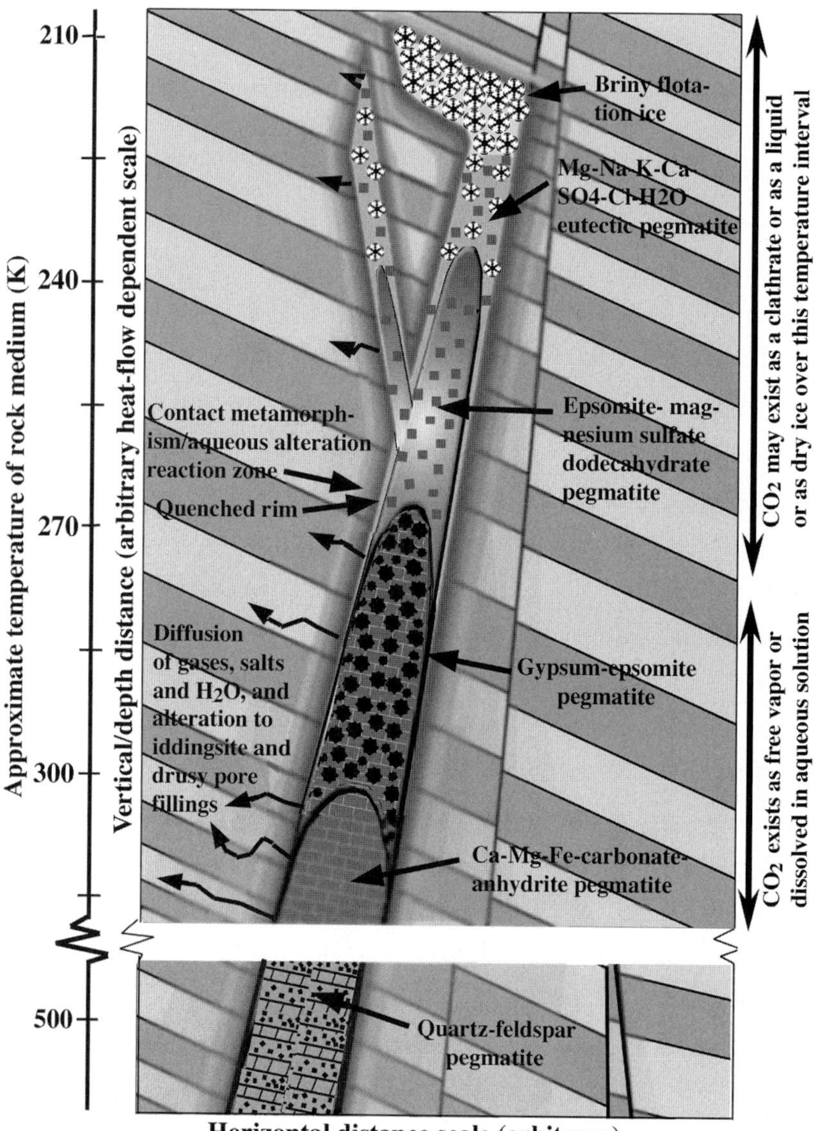

"Ice pegmatite." Quartz-feldspar pegmatites probably occur on Mars, as on Earth where hydrous fluids emanate from silicic igneous intrusions. On Earth the residual fluid after crystallization of the quartz–feldspar mixture will commonly erupt as a briny hot spring. On Mars, however, the brine may crystallize as a pegmatitic appendage formed of carbonates, hydrated sulfates, chlorides, ice, and clathrates. Here, an ice pegmatite forms a series of precipitates related to a fractional crystallization sequence (simplified here) predicted for a Martian salt assemblage. Some Martian meteorites may be samples of the aqueous alteration zone. The pegmatites themselves are not apt to survive the trip to Earth.

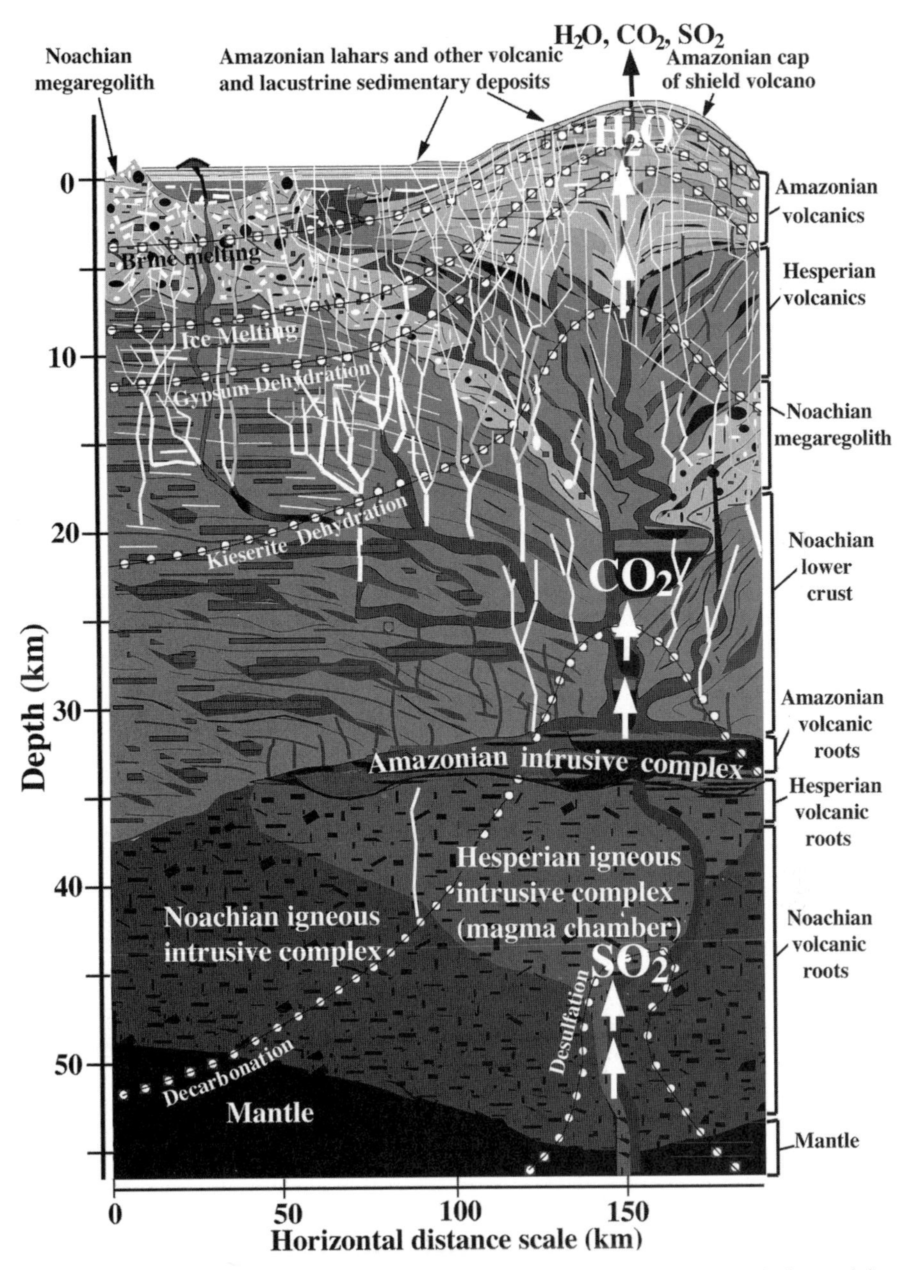

Schematic illustration of the formation of a volatile-rich upper crust (salts and ice depicted in white). A volcanic complex spans the eons but post-dates the early Noachian megaregolith. Young lake deposits related to volcanic activity bury the old cratered terrain. A series of isotherms represent mobilization of various volatiles. White veins and dikes represent pegmatite-like intrusions (icy where cold enough) associated with igneous and metamorphic activity.

the upper crust along with salts and other soluble minerals. Fluxes between reservoirs and among their components may include surficial climate-driven cycling of outgassed volatiles, emission of metamorphic fluids and vapors produced by geothermal heating of deeply buried crustal volatiles, active outgassing of juvenile volatiles from the mantle, slow downward burial and isolation of crustal volatiles that may have once participated in the climate system, and occasional impact-caused injection into the shallow hydrosphere of deeply sequestered volatiles.

While the lack of plate tectonics (at least for 4 billion years) means that recycling of the hydrosphere/cryosphere back into the mantle is nil, it is all but inevitable that deep crustal burial (especially in large sedimentary basins and perhaps in places like Tharsis) and aqueous chemistry in the groundwater system have proved to be effective means of sequestering H_2O and CO_2 and other volatiles into rocky minerals deep in the crust. Deep polar outflows of groundwater and CO_2 due to basal polar melting and intracrustal refreezing are expected to have a major influence on rock chemistry and may have sequestered large volumes of volatiles within the northern plains and regions surrounding the south polar cap.

6

Old Times, New Wrinkles

The hydrologic cycle during ancient epochs of the "big waters" are of obvious great interest. Gravitationally and geothermally driven global-scale subsurface groundwater transport and thermally driven redistribution of ice has been hypothesized by Steven Clifford (Lunar and Planetary Institute, Houston). An early period of recharge via heavy ancient rainfall and snowfall has also been considered by Robert Craddock and Ted Maxwell (Smithsonian Institution, Washington, DC). Transient, repeating episodes of a volcanically triggered humid, marine-glacial dominated climate has been proposed by Victor Baker, Robert Strom, me, and others.

Each research group has its own unique hypothesis and bias. Like geese imprinted by first out-of-shell experiences, these scientific biases are often based in part on where on Earth a geologist first planted his field geology boots during graduate school and what on Mars she first laid eyes upon. It is often difficult for some geologists to move beyond those formative experiences, but most of us do, in fact, eventually take a broader look at the issues and evidence.

Renewed Mars exploration, starting in the mid-1990s with *Mars Pathfinder*, *Mars Global Surveyor* (*MGS*) and, most recently, *Mars Odyssey* (*MO*) – and by press time a whole new international armada of spacecraft – has heralded new opportunities to test hypotheses that were built on a foundation of classical astronomy and *Mariner/Viking*-era exploration. Improved imaging by MOC at meter-scale resolution, Mars imaging by THEMIS at resolutions bridging that of MOC and *Viking Orbiter*, an accurate global topographic map provided by MOLA, satellite-borne thermometry and mineral mapping (TES and THEMIS), chemical element mapping (HEND and GRS), and gravity (Doppler tracking) and magnetic field (MAG) mapping have revolutionized our understanding of Mars, while geochemical analyses and imaging of rocks and soils by *Pathfinder* have provided a bit more ground truth to validate remote-sensing observations. Long-standing ideas, untestable until these new missions, have now been subjected to unprecedented scrutiny. New theoretical advances, spurred to a considerable extent by the new observations, have enriched our concept of how Mars operates as a system. The advances have been too numerous and diverse to summarize adequately, but I present a few examples of recent observations and models that tend to confirm, clarify, or modify previous concepts regarding ancient Martian climate, water, and ice. This chapter is about new findings regarding the processes occurring at the low-frequency/high-magnitude end of the Martian geoprocess spectrum: the big-picture events, mostly from very long ago. I save for the next chapter the possibly more mundane but more immediately relevant everyday processes dominating the present period and thought to dominate the long eons between mega events.

U CHANNEL

I shall consider the formation of glacial U channels on Earth and Mars, but first I explore why these forms are significant. Dr Eugene Wigner (1902–1995), of Princeton University, wrote in 1960 on *The Unreasonable Effectiveness of Mathematics in the Natural Sciences*. He had in mind the myriad occurrences in daily life and throughout the physical world where, recognized or not, pure mathematical elegance emerges in the most amazing and sometimes absurdly simple ways. The numbers of theory are usually simple numbers. When the numbers of nature match the numbers of theory, a powerful diagnostic and predictive quality emerges. Departures of observations from theoretical ideality are also important, as they say much about the possible inaccuracies or incompleteness of the theory.

Nature is not always as mathematically clean as Wigner would have it. It sometimes is, but multiple complex processes overprint and interact, and the result commonly looks like a mess too complex for simple theory to solve. Gravitation illustrates a rapid trend toward mathematical intractability of some simple systems as they become even slightly more physically complex. Kepler, Galileo, and Newton did well to place motions of two co-orbiting gravitating objects into mathematically precise terms of perfect ellipses and the spherical radial symmetry of the inverse-square law. However, there is no known analytical solution describing an incrementally more complex system, e.g., the motion of a massless third object when it is added into this system. Mathematicians believe this so-called "Restricted Three-body Problem" to be insoluble. Approximate short-term answers can be obtained by iterative computational methods (e.g., for calculating a trajectory to Mars) or numerical statistical methods (e.g., Öpik's theory of myriad massless particles perturbed by two gravitating objects). Understanding complex systems requires approximations of behavior and statistical approaches. Rather than tracking the paths of single gas molecules in a flask full of air, the air can be considered statistically, reliably, with the Ideal Gas Law, the Boltzmann Distribution of molecular speeds, Fick's Law of Diffusion, and such.

In geology, the overlapping influences of multiple planetary processes or the cumulative effects of many evolutionary steps of a single process can yield chaos. This is seen, for example, in the bewildering assemblage of rock formations and structures in geologic maps of Earth and Mars; the map reader is normally aware that those maps are themselves gross simplifications of an even less tidy reality. It is the role of geologists to make an empirical approximate tracking of a sequence of inherently complex or even chaotic events of planets and to develop theoretical

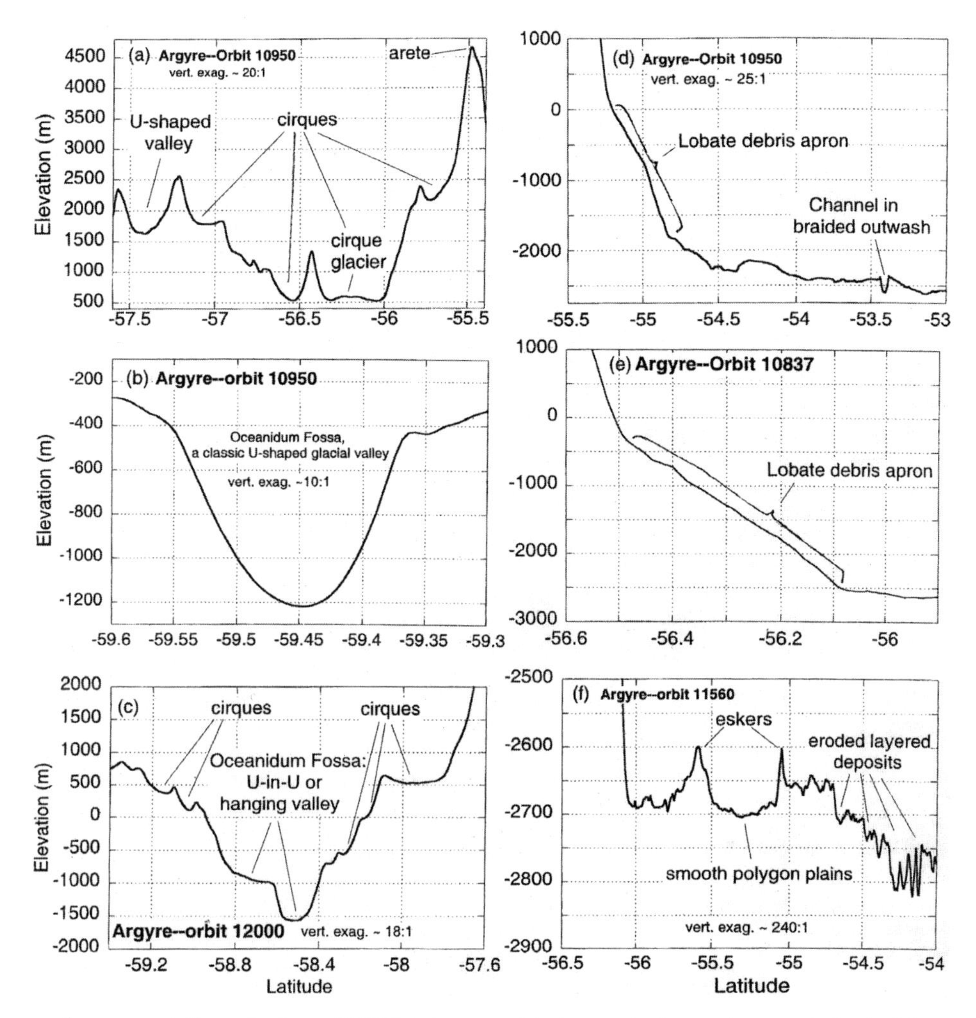

A representative selection of MOLA topographic transects shows characteristic glacial-type profiles typical of alpine glaciers and lowland glacial deposits. Orbit tracks 12000 and 10950 are overlain on *Viking Orbiter* photomosaics in Chapter 5. The ascending and descending series of sharp ridges separated by concave-up amphitheaters and valleys is a hallmark of alpine glaciation. The classic U-shaped valleys are especially significant. In general, such alpine glacial features are associated with vigorous erosion by wet-based glaciers. If this general relationship observed on Earth carries over to Mars, the implications are considerable, as it requires a mean annual surface temperature approaching the melting point of ice. The pressure at the base of a Martian glacier 2,000 m thick would be only 72 bars, enough to cause less than 1 K of pressure reduction in the melting point. Salts could lower the melting point much further, but it is difficult to avoid a situation where the climate was much warmer than it is today. This conclusion is consistent with evidence of surface melting and with esker-like ridges, which require basal melting. The formation of a massive ice sheet further implies a greater supply of moisture than is available today.

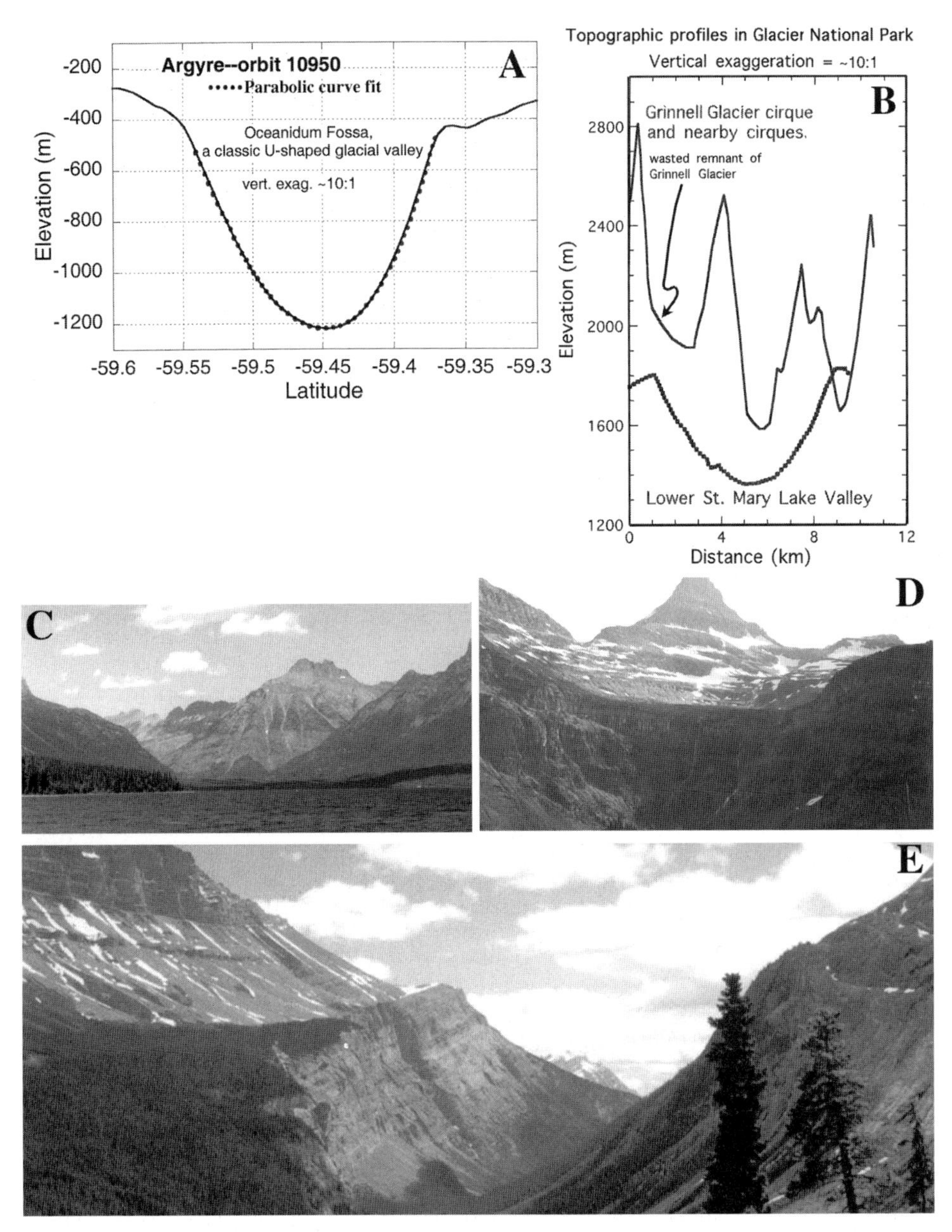

A Martian U-shaped channel – Oceanidum Fossa (A) – is compared here, at the same factor-of-ten vertical exaggeration, with (B) Lower Lake Saint Mary valley (Glacier National Park), a classic U-shaped glacial channel. More extreme slopes of glacial topography elsewhere in Glacier National Park are also shown for completeness of the comparison. Also in panel A is a parabolic curve fit – the classic and diagnostic form of glacial U channels – to Oceanidum Fossa. (C) Lower Lake Saint Mary valley. (D) Grinnell Glacier valley, (E) U-in-U valley in Banff National Park.

Two classic U-shaped valleys – actually, they are glacier-cut channels – in Glacier National Park.

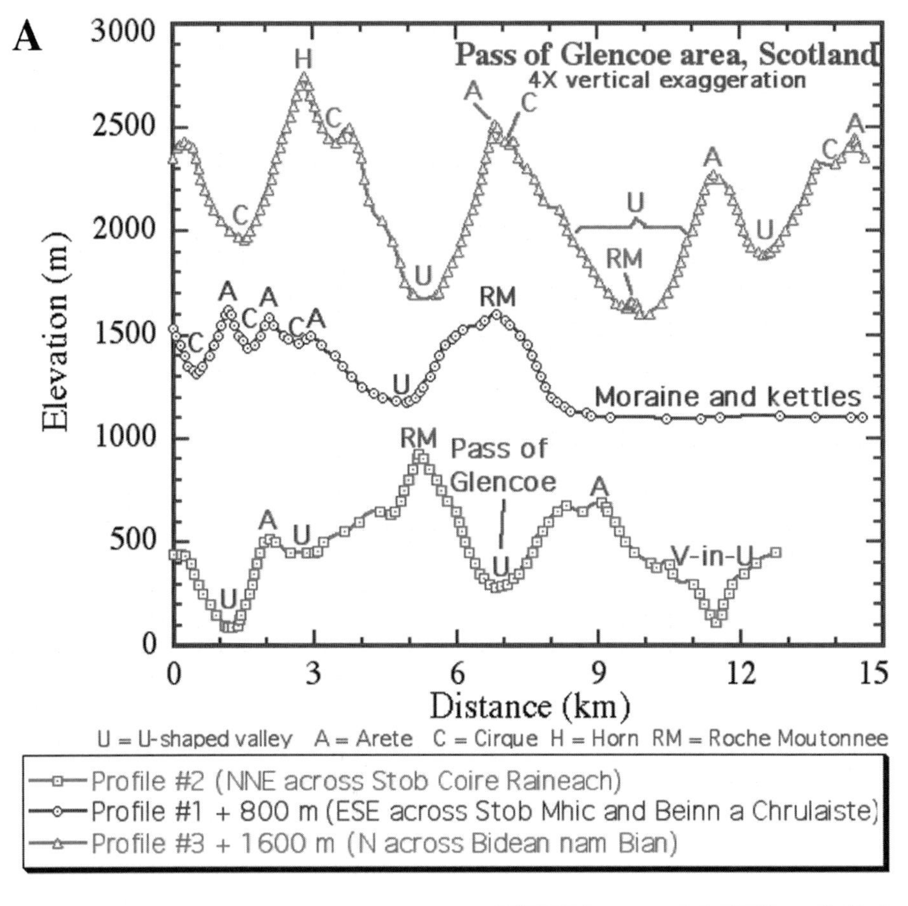

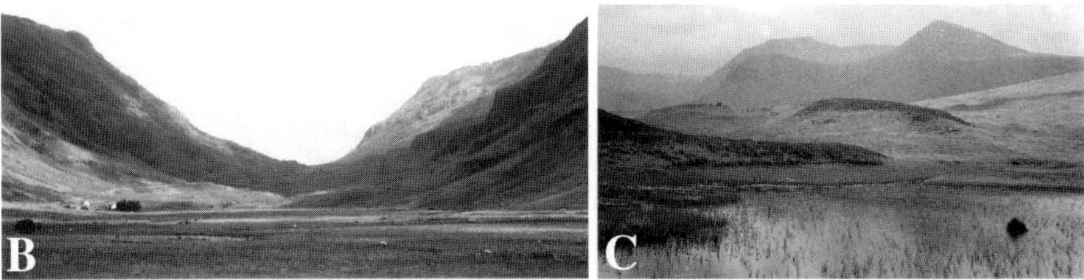

Topographic transects across the glaciated Scottish Highlands show abundant U-shaped valleys (glacial channels) and sharp ridges (aretes) and peaks (horns), no matter where the transect is drawn. Rounded peaks occur where ice completely overtopped the mountains. (B) Pass of Glencoe. (C) Kettles, moraines, and eskers in the lowlands; glacier-carved highlands loom in the background. The slopes and shapes of mountains and valleys in the Scottish Highlands are typical of (but the scale a factor of five smaller than) those in the Charitum Montes. The large-scale structure in map view of valleys is different, being controlled by pre-glacial fluvial drainage and strike-slip faults in Scotland and by craters in Argyre.

Stereo pair of the spectacular Grinnell Valley, Glacier National Park, viewed from the glacier, looking down valley. The natural lakes are due to erosional overdeepenings, a common attribute. The valley has an ideal parabolic cross-section, or one parabola inset into a slightly larger one.

explanations for individual events. Our purpose is to understand geologic histories rough-hewn by overlapping processes.

Physicists' common dissatisfaction with geology (and geologists) is due to a usual lack of the perfect beauty of geology's complex systems; the physicists would much prefer unphysical oversimplified systems, "spherical cows," as planetary scientist Torrence V. Johnson calls these simplifications. In reality, the simplicity and symmetry relished by physicists is always there, but simplicity hides like Orion in a star-spangled sky; sometimes we see it, and other times it may evade us. It is the composite of multiple hierarchical processes and multiple overlapping events, and our limited ability to track their cumulative outcome, that creates an impression of chaos in many systems, such as the seeming anarchy of breccias and of cratered landscapes. It is refreshing when a system appears so isolated and perfect that we can apply ideal theory to describe it, and when alternative explanations are few or absent. When a multitude of interactive objects work as a system, an elegance of symmetry born in fundamental physics of simple systems can reappear, e.g., in the grace of spiral galaxies. This re-emergence of simplicity from chaos happens in geologic systems, too. Despite prevalent complexity, geological nature abounds with geometrically simple, beautiful forms, such as the sinusoids of alluvial river channels and folded strata and the triaxial ellipsoids of beach-rounded pebbles and planets. Much of the beauty of these structures, and the scientific diagnostic quality of their beauty, is in their adherence to symmetry – a theme recurrent in science.

Peter Atkins, in *Galileo's Finger* (2003, Oxford University Press), described symmetry and its mathematical nature and physical meaning as a chief basis of the quantification of beauty; he considered this evolutionary advance in understanding to be among the 10 greatest ideas of science. The roots of our understanding of symmetry go back further, probably as long as sensory systems could sniff, feel, or see symmetry in the opposite sex; surely this appreciation was well advanced when Mark Antony set his eyes on the symmetric form of Cleopatra. While Atkins' application was mainly in cosmology, most unitary acts of nature carry with them some aspect of symmetry in the processes and their products, e.g., the ripples on a pond from a raindrop impact and the geomagnetic dipole generated by core convection.

Symmetry is produced by surface flow, impact, diapirism, diffusion, melting, crystallization, folding, faulting, and gravitational plastic deformation. Symmetry commonly is handed down from one process to another, and this is certainly the case in glacial flow and erosion. The graceful curving medial moraines and the bilaterally symmetric patterns of flow lines and surface flow speeds of glaciers are reflected also in the topographic forms of rocks etched by living ice (or formerly by long-gone ice) at the beds of glaciers.

Consider the glaciers that form U channels, such as the Lake Saint Mary valley, Glacier National Park. It is a marvelous case of order arising from the superposition and chaos of individual interactions of water molecules, falling snowflakes, snowstorms, avalanches, and heat waves. Glacier mass balance is set by the cumulative impact of these events. The accumulated ice flows down valley in an organized, sinuous mass movement. The glacier seeks the most expeditious route to maximize throughput of ice until each annealed snowflake finally finds its moment to melt. In one meter creeping past another, the mass of glacier ice plucks and drags billions of boulders and sand grains along its bed, scarring the rocks beneath in irregular jagged gouges, scratches, and chattermarks.

Despite all the complex inputs to natural glacial erosional systems – the rock structure, pre-existing mountain morphology, the chaos of meteorology, the physics of ice deformation, and so on – what appears throughout glaciology are functions of shape and process rates defined by the parabola or similar simple functions. There are few examples in geology that are simultaneously so visually bold and mathematically simple as the perfect grace of glaciers and the glacial valleys and basins they carve. Recognition of the geomorphic significance of the U channel dates back at least as far as W.J. McGee (1894), who wrote, with contributions from J.E. Hendricks, that "*Glacial cañons are ... U shaped rather than V shaped in cross section.*" McGee noted several other properties of glacial channels, and found that together these characteristics are diagnostic of glacial erosion. No other known geologic processes on Earth generates the same type of landform at that scale.

This view has been echoed through the generations. J.M. Harbor (1992), in a numerical simulation of the erosion of U channels, wrote, "*The U-shaped valley is one of the most well-known products of alpine glaciation.*"

K.R. MacGregor (2000), in a numerical simulation of the longitudinal forms of U channels, reiterated this essential point, "*Glaciated valleys are easily identified by U-shaped cross-sections... that contrast strongly with fluvial valley forms.*"

Researchers have shown that many U channels are, in fact parabolic or a simple composite of paraboloids. The shapes of U channels derive fairly directly from the mathematical functionality of glacial flow fields and time history of glacier thickness. While exceptions also abound, depending on the details of the glacier, parabolic longitudinal and transverse surface topographic profiles and surface and internal flow rates are the rule for glaciers. The dynamical aspects of accumulation, flow, and ablation are normally manifested in parabolic profiles of cross-channel glacial erosion rates. It is perhaps no surprise that the bed relief etched out by glaciers is also represented commonly by a parabola or a simple derivation from the parabola. At the scale of glacial channels of cirque basins, the glacial parabolic form is not mimicked by other erosional processes.

The Charitum Montes of Mars bear classic examples of U-shaped channels and cirque basins. While some are imperfectly rounded at their bottoms, others possess cross profiles having no geomorphologically significant departure from perfect parabolic forms. If these were on Earth, valley shapes in the Charitum Montes would be enough, by themselves, to suggest glacial processes. U channels, of course, are not unique to erosion by ice; erosion by water and lava can create similar mathematically beautiful forms because they involve fundamentally similar flow processes, but their channels occur on different scales.

The assemblage of U channels and cirque basins in the Charitum Montes is indeed as spectacular as any on Earth. But seeing that glaciation on Earth never produces one glacial landform in isolation, a critical geomorphologist would wish to see other types of evidence for glaciation located nearby, for example, lowland glacial deposits. Observing that other features are present that can be plausibly attributed to glaciation and are ordered in a logical spatial arrangement, the glacial hypothesis provides a robust explanation. Indeed, eskers, scour, outwash plains, and other probable glacial landforms were identified before the U channels – it is improbable that any process other than glaciation could have formed this landscape, though much room remains for specific interpretations. Some legitimate remaining questions regard other important issues, such as how Argyre's glaciers were similar to or different from typical terrestrial alpine glaciers and ice sheets, whether the glaciers were wet based or cold based, how long they existed, how widespread they were, what caused them to recede or disappear, and whether icy remnants are still in existence. Are the thermokarstic terrains, lobate debris aprons and lineated valleys flows, for instance, some of these remnants? These incompletely resolved or unanswered questions of geology, climate, and process pertain to almost everything we might want to know about these landforms and the environments in which they were produced.

The similarities are profound among the glacier-sculptured mountains of Argyre and those of classic glacial alpine landscapes on Earth, such as the Scottish highlands and Glacier National Park, but the differences are also evident and very informative. Key among these differences are scale and organization. The Charitum Montes exhibit cirques a factor of several larger than those common on Earth, and the organization of cirques and valleys is not like the dendritic structure prevalent on Earth. These differences are dealt with elsewhere in this book. For now, the take-home message is that the more we have learned of some landscapes on Mars, the more it appears unavoidable that they indeed have glacial origins.

The cameras of the *Viking Orbiters* and *Mars Global Surveyor* have produced a wealth of image evidence of bedrock sculpture in the Charitum Montes on all resolved scales. Spiny ridges and deep furrows marking the sides of mountains indicate an erosive medium that responded to gravity. High alpine amphitheaters and sharp ridge crests indicate an erosional process that preferentially worked on high altitudes, or steep slopes. The *Viking Orbiters* indicated features that lent impressions of having formed glacially, but it was difficult to use those images and do much quantitative beyond what is possible in two-dimensional map analysis. A breakthrough was presented by the *Mars Global Surveyor* mission, which provided not only improved image resolution, but also accurate topography of landforms, thus enabling quantitative assessments of their shapes.

SPECIAL THINGS I SEE IN U

Nonglaciologists and nongeologists may tend to think of solid masses of ice as a hard and brittle substance. Most of them probably also know about the flow-like nature of glaciers. In many regards, ice behaves similarly to water. Glaciers slip and slide downhill and can erode a coalescing network of channels. Full-blown alpine valley glaciation produces dendritic drainage systems or etches out pre-existing fluvial drainages. Glaciers on Earth commonly exhibit supraglacial, subglacial, and proglacial stream flow, and these leave their fluvial marks on landscapes produced by glaciers. However, this is roughly where the analogy between glaciers and streams ends.

Kilo for kilo, subglacial meltwater tends to be more effective at erosion and sediment transport than ordinary stream water, because in wet-based glaciers the water is usually hydraulically highly pressurized. More fundamentally, and more generally for all types of glaciers (with important exceptions), it is the dominant solid phase state of glaciers that is so significant. For being the same substance, the difference in molecular bonding between solid and liquid H_2O makes these phases very different as geomorphic agents. The rheology of ice is different from water in three ways, and these exert profound differences in the landforms that streams and glaciers generate. First, the effective viscosity of warm ice (close to its melting point and at shear stresses similar to those of glaciers) is typically about 17 orders of magnitude greater than that of liquid water; second, the viscosity of ice is far more sensitive to temperature than the viscosity of liquid water; third, liquid water is Newtonian (meaning that its viscosity is independent of the applied hydraulic stress), whereas ice is a strongly non-Newtonian power-law substance under most conditions (with effective viscosity strongly dependent on applied differential or shear stress).

Before I go on, it will be useful to remind readers that there is a difference between valleys and channels. A channel is the longitudinal depression produced along the "wetted" perimeter of the flowing condensed fluid, whether that fluid is water, ice, mud, or lava. A valley is minimally identical to the channel if no other processes that may widen the channel are active; however, with mass wasting, slumping, and runoff from the channel walls, or other processes, the valley may widen over time, such that the channel occupies a small part of the valley bottom. The channel also may incise vertically into the bottom of the valley, which continues to widen over time. Just for example, a stream may be 100 m wide and the water 2 m deep in the middle of the channel, but its valley may be 20 km across and 200 m deep from rim to channel.

A useful equation approximately describing the laminar flow of condensed fluids down an inclined plane is:

$$u = (\rho g \sin \alpha/2\mu)(h^2 - y^2), \quad (6.1)$$

where ρ is the density of flow, α is the slope, μ is the effective viscosity, h is the flow thickness, and y is the depth within the flow. This is an approximate formula because it assumes Newtonian viscosity – an acceptable approximation for a glacier if the shear stress remains roughly constant across its bed, as it commonly does; a student of glaciology will recognize other reasons why equation (6.1) is merely a rough approximation; the glass-is-half-full characterization is that equation (6.1) roughly works. A typical terrestrial valley glacier, 200 m thick at its centerline, much wider than it is thick, flowing down a slope of $\sin \alpha = 0.1$, with density 900 kg/m^3, would exert a basal shear stress of 0.18 MPa. If this glacier has a temperature of 270 K, its basal shear, according to lab data of William Durham (Lawrence Livermore National Laboratory), is $\sim 2.2 \times 10^{-9}$ s^{-1}. The effective viscosity, $\mu_{\text{eff}} = \sigma/\varepsilon'$, at the base of the glacier would be 8×10^{13} Pa-s (where σ is the shear stress and ε' is the strain rate). To a rough order of magnitude, the calculated surface flow speed along the centerline, according to equation (6.1) above, would be 2.2×10^{-7} m/s, 6.9 m/yr. This is tolerably close as a back-of-the-envelope approximation, but large valley glaciers normally flow an order of magnitude more rapidly than calculated. In fact, real glacier ice contains impurities dissolved in intergranular liquid water films or in very small amounts in the ice itself, and these soluble impurities make glacier ice softer than the very pure ice that laboratory experimentalists normally analyze. Wet-based glaciers, furthermore, normally have a component of basal sliding. Hence, a large valley glacier normally moves tens of meters per year.

Equation (6.1) roughly works for any laminar channel flow. It happens to work for glaciers and lava flows, too; departures of reality from the back of the envelope are small (speaking in order of magnitude language) and are reasonably well understood; this rough agreement is possible because glaciers and lava flows are laminar. For parametric values appropriate for a typical lazy river (depth 5 m, density 1000 kg/m^3, viscosity 10^{-3} Pa-s, slope $\sin \alpha = 10^{-4}$), equation (6.1) gives u = 12,250 m/s – which is above escape speed and is four orders of magnitude too fast for reason! River are not long enough to accelerate to such a speed, but with the physics embodied in equation (6.1), the Mississippi River would accelerate continuously from its source in Minnesota to 80 m/s in New Orleans, a couple orders of magnitude too fast. The physics behind equation (6.1) is obviously inapplicable to rivers. The failure of equation (6.1) to work with rivers is because all of them, even the laziest ones, are turbulent; they lose the energy constantly added from gravitational acceleration by turbulent frictional dissipation in eddies and along the wetted channel floor. In the turbulent dissipative process, they erode rocks and move sediment in the distinctive style of streams. Glaciers share aspects of this same style, but only if they include subglacial stream activity as part of their dynamics, and even then the high hydraulic pressurization of subglacial streams causes extremely turbulent or jetting flows.

However, all streams – even the laziest ones – flow far more rapidly than glaciers; a typical difference in speed is a factor of 500,000. We can imagine the common circumstance where a glacier directly feeds a river. The annual mass flux of flowing ice at some point along the glacier (not too far from the equilibrium line, which is where local annual accumulation equals annual ablation) is nearly identical to the annual mass flux of water just downstream from the snout of the glacier, if the glacier is close to hydrological balance. Since the stream is flowing 500,000 times more rapidly than the glacier, mass balance requires that the cross-sectional area of the river is 500,000 times smaller, averaged over the year, than the cross-sectional area of the glacier. The seasonality and diurnal cycle of glacier stream flow (where it may flow near peak discharge just 10% of the time) may mean that the glacier stream at peak flow has a cross-sectional area 50,000 times smaller than the glacier. A glacier 2 km across may feed a stream that has an active channel just 10 m across and less than 1/200 as deep. This explains another key distinction between glacier channels and stream channels; streams channels are narrow and occupy the very bottoms of much broader valleys, whereas glaciers are wide and in glaciated terrains are responsible for carving the broad valleys in which streams may flow after the glacier retreats.

Both streams and glaciers are normally sedimentologically or erosionally active over their whole channel surface. For a unit length in the channel's longitudinal direction, active glaciated segments erode bedrock or deposit sediment (as the local glacier dynamics dictate) across much greater areas than the rivers they feed. This brings us to the special issue of the U valley. Stream valleys flowing without the inheritance of glacially sculpted terrain almost invariably have the classical V-shaped valley cross section known to geomorphologists. Glacier-eroded channels and stream valleys still reflecting recent glacial erosion are almost invariably U shaped (in cross section) across the whole extent of contact (or recent contact) with glacier ice. Erosive stream channels are commonly also U-shaped, but not their valleys, unless they inherit a young glacial channel. Stream channels in fact may have much more complex cross-sections due to the vagaries of turbulence and effects of meandering and sediment deposition; glaciers just about always, with their laminar flow, carve exquisite U-shaped channels (also called U-shaped valleys, if their glaciers have disappeared and they are now under a fluvial regime).

The parabola emerges everywhere in the geometry and flow dynamics in idealized analytical solutions of simplified glacier flow geometries. The beauty of the parabolic form of glacier channels arises naturally from the ideal parabolic surface shape profiles and parabolic distribution of speeds across the surfaces of and in vertical transects within glaciers. While the diagnostic quality of U channels was recognized early in the development of glacial geomorphology in the late nineteenth century, the quantitative understanding of this special landform has emerged slowly. Never in doubt since it was first recognized, the marvelous symmetrical beauty and smooth profiles of glacial channels, and the bold expression of glacially sculpted mountains with which U channels commonly associate, tell a story of sculpture by ice. The erosional development of parabolic and other U-shaped glacial channels has been modeled by J.M. Harbor (1992), who found that glacial channels evolving from pre-existing V-shaped fluvial valleys and realistic models of changing glacier ice output can generate forms more complex than a parabola. However, it never stops amazing me to see the common occurrence of ideal parabolic channels, to see the

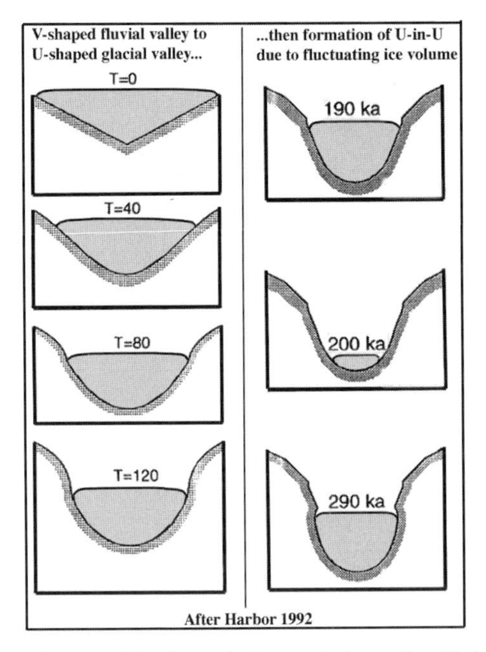

Computations by Harbor (1992) showed the evolution of a V-shaped stream-eroded valley to a characteristic U-shape of glacial valleys. The detailed shape depends on details of glacial history.

mathematically pure physical effect in channel shapes carved by ice that no longer is there. When we see the big U on Earth, it's saying, "*a glacier was here*."

Topographic cross-sections taken oblique to the down-valley direction in glaciated mountains typically will cross tributary valleys and cirques and high mountain peaks. The topographic profiles along any random transect are almost diagnostic of glaciated terrains. These profiles have a beautiful ascending and descending series of scalloped concave-up forms, each concave-up curve separated from the next by a sharp pinnacle, ridge, or other abrupt change in slope. The famous glacial "hanging valleys," formed where tributary glaciers once poured forth into trunk valley glaciers, and now where waterfalls grace the mountains, show up in random transects over glacial mountains. Fluvially dissected mountains, by contrast, reveal ascending and descending series of V's and other angular forms lacking concave-up relief elements. It is a simple diagnostic, and one that is instantly recognized by geomorphologists as separating glaciated from nonglaciated mountains. Unless there is a complex overprinting of glacial and fluvial erosion, it takes a mere glance to make a diagnosis of process.

The *MGS* laser altimeter has produced data indicating the general U-shaped cross profiles and ascending/descending series of concave-up forms, separated by peaks and abrupt slope breaks throughout the Charitum Montes. Set between sharply crested, scalloped peaks and high cirque-like amphitheaters of the Charitum Montes, some U channels have nearly perfectly parabolic profiles. There is little doubt that the Charitum Montes have been deeply eroded by glaciers. The imprint is there and obvious at a glance. It remains to be determined whether the alpine glacial erosional features imply wet-based glaciers, or whether cold-based glaciers can achieve the same result over longer periods of slow erosion.

NO JOHN F. KENNEDY!

There is glaciation as we know and love it here on Earth, and then there is glaciation the Martian way. The two are similar in many respects, but are dissimilar in as many ways. To exclaim the similarities and ignore the differences would do an injustice to science.

During a memorable 1988 debate between two Vice Presidential candidates, the youthful Senator Dan Quayle took to the floor against the opposing candidate, former Senator Lloyd Bentsen. Mr Quayle evoked the memory of President John F. Kennedy, cloaking himself in the honor and eloquence of that esteemed leader in a tried and proven political tactic, which nevertheless did not work for Quayle any more than his hosting of a child's spelling bee did. When his turn came, the more experienced Lloyd Bentsen shot back, with no minor indignation at Quayle's cloak awkwardly worn, *"Senator, I knew John F. Kennedy, and you are* **no** *John F. Kennedy!"* Though the harshly witty but honest response did not carry the election, it remains in the lore of American politics. We might imagine a fanciful debate between Venus and Mars, both looking up to the qualities of planet Earth. Venus adores her lifelong friend and neighbor, Earth, but Venus accepts and humbly announces their profound differences, leaving the equally profound similarities to be self-evident and to be exclaimed by others. Mars aspires to the greatness of Earth, but recognizes neither the preposterous indulgence of comparison of traits that are not quite comparable nor its shortcomings that are evident to all but Mars.

If we think of Martian alpine glacial sculpture much as I have written thus far, the preposterousness of an incomplete comparison of the Charitum Montes to Earth's alpine glacial terrains would go unchallenged and unanalyzed by me. I should confess that the differences are as glaring and as numerous as the similarities. Be assured, the differences would be considered, as they should be, by those who are qualified to judge. I shall now offer some critical contrasts, first in theory, and then in observation. (In fact, the intellectual process was the reverse.) The differences are as insightful as the similarities.

Alpine glacial erosion on Mars should be different from that on Earth in three ways, at least, and these three ways do not yet consider the possible role of exotic substances, e.g., CO_2:

1. The minimum scale of pre-existing valleys and other topographic "flaws" that are erodable and etchable into distinct glacial forms such as cirques and U channels is larger on Mars, also due to lower gravity and probably lower temperatures and more viscous ice on Mars (thanks to power-law rheology of ice).

2. The plan view of inherited topographic structure, and hence, of prior topographic control of Martian glacial valleys is different from that on Earth. On Earth, glacial valley control is due mainly to prior fluvial dissection, whereas on Mars the structural control is by impact basins and smaller craters.
3. The rate of glacial erosion is lower on Mars, and the timescale required for deep erosion is longer, due to lower gravity and probably lower prevailing temperatures (mainly cold-based glaciers) during the episodes of glaciation.

It should also be included here that cold-based glaciers are restricted in occurrence to Earth's polar regions and certain other areas, but they should be the rule on Mars. This difference has dramatic implications for the nature and rates of bed erosion. The three distinctions above are explained below, with the third one considered in some detail where gravity is concerned.

The rheology of ice is non-Newtonian, such that it has a high apparent viscosity at low shear stresses and a sharply lower apparent viscosity at high shear stresses. Ice deformation is sensitive to about the second, third, or fourth power of shear stress, depending on conditions. It is as though there is a yield stress (product of glacier thickness, density, slope, and gravity), such that small, thin accumulations of ice effectively do not flow and erode efficiently, but slightly thicker and broader masses may be very effective at erosion. The shear stress dependence almost guarantees that thicker, broader accumulations of ice will be needed to achieve flow and erosion, but further guarantees that with a sufficient supply of ice, it will eventually reach a point where flow will become effective. Glaciers will seek particular length scales on which to operate, etching out and moulding pre-existing landforms of that scale and tending to obliterate all those that are smaller. This length scale is larger on Mars by roughly the inverse of surface gravity times an additional factor due to low temperatures. Martian cirques and glacial valleys will be larger than those on Earth. On Earth, large hollows and defects may be etched out initially into a large cirque, for instance, but then flow instabilities will develop in the ice and accentuate erosion in certain spots and reduce it in others: a large glacier-filled cirque will self-divide into smaller, nested cirques. We see evidence of this on Mars where the cirques are, indeed, huge: 3–10 km across, compared to cirques usually 300–1000 m across on Earth. A factor of 2.6 of this order-of-magnitude size difference is explicable by gravity differences; the remaining factor of about 4 is presumably related to colder Martian ice.

The control of glacial erosion by fluvially eroded topography on Earth and by impact craters on Mars will produce a distinctive organization of glacial valleys and cirques on the two planets. On Mars we expect a predominance of piedmont and crater-wall glaciers; large craters will provide the relief that will be etched into cirques, whereas on Earth we expect dendritic systems of valley glaciers. A common glacier erosional form on Mars appears to be a succession of craters, side by side, which have been breached on one side, with ice spilling from one crater to another. On Earth, ice flows down one inherited fluvial valley and empties into another, with the dendritic structure of glacial U channels preserved and the valleys enhanced down to a minimum scale where glaciers can flow as separate entities. At smaller scales, fluvial channel structure is destroyed.

Glacial erosion rates on Mars, and particularly the effect of lower gravity there, can be assessed by adaptation of the erosion model of J.M. Harbor (1992). Harbor started with a simple erosion rule, where the erosion rate, E', is related to basal sliding speed, U_s, by:

$$E' = C_2 \times U_s \tag{6.2}$$

C_2 is an erosion coefficient = 0.0001, according to Harbor (1992). However, it cannot be constant. C_2 must be positively related to normal load (and also rock characteristics), hence, to surface gravity and ice thickness. I assume that C_2 is directly proportional to normal force, hence, that C_2 has a linear dependence on gravitational acceleration, g. The sliding speed is given by:

$$U_s = C_1 \times \tau_b{}^2/N_e \tag{6.3}$$

where C_1 is a sliding coefficient = 0.0012 m/Pa per yr.

$$\tau_b = \text{basal shear stress} = \rho_i g H_i \sin(\theta) \tag{6.4}$$

where ρ_i and H_i are the density and thickness of ice. The effective normal stress, N_e, is given by the difference of the weight of ice on the bed and the basal hydraulic pressure:

$$N_e = \rho_i g H_i - \rho_w g H_w \tag{6.5}$$

where ρ_w and H_w are the density and effective column height of water in glacial conduits extending down to the glacier's bed. It follows that, for low hydraulic water pressures, the erosion rate exhibits a simple proportionality to gravity:

$$E' \propto g^2 \tag{6.6}$$

Thus, on Mars, all other things besides gravity (38% of that on Earth) being equal, glacial erosion rates will be 14% as rapid as on Earth. However, all other things are not equal; e.g., ice thickness for a given throughput of ice and a given ice rheology will be greater on Mars due to lower gravity; and cold-based ice will not slide at all or erode in the same fashion. There is a hint here, however, that lower gravity on Mars means lower erosion rates. This suggestion implies longer durations of glacial erosion on Mars compared to Earth for a given level of glacial valley development. For instance, large glacial U channels can be eroded on Earth in 10^5–10^6 years, whereas on Mars it might require 10^6–10^7 years (if the U channels are wet based and the glacier is sliding). For cold-based Martian glaciers, it is evident that much longer periods would be required to form deep abrasional channels.

ROCKY ROADS OF ARGYRE

In northern Minnesota and Ontario – my favorite wilderness canoe country – the classical transportation corridors were lakes and streams, suitable for canoe passage, located in depressions etched out of the bedrock by glaciers or depressions (kettles) caused by decay of glacier ice relics that had been buried in glacial sediment. Overland transportation routes across the same regions of Ontario were often also formed by glaciers – they are the high-standing bedrock and sedimentary ridges, including moraines and eskers. The eskers of this area are stunning for their number, size, complexity, and relations to other types of glacial landforms. They tell important parts of the glacial history of the region, especially during that period late in the life of the Laurentide Ice Sheet when it was in a mad hurry to melt. The sand, pebbles, cobbles, and boulders of these eskers were originally derived from the bed of the ice sheet in the far north of Ontario; these granitoid and greenschist fragments mostly were transported in excess of 1,000 km by the ice, but their final hundred kilometers and final emplacement was by way of a vast system of subglacial meltwater tunnels. These tunnels carried confined debris flows, hyperconcentrated slurries, and stream bedloads of coarse sediment derived by erosive glacial plucking of the Canadian Shield. As the hydraulic pressure declined at the exit points of the subglacial floods or within the tunnels, the sediment was deposited. The ice eventually melted, and the glacial sediment was left as high-standing ridges, suitable for gravel mining and road building. Some of these eskers formed fans that once were deltas built into the vast Glacial Lake Agassiz.

Kargel and Strom (1992b, *Geology*) recognized the likely glacier eskerine or esker-like nature of Argyre's sinuous ridges, and the relations of these ridges to adjoining alpine glacial erosion and plains deposits, including what appears to be a vast lake deposit. The esker interpretation forms one of the key elements of the glacier interpretation of Argyre. Somewhat similar features occur elsewhere on the planet, including parts of the northern plains near Tim Parker's Contact 2. Kargel and a cast of coauthors drawn from the Mars and terrestrial cryospheric communities published in 1995 a broad-ranging paper on "Evidence of ancient continental glaciation in the Martian northern plains" (*Journal of Geophysical Research*). Although this paper concerned primarily a different area of the planet, we included a presentation of key characteristics of the Martian sinuous ridges, also summarized briefly in Chapter 5. That 1995 paper also included a table of predictions for what would be found with MOC and MOLA that was not so readily observable with *Viking Orbiter*. Included was the prediction that MOC would resolve boulders, sedimentary beds, and additional esker morphologies not observed by *Viking Orbiter*, and that MOLA would discern that Martian sinuous ridges sometimes cross low divides and have the characteristic range of esker cross profiles (sharp-crested, rounded, and flat-topped). All of these additional features have been found in the new data.

Perhaps most striking in the MOC images is the abundance of very large boulders (many 10–20 m across) in unmantled sinuous ridges. This discovery, in fact, exceeds my expectations to such an extent that some rethinking is needed. On Earth, glacier-deposited boulders are characteristically up to 3 m across and rarely larger. On Mars, just using inverse linear gravity scaling and a strength-of-rock argument, I would have expected abundant boulders up to about 7 m across. The fact that many boulders are 50–300% larger than predicted suggests that they were transported primarily via laminar debris flows, thus reducing violent boulder-on-boulder clashing, which would have comminuted the boulders. The author has seen similar size boulders in the deposits of catastrophic glacier lake outburst floods, but the transport distances in each of those cases was measured in a few hundred meters or a few kilometers, not hundreds of kilometers. The subglacial debris-flow model is consistent with Ken Tanaka's (and my) perspective on the northern plains "mud-ocean" deposits, which have been reinterpreted as primarily fast-spreading but mostly laminar debris flows more so than as turbulent flood deposits. The Martian sinuous ridges probably were emplaced as tube-filling (and tube-eroding) debris flows – a particular form of esker. Even so, the association of the bouldery sinuous ridges with bouldery plains and smooth plains of southern Argyre Planitia indicates that these esker-forming discharges formed complexes of diverse, but related, sedimentary units. As with the northern plains deposits, the hard work of mapping out and deciphering precise stratigraphic sequences and the full range of transport and depositional processes remains ahead of us.

DIRTY OCEANS: SMOOTH AND FILLING!

The Mars ocean model is the most controversial and potentially most important recent advance in our concept of Mars. Whether or not there ever was an ocean, there certainly are ocean-size sedimentary deposits precisely in the places where the "Mars oceanographers" would have oceans exist. Subsequent to the classic papers of Parker et al. (1989) and Baker et al. (1991), the most stunning achievement came with the MOLA topography-mapping experiment of *MGS*. James Head and others at Brown University reported in 1999 on "Possible ancient oceans on Mars: Evidence from Mars Orbiter Laser Altimeter Data" (*Science*, **286**, 2134–2137). This paper initiated a long succession of other observations, but the basic point reported in that seminal paper remains at the center of the new controversy: the Martian northern plains exhibit vast expanses of very smooth plains whose surfaces occurs at nearly a constant elevation and whose edges exhibit a change of slope and morphologic contact with surrounding, more rugged and higher plains. This contact closely corresponds, with important exceptions, to Tim Parker's "Contact 2," the interior of two hypothesized ocean strandlines and mostly identical with H.-P. Jöns' mud-ocean contact. The notable exceptions occur in areas where long-wave isostatic adjustments (uplifts due to tectonic processes and subsidence due to volcanic and sedimentary loading) of the crust and lithosphere have apparently taken place, presumably subsequent to plains emplacement. In their initial and subsequent analysis of MOLA and MOC data, the team led by Jim Head has backed up their initial conclusion that the new spacecraft data have provided tests for "*a large number of predictions ... consistent with hypotheses calling on large standing bodies of water in the northern lowlands in the past history of Mars....*"

Some of the most interesting new results have come from an analysis by M. Kreslavsky, J. Head and others of root-mean-square slopes and slope spectra: the lowest slopes and smoothest topography on Mars outside of the polar caps – in fact some of the smoothest, flattest terrain known on any planet – occurs in the northern plains interior to "Contact 2." The slopes and smoothness is comparable to that of oceanic abyssal plains on Earth.

One team's overwhelming evidence, as I learned in my work with Vic Baker and others, is another team's point of contention. Head et al. and Baker et al. have never claimed that the case for an ocean is rock-solid, but the case certainly demands some process at least comparable to ocean formation. It might not have been an ocean as beachgoers and surfers know and love on Earth. For one thing, it was probably literally ice-cold from the start, and it probably promptly froze nearly to the seafloor (Kargel et al., 1995). For another point, Ken Tanaka (USGS) and colleagues (including me) have pointed out that some features of the ocean deposits, and especially the morphology of the deposits and the stratigraphic nature of the sedimentary contact, are more consistent with a mud ocean and a massive series of thick, fluid debris flows. The difference in some regards is minor, since nearly everyone except Tim Parker believes that the ocean's liquid existence was transient and that it contained a large amount of sediment and, in any case, contained a huge volume of liquid water. Climatologically and geologically, there is quite a difference; a transient water ocean will tend to sublimate rapidly, with the water being redeposited where it is more stable. A mud ocean would rapidly form a nearly impermeable lag deposit, which could protect underlying ice to this very day. From a geomorphic process standpoint there is also a major difference, because a water ocean that was liquid for long would produce a wide array of ocean shoreline features known from Earth; despite Tim Parker's best efforts, he has not convinced many skeptics. A mud ocean would exhibit different types of contacts, much like those observed at or near Tim Parker's "Contact 2."

It sounds political to say so, but it also makes much sense that a large debris flow would probably deposit several components in rapid succession: a proximal (near-to-source) debris flow, expressed mudflows, segregated water bodies, salt deposits laid down by fractional crystallization and/or evaporation (all together forming a "mud-ocean complex"), and secondary H_2O and CO_2 ice deposits accumulated at cold spots across the planet due to recondensation of sublimated volatiles from the ocean. The midlatitude lobate debris aprons, lineated valley fill, and terrain mantling units, and the polar caps may be these secondary deposits related to the most recent major marine episode. The Mars science community has yet to map these mud-ocean complexes and secondary deposits in full detail, but I am confident that these complexes will be found in logical stratigraphic sequences, that multiple phases of mud-ocean emplacement will be discerned, and that each major mud-ocean complex will be correlated with broad ranging global climatic effects and glaciations.

MOLA data also reveal step-like or terraced platforms at the periphery of the interior lowland plains. The terraces are interpreted by Jim Head's group as due to coastal erosional influences or deposition along the Martian equivalents of continental shelves. This could occur only if there was a free-standing body of water subject to wave and current activity. Ken Tanaka and colleagues (including this author) have highlighted the lobate edge of these interior lowland plains, seen in MOC imagery at a higher resolution than that first observed with *Viking Orbiter* scenes. We have explained that, as H.-P. Jöns first stated, the lobate edges are characteristic of the properties of mud flows, debris flows, or possibly thick turbidity flows. The lobate edges are not consistent with the features of sedimentary blankets deposited by bed traction (e.g., current-driven sand grains bouncing over the bed surface) or pelagic accumulation (by slow settling in deep, quiet water) of fine clastic grains. The lobate edges are certainly not consistent with marine chemical evaporitic deposition, either. Thus, the major identifiable sedimentary units are most consistent with sudden deposition of massive mud flows, as first suggest by H.-P. Jöns in 1986, and as one might expect from the sudden and catastrophic origins of the outflow channels and chaotic terrains. Of course, Earth's oceans tend to fill up with sediment by numerous processes and many sources, and it is likely that the same is true of Martian oceans. Furthermore, some degree of sediment/water segregation would inevitably occur.

So the controversy goes on, but it is clearer than ever that the northern plains received huge volumes of sediment. This sediment originated deep in the highlands and at the edges of the highlands, was transported through the outflow channels and fretted canyons, and was deposited, as far as we can tell, from an aqueous (and possibly icy) medium. Whether it was an immense watery ocean with turbidity flows or a gigantic mud ocean, or involved the slow settling of suspended and chemically precipitated sediment in a quiet pelagic marine environment, or dominant glacial processes, we do not know. It probably involved all of these processes at various times and places. In any case it represents millions of cubic kilometers of eroded rock and (cumulatively) millions or tens of millions of cubic kilometers of water. The details of these deposits are not yet known to us but will provide the essential information regarding the specific depositional processes and climatic conditions represented by this most amazing hydrologically driven erosional/depositional system.

MOLA topographic data also show numerous ring features, each tens of kilometers wide, scattered across the northern plains in areas where ocean deposition is thought to have occurred. Invisible in ordinary imagery, these rings are clearly expressions of impact craters. The Brown University group believes them to be impact craters buried in oceanic sediment. Ken Tanaka (USGS, Flagstaff) believes that they may be craters developed on the surface of sedimentary units, but that eons of day-by-day ice-related creep may have caused their relaxation. In Head's view, the craters are developed on a Noachian ocean floor dating from 4 billion years ago or so, and that more youthful oceanic processes have buried the craters. In Tanaka's view, the sedimentary deposits are themselves extremely ancient, mostly dating back 4 billion years, but have been actively disturbed by permafrost processes in more recent times.

Unlike usual polar caps, but similar to some abyssal plains, the northern plains have vast expanses that are not just smooth but occur with nearly horizontal surfaces, suggesting deposition of a material to a level defined by an equipotential surface. Perhaps the ocean filled up with sediment, or highly fluid mud flows expanded across the northern plains and settled in thick blankets, thus partly filling, smoothing, and flattening the northern plains. The abyssal plains analog would suggest the possibility that immense turbidity flows spread across the former Martian seafloor, depositing blanket after blanket of silt and mud-capped beds of sand and gravel. On Earth, abyssal plains are mostly several thousand meters beneath the surface of the sea and cover tens of millions of square kilometers. On Mars, the smoothest surfaces occur at some of the lowest elevations in the northern plains, where deposits mantle underlying terrain over expanses of nearly 10 million square kilometers.

GRAND-DADDY OCEAN?

Possibly the biggest story that Mars has to tell has barely been read from the Martian record and transcribed into scientific language, although Tim Parker and colleagues are onto the story. Parker's Contact 1 and Contact 2 have nothing on this newly proposed super-ocean. Whether caused by an ocean or something else, there exists a ragged escarpment in both hemispheres of Mars, notably along most of the fretted terrain and almost entirely around Hellas and Argyre, and elsewhere, too. This ragged boundary is indeed a fretted boundary, with complex embayments and fjord-like indentations and island-like mesas. The nature of icy flows in the fretted terrain is described in the next chapter. The MOLA-derived topographic map of Mars, published recently as the *Topographic and Color-Coded Contour Maps of Mars* (USGS, 2003, Geologic Investigations Series I-2782), shows this escarpment consistently at an elevation of about 0 km. The main exception is in the fretted terrain of northern Arabia Terra, where apparently the same topographic escarpment occurs at a lower elevation of –1 km or less.

While 0 km is just a coincidence, since the Martian datum is based arbitrarily, it is an ironic value recalling Earth's sea level. Tim Parker would have this as the boundary of an exceedingly ancient ocean. I have quite a bit of difficulty with that concept, mainly for the simple question of: Where did all the water go? An ocean as deep as that would exceed by an order of magnitude the volume of the already-troublesome "Contact 2" ocean. However, the escarpment is there. It is real. It is more profoundly a global feature than the much-heralded Contact 2, which by now almost everyone agrees is a major, real feature needing to be explained (with different explanations depending on whom you talk to). The exceptional depressed elevation of the fretted terrain boundary in Arabia Terra could be due to gravitationally driven isostatic sagging of Arabia and nearby Chryse Planitia, due to loading of Chryse with deposits from the circum-Chryse outflow channels and chaotic terrain.

Imagine Hellas and Argyre filled to the brim, with Argyre spilling over to the north. Imagine the northern plains overflowing their brim and flooding far onto the highlands. It's an amazing concept. This highest escarpment might not be ocean-caused, but may be related to a groundwater table, sapping, ice extrusion, and other water-related processes (locally involving lakes and rivers at times) that have fretted the terrain to base level. Whatever the cause of this feature, it is one of the most profound and enigmatic global topographic and geologic features on Mars, almost right up there with the lowlands/highlands dichotomy. The future of Mars exploration indeed has some interesting problems to solve!

MARTIAN LAKES' CHRONICLES

There is ample evidence that long-term sedimentary deposition occurred in large lakes within craters, intercrater basins, and canyons. Thick layered sequences abound in many areas of Mars at all latitudes. The characteristics of these deposits in many cases indicate aqueous deposition, and the rhythmic layering suggests control by climatic oscillations – indicating easily several hundred thousand years or longer of oscillating water-mediated deposition.

Whether these represent hypersaline lakes, ice-covered lakes, cyclic ephemeral lakes, or some other special climatically modulated depositional environment is not clear. With little doubt, they represent an important component of the Martian hydrological story. Many have discrete solitary channelized inlets and outlets, and some have multiple valley-network-type inlets and appear to represent interior drainage (not evident outlet channels) where evaporitic salt deposition would probably have occurred. Some are in areas where glaciation may have been a local

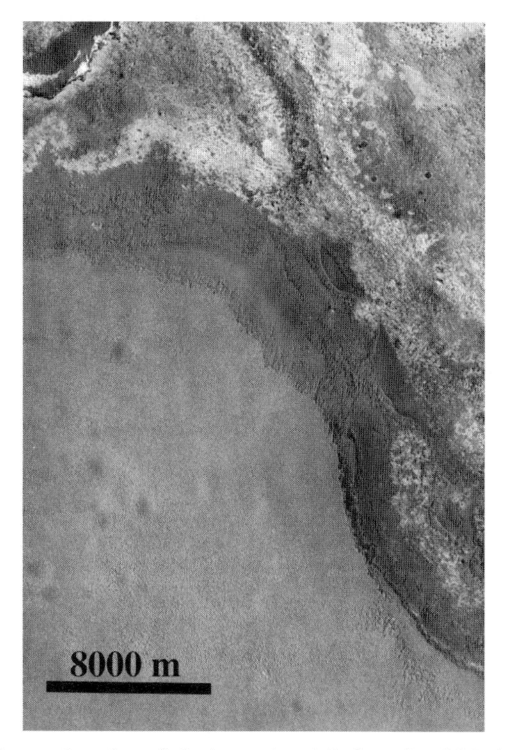

THEMIS image shows the edge of the hematite-rich deposit of Meridiani Planum. The hematite-rich deposit is the smooth material at lower left, which caps a 200-m-thick deposit of eroding sediment. This unit has been stripped away, leaving a more resistant and more rugged plains-forming rock unit at upper right. Modified from THEMIS press release 20031030.

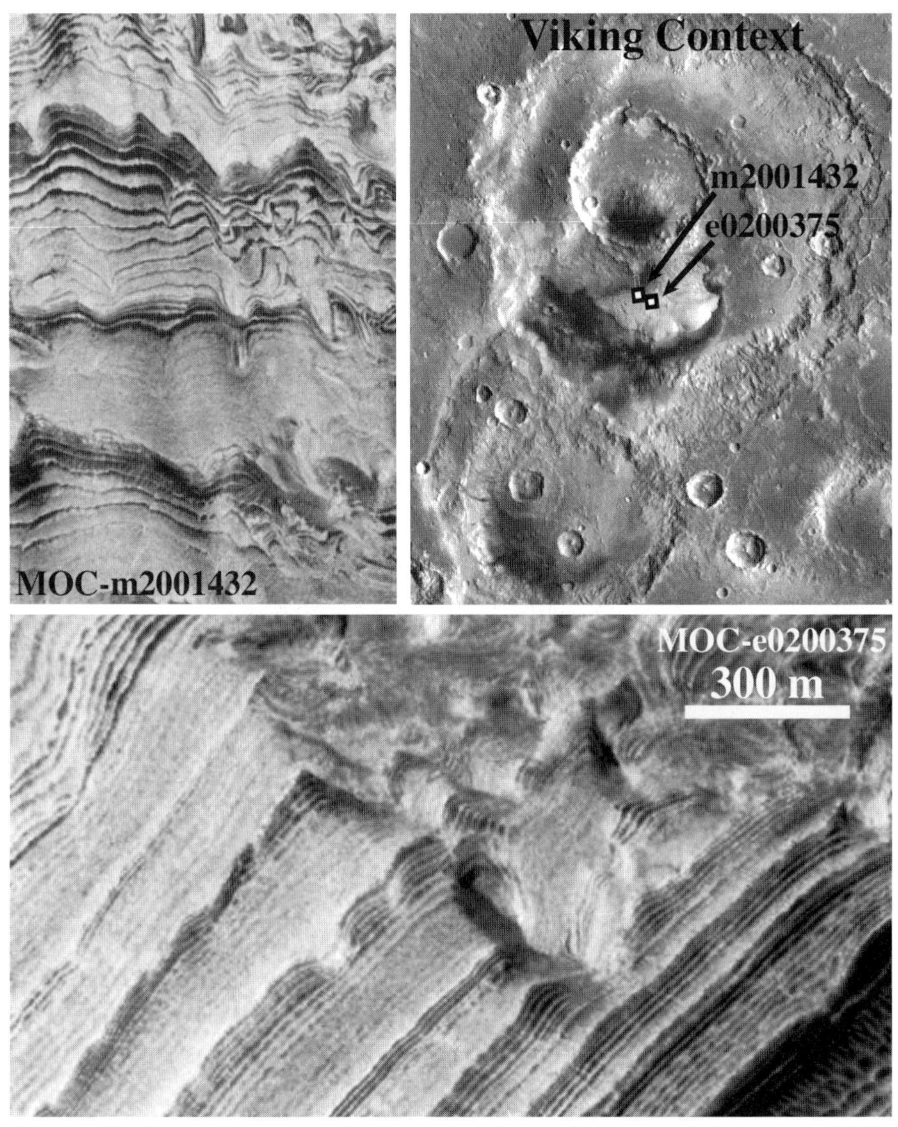

"Packets" of layers form a two-rhythm series of strata in Becquerel Crater, near 22°N lat., 8° long. About nine to eleven light-dark couplets exist per larger bedding unit. Such sedimentary rhythms are commonly achieved on earth through climatic modulation of deposition. Wind-blown dust or aqueous deposition is likely, but not volcanic processes. A rich depositional record is likely. Speculatively, the rhythmicity is similar to that recorded in some sedimentary rocks on Earth due to the 11-year and "double-sunspot" Hale cycle of solar sunspot and extreme ultraviolet activity. Such thick strata as these would require enormous influxes of sedimentary material in a matter of just a couple millennia to explain the entire sequence of strata. More likely, a long-term climatic modulation is involved, such as the 1.2×10^5 and 1.2×10^6 year cycles of obliquity variation.

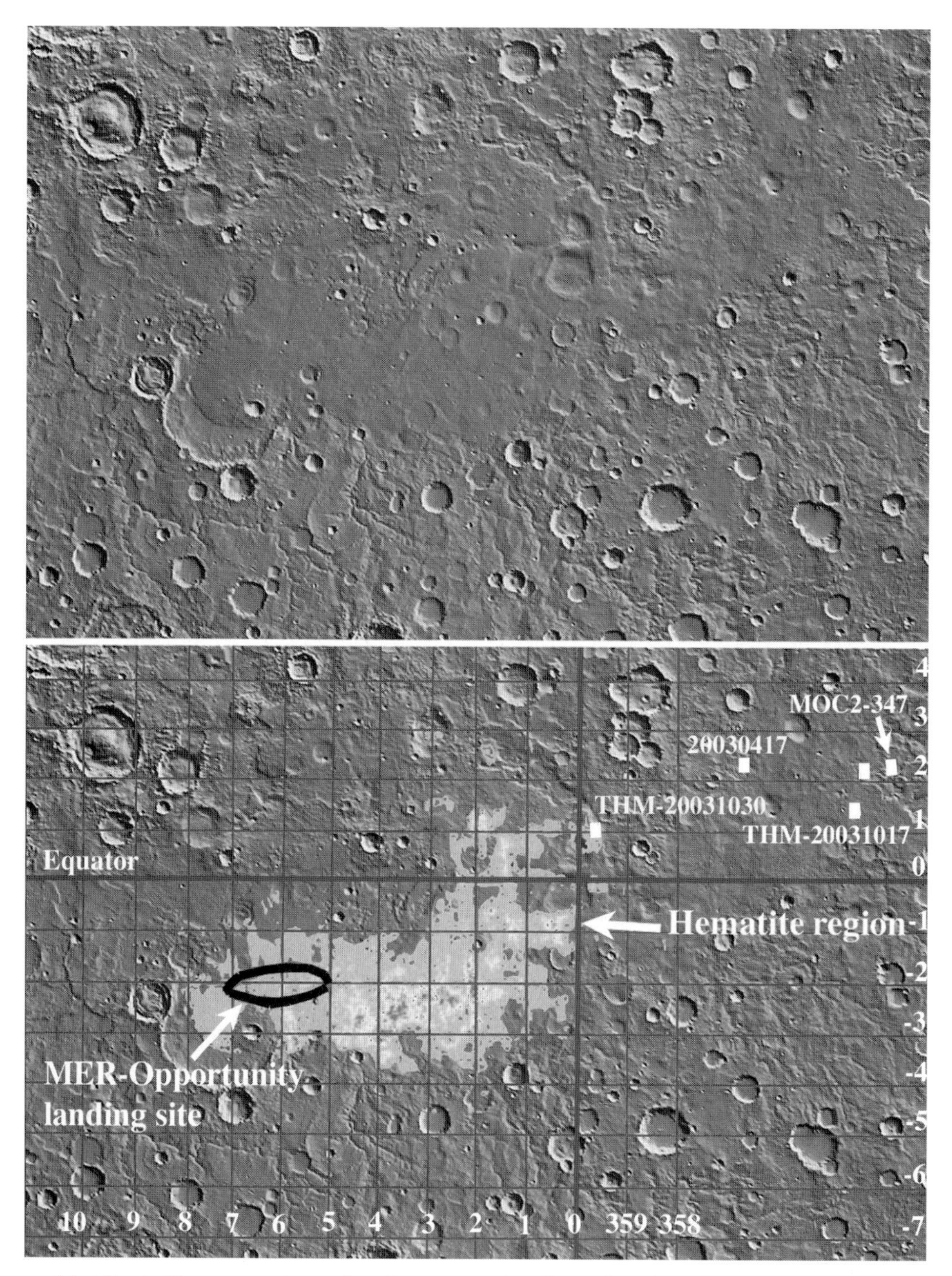

Meridiani Planum region of valley networks, hematite deposits, and the MER *Opportunity* landing site. Top panel is a shaded-relief map prepared from a MOLA digital elevation model; bottom panel shows the same product with superposed mapping of hematite from the TES team. See also Hynek et al. (2002).

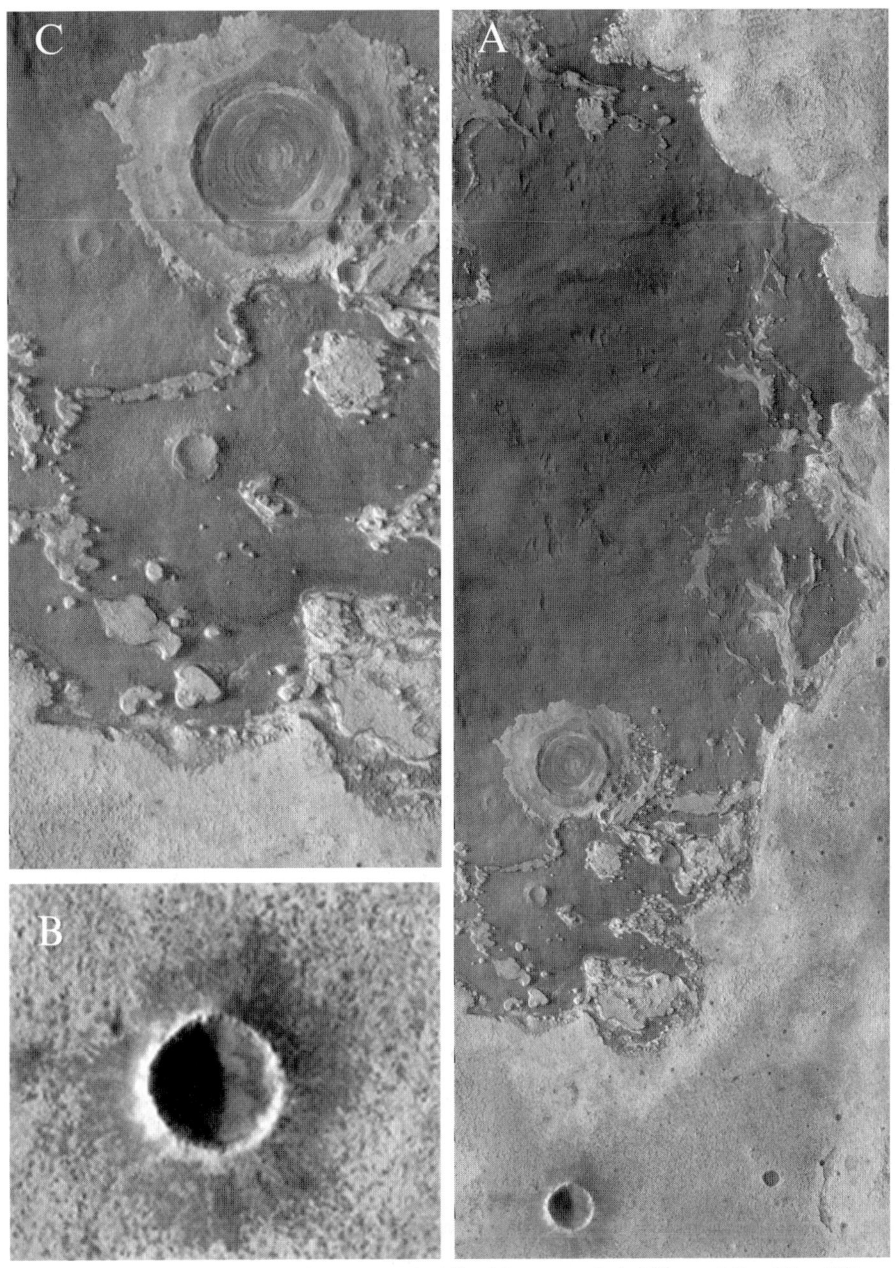

Layered sedimentary strata in Terra Meridiani have eroded differentially. The diffuse, radial ejecta pattern (lack of "muddy" texture) of the crater (panel B) indicates a lack of ground ice and groundwater at this time in this region. The crater near the top of panel C existed before these deposits were laid down; it filled with sediment layers, which now are being eroded. NASA's rover *Opportunity* will land in this region, though not at this specific site. THEMIS press-released image 20030417. 2.3°N lat., 356.5°W long.

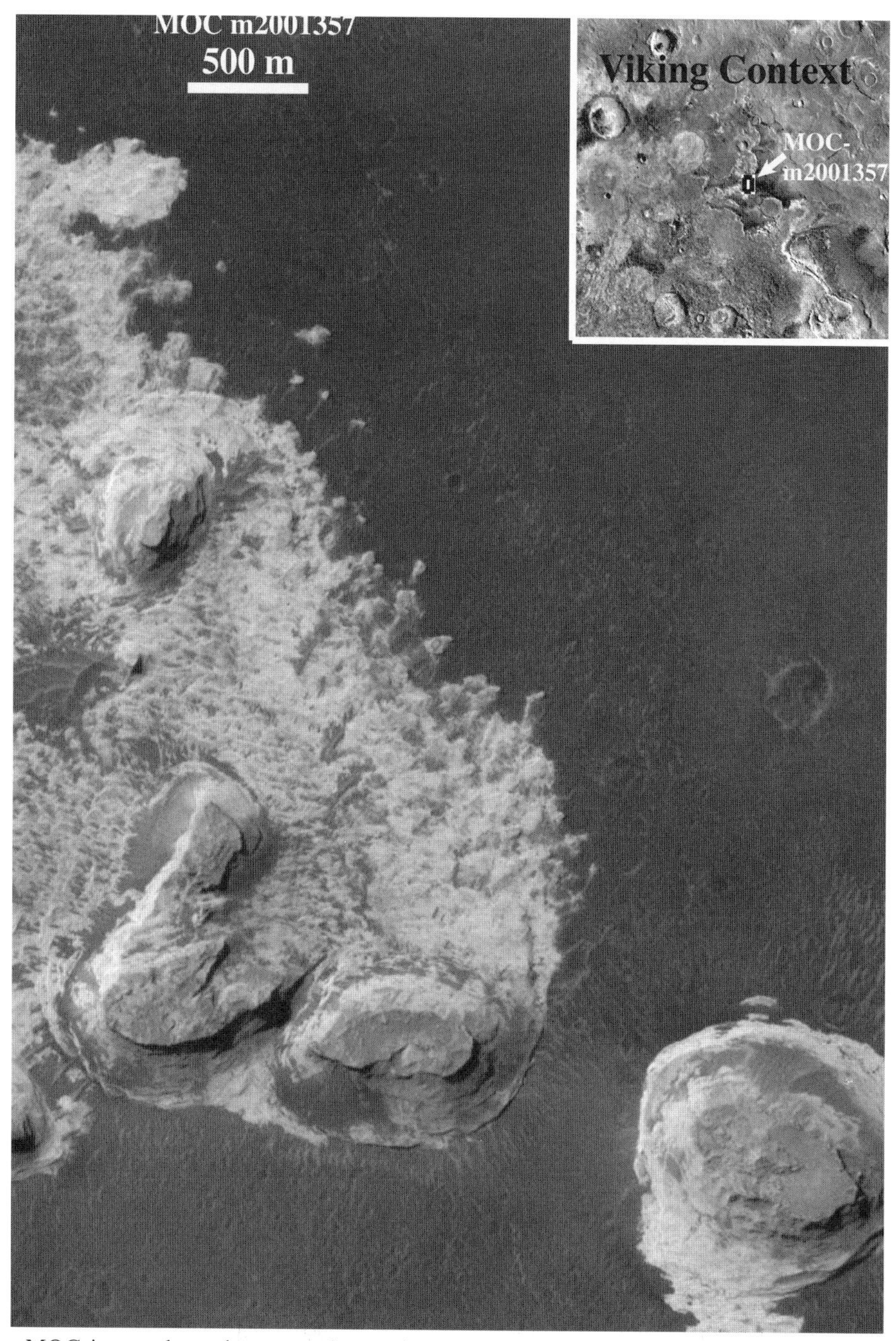

MOC image shows buttes at the northeastern edge of Terra Meridiani, a highlands terrain of eroded, light-toned sedimentary rocks. This landscape resembles Arizona's redrock country. 2°N lat., 354° long.

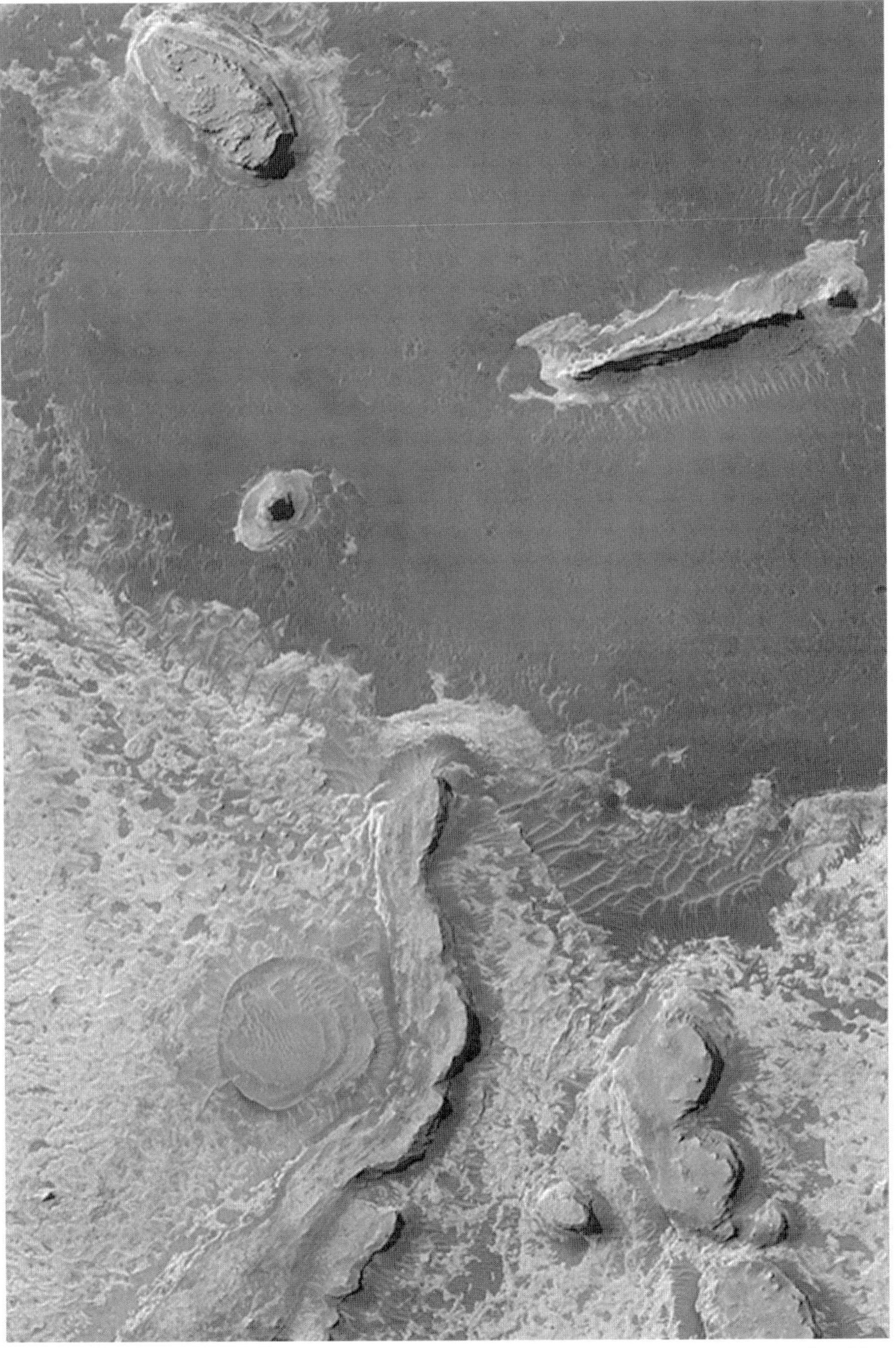

Mesas and buttes in Terra Meridiani are probably eroded sedimentary rocks. The depositional environment and type of sediment are unknown. 2.3°N lat., 353.6°W long. MOC Release 2-347, NASA/JPL/MSSS.

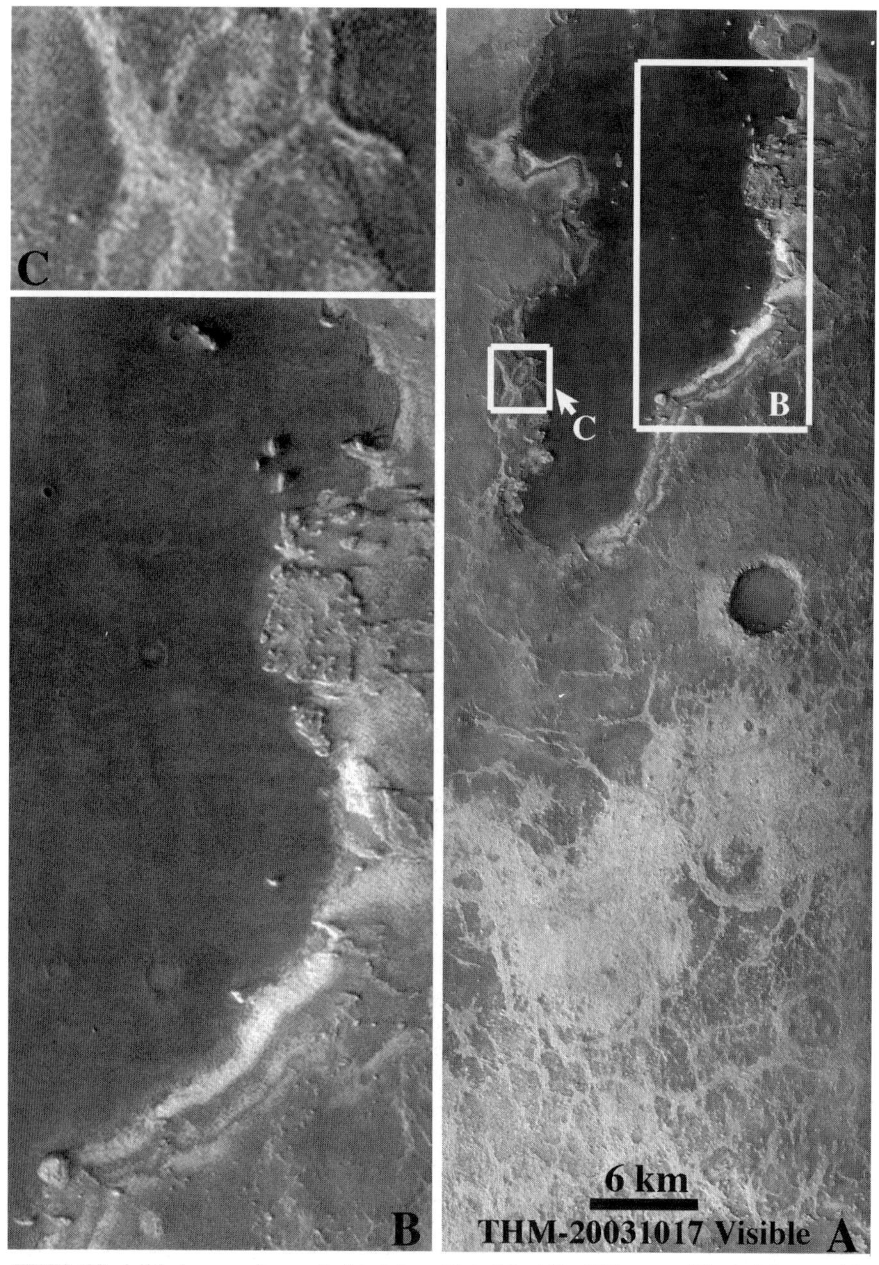

THEMIS visible image of a probable lake within Meridiani Planum. The bright speckles on the dark lake-like surface could be either icebergs (now frozen in) or small islands protruding through the frozen lake water. The lake margins show signs of erosion by either waves or melting of ice-rich shores. Panel C shows polygons that have been eroded by the lake. Light and dark materials along the lakeshore could be eroded layers or might be high-water lakeshore deposits (salts, sand, or mud). 1.5°N lat., 5.6° E long.

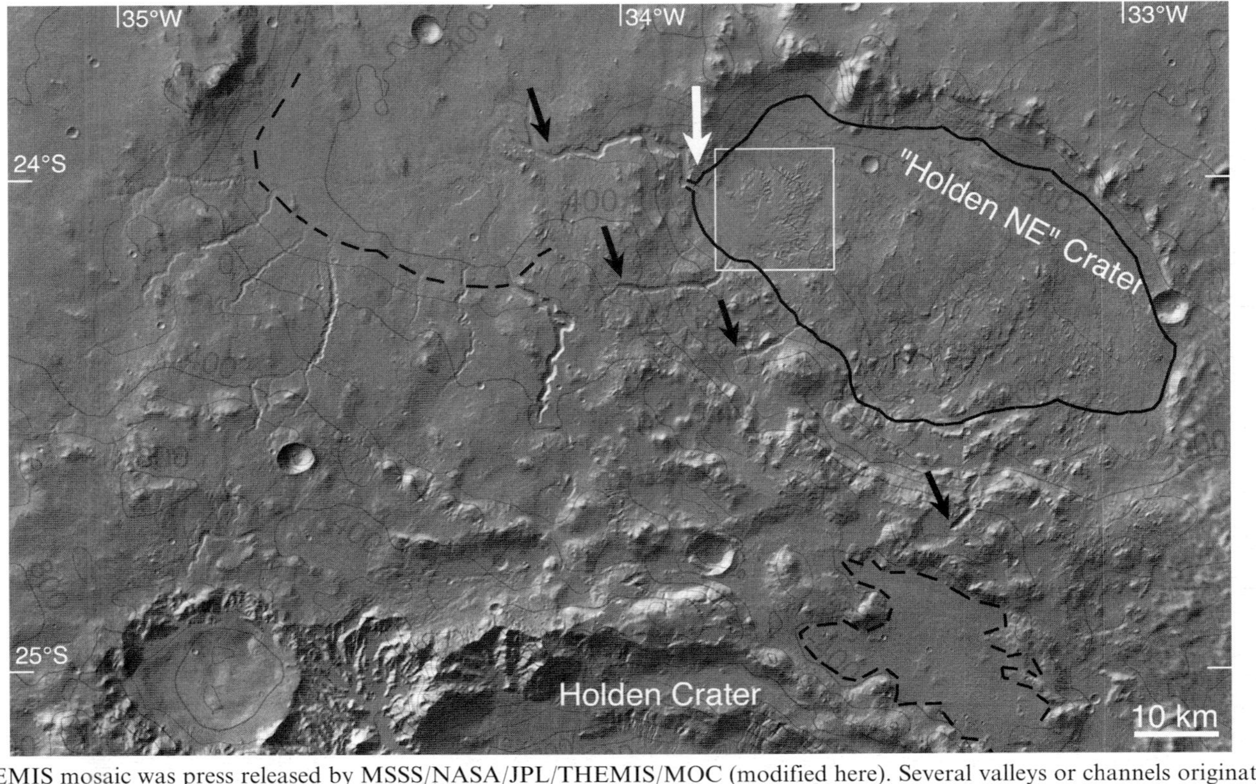

A THEMIS mosaic was press released by MSSS/NASA/JPL/THEMIS/MOC (modified here). Several valleys or channels originate in hilly terrain north of Holden Crater, coalesce, and terminate in a delta (white arrow) in a highlands basin ("Holden NE Crater"). Several other valleys/channels (black arrows) also terminate in this basin. The black-highlighted elevation contour ($-1,200$ m) forms a closed basin and marks a transition from stream erosion to deposition. An approach toward equilibrium in sediment supply, stream transport, and lake deposition is indicated. A black dashed contour (-280 m) shows a similar transition.

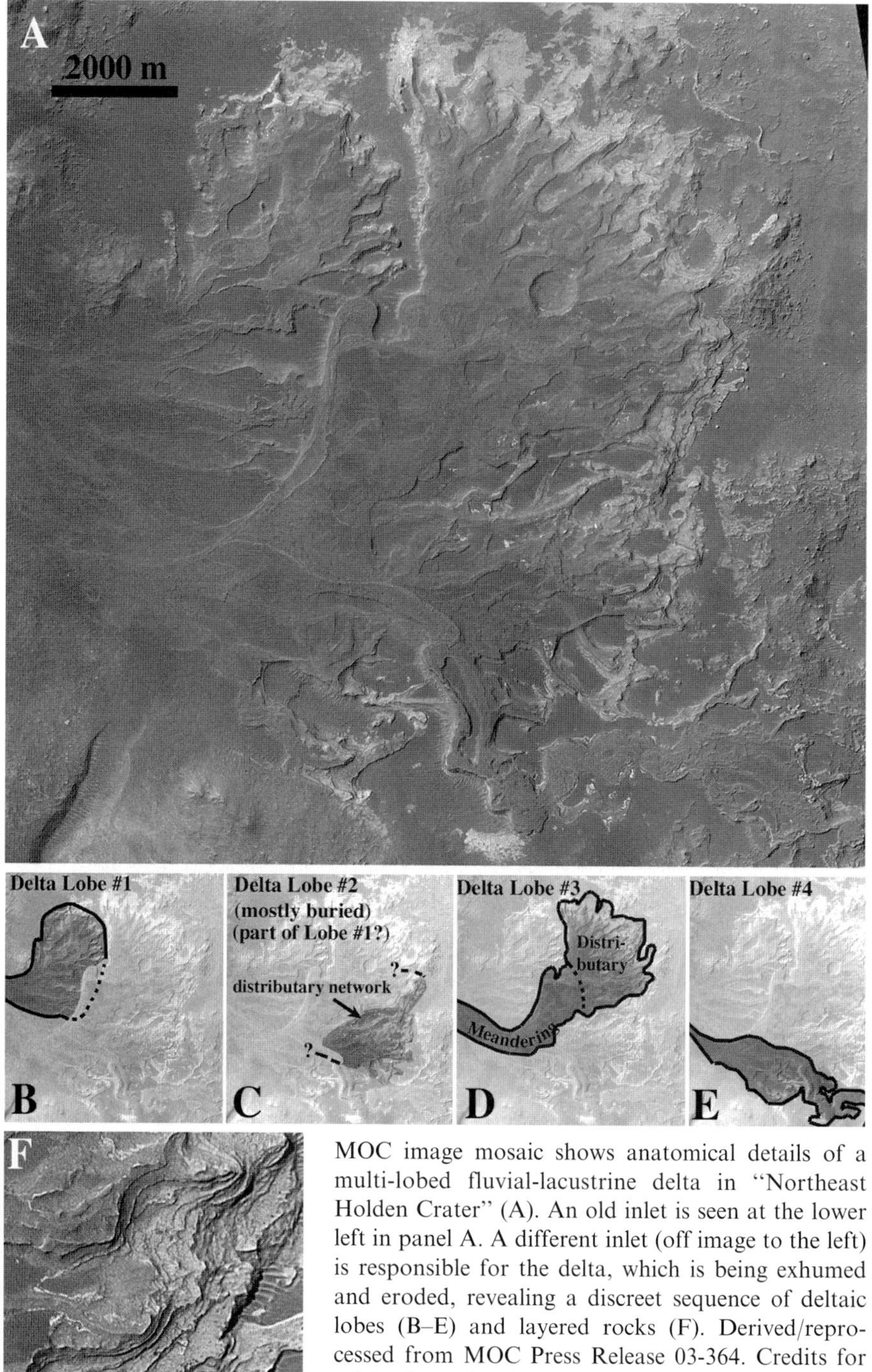

MOC image mosaic shows anatomical details of a multi-lobed fluvial-lacustrine delta in "Northeast Holden Crater" (A). An old inlet is seen at the lower left in panel A. A different inlet (off image to the left) is responsible for the delta, which is being exhumed and eroded, revealing a discreet sequence of deltaic lobes (B–E) and layered rocks (F). Derived/reprocessed from MOC Press Release 03-364. Credits for original mosaics: MSSS/NASA/JPL/MOC and MOC and THEMIS teams.

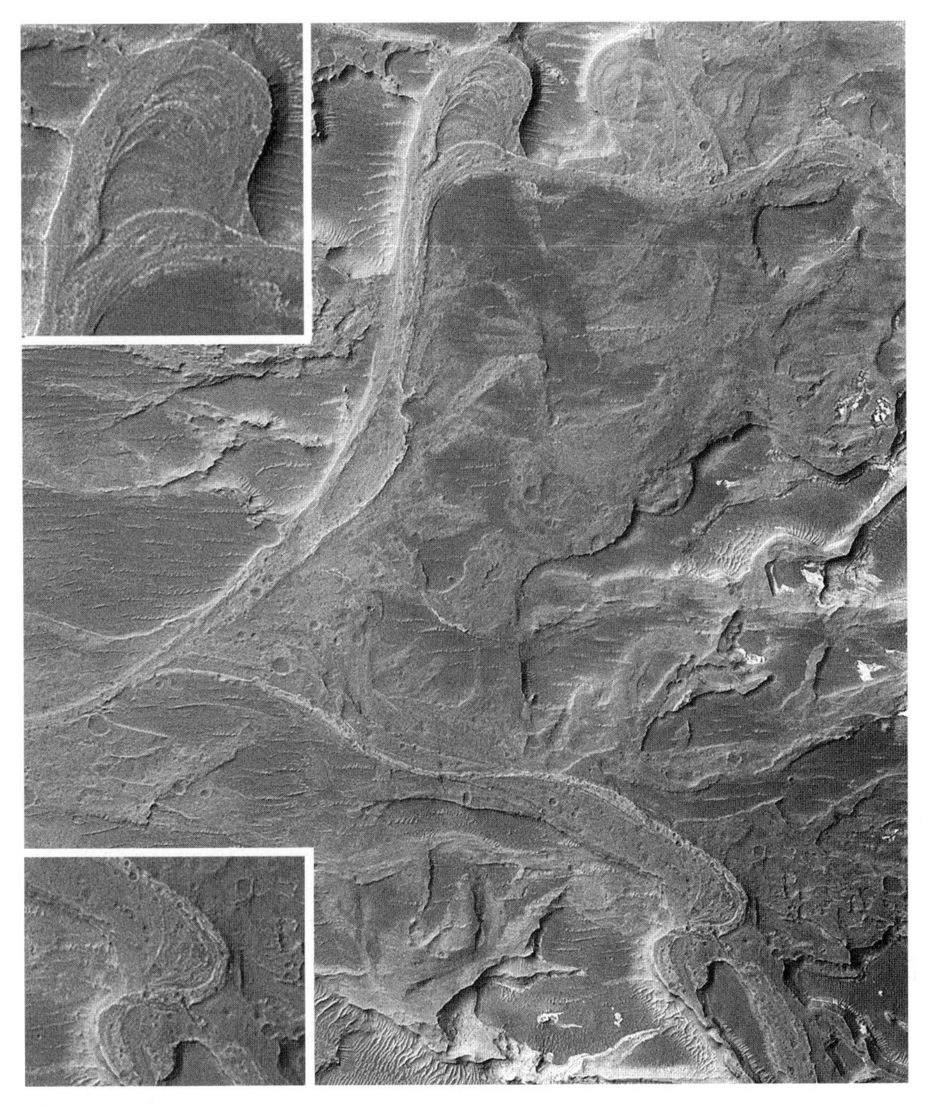

This compelling MOC image mosaic offers positive proof of sediment deposition by rivers of the delta featured on the previous image. The rivers were near sedimentological and hydrological equilibrium as a depositional system where bedload dominated. Devris flows and hyperconcentrated flows would not have behaved as indicated here. The riverbed meander loops (enlarged in the insets), are now ridges due to their erosional inversion (intervening fine sediment has probably blown away). The meander scars indicate meander avulsion (migration). The evidence is a "smoking gun," to use the words of a MSSS/NASA/JPL press release, that Mars had rivers that flowed for an extended period. A single stage of meander migration as recorded here requires 100–1,000 years. The delta probably was active for far longer than that. Pits may be due to sublimation of entrained ice. Cropped/enhanced from MSSS/NASA/JPL release MOC 03-364.

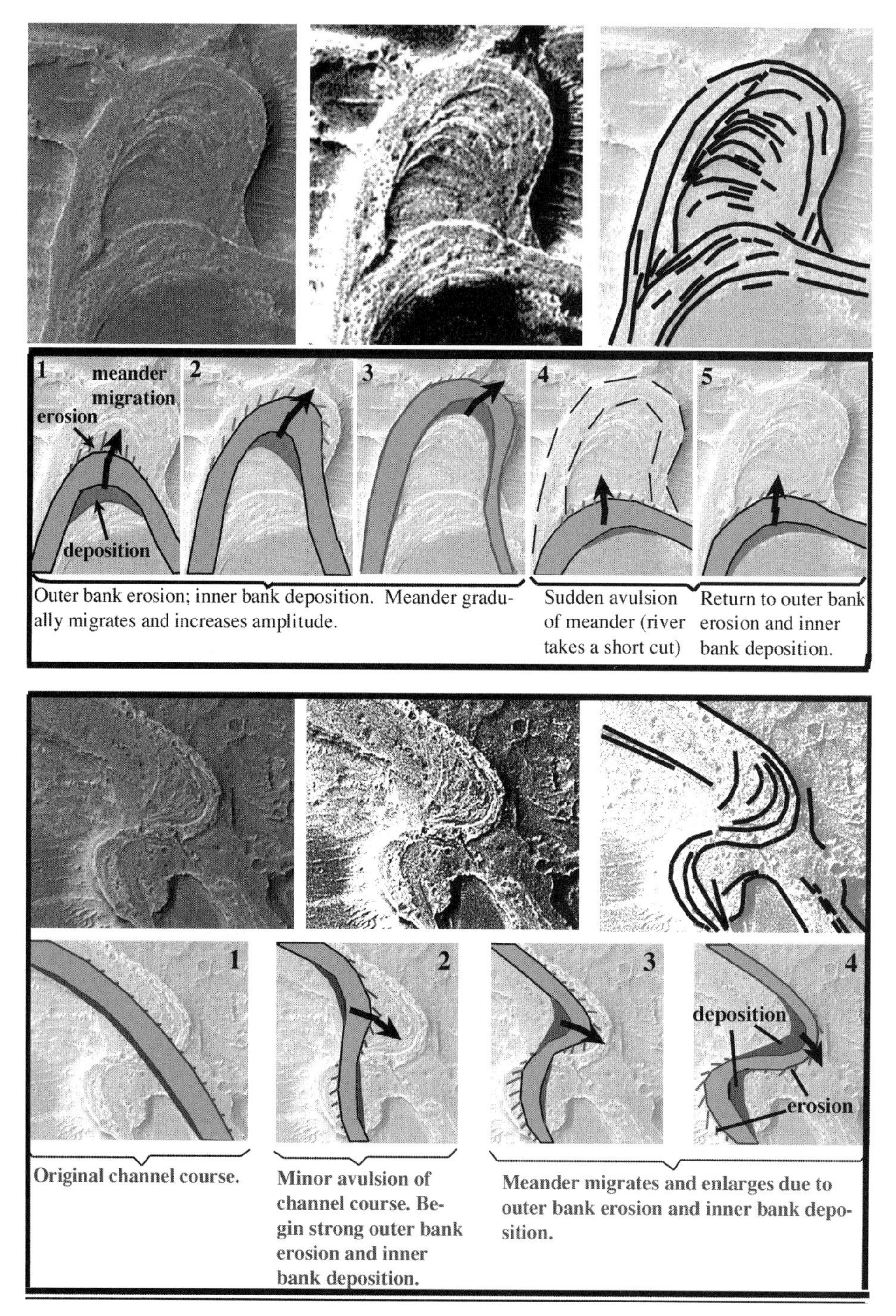

Reconstructed and interpolated sequence of meander development.

Meandering rivers in central Alaskan permafrost has left innumerable oxbow lakes, point bars, and other meander scars. Scene at left includes two pingoes (white arrows) formed where the soil has re-frozen after the river meandered away and left the terrain exposed to permafrost conditions.

climatic response to humidification; some crater-basin layered sedimentary complexes, such as that in Galle, occur proximal to rock glaciers or debris-covered glaciers on the crater rims. Like the geo-stories told by the Grand Canyon series of sedimentary rocks, the stories told by these Martian layered sedimentary deposits will one day help us to elucidate the details of Martian hydrologic and eolian processes, the sequence of events, and the causes and environmental conditions of hydrologic activity. Indeed, these deposits may make it easier to explore Mars geologically – and to discern the big-picture view – than the deposits of the northern plains.

With the importance of lake deposits evident, it is no surprise that the next steps in Mars exploration will feature detailed lander and rover studies of such depositional systems. One of NASA's Mars Exploration Rovers, Opportunity, is slated to land on the vast plain of Meridiani Planum, composed mainly of the iron oxide mineral hematite. This region includes layered sedimentary sequences eroded into spectacular buttes and mesas and spires not terribly unlike Monument Valley, Arizona (featured in Chapter 1). We are about to begin a new era of exploratory surveys to discern and map, as a field geologist would, the finer bedding structures, grain fabrics, and mineralogy in freshly exposed, well-preserved sedimentary rocks. Until this is done we lack definitive discrimination of one manner of sedimentary deposition versus another. I look forward to seeing, one day, current ripples, cross-bedding, scour marks, water-escape structures, sand layers, fine clayey laminae, desiccation cracks, and other indications of flow regime, flow direction, emplacement mechanism, and depositional environment in the lake, marine, and stream deposits of Mars. The final word from this historic era of Mars exploration may vindicate much of what was proposed by Henry Faul, Richie Williams, Henry Moore, Baerbel Lucchitta, Heinrich-P. Jöns, Tim Parker, Vic Baker, Jim Head, Ken Tanaka, and me. The early indications are that each of us got parts of the problem right; but all scientists are re-evaluating data and seeing new things that do not fit old ideas, as well as discovering great new evidence that adds new support. No doubt, the real Mars will prove to be more amazing than any of us has considered. The possibility remains that we will discover that Mars has been a Cold, Dry, Windy planet for 99.999% of its history, with just brief catastrophic excursions to transient wet conditions; personally, I am betting more on the 99% solution, with most of the interesting things happening during that 1% of wild and crazy times. No doubt the next months and years will be as fascinating and as full of surprises as the last several years have been.

FRACTAL EROSION

Floods, oceans (or mud oceans), lakes, glaciers, and icy permafrost are dramatic features of the Martian past, and they are probably closely related to one another. It snowed, it froze, it seeped, and it gushed on Mars, but did it ever rain? Did it drizzle, and did it ever pour 'cats and dogs'? That is quite another issue, though if it did rain, there is little doubt that it was related to oceans and lakes or transient volcanic or impact-driven emanations. My sense is that it never rained significantly on Mars since the start of the well-preserved geomorphic record of the planet (going back maybe 4 billion years), but this belief is being challenged by plausible new evidence to the contrary. I really am uncertain of the ultimate resolution of this issue, and I do not take a strong stance on it. But I have my view.

Erosion of landscapes by precipitation-fed runoff of liquid water is well known for the generation of self-similar ("fractal") erosional gullies and stream channels/valleys from the scale of the drainage basin down to some small limit where the rheology of water transitions from control by its low Newtonian viscosity (10^{-3} Pa-s) to control by the surface tension of water at the scale of plant roots and lithic fragments of soils, or to a limit caused by the effects of processes that may degrade small gullies.

Water-eroded channel and valley systems formed dominantly by spring drainage, including sapping valleys, may have far simpler structures that may or may not be fractals, depending on the size–frequency distribution and spatial distribution of spring sources. More often, there is a strong structural control of spring-fed drainage, and even if they do not, spring sources are few and far between; hence, these systems rarely form fractals.

For decades, almost ever since the small-valley networks on Mars were discovered, it was noted that very few valley systems exhibit something close to a fractal structure, and then only over a very limited range of scales. (I should note explicitly: The small gullies discovered in recent years are not a direct part of the small-valley network story, because there is a discontinuity in scale and formation time. Thus, the gullies do not actually have a central bearing on the debate over small-valley networks.) Most small-valley networks are morphologically very simple systems as observed and mapped from *Viking Orbiter* images; they typically consist of one major trunk valley and a few short tributaries, or perhaps one further order of tributaries. Furthermore, these systems as mapped from *Viking Orbiter* data do a poor job of covering surfaces; the sporadic distribution of preserved valleys would not allow efficient or complete surface drainage. Large, wide divides between tributary branches have no apparent dissection by smaller tributary valleys. It is as though Mars was very spartan about where it placed springs and how many it spread across the surface, but it does not appear that large amounts of water drained over broad areas of the surface on a sustained basis, as would have occurred from a rainfall-dominated epoch. The alternative mechanism of sapping by spring-fed sources seemed much more plausible.

Well, that was the consensus of most of the Mars community until late in the *Mars Global Surveyor* mission.

It was always my impression that the inefficient dissection of the surface by valleys may be due to post-valley-formation erosional muting by other processes, or mantling by dust and dust and ice; these processes would preferentially obscure the smaller valleys. This type of obscuration is evident across parts of the American West, where a general climatic drying subsequent to the Ice Age has caused some tributaries of drainages to become inactive; they have starting filling in and commonly are visible only at very low Sun angles. For Mars, this was never a very successful argument, because it was difficult to find evidence that would support (or refute) this conjecture. When something (such as the fine-scale component of fractal valley systems) is not observed even when looking for it, maybe that's because it was once there but has now gone, or perhaps it never existed. Given the process and climate implications of high-order, space-filling dendritic valley networks, the burden of proof for high-order space-filling valley systems rests on those who would propose them. This burden could never be carried, until now. ...

The sapping-only hypothesis survived intact as the favored hypothesis of the Mars community until the past year. New analysis of MOLA data has picked out subtle valley forms that are not readily evident in *Viking Orbiter* and MOC scenes, and other studies have pointed out other observations or results of models that now point fairly consistently toward a far more complex global network of small valleys and a more highly dissected surface eroded by fluvial processes. Brian Hynek and Roger Phillips (Washington University, report in *Geology*, 2003) have shown highland terrains that seem barely dissected by small valleys in *Viking Orbiter* scenes, but that MOLA data reveal it to be intensely dissected by up to sixth-order tributary valley systems; furthermore, the geometric integration of the valley systems and their coverage over the surface appear to be dendritic fractal valley networks caused by spatially distributed runoff, as would occur most plausibly by rainfall or snowmelt. Thus, several of the chief observational bases for sapping origins and against distributed precipitation-fed runoff have been overturned, at least in some areas of the planet.

Such studies have produced a strong new interpretive trend in the Mars science community toward climate models that involve ancient heavy precipitation across the highlands. Some of the most ardent defenders of sapping, such as Michael Carr (USGS, Menlo Park), have been convinced that the new school is the right school. Quoted in an article by Richard Kerr (*Science*, June 2003), Carr stated, "*It looks more and more as if you do need precipitation.*" Accepting the evidence from several recent studies, Oded Aharonson (Caltech) sums the trend, "*There's been a shift [toward] a somewhat wetter Mars, implying a more hospitable [ancient] climate.*"

This revolution in thinking must feel like a warm community embrace to Robert

Craddock and Ted Maxwell (Smithsonian Institution, Washington, DC), who have been perhaps the only consistent, ardent, expert defenders of a history of ancient, long-continued but declining rainfall on Mars. While Craddock and Maxwell may yet be fully vindicated if the new trend in thought persists, there are good reasons not to abandon the sapping hypothesis completely, as many well-preserved ancient and young landforms (including some small young gullies) have distinctive sapping morphologies. Of course, sapping occurs on Earth along with rainfall and snowmelt, and it is the precipitation that sustains the sapping systems with replenished groundwater. On Mars, a likely prevalence of ice and transient existence of unfrozen lakes and seas would favor a climatic regime where precipitation fell as snow, rather than as rain. Rain is a plausible alternative, though it would entail a considerably different paradigm than the one I and many other Mars researchers currently see as likely.

Surface runoff from snowmelt is supported by the observations and makes more sense to me theoretically. The sources of precipitable humidity would have been along the ice-vapor saturation curve (triple point and colder), and condensation would likely have been triggered by cooling from that point; with local atmospheric warming, the air would have found itself undersaturated and unable to precipitate. Any periods of rainfall would have been exceedingly brief, lasting perhaps days to months (depending on the size of the water body) adjacent to pondings of geothermally warmed floodwaters.

WASHED AWAY

Impacts on Earth are more frequent per unit area than on Mercury, Venus, and the Moon, and they are almost as frequent as on Mars, yet Earth retains a mere smattering of deeply eroded remnants of large, very ancient impacts, and a handful of really fresh-looking small impact craters. At small scales, *Apollo 16* astronaut John Young reported from the lunar surface, *"Even the craters have craters!"* The different appearances of Earth and the Moon caused no small amount of confusion among geologists a century ago, but the reasons are now very well understood. The moon lacks effective processes – aside from subsequent impacts – whereby craters can be eroded.

Most adults living in anything but the most urban setting remembers a storm, a drought, a flood, or some cumulative effect related to flowing water that caused a notable human-scale change to a familiar landmark – a once fertile corn field now gullied, a shallow lake dried up and mud cracked, a beach sand dune no longer there, a favorite fishing spot on a tree-lined riverbank now collapsed and the trees gone, or our first home inundated with floodwaters and inches of mud. These may be once-in-a-decade or even once-in-a-lifetime changes, but compounded over millions of years and combined with the day-in and day-out riverine transport of sediment to the oceans – huge changes to the landscape ensue. Mountains wear away, deep lakes fill up, and whole continents wash into the sea. Make a deep footprint impression on a muddy surface and it will probably not be there in a month or a year, depending on the climate. This is erosion and deposition on a watery planet. This is not how most worlds operate. On the Moon a deep footprint impression would last about 200,000 years before micrometeoroids stir the dust enough to obliterate it.

What a different world Earth is! More profound than obscuration of insignificant surface details, huge quantities of sediment are produced and transported to the sea. Stuart Ross Taylor and Scott McLennan, in an insight-packed book for the data-hungry scientist (*The Continental Crust: its Composition and Evolution*, Blackwell Scientific Publishers, 1985), give pre-human erosion rates about 4 km^3 per year, or about 30 m per million years of land surface lowering. Human activities have more than doubled this rate. It is spatially heterogeneous; places like the Karakoram Himalaya have a kilometer (and more) of erosion per million years, whereas in the semi-arid region around Meteor Crater, Arizona, half a meter (or less) of denudation since the impact 50,000 years ago gives an average erosion rate of less than 10 m per million years; and rates go down from there. Taking the average pre-human erosion rate, about half the crustal volume of the continents lying above sea level would be emptied into the sea in just 15 million years. The point is clear. Our world as we know it is fast eroding. It has always been this way; fortunately, our land world is not actually disappearing but is being continually reconstructed. Magmatic activity, deposition of eroded sediment near the continental margins, scraping and crumpling and folding of that sediment back onto the continental edges, uplift occurring with earthquakes, and slow elastic and plastic uplift due to gravitational isostacy all tend to counter the effects of erosion. A near steady-state of continental volume and relief has prevailed for billions of years.

The surface of the Martian highlands has been eroded by an average of around 1 km since heavy bombardment, as judged from the degradation of crater rims and amount of sediment infilling of basins. Locally, for example in the Charitum Montes, two or three times this average highland rate has applied, but in the Charitum Montes erosion has clearly been dominated by glaciation and probably was restricted to narrow intervals of time. This gives an average highlands erosion rate of about 0.25 m per million years, just 1% of the pre-human erosion rate on Earth's land areas. Another way to view this is that the Martian highlands have experienced about 4 million years' worth of erosion at typical terrestrial rates. Recent rates of erosion on Mars have been sufficient to wipe out around 90–99.9% (depending on area) of the small impact craters (400 m in diameter) that are expected based on extrapolation of the densities of large craters (>4 km in diameter).

Erosion on Mars has been far from constant. It is evident that Mars has had brief episodes of greatly heightened erosional activity superposed over a generally declining trend in erosion rate. The brief intense erosional episodes are interspersed between eons where comparatively little happens. Mars is now in such a low-activity period, but it is far from eventless; it is what I call the Neoglacial.

7

Neoglacial

NEOGLACIAL IS NOW

The Martian hydrologic and climatic regime and the myriad effects of water and ice on the solid planet will be critical for any future human explorers to understand. More than any other aspect, H_2O and its influence on Mars is what engenders a compelling dream to extend our presence there. It is also what will allow that dream to come true. The present condition was established over eons of hydrogeological activity, but the present disposition of water and ice is just as much influenced by current and recently active conditions and processes. Now I step forward from the long-gone distant ages to this recent time, the latest of the Late Amazonian. Hardly anything was known about this period and its chief processes until *Mars Global Surveyor* started returning thousands of high-resolution images.

There are some major caveats pervading this chapter. We really do not have a solid grip on absolute ages or even relative ages for small-scale young landforms (which includes most of those discussed here). We know or suspect that they are young by virtue of their lack of abundant impact craters and their crisp preservation. Furthermore, a landform may be very old but appear young, fresh, uncratered, and unweathered because it was mantled for eons by dust or ice, which may have only recently been stripped away; or it may consist of particularly durable rocks that take a lashing by the elements and still look pristine; or it may in fact be undergoing highly selective erosion or sublimation, thus bringing out fine details of internal structures and making it appear crisp, even though it is highly degraded. Finally, the populations of small Mars-impacting asteroids are poorly known; there is a distinct possibility that the crater size–frequency curve used for crater-based age dating may be completely erroneous at small crater diameters.

With meter-scale image resolutions now available, we have entered an intellectual *Terra Incognita*, and we are just feeling our way through some challenging chronostratigraphic issues. I take an approach to these issues that is admittedly not satisfactory: I almost ignore them, so as not to get caught up in minutia of something that is not understood by anyone. I take this approach fully knowing that accurate reconstruction of planetary climatic evolution is always dependent on knowledge of what happened when. Maybe in another three to five years the Mars community will have had time to sort through the issues and develop a new quantitative chronostratigraphic timeline for late Martian history. There are some promising possibilities based, for instance, on secondary crater ray tracing, as well as more conventional analysis of crater densities and landform superposition. Until then, we are working somewhat in ignorance of time, which for late Martian history encompasses thousands, millions, even tens of millions of years, during which time Mars has seen many climatic oscillations and dramatic geologic events whose relative sequences, much less their absolute ages, are unknown. Imagine, for instance, confusing the time sequences here on Earth of the Mesozoic age of dinosaurs, the Cenozoic age of mammals, the Ice Age, the construction of the Great Pyramids, and the assassination of Abraham Lincoln. Lacking such relative chronologic information, and being almost blind to important matters of process rates, we might be led to arrive at absurd answers to nonsensical questions, deriving histories and geologies that make about as much sense as the *Flintstones*. Indeed, I take the *Flintstones* approach and say that anything that looks really fresh and is uncratered probably is very young and all virtually the same age. The possible nonsense of this approach may be known in a few years, but for now it is the best we can do.

The observable processes and geologic record of Mars show that the Martian water cycle does not now resemble and probably never has closely resembled that on Earth, some comparable aspects notwithstanding. Currently, the relative magnitudes of subsurface water transport versus atmospheric water vapor flux seem to be quite different, possibly inverted, on the two planets. It is not that the Martian subsurface component of the hydrologic cycle is any more vigorous than on Earth – it may be a few factors less vigorous if gauged from the geothermal energy flux available to drive the system – but the atmospheric component is several orders of magnitude less efficient due to low temperatures. If rates of surface/subsurface recharge, hence, of surface/subsurface flow, are limited by rates of atmospheric transport of water vapor, then the present-day pace of Martian water cycling indeed is far more sluggish than Earth's. Regardless of how Mars operated in the distant past, the Mars we see now is one of more modest hydrologic behavior by all accounts. Current/recent hydrogeologic activity is more restricted spatially and in intensity compared to certain periods of punctuated activity in the distant past.

The present-epoch hydrologic cycle, though feeble by comparison to Earth's, is the main thing happening on Mars today and is highly consequential for the state of preservation of older features at middle and high latitudes and the formation of younger landforms. A few million years of the ongoing hydrologic regime is ample to render a complete facelift of the entire region poleward of $\pm 40^\circ$ latitude at resolutions that would be seen by a human standing on the surface. Current processes are shaping the parts of Mars – the upper few meters– that will greet and be used by the first human explorers.

We have seen from *Mars Global Surveyor* and the *Viking Landers* some effects of current-era hydrology and volatile transport happening on a daily and seasonal basis

– fresh morning frosts at Utopia Planitia and lowland ice fogs in Argyre Basin and the fretted terrain, orogenic ice clouds at Olympus Mons, and humidified summer air over the northern polar cap, and myriad CO_2 processes at both poles. Our time is one of creeping and cracking icy permafrost, waxing and waning and flowing polar caps, seasonal seeping icy cliffs and ledges, landslides and debris flows, ice sublimation, and collapsing ice caves. Rather than megafloods and enormous mid-latitude ice sheets and the other drama of the distant past, the real story now includes some more subtle roles of H_2O and CO_2 working separately or, in some cases, together. As Baerbel Lucchitta reported, following 2003's annual Lunar and Planetary Science Conference, "*Ice is practically everywhere!*"

Within the middle and high latitudes, Lucchitta's summary is not overstated. It is difficult to find places where there is neither a plausible link of landforms to ice nor an indication of ice in neutron spectrometer data. One paper after another of the latest research confirms what Lucchitta, Steve Squyres (Cornell University), and a few others had started saying against the grain of prevalent thinking, since the *Viking Orbiters* relayed their images. Indeed, the first indications that climatic influences have produced a global latitude-dependent surface geology came soon after *Mariner 9* images poured in. Larry Soderblom (US Geological Survey) wrote about a latitude-controlled debris mantle on Mars as early as 1973 (*Journal of Geophysical Research*, **78**, 4117–4122). Although the climatic and geologic mechanisms of mantling now are understood much better than they were 30 years ago, the more we have learned, the more this mantling concept has been affirmed.

Ice deposits and their effects on the surface are widespread. Most shocking to the last converts from the Cold, Dry, Windy Mars paradigm, H_2O seems to remain an active agent of surface modification right until recent times. Although the roles of liquid water are less pervasive than indications of ice, we do see many signs that water recently has been erupted onto the surface; we can be sure that liquid water is abundant at depth. However, far more significant than melting ice for geomorphic work being done today is sublimation (deposition and evaporation) of ice. The construction and destruction of icy permafrost by the ebb and flow of traces of water vapor in the thin Martian air drives processes that dominate ongoing geomorphologic processes at middle and high latitudes. The long-sustained accumulation of polar ices, just 60 micrometers each year, builds towering ice caps that creep and flow, melt and sublimate. The heating and cooling of permafrost causes ground ice to accumulate, crack (by thermal contraction), and sublimate. In some regions where condensed carbon dioxide may be trapped and pressurized beneath ordinary ice, with the fate of ice so goes those subsurface CO_2 reservoirs.

By contrast to what we see poleward of $\pm 30°$ latitude, there is very little surficial activity at the low latitudes, except for scattered eolian activity and slow erosion of dust deposits. Thus, the grandeur of the climatic record of the low latitudes is mostly a battered relic of the distant past. By contrast, the weathering and sedimentary overprint due to water and ice is so deep and pervasive at middle and high latitudes, and so recent, that most traces of ancient features discernible at low resolution are invisible when seen close up.

Water and ice have had both strong indirect and direct influences on the Martian

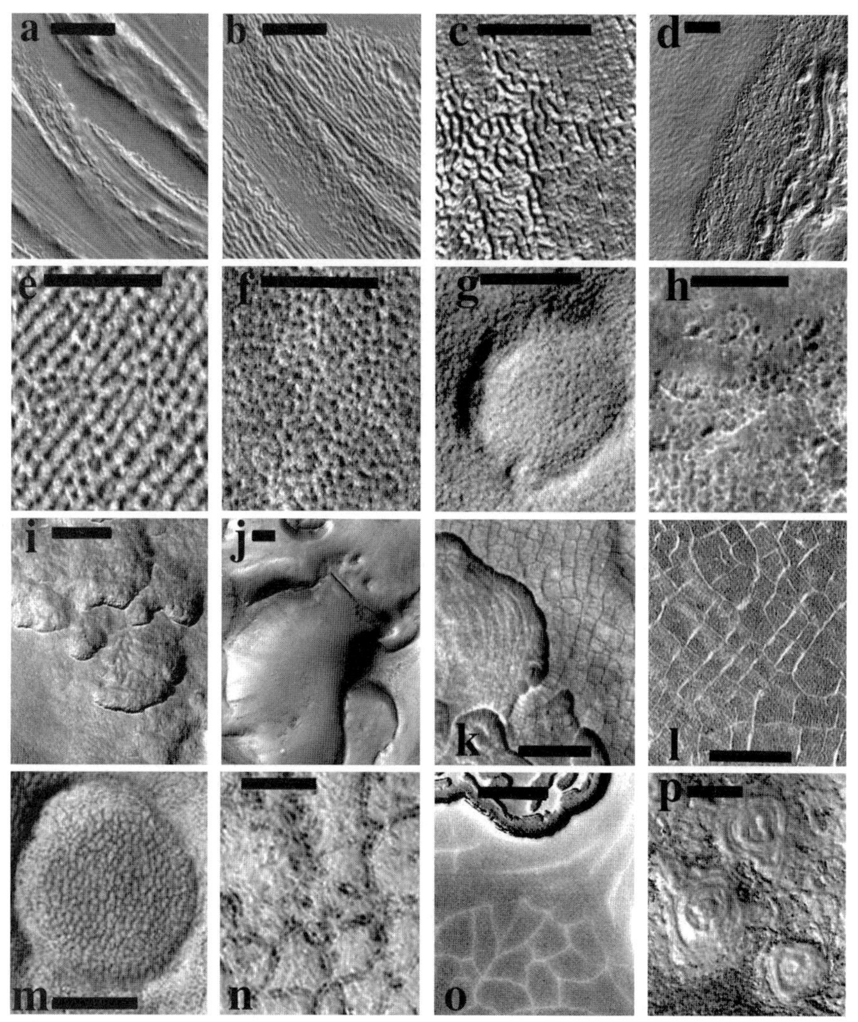

Ice features. North down, illumination from right, black bars 300 m long. (a) M1002174, lineated valley fill, 40.37°N lat., 329.76° long. (b) Lineated form, M1002174, (c) M0303561, "brain terrain" lineated valley fill, 40.06°N lat., 338.04° long. (d) M001981, lobate debris apron, 45.09°S lat., 241.99° long. (e) M0201054, northern plains knobby/lineated deposit, 71.24°N lat., 241.85° long. (f) M0306351, northern plains knobby deposit, 71.24°N lat., 156.02° long. (g) M1600879, knobby deposit mantling crater, Noachis Terra intercrater plains. 60.60°S lat., 37.15° long. (h) M1302002, intercrater pitted plains, such as mapped by Mustard et al. (2001), 38.96°S lat., 172.55° long (i) M0300502, scalloped pits, Malea Planum, 54.62°S lat., 314.16° long. (j) M0300139, highland chaos, 57.24°S lat., 349.94° long. (k) M0401631, polygons and scalloped thermokarstic pits in Utopia Planitia, such as studied by Costard and Kargel (1995), 44.33°N lat., 272.68° long. (l) M0701136, polygon network on large crater floor, cratered highlands, 60.11°S lat., 274.59° long. (m) M0700451, small polygons in unit mantling crater, intercrater plains, cratered highlands, 63.47°S lat., 336.93° long. (n) M0700797, defrosting or annealing polygons in crater-floor CO_2-frost deposits; cratered highlands, 67.93°S lat., 116.58° long (o) M1300618, polygons in sublimating dry ice, south polar cap "swiss cheese terrain," 84.92°S lat., 76.67° long. (p) M0305085, northern plains, possible Lyot crater ejecta, 49.75°N lat., 322.13° long.

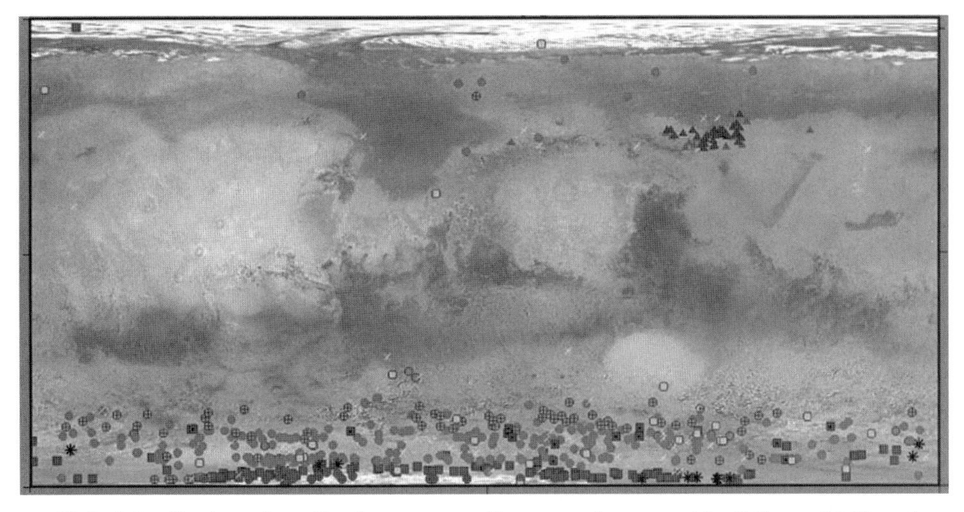

Global distribution of small polygons according to work reported by Seibert and Kargel (2001), annual meeting of the Division for Planetary Science, abstract 530. Symbol types refer to various morphological classes of polygons.

surface. Much Martian eolian activity – so long presented as an alternative to volatile activity – can be related to Martian volatile activity. The original formation of the fine sediment on Mars, as true also of Earth, is largely due to the erosional activity of water and ice. Sand and dust is trapped and released as volatiles undergo their sublimation and melting/freezing processes. Wind stirs up and redeposits detritus in nature's most perfect recycling program, forming dunes and layered dust sheets. Even free H_2O is not as well conserved as dust on Mars, because H_2O can react with silicates or can be photolytically destroyed and lost forever from the hydrologic cycle. The dust of ages turns to stone or is further pulverized, but once formed it always remains available to be disaggregated and recirculated and to tell a new story. This principle of "conservation of dust" applies to Mars, but not to Earth, where metamorphism and subduction destroys dust.

The creeping of icy permafrost, the gradual erosion of permafrost terrains in areas where ice becomes unstable, and the burial of terrains by sand and wind-blown dust all contribute to the degradation of old impact craters, volcanoes, and other elements of the landscape. While not every Martian sand dune can be attributed to sand produced by glacial grinding and liberation from sublimated ice, on Mars there might be a strong correspondence of many dune fields and dust sheets with glacial (past or present) and permafrost landscapes. It is difficult to avoid the conclusion that the glacial and periglacial deposits have entrained dust and sand, and as the ice goes the fine sediment also goes. There is a direct analog of this type of occurrence and process on Earth, where vast dust blankets ("loess") and dunes fields (e.g., the Nebraska Sand Hills) are made of wind-worked outwash from Ice Age glaciers. The dust and sand was generated by glacial abrasion of bedrock and comminution of subglacial till, and then were expelled by floods of meltwater; as Ice Age lakes dried

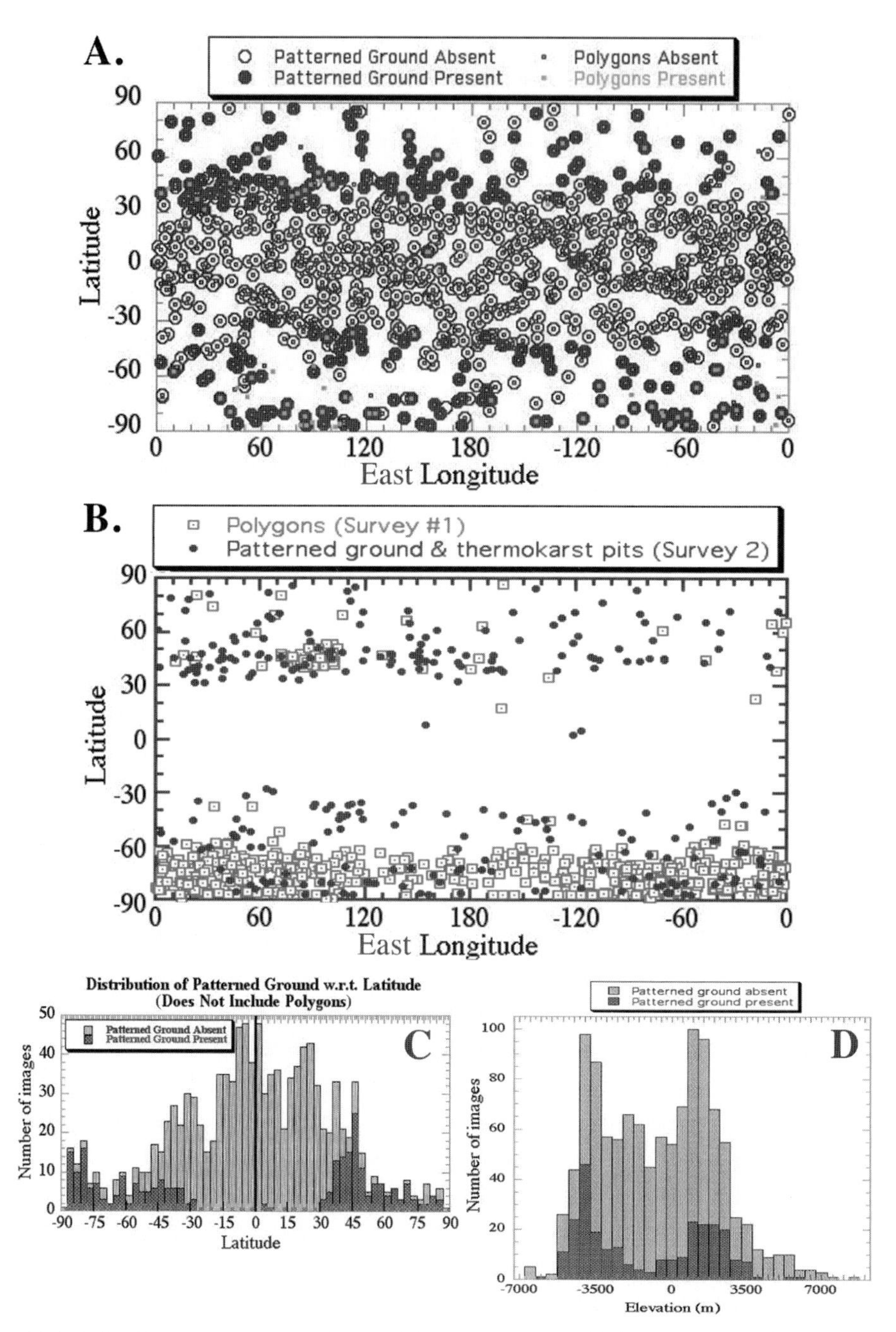

Global distribution of patterned ground (ice indicators). (A) Results of global survey by Joseph Restaino in collaboration with Kargel. (B) Comparison of ice indicators mapped by Restaino with a more complete survey of polygons by Nicole Seibert. Dependence on latitude (C) and elevation (D).

up, dust and sand was re-eroded. Fine sediment re-accumulated wherever conditions stabilized the dust; this was often where ground ice could accumulate in permafrost. In places such as Argyre and many other areas this seems a highly plausible framework in which to consider eolian activity.

Many of the myriad effects of ice on Mars were not directly visible to the *Viking Orbiters* and are blurred from a distance, forming what Steve Squyres and Michael Carr termed "terrain softening." With the close-up lens of *Mars Global Surveyor*, we see the details of terrain-softening landforms generated through the action of ground ice. These effects are pervasive across vast expanses of the middle and high latitudes, and in many areas the processes still appear to be either active or very recent. Indications of ice flow abound. Many of these icy flow features appear to be heavily degraded or etched by sublimation, and nearly all (except at the poles) are debris covered, but very few have evident impact craters superposed over them. There are powerful analogs of these features on Earth, especially debris-covered glaciers and rock glaciers. These are not perfect analogs but anyway are very insightful. These Martian features and their terrestrial analogs indicate some important workings of a planet undergoing climatic change to a less humid and perhaps colder state. The glacier-like landforms on both planets are generally receding and wasting; and although the specific nature and certainly the causes of climate change on Mars are different from those occurring on Earth, the consequences for the cryosphere are comparable. On Earth we see remnants of an icier time, not unlike that Mars exhibited on Mars.

It is not a well-dated period of time, but a plausible estimate is that many of the small-scale high- and middle-latitude features we see on Mars – for instance, the small gullies, the small polygons, small thermokarstic pits, and the polar caps – date from a period that may be roughly concordant with Earth's Late Tertiary and Quaternary, including the Miocene to Recent epochs. Another way to view it is that these Martian features were forming just in the period since the youthful Hawaiian Islands and Grand Canyon have taken form, and over the period since equine, pachyderm, and hominid species became recognizable as close relatives of their modern forms. It is quite possible that these Martian landforms were initiated long before that and have ancient precursory roots, but they have evolved into their present forms just over this geologically recent time. Indeed, very many of the sublimational forms, polygons, and small gullies of Mars are probably less than 100,000 years old, dating from when early *Homo sapiens* and *Homo neanderthalensis* cohabited Earth, a period roughly coeval with the period since the great Eemian Interglacial, and including the last great glacial maximum and the more recent period of human civilization.

The present epoch on Mars is one of occasional volcanic eruptions and small gushing floods of lava-melted water and mud, but more than that it is a period of sublimating permafrost and blowing dust and sand, accumulation and sublimation and slow flow of polar caps, and the occasional melting of icy remnants of a past era. It is surely not the most active time in Martian history, but this is a period of dynamic climate changes and responses. The present epoch may be representative of the eons between major MEGAOUTFLO disturbances, but another possibility is that the present epoch is a late-stage, tail-end response to the most recent big MEGAOUTFLO event. This is the latest of the Late Amazonian, the most recent chapter in the ongoing Mars story: the Neoglacial, and it is now. What comes next is partly what Humankind is apt to make of it: The Future.

PERMAFROST PATTERNS

MOC imaging reveals an amazing diversity of finely textured patterned terrain at middle and high latitudes. At resolutions of 1.5 to 10 m per pixel, the patterned terrains are totally unlike anything observed at the lower latitudes, where the surface may be rugged or smooth depending on the geology and age of the terrain. The texturing and patterning at middle and high latitudes is the fine-scale expression of the processes that are collectively termed "terrain softening" when seen at lower resolutions. The two commonest patterned ground forms include polygonal networks of cracks (polygons typically 10-200 m across) and equi-spaced 10-m-scale mounds. The polygons referred to here are distinct from the giant polygons 5-10 km across formed of troughs about 1 km wide.

Most Martian polygons consist of evident cracks or troughs or simply bright or dark lineaments of uncertain topographic nature. Some ("south polar type," which occurs mainly south of 60°S latitude up to the edge of the permanent south polar cap) consist of ridges, ridges-within-troughs, rimmed troughs, or light-and-dark albedo lineaments interpreted as partial frost coverings of ridges and troughs. Others ("Utopia type" named for the only major geographic occurrence) consist of isolated pits arranged along lineaments that are further organized into polygonal patterns.

When categorized by geometric organization rather than their topographic expression, two types are dominant. "Random orthogonal polygons" have approximately random lineament orientations with orthogonal intersections produced by late-forming cracks curving into older cracks at right angles; "oriented orthogonal polygons" have truly straight lineaments having preferred orthogonal orientations (forming a rectangular pattern) or one set of small-circle cracks with a second orthogonal set (common in the bowls or on the rims of impact craters). There are also cases of non-intersecting parallel systems of cracks, and randomly oriented cross-cutting cracks.

The most striking association of Martian polygons is with latitude. Polygons are virtually absent from the equator to about 30° latitude (north and south), but common poleward of that. In the band between $\pm 30°$ and $\pm 45°$ north and south of the equator, these features primarily occur among highly degraded, eroded (probably sublimated) landforms, or otherwise occur in terrains that appear to have lost ice (Utopia Planitia having the prime examples, where scalloped pits further attest to thermokarstic ice-loss processes), whereas poleward of $\pm 45°$ most polygons seem relatively fresh or may even be actively forming.

The dekameter-scale patterned ground is normally spread across all elevations or can be concentrated at high elevations in the mid- to high-latitude regions where it occurs. This is a generic grouping based on the typical 10- to 20-meter scale, barely resolved in MOC images, of textural elements comprising it. While some may simply be the smaller, poorly resolved examples of polygons already described, most appear to be a couple of different types of landforms related to the polygons only by being ice-related. One form is called hummocky patterned ground, and the other type is called scaly patterned ground. These are the classic expressions of 'terrain-softened' southern-hemisphere, mid-latitude landscapes.

Hummocky patterned ground occurs on mantling units that drape ridges and hills, crater rims, crater floors, and all terrain high and low. This type has a highly distinctive morphology everywhere it occurs. A type example occurs widely in the highlands south of Argyre's Charitum Montes. This form shows little preferred structural orientation, with almost uniform spacing of hummocks; it is almost isotropic in the plane of the surface, and shows no strong deviation of form or scale with slope or slope aspect.

A second and less common form of dekameter-scale hummocky terrain, but no less distinctive than the first form, occurs on lobate debris aprons and other icy flows. This type has a 'scaly' texture of cracks or closely spaced hummocks oriented transverse to the inferred flow direction.

Polygonal terrain was anticipated, and though major questions remain it is not particularly enigmatic. Baerbel Lucchitta (US Geological Survey) and others had seen hints of small-scale polygonal terrain in *Viking Orbiter* images, and a polygonal network of troughs was seen on the ground at the *Viking 2* lander site, and now with MOC we have synoptic views almost as detailed as those obtained from the landers. Wade and deWys had predicted permafrost polygons on Mars as early as 1968, so the recent discovery that they are common and widespread has come as a surprise only to researchers who had still been entrenched in the Cold, Dry, Windy Mars paradigm.

PERMAFROST PROCESSES

While some relationship of Martian polygons and other patterned ground to a condensed form of H_2O is not now doubted, this relationship admits a wide range of specific origins. The link could be to liquid water, ice, or some major hydrated mineral phase. This compelling link is shown by the morphology, global distribution, and local distribution of these features, and by the similar latitude-controlled distribution of hydrogen, as seen in neutron spectrometer data. A fairly well accepted explanation is that most polygons and other patterned ground on Mars is related to icy permafrost, though some polygons may be mud desiccation cracks (e.g., dried lake beds) or simple thermal contraction cracks. The latitude control of Martian polygons and other patterned ground is good evidence that these features are mainly connected to water ice, because these mid- to high-latitude zones are generally where climate models indicate that deep ground ice should survive stably, indefinitely. According to terrestrial analog studies, it is thought that most Martian polygons represent some type of shrinkage contraction cracking operating with various possible crack-filling processes. This is about as far as the consensus goes.

Desiccation cracking of mud and silt offers one explanation for some polygons, but most polygons on Mars are probably not of this type. Some polygons and other forms of patterned terrain may be due to freeze–thaw driven heaving, with or without thermal contraction cracking, that can sort sand, silt, and dust from pebbles, cobbles, and boulders in polygonal patterns. A broad range of other patterned ground occurs on the surfaces of icy flows (debris-covered glaciers or rock glaciers) and may be due to ice sublimation controlled by periodic structures, such as crevasses.

Small polygons on Mars are probably primarily permafrost ice features, although some may be mud desiccation features. Mud desiccation makes most sense where the polygons are strictly limited to deposits contained in depressions and having evident inlet channels and layered sedimentary deposits exhibiting onlapping relations (embayment) with higher ground, terraces, or other indications that the site was where water once ponded. These situations constitute a minority of occurrences of polygons; even these could be due to icy permafrost, because permafrost polygons on Earth have a special tendency to form in fine-grained sedimentary deposits of fluvial or lacustrine origin. Polygonal crack and ice-wedge networks, sorted stone circles, and other patterned ground is common in Earth's permafrost regions. Common on Mars are general associations of polygons with low-lying terrains regardless of other evident hydrological connections to floods or pondings; ice is most likely responsible. On Earth, polygonal networks of ice wedges formed by freeze-thaw are usual. These polygons have cracks with V-shaped cross-sections, widths of 1–3 m near the surface, and depths about 1–6 m; the cracks taper downward and are arranged in polygonal (often random pentagonal or oriented rectangular) networks; the polygons are typically 10–200 m across. Ice wedges initiate by tensional cracking of ice-bonded permafrost during winter cooling. They fill with meltwater during the summer and then the water refreezes. The ice-filled cracks recrack and grow wider by a few millimeters each year. After hundreds of annual cycles of winter thermal cracking, summertime thaw and infiltration of liquid water into the crack, and subsequent freezing and recracking, classic ice-wedge polygons are formed. As they grow and the water freezes and expands, a classic ice wedge will push adjacent silty-icy sediment aside, thus heaving the ground surface and producing low rounded rims. The ice wedge may be expressed as a trough or as a medial ridge within a rimmed trough.

Aaron Zent (NASA Ames) has emphasized permafrost processes involving thermal fluctuations, liquid water transport, and volume changes in permafrost containing grain surface-stabilized films of so-called "unfrozen water" at temperatures between 240 and 273 K. As I point out below, water transport mechanisms driven by daily and annual thermal fluctuations and unfrozen water or vapor-phase transport could produce hydration/dehydration of hydratable salts and clays in soils, as well as cause the segregation of ground ice. Stresses and cracking processes similar to those caused by bulk freezing and melting or wetting/drying cycles potentially could, over time, drive formation of similar polygonal crack morphologies, but without the bulk melting point ever being reached. Indeed, it should be difficult (without detailed field scrutiny) to distinguish, based on purely on imaging, permafrost melt-caused polygons from other types of permafrost polygons and desiccation polygons by morphology alone. These features are not unambiguous indicators of summer melting, though they seem to require considerable H_2O transport.

Salt hydration and dehydration can produce a complex overprint of dekameter-scale polygonal contraction cracks and compressional tectonics. Winter wet-season conditions observed on Searles (Dry) Lake (Mojave Desert, southern California) and nearby Bristol Dry Lake produce hydration reactions and volume increases in the salt minerals and phyllosilicates. The brine evaporates during the spring, leaving a salt crust overlying still-wet saline mud. As drying continues, the salt crust fractures like typical mud cracks, but quickly tension forces give way to compression, as the brine is wicked up and fills the cracks with salt precipitates. The growth of salt crystals is sustained by a continued capillary transport of brine to the evaporative surface. The salt crust folds and buckles, forming polygonal networks of small wrinkle ridges like those of the lunar maria (but 10^4–10^5 times smaller) and overthrusts of one polygonal plate shoved over another. This process thus generates polygons, but there is no need for temperatures to exceed 273 K, since it can operate at lower temperatures through the liquid-stabilizing influence of solutes.

Hummocky patterned ground on Mars is probably closely related to permafrost "stone circles" and other forms of "sorted terrain" on Earth (which Troy Péwé used to call "sordid terrain," in reference to the enigmatic origin of terrestrial examples). Most investigators usually attribute sorted permafrost terrain to freeze–thaw driven heaving and slow convection of permafrost in material composed of multiple grain sizes. Freeze–thaw cycles cause formation of thermal contraction cracks and ice wedges, but, more important in these terrains, it churns the sediment, which then separates by grain size in circular or polygonal patterns. When both coarse pebbles/boulders and fine sand/silt are present, sorting produces stone circles or polygons (on low slopes) and stone ellipses or stripes (on steep slopes). The more common ice-wedge polygons are favored by fairly homogeneous fine-grained sediment. Sand-wedge polygons may well be an intermediate form.

Sorted stone circles and permafrost hummocks form in extreme polar environments where melting is minimal, and this alone should caution us from drawing too definitive conclusions regarding Martian processes. There may be means of generating this morphology without melting. Annual cycles of hydration and dehydration of soils, particularly hydratable salts and clays, might drive a similar process and geomorphic response. Theoretically this process can be just about as effective as water freeze–thaw in driving sediment heaving and size sorting. For soils rich in hydratable sulfates, 10% (volume) of epsomite for instance, an annual cycle from $MgSO_4{\cdot}7H_2O$ to $MgSO_4{\cdot}12H_2O$ and back would exert a volume change of $\pm 4\%$ during soil hydration. This is about the same as for expansion of water-saturated sediment caused by freezing.

Scaly patterned ground, which is normally associated with icy flows on Mars, resembles seracs of terrestrial glaciers. Seracs are formed where crevassed structure and ablation causes widening of the fractures or partial disarticulation of the glacier. The sizes of icy Martian "scales" is similar wherever they occur and is similar to that of terrestrial glacial seracs. In general, the Martian scales have a subtler vertical

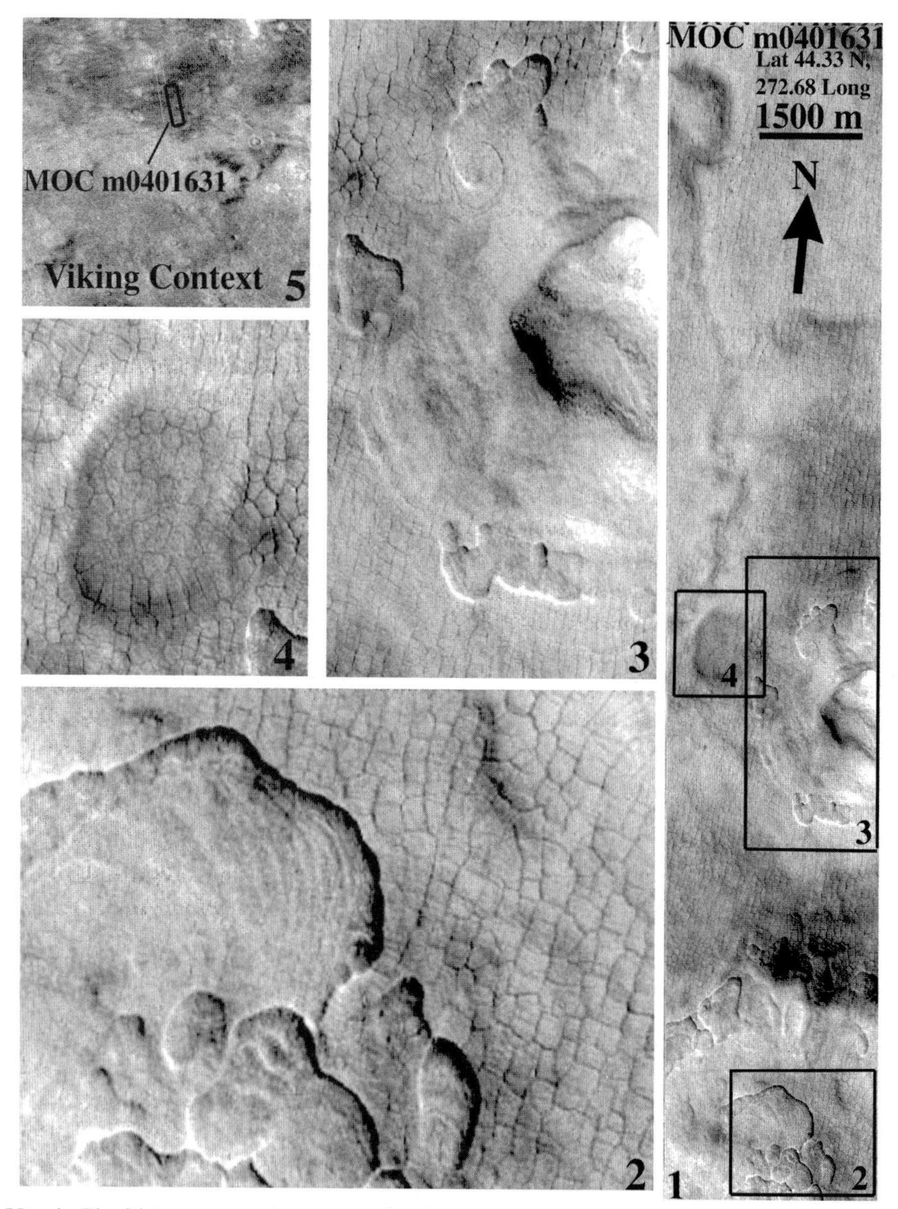

Utopia Planitia's nearly unique type of polygons are associated with equally rare types of scalloped, terraced depressions and lobate, bulbous mounds. These features were observed at much lower resolution by *Viking Orbiter* and were interpreted by Costard and Kargel (1995) as an icy permafrost assemblage, including ice-wedge polygons, alas lake basins (scalloped pits), and possible mud volcanoes (the mounds such as that in panel 3). With improved resolution now available, this interpretive model remains a viable candidate, but some domical structures (such as that in panel 4) may be mud diapirs that have not produced eruptive deposits, and polygons appear to be desiccated.

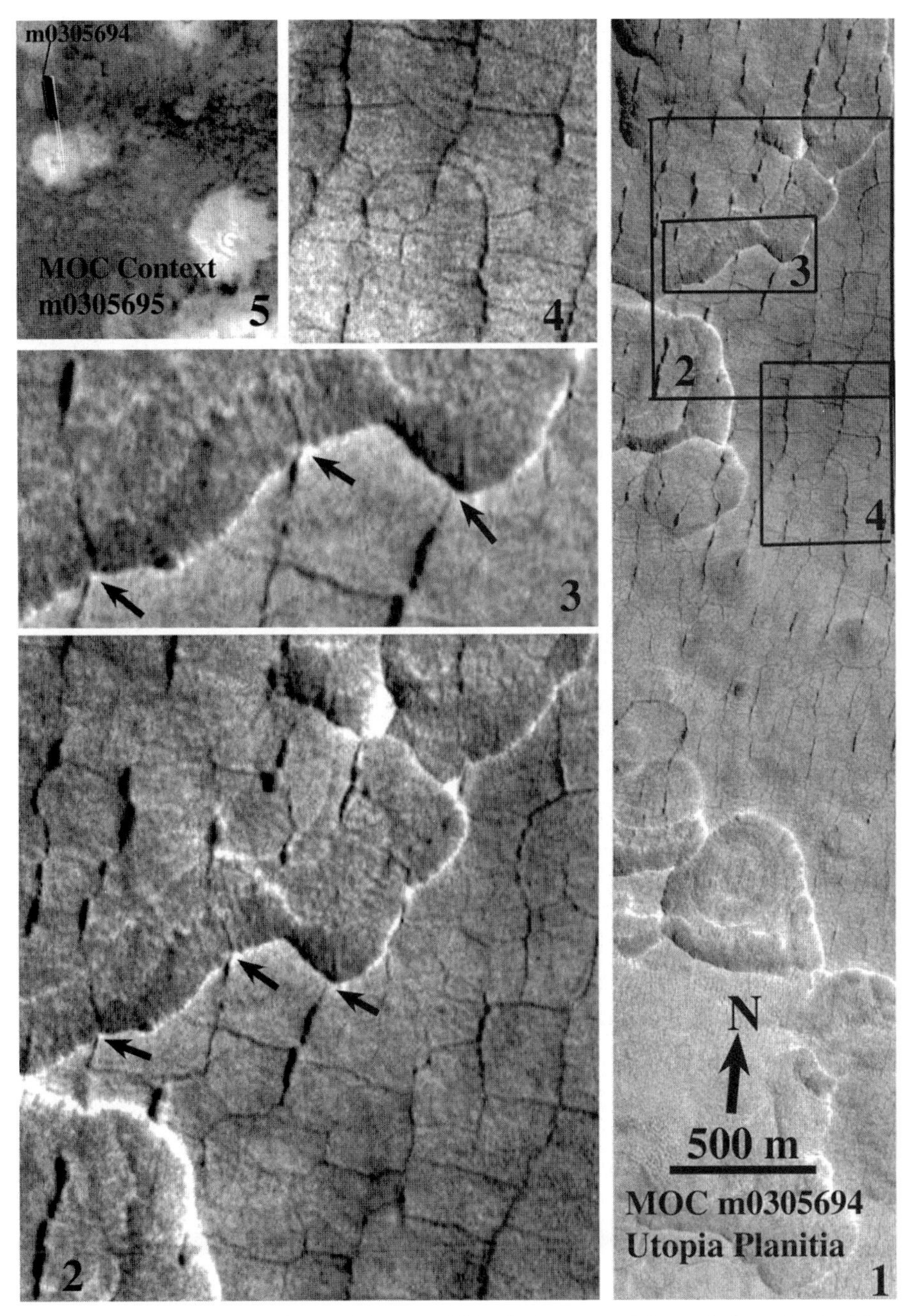

Polygonal cracks in Utopia Planitia are within the interior plains unit hypothesized by Parker et al. as having formed in the younger ocean. They could be sublimated ice-wedge polygons or due to mud desiccation or dehydration of highly hydratable salts, such as magnesium sulfate hydrates.

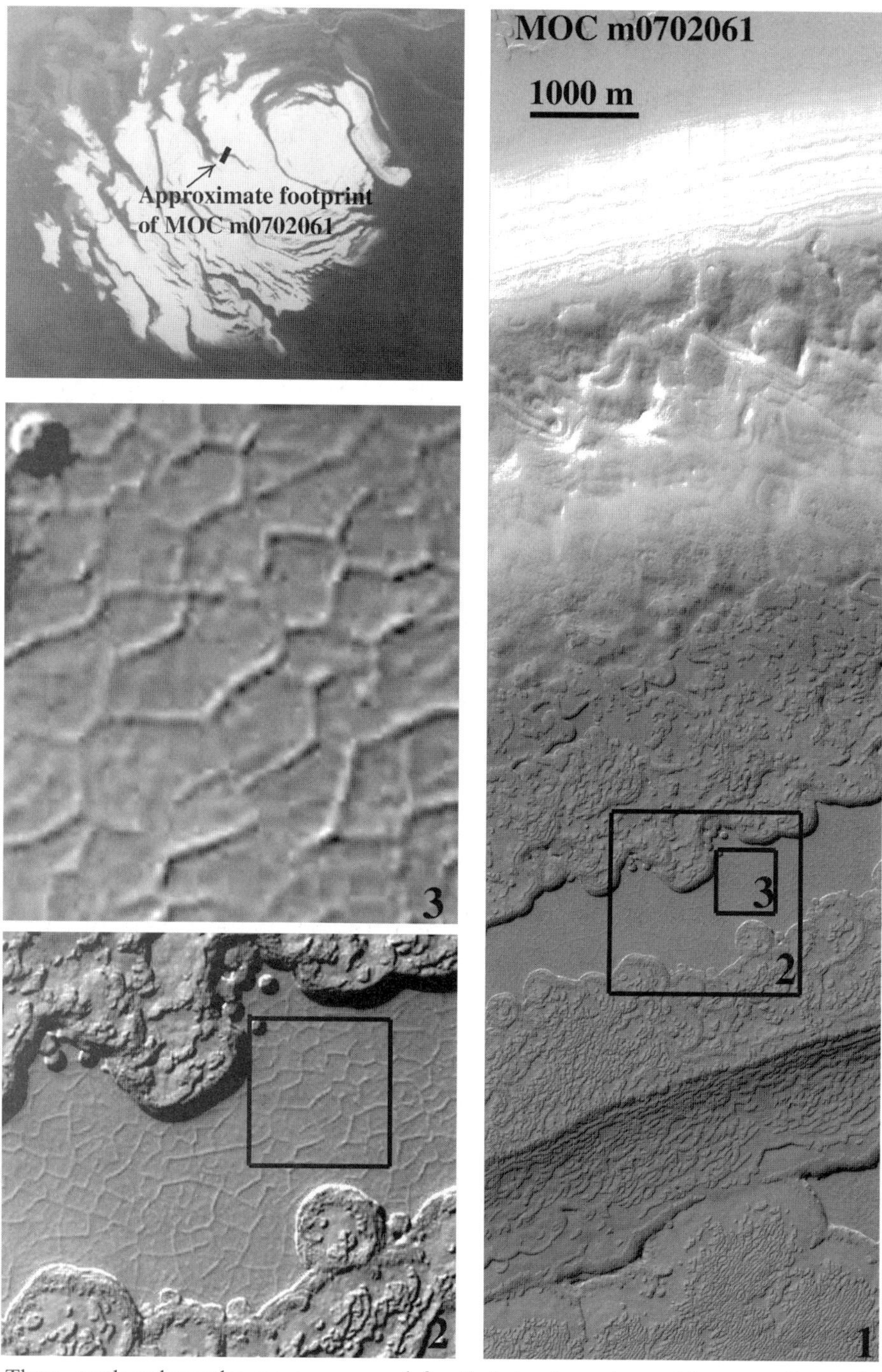

These south polar polygons are unusual for their tendency toward hexagonality and occurrence in dry ice.

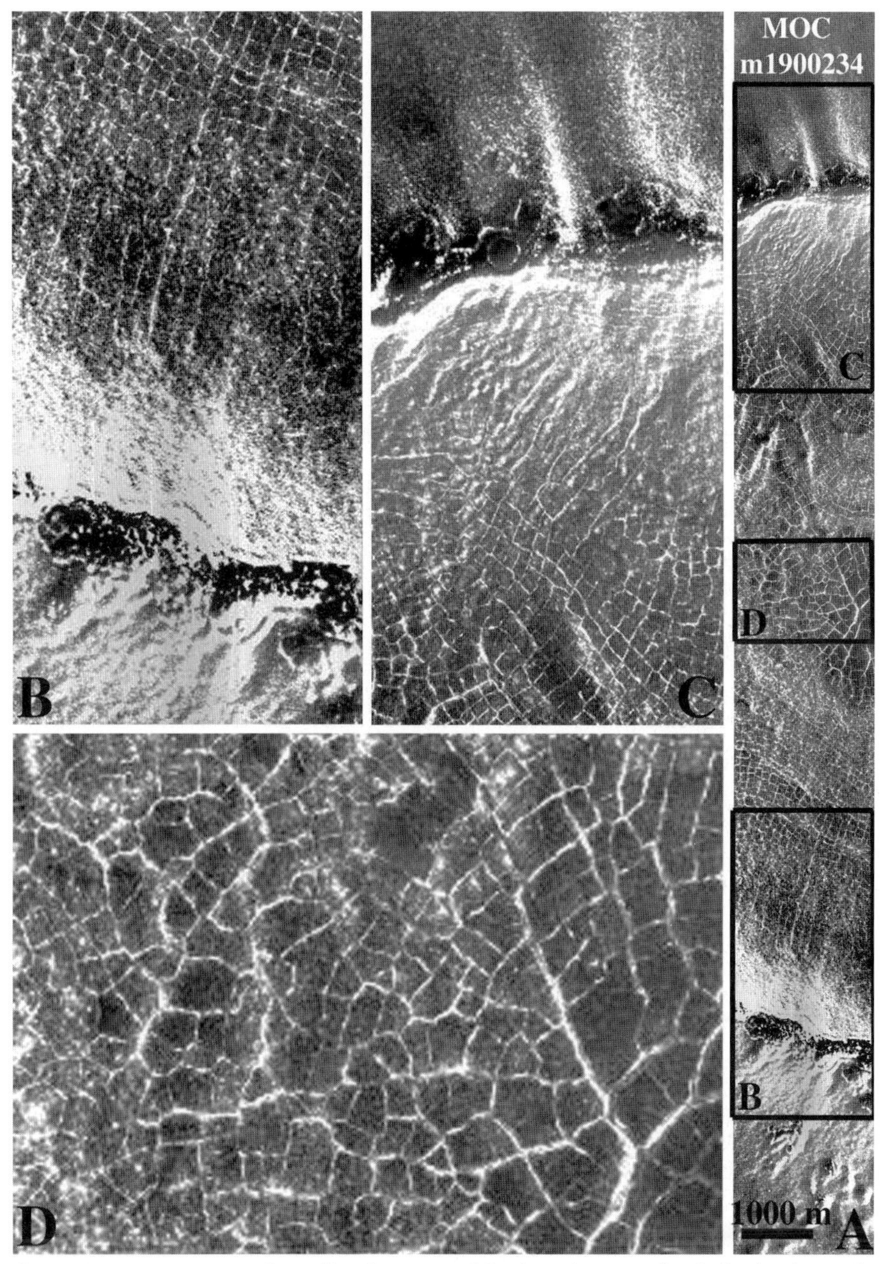

A common occurrence of small polygons at Martian circum-polar latitudes is on the floors and interior walls of impact craters, as made visible here by retention of late-season frost (A). Polygons occur in a variety of geometries, including parallel near the rim (B and C), random orthogonal at the nearly horizontal bottom of the crater (D), and oriented orthogonal (rectangular grid) at positions intermediate between the crater bottom and the rim. At this locality, polygons are undeveloped outside the crater.

Hexagonal columnar pattern of cooling cracks in basaltic lava, Devil's Postpile, California. Small grooves and chattermarks in the lower panel are due to glacial abrasion.

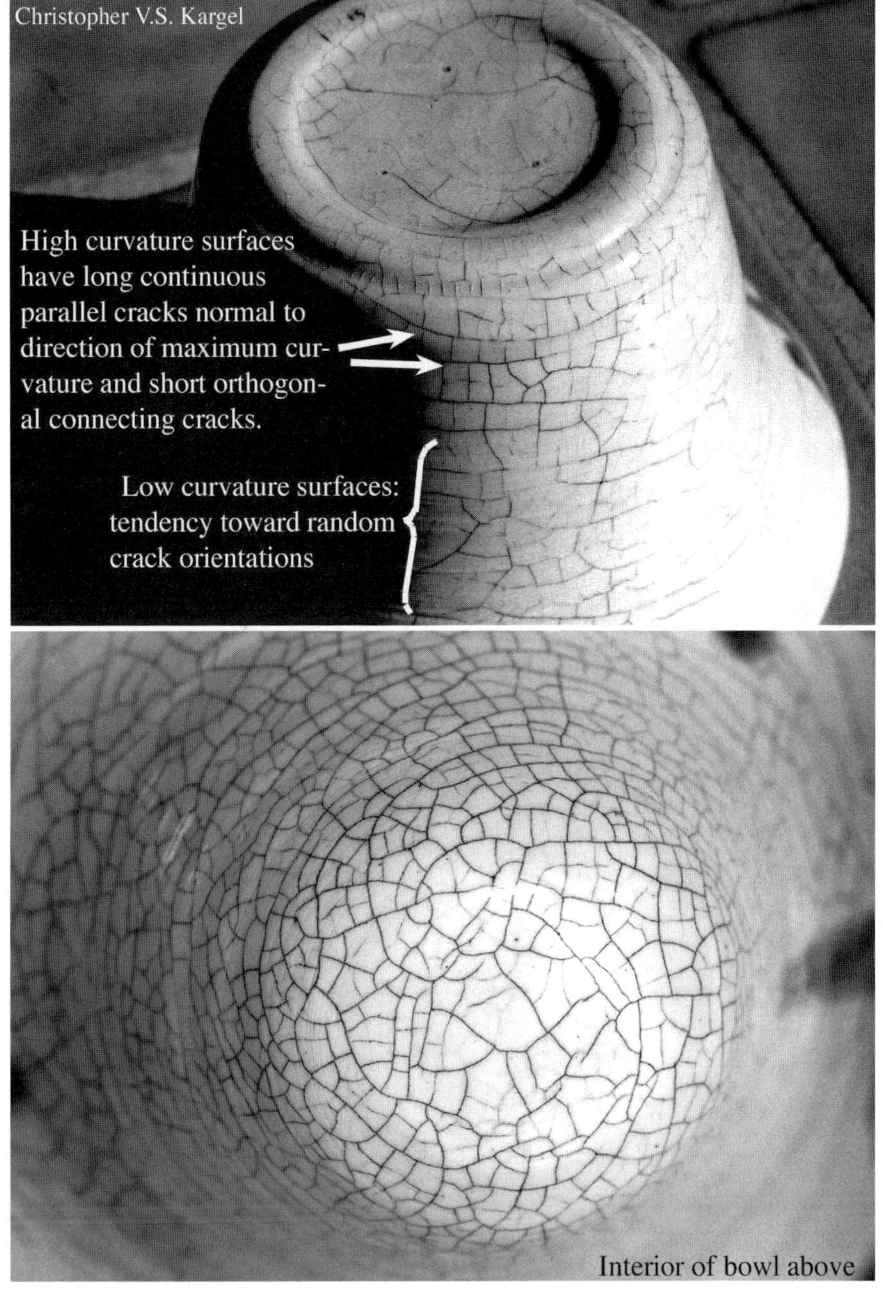

The materials and conditions of formation of these cracks were different from those of Martian and terrestrial permafrost cracks, but the tension cracking in both cases involves the same physics of the rupture of brittle membranes due to contraction cooling stresses.

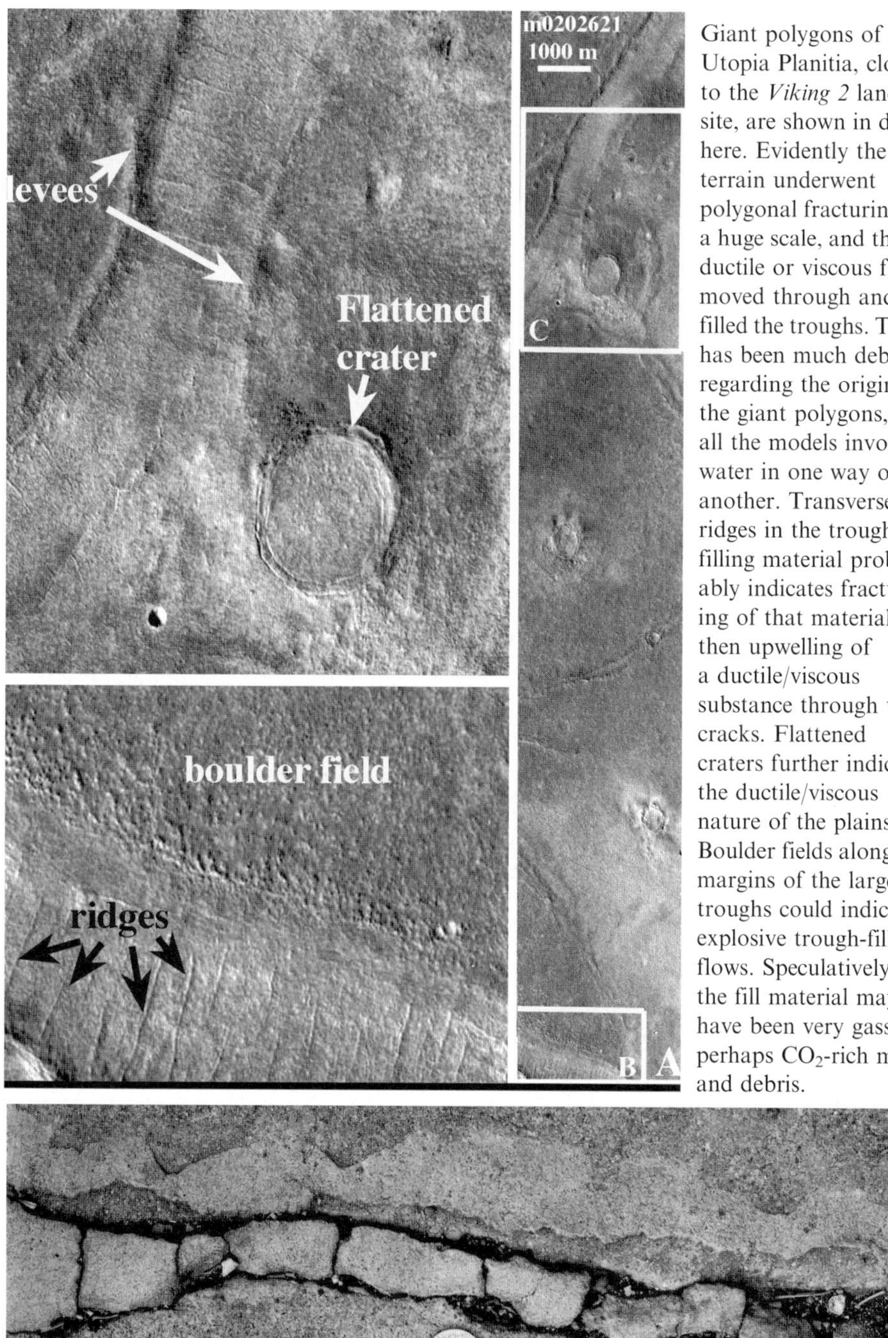

Giant polygons of Utopia Planitia, close to the *Viking 2* landing site, are shown in detail here. Evidently the terrain underwent polygonal fracturing on a huge scale, and then a ductile or viscous fluid moved through and filled the troughs. There has been much debate regarding the origin of the giant polygons, but all the models involve water in one way or another. Transverse ridges in the trough-filling material probably indicates fracturing of that material, then upwelling of a ductile/viscous substance through the cracks. Flattened craters further indicate the ductile/viscous nature of the plains. Boulder fields along the margins of the large troughs could indicate explosive trough-filling flows. Speculatively, the fill material may have been very gassy, perhaps CO_2-rich mud and debris.

Cement fill in a sidewalk crack has re-cracked, with secondary cracks orthogonal to the primary crack.

Ice-wedge polygons and permafrost thaw lakes, Arctic Coastal Plain, Alaska. Individual ice wedges here have a surface expression consisting of parallel raised rims (produced by compression of soft sediment during wedge growth) and a trough over the ice wedge itself. Enhanced solar radiation absorption in the lakes causes accelerated permafrost thaw. The resulting erosion is controlled by ice-wedge tectonics.

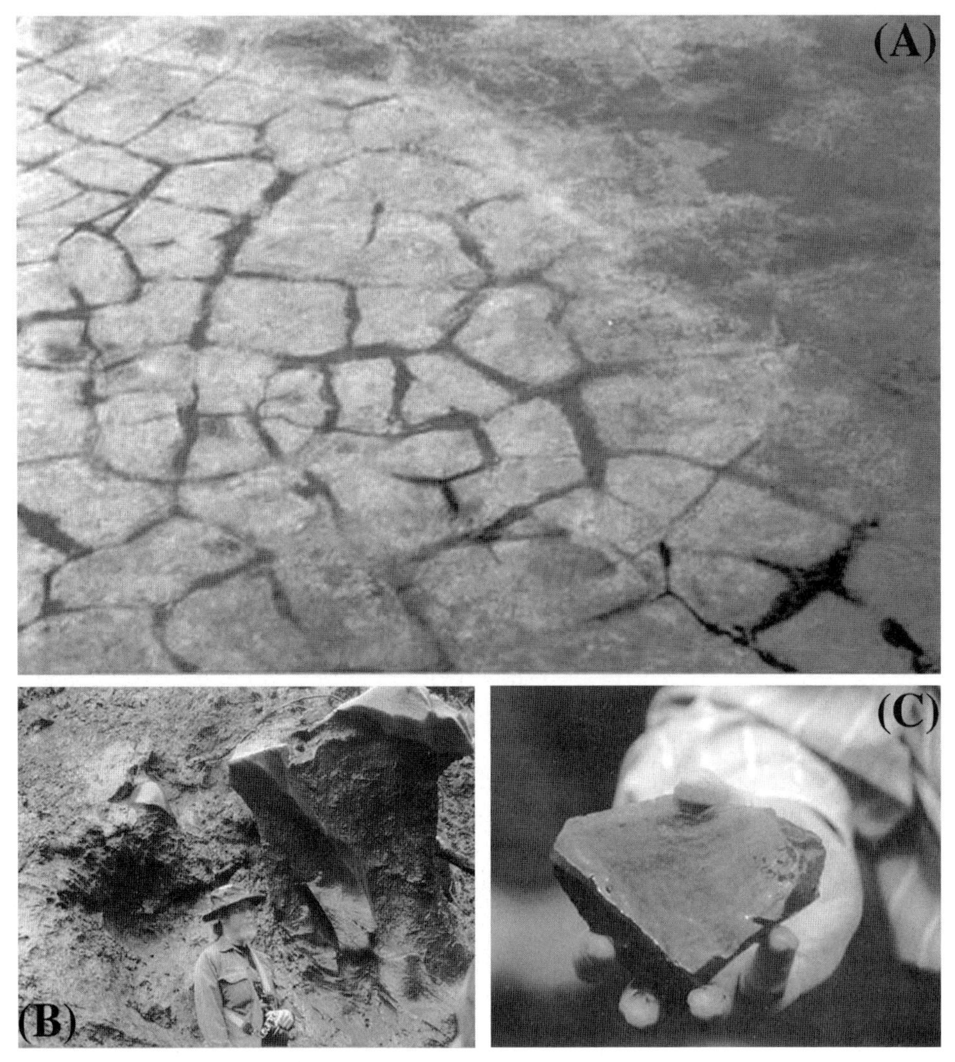

Ice-wedge polygons or other types of patterned ground occur through most of Alaska's permafrost. (A) Random orthogonal ice-wedge polygons (Arctic Coastal Plan, Alaska) were produced by centuries of annual freeze-thaw cycles, including wintertime contraction cracking of ice-cemented lake silt, spring and summer thawing of the active layer and meltwater drainage into the crack, autumn freezing and expansion of the infilled water and heaving of the sediment to the side to accommodate the volume increase, and then repetition of the cycle. The ice wedges occur beneath troughs, which here are preferred sites for vegetational growth. (B and C) Near Fairbanks, hydraulic mining of frozen loess exposes ice wedges, Pleistocene mammal fossils, and other structures. Here, ice-wedge ice is dark from decayed, malodorous organic material, roughly 38,000 years old. Jeff Moore (NASA Ames) for scale.

Sorted stone circles are well developed in a small permafrost lake basin in the Delta River Valley, Alaska. Still enigmatic, they were termed "sordid circles" by the late Troy Péwé (top panel). Sorting of sediment is thought to be related to active-layer convection caused by seasonal freeze–thaw.

Desiccation polygons in hydratable clay deposits from the 1993 flood in the Gila River, Arizona.

Desert playas – where ephemeral water and transient bursts of wind go together to make for a very dusty place. Here at a small Four Corners area unnamed play in Arizona, dust devils whip up a storm of swirling salt and clay (A). In panel B, the flatness of the playa is evident, as is a former shoreline terrace in the background, formed when the playa held more water than it occasionally does these years. (Beware the mirage in panel B, which makes interpreting it a bit complicated.) In panel C, evidence of water's role is seen in the meter-scale larger desiccation polygons and the smaller secondary polygons inset into the larger structures. The playa sediment mineralogy also attests to a role of water: mostly clay (primarily hydroxylated phyllosilicates formed by aqueous chemical weathering) and salts formed by aqueous chemical weathering and aqueous precipitation. When the sediment surface dries and heats in the summer sun, turbulent atmospheric vortices take over as the chief active process. The Mars science community would like to investigate a similar landscape on Mars with a well-instrumented rover. I expect that such terrains will soon be explored, but we will find some critical differences from what we commonly observe in such terrestrial hot and temperate deserts. On Mars the salt precipitates will be more highly hydrated: minerals such as hydrohalite ($NaCl{\cdot}2H_2O$) instead of halite ($NaCl$), mirabilite ($Na_2SO_4{\cdot}10H_2O$) instead of thenardite (Na_2SO_4), magnesium sulfate dodecahydrate ($MgSO_4{\cdot}12H_2O$) instead of epsomite ($MgSO_4{\cdot}7H_2O$). Ice will be a major chemical precipitate along with salts, and where the sediments were buried at least a few decimeters deep and climate remained cool and humid enough, ice should persist.

Halite deposits of Bristol Dry Lake, Mojave Desert, California, are broken into polygonal plates which are undergoing compressive tectonics. The sediment is still damp beneath the salt crusts. Alluvial fans, such as that in the background, completely surround the closed-drainage, evaporative lake, which sometimes fills with shallow water in the winter. The polygonal thrusts are dissolved and recreated each year.

expression than seracs of vigorous terrestrial glaciers, which may be attributed to recent reduced flow activity and erosional muting of the serac structure. All of the serac type structure is debris covered, as indicated by the dirt-red color and albedo.

Other forms of patterned ground have slightly different latitude dependencies than the polygons (but closely overlapping with them). The case is still compelling that some type of climatic control is forcing the responsible processes, with the differences in distribution attributable to deep ice versus shallow ice, sublimating ice versus melting ice, melting ice versus dehydrating salts, etc. There are good prospects that the specific phase changes responsible for each morphology can be determined by careful assessment in terms of climate models, scale and form analysis, and correlation with neutron spectroscopy data. These will be important features to examine in detail, including by rover traverse, because the substances involved in their hypothesized formation – ice and salts and clays – will provide important, versatile, and widely accessible resources to support establishment of a viable human presence.

The middle/high latitudes where small polygons and other forms of patterned ground occur are approximately where ground ice is predicted to remain stable close to the surface, and also is where the neutron spectrometer data indicate that ground ice exists within the upper meter. Thus, any proposed fundamental connection to carbon dioxide or wind, or any other substance just does not make sense. We can draw two powerful conclusions: (1) H_2O in the form of water ice and/or other major reservoirs of condensed hydrogen is fundamentally the responsible agent for most types of Martian polygons and the other forms of patterned ground, and (2) more or less the present climate is relevant to their formation. However, the common occurrence of features affected by the sublimation of ice from the upper layers (generally tens of meters) in the latitude zones between $\pm 30°$ and $\pm 45°$ on both sides of the equator is good evidence that the Martian climate has recently changed – it has become warmer or drier, or both, such that relative humidity has declined, causing either the sublimation of ice or the dehydration of hydratable phases. These are basic conclusions that seem as firm as anything we can say about Mars. Beyond these statements, this set of observations and inferences allows for a wide range of interpretations and implications, with uncertainty shrouding the exact phase changes involved. Whether surface melting is occasionally involved in loss or replenishment of ice, or ancient dried lake beds are indicated by some polygons, whether carbon dioxide plays some accessory role, or orbital obliquity changes or long-term epochal climatic evolution is involved – all these intriguing ideas await further observations and modeling.

The similarity of Martian patterned ground to that on Earth is striking, and it seems improbable that glacier ice and icy permafrost are somehow not involved. However, similar morphologies can form by several volume-reduction and tensional shrinkage cracking mechanisms. These include thermal contraction cooling (seasonal cooling of icy permafrost, and lava cooling), loss of pore water in drying mud (desiccation), loss of excess pore water in supersaturated sediment (including subaqueous synaeresis), and loss of crystallographic water in hydrated salts or other hygroscopic minerals (dehydration). In addition, tension cracking of surficial material without shrinkage can occur by gravity sliding, stresses imposed by piercement (in sediment affected by ascending salt domes or mud diapirs, for instance), and other tectonic processes. Since these various processes would carry different environmental implications, we seek observations that can constrain their genesis.

Various forms and causes of shrinkage cracks are not always readily distinguished from one another. Mud shrinkage cracks typically (but not always) have curled rims. Permafrost contraction polygons, while generally of the same scale as the Martian polygons, commonly (but not always) have rounded, heaved rims. Rim features, however, are not always produced, and if they are, they can be readily destroyed by weathering and erosion. The usual lack of rim features thus offers fewer diagnostics. Permafrost polygons commonly exhibit a medial ice wedge, which can be expressed as a medial ridge running down a trough, but mud cracks occasionally possess a medial clastic dike of intruded mud or sand that can appear as a medial ridge. Scale is a useful criterion but not in itself diagnostic. Permafrost polygons occur principally at scales of several meters to tens of meters, and desiccation polygons ordinarily occur at scales of centimeters to a couple meters, but there are exceptional occurrences of desiccation (or dehydration) polygons 100–200 m across. The significant overlap of scale of polygons formed by drying and those caused by seasonal thermal contraction thus prevents scale from use as a diagnostic.

The most compelling diagnostic – though not readily discriminating among desiccation and permafrost polygons – is the geometric integration and intersection characteristics of crack networks and the topographic/tectonic associations. Tectonic tension cracks are usually easily identified by the orientation of cracks (e.g., parallel, radial, or concentric) and association with a tectonic cause and a discrete tectonic feature. Lava cooling cracks are ordinarily organized in ideal hexagonal networks not seen commonly in other types of shrinkage crack networks, and of course they occur in volcanic materials. Mud desiccation and seasonal cooling cracks (including ice-wedge polygons) ordinarily form quadrilateral or pentagonal polygons, where cracks progressively propagate and intersect at right angles (T-junctions) by curving into one another. Dried mud tends to occur in fine-grained sediment accumulated in lake basins, river channels, or other areas of ponded and then evaporated water. Ice-wedge and sand-wedge polygons may occur in similar materials but can also occur on gentle hillsides.

Drying and seasonal thermal contraction thus produce similar forms of polygons, and no property seen in images is fully diagnostic, but in aggregate the properties and associations can be fairly compelling in distinguishing the two processes. Sometimes compelling is an association of permafrost polygons with other ice-related features (and ice itself) and desiccation polygons with water-formed features; however, ice- and water-worn features often occur together. The regional and global distribution (and climatic associations) of these polygon types turns out to be the most compelling indicator of origin, short of having subsurface information.

On Earth, ice-driven creep, thermal contraction cracking, seasonal melting and meltwater drainage and sediment infiltration into cracks and crevices, and heaving and settling caused by seasonal freeze–thaw in the surface layer also tend to disturb permafrost. These processes sort materials by grain size, fracture the terrain in various patterns, heave and thrust the surface layer, create subsurface voids, and impregnate any voids in the subsurface with segregated massive bodies of ice. Sorted stone circles, stone stripes, ice-wedge polygons, pingos, lobes, ridges, troughs, mounds, pits, and other sorts of peculiar features and patterned terrain is the characteristic result. Destabilization of massive layers, lenses, nodules, and dike-like wedges of ice in permafrost generates sink holes, shallow lake basins, and organized or patterned arrays of scalloped or nested pits and polygonal troughs. Though wind and other processes can generate patterned terrain features (sand dunes and erosional yardangs for instance), the variety of features formed in permafrost and their specific characteristics readily identify these terrains. There is nothing else on Earth quite like the assemblages of permafrost landforms. While certain permafrost landscapes are very widespread and distinctive, there is no single diagnostic permafrost assemblage.

A particularly important permafrost landscape consists of the thermokarstic thaw lakes and ice-wedge polygons typified by those of the Alaskan Arctic Coastal Plain. In Alaska, meandering rivers draining from the Brooks Range and other uplands have deposited a thick wedge of primarily silty sediments, forming a low-relief coastal plain. Unlike coastal plains along tropical and temperate coastlines of passive plate tectonic margins, the Arctic Coastal Plain has mean annual surface temperatures well below freezing; consequently, the water-saturated sediment is, in fact, ice-saturated to a depth controlled by the geothermal gradient and the pressure-melting curve of ice. There is a shallow surface "active layer," typically a meter thick, that undergoes annual freeze–thaw cycles. Intense summer heating melts the past winter's snow; lakes overfill, and rivers coursing over the landscape permeate every accessible void with liquid water until an impermeable, perennially ice-cemented layer is reached. By the next winter, the landscape is frozen hard. Thermal stresses due to winter cooling causes the ice-cemented sedimentary matrix to fracture. These fractures represent voids into which the next summer's meltwater percolates. The water-filled icy cracks then refreeze during the next winter, and as the water expands it pushes the adjoining sediment away from the crack, commonly causing it to heave at the surface. The ice-filled crack often then refractures due to thermal stresses, and the cycle continues. The crack thus widens and fills progressively with ice. The ice filling is termed an ice wedge.

Similar cracks can form in less wet or even dry permafrost environments, such as the Antarctic Dry Valleys, where the filling substance is commonly sand or a mixture of snow and sand, though the dynamics of permafrost sand and ice wedges are different. Despite different formation mechanisms and products of the two extreme end-members, there appears, however, to be a continuum from the icier to the sandier polygon end-members. Even in the extreme cases, it appears that seasonal thermal contraction cooling of ice-impregnated permafrost is a basic underlying cause. The other principal polygon-forming tension cracks, aside from lava cooling cracks, also involve water (liquid water or water of hydration). Thus, except where lava cooling can be implicated, polygonal terrains in general are indicators of H_2O.

Tension fractures normally do not occur in isolation but rather in polygonal networks. Ice wedges can be arranged in cross-cutting patterns as younger cracks cut across older healed cracks; or they can form the orthogonal patterns most characteristic of mud desiccation cracks. In orthogonal mud cracks and ice wedges, each crack relieves the tensional stress only perpendicular to the crack; as one crack propagates and approaches an older one, it enters a region of stressed sediment where the older crack may already have relieved stress in one direction; and so the younger crack curves into and intersects the older one at right angles. Other geometries of crack networks are possible, including rectangular networks or systems of parallel or *en echelon* cracks, which are especially prevalent where the sediment has an anisotropic or unidirectional stress field even before cracking commences. This commonly occurs on sloping ground or on substrates that are curved. A very common pattern involves orthogonal systems of large primary tension cracks that break the sediment into large polygonal plates, which are then cut by finer networks of smaller secondary cracks. This can signify repeated episodes of tension stress formation (e.g., repeated wetting/drying episodes for desiccation cracks, or equivalent cold-climate cycles for ice-wedge polygons) or prolonged episodes of tensile fracturing where the first episode of cracking relieves some stress, which then builds up again. Populations of fine-scale secondary and tertiary cracks superposed on larger primary ones can also be due to cracking on multiple scales in layered media.

Insights into tension-cracking behavior can be gained in simple table-top experiments with mud dried under a hot lamp, and in studies of crackle glazes in common ceramics. The influence of substrate geometry, especially curvature and proximity to strong, rigid boundaries, is a critically important variable, as is the thickness and shrinkage properties of the tensed material. We see the roles of such parameters in natural desiccation cracks and ice-wedge polygons on Earth, and we see it on Mars, for instance in the geometry of cracks in craters.

As yet it is not clearly evident from the scrutiny of most Martian crack networks (work by Seibert and Kargel) whether desiccation of mud or thermal stressing of icy permafrost was principally involved, though the global distribution of polygons is similar to that of near-surface ice. These two classes of polygons are difficult to distinguish without direct field access or detailed geophysical means to ascertain the subsurface structure of the polygon systems. There are particular morphological attributes that can distinguish them (e.g., curled rims of mud desiccation cracks), but it requires very fresh, undegraded forms, and in any case these diagnostic features are not always present even when the features are fresh.

I tend to think that permafrost processes, rather than sediment drying, are chiefly involved. This supposition betrays a desire to explain all polygons with one process, an approach that definitely would not work on Earth. This is not true in all cases on Mars, either, since some Martian polygons occur where ice and water is not and probably never has been present; those appear to be very special cases (occurring in dry-ice polar deposits and in low-latitude volcanic plains) and have a unique morphology; contraction-driven tension cracking appears to be the cause of all Martian polygons. Seasonal thermal contraction cracking is probably the underlying cause of most Martian polygonal cracks. Unexplored is the possibility that some cracks on Mars, and in certain evaporite lake basins on Earth, may be formed by dehydration and shrinkage of hydratable salts and other water-rich minerals. It is also possible that on Mars, where competing geomorphic activity is less than on Earth, ordinary heating/cooling expansion/contraction cycles of non-icy but slightly cemented sediment can cause polygons to form; this seems improbable, since the polygons are distributed preferentially almost the same as the global distribution of ice. Thus, some connection with ice seems almost assured in all but a few exceptional instances.

Roles of liquid water as well as ice are widely considered to be essential in the formation of most terrestrial permafrost landforms. This is also probably true on Mars, although some terrestrial permafrost processes occur in the absence of a liquid phase, which is apt to be true also on Mars. This interpretation, of course, presents a challenge – ever since the first remote temperature measurements were made in the decades before the dawn of spaceborne observations, Mars has not been a place warm enough to allow melting of ice in areas humid enough to permit ice to exist at the surface. Liquid water can exist in the subsurface many miles beneath the surface, or closer to the surface where igneous activity or hot springs occur, and occasional breakouts of liquid water could occur at hydrothermal springs. However, there is no chance that under today's environment liquid water could produce the polygons and patterned ground observed so widely on Mars. These features are very young, and so it would appear that the Martian environment has changed very recently. Consistent inferences are obtained from other classes of features, such as gullies.

PERMAFROST COCKTAIL

My first personal experience with a permafrost landscape and ice-wedge ice was in Alaska in 1993. I had organized a workshop on the Martian northern plains, and an invited guest of honor was the distinguished geomorphologist, the late Troy Péwé, an expert in cold and arid lands. Upon my arrival, Troy offered up a glass of his infamous permafrost cocktail. The slightly aqueous grain-alcohol-based solution was cooled by 38,000-year-old ice-wedge ice with dark inclusions of foul-smelling organic/bacterial detritus. So offered by the reigning King of Cold, I could do nothing but chug it and then listen in mellow contentment, over dinner, as Troy recited Robert W. Service's *The Cremation of Sam McGee* entirely from memory.

Dutiful after recovering from that night's work, the workshop participants later watched as commercial gold miners outside Fairbanks used powerful hydraulic jet sprays to melt away a thick blanket of permafrost-cemented dust ("loess") to collect flecks of gold that had been hoarded for 38,000 years by filthy-rich bacteria and ice-entombed remains of bison, camels, horses, and other creatures. The frozen remnants of thousands of mammals from the loess have no opportunity to tell their many stories that must include countless escapes from predatory attacks, bitter winters, blooming summer meadows, and mating competitions won and lost on the tundra plains.

The light-gray talcum powder-like loess has its own story, the bare framework of which is told partly by the glacial landforms arrayed around the plains and in the nearby mountains, and partly by analogous landscapes presently forming further north. The loess was originally produced at the beds of nearby valley glaciers and mountain ice caps. Flowing for millennia, the ice plucked pebbles, cobbles, and boulders off the bedrock. Dragged over the bed, the bedrock and entrained grains were scraped and polished to optical quality. Optical grinding produced a million trillion trillion individual grains ranging from one micron to a few tens of microns in size (called "glacial flour"). As the glaciers melted each summer, torrents of meltwater rushed across the glacier bed. The coarser sand, pebbles, cobbles, and boulders settled onto the beds of subglacial rivers, forming eskers, and onto the beds of braided streams in front of the glaciers. The finer sand, silt and clay-size sediment was flushed completely beyond the glaciers' snouts. The discharged turbid water ("glacial milk") eventually laid down sand bars of braided streams, and the finest flour was deposited further downstream in muddy beds of sluggish streams and lakes, which watered tundra vegetation. The glaciers eventually receded; the volume of discharged meltwater declined; and the lakes dried or drained. The summer sun-baked, mud-cracked lake plains were whipped by windstorms and dust devils, raising an endless train of dust clouds. The dust eventually found places to settle where meteorological conditions allowed; those places included the tundra plains near Fairbanks. The dusty silt was stabilized in place by shallow summer thaw lakes and permafrost ice. Seasonal cycles of freezing and thawing opened thermal contraction cracks each winter. By summer the cracks filled with meltwater, which then froze during the next winter, wedging the cracks wider before refracturing under deep freeze. This freeze–thaw process continued over the millennia, and this process constructed polygonal networks of ice-wedge-filled cracks. Summer thaw and lush summer growth under the midnight Sun attracted animals, which lived and died and became entombed on those icy and briefly seasonally muddy plains. A somewhat similar physical and biological story is unfolding even today just 500 km north along the Alaskan Arctic Coastal Plain, which offers some of the most marvelous partial landscape analogs for Mars.

The story of the loess went untold, in deep-frozen preservation, until it was disturbed by the gold-mining operation. As the permafrost washed away, hydraulic erosion exposed wedge-shaped masses of debris-darkened ice, which came crashing down the cliffside, carrying with them their iced archives of prehistoric times not really that long ago. Those were the final days of a truly wild and unpeopled America, a time just before the last great ice sheet and mountain ice caps climaxed across the northern expanses of North America. Meanwhile, as the loess gathered in Alaska, Europe was populated by Neandertals, exactly as modern humans began a 10-millennium-long expansion through Europe, starting in the Middle East. *Homo sapiens* mastered Europe, either co-opting the cold-adapted *Homo neanderthalensis* or causing their extinction. Then the full glacial maximum hit, straining the ability of mammalian and particularly human life to survive, and placing a premium on intelligent human adaptability and ingenuity. The next time the glaciers melted back, civilization burst forth across the Old World, and North America became populated by this clever human species. Overcome by land quest and newly empowered by Ice-Age survival skills and hunting technologies, the tried and proven survivors of global climatic upheavals during Earth's Ice Age went on to build 10,000 cultures and the dream and technology to go onward to Mars.

It is perhaps fitting that as we venture forth to Mars, we will be tested in a hostile permafrost desert terrain not too unlike the icy, desiccated landscapes across which *Homo sapiens* spread across Europe, the Mongolian steppe, and into North America. Ice and drought were some of the chief obstacles challenging human survival across most of the globe. Once overcome, nothing could limit human aspirations. On Mars, ice will be the resource challenging humans to find ways to exploit to help us to overcome an ultimate drought like none ever experienced on Earth. Aiding our survival, much of what we find on Mars is already familiar to Earthlings.

ICY FAMILIARITY

In addition to similar forms of permafrost, Mars also exhibits glaciers that, in many respects, are familiar. A glacier is defined here as a perennial mass of ice that exhibits morphological signs of flow under the stress of its own weight. This general definition is probably the simplest and most common one in use, and also the broadest; one can certainly argue about details and marginal cases that may or may not be glaciers, but I see no need to. This definition does not indicate mode of accumulation or ablation, but growth of any glacier requires accumulation from all sources to exceed ablation due to melting and/or sublimation and other loss processes. Many glaciers exist close to a state of hydrologic balance. Every terrestrial glacier exhibits one or more zones where ice accumulates averaged over the year (generally but not exclusively near the source), unless it is on a path to its final demise; and they exhibit one or more zones (generally but not exclusively at the terminus) where ice is lost (ablation zone). Integrated over the entire glacier, the net annual balance may be negative, zero, or positive. Mass balance and accumulation and ablation zones generally fluctuate from one year to another. Any healthy glacier exhibits a longitudinal topographic surface profile such that gravity drives ice from sites of net accumulation to sites of net ablation. The classical analytical and empirical description (a good approximation) is that glaciers exhibit parabolic longitudinal and transverse surface profiles needed to maintain constant local dynamical mass balance due to the sum of accumulation, flow, and ablation.

Viking Orbiter images convinced many Mars researchers that some form of ice flow process is responsible for several classes of striped and lobate features, including lobate debris aprons, lineated valley fill, and concentric crater fill, so named according to the topographic bounds of these deposits. These are probably similar phenomena, here termed synonymously "icy flows" and "glaciers." This broad group of features has the definite appearance, even with casual inspection, of having flowed. Their icy and flow natures are justified below. The figures presented here should be sufficient to convince readers that debris-rich icy flows – glaciers – exist on Mars. These examples look very Earth-like in many regards. However, as succeeding examples will show, their flow-like nature is justified, but most are not very Earth-like in several other respects.

Mapping of the global distribution of these features completed in the 1980s by Steven Squyres (Cornell University) and Michael Carr (USGS/Menlo Park) showed that the icy flows are overwhelmingly located at middle latitudes. The interpretation offered then was that (1) at lower latitudes ice is absent due to the dry equatorial climate, (2) at higher latitudes ice is present but too cold and hard to flow, and (3) at intermediate latitudes ice is both abundant and warm and soft enough to flow. Certainly some kind of climatic control is evident. There also evidently are strong geologic controls, because they do not occur *everywhere* at these latitudes. These mid-latitude flow features are particularly common and well developed in parts of Argyre, the "Hellas Montes", the northern fretted terrain, and locally on crater rims. They are certainly affiliated with steep relief as much as with middle latitudes, and in some regions, especially the "Hellas Montes", there is a preferential occurrence on pole-facing slopes.

Theoretical calculations of ice stability over the last quarter century by Jack Farmer and Doms, Michael Mellon, and many others have shown that ground ice

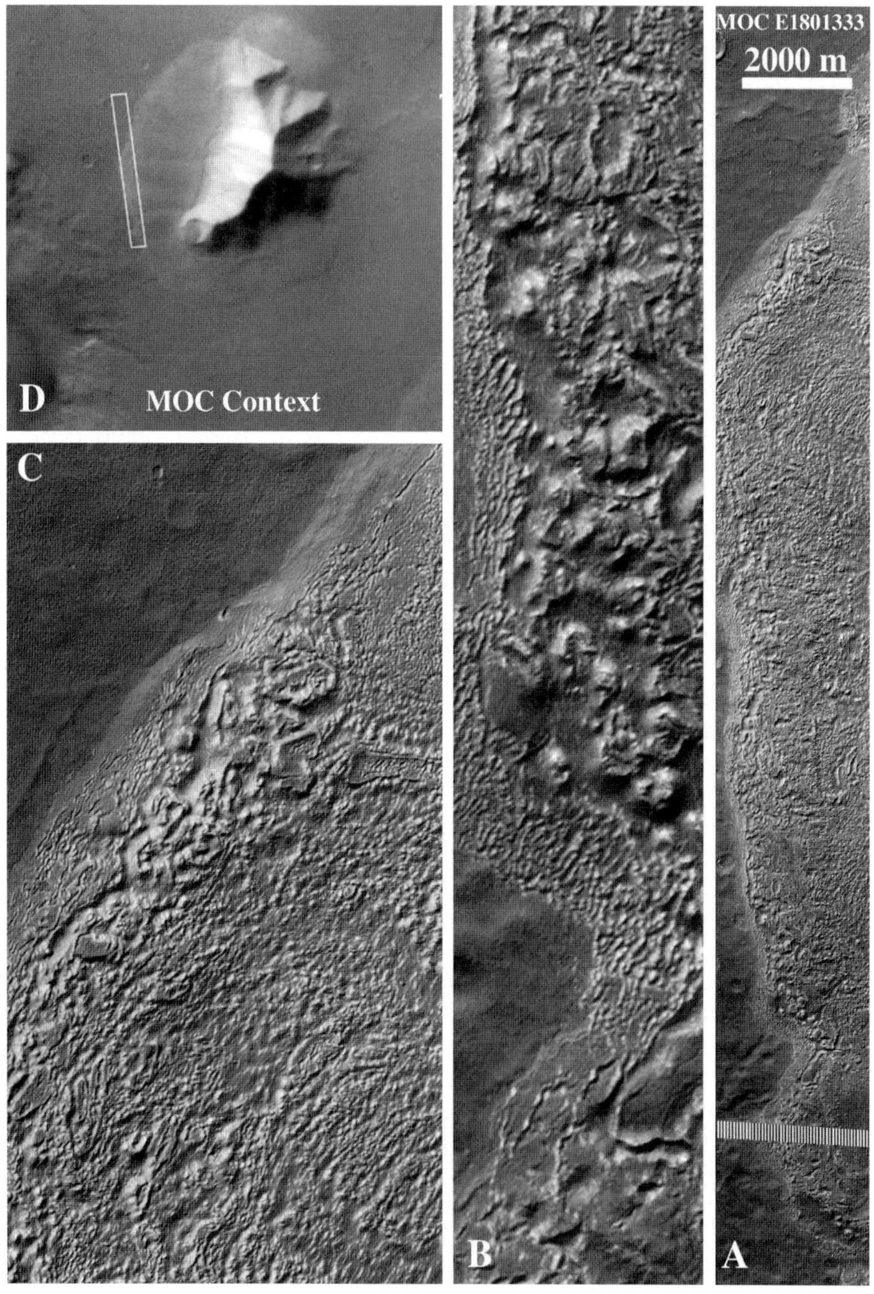

Typical of the "melting mountains of Hellas", the lobate debris aprong of this one shows compelling indications of flow, including longitudinal striping and transverse corrugations, which have been etched into sharp relief by ice sublimation. Panel B shows cratered cones, possible mud volcanoes, near the flow terminus. These features make the "ice cream model" proposed by Greg Cardell seem almost tasty.

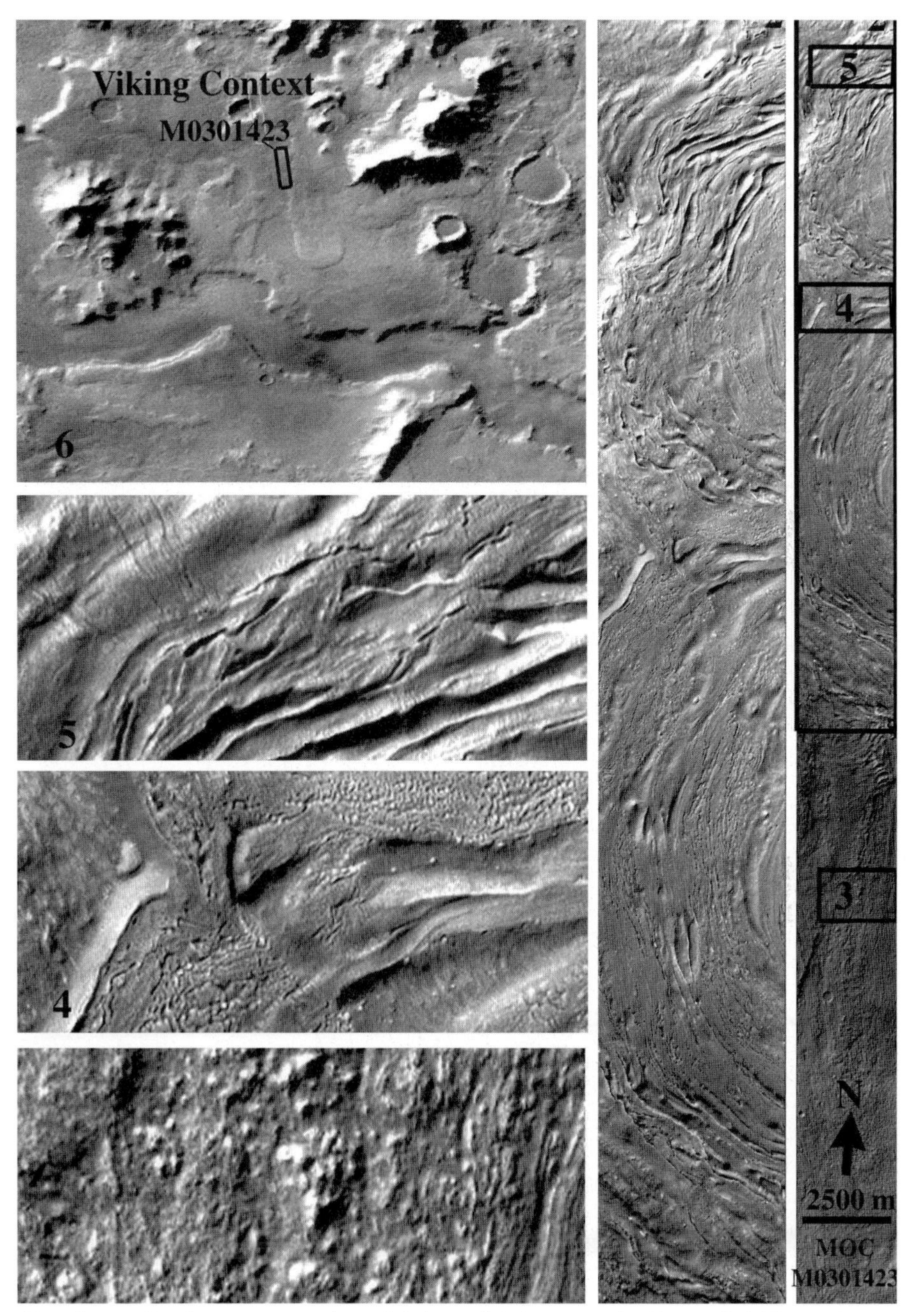

Terminus of a large tongue-shaped lobate debris apron in Promethei Terra shows evidence of flow in its lobate form (panels 1 and 2), in crevasses (panel 5), and in compressional folds (panel 2), which have been eroded by sublimation into grooves and ridges (panel 5). Pits (panel 4) further indicate ice sublimation, while knobs and ridges (panel 3) are moraine-like and suggest deposition.

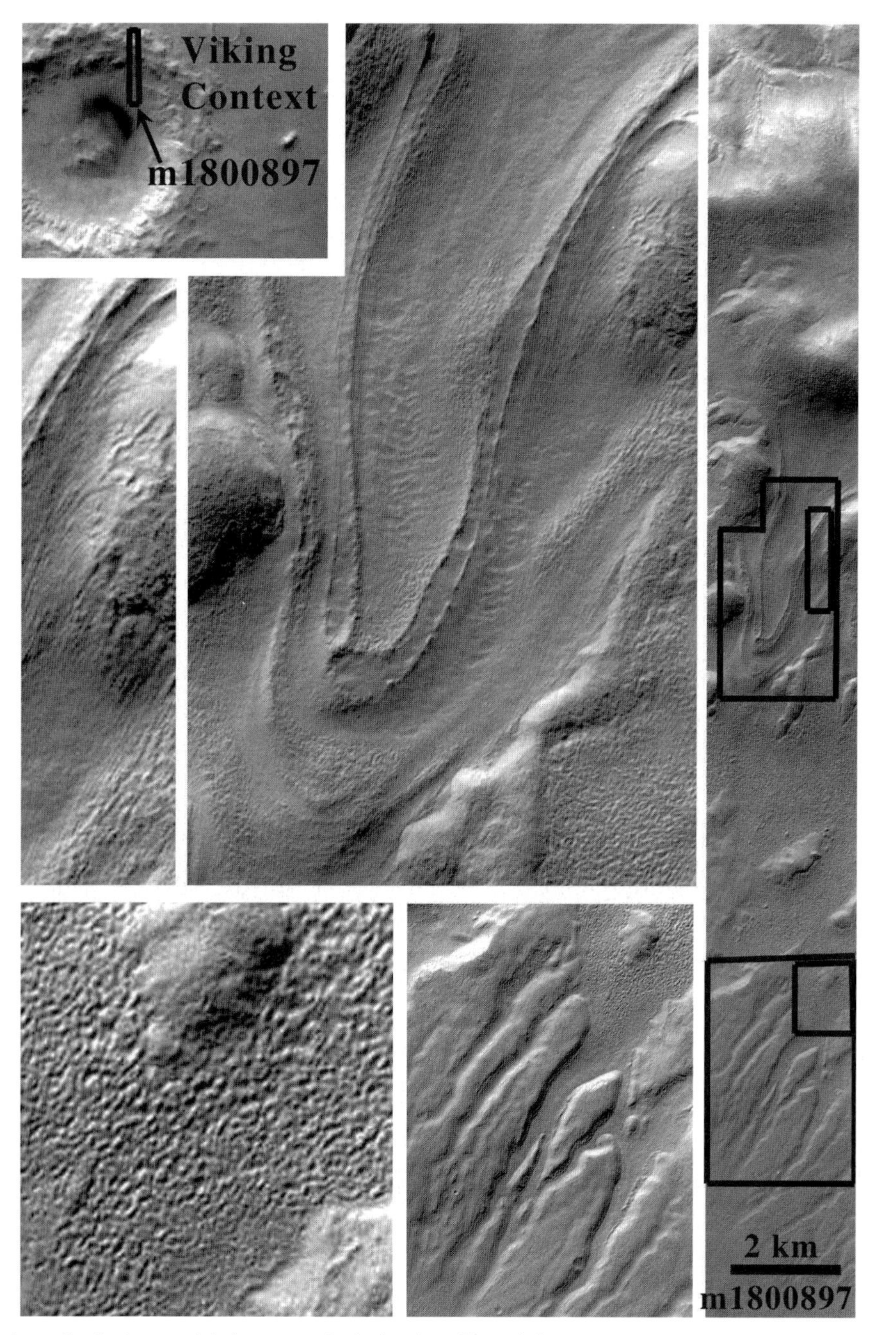

A rock glacier or debris-covered glacier has flowed from a source near the crest of a crater rim toward the floor of the crater. Associated with the flow are ice ablation features and gullies or erosional scars due to debris flows. This flow occurs on a large crater east of Hellas, where many mountains have dirty glaciers.

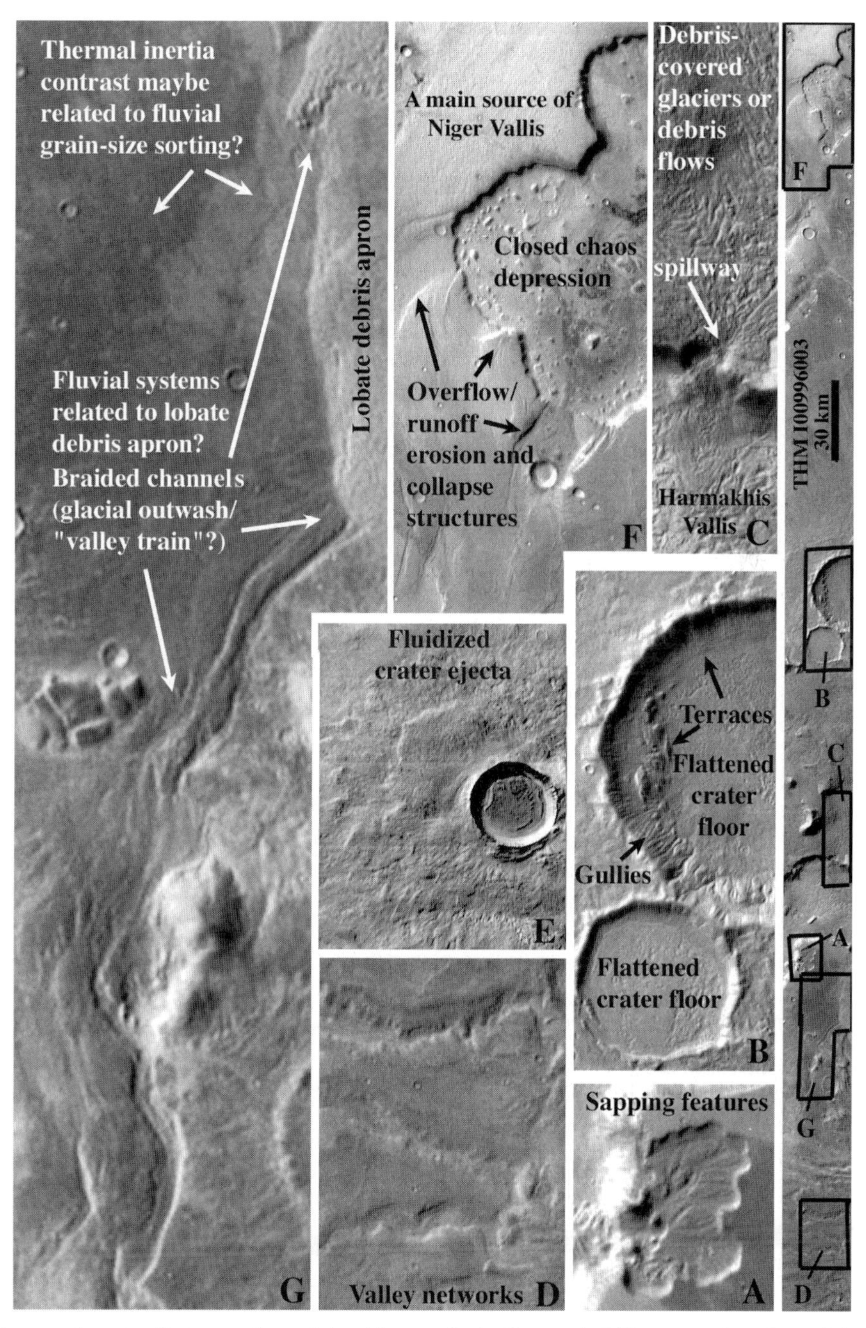

Ice- and water features abound in this terrain in Promethei Terra at the edge of the Hellas Basin, near where Harmakhis Vallis enters the basin. This terrain resides within the bounds of the circum-Hellas 0-km contour, which roughly coincides with an escarpment that in places is fretted. –37.2°S lat., 264.5°W long.

should be stable poleward of about $\pm 30°$ latitude, but unstable at lower latitudes. This is closely consistent with observations of the lower latitude cutoff of the icy flows, a fact that formed the first compelling indication that these flow features are icy. Skeptics, however, during the *Viking* era cautioned that wind-deposited dust and sand may compose these features and give rise to textures that seem as though they were created by flow of ice and mantling of terrain by ice-rich debris blankets. In some cases, this was true when the best *Viking* imagery was examined. Though we had global image coverage of Mars from *Viking*, it primarily lacked the details necessary to provide any but the most basic morphologic characterization. These types of disagreements were usually impossible to resolve. As handicapping, useful topographic control was almost lacking until MOLA orbited Mars.

Perhaps more debilitating in the 1980's than the limited data was inadequate understanding by most members of the Mars science community of analogous ice processes and ice-caused landforms on Earth. The Mars community during the *Viking* era collectively had good training and experience in wind, impact, and volcanic processes, but it was ill prepared to recognize ice processes and ice-modified landscapes. A few of the more prominent and influential scientists were unwilling even to listen to the evidence pointing to glacial and permafrost processes. There was the further limitation, burdening all of us, that Mars is not Earth; conditions there are not those familiar to Earth-trained observers, and great uncertainty shrouds our understanding of past climate and process variability; nobody can do a good job of predicting the details of Martian ice-related landforms and landscapes, though we may recognize them once they are seen. With such limited types of data, mainly remote-sensing data, and with so many uncertainties regarding a different world, we should expect a wide range of interpretations of Martian landforms and landscapes. The *Viking* data made for a compelling case for icy landforms and ice-flow processes but did not offer unassailable proof. As just one observer, I must conclude that the icy and glacier-like nature of the Martian flows is confirmed beyond reasonable doubt by MOC and THEMIS imagery, neutron spectroscopy, and older data. However, we are probably not yet near the end of the upheavals of paradigm, as Mars knowledge remains thin in variety despite terabytes of data. One would worry at this point about the collective objectivity of the Mars community if the consensus became too cozy. Not to worry, the consensus is not so tight.

For some researchers, the threshold burden of proof for the icy constitution of the Martian flow features, over all other possible bizarre explanations, came from neutron and gamma-ray spectroscopy. Those data, presented by William Boynton (University of Arizona), William Feldman (Los Alamos National Laboratory), and others on the Gamma Ray Spectrometer team, have shown that ice is abundant and widespread at the middle and high latitudes on Mars, just as the theoretical models predicted (such as the work of Michael Mellon) and as the morphological observations suggest. Seeing is believing, and these data are the closest we have come to seeing buried ground ice on Mars; the consistency with morphology is nothing short of stunning.

The *Mars Odyssey* neutron/gamma ray spectrometer provides two completely independent and complementary means of mapping the shallow three-dimensional distribution of hydrogen. Neutrons generated by cosmic ray interactions with ice and rocks bounce around in the regolith and are emitted into free space, where the orbiting neutron detectors determine their energies (speeds). Neutrons that bounce off heavy nuclei, such as silicon, magnesium, iron, and oxygen, rebound almost elastically as though off a fully elastic and immobile wall, and so tend to be emitted from the regolith at high energies. Neutrons that tend to encounter and rebound off nuclei of about the same mass as a neutron (particular hydrogen nuclei in water molecules or any other form of condensed H_2O) tend to lose their momentum and speed when they collide head-on, like a billiards ball colliding squarely with a stationary ball of equal mass; thus, large amounts of hydrogen eventually will reduce the average speed of neutrons until they move about at characteristic thermal speeds. The fraction and number of high-speed and low-speed neutrons thus provides an indirect but powerful means to measure the near-surface abundance and distribution of ice, hydroxyl, and water of hydration.

The gamma-ray spectrometer operates off a different principle; the thermal neutrons produced by interactions with hydrogen can be absorbed by hydrogen nuclei, and in the process a gamma photon is emitted at a characteristic energy, which can be measured for a direct determination of the hydrogen abundance. The summertime residual north polar cap provides an ideal calibration target, where it can be safely assumed that the ice abundance is nearly 100%.

Allowing for the fact that some deeply sublimated icy flows ought not and do not show neutron data indicative of ice in the upper meter, the correspondence between moderated neutron flux (indicating hydrogen; hence, H_2O) and icy landscapes is strong. This key result from *Mars Odyssey* has basically unraveled the arguments of a few remaining Cold, Dry, Windy Mars adherents and has overturned some upstart hypotheses, such as that liquid carbon dioxide can somehow explain Mars, including these flows, without water and ice.

No doubt, Mars has icy glacier-like flows and icy permafrost. The Mars community's change of perspective is a revolution in thought. Revolutions in science, as with politics, are not normally steady and homogeneous, but chaotic. We lack the cozy consensus that I would fear at this point, as huge ambiguities must be acknowledged in the simple issues of whether melting has worked with sublimation of icy Martian flows, about the timescales and rates of flow and accumulation processes, concerning the mechanisms of accumulation, and regarding the roles of epochal climatic upheavals versus astronomically forced climate change of lesser magnitude. Thus, Mars conference halls remain a dynamic and energized forum of discussion and debate, but the debates are moving on from where they were a few years ago.

The glacier definition above says nothing about the amount or distribution of entrained or surface debris, nothing about melting versus sublimation or calving as the cause of ablation, or about the relative magnitudes of various mechanisms of accumulation (ice contributed by snowfall, snow avalanches, subsurface freezing groundwater or percolation and freezing of surface water, surface/subsurface frost condensation; and rock matter contributed by air fall, rock avalanches, and bed erosion). Temperate glaciers on Earth normally contain substantial meltwater, and on Mars they might also if the water is Na–Mg–Ca–Cl-rich brine.

Strictly applied, the definition of glacier includes cases of ice flows due to purely periglacial phenomena, particularly rock glaciers. The definition does not really specify the ice composition; "ice" can be taken in the planetological sense, which on Mars may include CO_2 ice and CO_2 clathrate hydrate, especially in the polar caps, in addition to H_2O ice. Fluid phases are important components of many glaciers. Temperate and polythermal glaciers on Earth include a component of liquid water in the form of supraglacial lakes and streams, englacial plumbing, and subglacial lakes, streams, and saturated till. Polar glaciers and ice sheets, unless they are polythermal due to pressure melting, are mostly frozen throughout, though they may contain intergranular brine films and seasonal surface water. All terrestrial glaciers contain air bubbles, but at depths greater than about a kilometer in polar ice sheets air is compressed and reacted to form about 0.04% air clathrate crystals. Water ice, however, is the primary volatile of Earth's glaciers and those at temperate latitudes on Mars. At high Martian latitudes, prodigious quantities of CO_2 clathrate are likely, and clathrate may be significant at deep levels in the middle latitudes, too; even dry ice and liquid CO_2 are possible constituents.

Rock glaciers attract more than their share of controversy among terrestrial cryosphere experts, but at least the experts agree that some rock glaciers are strict periglacial phenomenon that do not involve net annual accumulation of snow. It is a matter of semantics whether periglacially formed rock glaciers are a subclass of glacier. My definition of glacier certainly encourages the broader grouping and allows a greater range of glacier types than usually described for Earth. Just considering terrestrial glacier types thus defined, there is a continuum between perennial snow fields, snow-fed glaciers, debris-covered glaciers, and rock glaciers, according to many observers. In terms of morphology, there definitely is a blurring of distinctions among these types of icy flows, and I suggest below that these terrestrial glacier forms are just a good start for the kinds that occur on Mars. I ask the reader who may prefer a more limited definition of glacier to tolerate a broader consideration of icy flows in general, a term that I will use synonymously, with more specific types of glaciers referred to with appropriate modifiers. This definition then would include the two chief alternate interpretations favored, for instance, by Timothy Pierce (University of Texas) and David Crown (Planetary Science Institute). Those researchers studied the shapes and other morphologic attributes of Martian flows and found that the creep of icy rock glaciers or ice-rich landslides and their degradation by partial sublimation or melting of ice explains the observations (*Icarus*, 2003). Indeed, ever since their first discovery, there has been a consistent recognition of the roles of both rock debris and ice in the accumulation, flow, and degradation of these landforms, and a role of climate in controlling their locations.

There is good evidence that most Martian icy flows are not very typical of those on Earth and may be exceedingly odd beasts. There are, however, many aspects of the Martian icy flows that are similar to terrestrial glaciers. Some Martian icy flows are dead ringers for rock glaciers and debris-covered glaciers on Earth. These are the exceptions, and for a general explanation of Martian icy flows we probably have to extend our imagination to possible unearthly glacier phenomena. Before becoming

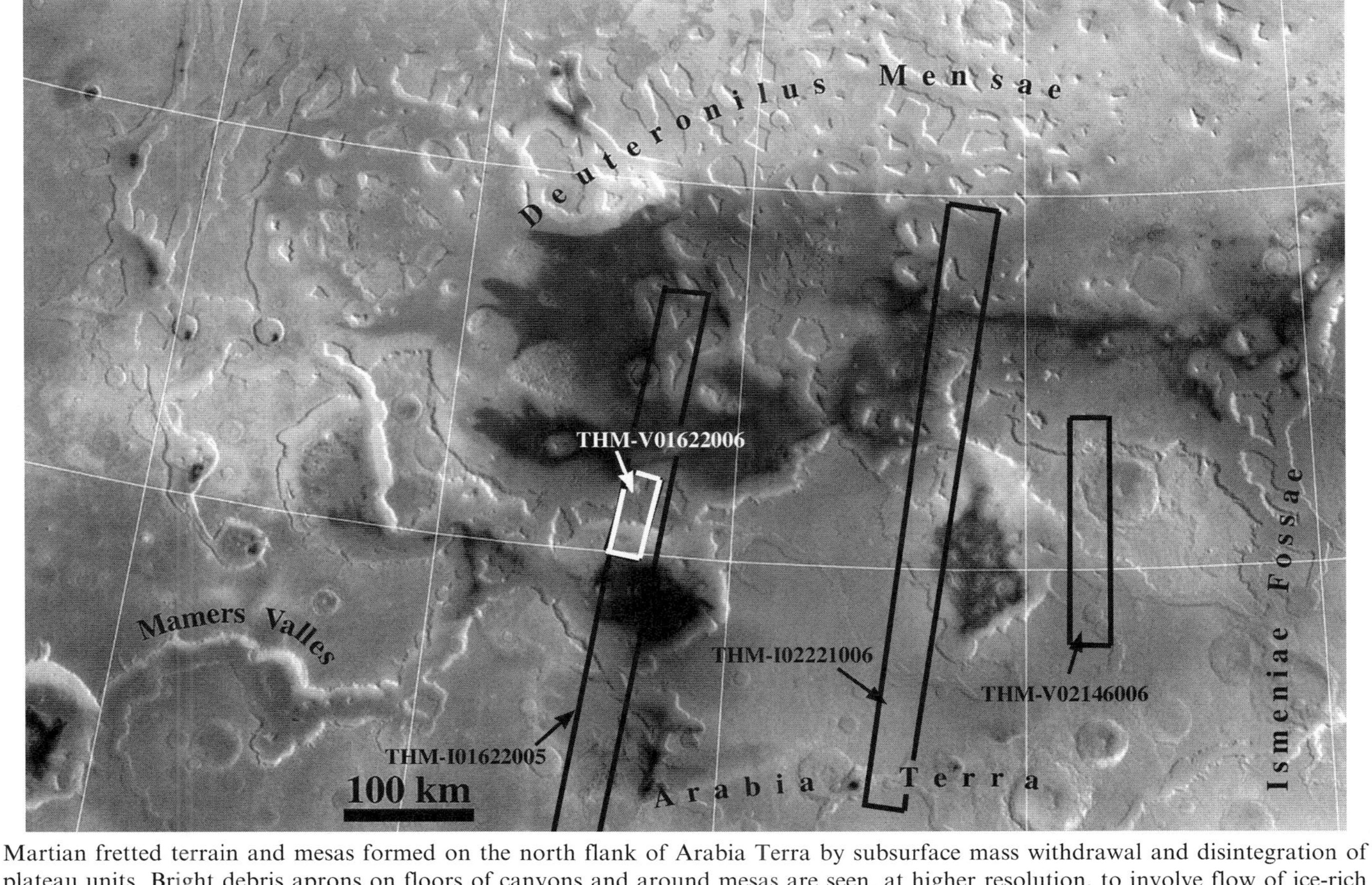

Martian fretted terrain and mesas formed on the north flank of Arabia Terra by subsurface mass withdrawal and disintegration of plateau units. Bright debris aprons on floors of canyons and around mesas are seen, at higher resolution, to involve flow of ice-rich debris. MOC MC-5. Figures on pages 286–290 occur in this region.

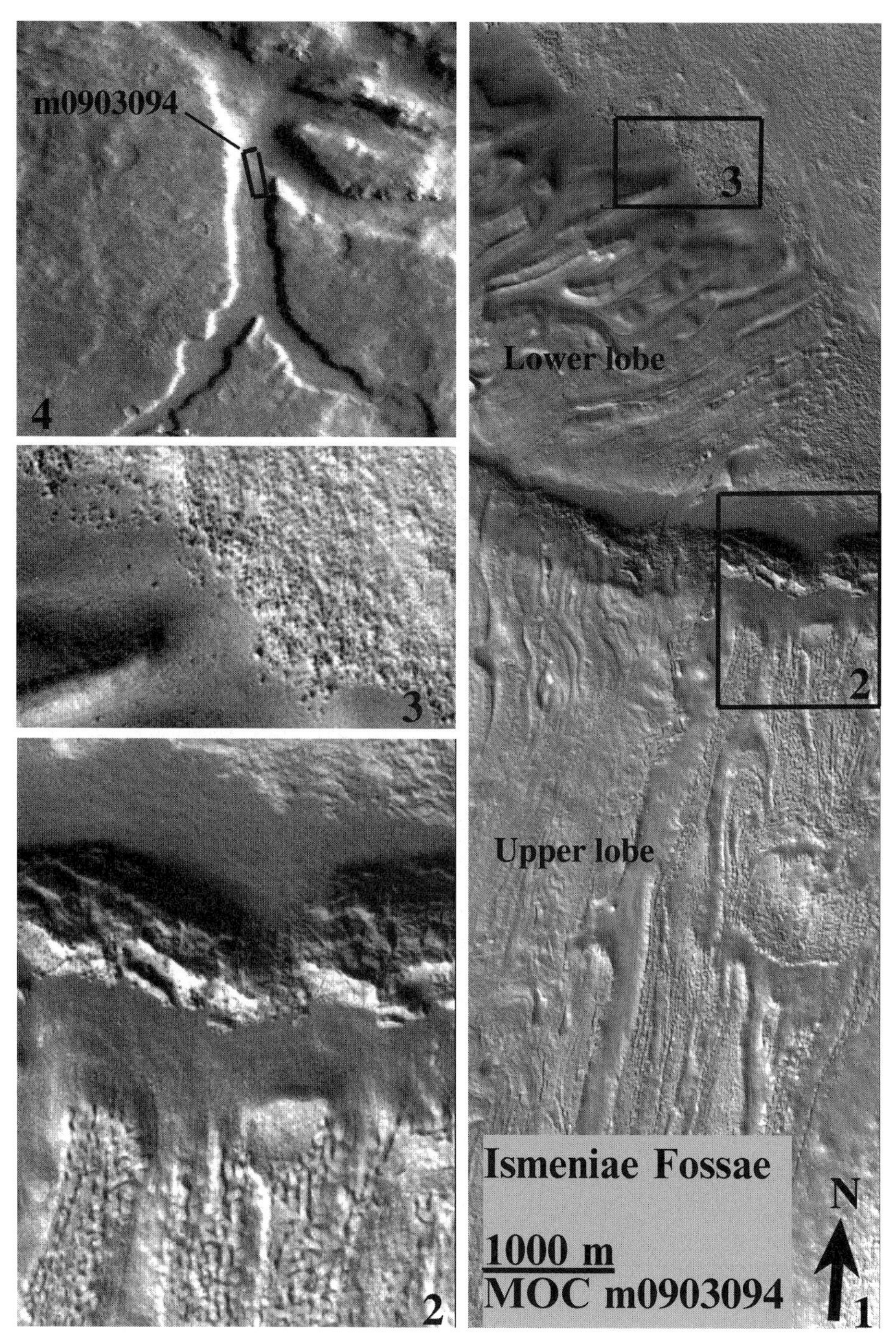

Two overlapping icy flows in a canyon of Ismeniae Fossae are deeply etched by sublimation. Bouldery deposits – probably end moraines – occur at the flow termini.

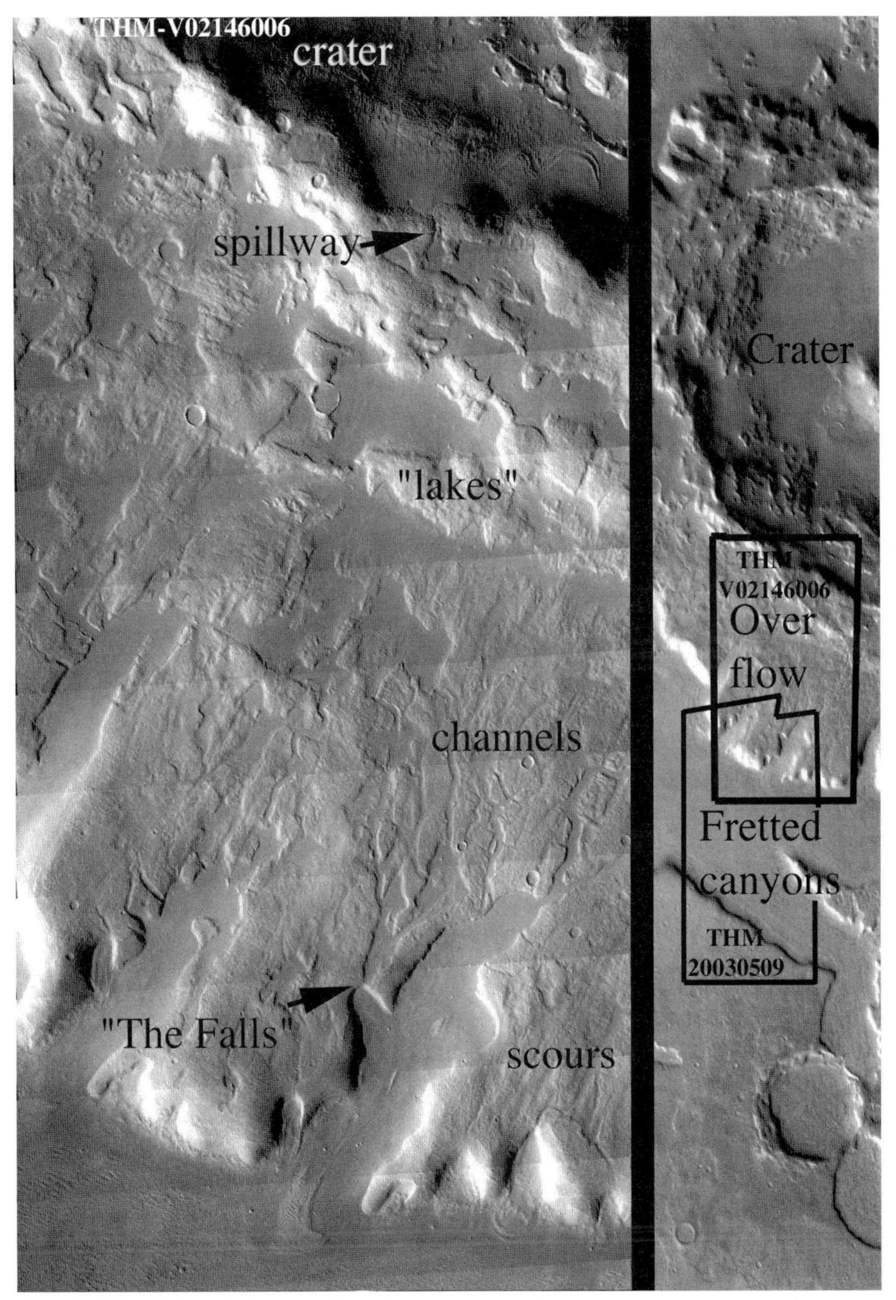

Fretted canyons and lineated valley fill (debris-covered glaciers) of the Ismeniae Fossae. The erosional scours suggest – as unbelievable as it may seem – that the crater literally filled to overflowing from some unknown source and spilled its erosive fluid – presumably water and ice – into the canyon. "The falls" is an area with interconnected channels and hanging valleys. It has hallmarks of classical glacial erosion.

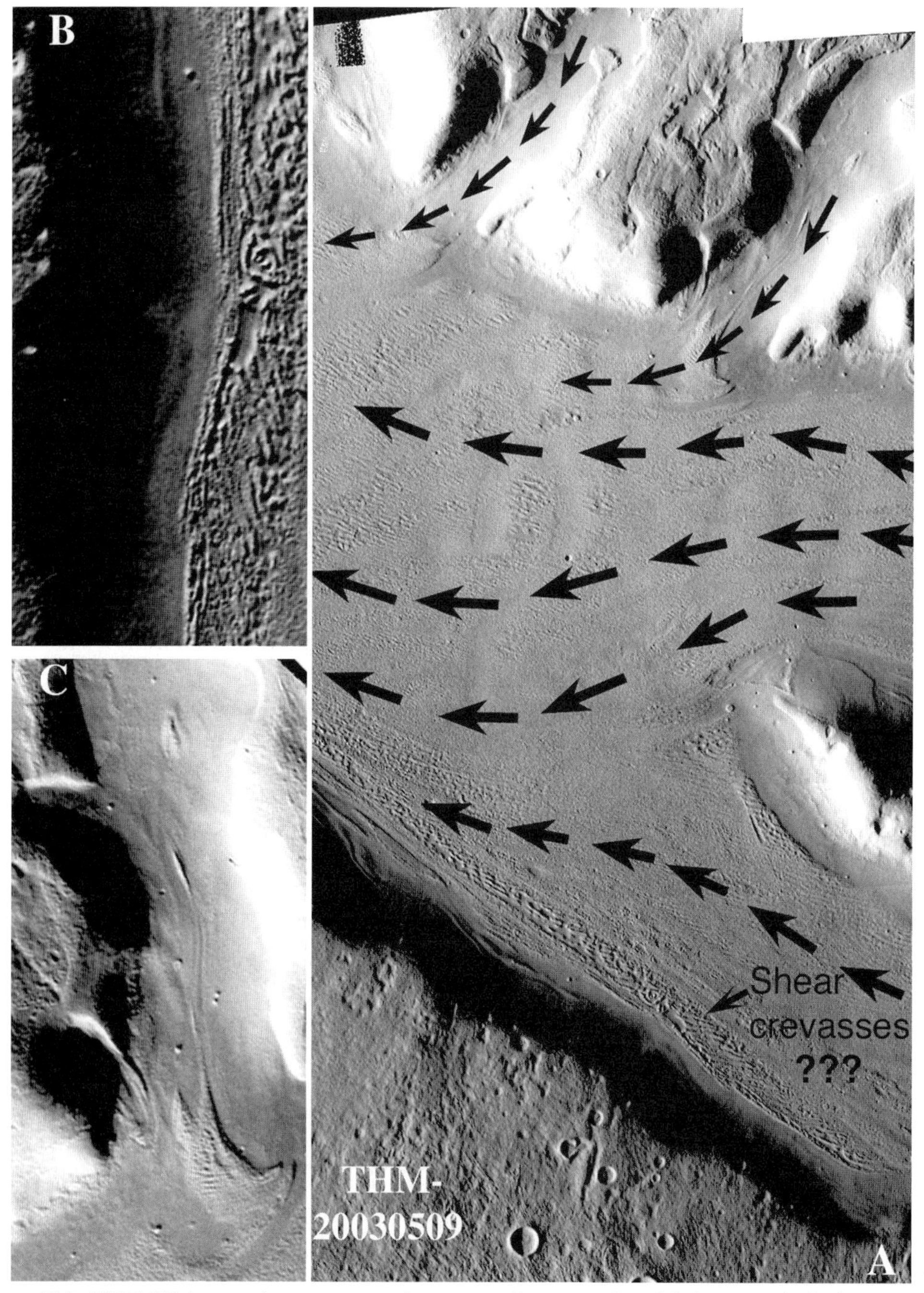

This THEMIS image shows an amazing scene of present-day debris-covered glaciers and glacial erosion by more massive ancient glaciers in Ismeniae Fossae. A marginal shear zone (B) and hanging valleys (C) are evident.

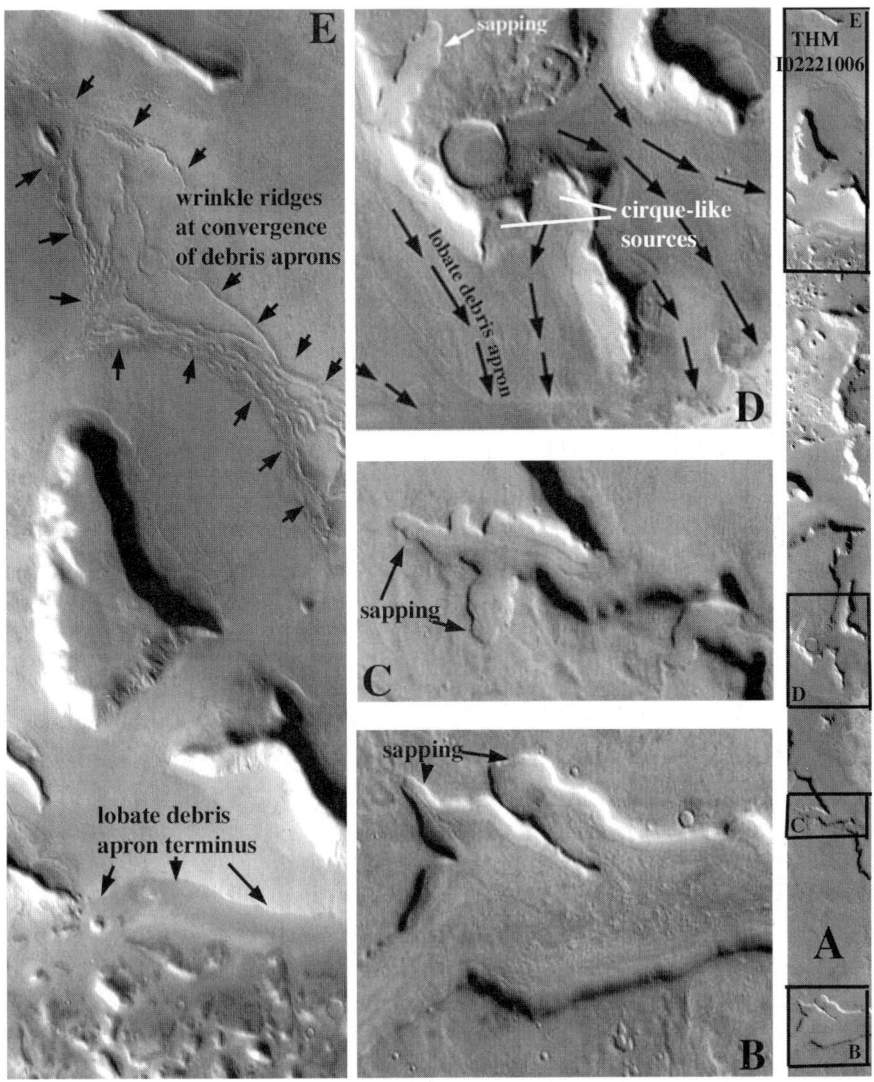

Fretted terrain and mesas of Deuteronilus Mensae show a variety of flow and sapping forms. The sapping appears to involve undermining, collapse, and headward growth due to extrusion of ice from beneath the plateau, or perhaps spring sapping with rapid freezing and glacier-style creep of material away from cliffs and down valleys. Panels B, C, and D show some typical amphitheater-headed sapping tributaries. Panels D and E show some typical lobate debris aprons, including a collision zone in panel E where lobate debris aprons have merged from different directions, creating a zone of compressional wrinkle ridges. The plateau rocks have a Hesperian age, according to impact crater densities, but the lobate debris aprons and debris-mantled floors of sapping valleys are almost uncratered at this resolution, indicating an Amazonian age. In fact, even at far higher image resolutions provided by MOC, craters are few, indicating a very late Amazonian age of at least the latest reactivation of flow activity.

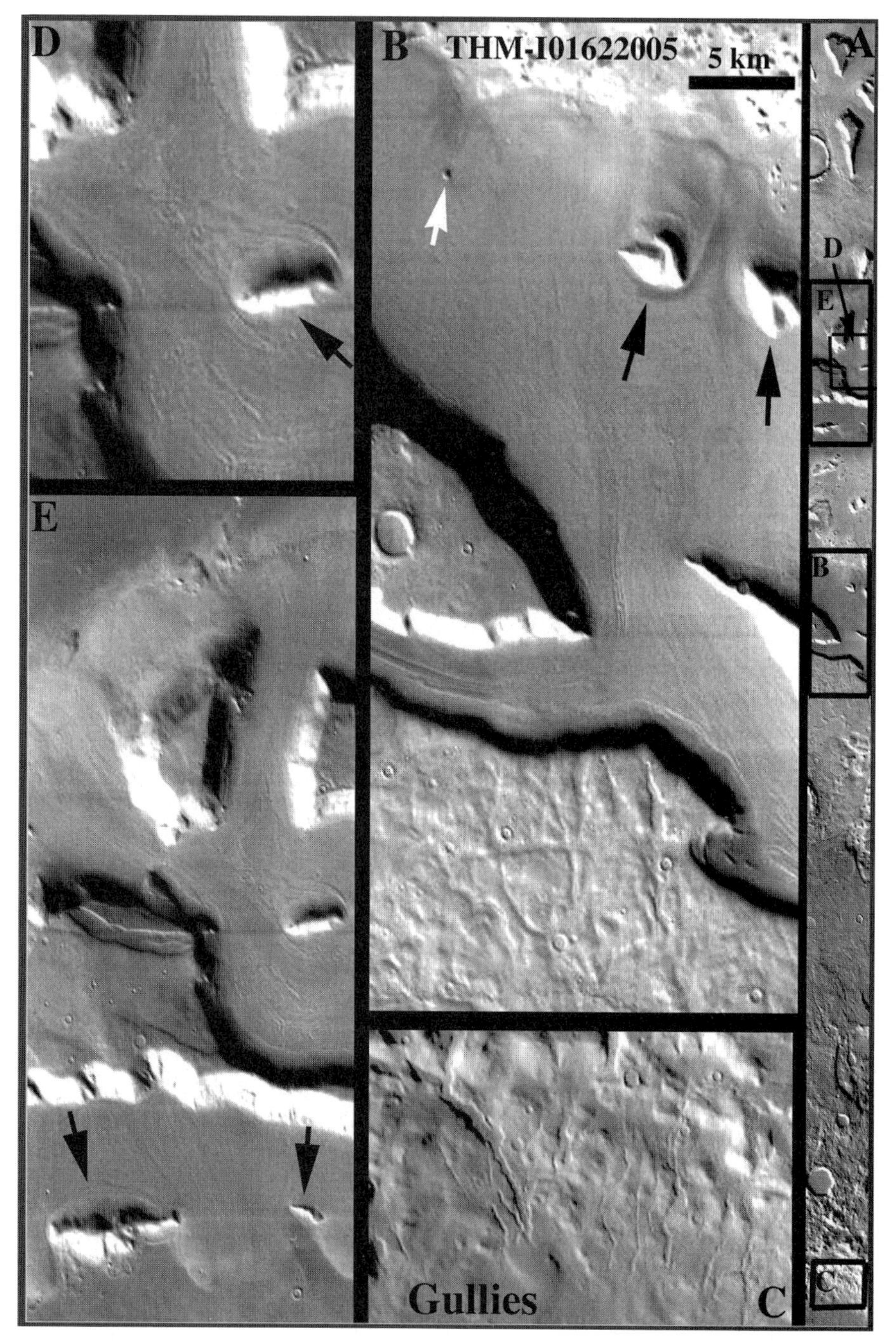

West Deuteronilus Mensae exhibits innumerable dissected plateaus and mesas and well-developed flow structure of lobate debris aprons. This infrared image (I01622005) shows flow diversion around obstacles.

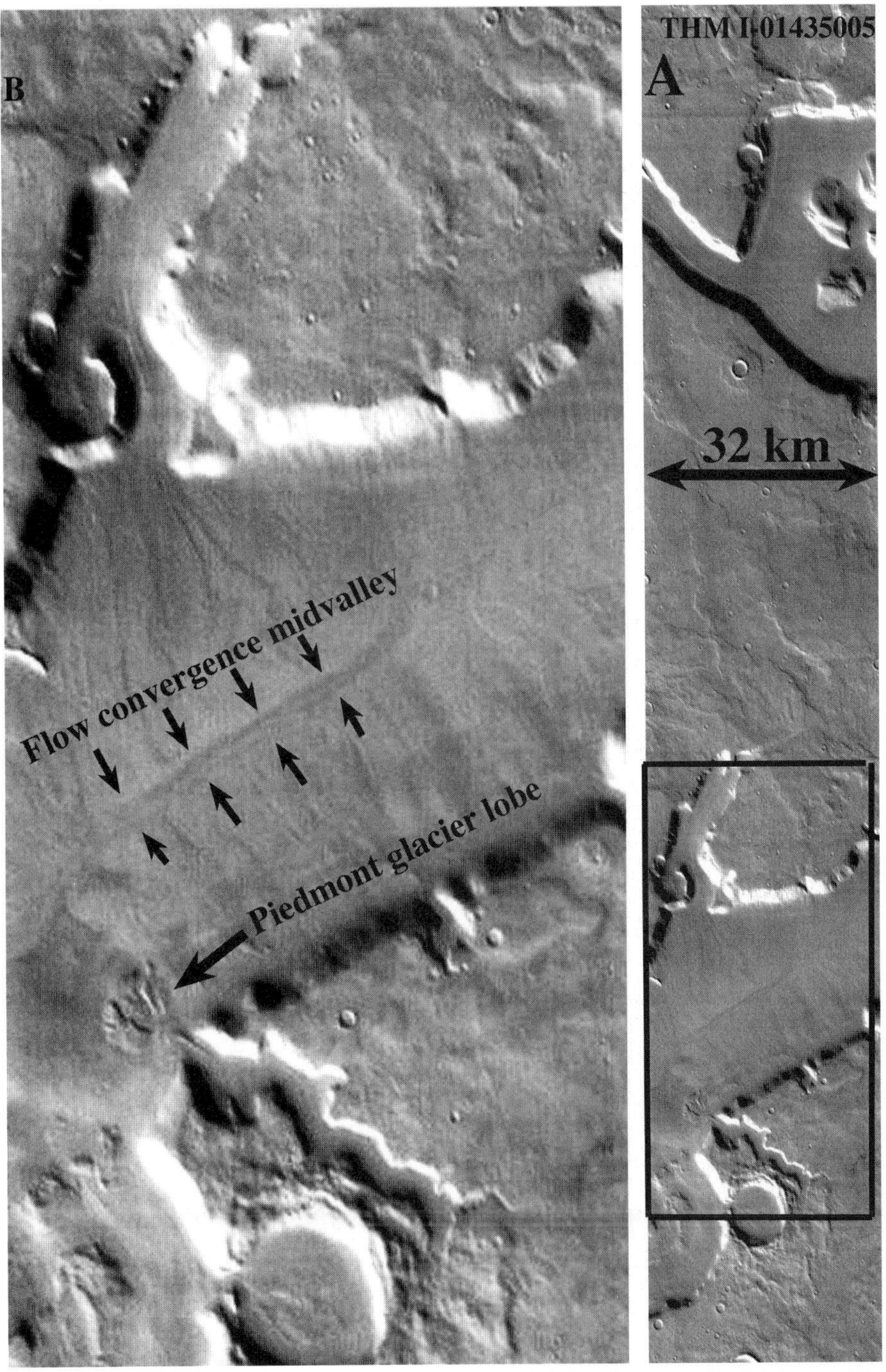

Mars Odyssey THEMIS image of fretted canyons with icy flows or alluvial fans, with sapping sources.

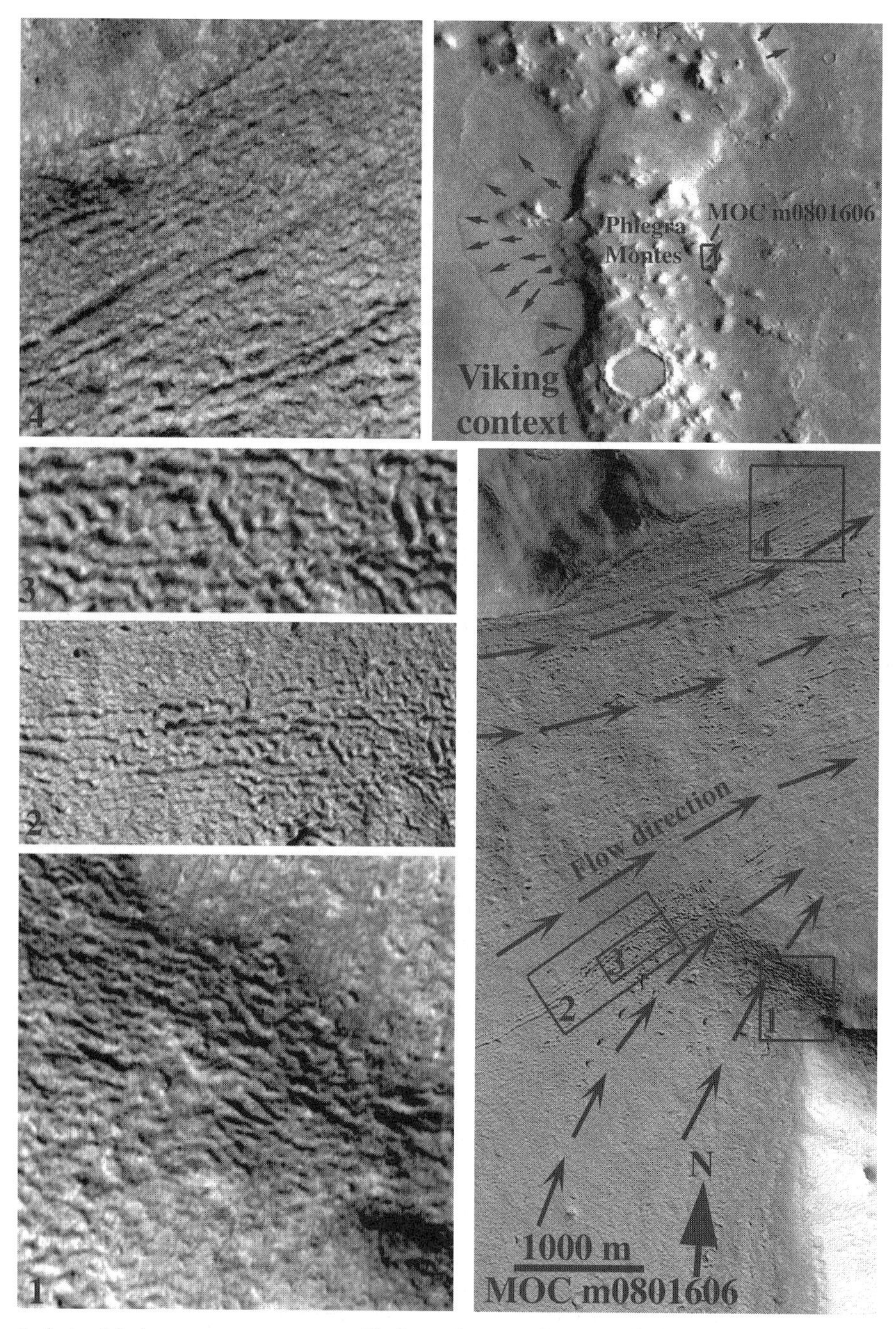

Lobate debris aprons appear to spill through a gap between massifs. The deposits are lineated with pits, troughs, ridges, and hills. The general fabric is interpreted as flow lines etched by sublimation. Where the icy flow crosses an escarpment, it appears to form heavily crevassed ice falls (panel 1).

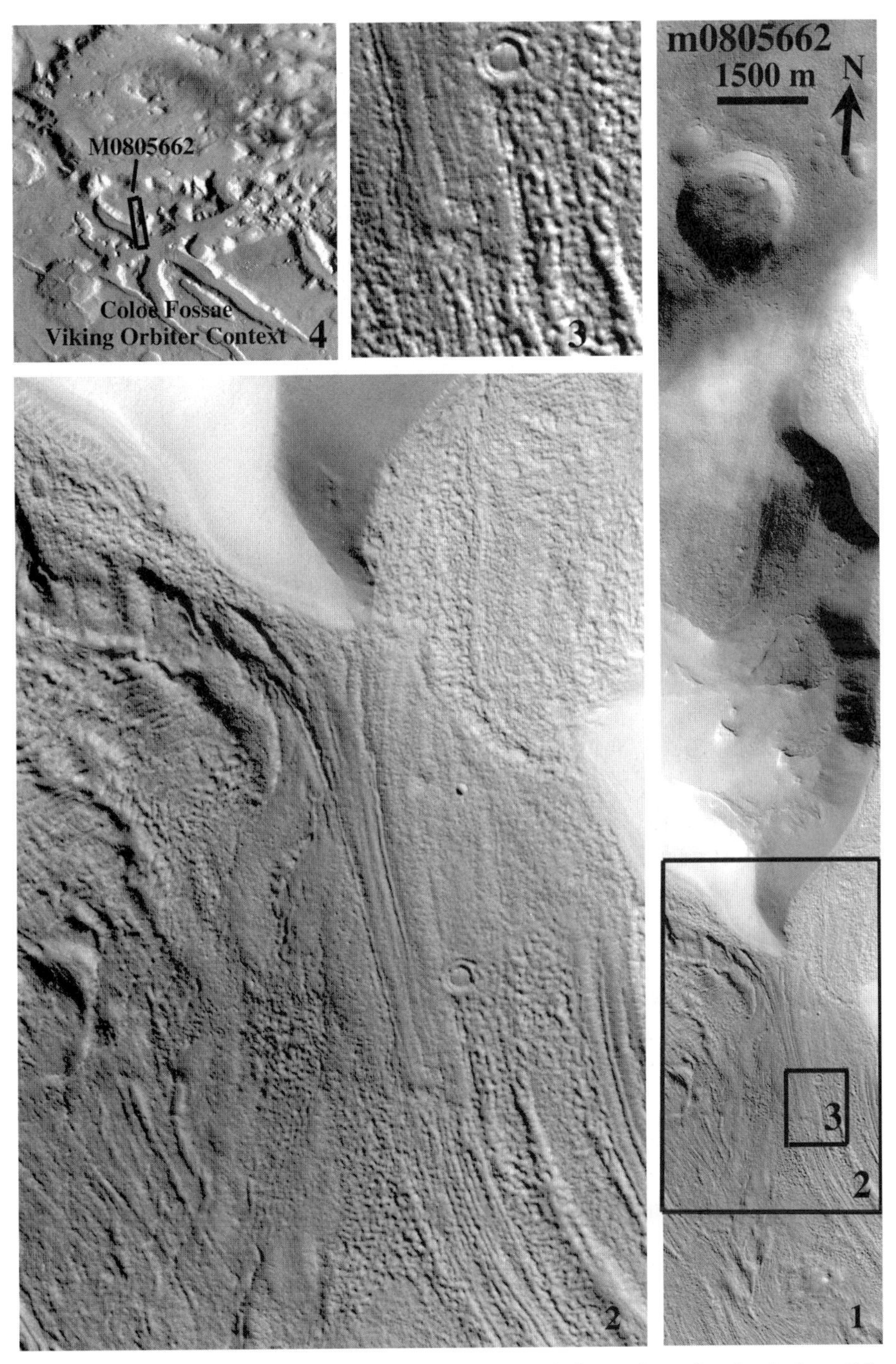

Icy flows in two canyons of Coloe Fossae merge, with formation of a classical medial moraine structure.

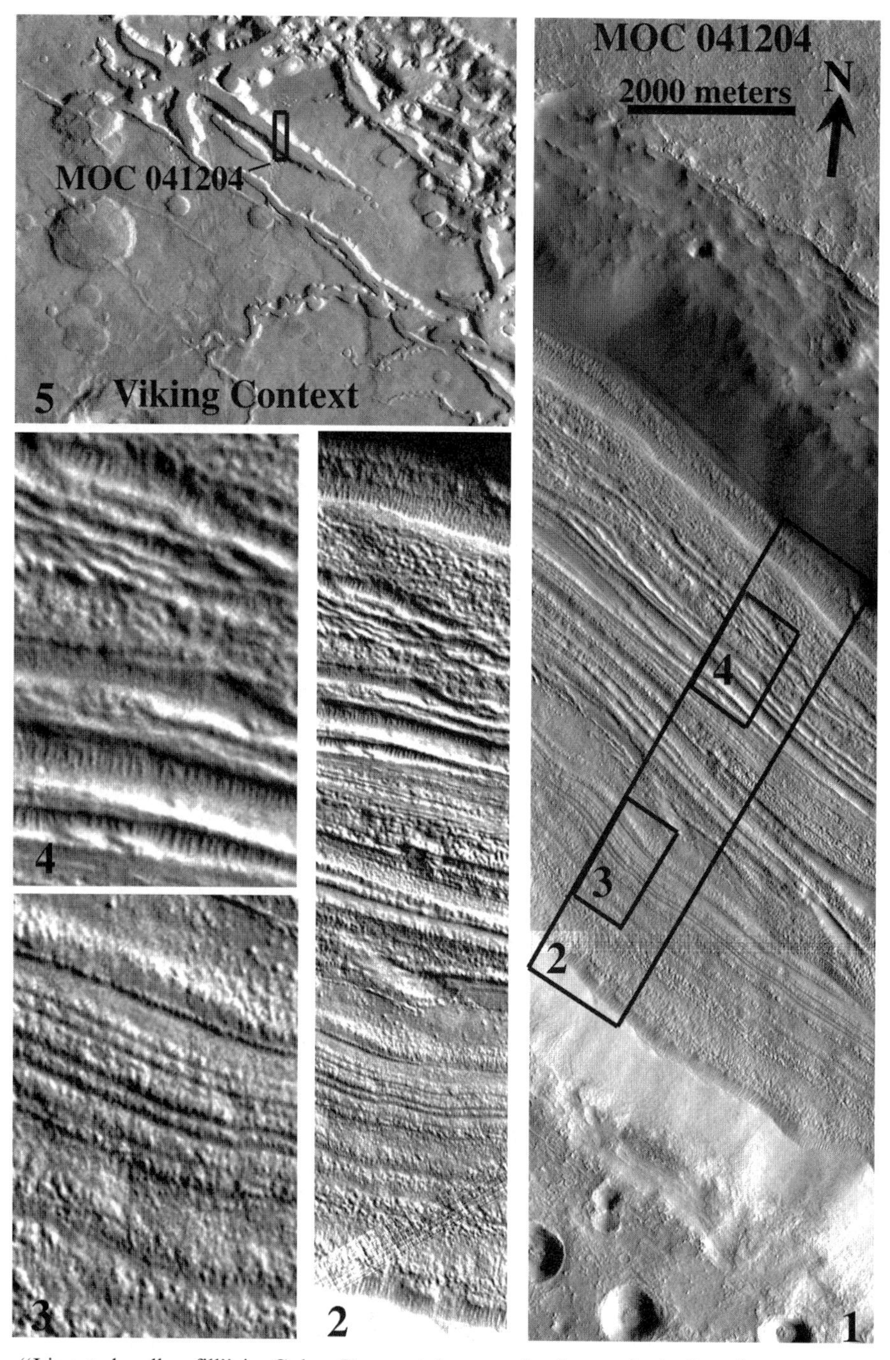

"Lineated valley fill" in Coloe Fossae takes on the forms, including fine structures, ideally analogous to the structure of many heavily debris-covered glaciers, such as Baltoro Glacier, Kashmir.

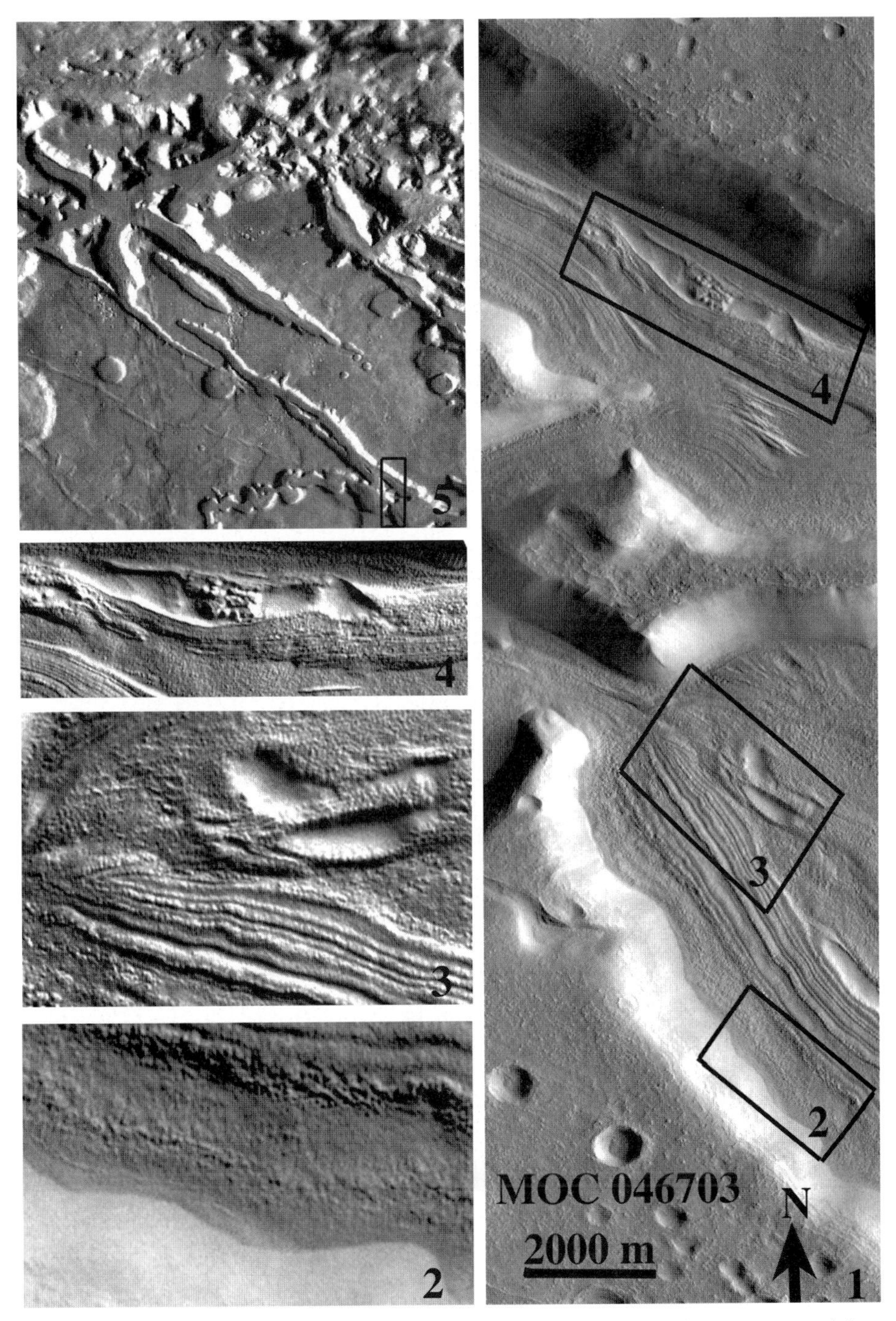

Icy flows in canyons of Coloe Fossae have deeply etched folded strata exposed in classical anticlinal and synclinal structures.

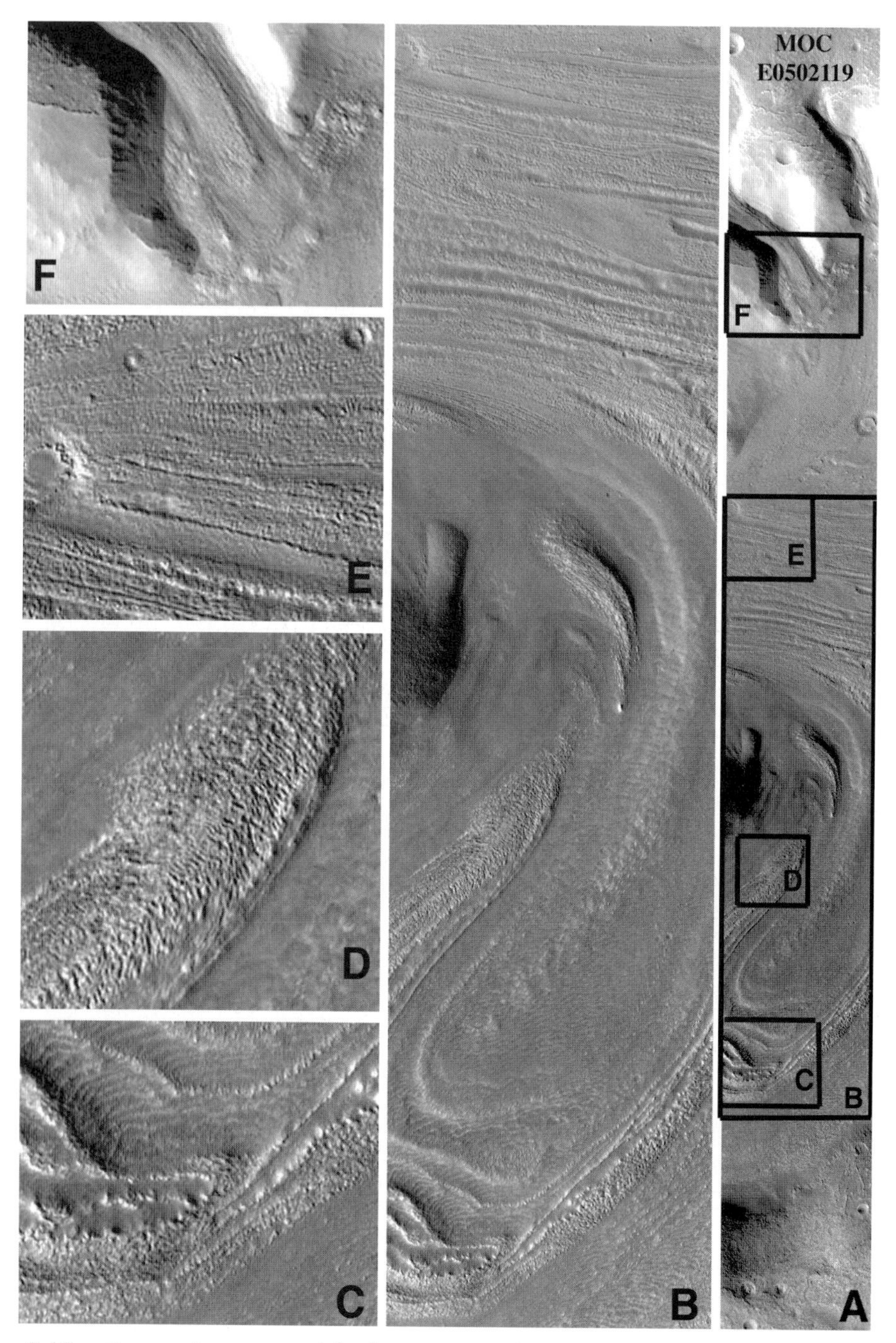

Sublimation erosion may explain the exposure of internal structures and ice tectonic features such as layers, crevasses, moraines, and shear zones. Highly eroded impact craters indicate a great age of this flow.

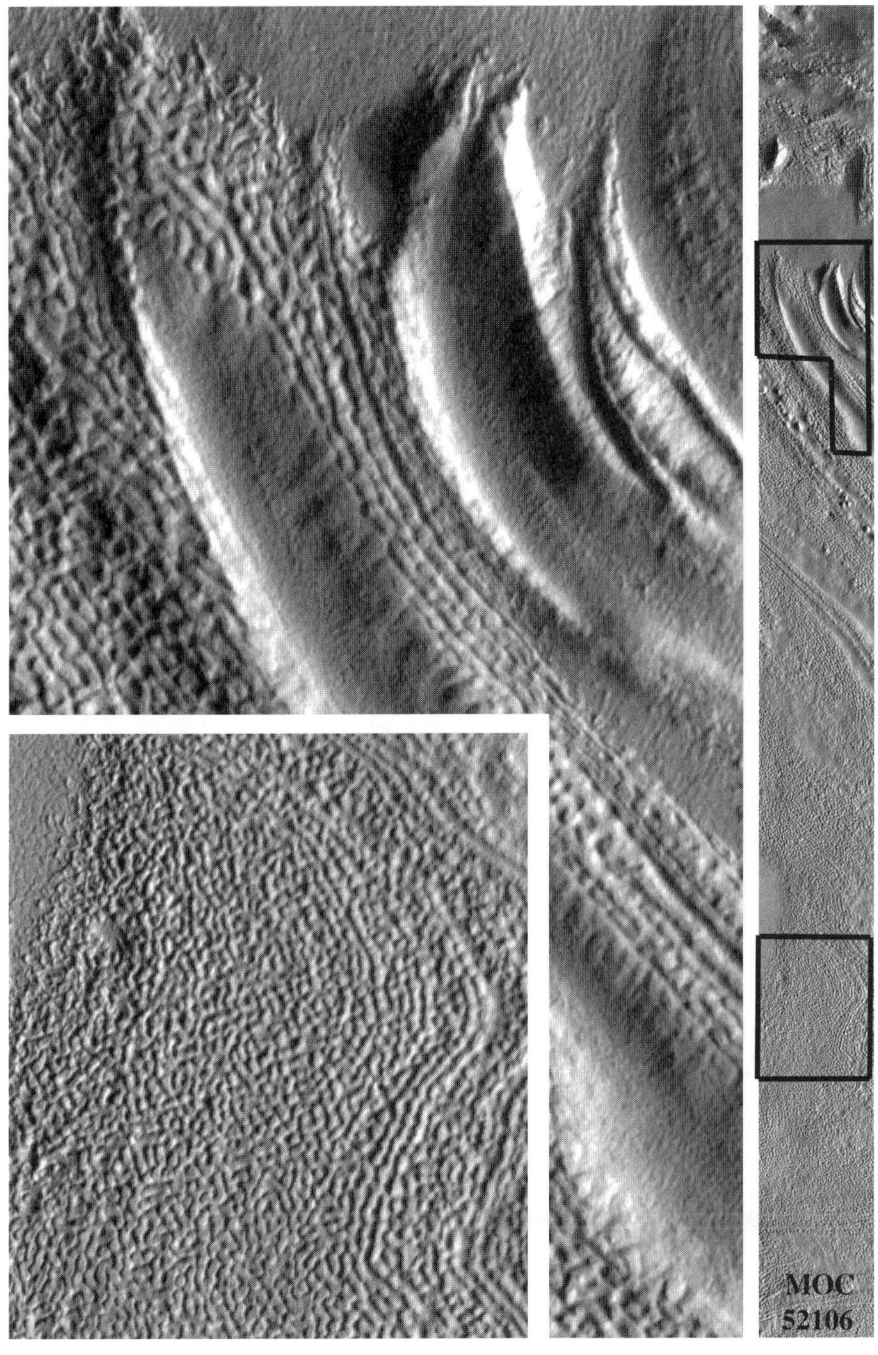

These textures – common in Martian lineated valley fill – are rare in terrestrial glaciers but are similar to some fine textures of the Beacon Valley (Antarctic Dry Valleys) rock glacier.

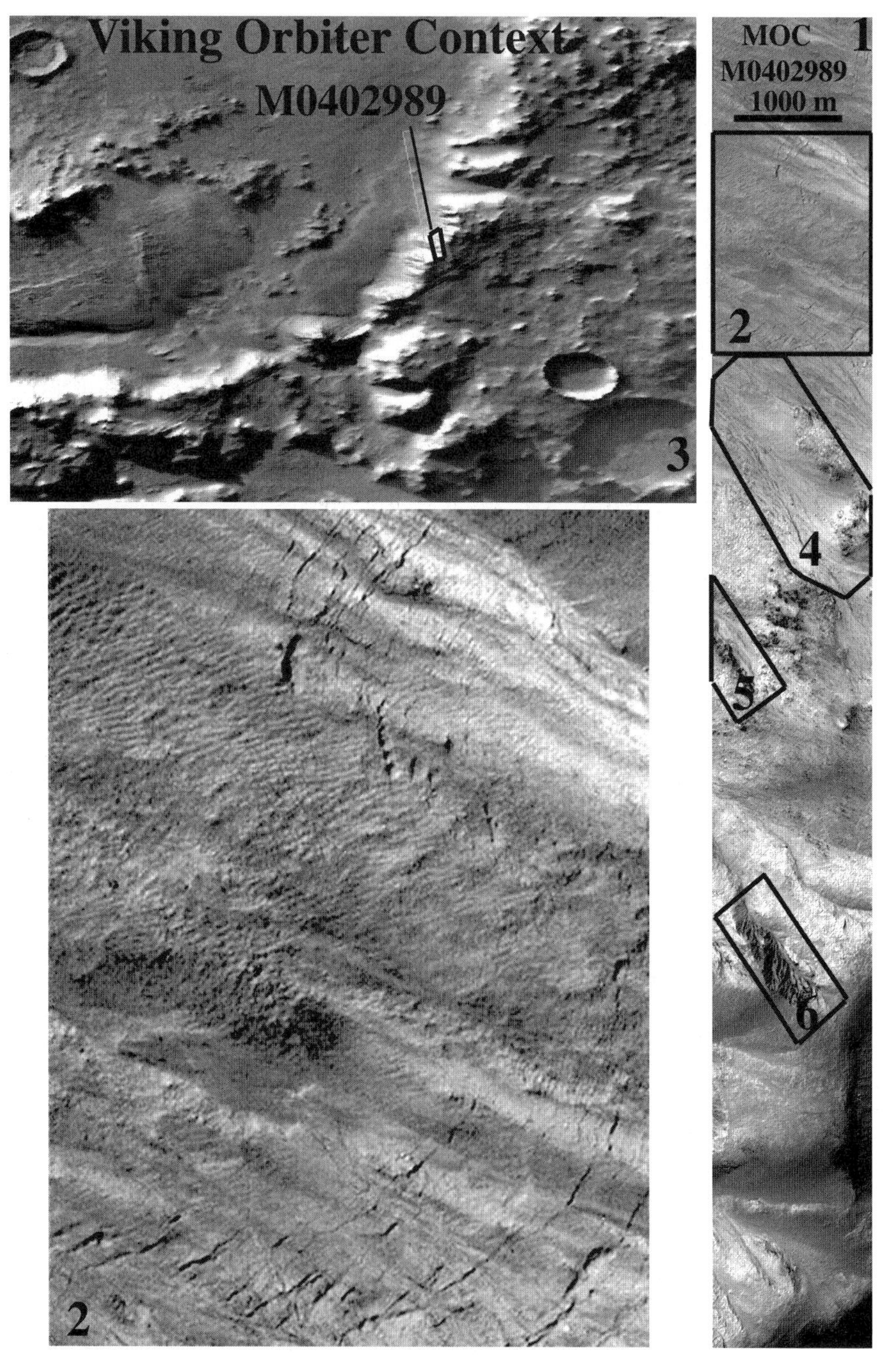

Crevasse- and gully-riddled surfaces of glacier-like lobate debris aprons on the southeastern rim of Galle. Crevasses indicate recent flow activity. Boxes 4, 5 and 6 are gullies shown in detail on next page.

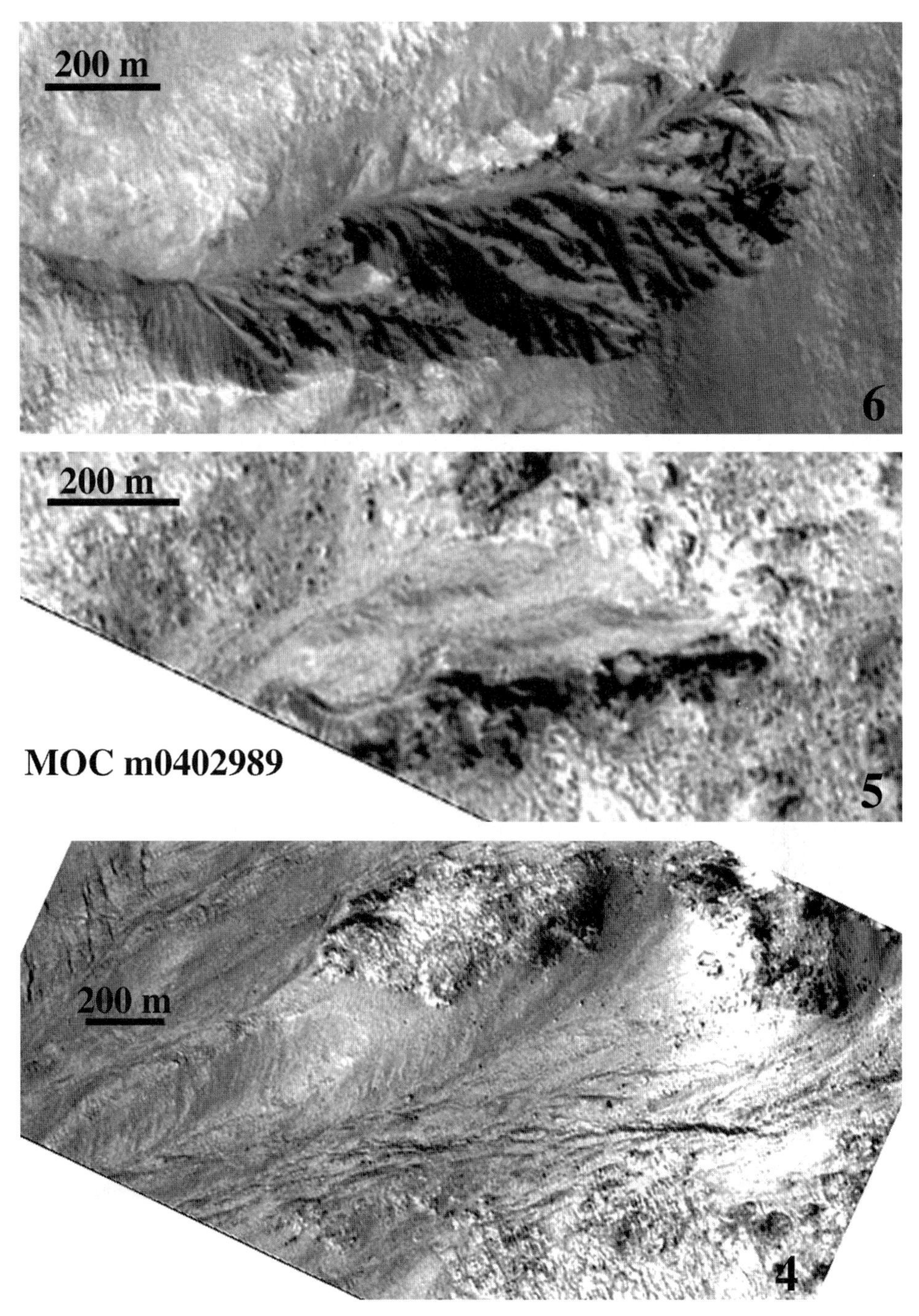

Deep erosional scars – probably the gully sources of debris flows – incise the glacier-like deposits on the southeastern rim of Crater Galle, in eastern Argyre.

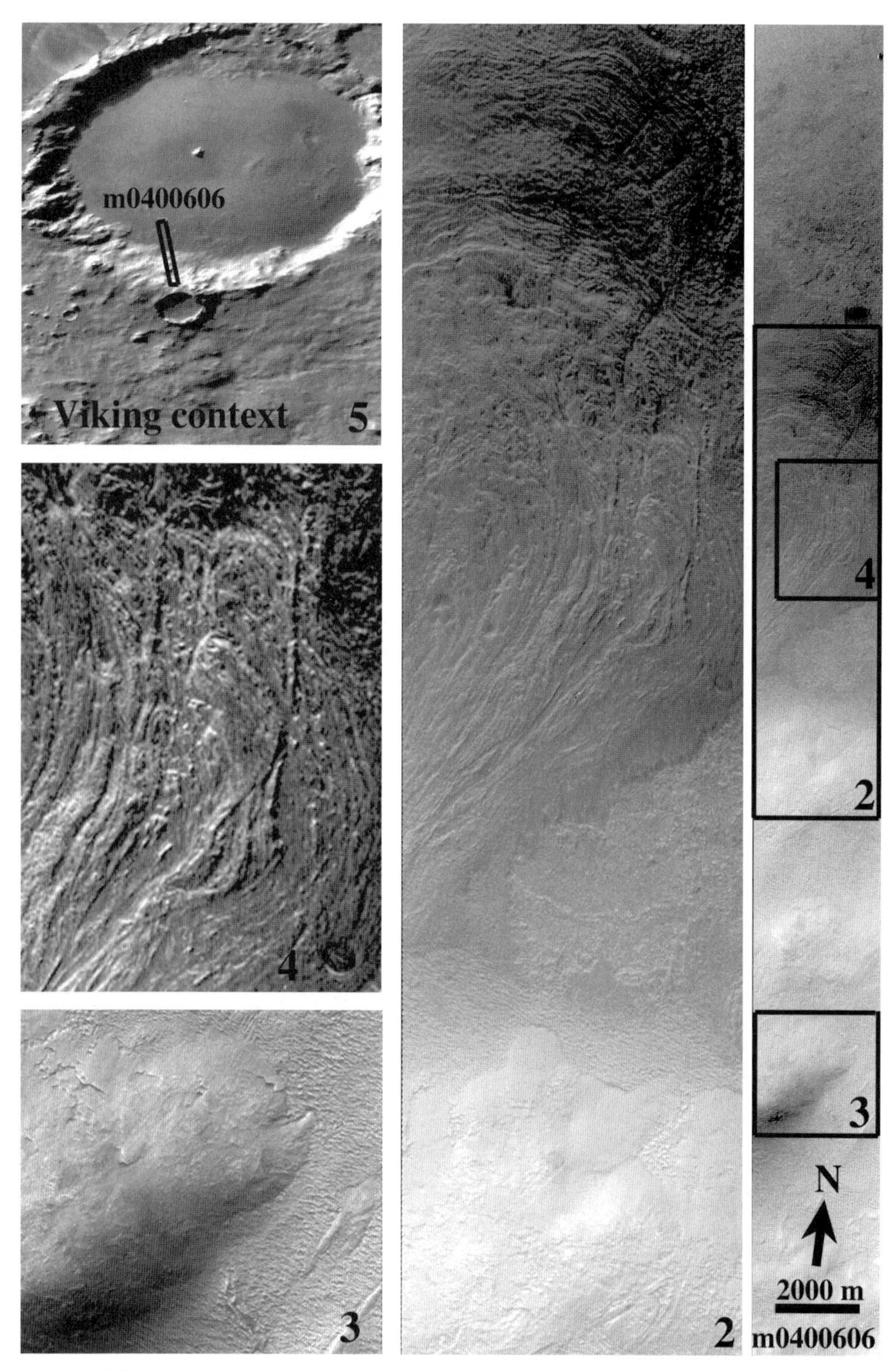

Crater Wirtz, east of Argyre, has glacier-like deposits on the crater wall and piedmont lobes on the floor.

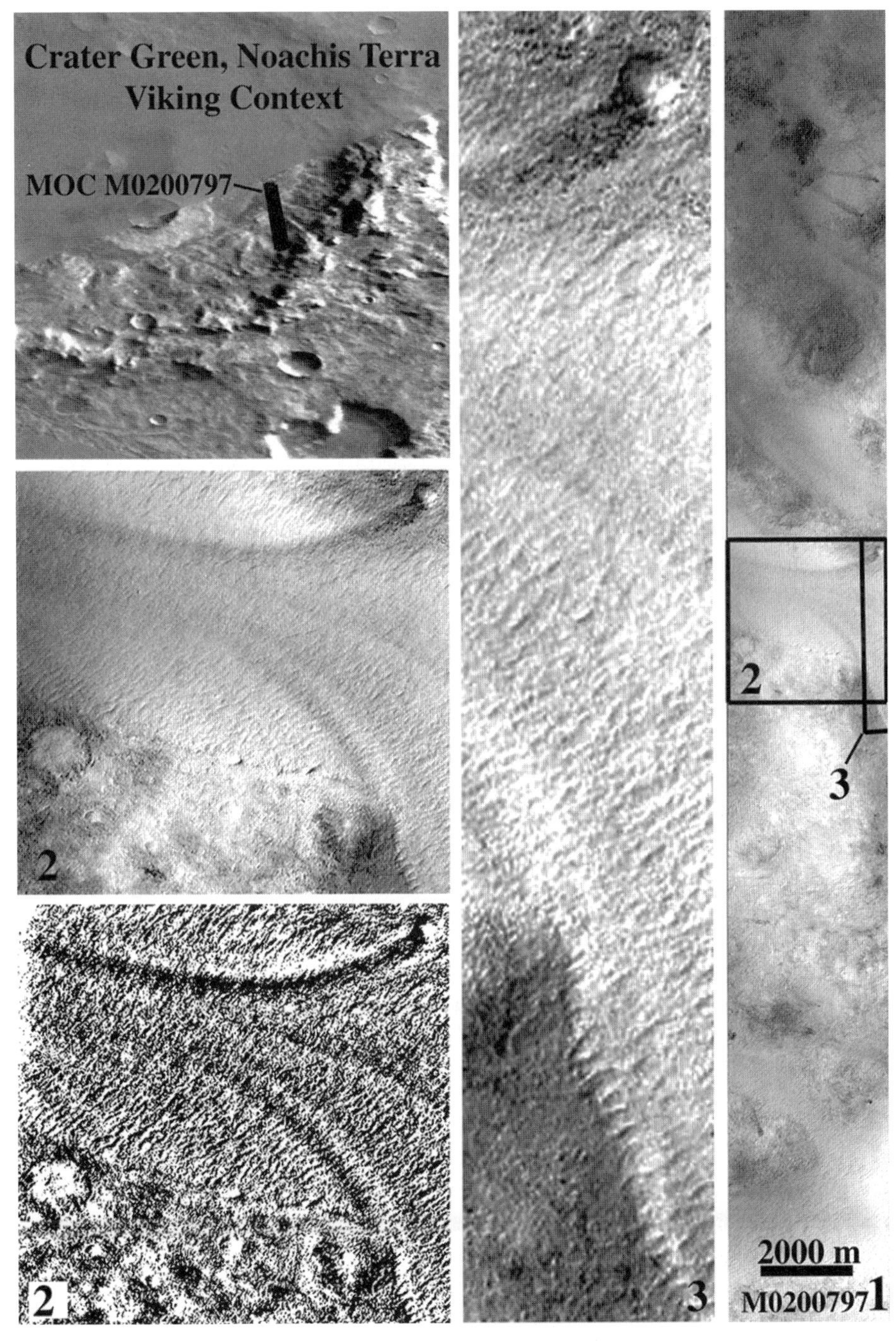

These deposits on the inner wall of Crater Green (Noachis Terra) appear to be highly crevased, serac-riddled, and somewhat sublimation-degraded glaciers with medial moraines of debris. Two panel 2's use different contrast stretches.

too otherworldly, we are well warranted to consider the textures and deformation processes of the Martian icy flows in the context of those of terrestrial icy flows (glaciers, debris-covered glaciers, rock glaciers, and creeping permafrost), as there are many compelling similarities.

Within the ice accumulation zones, glacier surfaces exhibit subtle flowlines and swarms of sharp crevasses, but far more internal and surface structure is produced with little visible manifestation of it. The coupled brittle and ductile deformation is apparent at every scale and stratigraphic level within the ice, at its surface, and on its bed, wherever these are seen. The brittle and ductile deformation of glacier ice affects the crystal structure and shapes of individual ice grains, the ice fabric represented by the aggregate of grains, and the larger structure and overall shape of the flowing mass.

The surface form and internal structure of glaciers controls the movement of heat and formation and movement H_2O liquid and vapor phases. When glaciers ablate by melting or sublimation, the inner anatomical structure is etched like a bass relief, bringing out the layered stratigraphy, folds and rumples, medial moraines and sutures of icy flows contributed by tributaries, and the faults and crevasses. These structures are best revealed where ablation is deep but incomplete, where thermokarstic forms have eaten selectively deep into the icy flow. Debris-rich or debris-mantled glacier ice ablates in especially interesting patterns due to the thermal insulating role of thick accumulations of debris and the energy absorption due to decreased albedo. Production of local surface slopes and differential debris accumulation (with differential solar energy absorption and thermal insulation) of debris-mantled glaciers and rock glaciers can bring internal structures into sharpened relief as the ice ablates. Initial supraglacial ponds can become deeply burnt into the glacier through the enhanced absorption of solar energy by lake water and transmission of the heat to ice through rock debris. Topographic relief may become inverted over time; for instance, formation of a thick debris layer on the basin floor of a drained supraglacial lake can, over time, cause the lake floor to become elevated above surrounding ice that has less debris.

This concept of erosional etching and accentuation or inversion of structure is not one related strictly to icy masses. Geologists see it all the time. We see it in the polished and acid-etched Widmenstätten structure of intergrown metallic crystals of iron meteorites. We see it in carbonate and other sedimentary rocks that are slowly etched by surface and groundwater, which brings into relief the internal structures of layers, folds, faults, clasts, and fossils, all of which help to tell the rock's story. It is no different for glaciers because of the selectivity of ablation. Martian glaciers, too, would be prone to ablative etching, though the etching may progress differently when sublimation is the key process on Mars versus melting on Earth.

Ice fabric anisotropy is caused by the ductile stretching and recrystallization of ice grains. Layered structure forms with seasonal variations in air bubble and debris content. Stretching of the ice as it flows produces a boudin ("sausage") structure, where layers pinch and thicken. Tributary glaciers drag trails of rock debris into the main trunk glacier, forming medial moraines. Since glacier flow is laminar, the tributaries produce a three-dimensional suture zone that is preserved as the ice flows down valley. Rock avalanches add surface debris in patches, which may be stretched into discontinuous medial moraines; if the avalanches occur within the accumulation zone, the debris may form a buried lens of debris. Buried layers and other structures gradually distort with flow and eventually emerge at the surface in the ablation zone. Medial moraines and other surface debris may end up in a giant "train wreck" of contorted, folded moraines or continuous cover of thick debris over the ice at the glacier terminus, such as seen in the Malaspina and Bering Glaciers of south-central Alaska. Bedrock obstacles or changes in slope produce compressional folds and buckles that may be barely discerned at the surface and invisible in the subsurface until revealed by ablation. The spectacular waveform known as "ogives" represent on Earth an annual wave of oscillating debris content, fracture intensity, and ice amplitude produced where ice pours over a cliff in icefalls. Extensional fractures and open crevasses, and shear zones of *en-echelon* S-fractures and strike–slip displacements produce additional structure. Meltwater channels on the surface, beneath the ice, and within it produce troughs and englacial conduits. Except for certain types of bedrock-controlled waves and icefalls, all of these structures move with the ice and become distorted by flow; buried structure eventually are exposed at the glacier's surface. Exposed structure is enhanced to sharpened relief or inverted, and debris is concentrated at the surface as ablation proceeds.

On Mars, the general fine-scale appearance of the "lineations" of lineated valley deposits and some lobate debris aprons is that of sublimation-etched flow structures, including medial moraines, linear zones of similar stress and fracture characteristics (e.g., shear zones), and compressional folds. In some cases, the lineations form structures that resemble the eroded synclines of terrestrial mountain fold belts. Martian icy flows exhibit signs of considerable deformation, and in most cases also selective sublimation. In numerous instances of lobate debris aprons, many examples are evident where flow paths of materials have been deflected by underlying and partly protruding obstacles. While the annual Martian cycle of accumulation and ablation has much too small a magnitude to produce ogives, longer climate cycles, such as the obliquity cycle, might have an influence in generating some of the periodic surface rumples and waves observed on the Martian glaciers.

Martian icy flows generally have convex-up surface profiles and markedly convex, lip-like edges. These are hallmarks of flowing solids and viscous liquids and composite mixtures of liquids and solids, such as salt piercement diapirs (salt flows), lava flows, debris flows, and especially glaciers. In rheological terms, the convexity is an indication of non-Newtonian behavior. These flows obviously are not lava flows or salt flows. The sublimation structures and the global distribution of the flows offer compelling evidence that ice is the major constituent. A logical inference, considering the rheology of ice, is that ice is fundamentally responsible for the flow morphologies. Alternative suggestions that these are somehow related closely to terrestrial alluvial fans or CO_2-driven flows of some type just do not explain the major observations regarding morphology, topography, and global distribution.

Alluvial fans have a different type of profile, generally cone-shaped or concave-up, and where two alluvial fans coalesce they form shallow V-shaped junctions, completely unlike what we normally see in lineated valley fill, lobate debris aprons, and concentric crater fill. Alluvial fans, furthermore, would lack the ice-tectonic and sublimation features. There are certainly alluvial fans on Mars, but they are not the flow forms that we can observe. Debris flows produce some features similar to what is observed, but not the full suite of them, and in any case debris flows require a more seriously modified climate than that required by glaciers. The conservative assessment must be that these landforms are icy flows: or glaciers by the definition above.

As glaciers flow down valley and reach climatic conditions such that net annual ablation occurs, the flow lines of ice bring ice up toward the glacier surface, where layers, folds, boudins, crevasses, subsurface sutures, and all these structures are exposed by ablation. These structures variously channel and bar the flow of energy, meltwater, and water vapor and so become better revealed or even self-accentuating. The ice dynamical structures revealed or enhanced by ablative etching are extremely informative about glaciological processes. Within the ablation zone, icy flows exhibit hummocks, troughs, pits, and ridges arranged in curvilinear, parallel, and en echelon patterns related directly to flow processes. They key to identification of these relief elements as due to flow is their spatial organization relative to one another, the surface slope of the flow, and adjacent topography. The form and spatial organization of similar features, whether on Earth or Mars, indicate flow directions, ice sources, shear and extensional failure, medial moraines, compressional ridges and folds, and thermokarstic structures etched out along these ice-tectonic features. There are many Martian icy flows that are indistinguishable from rock glaciers and debris-covered glaciers on Earth.

Also of interest are indications of brittle behavior. Crevasses are hallmarks of terrestrial glaciers. They are not ubiquitous but are very common in lineated valley fill, concentric crater fill, and lobate debris aprons. Spectacular instances of brittle behavior occurs in some coalescing lobate debris aprons; the site of convergence includes curled flow edges and overthrusting of one flow over another. In the high alpine lobate debris aprons of Argyre, Noachis Terra, and the Hellas Montes, knobby terrain arcuate swarms of nested crevasses occur locally, resembling the more ubiquitous crevass fields of terrestrial alpine glaciers. In general, it has to be stated that fresh crevasses and other indications of recent brittle flow are less common than among terrestrial alpine glaciers. This is presumably an indication of lower flow vigor on Mars relative to competing processes that tend to obscure the crevasses by burial, flow closure, or sublimation. It may indicate that many Martian icy flows are in a state of comparative dormancy, a possibility that is supported by the complete debris cover of most Martian icy flows and the deep sublimational etching of many of them.

What we find, in sum, are that these flow features consist of a material that is capable of flowing, fracturing, melting, and sublimating. All of these phenomena do not always occur in the same flow feature, but they sometimes do. The choice of possible materials that can do all of these things at surface and near-surface

COLOUR SECTION

Plate 1

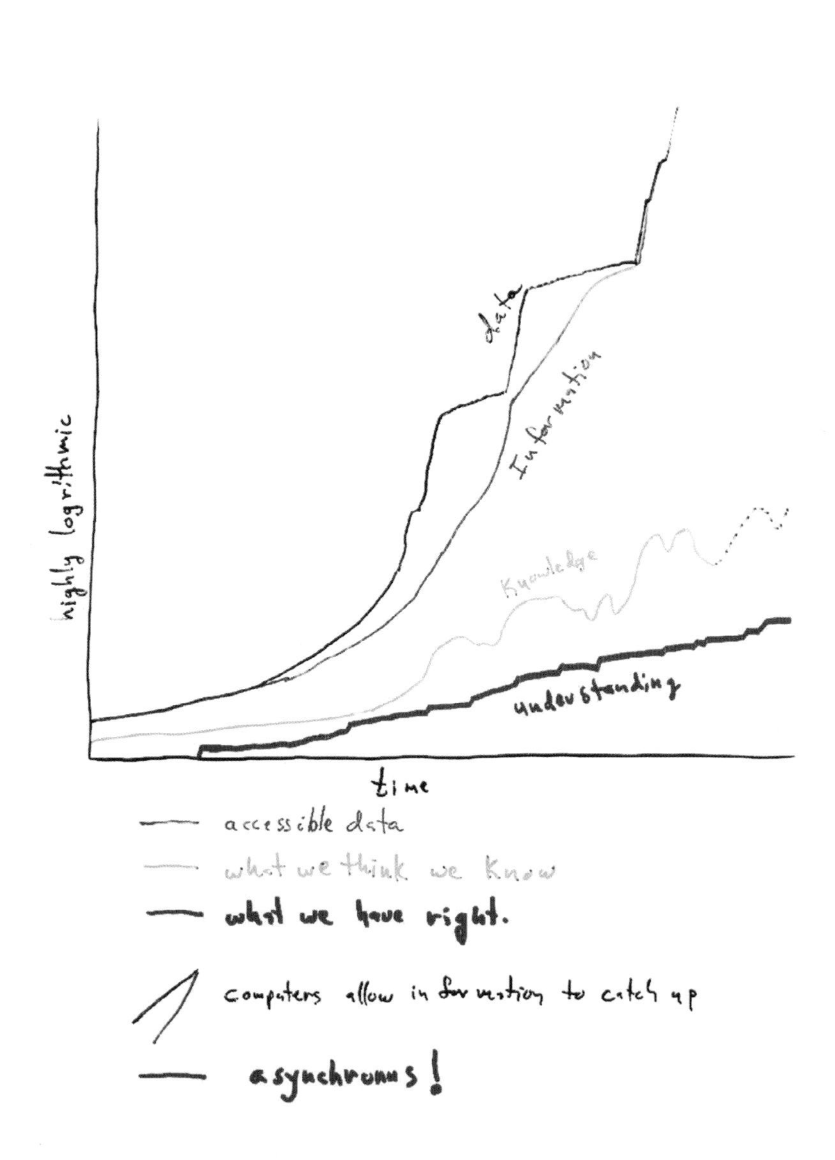

Hugh. H. Kieffer (US Geological Survey, *emeritus*) produced this graphic in about 1990 before people used Powerpoint and when geophysicists wouldn't be caught dead using a 35 mm slide; so they scrawled their important ideas, usually illegibly, on overhead transparencies, usually in real time during presentations. Over the last several years I have seen this graphic about three times at opportune moments, usually about the time a room full of people became intoxicated on the vast new quantities of data that had been pouring in. The crowd always sobered to the fact that knowledge always lags bits and bytes by a large degree; the vague, uneasy thought then struck that our understanding lags perceived knowledge.

Plate 2

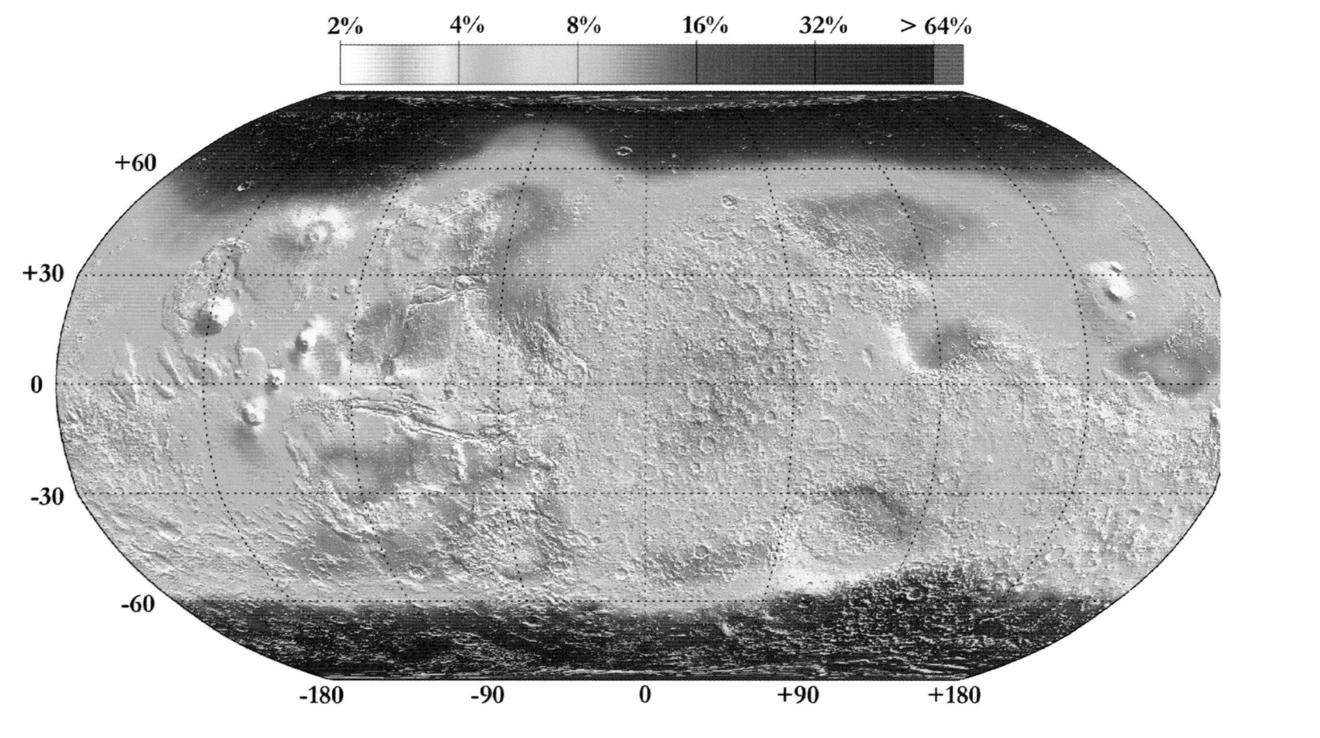

Neutron flux data have been inverted to hydrogen abundance (here represented as H_2O mass percent) in the top several decimeters of Mars. The neutron data have been calibrated to H_2O abundance using the north polar residual cap (assumed to be 100% H_2O), and taking a plausible value for H_2O abundance at the *Viking 2* landing site. Further refinements in terms of multi-layer models will be possible when gamma-ray data are also used in the deconvolution analysis. Data are courtesy of William Feldman (Los Alamos National Laboratory) and the *Mars Odyssey* Gamma Ray/Neutron Spectrometer team.

Plate 3

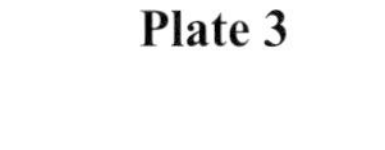

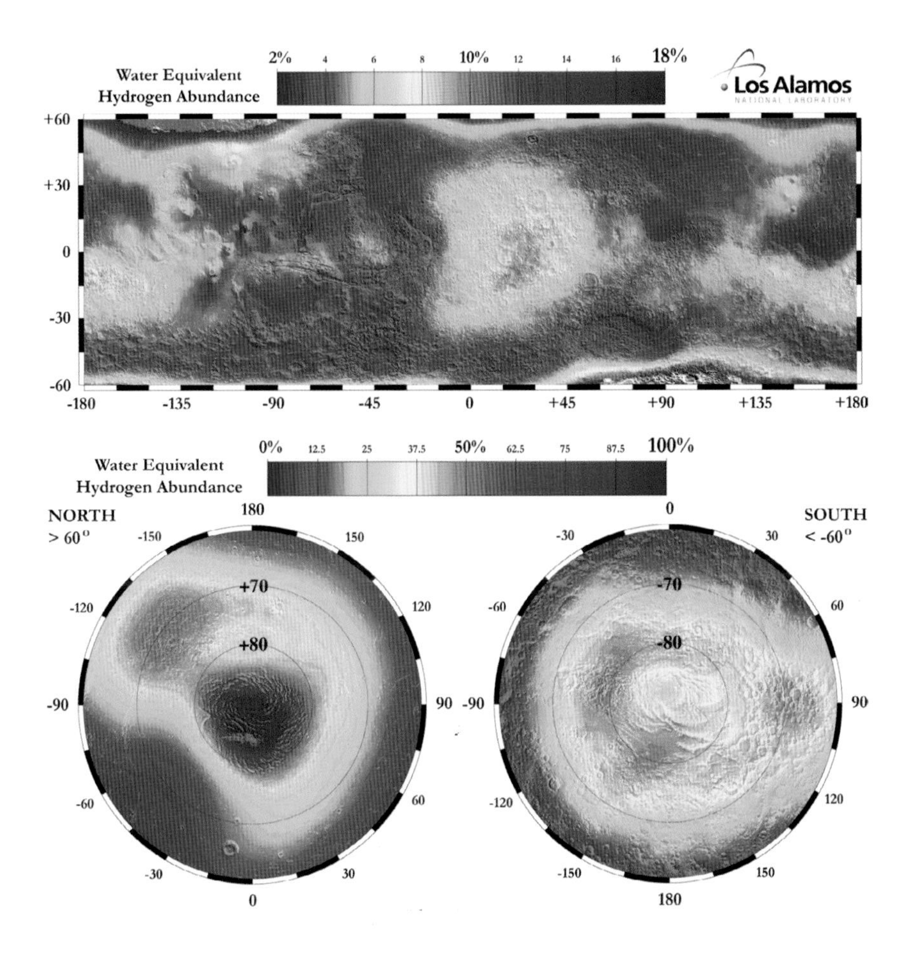

Ice abundance maps produced by and provided courtesy of William C. Feldman and the gamma ray/neutron spectrometer team and the Los Alamos National Laboratory. The data used here are virtually the same as used for Plate 2 but the color scaling is different to bring out details in different regions. The top figure enhances the variations seen at low latitudes, particularly the H-rich tropical zone in and near Arabia Terra, which is noted also for its hematite deposits and its unusual abundance and exquisite geomorphic expression of lineated valley deposits (dirty glaciers) along the fretted boundary with the northern plains. The bottom figure shows summertime polar ice abundances (after withdrawal of the seasonal CO_2 caps in each hemisphere). Note the different color scale bars in the top and bottom figures. As in Plate 2, appearances of fine gradations of ice abundance between regions may be artifacts due to the roughly 300-km spatial resolution of the *Mars Odyssey* neutron data. The "hole" at the south pole is due to coverage by residual summertime CO_2, which would bury any ice that may be underneath. A similar residual dry-ice cap does not exist at the north pole, so its H_2O abundance approaches 100%. For further information see: Feldman W.C., T.H. Prettyman, S. Maurice, J.J. Plaut, D.L. Bish, D.T. Vaniman, M.T. Mellon, A.E. Metzger, S.W. Squyres, S. Karanatillake, W.V. Boynton, R.C. Elphic, H.O. Funsten, D.J. Lawrence, and R.I. Tokar, The global distribution of near-surface hydrogen on Mars, J. Geophys. Res.–Planets, submitted July 2003.

Plate 4

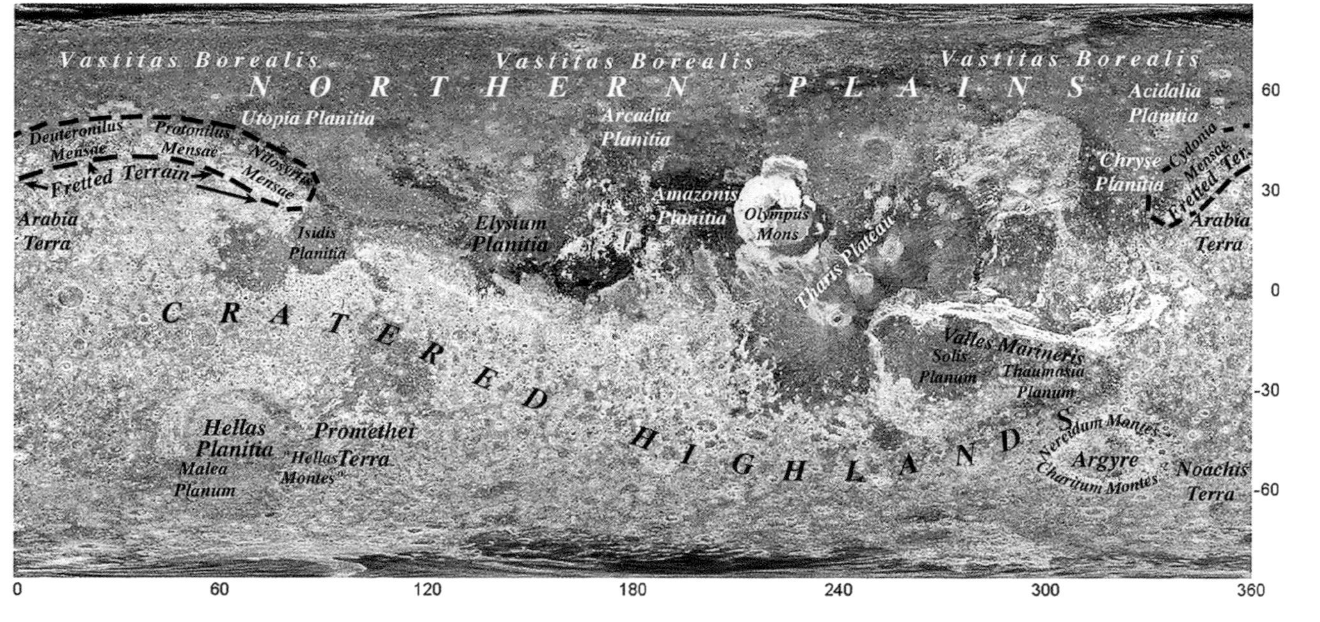

Global roughness map courteously provided by Mikhail Kreslavsky and James W. Head, III (Brown University), then annotated with names of most large-scale features mentioned in the text. The roughness map, computed from MOLA data, is a red-green-blue composite image. Blue is surface roughness at 0.6 km wavelength, green is roughness at 2.4 km wavelength, and red is roughness at 19.6 km. For each channel image intensity (brightness) correlates to roughness: brighter is rougher. Thus, a bright, vivid blue terrain is rugged on a scale of 0.6 km and smoother at longer wavelengths; a bright red terrain is rugged at 19.6 km wavelengths but smoother at shorter scales. Valles Marineris is bright in all channels, indicating ruggedness at every scale examined. The relatively bright reddish Charitum Montes and Deuteronilus Mensae are rugged at a scale of 19.6 km, but fairly smooth at short scales, consistent with morphologies indicating ice flow. There is a pronounced latitude control on roughness; darker shades at high latitudes indicate a generally smooth surface, with a small peak in roughness near 2.4 km (green). The transition from rough low latitudes to smooth high latitudes occurs between 45° and 60°, which Kreslavsky and Head interpret as due to (1) terrain degradation caused by repeated seasonal condensation and sublimation or (2) ice flow. See also Kreslavsky, M.A. and J.W. Head, III (2000) *J. Geophys. Res.*, **101**, 26,695 – 26,711.

Plate 5

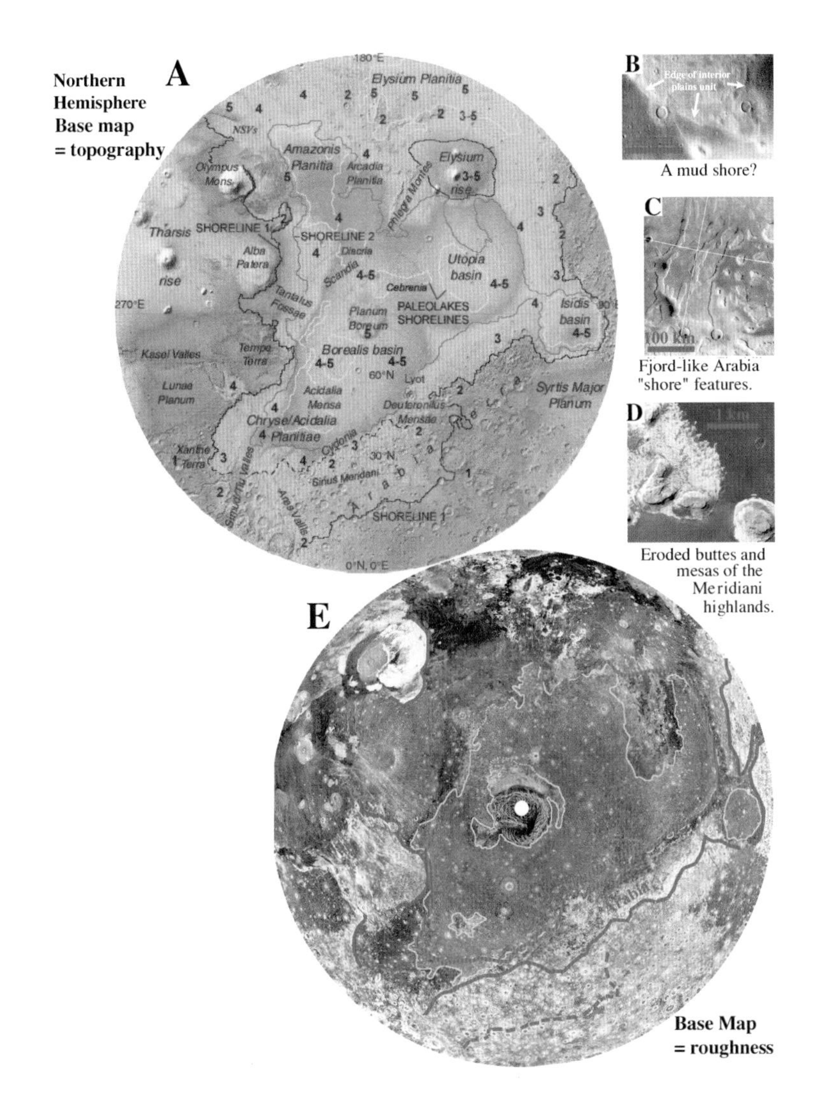

Ocean on Mars? (A) Alberto Fairen (Universidad Autónoma de Madrid), James Dohm and Vic Baker (University of Arizona) compiled a vast number of observations and proposed, following Tim Parker and colleagues and Baker et al. (1991), a series of ancient ocean shorelines and a wide range of coupled geologic events (numbered in sequence from oldest, 1, to youngest, 5). (B, C, and D) Examples of putative shoreline features. (E) Map of topographic smoothness (dark = smooth, bright = rough, with color channels based on mean slopes at horizontal scales of 0.6 km (blue), 2.4 km (green) and 9.6 km (red) from Mikhail Kreslavsky and James W. Head, III (Brown University). The red and pink solid lines are the Arabia and Deuteronilus contracts (possible shorelines) simplified from Steve Clifford (Lunar and Planetary Institute) and Tim Parker (JPL). Dashed lines are extended according to the interpretations of Fairen et al. (2003). Yellow is the outline of the Vastitas Borealis Formation, based on slope characteristics, from Kreslavsky and Head (2002a). According to Carr and Head, water volumes corresponding to the Deuteronilus, Arabia and Meridiani "shorelines" are, respectively, 1.9 x 10 , 8.7 x 10 , and 2.2 x 10 km , and corresponding to global equivalent water depths of 130, 600, and 1500 m. (Note: Meridiani shore not shown.)

Plate 6

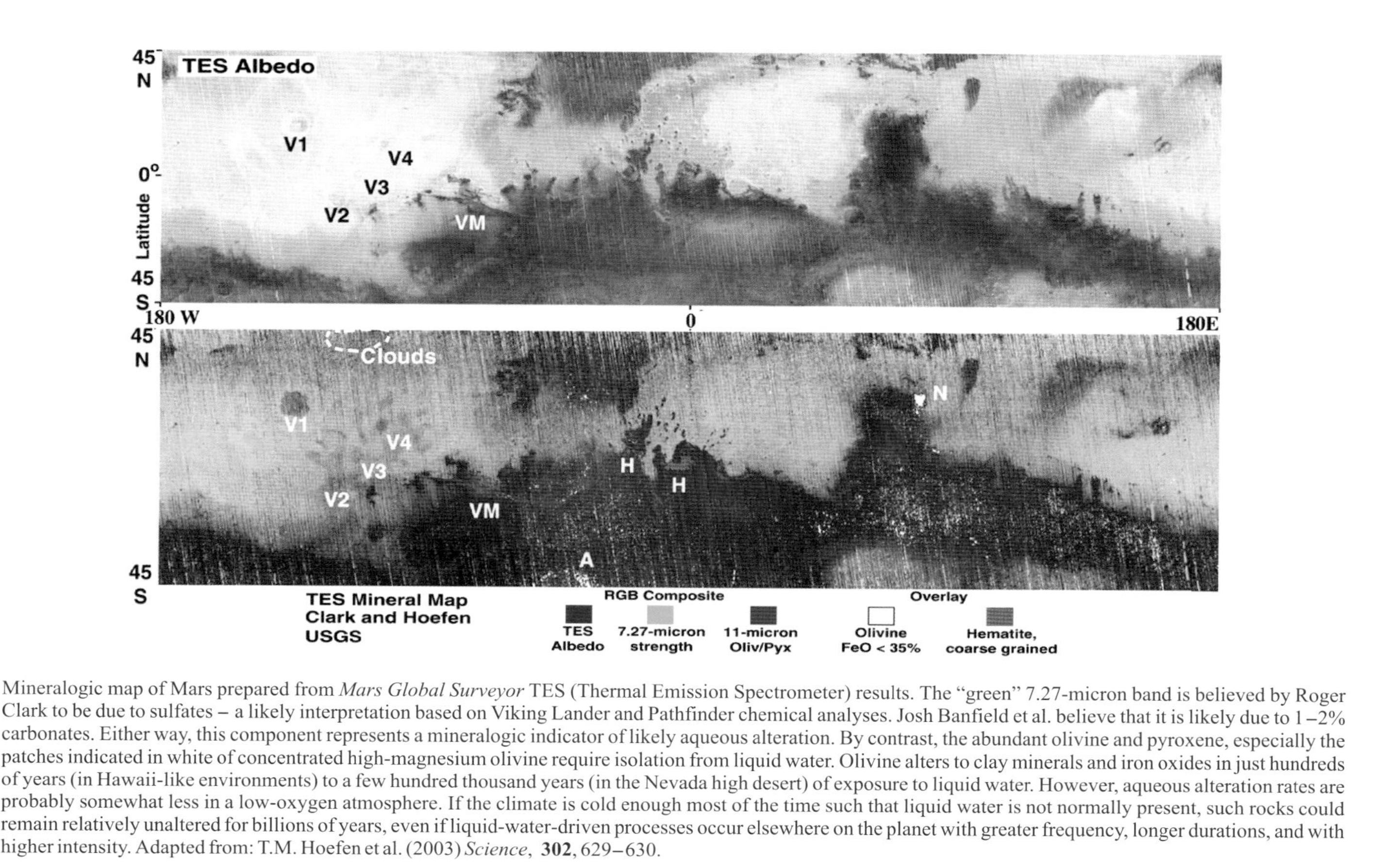

Mineralogic map of Mars prepared from *Mars Global Surveyor* TES (Thermal Emission Spectrometer) results. The "green" 7.27-micron band is believed by Roger Clark to be due to sulfates – a likely interpretation based on Viking Lander and Pathfinder chemical analyses. Josh Banfield et al. believe that it is likely due to 1–2% carbonates. Either way, this component represents a mineralogic indicator of likely aqueous alteration. By contrast, the abundant olivine and pyroxene, especially the patches indicated in white of concentrated high-magnesium olivine require isolation from liquid water. Olivine alters to clay minerals and iron oxides in just hundreds of years (in Hawaii-like environments) to a few hundred thousand years (in the Nevada high desert) of exposure to liquid water. However, aqueous alteration rates are probably somewhat less in a low-oxygen atmosphere. If the climate is cold enough most of the time such that liquid water is not normally present, such rocks could remain relatively unaltered for billions of years, even if liquid-water-driven processes occur elsewhere on the planet with greater frequency, longer durations, and with higher intensity. Adapted from: T.M. Hoefen et al. (2003) *Science*, **302**, 629–630.

Plate 7

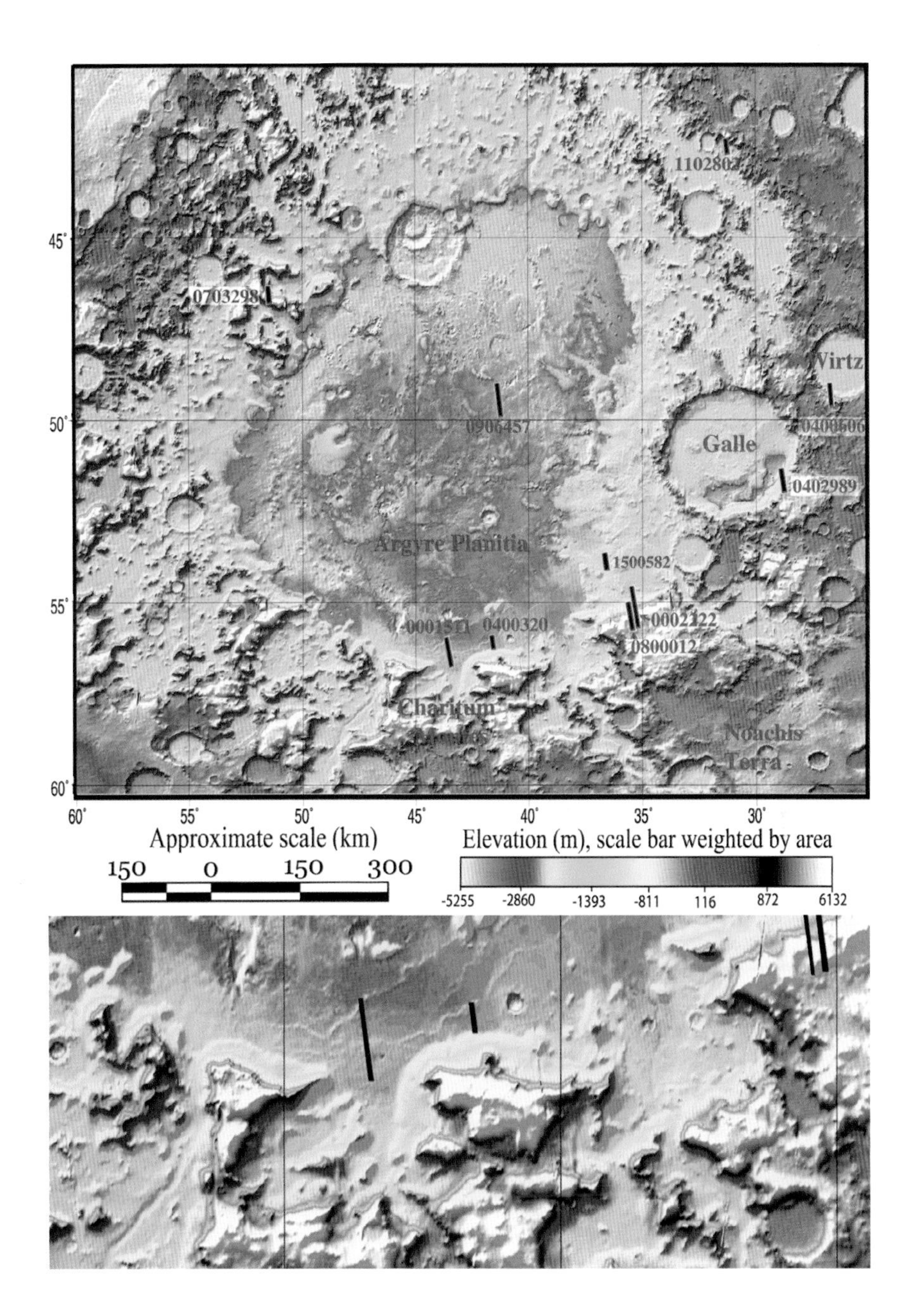

Color-coded topography of Argyre rendered onto a hillshade relief map. Note the strongly nonlinear scale. Selected images presented and discussed in the book have their footprints indicated on the map. Local relief reaches 8 km. Computed from original MOLA data by Jim Skinner.

Plate 8

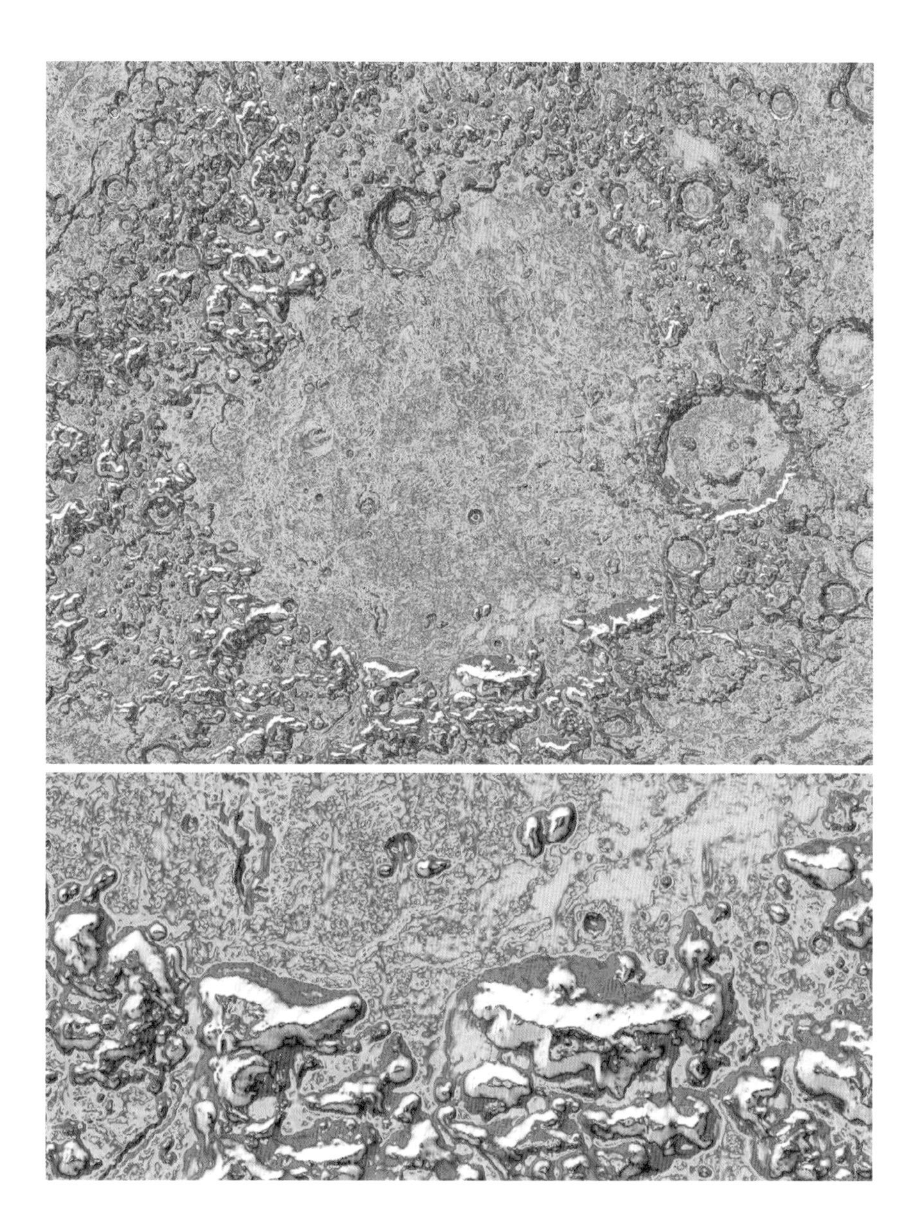

Slope map of the Argyre impact basin computed from original MOLA topographic transects. (A) Color-coded slopes are overlain on a hillshade product derived from a digital elevation model. Whites represent mountain slopes exceeding 20° in many places. Lower slopes represented by the color spectrum range from a few tenths of 1° in much of Argyre Planitia (the broad violet area is probably a former glacier-fed lake), to the 4°–8° typical of the lobate debris aprons surrounding the large massifs. (B) Enlargement of southern Argyre. Note the channels cutting through the mountains and the esker-like ridges on the plains. Computed, rendered, and provided courtesy of Jim Skinner.

Plate 9

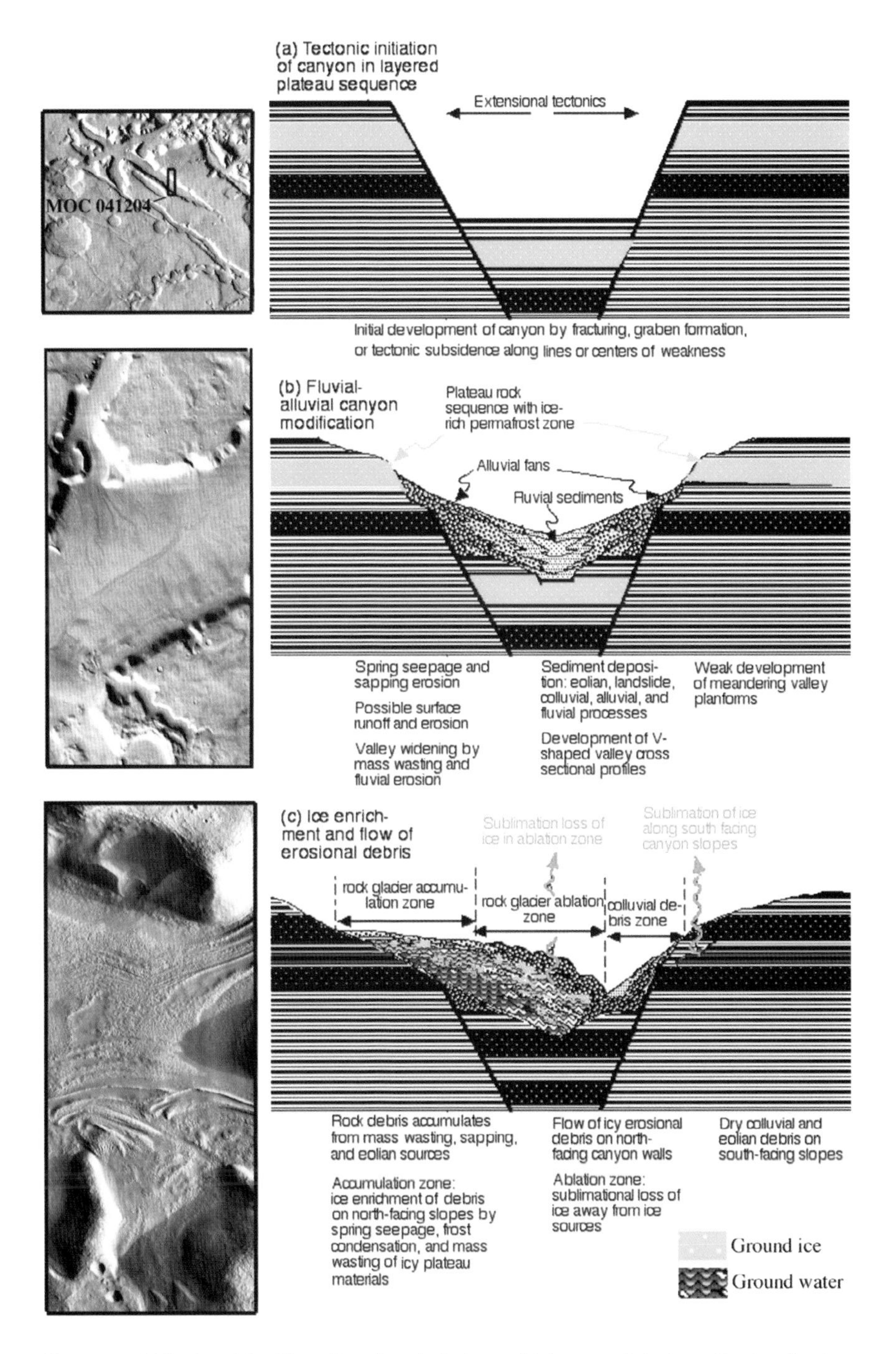

Concept model for the origin of fretted terrain rock glaciers or debris-covered glaciers with subsurface ice.

Plate 10

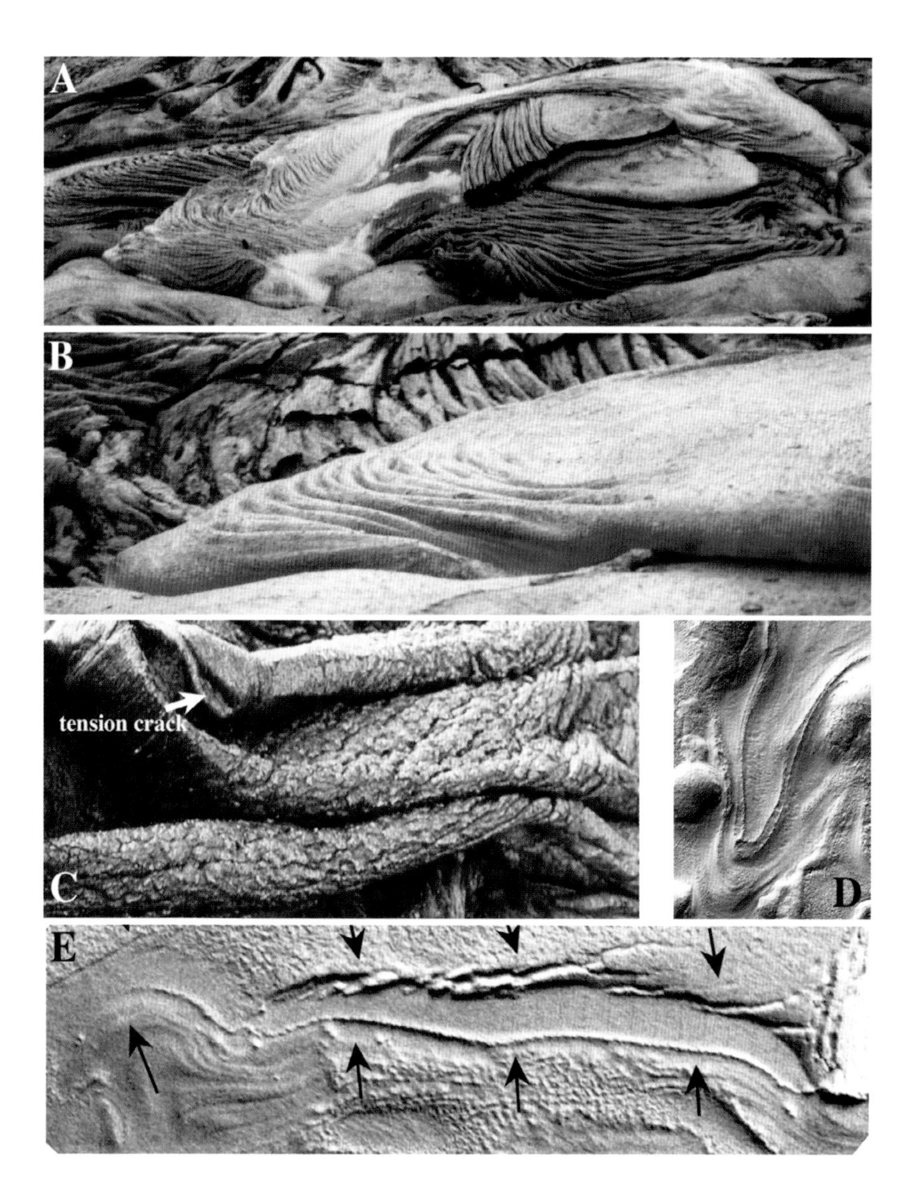

Great morphologic analogs do not necessarily imply the same phenomenon. Then think again; maybe it does! Panels A, B, and C, show fresh, hot lava at Kilauea undergoing compressive folding related to a hot, low-viscosity flow interior moving faster than the partly chilled outer skin. The rheological contrast causes flow dynamics that produce folds, called "corda", a typical structure in pahoehoe ("ropy") forms of basaltic flows. Note in panel C that tension cracks have formed where the lip of a fold has curled too tightly. In panels A and B the red-hot lava is flowing across other flows, some of which are concurrently active. It seems that similar structures may occur on much larger scales in the icy Martian flows shown in panels D and E. Obviously these materials are not the same material. They could be an analogous phenomenon: flow seems to involve a more fluid interior deforming a more viscous, semi-brittle skin.

Plate 11

"*Chaos*" by Michael Carroll. This painting derives critical insights from several terrestrial analog sites shown in this book. Michael Carroll's astronomical art has graced hundreds of murals, magazines and books throughout the world. Carroll, a cofounder and Fellow of the International Association of Astronomical Artists, is also a science journalist. He lives with his wife, son, daughter, five seahorses and a cat, turtle and god in Littleton, CO. His work can be found at *www.spacedinoart.com.*

"*Volcano*" by Michael Carroll. This painting captures some of the most universally accepted water- and ice-related processes on Mars: volcanic interactions with ground ice. Subtleties of this painting suggest that the subsurface may be extremely enriched in ice, as neutron flux data have clearly indicated.

Plate 13

"*Argyre*" by Carroll. This painting first appeared in an article by Kargel, J.S. and Strom, R.G. (1992a, December) Ice Ages of Mars, *Astronomy*, pp. 40–45. The painting derives some of its insights from Alaskan glaciers observed by the artist. It also successfully portrays the author's interpretation of alpine glaciers believed to have formerly poured into a basin-floor lake extending across southern Argyre Planitia. The painting uses modest slopes typical of the Charitum Montes (Mars) and certain terrestrial alpine regions, such as the Cairngorms (Scottish Highlands), without resorting to the exaggerated relief common in Mars paintings and common in many other alpine glacial landscapes on Earth. Nevertheless, the Charitum Montes attain peak elevations 8,000 m above nearby basin areas. The massifs are huge, and so were the glaciers that apparently once shaped them.

Plate 14

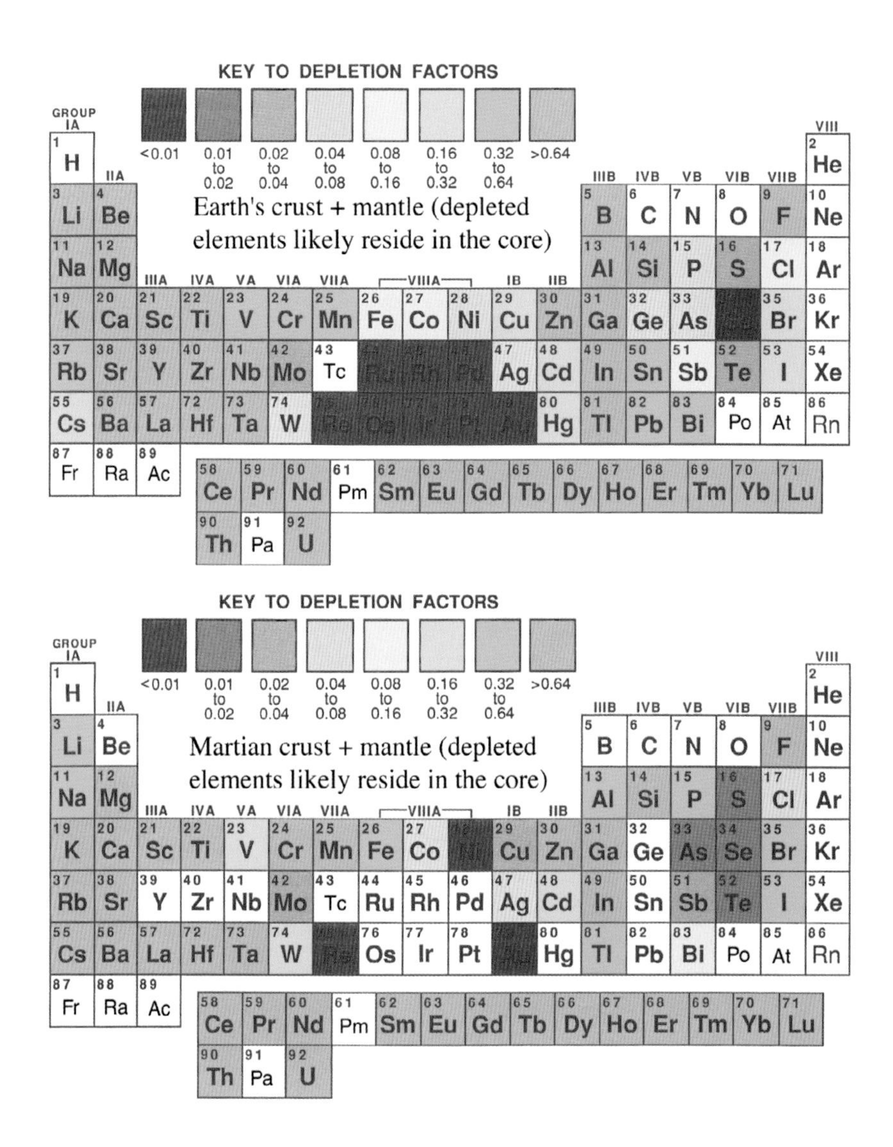

The compositions of the rocky parts (plus the hydrospheres) of Earth *(top)* and Mars *(bottom)* are indicated on standard periodic tables. Depletions (discussed further in Chapter 1) are probably related to sequestration into the planets' cores, but for some elements another possible cause could be impact blowoff of massive primordial oceans, including their salt-forming elements such as sulfur and chlorine. Elements that are not color coded either are not usefully portrayed on this diagram due to their specific geochemical natures, are not naturally occurring, or do not have sufficient chemical analyses to warrant plotting. Top figure for Earth is modified after Kargel, J.S. and J.S. Lewis (1993). The composition and early evolution of Earth, *Icarus*, **105**,. 1–25.

Plate 15

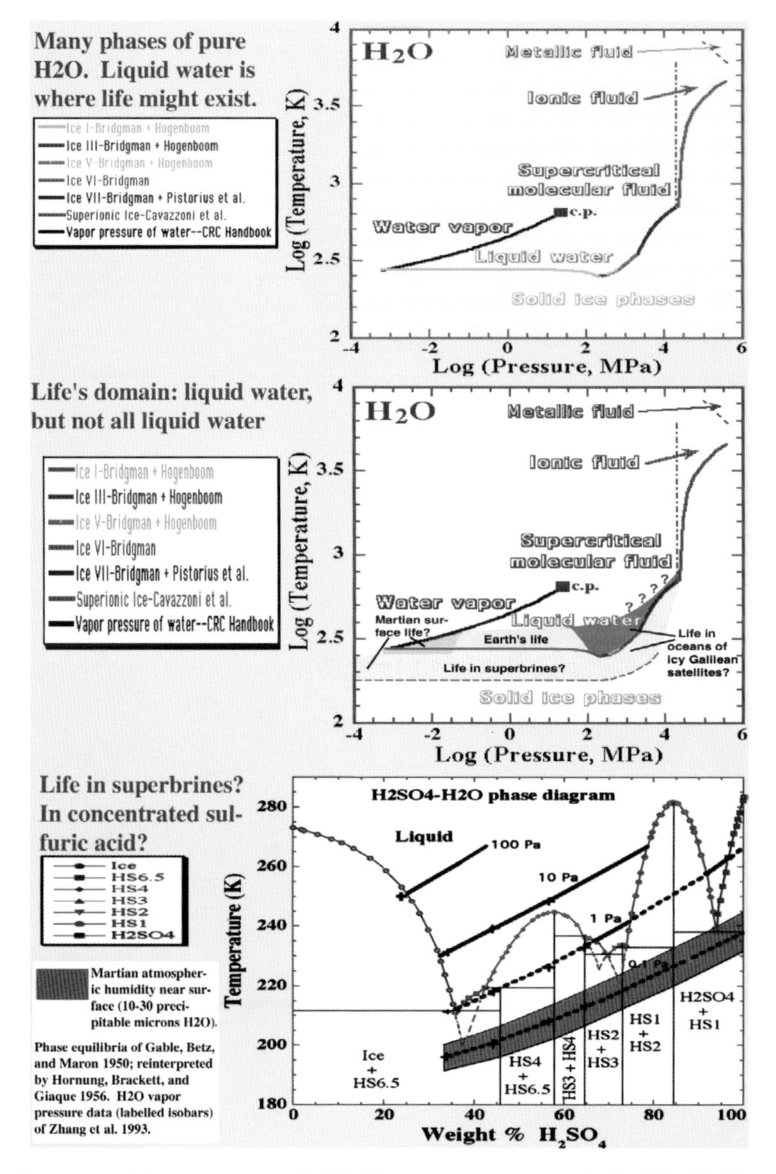

Liquid aqueous solutions are necessary for life. The pressure-temperature limits of Earth's life are set by the upper and lower pressure limits of Earth's hydrosphere, and the upper temperature limit of stability of certain large organic molecules. Only the latter limit is apt to apply to Mars, which may have far colder brines than any on Earth. In fact, aqueous sulfuric acid solution may exist stably on the Martian surface at low and middle latitudes. Eutectic sulfuric acid solution (93% sulfuric acid + 7% water) could exist in ponds, seeps, springs, marshes, or permafrost soil layers, stable as liquid without evaporating during the warmer months, but apt to freeze at night and during the winter. The more water-rich eutectics (including 35% sulfuric acid) could be erupted as spring water around degassing volcanoes, but the water would slowly evaporate. Whether life could exist in such concentrated acid is doubtful, although some acidophile terrestrial aquatic microbes come close.

Plate 16

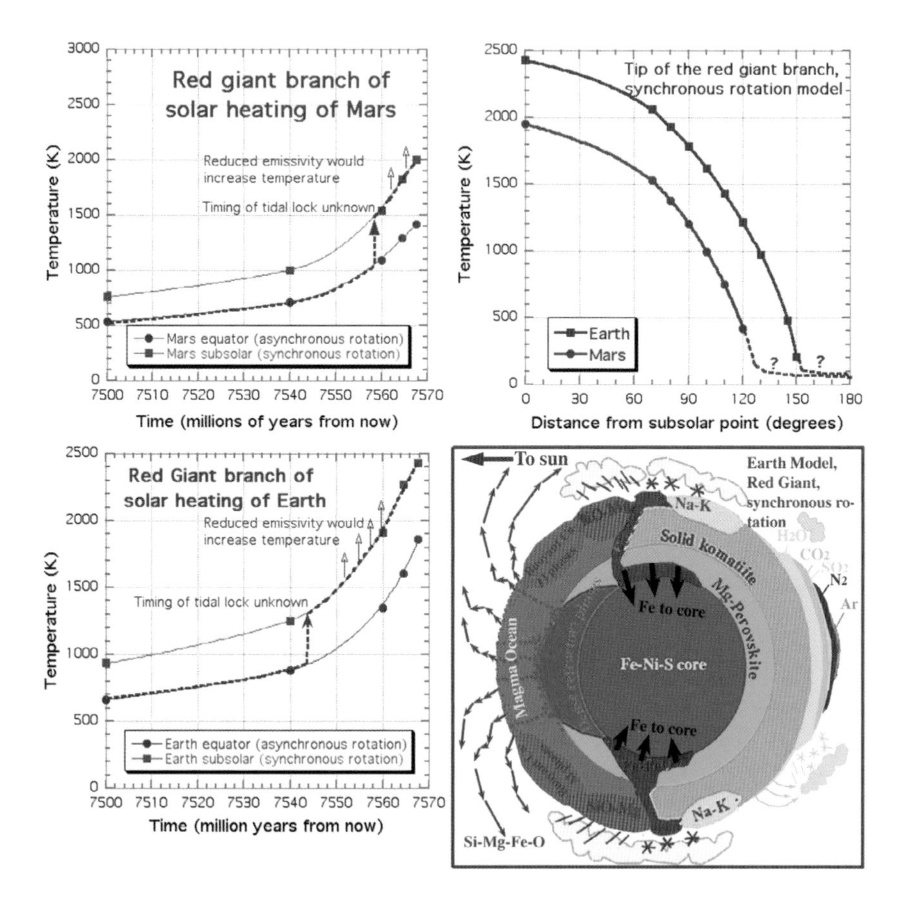

Model results illustrate the fates of Earth and Mars during a red giant phase of the sun's future evolution. Results first presented at the 2003 meeting of the Division for Planetary Sciences, American Astronomical Society. Left two figures show the maximum surface heating for Mars and Earth during a 68-million-year progression of the red giant sun. Each plot has two curves representing the cases of tidal rotational locking (like the Moon has to Earth). Upper right shows variation in temperature across the surface of Earth and Mars (distance from subsolar point) for the cases of tidal locking of rotation. Lower right is a diagrammatic depiction of Earth's possible cross-sectional structure. The view shown is cut from pole to pole and through the subsolar and antisolar points of a tidally locked Earth. Imagine the magnesium rain and sodium glaciers!

Plate 17

Viking Lander 1 images, from megaflood-ravaged plains of Chryse Planitia. Courtesy of NASA.

Plate 18

Viking Lander 2 images of what may be an impact crater ejecta blanket (crater Mie) on Utopia Planitia, but other processes may explain this landscape and the rock distribution: giant debris flows related to distant outflow channels, glacial deposition of ground moraine, or periglacial heaving due to freeze–thaw. The rocks appear to be mainly volcanic. Panel D shows early morning frost. Images courtesy of NASA.

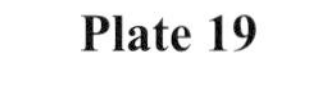

Plate 19

Mars Pathfinder rover *Sojourner* was hemmed in during its short traverses by an abundance of large rocks and by a need to communicate with Earth via line-of-sight radio transmissions to and from the mother landing craft (the *Sagan Memorial Station*). The high rock abundance is related to the landing site being situated directly on the deposits left by the megafloods that formed giant Ares Vallis. The rocks were determined by the rover's alpha-proton X-ray spectrometer to be andesites and basalts (volcanic rocks). The soil is similar, except that it contains a higher abundance of the salt-forming elements sulfur and chlorine.

Plate 20

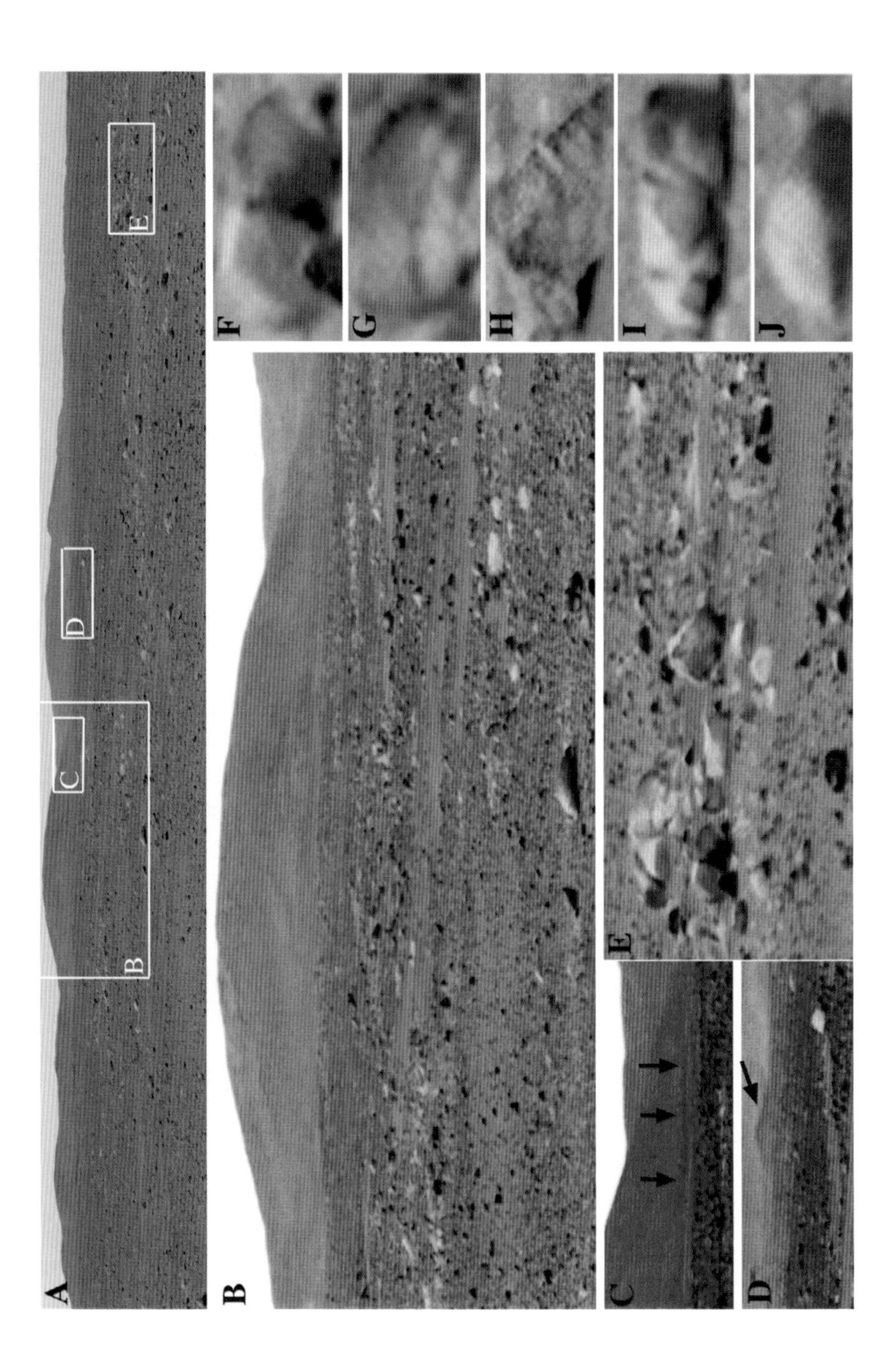

Spirit's PANCAM gave a great view of the "Eastern Hills," which could be the long-term destination of the rover if it has the good fortune to live longer than engineers had planned. The hills are about 2–3 km away at their nearest point. This scene spurs many questions and raises many opportunities for exploration. (B) What process accounts for the irregular distribution of cobbles? Are these small ridges of overlapping crater ejecta blankets, related to ancient flood deposition, or caused by recent heaving and sorting of sediment by volume changes due to permafrost freeze–thaw or salt hydration and dehydration? (C) The base of the hills is marked in many places by a terrace (arrows). This is seen also from orbit in MOC images. Is this terrace a sedimentary outcrop? A wave-eroded beach terrace of the ancient Gusev Crater Lake? (D) A conical relief form (arrow) is seen in the distance at the base of the hill. Is this a volcanic cinder cone? If so, and if these volcanic materials are found by inspection to overlie the Gusev sedimentary rocks, then a future mission carrying an appropriately designed mass spectrometer could determine the age of the volcano and thus derive a younger limit to age of the Gusev lake. Impact craters have probably excavated older underlying volcanic rocks deposited by ancient activity of the giant nearby shield volcano Apollonaris Patera, and so an older age constraint also could be produced. (F–I) Many rocks in the area near the lander are complexly structured, indicating that they have a complex history. Although closer inspection is needed to confirm these structures and determine their significance, we see here rocks that appear to be multi-toned, layered or cut by dikes, and composed of multiple rounded fragments that have agglomerated together (a conglomerate). Unlike any landing site seen before, this site also bears many rocks that are well rounded (J), as can be produced by rolling along the bed of a stream channel.

Plate 21

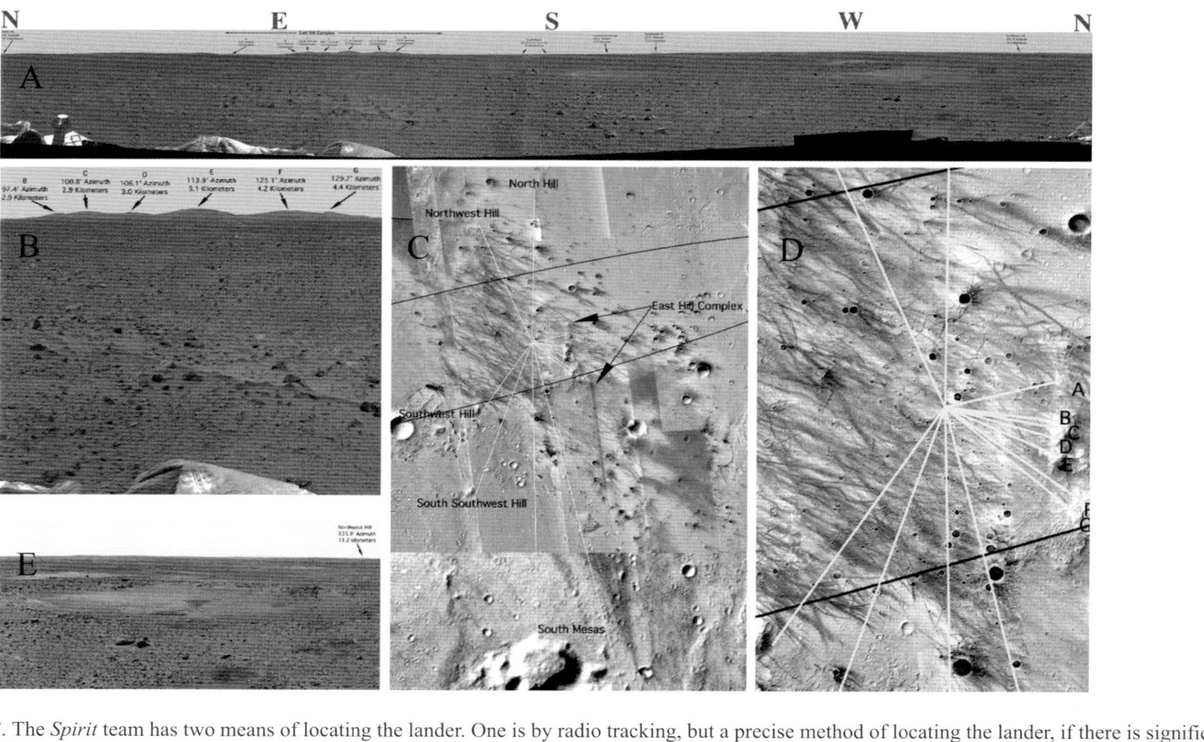

"Spiritual vision". The *Spirit* team has two means of locating the lander. One is by radio tracking, but a precise method of locating the lander, if there is significant nearby topographic relief, is to use satellite imaging in conjunction with peak matching in panoramic scenes. (A) 360-degree panorama (almost true color). (B) The East Hill Complex (70–100 m high) is the boldest feature on the horizon. The peaks in the complex are about 3.1 km away. (C) Sleepy Hollow (center of scene), just a few tens of meters away. Northwest Hill (marked) is not visible in this rendition. Its hazy summit, 11.2 km distant, barely protrudes above the horizon. (D) Horizon features visible to *Spirit* and used to pinpoint its location are indicated. (E) Blow-up showing the East Hill Complex. Courtesy of NASA and the MER team.

Plate 22

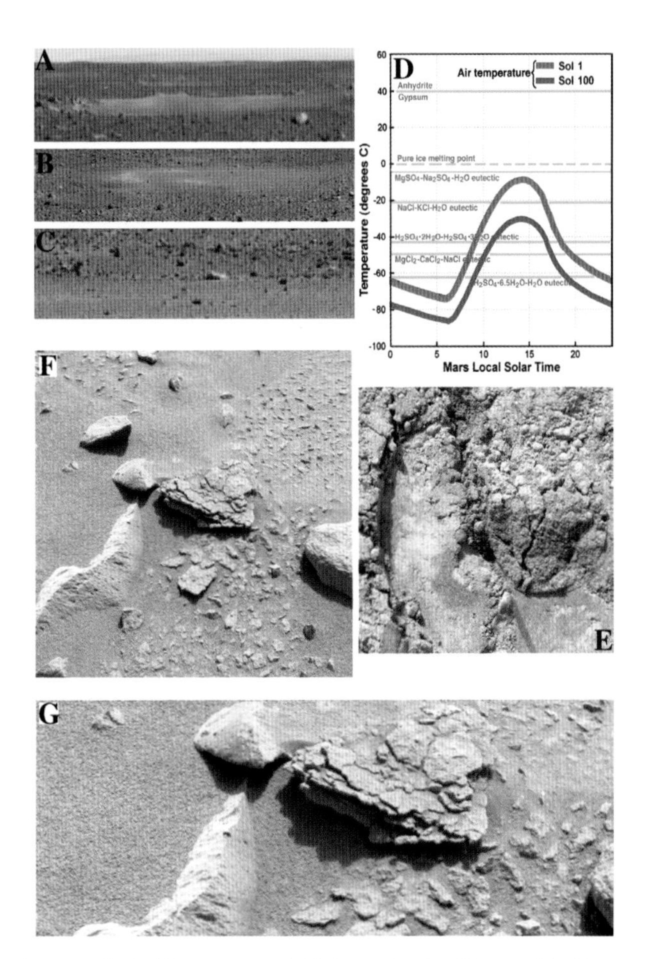

Small depressions near the *Spirit* rover are probably impact craters (A, B, and C). However, they have been heavily modified, as indicated by their eroded rims and sediment-filled bowls. Is the infilling and erosion due to wind processes or perhaps water processes? Erosional and depositional activity due to liquid water is possible even under today's climate if the water is highly impure. Perhaps the most effective and stable brine would be sulfuric acid-water, which includes eutectic compositions stable during the daytime both in terms of temperature and relative humidity (D). The MER team has reported predicted daily temperature oscillations (D) well within the range where some highly impure saline or acid solutions would readily melt. Such activity is suggested – weakly – by the possible eroded edges of the craters (C), the flat, rock-inundating sediment blankets in some of these deposits (F and G) and the eroded, crumbling rocks, such as "Flaky Mimi" just right and above center in G. However, nothing shown here proves unusual water activity, and in fact olivine abundances in rocks and soils suggest very slight or no aqueous chemical alteration that would be expected if acid brines flooded these areas. Even so, the soil clods and crusts dug up during the rover's wheel spins and the same crusts evident when the Mössbauer spectrometer compacted the soil (E) offer compelling evidence of some slight salt cementation and aqueous activity. Thin unfrozen films of briny solutions could explain these crusts and might also, over time, contribute to some physical weathering of rocks such as Flaky Mimi. More substantial seasonally unfrozen permafrost or marshes of sulfuric acid solution or other superbrines undoubtedly exist on Mars in areas of volcanic outgassing and spring activity, but probably not here. Panel D is after Kargel and Marion (2004), whereas the other images are courtesy of NASA/JPL/Cornell and the instrument teams.

Plate 23

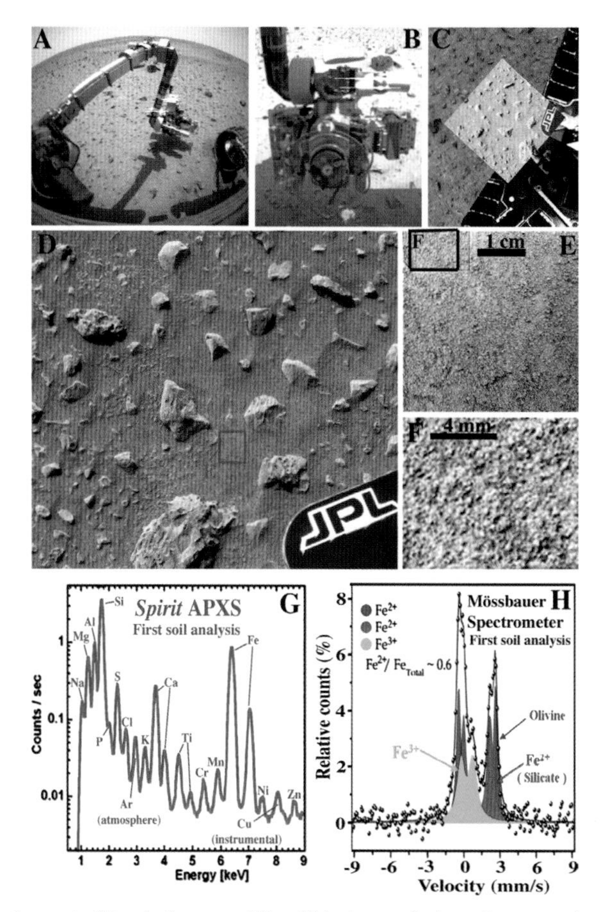

Spirit's first deployment of its robotic arm and "hand" (a cluster of science instruments) was captured by one of the rover's hazard avoidance cameras, with the shallow dust-filled depression known as "Sleepy Hollow" in the distance (A). The four arm-mounted science instruments (B) include a Microscopic Imager designed to simulate a geologist's powerful hand lens, a Rock Abrasion Tool designed to remove surface dust and weathering rinds and to measure rock hardness, a Mössbauer Spectrometer for determining iron-bearing mineral types and composition, and an Alpha Particle X-Ray Spectrometer for determining elemental abundances in dust, sand, and rocks. Besides these instruments, the rover also has nine cameras, including a 14-band visible/near infrared camera and navigation and hazard avoidance cameras. Panels C and D show the context for the image returned by *Spirit*'s Microscopic Imager (E and F). The first quick analysis of elemental composition of this same soil (G) showed a familiar suite of elements known from *Viking Lander* and *Mars Pathfinder* analyses of rocks and soils. The very high peaks for S and Cl indicate the presence of sulfates (probably salts containing Ca–Mg–Na–K–SO_4, and possible $FeSO_4$, and possible sulfuric acid) and chlorides (probably mainly NaCl salt and possible K–Ca–Mg–Cl components, and possible $FeCl_3$ and HCl acid). Most of the other elements are contained in these salts, unsilicated metal oxide minerals, and silicates. However, Ar is in the atmosphere and Cu is primarily in the APXS instrument itself. Zinc is the only element detected here that has not been detected previously. The first Mössbauer Spectrometer analysis of this same soil revealed minerals including both divalent (ferrous) iron and trivalent (ferric) iron. Ferrous iron is contained dominantly in silicates, but some is also possible in ferrous sulfate and ferrous chloride salts. A positive identification of ferrous-iron-bearing olivine was unexpected and is consistent with partial (or no) aqueous alteration. A more detailed interpretation of these and other results is offered in Chapter 11.

Plate 24

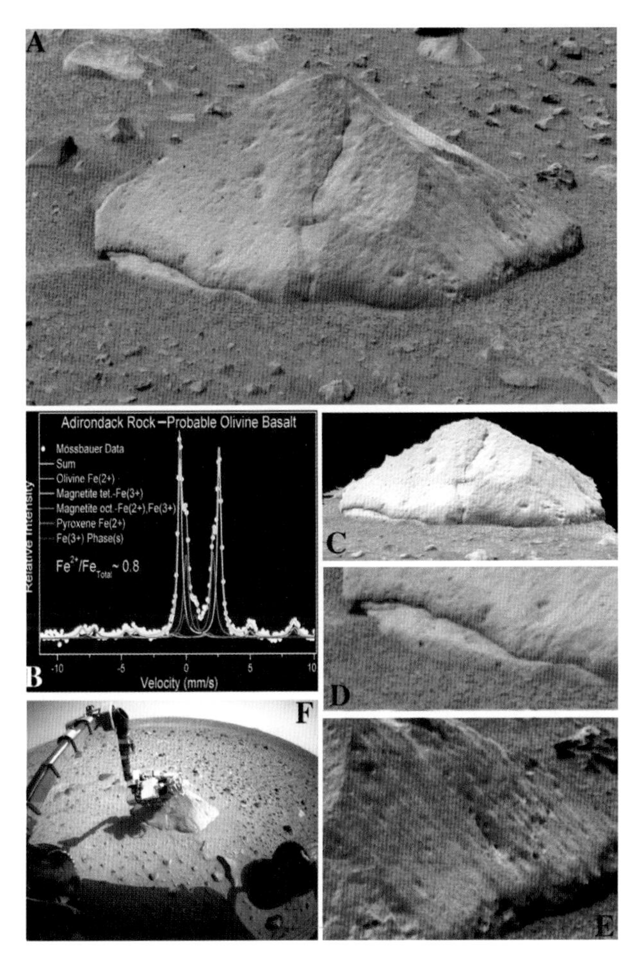

Spirit examines Adirondack, a faceted and fractured boulder measuring about 20 cm high. The rock has been heavily weathered, as indicated by its facets, pits and striations or grooves, the rounded fracture and the overhanging piece of rock along the fracture. Although the striated, faceted surface of the rock look just like glacial erosion, a more likely explanation is erosion by wind-driven sand-blasting. Stream-worn boulders are rounded and not faceted, because they periodically roll or flip over as they are progressively worn down by innumerable impacts of smaller current-driven particles. Desert wind-shaped boulders (called ventifacts) acquire faceted or cuspate shapes because they are largely stationary and eroded by sand blasting caused by one or two or more prevailing winds. Subglacially eroded boulders are faceted because they are scraped along bedrock and periodically flip over. Microscopic imaging can provide a diagnostic test, because wind erosion will show up as myriad microscopic pits and chips caused by impacting sand grains, whereas glacial erosion has long scratches and periodic "chattermarks" produced where silt, sand, and pebbles became trapped between a facet of the boulder and bedrock. They Mössbauer spectrum reveals the presence of iron-bearing olivine as the major mineral, plus pyroxene and small amounts of iron oxide minerals such as magnetite. The MER Mössbauer team interprets this rock as an olivine basalt, which on Earth is a common type of volcanic rock; however, it could be an ultramafic rock originally derived directly from the mantle, perhaps emplaced as impact ejecta. The abundant olivine indicates a limited amount of aqueous weathering. Olivine survives in central Nevada, for instance, for about 100,000 years before altering to other minerals. Data courtesy of NASA/JPL/Cornell and the imaging and Mössbauer instrument teams.

Plate 25

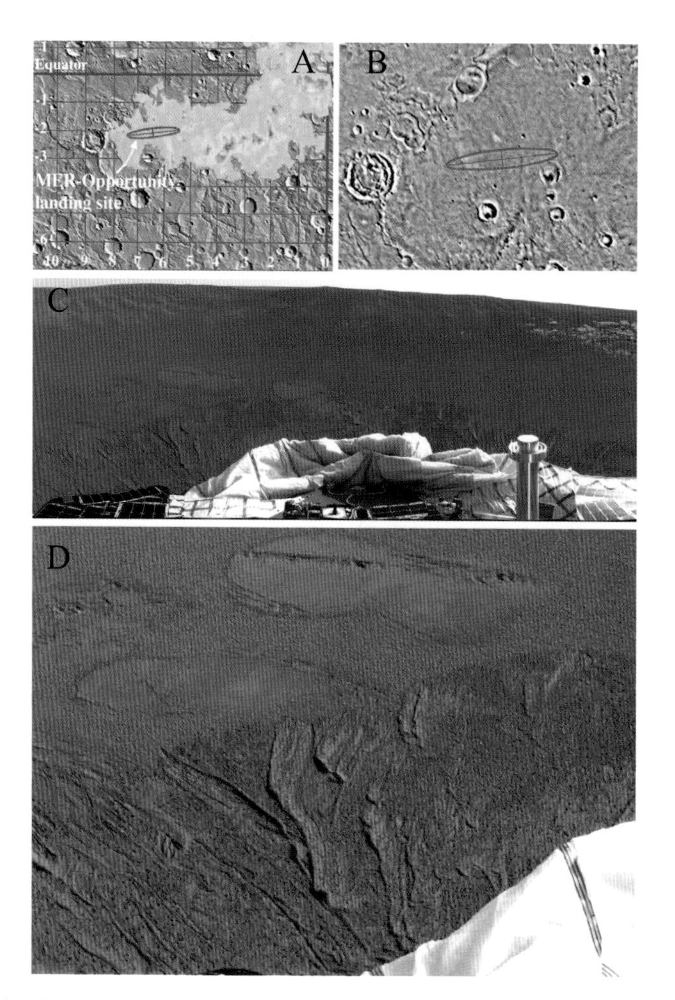

Landing site of *Opportunity* is deep within a Latvia- (West Virginia-) size deposit of gray crystalline hematite (Fe O), represented by the color-coded map in panel A, which shows the estimated abundance of hematite in the upper soil and rock layer overlain on a topographic shaded-relief map produced from laser altimetry. The map of hematite distribution is from analysis by the *Mars Global Surveyor* TES (Thermal Emission Spectrometer) team. There are other hematite concentrations on Mars, but this is by far the largest such region. In total these areas comprise about 0.05% of the planet, so this is really a rare locality. The deposits are of great interest because rich concentrations of hematite usually form on Earth under aqueous, acidic conditions. The region lies in a regional depression and consists of layered rocks, which locally (far to the east of the landing site) are associated with spectacular mesas, buttes, and pinnacles of layered rocks resembling Arizona's Monument Valley and the red banded iron-formations of Western Australia. Here at the landing site the terrain is about as flat and smooth and rock-free (safe) as any site on Mars. The lander by dumb luck bounced into a small crater about 20 m across. Near the rim of the crater are exposed layers of layered bedrock, a gift to geologists. The soil in the crater is unlike anything previously seen on Mars. As the airbag-cushioned lander bounced and rolled to a stop, it made impressions on the surface. When the airbags were retracted it left scrapes and ridges of fine sediment. The unique soil and rocks in the crater will be the first targets of analysis, and then the rover will drive out of the crater. Courtesy of NASA/JPL/Cornell (panels B, C, and D) and NASA/JPL/ASU (TES data, panel A) and the NASA/MOLA team (shaded relief map in panel A).

Plate 26

The Mars *Opportunity* rover has obtained these spectacular paroramas of rock outcroppings in a small crater, which, by the good fortune of this mission, it landed in. The top image is roughly natural color although contrast enhanced; the bottom two scenes are false color. The outcroppings give an impression of being massive, but in fact they are about a quarter meter high. Thin laminations – about 2 mm thick – have been found to be made largely of sulfate salts. Included are large amounts of jarosite and unidentified calcium and magnesium sulfates. Eroding from the outcrops are spherules (blue in panels B and C), 2–3 mm diameter, made largely of the iron oxide hematite. The polygonated outcrop at first struck me as most likely having been shattered by impact. While this is still a possibility, it now appears most likely to be solution weathering of salts. The hematite spherules are probably concretions, formed under aqueous conditions, according to the MER team. Images courtesy of NASA/JPL/Cornell and the MER team.

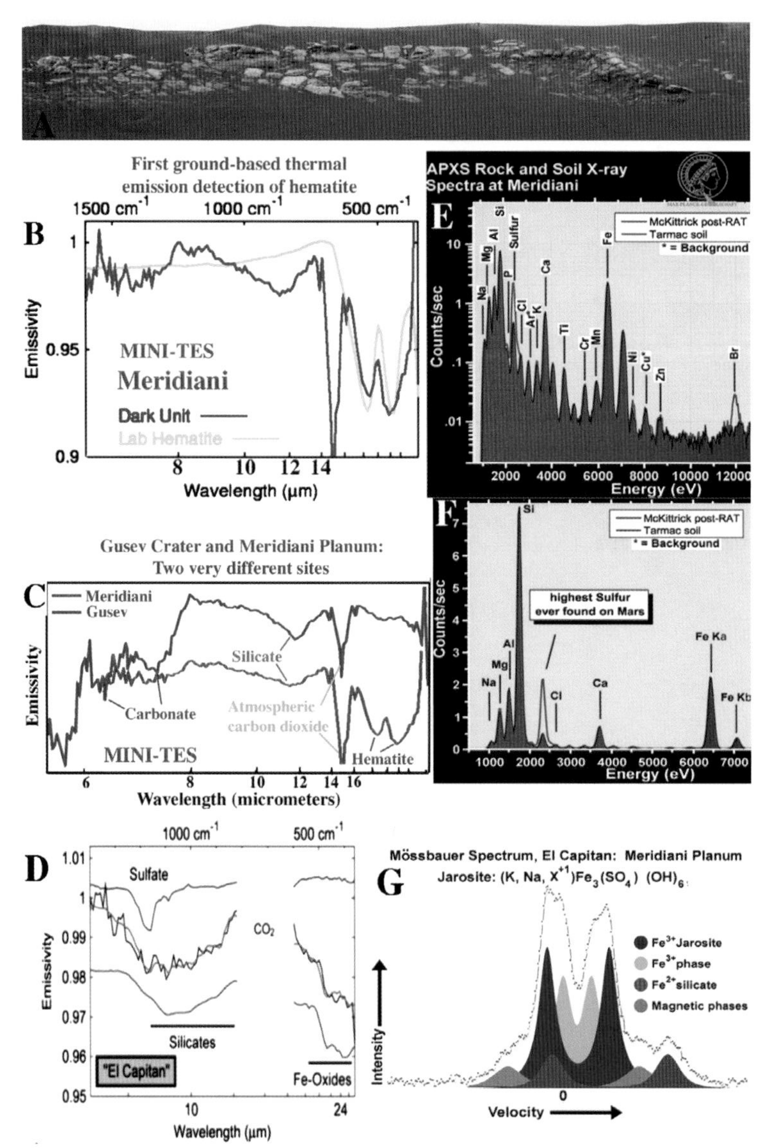

Opportunity has now completed its exploration of a little rock outcrop in its little crater. A sampling of chemical and mineralogical analyses presented by the MER team is shown here. Enigmatic spherules, about the size of B's (2–3 mm diameter) are weathering out of the layered rocks. The spherules are made substantially of the iron oxide hematite, which also constitues a fraction of the dark soil in the crater and beyond. The rock outcrop, aside from the spherules, contains relatively little hematite, but instead is made dominantly of a variety of sulfate salts, including jarosite and unidentified Ca–Mg–rich sulfates. Carbonate is far less abundant (or absent) at Meridiani compared to Gusev Crater. This is consistent with aqueous acid chemical precipitation or alteration of these rocks, as explored in Chapter 11. Data courtesy of NASA/JPL/Cornell/DLR/ASU, and the MER team. See the MER website (http://marsrovers.jpl.nasa.gov/gallery/press/opportunity/) for the *Opportunity* mission for further details.

Plate 28

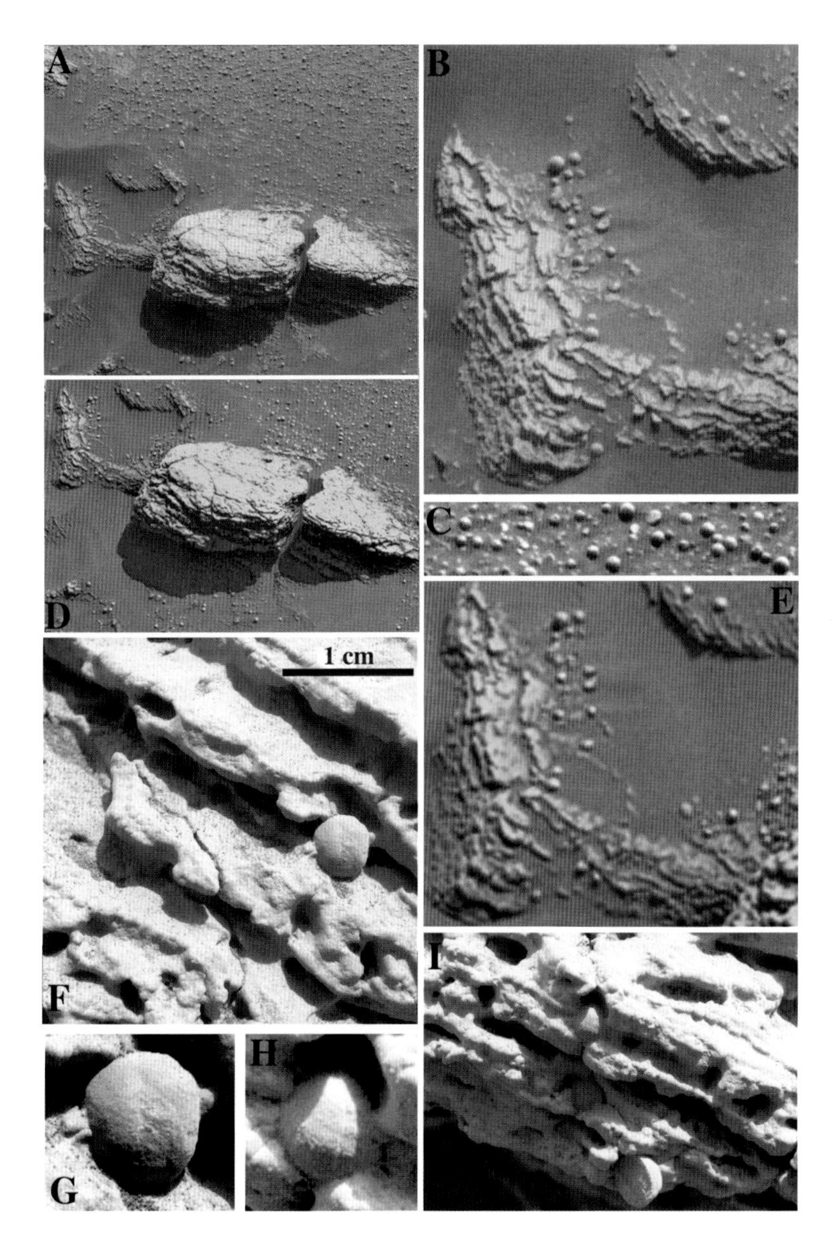

Opportunity's fortuitous discovery of layered fine-grained rock outcrops containing granular spherules has dominated the early weeks of exploration. The spherules have a composition distinct from the laminated fine-grained material and from the hematite-rich dust. The laminations shown in panels A and B are each a few millimeters thick, comparable to the diameter of the spherules. The laminated rocks have an appearance typical of water-lain rocks, such as annual silt layers known as varves, annual chemically precipitated layers produced in saline lakes, and hydrothermal spring deposits. However, wind and turbulent ash and ejecta clouds can also produce layers. Further analysis is needed. Courtesy of NASA/JPL/Cornell and the MER teams.

Plate 29

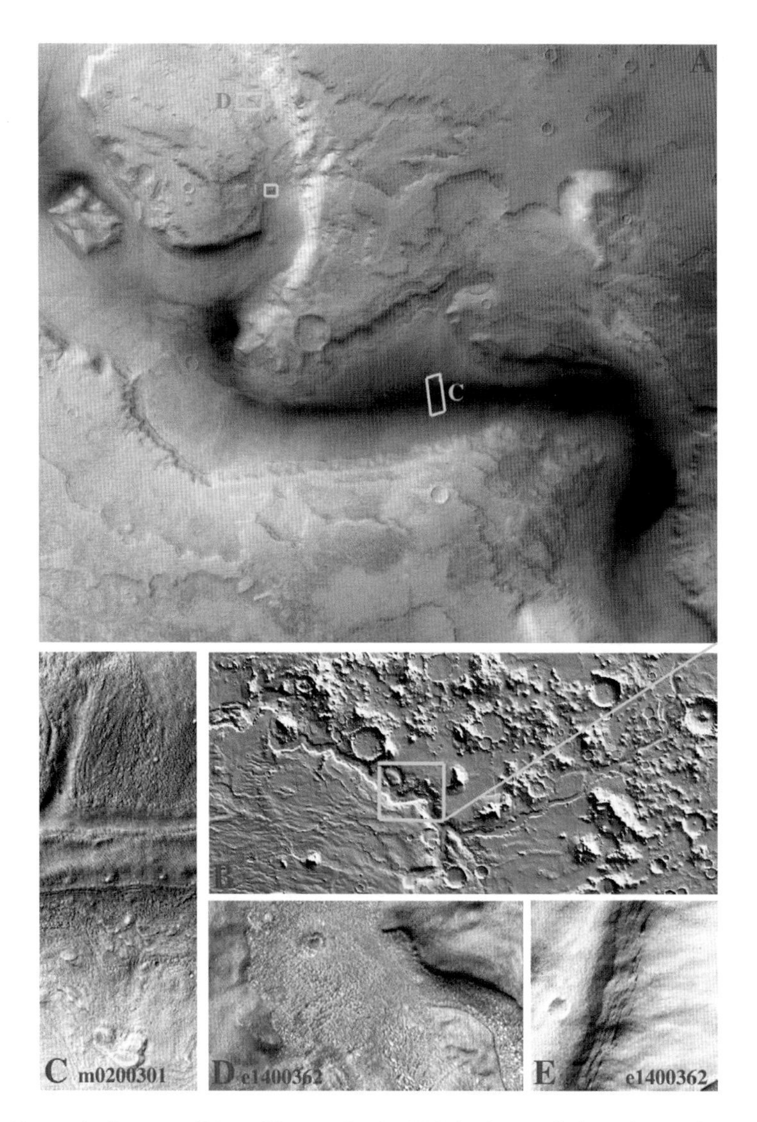

Mars Express, having entered Mars orbit December 25, 2003, has begun a firehose-deluge of spectacular data, including stereo, full-color images, ground-penetrating radar, and spectroscopy capable of discerning chemical constituents of the atmosphere and surface. (A) Color image of a small portion of Reull Vallis, a major channel that emptied into the Hellas Basin. (B) A MOLA-derived shaded-relief image shows the topography of Ruell Vallis and adjacent volcanic plains to the south and rugged mountains of Promethei Terra to the north, as well as the context for the *Mars Express* scene. A channel-floor deposit is evident in the *Mars Express* image, as seen also in spot coverage by MOC with finer detail (C, D, and E). The deposit has characteristics of partly sublimated, debris-covered ice or icy debris flows. The adjacent mountains are heavily mantled by icy debris aprons, thought to be debris-covered glaciers, as described in Chapter 7. This set of images shows how data from different satellites and instruments can be used together to gain improved context and details at all scales, thus improving interpretability. *Mars Express* image courtesy of ESA, the *Mars Express* team, and Gerhard Neukum (DLR). MOC data courtesy of NASA/JPL.MSSS, and MOLA data courtesy of NASA/JPL.MOLA team.

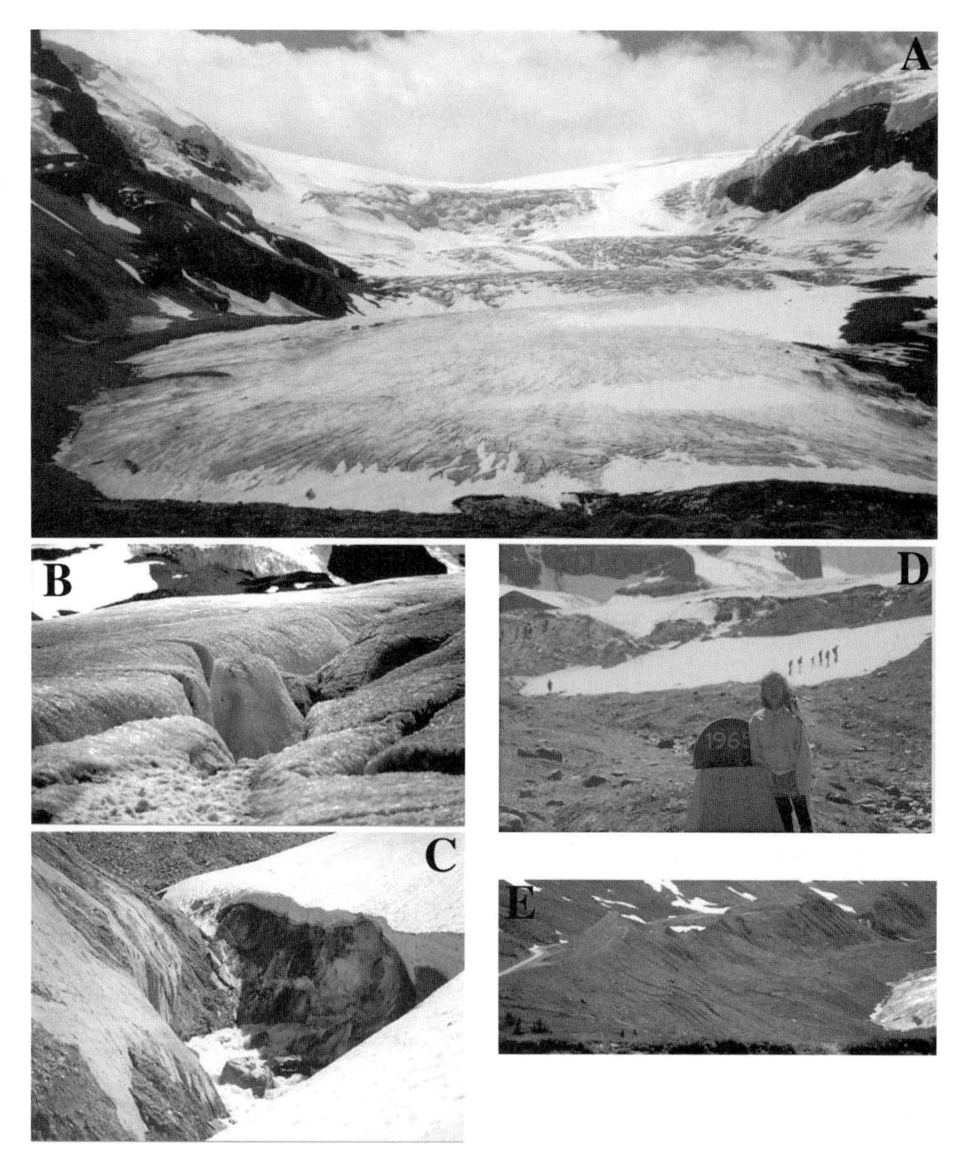

(A) Athabasca Glacier, Jasper National Park, Alberta, in 1990. (B) Crevasses. (C) Subglacial stream at the glacier snout. (D) Marker locates glacier terminus in 1965. (E) High lateral moraine from late nineteenth century.

Alaskan glaciers exhibit crevasses, as seen in some Martian glaciers, and ogives, not on Mars. Wave-like ogives occur in terrestrial glaciers beneath ice falls, with one wave per year. Martian glaciers are far too sluggish to develop ogives, but crevasses are wide and well developed in cases. The point of closure of annual crevasses and transition to ogives is indicated by the arrow.

Large valley outley glaciers in east-central Greenland. Although snow-covered and thus not looking like Martian debris-covered glaciers, these glaciers have developed some of the fine-scale textures common in lineated valley fill, such as the evenly dissected medial moraine, enlarged (bottom panel, upper left). These valleys also show, like glacial valleys on Mars, dominant tectonic control, rather than the antecedent fluvial control seen in most temperate glaciers on Earth.

Mer de Glace, French Alps. (A) Ice Falls. (B) Meltwater-eroded crevasse. (C) Typical crevasse; note human figure for scale. (D) Ogives produced at ice falls and flowing down-valley. These will not occur on Mars, because they represent annual periodicities, and on Mars the annual flow displacement is negligible. (E) Close view of ogives rounding a bend in the valley. (F) Crevasses have planview pattern that is concave up-glacier (to the right). Crevasses are expected on Mars because ice is brittle at Martian temperatures. The sinuous white ribbon is an alpinist's trail. (G) The author.

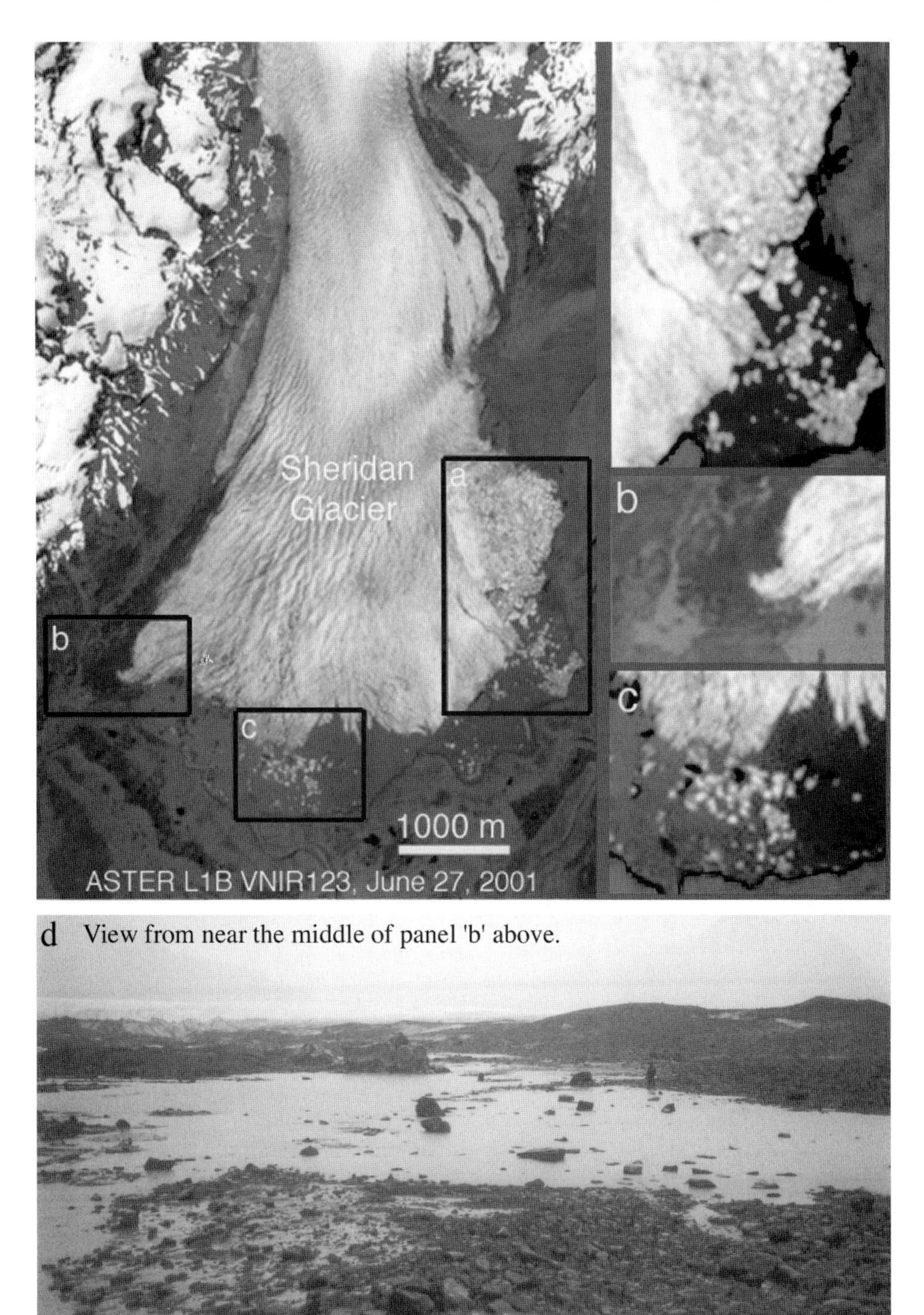

Sheridan Glacier, Alaska, has a debris-covered terminus due to down-glacier transport of landslide debris.

The debris-covered terminus of Sheridan Glacier, Alaska, has nearly stagnated. The preceding figure shows this part of the glacier in panel B. Rock debris plugs conduits, allowing meltwater to accumulate on the glacier surface and etch a swiss-cheese structure of water-filled moulins and caverns patterned around ice-tectonic structures. (A) Crevasses etched by water. (B) *En echelon* pattern of sigmoidal crevasses etched by water. (C) Disordered crevasses etched by water. (D) Water-filled Moulin. (E) Moulin shapes ranging from precisely circular to elliptical tubes are common. The ellipticity of this Moulin (same as in D) is probably caused by ice tending to close the Moulin, but anisotropic thermal erosion of oriented ice grains is possible. Like proximal hot springs and geysers, systems of moulins are commonly hydraulically linked, although separate clusters of nearby moulins may exist in hydraulic isolation. Look for possible Martian moulins as sediment-filled tubes.

Sheridan Glacier. (A) Glaciers exhibit a rich tectonic fabric not unlike that of high-grade metamorphic silicate rocks, such as gneiss. The mineralogic-scale causes of the fabric are not all the same; for instance, a crucial part of a glacier's fabric is its trapped air bubbles, which do not have a close analog in gneiss. (B) A sheared-to-shreds "mylonitic" fabric, like that of gneiss, is very common in glaciers, as are folds, such as seen here. (C) The rheological behavior of bubble-rich and bubble-free and rock-rich and rock-free layers is different; when glaciers flow these layers stretch differentially. The result sometimes is boudinage ("sausage structure"), where some layers alternately pinch and thicken. It is a tectonic phenomenon familiar in gneiss.

Debris mantling of the Root/Kennicott Glacier, Alaska, is nearly complete in the lower several kilometers but is not as thick as it appears; in places, about a meter of rock debris blankets all the ice, but elsewhere a monolayer of sand and gravel obscures all but a few ice exposures. Thick or thin, the debris has a strong influence on ice ablation and formation of thermokarst structures. Thin debris melts ice faster; thick debris insulates it and reduces melting.

A dendritic, thermally eroded gully pattern is etched into the dirty ice face of a hummock of the debris-covered Root Glacier, Alaska. The debris cover is about 0.5 m thick, seen at the top of the ice hummock in the enlarged view, lower panel. An ice cave, formed by a subglacial stream, is also evident.

Some terrestrial glaciers are very poor analogs for Mars, including glaciers with highly contorted medial moraines due to unsteady or unsynchronized flow of multiple tributaries. (A) Bering Glacier, Alaska, during its 1993 surge into a proglacial lake. While lake calving and wet-based conditions conducive to glacier surges might have occurred on Mars in the distant past, they certainly do not occur today. (B and C) Malaspina Glacier is the archetype piedmont glacier, with tightly compressed contorted moraines.

Rock glaciers in Wrangell Mountains near McCarthy, Alaska. Both of these could be rock-enriched remnants of ice glaciers, since they are associated with high alpine cirques. But they could represent ice-enriched talus or old glacial moraines that now are creeping by periglacial processes. Either way, by the classical definition of a glacier, these are glaciers.

Part of Baltoro Glacier, Pakistan, near K2. Most of the glacier section shown here is heavily debris covered, but here the surface covering is snow in this January ASTER scene. The relief of medial moraines has been enhanced and the internal structure of the glacier etched by melting. Crevasses, ogives, and other features have been formed and accentuated by flow and melting. Although ogives are not part of the Martian glacier story, crevasses and medial moraines are. Similar glaciers occur on Mars in the fretted terrain; however, that region has no cirques, aretes, horns, and other classic alpine glacial features.

conditions on any planet is not large, but ice is one, probably the only one, that would work on Mars. Furthermore, these features are distributed at latitudes either where ice currently occurs at shallow depths (observed in neutron spectrometer data and indicated by theoretical calculations) or where it likely exists somewhat more deeply buried (theoretical calculations).

There can be no reasonable doubt: these are icy flow deposits, and the primary ice is ordinary water ice. They are glaciers.

My interpretation, benefiting from new data, is slightly different from that of Squyres and Carr. The absence of glaciers at low latitudes is indeed because that region is desiccated, but at high latitudes ice is abundant and is flowing sluggishly in many places, but the flow indicators are hard to see because the region is generally one where ice is accumulating and burying the landscape. At middle latitudes, ice is sublimating, thus exposing the inner structure of the flows and making them easier to see. In my view, dust and fine sand are present and co-deposited with ice; Martian glacier ice is dirtier than terrestrial ice because the dust flux of the atmosphere is not that much less than on Earth but ice precipitation rates are orders of magnitude less. However, the main reason that Martian glaciers are so heavily charged with debris is that they are very strange glaciers – their accumulation is largely from the subsurface and by frost condensation just beneath the surface. Most terrestrial cryosphere experts would feel comfortable if the term "rock glacier" were applied.

We should not feel too comfortable yet or think that we really understand Martian glaciers. Most are really weird in some ways.

ICY UNFAMILIARITY

There are some specific features of Martian glaciers and their valleys that are not seen or are rarely observed on Earth. We see circular structures that may be viscously relaxed or glacially strained impact craters, for instance. This type of difference between Earth and Mars is important, as it attests to differences in key timescales, but it is not a puzzling distinction. The self-consistent interpretation reached above, based on texture and integrated form, is one of flow deformation phenomena and ablation responses that are strikingly similar to what we see on Earth. That basic point is well grounded. However, Martian glaciers have some real puzzles to explain.

New evidence is overwhelming that most Martian icy flows, including all of those in the fretted terrains, have not quite earthly modes of formation. The fretted terrain presents some fascinating landscapes. Webster's defines the verb *to fret* as *"3. To make a way by gnawing, corrosion, wearing away, 4. To become eaten, worn, or corroded."* And the noun form of *fretting* is *"1. An interlaced, angular design; fretwork. 2. An angular design of bands within a border."* Without becoming too geologically detailed, it is hard to find a better word and definition to describe the fretted terrain, which was first described from *Mariner 9* images and named by R.P. Sharp in 1973. As we have learned recently, this terrain type represents a superposition of extremely young (late Amazonian) ice-flow features (debris-covered glaciers) in extremely ancient (Noachian and Hesperian) fretted canyons. The glaciers and canyons are clearly related. It may be that the glaciers have been forming and developing slowly, episodically since the incipient formation of the canyons, but that the most recent episodes of flow are very recent.

An odd thing about Martian glaciers (the polar caps excepted), and something that any successful model will have to explain, is that most have no clearly evident accumulation zones that are just like those on Earth. They are completely debris-covered, and the fact that most are so deeply textured across their whole surface indicates that we are looking at their eroded inner structures, or at least partly sublimated surfaces. This would not be so odd if that was the only anomaly, because terrestrial rock glaciers are somewhat similar to this, and many dying snow-fed glaciers are not too different. I do not doubt that some Martian glaciers are of these varieties. Several models of Martian glaciers are presented below to explain certain problematic types of glaciers.

In any consideration of Martian glaciers formed under anything close to the present climate, it should be noted that the annual cycle of accumulation and ablation – while critical for determining long-term mass balance – is not going to have dynamic significance for flow. One year's worth of ice might be less than a millimeter, and if a glacier or ice cap flows at even a few centimeters per year, it's really clipping along. For dynamical considerations, the equivalent of Earth's mass-balance year might instead be the Martian obliquity cycle. The complete absence of melting in most cases makes for one of the most unique aspects of Martian glaciers, because even in the cases of hard-core polar glaciers of the Antarctic Dry Valleys, seasonal surface melting is significant, even where sublimation is dominant.

Glaciers of the fretted terrain

There are key hemispheric differences among the icy flows; in the north, the lineated valley fill (striped valley deposit) morphology is dominant in the fretted terrain, whereas lobate debris aprons are the most usual icy flow feature in the south. Why there should be this profound difference in icy flows between the two hemispheres is not clear, but this is certainly not the only hemispheric dichotomy. Geology in general is dichotomous and regionally variable, and presumably the icy flows are in some way reflecting these differences. We see regional differences in glacier style on Earth, too, and it is not particularly a puzzle that regional variations of glaciation would occur on Mars. Ice accumulation on Mars may include both subsurface and atmospherically driven processes, with one or the other dominating depending on latitude, local topography, and geology. Generally, Martian glaciers seem to be variations on weird glacier themes.

A large segment of the northern fretted terrain occurs along the lowlands boundary with Arabia Terra, which is a geophysically/geologically unique part of the cratered uplands in many ways. This terrain is as low as the fringes of the northern plains; its crust is thin, its lithosphere is thin, it lacks the magnetic anomalies that pertain to much of the rest of the highlands crust, it merges with the strange hematite deposits of Meridiani Planum, and it has a more siliceous composition than the rest of the highlands. Geophysically and geochemically it looks like honorary lowlands, except that it has a high density of large craters. The craters are exceptionally

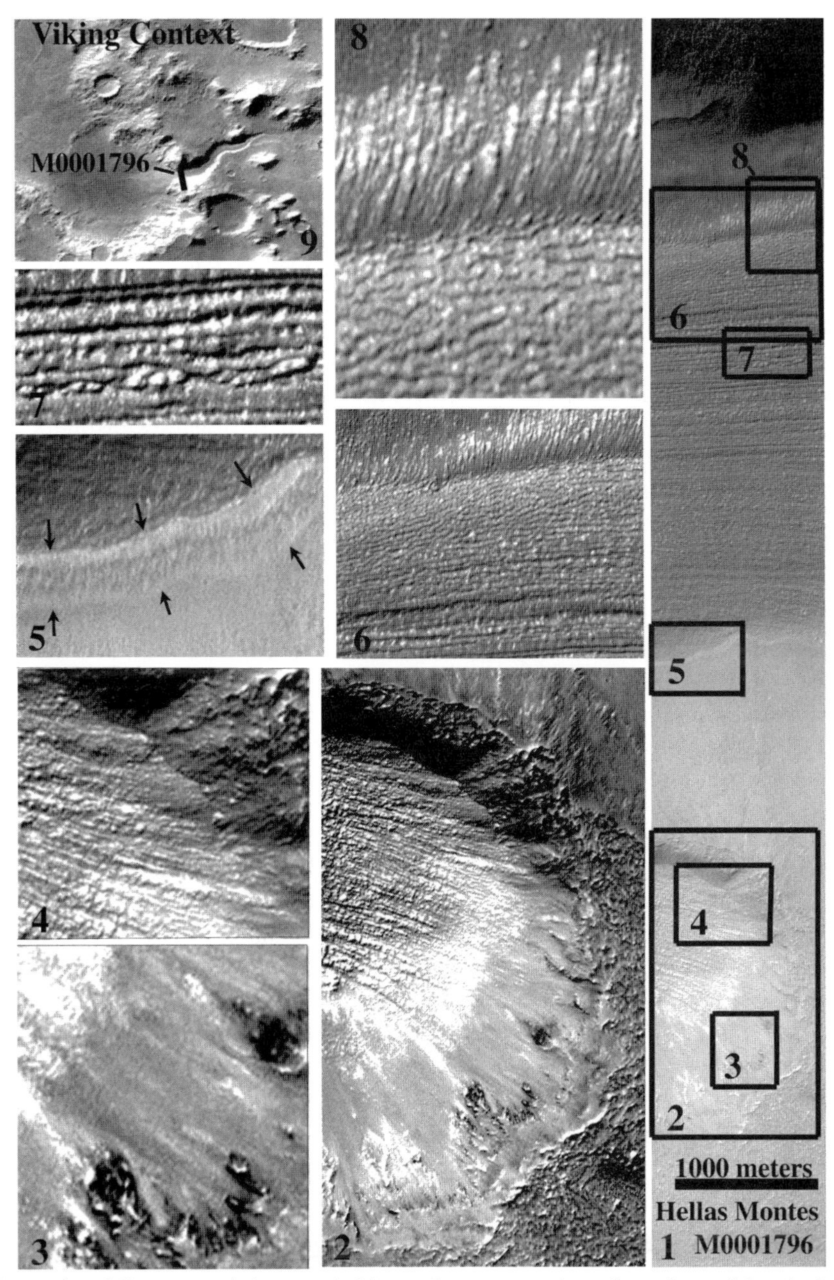

A sapping falley network is occupied by and appears to have been formed by glaciers. Termed "striped valleys" by Nathalie Cabrol and others, these icy landforms appear to have revealed their internal structures by sublimational etching. Close inspection reveals a hierarchy of sapping type features. One possibility is that these features form by tectonically controlled stream sapping and freezing of water and rock debris. The frozen rocky mass then creeps down valley over the eons.

flattened, and many are dissected by channels or disarticulated by fault valleys and other tectonic gashes forming the fretted terrain. However, fretted terrains exist widely along the edge of the cratered highlands and similar terrains circumscribe Argyre and Hellas.

Arabia has, like many terrains on Mars, and more so than many cratered highlands terrains, an unusual paucity of small craters considering the number of remnant large craters. In general, preferential erasure of small craters indicates a resurfacing process, such as sediment deposition, volcanic lava flow emplacement, and glacial or fluvial erosion, that can wipe the slate clean of small scales while leaving the largest craters. In Arabia, even though the indications of glacial flow abound in the fretted canyons, the morphology of the plateau surfaces indicates that they are untouched by glacial erosion. Southern Arabia merges into Meridiani Planum, where the hematite terrain exists along with layered strata and smooth plains filling in the low-lying areas, as though this region was once where shallow marine embayments occurred. If there were repeated marine incursions, sediment deposition and ice deposition could explain these features.

The fretted terrain appears as though large crustal blocks are slowly gliding downslope over a ductile substrate toward the northern plains. There are many aspects shared with fluvial sapping valleys, but the sapping fluid seems to be ice! Between fairly rigid crustal blocks are the fretted canyons and debris-covered glaciers on the canyon floors. Lacking in the fretted terrain are high-alpine cirque basins at the sources of glaciers, aretes, and other classic alpine landforms. The drainage structure of the lineated valley fill of the fretted terrain is controlled by fault canyons (graben) and extensional fractures and apparently not at all by pre-existing fluvial valleys, as occurs with most of Earth's alpine glaciers.

The only terrain almost like it are some adjacent highland units, including chaotic terrains, which have given rise to the circum-Chryse outflow channels, and the knobby terrain east of Hellas. The source regions of the Chryse outflow channels may be just a more extreme version of Arabia's fretted terrain – the discharges of ice and water and debris occurred in Arabia more slowly, in more controlled fashion, but similar processes may have been involved, with wholesale undermining of plateau units by massive volatile discharges. In Arabia it seems that the discharges continue. This hearkens back to Lucchitta's glacial model of the outflow channels.

In meeting Hugh Kieffer's mandate that we try to "think like Martians," I offer some further wild thoughts on Martian glaciers. Some are fairly well developed hypotheses, some are half-baked (or at least half melted), but all are serious and worthy of consideration.

Glaciers formed below a low troposphere at the edges of ice-filled basins

With some exceptions, the tops of most Martian glaciers are associated with the bases of mountains and cliffs, rather than the peaks and cliff tops. The lobate debris aprons in Argyre, for example, fringe the bottoms of the high peaks of the Charitum Montes, showing a fairly consistent maximum altitude of 0 km, i.e., 2 km above the bases of the mountains, which rise a further 3–5 km above the tops of the glaciers. Likewise, the lobate debris aprons and lineated valley fill in the fretted canyons at the fringe of the northern plains and east of Hellas are topographically constrained to the lower mountain elevations. One possibility is that these basins contain large reservoirs of ice, an idea that has much independent support. These ice deposits may serve as major regional sources of water vapor and may drive local hydrospheric cycles, including snowfall or frost condensation on nearby elevated terrain. Local humidification of the atmosphere – either under the present environment or some other recent climate – may have caused growth of snow- or frost-fed glaciers up to the tropospheric limit, above which the global dry Martian atmospheric system dominates and prevents snow/frost accumulation. The 100% debris cover of these glaciers may be an indication of an active dust transport regime on Mars – any accumulated ice is very dirty and dark. This may sound very un-Earthly, but it is really the cold-planet analog of Earth's maritime mountain glaciers. One difference is that the "ocean" on Mars is frozen. Another might be that the Martian troposphere extends only partway up the mountains and cliffs; on Earth even Mount Everest is sometimes beneath the tropopause (the boundary between the troposphere and the stratosphere) and thus receives snow. The occurrence of striae of likely glacial origin right up to the summits of the Charitum Montes suggests that the tropopause was once higher, but with the sublimation of ice in Argyre the climate has dried and the glaciers have receded. This model is readily testable by supercomputer-driven global circulation models, but it has not yet been tested.

Expressed glaciers

The flowlines and other flow direction indicators of Martian icy flows suggest large-scale flow variously (1) transverse to major valley walls and crater rims and the edges of mesas, or (2) longitudinally down valleys (parallel to valley walls). Some instances where the lineated valley deposits appear to have flowed longitudinally down valleys offer striking analogs to terrestrial debris-covered valley glaciers. More commonly, the flowlines and other structures seem to indicate that accumulation occurs at valley walls and the flow is dominantly transverse to them. Often, both motions are indicated, with flow initially directed away from sources at valley walls but then swept down-valley in gracefully arcing flowlines. Baerbel Lucchitta was among the first to note this from *Viking Orbiter* imagery. Michael Carr added to this perspective in describing his model for formation of small-valley networks. This model and the next are adapted and further developed from the ideas of Lucchitta and Carr.

Many morphologic indications point to subsurface sources of ice feeding these flows at the bases of valley walls and other major cliffs and mountains – a very peculiar type of icy flow. The appearance in some cases, particularly in the northern fretted terrain and the knobby terrain east of Hellas ("Hellas Montes"), is that the weight of the plateaus and mesas is expressing a buried layer of ductile ice into the valleys; the ice flows into and then down the valleys. The crust appears to be literally oozing ice. Where the plateau materials are heavily dissected into isolated blocks, the icy flows are apparently expressed transversely/radially from the bases of the isolated mesas. In some cases, particularly the Hellas Montes, the appearance is that the mountains themselves are melting and oozing away along with their buried layer of ice. I call these the Melting Mountains of Hellas. If these appearances are correct, it suggests that these areas of Mars have a crustal stratigraphy that includes massive amounts of ice, perhaps a buried layer of ice hundreds of meters thick. The melting mountains may be composed largely of ice – we may have to think of gigantic blocks of ice ejected by the original Hellas mega-impact into a glacial ice sheet, or something equivalently bizarre.

According to this interpretation, ice is a major crustal rock type in certain provinces of Mars, but it is a viscously deformable one that sometimes behaves more like glaciers. The suggested mode of accumulation is, of course, very distinctly unlike Earth. The "ice-express" model can explain the lack of evident surface accumulation zones; these glaciers have subsurface accumulation zones, but they are hidden from view in the subsurface. The exuded ice is heavily charged with rock debris, and so no clean, snowy accumulation zones exist for this type of glacier.

At some critical threshold of crustal ice abundance, surface temperature, heat flow, and surface slope (driving stress), the mechanical integrity of the crustal plateau and underlying icy materials would be violated. Ductile failure of the icy layer and brittle failure of the overlying rocks would ensue. Ice-rich material and ground water would be erupted through fissures, and as the fissures widened into valleys, icy flows and ground water would be expressed from the ductile layer into the valleys. The progressive disintegration of the plateau would leave isolated upland blocks, whose weight may continue to express the ductile medium beneath. However it formed, it seems to be very old ice dating from the earliest Noachian or earlier in Martian history from a time when the hydrology of the planet allowed segregation and burial or crustal storage of huge volumes of ice.

Ice cream model

The "melting mountains of Hellas" came up at a conference dinner recently. With the Red Sox and Yankees playing in a historic baseball playoff game in the background, Greg Cardell (JPL) offered the Melting Ice Cream model from the Fifth Inning through the Tenth. I embarked on this discussion with him and Frank Carsey (also JPL), but then I thought to pull back as an observer on the fascinating thought processes of two scientists in a discussion so characteristic of science geeks such as ourselves. This "fun" hypothesis may, in fact, include a nugget of scientific insight. There is at least a superficial resemblance of the melting mountains of Hellas to melting scoops of ice cream, the melting scoops surrounded at their bases by low-sloping girdles of icy flows. As Cardell explains it, the morphologic similarity may have some rheological sense behind it. Good brands of ice cream, as Cardell states, are multicomponent, multiphase substances, including intermixed solids, liquids, and gases. As ice cream melts, it yields in part by releasing gases, and in part by fluid flow. The melting mountains of Hellas may undergo a similar process. If so, the Ice Cream model would suggest that the rocks composing these wasting massifs might not be composed of rocks as we know and love them here on Earth, but may be more complex mixtures of ordinary lithic material and frozen volatiles, such as gas-bearing clathrate hydrates. Regardless of the applicability of the Ice Cream model or any other model, the glaciers in this region of Hellas and around Mars include some un-Earthly types.

Glaciers fed by subsurface seeps and springs

This model is somewhat similar to the ice extrusion model, but dynamically it is very different, because the accumulation of ice occurs by subsurface spring discharge into debris. The discharged water freezes, and eventually induces the mass to start flowing. These are basically spring-fed rock glaciers, which can grow on Mars to supersize because of the absence of snow-fed glaciation, rainfall, and other processes that tend to limit the ages and growth of rock glaciers on Earth. Though the Martian glaciers may develop at a small fraction of the rate they may grow on Earth, Mars has orders of magnitude more time to develop them.

Glacier indications of process and recent climate

Considering the models above and issues of chronology, there is plenty room for debate about the temperature conditions and time over which these icy flows have formed, whether they continue to flow, and whether major epochal climate change has been involved. There is little question that most lineated valley deposits have not seen vast quantities of water flowing over their surfaces, along their edges, and from melting termini – we simply do not see the river beds of such outwash or the supraglacial stream channels that scar many temperate terrestrial glaciers. Rarely, some lobate debris aprons in the southern hemisphere exhibit what may be outwash patterns of braided channels. More common in the southern hemisphere – but rare in the north – are gullies that taper down-slope. Gullies are considered more generally later in this chapter. This set of observations suggest that Martian glaciers have not been so affected – or not at all affected in some cases – by melting. Their ablation processes have been dominated by, or in some cases restricted entirely to, sublimation. Thus, we can rule out a really Earthly model where a warm ocean supplies moisture needed for glaciers.

Ice-water ocean-driven glaciers

If the Martian glaciers are largely relict forms, growth due to atmospheric humidification by an ice-cold ocean – one that was equilibrated at the ice point – is possible, because then precipitation and ablation would likely occur under conditions where no liquid phase is present. This ice-water maritime model can readily explain the presence of tapering gullies due to eruptions of geothermally heated water coming from within or beneath the glaciers.

POLAR CAPS: GOLLY, THEY'RE GLACIAL

No surprise to glaciologists

Of course they flow! Even if lately there is a creeping acquiescence in the Mars science community that mid-latitude icy flows are, indeed, icy and glacial, there remains a widely held interpretation of Martian polar caps that they are totally rigid and static deposits of dust and ice. That Martian ice caps appear to flow somehow still unsettles many Mars scientists, but terrestrial glaciologists who have considered the Martian polar caps are in universal agreement. When I asked about theoretical evidence for polar flow, Christine Hvidberg, a Danish glaciologist, remarked "*Of course they flow!*" with a polite smile not quite hiding an air of incredulity that it could be any other way. She was of course 100% right. Those who are still convinced otherwise deserve a heap of incredulity for turning a blind eye to ice physics and for their disregard of nature's unstoppable whim and design. The real issue is the uncertain significance of flow rates relative to other processes; the specific dynamics of flow; and which features are due to flow. There remain legitimate questions about whether the polar caps exist long enough at the poles, amid climate oscillations, to establish a glacial-type regime of accumulation, flow, and ablation, although this questions seems to be close to resolution in favor of a glaciological regime. It is thought that the residual bright polar caps are areas where accumulation is occurring – at least ice accumulation, even if dry ice is undergoing net annual sublimation. The polar caps accumulate there and then flow to the dark lanes and the edge of the residual cap, where ablation balances accumulation. A complex flow field is implied, but the details are not at all understood.

The theoretical argument for Martian polar flow is the same argument, but even more conclusively in favor of flow, given above for mid-latitude glaciers: with sufficient precipitation, ice will pile up until it flows. Ice is not so stiff that it will sit passively for ever. Crater counts presented recently by Jeff Plaut have shown that the south polar cap's ices are around 14 million years old, or alternatively, that the ices accumulate at an average rate of 60 m per million years. Either way, the layered polar sequences are millions of years old. That is enough time even for cold ice to deform glacially, though questions remain about the total magnitude of glacial strain. This argument gains added strength when the ice sheet thickness is large and the geothermal gradient is significant, because of course ice softens dramatically as it warms up near the base of a thick ice sheet. It will get only so warm before flow occurs at a fast enough rate to transport the ice to lower latitudes and elevations where it can sublimate, such that the glacier ablates at a rate approximately balancing accumulation. If the glacier is too thick for the accumulation rate, then its basal shear stress and temperature will both be too great and its effective viscosity too low for mass balance; it will flow more rapidly than a balanced glacier, will expand beyond where the terminus of a balanced glacier should be, will become thinner, and will ablate faster than the balance value, until the glacier finally stabilizes with the balanced dimensions and flow regime. Oscillatory behavior is possible, where cycles of slow flow and thickening alternate with periods of fast flow and thinning. Both steady-state and oscillatory behavior are known from the terrestrial polar ice sheets (past and present).

One thing is certain: glaciers and ice sheets are naturally regulated such that flow deformation transports ice from areas of accumulation to areas of ablation; the glacier flows at just the right rate such that, over the long term, lacking large climatic changes, their mass balance is nearly zero. Of course, climate is dynamic, and so glaciers are almost continually readjusting their flow regime so that accumulation and ablation tend toward a balanced state.

Many aspects of the polar caps – especially the south polar cap – are completely beyond the experience of Earth-educated glaciologists. It is not surprising, then, that

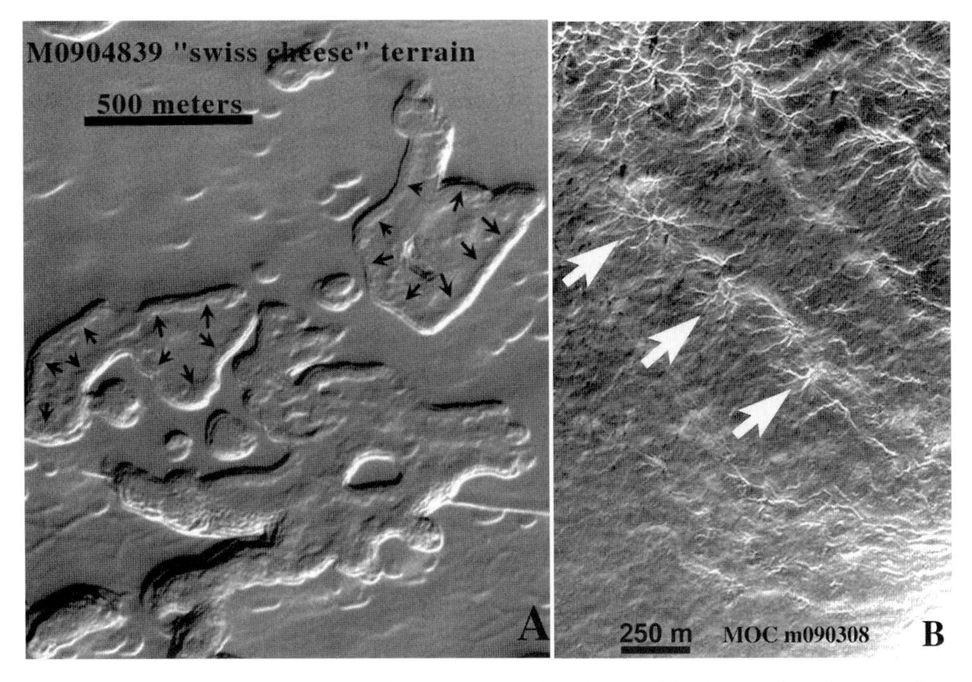

Two examples of enigmatic landforms and terrains observed in the south polar cap of Mars with MOC images. (A) One variety of "swiss cheese" terrain has been observed to change significantly in just a few years of observations by MOC. The pits tend to widen and the remnant divides and knobs between pits narrow, with typical changes on the scale of a few meters per year. Such rates are so rapid that the only possible composition of this material is dry ice. Thus, "swiss cheese" equates with CO_2. Sublimation can widen a pit on all sides because, so near the pole, the summer sun just swings above the horizon and shines on every surface no matter what the slope aspect. There commonly exists, as in this example, a moat or trough at the edges of the pits. This may be related to "indirect radiation" scattered off the steep bluffs bounding the "holes" in the swiss cheese. Summed with the direct solar beam, Karen Lewis, working with me and other colleagues, has found that dry ice could undergo enhanced sublimation, thus forming a trough. A similar but weaker effect has been found on Antarctic glaciers. (B) "Spiders" are radiating gully-like and ridge-like forms found on the south polar layered deposits. These forms are not at all understood. Speculation has centered on some type of degassing and erosion due to vapor-flow erosion or liquid eruptions. Two models include (1) "solar greenhouse" models (advocated by Hugh Kieffer and Tim Titus, US Geological Survey), and (2) subsurface heating and CO_2 mobilization due to geothermal heat. The vapor could be emitted due to basal dissociation of clathrate hydrate or melting of dry ice. The common preferential occurrence of spiders along bedding planes may indicate venting along those planes or solar greenhouse heating that is accentuated along slopes related to bedding plane outcrops. The debate mirrors exactly the models put forward to explain the geyser-like plumes on Neptune's moon, Triton, where nitrogen instead of carbon dioxide is thought to be the active volatile.

a variety of cute names have been applied to features never seen before, a tradition started with the *Mariner 9* mission to Mars, continued during the *Voyagers* exploration of outer planet satellites, and now extended to the new discoveries of more unfamiliar landscapes on Mars. The "swiss cheese terrain," "spiders," and "dalmatian spots" all refer to utterly bizarre forms of sublimation-formed landscapes. The uniqueness of these terrains is believed to be related to the prevalent frozen CO_2 of the south polar cap – a substance never seen in detail as a geological material until now. One of the most un-Earthly aspects of the polar caps is their large-scale quasi-spiral structure of debris-rich scarps and troughs. This is not a characteristic known from Earth's polar ice sheets or any ice caps and mountain glaciers. Both Martian polar caps exhibit this structure, the northern cap most perfectly so. It affects both the perennial ice deposits and the outlying polar layered deposits. It is along these erosional troughs that we can best observe the internal layered structure of the polar caps, and it is those layers that provide some of the best structural indications of flow.

Evidence of flow

The south polar cap exhibits multiple indications of flow. The flow indicators are best seen where strata are in evidence along the trough walls (the dark lanes of the quasi-spiral structure). This because these are places where strata are being eroded and exposed by sublimation. The fresh frost-covered areas hide their layered structure. Along the trough walls we see folds, faults, and what are called boudinage – "sausage structure," where layers alternately pinch and thicken due to ductile flow. The folds and boudins require a rheologically layered structure – that is, where the viscous and other mechanical properties of layers provide mechanical contrast. In addition, plate flexure responses add to a rich phenomenology of a rheologically highly structured ice cap.

Folding and faulting

Folding requires mechanical layering partly for the same reason that enables a sheaf of paper to be buckled or bent, whereas the same mass in a block of wood would resist folding; with a compressing vice, the wood would fracture before it would buckle. Indeed, we see both folding and faulting in the Martian south polar cap. The rheological layering of rocks allow shearing in weak, "incompetent" beds to occur between thin, more "competent" beds, which are free to slip along the incompetent layers and are thin enough to be bent by intracrystalline strain, intergranular shear, and recrystallization. For this reason, we see on Earth folded thinly bedded sedimentary and thin-bedded volcanic rocks, and we see folding of metamorphic rocks that were once well layered, but we do not see folding in massive, homogeneous sedimentary rocks or massive igneous rocks. Folding in the Martian south polar cap must be related to rheological layering as well as stresses sufficient to induce buckling and bending of the ice. It is thought that alternating layers of different ice types or variable inclusion of dust within ice is responsible for the rheological contrast needed for development of folds.

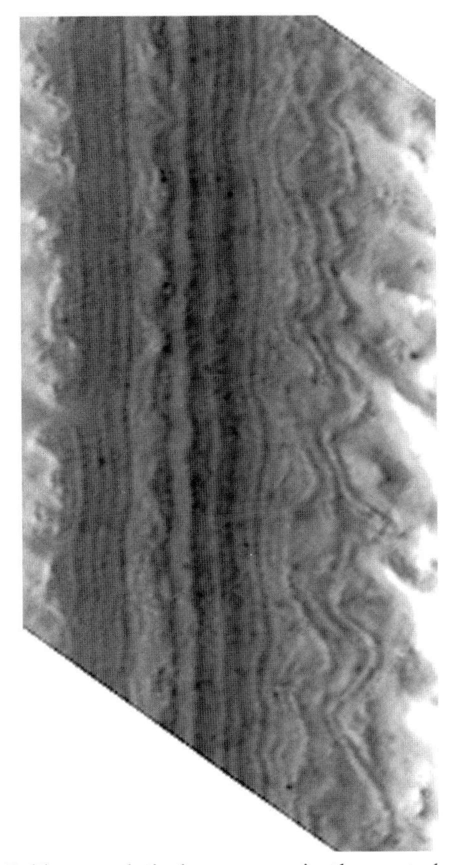

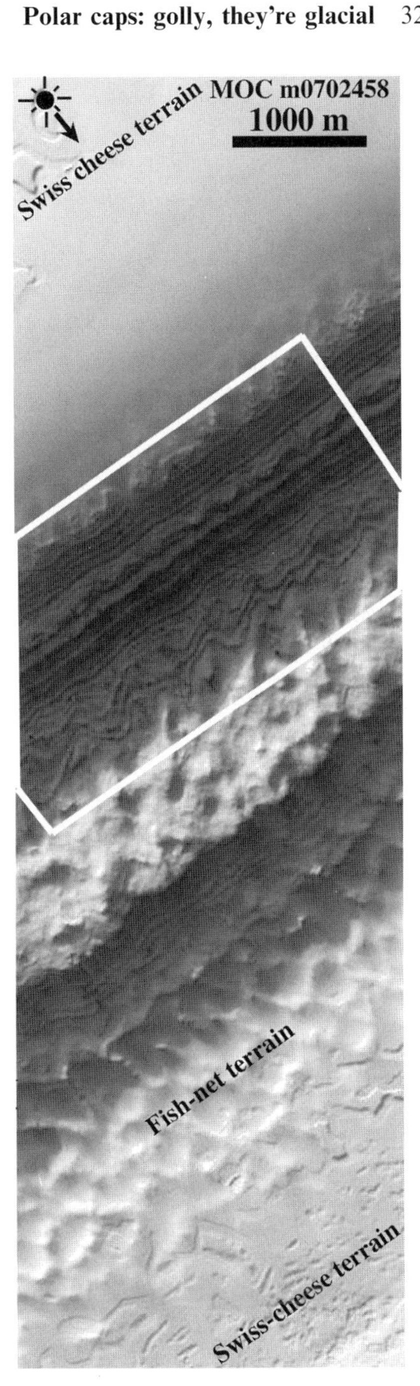

Folds are relatively common in the central core of the perennial CO_2 cap. These are exposed where sublimation has revealed the internal structure of the polar caps. (A) Folds occur in areas where CO_2 remains stable around the year, including terrains such as the "swiss cheese". Here it appears that a sequence of folds is exposed along a folded bedding plane, and also in an eroded outcrop where multiple folded layers are exposed. A net like terrain of shallow depressions also occurs; it has the appearance of boudinage, in this case perhaps boudin casts. (B) Closer view of the folds shows that there exists a sequence of unfolded strata above the folds. Ken Tanaka (USGS) has offered an alternative interpretation that the fold-like layers could be draped over underlying rumples without involvement of flow or compressional tectonics. The occurrence of fold-like strata only near the CO_2-rich central region of the polar cap supports folding of layers including soft dry ice.

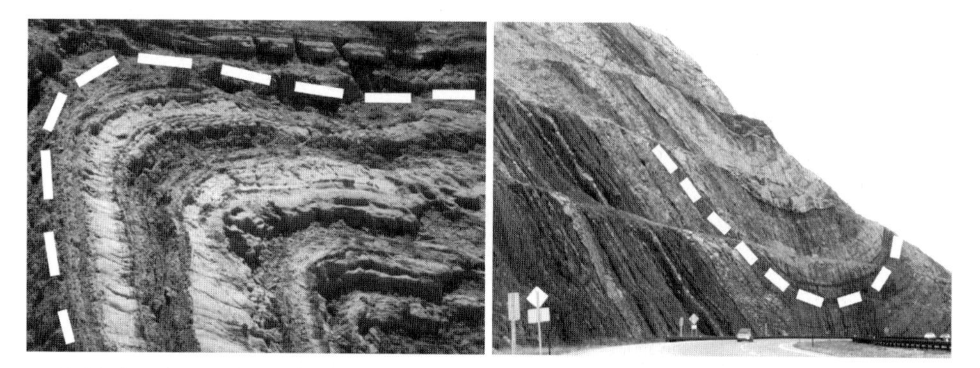

Folded rocks are always layered and include high-ductility beds, slip planes, or very thin bedding of stiff rocks. In the case at left, a Southern California sequence of interbedded shale, siltstone, and gypsum is tightly folded. In the case at right, these Appalachian folds in Pennsylvania were composed of a similar sequence of shale, siltstone, sandstone and coal. Folding in both cases was due to plate tectonic motions. Folds are rare on Mars for lack of plate tectonically driven compressional stresses sufficient to fold the layered rocks, which abound. The south polar cap is locally folded, implying lateral flow processes and rheologically layered ices that are susceptible to folding. For those lacking geologists' vision, the folds are highlighted with the white dashed lines.

Boudinage

This is a completely different process from folding, but it also requires an alternating sequence of weak and strong layers; one analytical model requires the layers' power-law stress exponent, commonly designated n, to differ, whereas other models just require a contrast in viscosity. Boudinage occurs where layered rocks are stetched. In the deforming layered sequence, the stronger layers form the boudins (the "sausage links"), and the softer, more ductile layers form a matrix that wraps around the boudins, fills in the gaps between boudins, and deforms in whatever manner is needed to accommodate the less drastic deformation of the boudins. The boudins may fracture and thus form block-like structures, or if they are somewhat ductile they may pinch out at their ends like stretched salt-water taffee.

In ideal theory, boudins form only beneath ice divides (where ice piles up and flows radially outward in all directions) and deep within comparable crustal spreading centers in layered crustal rocks. They occur, for instance, where thick sequences of sedimentary strata are piled up and over-thickened by tectonic plate collisions and then, when the compressional tectonic forces are relaxed, spread out laterally, somewhat as pancake batter spreads radially on a pan. Pancake batter does not form boudins, because it is rheologically not layered (it is made of one uniform mixture, lumps excepted). Data-hungry experimentalists with well-stocked refrigerators and workshops have produced boudins from vertically compressed and laterally spreading layered sequences of cheese and putty. In nature, boudinage is commonly found in metamorphic gneisses and schists and in sedmentary sequences from mountain belts where thick layered sequences have spread out under their own weight. The softeness required for this type of deformation is provided by

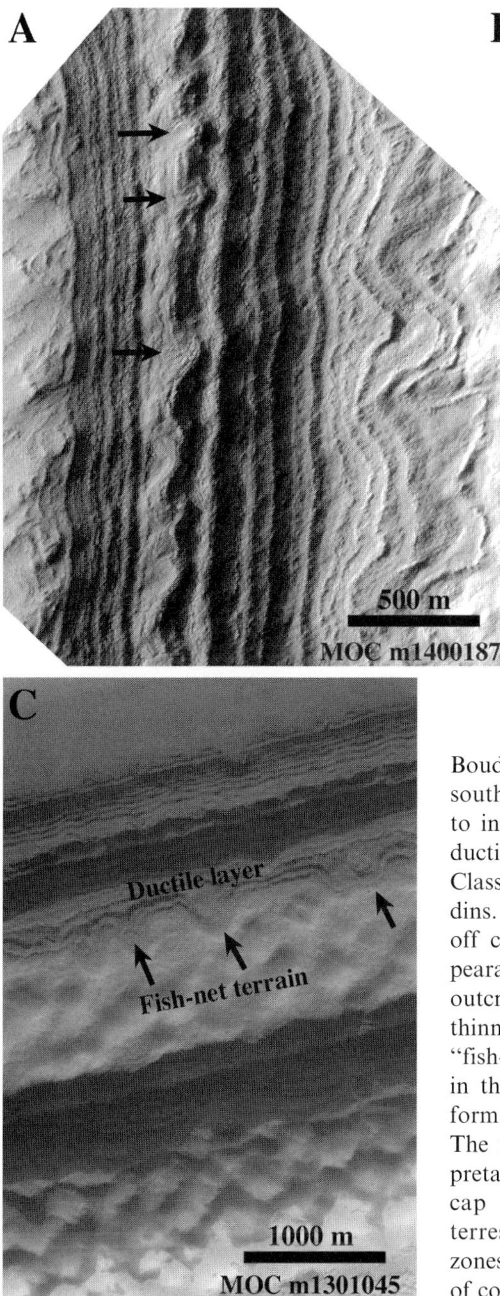

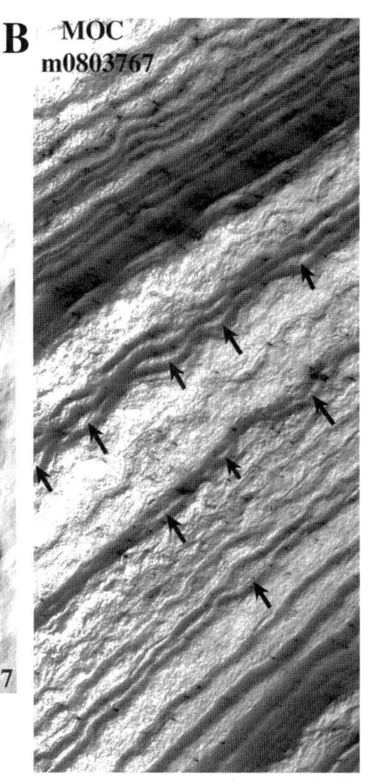

Boudinage in the perennial CO_2 zone of the south polar cap. (A) A ductile zone appears to intrude (arrows) into thinnings of a less-ductile bed to the right of the intrusions. (B) Classic pinch-and-swell structures mark boudins. The boudins (arrows) appear to pinch off coompletely at this resolution. (C) Appearances of ductile intrusions viewed in layer outcroppings correlate with thickenings and thinnings in map-view of a bedding plane (the "fish-net terrain"). Appearances may deceive in these eroded surfaces, but these do conform to classic examples of boudin structure. The implications, if this is a correct interpretation, are profound. Mars' south polar cap has behaved very much like some terrestrial glaciers and crustal spreading zones of relaxing bulges or overthickenings of continental crust. Boudinage abounds, for instance, in the layered sedimentary rocks forming North America's only triple divide in Jasper National Park, Alberta; this is where surface streams flow to three different oceans, and where hot, deep crustal rocks at one time spread out laterally under the weight of a thick dome of crust.

Boudinage may be a fully ductile pinch-and-swell phenomenon (A), or a semi-brittle process (B and C). The process is theoretically scale-independent and occurs on whatever scales of layering are present. These examples are from the Catalina Schist near Tucson, Arizona. Courtney and Bé Kargel provide a scale.

geothermal heat deep within the crust; hot rocks under steady stress behave ductily in a similar manner to a glacier.

Rheologically homogeneous glaciers would not form boudins. However, glaciers and ice sheets commonly are layered, and they sometimes form boudins. The ideal, classical theory of boudin formation would require icy boudins to form where layered ice spreads under an ice divide at the summit of the accumulation zone. The rheological layering can be provided by differential content of dust or rock debris (providing either acids and salts that soften the ice or a debris load that hardens it) or differential ice-grain fabric (grain size, shape, and orientation) or air bubble content. In contradiction to ideal theory, I have observed boudins in the marginal shear zone of the Sheridan Glacier, Alaska; this find only goes to show that nature often laughs at ideal theory. However, observations are invariably in accordance with the theoretical need for rheologically contrasting layers.

Plate flexure

There are several models accounting for the development of the polar spiral-structured scarps and troughs. One of the most successful is David Fisher's accublation model, which involves a combination of local net annual sublimation and nearby areas of net accumulation, with effects of ice flow also included. Modelers are in unanimity that there is a tendency for the troughs to close by glacial flow. Something – presumably sublimation – keeps them open, or they would disappear on timescales short compared to the age of the caps. It is apparent that something like accublation is happening, but that does not appear to be the whole story.

I have reported that the spiral-structured troughs have topographic forms and associated fine-scale tectonic features consistent with plate flexure similar to that seen in Earth's oceanic lithosphere around major loads, such as the Hawaiian Islands and in the vicinity of oceanic trenches. This is not a phenomenon recognized in Earth's polar ice sheets and glaciers. The large-scale topographic signature pointing to flexure is what is known as the universal flexure curve, which fits the polar troughs very well in most cases. In this model, the weight of large polar mesas (the high sides of the troughs) presses down on an underlying elastic sheet (forming the surface on the low side of the troughs) resting on a viscously deformable layer.

Some compelling observations indicating that plate flexure occurs in the Martian polar cap are associated fine-scale tectonic features, including areas with small extensional fractures and other areas of compressional ridges somewhat as plate flexure would cause due to bending stresses. There is, in fact, a fairly bewildering zoo of tectonic features clearly identifiable as due to responses to the icy mesas or scarps. In some cases, the features are exactly as one would predict if the troughs indeed are closing, with an icy mesa sliding down-glacier over the surface of an underlying icy mass composed of an elastic sheet on a viscous layer. This is not a phenomenon familiar to terrestrial glaciologists, who are accustomed to seeing ice sheets behaving as units, rather than having multiple ice sheets overrunning one another, as this interpretation implies. The key difference may be that the Martian polar caps are composed of multiple interbedded ices of different types, which give rise to this complex layered flow behavior, as well as to folding and boudinage.

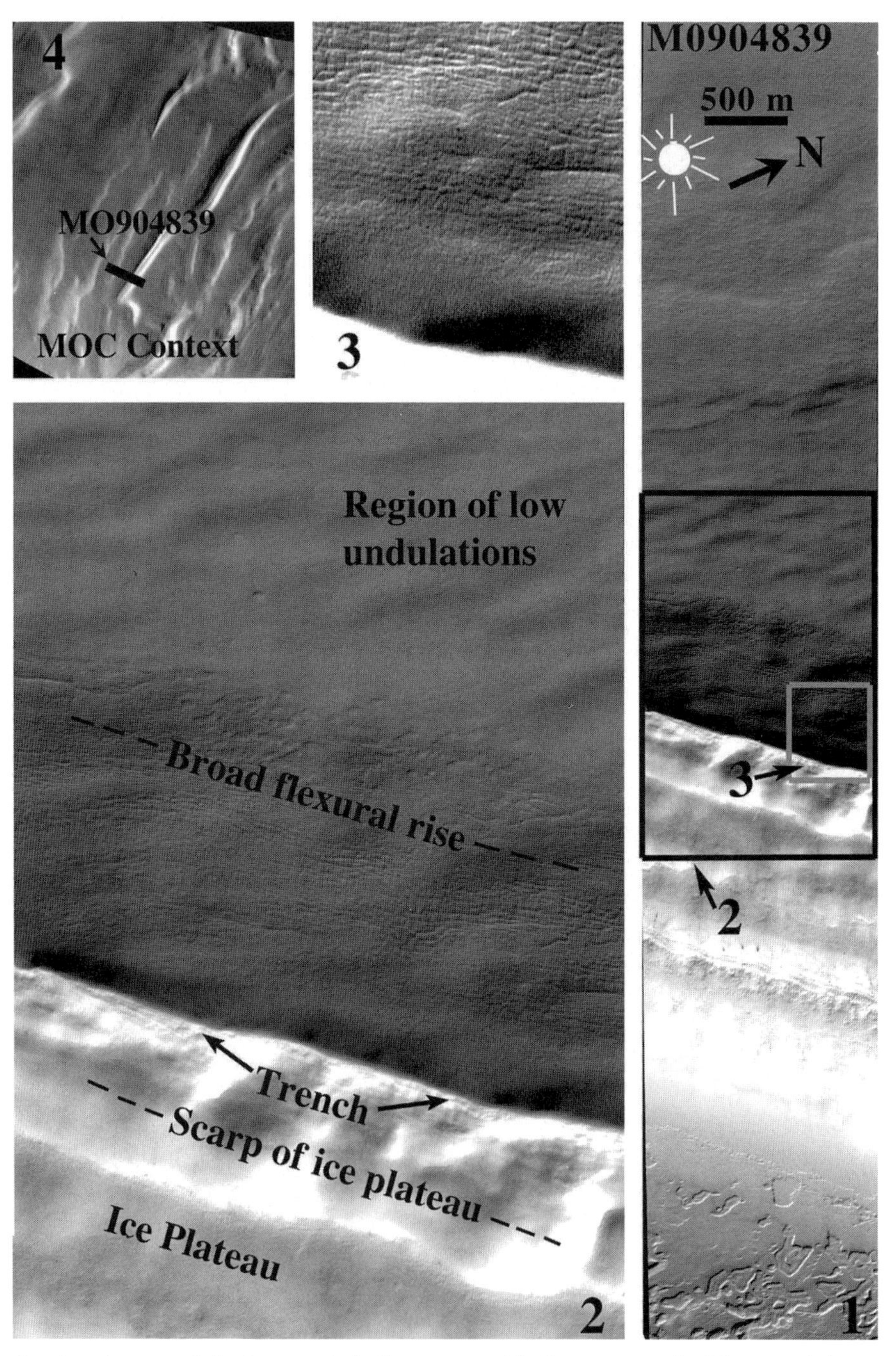

South polar trough is characteristically topographically asymmetric. The trough has a topographic profile and morphologic features consistent with plate flexure. This interpretation implies a ductile substrate.

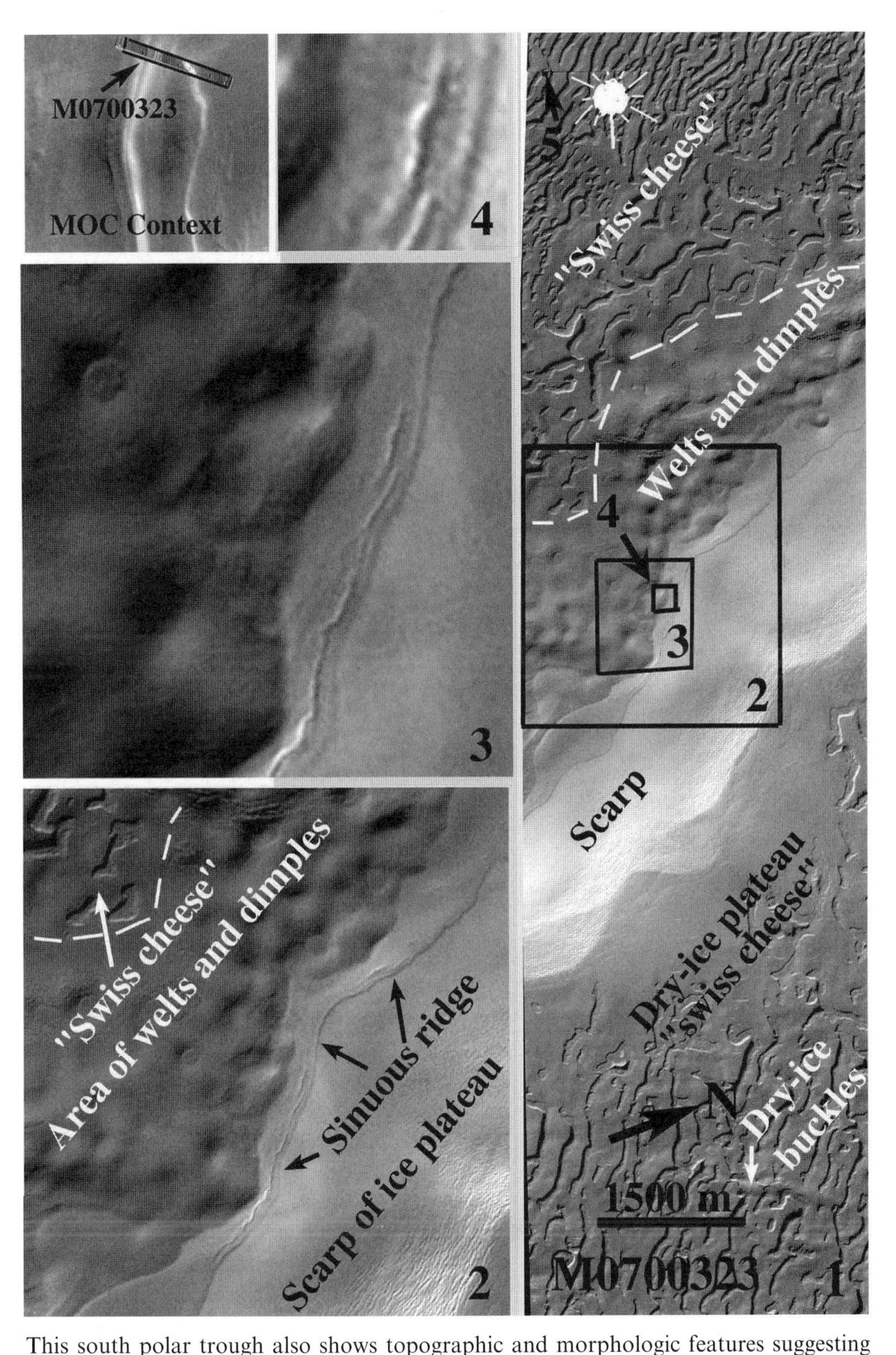

This south polar trough also shows topographic and morphologic features suggesting plate flexure, but in this case, it appears likely that the upper plate is actually advancing over the underlying plate.

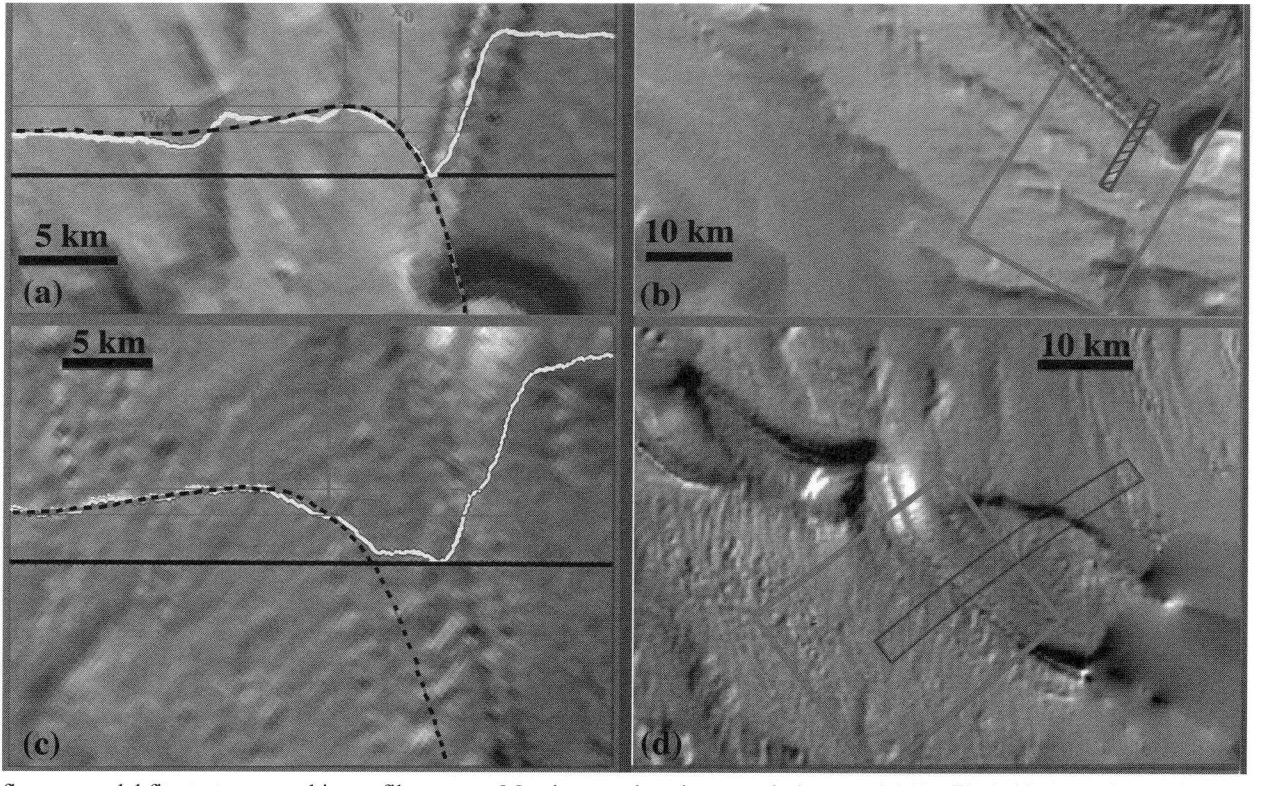

Plate flexure model fits to topographic profiles across Martian south polar troughs/scarps. (a) Profile (white curve) near image MOC M0904839 (see page 332). The dashed black curve is a model fit assuming plate flexure. (b) Shaded-relief/gray-scale topographic map including the region of profile in "a" and MOC m0904839. (c) Profile and flexure model for region near MOC M0700323 (page 333). (d) Shaded relief/gray-scale topographic map including region of profile in "c" and MOC M0700323. Hillshade maps courtesy of Jim Skinner. See also Kargel (2003).

Rheological contrast of Martian south polar ices

Vertical temperature and stress gradients can contribute to rheological layering in ice sheets on Mars, as they do on Earth. These gradients can produced ice that behaves in a brittle-elastic fashion near the surface, and as a softer, more ductile material at depth. While this behavior could contribute to the deformation behavior of the Martian polar caps, if the gradients are steep enough, it cannot explain folding and boudinage, which require *interlayering* of alternating hard and soft materials, rather than a simple downward gradient of effective viscosity. Different ice compositions is perhaps the easiest way to produce strong rheological layering of the Martian polar caps. The choice of ices include CO_2 ice, water ice, and CO_2 clathrate; these have increased effective viscosity in this sequence, according to the laboratory work of William Durham (Lawrence-Livermore National Lab) and his colleagues. According to a review paper (in the book *Solar System Ices*, Kluwer Academic Press) written by Russell Ross and me, the CO_2 phases (dry ice and clathrate) both have very low thermal conductivity (much less than that of water ice), which can provide warm temperatures and soft ice deep in the polar caps, as needed for effective strain, and can provide rheological contrast if interbedded with each other and/or water ice.

North and south polar caps poles apart?

The northern polar cap exhibits minimal direct geomorphic evidence of flow. It may not flow significantly; if so, this indicates that it has not yet reached a critical mass and shear stress such that flow is occurring at a rate fast enough to balance ablation. The evidence that the south polar cap flows, and that the quasi-spiral structure may have something to do with flow, however, suggests that the northern polar cap (with its quasi-spiral structure) does indeed flow. More compelling, the overall quasi-parabolic surface profile of the northern polar cap further suggests that it has adjusted its form to establish a rough dynamical balance between accumulation, sublimation, and flow. Going back to basic ice physics, it would be highly unexpected if somehow the north polar cap is not actively deforming. Ice, even at Martian polar temperatures, is simply too soft to be unyielding on timescales of hundreds of thousands to millions of years. The apparent planar nature of the layers in the north polar cap, in contrast to the obviously deformed strata in many parts of the south polar cap, is not a strong argument that the northern polar cap does not flow. Most of the Greenland ice sheet and other ice caps expose almost planar strata – but in fact they have been strained by stretching by factors of several tens. If the deformation occurs with processes of pure and simple shear without secondary flows and vertical perturbations, the layers can retain parallelism and quasi-horizontality. In any case, terrestrial glaciologists and experts in materials properties are not impressed by arguments from geologists who claim that somehow the massive northern polar cap does not flow. The thousands of polar images now available admittedly make it much harder to point to specific structural indications of flow in the northern polar cap, unlike the southern polar cap. My interpretation is that the north polar cap is compositionally and rheologically more uniform and its flow patterns more homogeneous than in the south polar cap.

Inventing words

Glacier modelers are well equipped with the analytical tools and data needed to show that, even at Martian temperatures, piles of ice 3 km high must inevitably flow as glaciers at rates that would be of interest. It is not an issue with them, and so the terrestrial glacial experts respond with no surprise at all when shown evidence of deformation. Some Mars experts are uncomfortable with concepts of Martian polar flow and are developing creative alternatives, but they seem to struggle against ice physics. The amazing thing about glaciers is that, with a long-continued supply of moisture and the benefit of geologic time, they adjust their profiles until the processes of accumulation, flow, and ablation reach some sort of dynamic balance, which is ever shifting with climatic changes. Glaciers will find the right timescale and process rates until something close to balance is achieved. The Mars community is beginning, at a glacier's pace, to see the polar caps in glacial terms. For example, after meeting the first time in Houston a few years ago, the second Mars Polar Conference was held near the ice caps of Iceland, and the third was held in 2003 near Jasper National Park's alpine glaciers. Certainly there remain huge gaps in our knowledge, and there remains enormous room for interpretation and recasting of models, but such massive ice caps that don't flow are improbable.

The polar caps represent areas where annual condensation exceeds or equals annual sublimation, or at least where sublimation has not yet caught up with past accumulations. While this description holds true, there is an abundance of new evidence that the polar caps are sluggish analogs of Earth's polar ice sheets (minus the fringing oceans). That is, they flow. They are glaciers. There are critical attributes of their dynamics that are un-Earthly, but to a first order a first lesson in the behavior of Martian polar caps has to start with the generalized behavior of terrestrial plateau ice-cap glaciation. What Mars specialists have long termed the "permanent" or "residual" polar caps – the bright deposits remaining after summer sublimation season – appear to be the accumulation zones. The dark lanes or troughs and surrounding layered deposits represent the ablation-exposed strata of the ablation zones. The quasi-spiral structure of dark lanes represent a peculiar sort of ablation zone generally not seen on Earth and indicates a very complex accumulation and ablation field, as well as a complex flow field.

Average accumulation rates can be assessed using cratering intensity data. Scientists such as Jeffrey Plaut (JPL) and Ken Herkenhoff (USGS) have spent a good part of their careers counting craters on the surface of Mars and assessing their data in terms of terrain ages and erosion and burial rates. In his latest assessment based on crater counts in MOC and THEMIS images, Plaut recently showed that the south polar cratering record can be explained by ice accumulation at a rate of 60 m per million years, or 60 microns per year. Such accumulation rates are about 1/2,000 that typical of Earth's polar ice sheets, a fact that can be explained by the lower temperatures, lower atmospheric partial pressures of water vapor, and lower precipitation rates on Mars. This low modeled accumulation rate is a good indication that H_2O is the dominant volatile composing the Martian south polar cap. This typical accumulation rate implies that most of the ice composing the polar deposits is millions of years old and, therefore, should record waxing and waning rates of deposition and inclusion of variable dust and perhaps clathrate layers over many cycles of obliquity, consistent with what has been inferred from the layering. At this typical rate of ice accumulation, polar glacier flow rates needed (along with sublimation) to balance accumulation would be on the order of 6 mm per year.

A typical winter's CO_2 snow accumulation over the polar caps is about 1 m of fully dense dry ice. Another implication of low long-term accumulation rates is that over the long term there is essentially no accumulation of dry ice, even though there appears currently to be a layer of dry ice at the south pole amounting to at least 20–30 m thick (the "swiss cheese" terrains). A 200-m-thick surficial layer of dry ice, as suggested by certain ice tectonic features, may build up over a period of 10,000 Mars years (about 20% of an obliquity cycle) due to a net positive mass balance amounting to 2 cm of dry ice per Mars year. This amount then presumably is lost during the remainder of the obliquity cycle, meaning that no dry ice accumulates over the entire course of an obliquity cycle. However, some CO_2 may be incorporated in the main mass of the polar caps in the form of clathrate.

Of course accumulation is not the only process occurring. Ablation is also clearly playing a major role in causing the polar caps to look the way they do. David Fisher (Canada Geological Survey), by extending to a glacial realm a polar stratigraphic model first proposed by Alan Howard (University of Charolotte, North Carolina), has proposed the *accublation model*. Accublation is the composite effects of ice accumulation on relatively flat-lying sections of the polar cap (between troughs and on pole-facing scarps), flow down the polar cap's surface gradient (especially flow into the steeper scarp toward the equator), and sublimation off the scarp faces. This process required a new descriptive word because it is a uniquely Martian process never considered by terrestrial glaciologists. The accublation model succeeds in partly explaining the spiral structure by showing the special stability of steep scarps facing the south-southwest in the northern cap and the north-northwest in the southern cap. Fisher proposes that accublation establishes local flow fields superposed over a deep flow field prevailing at levels well beneath the bottoms of the troughs. Whether the scarps are stably positioned, migrate toward the polar cap summit or toward the polar cap edge, and whether they grow wider and deeper or shallower and narrower depends on the state of local balance or the sign of imbalance of local and deep flow, sublimation, and accumulation. While terrestrial ice sheets have local complexities comparable to accublation, there is no example on Earth where a series of these systems operate together as they seem to do in the Martian polar caps.

There is little doubt that accublation is a key Martian polar process, but it does not explain the initiation of the scarps and troughs. Additional physics and structural glaciology are needed for that aspect. One possibility is that small initial perturbations in the surface relief grow and self-assemble into the spiral system. More likely, in my view, is that the small perturbations arise by intrinsically periodic structural dynamics of flowing layered media; boudinage and folding are two such processes, with boudinage the ideal theoretical result of self-deforming, pure-shearing domes of interbedded brittle and ductile materials. Such internal deformation could have surface manifestations that are accentuated by accublation.

The spiral-structured system of scarps actually form asymmetric troughs rather than simple stair-stepping scarps. The steeper and higher scarp faces the edge of the polar caps, while the gentler, lower scarp faces toward the summits of the polar caps. As such, the surface topographic gradients produce a shear stress that tends to close the troughs. The modeled rates of closure are quite rapid and require a process that keeps them open, unless we happen to be seeing their transient existence. Ablation is the obvious (and observationally well-supported) trough-widening process that counters flow closure. Accublation considered by itself would cause the upslope migration of troughs as the down-slope scarp face burns in toward the summit and the summit-facing scarp accumulates ice. A stable system of troughs could have them flowing toward the caps' edge as fast as they burn in toward the summit, with the slopes and trough geometry maintained by flow closure.

CREEPY, OOZY, ICY

Go with the flow

The best insights into the rheology of a flowing substance is obtained by watching it flow; the next best is by studying the already-deformed product of flow frozen in time. The spatiotemporal characteristics of the flow reflect the material's physical-chemistry, embued by the electronic configuration and bonding, and thus offers insights into the likely composition and phase state, if unknown. We all do this unconsciously in everyday life. Among the first things we note about an unknown substance is whether it is hard, soft, watery, slimy, muddy, gooey, waxy, slushy, clayey, plastic, brittle, or elastic. We note whether a substance changes deformation processes depending on the shear stress imposed, and we marvel at substances, such as silly-puddy and oobleck, that change behavior in unusual ways. We identify substances partly on the basis of their rheologies, a complex inference made possible by the well-defined deformation properties of the substances and learned by lifelong, everyday observation. We pour, squeeze, shake, sip, slurp, or step carefully into common substances and note their deformation behaviors. It is a primordial method of analysis. We fear quicksand for its rheological properties. We disdain greasy fingers for the potential toxic properties of the grease, the dangerous substances that may adhere to our skin due to the grease's rheological properties, or the contamination of objects we touch. We cook, build, and live based on rheological observations, and sometimes die if we fail to make an accurate rheological assessment of hazard. From an early age we predict the motion of rivers, gauge the volume of water poured or dripped, walk moist ground with or without trepidation, and bake cookies and pancakes based on an understanding built on lifelong observation of complex fluid responses to gravity or externally applied differential stress.

It should not be surprising that Earth and planetary scientists learn about their subjects often through rheological measurement or modeling or casual observation – we obtain insights into the composition of lava, assess hazards due to debris flows and unstable slopes, and model the ascent of salt diapirs and the convecting mantle. Glaciers are dominated by a rheologically interesting substance – ice – that deforms in nature on timescales long compared to everyday experience but short compared to the human lifetime. It is a geological material – a rock by definition – that is among the softest and most dynamic solid materials encountered by earth scientists. My love of glaciers is due to the fact that they show geology in action in such vivid ways. It is the same seduction used by Pele, the Hawaiian goddess of fire and the underworld, to lure her geologist lovers ever a bit closer and then onto the congealing surfaces of active lava flows, until the soles of their boots melt and extrude beneath their burning feet, the pain unhindering for the base pleasures of walking on a molten world.

Ironically, many Mars scientists have considered ice to be a purely hard and brittle substance at Martian conditions. They miss the pleasures of glacial seduction, and – no joking now – fail to see geology practically in action for lack of understanding the physics of ice. It is like the geologists and geophysicists of the 1920s, who dismissed Alfred Wegner's continental drift hypothesis, and failed to recognize all manner of geologic processes staring them in the face, for lack of understanding the ductile nature of hot rocks subjected to long periods of continuous differential stress.

It is true that ice will respond to short-term application of high differential stress by brittle failure, as it does when struck by an ice pick or gripped between the molars tightly and suddenly to crushing strength. However, rheology is a temperature- and stress-dependent property for all substances; ice, like almost all other solid substances including silicate rocks, becomes soft and viscously deformable on long timescales with rising temperature and long-term application of low differential stress. Apply the pressure of the molars more gently to a small bit of ice, and you can feel it deform by a combination of ductile and brittle deformation, though the product of this thought- and thirst-quenching experiment promptly melts. Due to low temperatures on Mars, it was thought by many Mars scientists that Mars cannot have glaciers because ice is so much harder than on Earth, about four orders of magnitude stiffer (all other things besides temperature being equal) at typical mid-latitude Martian temperatures compared to temperate glacial ice on Earth. Furthermore, the partial pressure of water vapor, hence, the water-carrying capacity of the present atmosphere and potential ice accumulation and glacier flow rates, are about an average of four orders of magnitude lower on Mars than is typical on Earth.

Ice extrusion model of Martian glaciers, reconsidered

The concept of unflowing, stone-like ice on Mars, and at the same time the idea of flowing ice, is understandable when one considers the deformation map of ice (the rate of deformation plotted as a function of temperature and applied shear stress). The driving shear stress (expressed in Pascals or MPa) for a glacier or ice cap is provided by the weight per unit area and surface slope of the ice mass ($\rho gh(\sin \alpha)$). The deformation map shows that ice, like other substances, has an effective viscosity (ratio of deformation rate to applied differential stress) that increases exponentially with inverse temperature. Cold ice is a lot stiffer than warm ice. At a 200 K typical temperature of Martian mid-latitude glaciers ice is, for a given differential shear stress in the range of 0.05–0.15 MPa (like that typical of Earth's glaciers), four orders of magnitude stiffer than typical terrestrial glacier ice at 270 K. At Martian polar latitudes, ice is even colder and stiffer. However, Mars has an immensity of geologic time to work with, and competing processes are generally less vigorous, so a slow and protracted creeping motion can result in cumulative morphologic effects that are visible through all the other geologic signals and degradation processes.

Furthermore, ice, like most solids and unlike most liquids, is a strongly non-Newtonian substance. Ice does not simply behave like extra-stiff liquid water, which is Newtonian. That is, the deformation rate of ice is *not* directly proportional to applied differential stress, but rather to the fourth power of it (or second or third power, depending on temperature). We call this a power-law substance. Double the shear stress (double flow thickness or surface slope), and decrease the effective viscosity by a factor of 8, thus causing the stressed ice mass to flow 16 times as fast. Triple the shear stress, and the ice flows 81 times as fast. It turns out that the effect of lower surface gravity on Mars may roughly balance a greater thickness of ice of lobate debris aprons (500 m) compared to the usual 200-m thickness of piedmont glaciers (a close geomorphic analog) for similar surface slopes (6°). Hence, shear stress of Martian lobate debris aprons is probably not that different compared to Earth's piedmont glaciers. Thus, Martian lobate debris aprons may flow a whopping four orders of magnitude more sluggishly than piedmont glaciers do on Earth. This could be possible, with the glacier staying roughly in balance, if the accumulation rate is four orders of magnitude slower than on Earth; that may be possible if temperatures are just about as cold and water vapor pressures as low as they actually are. That's an order-of-magnitude back-of-the-envelope approach; the devil may be in the details, but at least to a rough approximation, the Martian glaciers make perfect sense within a climatic regime approximately like today's. That is, until the meltwater features, which should not exist, are considered.

Healthy and vigorous fast-moving lobate debris aprons on Mars, under present-day conditions, may have surface displacements of order 5 mm per year instead of 50 meters per year for a typical large piedmont glacier on Earth. Nevertheless, 1 million years of this characteristic flow rate may result in a surface flow displacement of 5 km, and a good long runout distance of order 50 km may be attained in 10 million years. Flow speeds are not actually linear either with size or with distance from source, but this gives an idea that very low glacial flow speeds can, on Mars, produce immense glaciers over immense periods of time. Such slow movers on Earth would not survive climatic oscillations; and if somehow they did, they would not survive the overprint of other geologic processes; on Mars, except in areas of especially dynamic volcanism or eolian activity or tectonic formation of canyons, lobate debris aprons and other glaciers could easily survive this length of time and preserve evidence of their flow activity. They would, however, experience many climatic oscillations due to obliquity changes, and this should be recorded in their internal structure. They should be layered in some fashion, or have cycles of deposition overprinting cycles of sublimation; all of these cycles would be stretched by flow. A 10-million-year-old flow may contain 100 layers. The layers seen in glaciers in Argyre and elsewhere thus are consistent with long periods of accumulation and slow flow.

Lineated valley fill, according to the ice-extrusion model, may be dynamically very different from lobate debris aprons or an idealized snow-accumulation glacier flowing down a steadily inclined mountain slope; a better analog in terms of shear stress distribution, but not accumulation mechanism, would be a Himalayan glacier flowing first down a steeply inclined mountain face and then into a low-sloping valley. The driving shear stress at the point of extrusion may be of the order of 3 MPa, an astonishing value determined by kilometer-high rocky plateau materials ($\rho = 2{,}500\ kg/m^3$) sloping at $\sin \alpha = 0.3$. At such a high value of shear stress, the effective viscosity of ice at 210 K (assumed to be heated slightly by geothermal flux) may be $\sim 1/15$ that of typical free-flowing glacier ice on Earth. The ice layer, which could be hundreds of meters thick below the plateau surface well back from the canyon edge, may be pinched to just a meter or two thick at the point of extrusion, where it squeezes out at a flow rate perhaps as great as decimeters per year in channelized flow. The ice is squeezed into the valley, where suddenly the differential pressure due to the weight of the plateau is relieved. The ice experiences a sudden surge in viscosity to that more typical of lobate debris aprons. The ice piles up at the edge of the extrusion zone until it reaches a critical thickness, at which point it slowly flows away, carrying with it compressional ridge structure formed at the extrusion zone, or longitudinal lineations related to channelized extrusion.

The ice extrusion model would work only if the ice flow field beneath the plateau is channelized to some extent, an inference that draws support from the vague network of linear collapse zones in the fretted plateaus. This process would be closely analogous to stream sapping, in that bits and chunks of the plateau rocks would be drawn off with the ice, thus resulting in headward growth of the valleys, as morphologic evidence suggests.

Creepy mechanisms

The flow deformation of Earth's glaciers occurs throughout the ice mass and may or may not include a component of bed sliding or bed deformation. Bed sliding and bed deformation generally occur only if the glacier is melted at the bed, since shear stresses are generally much less than what is needed to break the ice–rock bond and cause brittle low-angle faulting-type sliding. Glaciers that marginally reach the pressure-melting curve along a fraction of their bed also undergo regulation flow, where pressure-melting on the up-flow sides of bed protuberances generates water, which flows to lower-pressure zones in the lee and refreezes in accordance with the phase diagram of ice. The internal deformation of ice occurs by multiple mechanisms, including grain-boundary sliding, internal crystallographic dislocation creep, and recrystallization, with each mechanism dominating under distinct regimes of temperature, impurities content and type, and applied shear stress (determined by gravity, slope, icy flow thickness, and flow density).

Permafrost, including rock glaciers and presumably the surfaces of debris-covered glaciers, may deform by surface freeze–thaw driven creep of the active layer in a process known as gelifluction. The seasonal freezing causes heaving normal to the flow's surface, whereas the thawing causing a vertical settling; the sum of the motions is an annual down-slope ratcheting motion and, on longer timescales, a tractor-tread type of down-slope translation, where the surface layer creeps to the flow's terminus, thereby exposing more icy rocky matter to freeze–thaw driven flow). Rock glaciers may also deform by glacier-like internal ice deformation. Permafrost composed primarily of ice-impregnated fine-grained sediment may deform largely by liquefied summertime flow of the water-saturated active zone (solifluction). All of these processes may be active within the same glacierized alpine environment, resulting in a composite icy landscape affected by multiple glacial flow processes (bed deformation, bed sliding, and internal ice deformation) and periglacial flow (gelifluction, solifluction, and internal ice deformation). The particular dominant flow processes may vary between adjacent features, and they may be difficult to determine for any individual feature without detailed measurement and analysis. Thus, arguments rage within the terrestrial cryosphere community about the deformation processes of these features.

Debates over the origins and specific flow processes of rock glaciers, which we can walk on and drill and observe surface changes from year to year, is the surest hint that we are far from uncovering the truth about Martian icy flows. While the back-of-the-envelope estimations, above, of glacial creep by internal viscous deformation of ice and accumulation by slow atmospheric precipitation or subsurface ice extrusion offer self-consistent explanations for Martian icy flows, these are merely some plausible solutions. They may equally be relict flows, no longer very active now, from a former warmer, wetter climate; perhaps they are still-very-active (by Martian standards) but wasted remnants from the latest transient MEGAOUTFLO oceanic episode. We shall certainly have to go to these places on Mars – with people or capable machines – to solve many riddles.

When ice is like lava

Remote sensing is not apt to solve some Earthly ice-flow controversies and is not apt to discern the origin, history, and suite of processes of Martian icy flows, but it is worth considering some of the insights that remote sensing can provide for Earth's glaciers (even if we know the answers from ground-based studies). Here, I take the single case of the Murtel rock glacier (Switzerland), which has been a topic of intensive field study and the subject of divergent interpretations as to origin. This rock glacier is neither unusual nor necessarily representative, but it is a specific example of a flow where knowledge can be gleaned from orbit that can be confirmed from the ground.

Now lacking a clear and substantial snow accumulation zone, Murtel rock glacier is heavily rock covered and contains both rocky and icy lenses or horizons. But does the Murtel have classical origins as a snow-accumulation glacier, perhaps having suffered a period of drying or warming that has reduced its supply of ice, but having a compensating increase in thermal insulation due to accumulated surface debris? Or is it an ice-impregnated, ice-mobilized deposit of rock debris having purely a periglacial origin, with ice supplied by percolation and freezing of snow melt, rainwater, and spring discharge into the rubbly mass, and entrainment of snow avalanches between debris layers? Remote sensing and photogeology does little to resolve the controversy over the Murtel rock glacier's origin, but it offers much to reveal the present rheological behavior and structure of the flow; if we had no independent information on the climatic regime of this flow, remote sensing would also offer useful insights about that. The evident fact is that ice is stable or at least metastable in the Murtel, but the supply of snow is not such that it has net annual surface accumulation. The Murtel is definitely a rock glacier and a permafrost phenomenon, but it can be considered a glacier by my definition above. It is on the hairy edge of existence; without its burden of rock, it could not exist where it does under the present climate.

Aerial photogrammetry of Murtel rock glacier done by Andreas Kääb and colleagues (University of Zurich) has produced a dense network of surface displacement vectors from imaging not much more detailed than full-resolution MOC images of Mars. We cannot do this for Mars because surface displacement speeds are too low to detect by normal imaging methods, but millimeter-wave interferometric imaging should be able to detect and map the surface flow displacement field, allowing the same sort of analysis of flow motions possible for studies of glaciers here on Earth.

I have assessed the flow displacement field of the Murtel by curve fitting cross profiles of longitudinal flow speed provided by Andy Kääb. It was discerned that roughly half of this rock glacier's flow displacement (varying along the flow's length) is due to bed sliding. Thus, this rock glacier, unlike many of them, could be eroding its bed, with bed erosion as well as rock avalanches contributing to the heavy rock debris load. A component of flow due to internal deformation of ice indicates a substance that appears to behave in Newtonian rheological fashion (like water, but 16 orders of magnitude more viscous). This result was a bit surprising, since ice is a non-Newtonian substance, but I speculate that it could reflect a large role of liquid water along grain boundaries in controlling intergranular ice deformation.

Some powerful insights into terrestrial and Martian icy flows can be derived from a very basic observation and a possibly surprising analog. The surface morphology of many icy flows on both planets is characterized by transverse rumples like the corda of lava flows. Corda-like rumples are common features of rock glaciers and are also observed in the terminal compressional zones of debris-covered glaciers, especially "piedmont glaciers," where alpine glaciers empty onto lowlands from a narrow valley or canyon and spread out laterally in a decelerating, compressing flow. The causes of volcanic corda are well understood in terms of compressional flow dynamics affected by sharp rheological gradients near the surface: extremely viscous but not brittle in the surface layer, less viscous in the subsurface of the flow. Pahoehoe flows observed in action exhibit progressive buckling and wrinkling of the colder, partly congealed, and more viscous upper layer; hotter, less viscous lava is extruded at higher flow speeds at the toe of the flow, while the crust is crumpled and partly restrained by attachment to the immobile edges of the flow lobe.

Although the substance and cause of rheological contrast is different from those for lava, the same mode of flow dynamics is inferred for transversely rumpled rock glaciers and debris-covered glaciers. This can best be explained for icy flows if there exists a highly viscous, ice-impregnated, rock-rich surface layer that is more resistant to flow deformation compared to an icier and less viscous layer below. A downward positive thermal gradient may contribute to rheological contrast for very thick and slow-moving ice masses, particularly large polar-type glaciers where low-magnitude mass balance budgets and slow deformation of very viscous ice allows long flow timescales and a rough approach toward thermal conductive equilibrium.

The surface rumples of glaciers are a special form of buckle-folding of layered media, a process that is understood in its more generalized form by tectonicists. Glaciers that lack appreciable surface debris generally lack corda-like rumples, a fact that accords with the inability of unlayered massive media to fold. Thinly bedded sedimentary and volcanic rocks, for example, can fold, while massive unbedded sedimentary rocks and massive plutonic rocks cannot fold – they fracture and fault before compressive stresses could reach a point where folding would otherwise be possible. However, some glaciers that lack debris cover are in fact folded, with the layering apparently contributed by small variations in debris and bubble content related to seasonal variations of accumulation and melting.

There is a known empirical and simple theoretical parameterized relationship in lava flows between the wavelength scale of corda and the composition and effective viscosity of the softer interior lava. Basaltic lava produces corda just centimeters in wavelength, whereas far more viscous dacite and rhyolite flows have corda measuring roughly 100 m from crest to crest. Application of the same empirical formulae to terrestrial rock glaciers and debris-covered glaciers indicates an effective viscosity corresponding to the rheology of highly siliceous lavas (and ice). Applied to Martian lobate debris aprons and other icy flows (where corda are evident with wavelengths of order 100-200 m), a deformable substance is indicated roughly with the rheological properties of water ice.

What insights regarding terrestrial glaciers and rock glaciers obtained from photogeology reflect the reality observed from ground observations? In the case of the Murtel rock glacier, it was unknown to me during my analysis that very similar conclusions had been obtained already by Andreas Kääb (University of Zurich) and colleagues using completely different, *in-situ* observations involving drill core and drill-hole analysis. The Murtel indeed exhibits a layered structure of debris-rich and ice-rich material, including a particularly stiff upper layer of debris-choked ice, which would allow surface compressional folding such as observed. The rock glacier also exhibits, as the remote-sensing analysis revealed, a major bed sliding component of deformation and a nearly Newtonian rheology (with viscosity independent of shear stress), unlike ideal laboratory ice, which has a strong stress-dependent apparent viscosity.

We still lack the capability to discern the flow displacement vector fields of Martian icy flows, but other details are readily observed in MOC and THEMIS images. Corda observed for some Martian icy flows indicate that these features are indeed ice-rich and probably have icier, less viscous interiors. A similar conclusion regarding the fundamentally icy composition was reached by analysis of the topographic surface profiles of the lobate edges of lobate debris aprons. This conclusion is supported also by the fact that these landforms are distributed in the regions where ground ice is theoretically stable and is within or near the regions where abundant hydrogen has been observed by the *Mars Odyssey* neutron spectrometer. These findings do not, however, solve the question of whether these are dying debris-covered snow-accumulation glaciers (relics of MEGAOUTFLO events perhaps) or ice-mobilized permafrost features; nor do they specify where along the ice saturation curve (in water vapor pressure–temperature space) the Martian icy flows formed and evolved, or when they first formed. The Murtel rock glacier is flowing perennial ice (a glacier by definition), whose ice probably originates from multiple sources, snow perhaps being less important than other sources.

Some Martian glaciers: not all that weird?

Glaciers are remarkable self-regulating systems that adjust their profiles and shear stress in order to cause deformation at just the right rate (to first order) to adjust the high-altitude accumulation of ice with the ablation of ice at lower elevation. If conditions change (altering the supply of snow or the rate of ablation) the profile of the glacier will adjust. On Earth climate is always so dynamic that glaciers are continuously adjusting; they may attain an approximate state of balance, though it is fleeting. To a cruder approximation, if we can throw around factors of several in the accumulation and ablation terms, glaciers are generally near a state of balance, even when they are growing or wasting.

On Mars any climatic regime that caused large accumulations of ice (no matter how slow the accumulation) would similarly find the right physical dimensionality of the ice deposit until the mass flows under its own weight at a rate whereby sublimation (plus any melting) roughly balances accumulation. If ice accumulates anywhere at all (and obviously it does at the Martian poles), and it accumulates by any mechanism, the only reason this approach toward glacial equilibrium would not be inevitable is if the supply of moisture was abruptly cut off before accumulation, flow, and ablation could reach a near balance. Thus, when we see that water vapor fluxes on Mars are roughly four orders of magnitude lower than on Earth, and effective viscosities are roughly four orders of magnitude greater, we ought to start thinking about glaciers near a state of balance, and specifically glaciers where the atmosphere provides a significant component of accumulation. Hence, at least some Martian glaciers may not be all that strange, just very slow with the flow.

SO, IT FLOWS. WHY THINK "ICE"?

A great variety of young features exhibit clear signs of brittle or ductile flow deformation, sublimation, or melting in the near-surface environment. Some of these features exhibit all of these behaviors. There are very few natural substances that can exhibit, at ambient Martian surface conditions, brittle, ductile, sublimation, and melting behavior. The only material that might plausibly do so across the middle and high latitudes of Mars is H_2O and ice-rich debris. The absence of such landforms at low Martian latitudes, and the prevalence of sublimation-related features at roughly ± 30–$45°$ latitude is ideally consistent with the predicted behavior of ice (stable tens to hundreds of meters deep but not nearer the surface at these latitudes), according to models by Michael Mellon (NASA Ames) and others. At the poles, the choices are a little broader, since CO_2 ice and CO_2 clathrate can be added to the short list, but the melting would have to be subsurface.

Some geologists have disputed the ice interpretation of these flow features – preferring instead to explain them with dust or sand dunes, for instance. Experts who have a close familiarity with terrestrial cold-climate ice-related glacial and periglacial features have never had much doubt about an association with ice ever since their discovery. The fact that small young gullies have a similar global distribution and often a direct association with icy landforms, supports an interpretation of gullies that relies in some way on ice. If there could have been any logical basis for skepticism, the discovery of near-surface ground ice by the neutron spectrometer on board *Mars Odyssey* should by now have dispelled that doubt. There is no longer any rational basis to suppose that periglacial-type features on Mars could be unrelated to ice. So now debate moves to a higher level. Exactly how is ice related to their origins? Are other ices besides water ice involved? How old are they? Are these features still actively forming? What is their relationship to recent climate history? Is there a relation to epochal hydrogeologic events of the distant past?

NEOGLACIAL GULLIES

Most Martian water-formed features and landscapes are heavily weather-beaten by wind and impacts and other ravages of time, but one class of feature stands in remarkable contrast for its youth and fresh expression at every scale of observation. Small gullies discovered only a few years ago attest to the activity of surface water in very recent times; perhaps they are still actively forming, but most likely their formation is episodic, occurring mainly at times of high obliquity. The small gullies are important not so much for the magnitude of their geomorphic imprint – their imprint in fact is fairly small – but for the recency of hydrologic activity.

The morphology of small gullies varies. Some gullies may be small glacier-cut valleys or troughs bounded by lateral moraines, and a few may be due to dry landslides, but the overwhelming majority of these features evidently are debris flows and erosional trenches cut variously by spring discharge and snowmelt. In general, their morphology and occurrence is consistent with the types of fluvial and mass-wasting landforms often found in arctic deserts and other polar and alpine terrains. The sources of many gullies are located just beneath or between ledge-like rock outcrops, as though the sources are springs; however, there are innumerable cases where gullies can be traced to the crests of crater rims or to mountain peaks. Most of the gullies form quasi-parallel systems of single-stem trenches and almost always taper rapidly downstream. The young gullies are only hundreds of meters to a few kilometers long and tens to hundreds of meters across, and are typical of the sizes of common gullies on Earth.

Debris flows are recognizable by their sources normally in very steep terrain, and by their down-slope tapering leveed channels that merge into debris sheets or cones. These were presumably caused by liquefaction of water-saturated rock debris where spring discharge or snowmelt and icemelt caused saturation of talus shed from steep

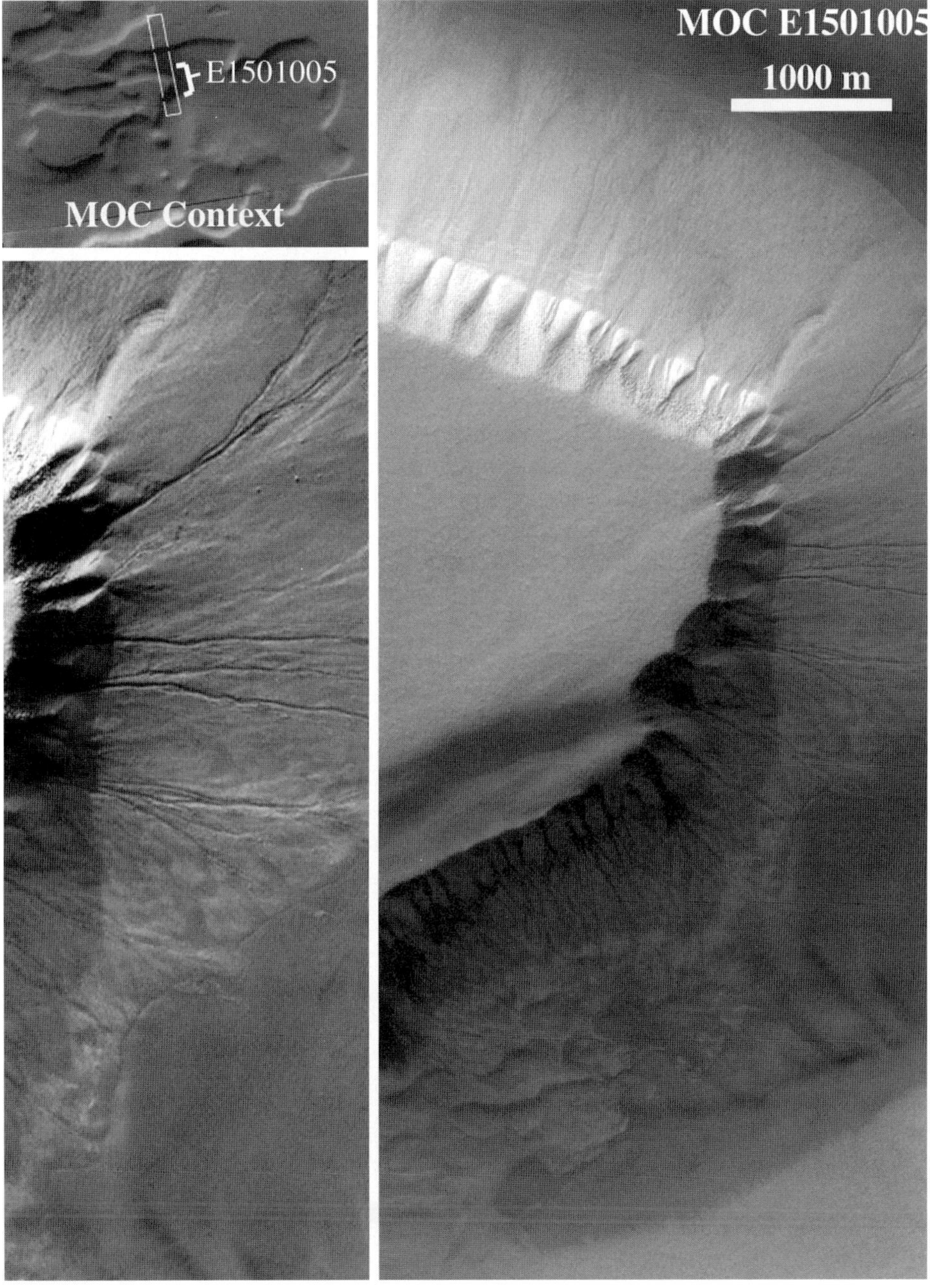

The south polar cavi were formed mainly by sublimation of massive circumpolar ice deposits; there may have been some involvement of volcanism and geothermal heating. Additional major processes include melting and debris flow emplacement. Here is evidence of debris flows, but downslope these have coalesced and form coherent viscous flows. Two discreet overlapping generations of flows are evident.

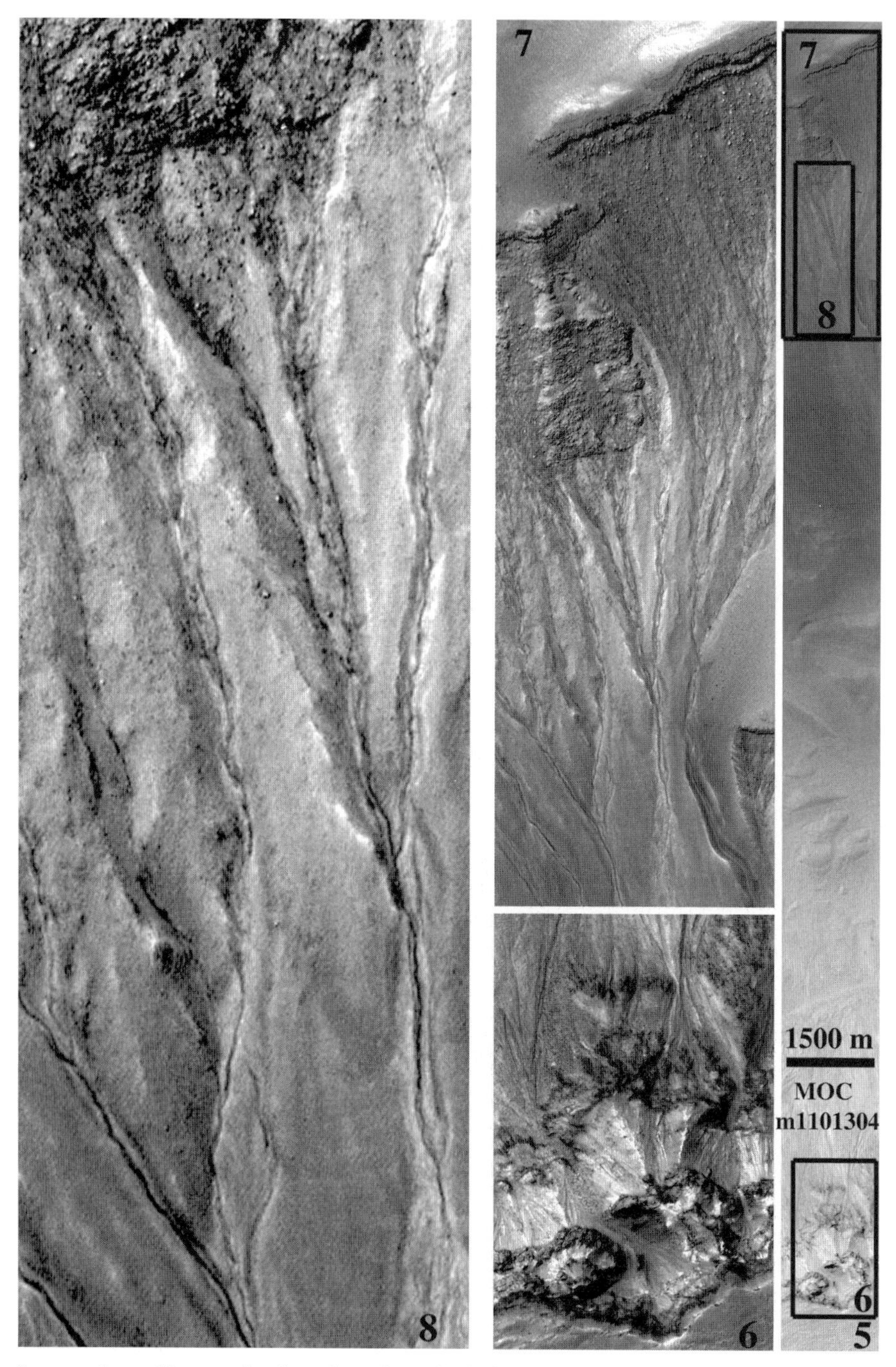

Spectacular gullies are developed on the pole-facing slopes of south polar cavi. The cavi terrains consist of giant collapse pits – like gigantic sink holes – developed in south polar volatile-rich materials.

Alluvial fans, gullies, and debris flows in the central peak region of the large Hale Crater.

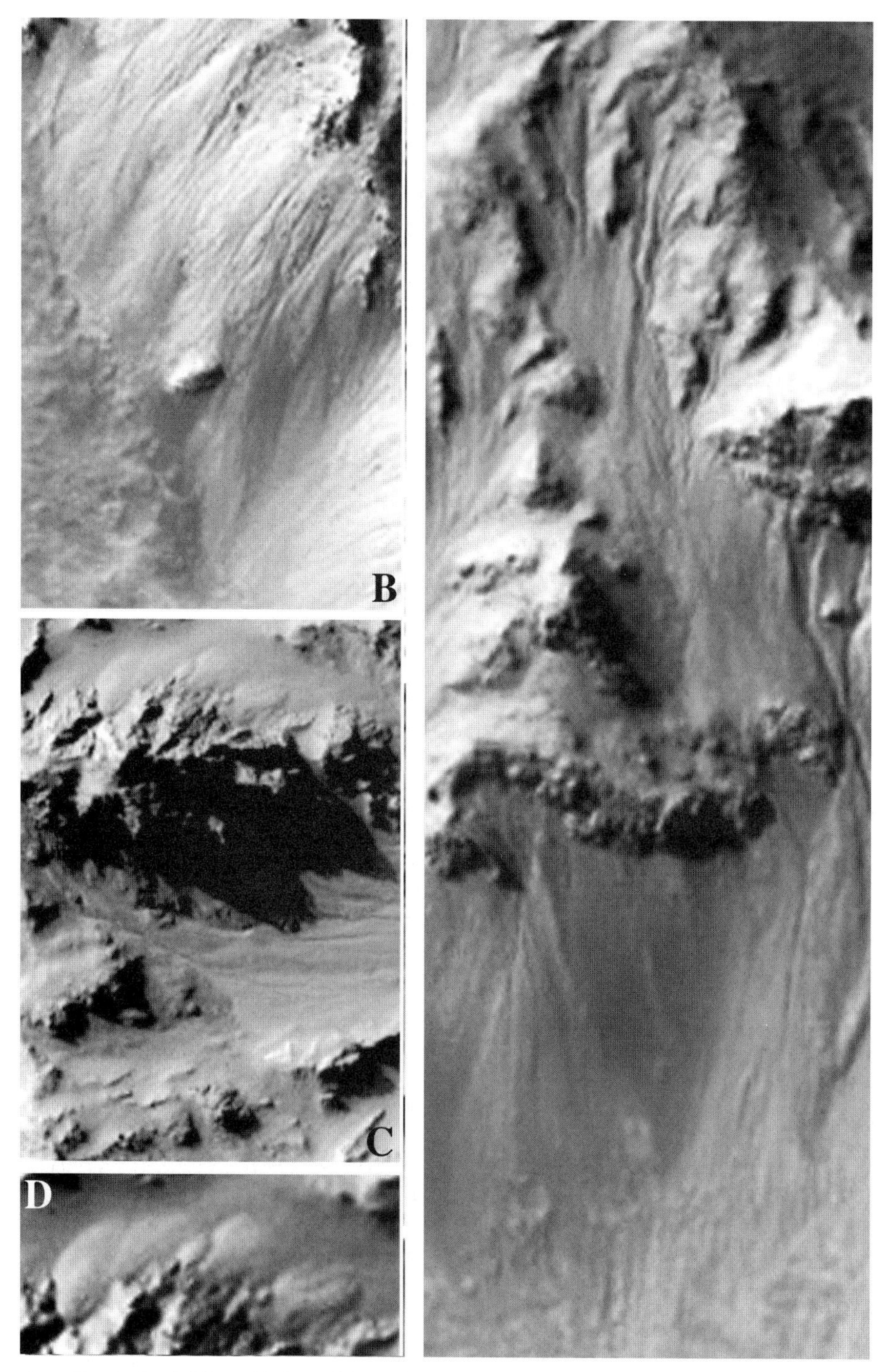

Details of alluvial fans (A, B and C) and possible small rock glaciers (D) in Hale Crater.

Alluvial fans, Banff National Park, Alberta, Canada. Morphology is highlighted by snow cover. Leveed channels and a slightly lobate terminus of the flow is visible at left in B. Note the overall conical shapes. Note trees for scale, around 2–12 m tall near treeline, in B.

Channelized, leveed debris flows in talus. (A) Mojave Desert, California, where almost no snow falls. (B) Crater Lake, Oregon, where snow and snow avalanches are important. Talus aprons and debris flows are often associated with arid, semi-arid, and polar desert environments. The chief requirements are for enough fine-grained debris to help to retain moisture; enough moisture to saturate the debris on occasion; slopes steep enough for flowage; and a near absence of physical restraints to flow, such as large plant roots. Talus development at Crater Lake has proceeded rapidly; in the 8,000 years since Crater Lake formed, 20 km of precipitation – mostly snow – has fallen.

A butte in Arizona's Monument Valley owes its existence to a sequence of three mechanically and sedimentologically distinct rock formations, as described in Chapter 1. The lower formation produces the butte's gullied girdle of badlands. The scale is given by the large electric utility tower at lower left.

Varied badlands erosional styles in Badlands National Park, South Dakota.

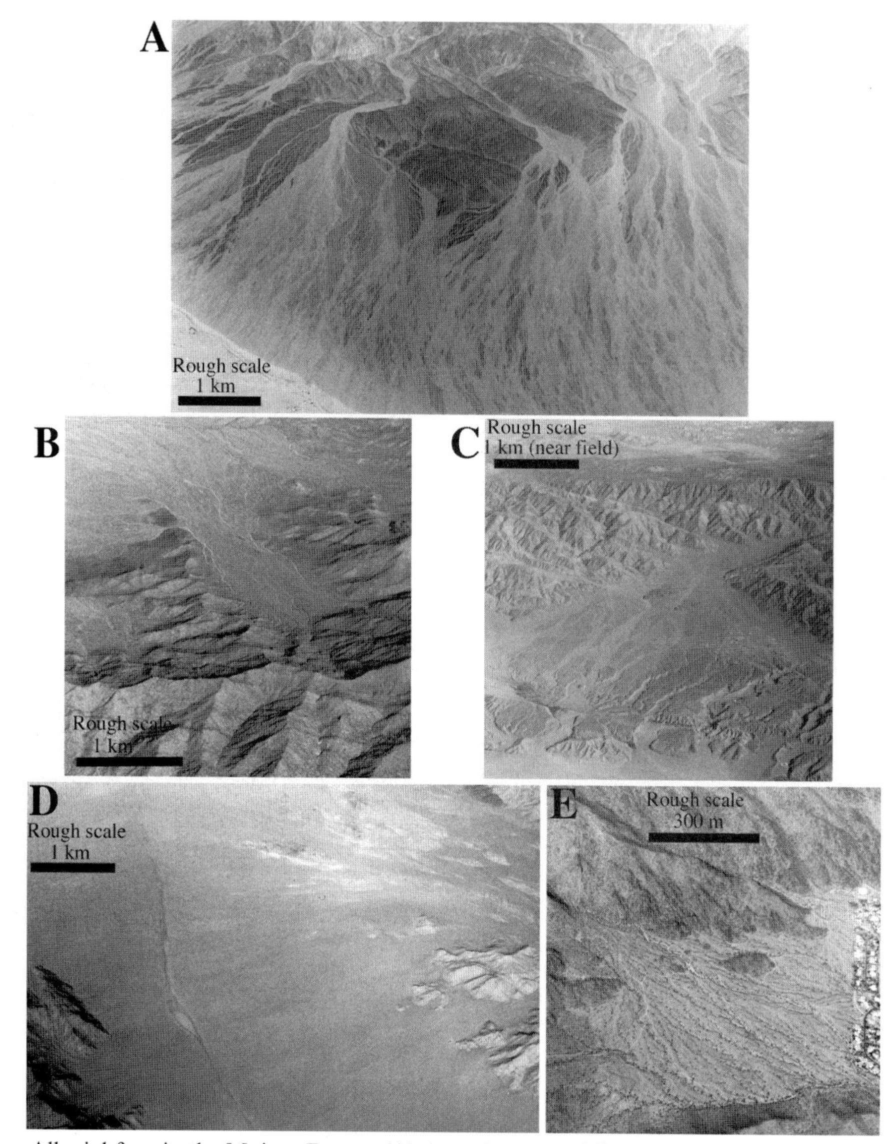

Alluvial fans in the Mojave Desert. (A) A coalescence of fans merges at lower left with another fan shed from another mountain out of the field of view. Active portions of the fan continually shift as sand, pebbles, and boulders build up and block certain channels, as fresh debris flows force active channels to be diverted, and as storms rip out new channelways. (B) The source of a fan in a fluvially dissected mountain. The V-shaped valleys are characteristic. (C) An alluvial fan is being incised due to a sudden change in the sedimentologic balance of sediment supply and transport; possibly the cause was faulting (uplift of the mountains or subsidence of the valley floor). (D) Coalescence of two alluvial fans. (E) Development encroaches far up a fan, a potentially dangerous situation in the face of flash floods and debris flows. Mars has excellent examples of alluvial fans, but they are not as well developed as in the Mojave Desert.

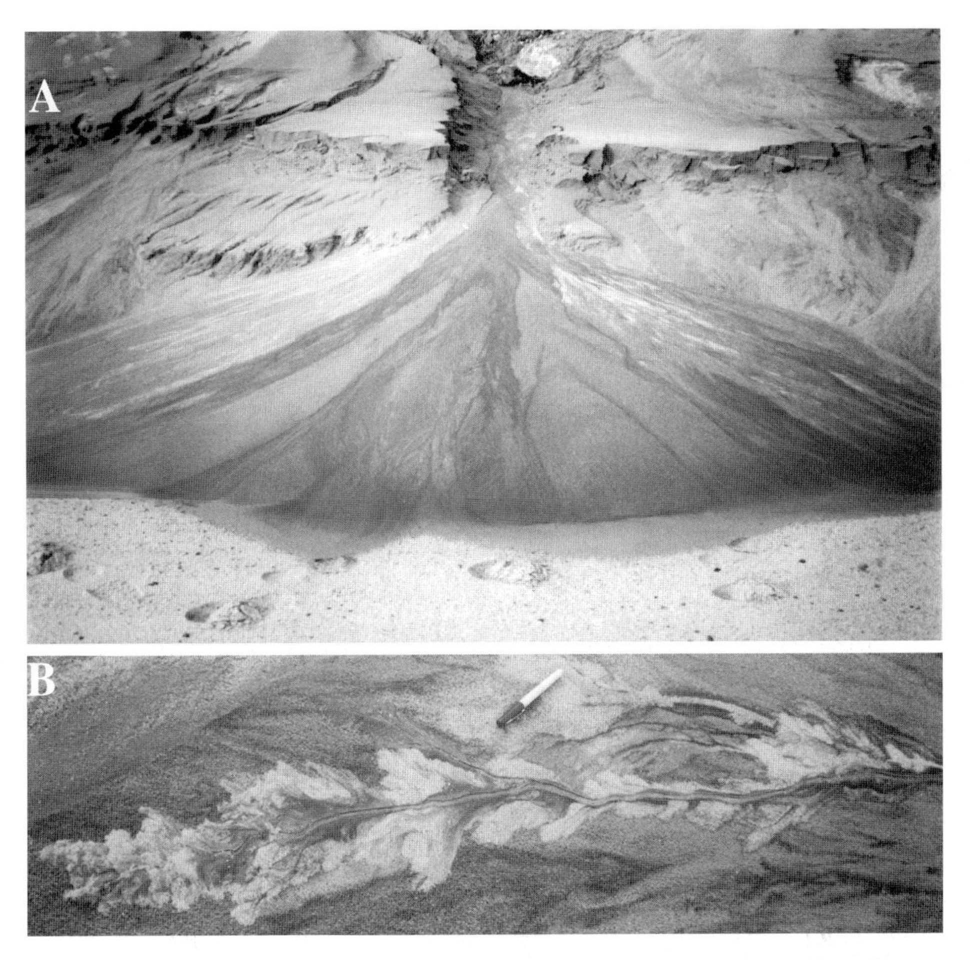

Sand flows, like debris flows, produce structures similar to alluvial fans and lava fans. (A) Sand fan produced by rapid water drainage. Footprints show the scale. (B) An individual sand flow exhibits a leveed channel and over-bank crevasse splays, like flood channels, debris flows, and lava flows. The pen is shown for scale.

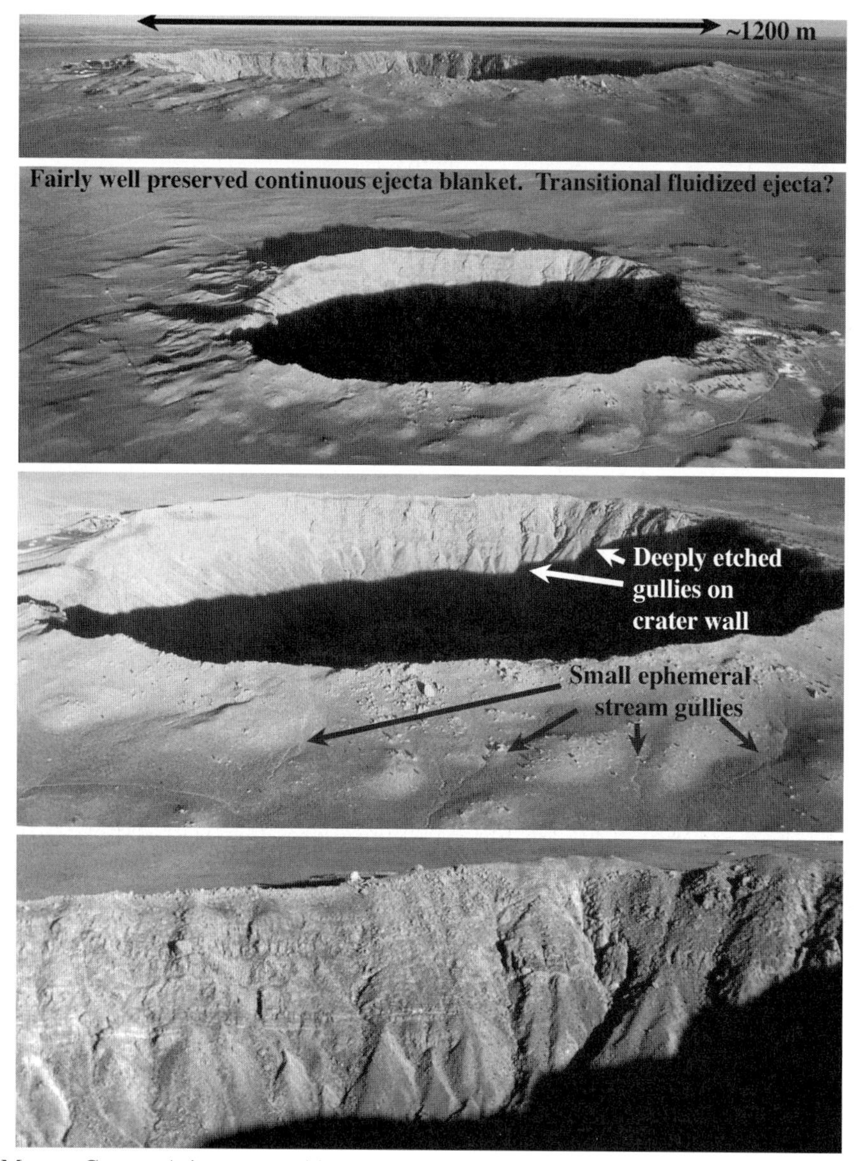

Meteor Crater, Arizona, provides an excellent measure of gully erosion rates in an impact crater. The crater is well dated at about 49,000 years, during which not more than 10% of the crater's volume has been infilled by sediment mainly derived from gully erosion of the walls (especially clastic sediment with some small chemical component) and wind-blown sand and silt derived from poorly consolidated impact-shocked sandstone. Since its origin, a cumulative total of about 12,000 meters of rain and snow-melt have fallen and eroded about 10–20 meters of rock from the walls. That amounts to about 1 m of eroded rock/1,000 m of cumulative precipitation. These are high rates because of the steep slopes and lack of vegetative cover. At Meteor Crater rates, gullies on Mars may take 50,000 years to 5 million years to form.

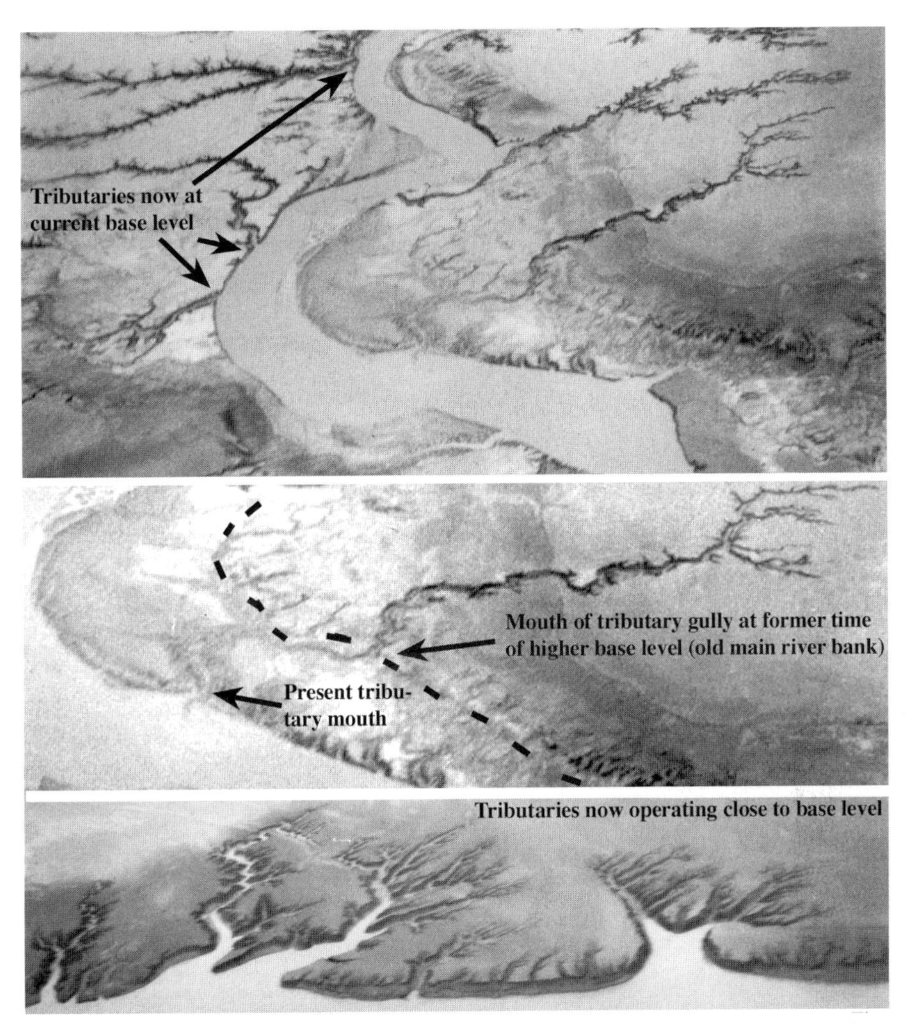

Low-order dendritic gully-like channels developed on flood plain of the Copper River Basin, Alaska. The formation of this landscape was not observed, but it is certainly a rapid process. Most likely here is what happened: Glacier-dammed lake Van Cleve drained suddenly. A flood coursed over the Copper River flood plain, depositing and redepositing abundant clay, silt, and sand (there was much evidence of this). The lake level and flood discharge dropped precipitously. As the lake emptied, water drained rapidly from the flood plain into the main river channels of the braided Copper River, including the channel pictured here ($\sim$30–50 m across). Stream flow and liquefaction carved the gullies. The total time to form this landscape was probably a few days, possibly a couple of weeks. The flood plain is remade every year. The liquefaction process suspected of occurring here is supported by the occurrence of "quick sand" at Van Cleve lake, and it can explain the gully morphology by headward growth, like sapping channels. Varied signs of sediment liquefaction are common on Mars, including ordinary debris flows. Evidence of shifting hydrological base levels are also common on Mars, as they are across many landscapes on Earth.

mountain slopes. In cold arid terrains of Earth, the location of debris flows may be controlled not just by places where talus may accumulate but also where snow may accumulate preferentially in shady areas or by snow avalanches; snow avalanches are not likely on Mars, but the shady locations are definitely favored by the Martian debris flows. The fresh, young gullies on Mars are overwhelmingly located at middle and high latitudes and on pole-facing slopes, such as on crater rims, central peaks, and other steep terrain. Stream gullies may lack such prominent levees and may (but not always) have sources at local drainage divides, but often it is difficult to discern debris flows from stream gullies.

One of the more remarkable and common associations is that the gullies very frequently occur on or very near features that are here interpreted as glacier-like deposits or ice blankets. Some of those not occurring on such icy landforms occur in cirques that may have been eroded by glaciers no longer in existence. Just as important, since they occur on pole-facing slopes preferentially, within their middle latitude band of formation, it is foremost not energy starvation but moisture starvation that prevented them from being more widespread.

While small Martian gullies are a very big story and provide one of the most powerful constraints on recent climate evolution, they constitute a small part of the last chapter of Martian history. Martian gullies of all sorts are much smaller and more restricted in occurrence than the ancient, degraded remnants of small-valley networks. The small-valley networks are "small" only by comparison to outflow channels, but nevertheless they are characteristically tens or even hundreds of kilometers long and one to three kilometers wide, and are organized into vast low-order dendritic tributary systems covering thousands to tens of thousands of square kilometers. The small gullies represent an eroded volume several orders of magnitude less than that of the small-valley networks.

The spatial distribution and common glacier association of small gullies indicates a pronounced influence of ground ice and surface ice across the middle and high latitudes of Mars. The much larger and older small-valley networks indicates a wider geographic distribution of water sources, but nothing as widespread as would occur from intense, widely distributed rainfall. Furthermore, there is an appearance of a discontinuity in age and scale between the small-valley networks and the small gullies, and different processes are also likely. While I could not conclude that there was an intervening episode when little of hydrologic interest occurred, that is a possibility. Regardless of details, the available observations indicate that the climate of Mars is dynamic. (Little can be said about the first half billion years or so predating the preserved remnants of small-valley networks, though degradation of large craters indicates a very vigorous phase of erosion far exceeding that accounted for by the small-valley networks.)

MARS' LITTLE ICE AGE

Mars appears to be an Ice Age planet of glaciers and permafrost across much of its surface through much of geologic time, though the geographic extent of icy and wet conditions have waxed and waned through time. It is insufficient that only surface temperatures have oscillated, but humidities, too (and probably the geologic/meteorologic sources of humidity), have shifted dramatically over time. Though such changes have been dramatic, Mars is not – and possibly has never been – a planet as warm as Earth, or with anywhere near the atmospheric flux of water vapor of Earth. It is, however, a place where time-integrated subsurface transport of water and subsurface/surface activity of ice has been considerable. Compared to global rates of other processes, such as eolian, impact, and volcanic activity, ice processes remain at least as significant, and water processes, too, are locally important.

Ephemeral stream flow due to melting snow, thawing permafrost, or spring discharge (perhaps episodically for hundreds of thousands of years) is now the most widely accepted explanation for small Martian valleys and gullies. Other ideas that rainfall runoff generated Martian valleys and gullies are not well supported by observations, as discussed in Chapter 2. The rainfall "backbone" of NASA's favorite poster-child of the day – the Warm, Wet, Early Mars paradigm – is very weak, insofar as the observable geologic record of Mars speaks, unless this record is read to the very earliest, indistinct record spoken by crater statistics and the very crudest remnants of geologic features. Another model with extreme climatic and geomorphological implications maintains that such features are formed by liquid outbursts of carbon dioxide, which flashes to the vapor state and drives sediment-laden turbidity currents. There is no compelling evidence that either rainfall or liquid carbon dioxide is responsible.

An alternative idea that I find very attractive – almost inevitable – is that large amounts of solids and gases are dissolved in liquid water, including salts and carbon dioxide. CO_2 would assist explosive outbursts of aqueous groundwater solutions and salts would depress the freezing point, thus making the briny liquid solutions stable closer to the surface and enabling greater flow distances across the surface. More important than the freezing point depression is the reduction of water vapor pressure of brines at the ice point, a fact that increases the elevation ranges where liquid water could be stable on and near the surface under present atmospheric pressure.

There are local instances where possible evaporitic salt deposits occur at the termini of gullies, but generally the expected salt cones or layers due to salt precipitation from freezing brine flows and seeps are not in clear evidence. This does not mean that they were never there, as they may be there but are mostly obscured from view, or perhaps the salts have a tendency to crumble and blow away when they dehydrate. In any case, salts and CO_2 can buy liquid stability for 20–30 K below the normal freezing point of water (somewhat more for unexpected special eutectic mixtures of magnesium—calcium–sodium chloride-water). This amount of freezing-point depression is of great interest to Mars researchers trying to explain evidence of recent water activity on Mars. However, it really is not sufficient, because the gullies tend to form at high latitudes and on slopes that face predominantly toward the poles; they form preferentially where it is extremely cold, apparently because that is precisely where shallow ground ice is stable. Warmer conditions without according increases in atmospheric water vapor content means that there is no ice to melt, regardless of geothermal heat flux or presence of salts. The basic problem remains that these gullies, like so many other Martian landforms, are fundamentally water-formed features in seeming contradiction to the present cryogenic environment.

ICY TIME

The Martian Neoglacial epoch is not a well-defined or well-dated period of time; nor is it a proper chronostratigraphic unit, but just my term of convenience referring to the very Late Amazonian time of formation and degradation of glacial and periglacial landforms. The ages of the Martian glaciers are not known, but in general they appear to be among the youngest features on the planet, along with small gullies, south polar dry-ice terrains, and some dune fields.

Some Martian glacier-like flows are extremely sluggish; they appear to be in the low millions of years old according to crater-counting chronologies presented by William Hartmann (Planetary Science Institute, Tucson), although some with a sparse scattering of small impact craters superposed over them may be tens of millions of years old. Hartmann concedes that cratering ages of just a few hundred thousand years are possible. More common, craters are absent or occur one to a flow, suggesting effective resurfacing ages typically a few million years or less. The young ages suggest that their flow deformation may substantially date to a very recent period of high orbital obliquity. Jim Head (Brown University) points specifically, in fact, to a period just 350,000 to 2.1 million years ago when frequent high obliquity may have caused periods of accelerated flow.

It is not too radical a thought – I would bet my bottom dollar on it – that some of these icy flows are even still sluggishly, actively flowing. Consideration of basic ice physics makes it impossible, if these flows are still as icy as they appear, that they are not active, because ice is a substance with a finite viscosity, and the driving stresses are significant with respect to flow on short geological timescales even for Martian temperatures (though three to five orders of magnitude longer than for typical terrestrial glaciers). Indeed, my statement here is far too conservative according to finite element dynamical flow modeling reported at many meetings by Elizabeth Turtle (University of Arizona, Tucson). She finds that the Martian lobate debris aprons should exhibit substantial flow deformation on timescales of just 1 million years assuming they are frozen at their beds. There is actually an embarrassment of high deformability. Geology urges that we somehow find mechanisms to put the brakes on these dynamical models. There are good reasons why Turtle's models may overestimate flow rates, but in any case the basic point is true that these features *must* be flowing even under today's climate.

Many crevasses of Martian icy flows are heavily degraded, presumably by sublimation, forming in some cases gaping shallow troughs and in other cases seracs typical of heavily melted and disarticulated crevassed glacier ice on Earth. Because crevasses will expose ice without an insulating debris cover, they will sublimate and erode rapidly if they are in an environment that allows net annual sublimation. Degraded crevasses may be only hundreds to tens of thousands of years old, i.e., geologically like yesterday, but as a rule they are not forming in large numbers today. The rare, exceptional fresh-looking fields of narrow crevasses in some of the lobate debris aprons in Argyre and Noachis Terra, however, indicate that some of these icy flows have been active, albeit weakly, within the past few decades or centuries; sublimation would have widened or subdued the crevasses if exposed for longer than that at their latitudes and with their slopes and aspects.

Basic ice physics says that lobate debris aprons are still in ductile motion, and their surfaces occasionally exposed to crevasse-forming tensional stresses, even under the present environment. Ice, even at Martian temperatures, is just too soft to stop all flow for ice masses hundreds of meters thick and sloping at several degrees. However, when considering the age of an icy flow, whether on Earth or Mars, we have to think about whether we are considering the flow's inception, the latest major episode of vigorous flow deformation, or the latest episode of erasure of craters and crevasses and other superposed features due to ablation or viscous creep. The inception of these flows almost surely goes back many millions of years, possibly hundreds of millions, whereas the flow deformation and sublimation is probably a continuous process that occurs at an uneven pace modulated by climatic oscillations.

It is worthwhile to consider the ages of these Martian flow features in a timescale of terrestrial significance. A plausible estimate is that many of the small-scale high- and middle-latitude features we see on Mars – for instance, the small gullies, the small polygons, small thermokarstic pits, and the polar caps – date from a period that may be roughly concordant with Earth's late Tertiary and Quaternary, including the Miocene to Recent epochs. Some may be much older, dating concordantly with the Mesozoic age of dinosaurs, but by and large we would see many more small craters (tens to hundreds of meters in diameter) if this was usually the case. Another way to view it is that these Martian features were forming mainly just in the period since the youthful Hawaiian Islands and Grand Canyon have taken form, and over the period since archaic horses, camels, and elephant species became recognizable as closely related to their modern versions, and since the first hominids strode the Earth on two feet. It is quite possible that these Martian landforms were initiated long before that and have ancient precursory roots even billions of years old, but they have evolved into their present forms just over this geologically recent time. Indeed, very many of the crispest flow and sublimational forms, polygons, and gullies that I write about here are probably less than 100,000 years old, dating from when *Homo sapiens* and *Homo neanderthalensis* cohabited Earth, roughly corresponding to the times of the last few great glacial advances or interglacials on Earth. To think that all this activity has somehow now ceased would probably not be accurate, though process rates may vary substantially over the 10^5-year Martian obliquity cycle.

ICE, NOW AND THEN

The Buddha wrote of physical existence at one moment conditioning the regeneration of a slightly altered existence in the next moment: cause and effect in one indivisible continuum of space, time, energy, matter, life, and death. It is certainly everyone's understanding over their lifespan. Modern science has arrived at exactly the same conclusion, with our present and future being linked moment to moment, like an ever-flowing and ever-changing river, back to the Big Bang and forward to whatever unknown fate awaits. (I ignore here the infinitesimal quantum unit of time.) The physical relationships of one continuum entity to its transmuted entity at another time can always be discerned with enough knowledge; science has shown this well.

Lacking sufficient knowledge, physical existence in one parcel of space-time can seem unrelated to that at another parcel of space-time, as when we board an airliner at New York JFK, take a long nap, and wake up while landing in London Heathrow, or if we were to take off from JFK in a light-speed spaceship and land in the same place a few millennia later. Short of hitting worm holes or other discontinuities, the direct and continuous relationships are always there, just perhaps are unknown. At some large level of apparent disparity, we are perhaps better off considering them to be almost different universes, as when comparing Crocodile Dundee's Australian life with his New York life. Some fundamentalist Christians do not recognize the links that science so well considers between human beings and the great apes and older life forms, just as Charleton Heston's character in *Planet of the Apes* at first failed to see the link between twentieth-century New York City and post-apocalyptic New York until the buried torch of Lady Liberty shed some needed light on an unwitnessed chain of tragic events. So in Argyre, what might be the links between today's lobate debris aprons and the esker-forming ice sheets of the early Amazonian? What of today's polar caps and the ancient ice sheets and the Chryse outflow events? Do we dare draw any link, or are they effectively different cryospheric responses, an early massive response and a later milder response, to two completely different climatic regimes and underlying hydrologic triggers? We lack knowledge of the full sequence of events, so we are left guessing about the gaps.

The morphologic characteristics of the glacier-like features in Argyre and their relationships with surrounding glacial-type terrain may suggest a genetic and temporal relationship to the ancient episodes of glaciation inferred from the presence of eskers, cirques, and other glacial landforms. I suggest that the most likely relationship is fairly direct; the present-day lobate debris aprons may represent the wasted or receded but still active remnants of a vast ice sheet that once dominated this region. This speculative relationship in some cases is as direct as that of present-day wasted remnants of Himalayan and Alaskan glaciers and rock glaciers relative to the mighty Pleistocene mountain ice caps that once completely engulfed whole mountain ranges; the present-day glaciers and rock glaciers continue to be nourished by the water cycle, but with reduced mass balance terms. In these cases, the ice contained therein is younger than the ancient ice caps, but it is likely that there has never been a complete interruption of glacial activity. In other cases, such as the ice caps of the high Canadian arctic and the Beacon Valley (Antarctica) rock glaciers, the relationship might be even more direct – wasting, barely active remnants of ice might date back all the way to the ancient periods of ice cap glaciation; this is apt to be true of some thermokarstic deposits on the floor of Argyre, surrounding the south polar cap, in and east of Hellas, and in Utopia Planitia. Finally, in other cases, the ice is gone completely, and what is left is just the rocky debris (moraines, eskers, lacustrine plains, and such) left behind by a mightier glacial past. In this concept, the Neoglacial permafrost and debris-covered glaciers of Mars are mostly analogous to the modern-day and Holocene glaciers and permafrost of Earth, while icy remnants and ice-free glacial landscapes (such as a the eskers of Argyre Planitia) remain in evidence of the greater glacial past (analogous to Pleistocene glacial/periglacial landscapes in places such as Minnesota, Ontario, and Scotland).

While there are many parallels to be drawn between Earth's cryosphere and that on Mars, the process of accumulation, the time scale of formation, and the duration of the mass balance cycle of extant Martian glaciers, however, appear to be quite different than that on Earth. Many other Martian landforms are interpretable as resulting from glaciation. In and around the Argyre basin, these include apparent alpine glacial sculpture (aretes, cirques, and U-shaped valleys) in the Charitum Montes, and lowland glacial landforms (moraines, esker-like ridges, and glaciolacustrine plains) in southern Argyre Planitia. Extant debris-covered glaciers are cold based and are variously weakly active or in a state of sublimational decay under present environmental conditions. Relic glaciers and ancient glacial deposits, however, show signs of surface melting and wet-based erosion and sedimentation.

Eolian processes appear to have acted on glacial deposits concurrently with and subsequent to glaciation to produce some of the most magnificent dune and dust-streaked landscapes on Mars. Large climatic swings may be implicated by the observed record of glacial landscapes, including what may be obliquity-driven climate oscillations and transient high-magnitude shifts of the global climate system due to thermotectonic mobilization of volatiles. Although glaciation explains the observations made here, specific glacial processes, the role of glaciation in the global system, the prevalent glacial climate, and even the composition of ice are subject to differing interpretations.

CLIMACTIC CLIMATE

Ancient Martian megafloods, monster ice sheets, and icy oceans get rave reviews for their hold over the planet. Their effects have dominion over the Martian eons and the upper 10 km but today's severe climate and more modest hydrology is the climactic episode of Mars in one respect. Today Mars is preparing for the arrival of humans. It may be egregiously anthropocentric to say so, but for the humans who soon will venture to Mars, there can be no overstating the significance of the present era's hydrology for the future ability of humans to sustain themselves and grow in our presence. If the present frigid climate and feeble hydrology touches just the Martian façade, it is precisely that upper 10 m that will be of greatest material significance for us, as surface-dwelling creatures.

Sublimating glaciers, seeping ice and small gullies, and cracking permafrost are the stories of these climatically dynamic times. Liquid water or brine was the agency responsible for the young ice-associated melt features; if these are as young as appearances suggest, then we must conclude that there has been some degree of recent climate change, because surface melting is not quite possible today. Warm enough temperatures are observed every summer in some areas, but those areas are where ice is absent. We may infer that today's Martian climatic regime is sufficiently dynamic to have allowed recent warm, wet episodes in places where ice exists near or on the surface. We see many additional indications of recent climate change, such as the polar layered deposits and sublimating south polar dry ice terrains, which reinforce the idea of rapid climate oscillations. Recent climate appears to have fluctuated up to and far below the melting transition of ice, as well as across the vapor–solid equilibria of polar dry ice and mid-latitude water ice.

Fluvial activity due to surface melting probably has not been common on Mars in long-term stable equilibrium with the climate anytime since the first part of the geologic record was laid down over 4 billion years ago. It could not have been sustained, at least not everywhere, or else we would not see the relics of heavily cratered surfaces. Fluvial activity remains a rare phenomenon, taking place only when and where conditions are just right. However, the cases where small gullies and proglacial drainages are observed are very important for understanding the limits of recent Martian climate oscillations. These brief warm, wet epochs indicate a climate system that recently has allowed, at the edge of the Martian envelope of change, melting snow, melting glaciers, and flowing rivers. Present data are quite compelling that this has been the case.

It is commonly supposed by Mars researchers that it would take a huge perturbation of conditions – for example, an excursion to a mean annual global temperature of 273 K (60 K warmer than present) – to unlock the vast stores of frozen water. This is not so; a global mean of 230 K (20 K warmer than now) would be warm enough, since summer diurnally averaged temperatures at middle latitudes then could exceed 273 K in areas where surface ice exists, thus thawing the permafrost active layer and unleashing torrents of water from ice-rich sediment and debris-covered glaciers.

We do not have to go far to find a suitable climate-change mechanism to explain the young water-related features. Mars has long been known to undergo large periodic oscillations in its orbital parameters (such as eccentricity) and the configuration of its spin axis relative to its orbit plane (obliquity). Earth's obliquity is responsible for the strong seasons that most parts of Earth experience, and periodic oscillations in obliquity, the precession of the spin axis, and changes in orbital eccentricity are responsible for much of the natural climatic variability that has led to waxing and waning of glaciers and ice sheets, advance and retreat of forests and grasslands, and other effects. Even human evolution and technological development can be tied to these astronomical changes of the Earth's distance and attitude with respect to the Sun. There is, of course, much more that controls Earth's climate, and the same doubtless is true of Mars. Martian orbital and obliquity oscillations are of a much greater magnitude than those affecting Earth, and so periodic climate changes are expected to be of greater magnitude than those on Earth. On our planet, perhaps the largest long-term control on climate is the location in latitude and the relative arrangement and integration/dispersal of continents and oceans; obliquity and other astronomical forcings play as an accessory, though one with a major influence in modulating climate on the timeframe of 10^4–10^6 years. In contrast, on Mars at least one of the biggest climate stories – the one we are certain of – is obliquity. In the time since the first late Miocene hominids walked the Earth, there have been five dozen of these obliquity-forced climate fluctuations on Mars, and one wild chaotic swing to extreme obliquity and seasonal variations three times as great as what Earth's middle and high latitudes experience. Each high-obliquity episode, repeated every hundred thousand years, lasts a few thousand years, with moderate and low obliquity prevailing the rest of the time.

We appropriately think of those high-obliquity periods as a warmer, wetter Mars, where the icy latitudes experience warm summers. Warm summers are balanced by the long season of short winter days and, much more so, the long winter nights, which are somewhere on the far side of harsh. This portrait of a recent Martian environment lacks any large, widespread source of abundant winter snow accumulation. There are no snow avalanches anywhere, ever, and no spring floods from melting of the past winter's snow. A big, bad winter might have all of one millimeter of accumulation in the most serious Sun-shaded mountain heights. We, in my home state of Arizona, defend the habitability of the desert by claiming that it's a dry heat making it an enjoyable place to live, as we invest in lip-balm and evaporative cooler stocks. Mars, too, cyclically experiences what may be called a dry heat (if you were a penguin). Certainly forget about any warm, sandy beaches recently, or, for that matter, in the past 4 billion years. Sublimation – both condensation and evaporation of ice, but lacking a liquid intermediary – is king today. Significant melting can occur only where accumulations of ice up to tens of meters thick have built up during the long periods of low obliquity, or where there are remnants of the ancient mega-ice deposits (relics of frozen lakes and oceans and ice sheets).

Free-flowing, alpine glacier-fed braided rivers and slow seeping pockets of ice are by far the exception on Mars, having occurred rarely in the recent past. But they have occurred, rarely washing the surface and doing fast geomorphic work during those short-lived episodes of brief seasonal melting. Prevalent in this present era at the middle and high latitudes are vast icy plains laid down originally hundreds of millions to billions of years ago, but remaining icy today due to local humidities close to saturation conditions; we see also icy plains slowly losing their ice, and debris-covered glaciers slowly creeping down their valleys at possibly as much as a few millimeters per year, if that. It may not seem so dramatic as the Mars of the ancient past, but there is little else happening, and so these water and ice processes dominate landscape formation across vast expanses of the planet. Certainly these processes are remaking the upper meters across half the surface, leaving ruins of the bigger ancient dramas of lakes, oceans, and ice sheets.

How warm is warm? How wet has wet been during those repeated excursions to high obliquity? The geologic evidence points toward cyclic shifts to a warmer, wetter climate not much more severe on average than the Arctic Coastal Plain of Alaska. The Taylor Valley and other Antarctic Dry Valleys is perhaps a closer analog. During those comparatively warm episodes the annual Martian temperature cycle brings extreme winter cold but also periods of sustained summer warmth across the middle latitudes. Temperatures exceed the melting transition for weeks on end around the summer solstice, such that significant meltwater is produced from any bodies of exposed or lightly debris-covered ice. Atmospheric pressure also probably increases modestly at those times due to sublimation of one polar cap. This increased pressure keeps any water from immediately boiling and prevents evaporation long enough for the water to move sediment and carve gullies. The down-slope tapering of short gullies and the downstream fading of larger river channels may indicate that atmospheric pressure was not too much more than it is at present.

While envisioning the Arctic Coastal Plain of Alaska or the Dry Valleys of Antarctica may help one to arrive at a mental picture of what Mars may have been like during the warmest years of the recent epoch, these terrestrial models are probably misleading in one key respect. The Earth has a vast, warm ocean that supplies prodigious precipitation. The equivalent hydrologic reservoirs on Mars in recent times are the polar caps and relic deposits of ice in the major basins. These are, in effect, like a very cold and solid ocean; they can provide abundant moisture to drive glaciation and periglacial ice processes, but only at low transport fluxes over long periods of time. Vapor pressures across large expanses of these reservoirs – even during the warmest years– are limited by the triple-point pressure of water. More often, vapor pressures of large sublimating ice masses are probably far lower. Any moist air mass that is somehow heated to warmer temperatures would be undersaturated in water vapor. Thus, it would never rain on such a Mars, and it would normally not snow very hard. Winter accumulations of snow or frost may measure more like one to a few millimeters – in the snowiest years and most humid places. Thus, annual cycles of winter snow accumulation and summer melting would not have any geomorphic effect unless sustained summer melting could be preceded by millennia of net annual accumulation.

A climatic oscillation from periods of seasonal melting to periods of perennially hard-frozen conditions where snow could locally accumulate may have been the prevalent Martian regime across many areas at high Martian latitudes for the last tens of millions of years. Cumulative warm, wet conditions during this recent epoch may have amounted to perhaps 1% of the total Late Amazonian epoch (10% of the years, occurring during high-obliquity phases, and 10% of each warm year).

WHEN EQUILIBRATION DISTURBS THE BALANCE

The rates of geomorphic processes on Mars are not well known. There is a definite sense that the pulse of Mars presently – and possibly through 99.9% of its history – is far slower than that of Earth; if it was not, we would not still see hundreds of thousands of impact craters on a planet where the impact rate is only a factor of a few greater than on the Moon (Strom et al., 1992) and Earth. There are good reasons to suppose that Martian surface modifications would occur more slowly than on Earth; the atmosphere is thinner and colder. That is fundamentally it. However, Mars is an active world. The activity requires energy, and it is worth reflecting on the sources and magnitude of energy flows and how energy is coupled to the surface.

Geomorphology is about recovery of solid planetary surfaces from a disturbed state. It is about the reshaping and reconstitution of solid matter and the record of processes inferred from the three-dimensional forms that are developed in solid planetary surfaces due to a drive toward re-equilibration from a disturbed state. The route taken to local re-equilibration is dependent on conditions, and it defines the geologic process. The product of time and process-rate defines the cumulative amount of geomorphic work done and the magnitude of modification. The process by which geologic systems attain their new equilibrium and the cumulative amount of work done defines the type of landforms produced. The interplay and sequence of multiple related and spatially associated processes in complex integrated systems define the landscape. Tracing landscape and landform back to process, it always comes back to disturbance, re-equilibration, energy flux, and time. Important determinants of the landscape produced are thus the sources and amounts of energy introduced to the system, the media by which the energy is transferred to and through the rocks, the efficiency with which available energy is used to grind and abrade rocks and transport grains, and the areal distribution of energy use. Herein are some differences, as well as striking similarities, between Earth and Mars.

The perpetual drive to attain physical order by equilibration at a lower and more homogeneous energy state drives geologic processes deep within planets. The mantle of a terrestrial planet, unsettled by heat liberated from natural radioactive atoms seeking their own more-stable states in their atomic worlds, softens, partially melts, and goes into convection. The mantle's churnings and oozings result in the progressive chemical differentiation of active planetary systems: partial melting causes segregation of mineral phases, dense matter sinks to the core, and buoyant liquids separate from a refractory mantle residue and form a crust. Whether expressed in volcanism, plate tectonics, diapirism, regional uplift or subsidence, or global compression or extension, mantle disturbances on each active planet produce differentiated crust and cause the tectonic disruption of the crust and planetary surface. Volcanoes are built, mountains rise in fold belts and thrust sheets, basins subside, and crust is rifted apart. The atmosphere, unsettled by differential solar heating, has its own continuous drive toward equilibrium, producing planetary climate and weather with its own vaporous overthrusts and cyclonic convection systems and with its own peculiar dynamic and reversible differentiation that we call snow and rain and evaporation.

As planetary mantles and atmospheres seek equilibrium, a seeming contradiction is that adjacent parcels of matter are set into contrasting energy states – warming or cooling, shifting laterally, heaving, thrusting, or sinking, changing pressure, and changing phase. This activity helps the planet as a whole to settle into a more ordered and lower-energy state through release of gravitational potential and thermal energy, but this activity also sets up unstable conditions along the way. Instability pervades active planetary interiors and surfaces, like the small disturbances that follow a cascading waterfall over its entire course. The cascading energies of the mantle and atmosphere require an energy source to sustain the disequilibrium that drives the cascade. Lacking such an energy source, the mantles of planets even as large as Earth would soon settle into cold dormancy, and there would be nothing to counter the erosion and infilling that would progressively level the planetary surface in just a few hundred million years. Lacking these energy sources, there would be no geology and no life. It would all be cold, hard, lifeless existence. Eventually it will be this way, but not yet.

Thanks to instability that otherwise we may lament, Earth and presumably many other planets are not lifeless. Thanks to volcanic eruptions, seismic activity, and the inexorable attack of ugly weather and a corrosive atmosphere. The effects of instability pervade the Earth but are most notable where major physical and chemical discontinuities exist. Diverse manifestations of clashing instabilities are perhaps nowhere greater than at planetary surfaces. This is also where surface-dwelling organisms, such as humans, observe these phenomena. The chemophysical clashing of planetary Fe–Ni cores with their silicate mantles may be nearly as great, but it is unseen by living things, except through our technology. It is at the solid surface/atmosphere interface where we see erosion and sedimentation by floods, sculpture due to glaciers, the drama of volcanism, shifting sands and scouring winds, and seismic upheavals of crust. Too bad for the individual who gets in the way, but that is but a sad little detail.

The activity of Earth's deep interior drives the surface topography that then drives the interactions of atmosphere, hydrosphere, lithosphere, and biosphere. Glaciers and alpine meadows, rain forests and coral reefs, coal seams and copper veins, cattle ranches, cities and the seven seas are all put in their places by the deep activity of the Earth, though none of that would be possible without the Sun. The whole global system and every component is described by physical and life scientists always in terms of instability, the processes sustaining instability, and the thermodynamic trend toward alleviating it and kinetic obstacles preventing re-establishment of equilibrium.

It is ironic that the cause of landform-generating instabilities on any active planet is the drive toward a greater planetary equilibrium, but this is an irony shared throughout the physical and biological Universe. Planetary surface materials are placed at high gravitational potential and high chemical potential energies, and then gravity and chemical thermodynamics run these systems back downhill, doing work, forming new substances, and creating new landforms. The cascading energies result in the grinding-down of bumps and infilling of holes on planetary surfaces. Minerals are altered, rocks are broken, grains are redistributed, and rock masses are sculpted along the way. Geomorphology is about these build-up and run-down processes and the continuum from disequilibrated terrain to new landscapes produced by modification or redistribution of old rocks. The metamorphosis is powered by energy from the deep planetary interior (radiogenic heat and gravitational potential energy long stored in an incompletely differentiated planet), by the Sun, and by the occasional impact.

Out of the nebula, planet Earth and our sibling worlds arose spontaneously through gravitational instabilities. Ultimately, the energy trapped during Earth's accretion, and that later liberated by radioisotopic decay, drove outgassing and mantle dynamics, and generated a globally structured surface with bumps and holes and other scars and bruises, with heterogenities ranging from the continents and ocean basins down to fault scarps and metamorphic minerals. Internally produced surface irregularities now organize pools of water and accumulations of ice, and control evaporation and precipitation of water and ice. It is through this complex hydrologic link that the geodynamics of Earth controls the deposition of an isotropic outflow of solar energy via gravitational potential energy carried with and deposited by precipitation. That potential energy, coupled with that of the rocks, drives the erosion of structured river valleys and cirques and deposition of deltaic plains and marine sediments. From uniformity arises hierarchically structured form. Clever designer or not, it is certainly an awesome design.

Planets differ in how they utilize internal energy to produce surface structure and how (or if) they regulate the surface deposition of solar energy and utilize it to alter the surface. Mercury lacks an appreciable atmosphere, so radiant solar energy is deposited one hemisphere at a time in accordance with its spheroidal geometry and rotational characteristics. On such dead worlds the trigonometry of direct solar insolation rules. Traces of volatiles, Sun-baked from the rocks and released by impacts, seek the poles; that and the churnings of the last dregs of a molten core is roughly the present chapter of Mercury's story.

Venus has a dense atmosphere but lacks a major surface-condensable volatile and thus lacks a massive volatile surface reservoir that could soak up and transport solar energy and transmit it to the rocks. Instead, the atmosphere of Venus absorbs and stores solar energy for a period longer than global wind transport times, and so the atmosphere homogenizes the delivery of solar energy to the surface, which remains almost isothermal. Though temperature decreases with altitude, and this has some surface effects, surface winds are feeble; the Sun accomplishes little work on the surface. The Venusian lowlands are so hot that trace volatiles, such as tellurium and lead, are sublimated out of the rocks and condense on mountain peaks. However, those volatiles are not abundant enough to do more than paint the mountain peaks. Venus' story is of internally driven volcanism and tectonism, but the surface heterogeneities so formed are poorly coupled to atmospheric processes.

More interesting than those planets, in my opinion, are worlds where a partly condensable, massive component of a restless atmosphere coexists in perpetual combat with a restless, active solid crust, as is true for Earth, Mars, Triton, and perhaps Titan.

COLD FACTS

Each planetary surface has a different ensemble of surface modification processes driven by distinct surface compositions and environments. On each planet destruction of relief and landforms by erosion and burial tends to be balanced by whatever processes generate relief and landforms.

On the Moon, primordial large-scale topographic relief was produced by crust formation processes, and major basin-forming impacts added to that relief. The chief means of erosion and surface-leveling on the Moon once occurred partly by volcanism, but now it is solely due to incessant chipping-away by innumerable small impacts and the random-walk-type ballistic diffusion of rocky ejecta. Impact generation and destruction of relief has nearly attained a steady state in the lunar highlands.

On Venus and Jupiter's moon Europa cratering is minor and on Io it is completely insignificant on the time scales of tectonic and volcanic generation of relief; wind erosion is nil on Venus and absent on Io and Europa, and there is no hydrologically driven erosion. Thus, relief on these worlds increases until volcanic burial and gravitationally driven tectonic subsidence, spreading, and collapse of steep relief features starts to balance the construction of new relief. It is not clear that Europa has reached any state that is near geological equilibrium, but Venus and Io certainly must be at such a point, where new features on average destroy old ones.

On Earth and Mars, all of these processes occur, but it is clear that additional, much more effective means of planetary surface flattening exist. On Earth, volatile-driven erosion of highs and infilling of basins competes with internally driven processes of vertical and horizontal continental accretion, including addition of new crustal matter and crumpling and uplift of old rocks. Mars lacks the plate tectonic processes of Earth, but it is basically a similar story of volatile-driven resurfacing processes competing alongside volcanic, tectonic, and impact processes.

As anyone who has manually sanded metal or wood knows, erosion of well-bonded substances is energy intensive. The quantity of energy available to do geologic/geomorphic work and the agents that transport and transmit energy around the globe are of course different between Mars and Earth. This difference relates to the heliocentric distance and greenhouse properties of the atmospheres (controlling surface temperature), but is greatly amplified by the strong temperature dependence of the vapor pressure of water and ice – both the IR-absorbing greenhouse properties of water vapor (substantial on Earth, trivial on Mars) and the depositional rate and form of precipitation depend on water vapor pressure.

It is worthwhile to consider the sources of energy that can drive relief-forming and relief-attenuating processes. First, I shall consider Earth.

The total energy available to Earth from internal heat sources is about 1.2×10^{21} J/yr provided primarily by radioactive isotopic decay and by the delayed release of primordial heat left from accretion and early global differentiation (Kargel and Lewis, 1993). This energy partially melts and softens the mantle, drives mantle convection and mountain building, causes volcanoes to erupt and geysers to spew forth, and sustains conductive heat flow through the crust. A minuscule but crucial part of this geothermal energy ultimately is consumed in lifting crustal rocks and forming surface relief: roughly 4.6×10^{17} J/yr of work is done in lifting continental rocks, raising oceanic sediment to continental heights, and constructing volcanic relief to compensate for erosion (this does not include the far greater energy used to melt rocks).

By contrast, the total solar energy incident at the top of the atmosphere of Earth is around 5.5×10^{24} J/yr. Of this, 1.2×10^{24} J/yr – three orders of magnitude more than geothermal energy – is used to evaporate and circulate water through the atmosphere. The energy consumed in evaporating water – over 2×10^{6} J/kg – is released to the air and eventually radiated to space when water vapor condenses in clouds. A small additional amount of solar energy is used in lifting water into the atmosphere – 70,000 J/kg to lift it 7 km, a portion of which is retained, as gravitational potential energy, by precipitation that falls onto elevated land surfaces.

Most surface-water hydrology is driven by just 0.014% of the total solar energy incident at the top of the atmosphere, about 7.5×10^{20} J/yr, that is stored as gravitational potential energy of water and ice that accumulates on land surfaces without soaking into the ground. Surface water runoff and flowing glaciers release this energy at continental surfaces, specifically at stream beds and glacier-covered valley walls, where over time the H_2O does much geomorphic work as it flows toward the oceans. This natural hydropower energy – almost the equal of all internal heat flow of Earth – carves river channels and glacial valleys, levels mountains, and carries erosional debris to the oceans. This energy flattens the continents at a rate that competes with natural continental construction due to deep tectonic forces and magmatism.

Earth exhibits a natural regulation of competing "constructive" and "destructive" energies. Geologic constructive activity, such as volcanism and compressional mountain building due to plate collisions, is episodic to some degree, and so then must be the opposing destructive responses. The two are not completely in phase, but overall there is a dynamic balance of construction and destruction. If continental construction and "freeboard" relative to sea-level increases or decreases, the stored potential energy of surface-accumulated precipitation also increases or decreases, and with it the rate of erosion increases or decreases until a new balance is achieved. The erosional response is possible because the erosive shear stress exerted by flowing stream water and flowing glacier ice is related to surface slope, and furthermore, the rate of precipitation is also related to topography.

The long-term rate of clastic sediment transfer from Earth's continents to the oceans, according to Taylor and MacLennan, is about 1.8×10^{13} kg/yr. There is an additional 5×10^{12} kg/yr of dissolved chemical sediment involving other energy terms, but this chemical erosion is made possible by exposure of fresh rock due to physical erosion. The total erosion rate of continents thus is about 2.3×10^{13} kg/yr. Since water- and ice-borne processes are the major cause (wind, waves, and landslides being minor causes), the average potential energy expended in erosion of continental material and transport to the oceans is about 3.3×10^{7} J/kg, including the "wasted" energy dissipated by viscous forces in rivers and glaciers. This quantity compares to 64,000 J/kg for industrial grinding of granite pebbles to 60-micron particle size; it takes more energy to transport bed load to the sea, for example, about 5×10^{6} J to transport each kilogram of 60-micron particles, 2,000 km as bed load (approximated as the work to lift each particle a height of 60 microns in each of 3×10^{10} steps needed to reach the sea). Larger particles will, in aggregate, cost less energy to transport as bed load, and transport of suspension load is highly efficient, but the point is that total hydropower energy far exceeds that needed, in strict accounting, to abrade rocks and transport sediment to the sea.

The deposition of gravitational potential energy with Earth's land-based precipitation is, coincidentally, similar in magnitude to the total geothermal energy available to drive internal geologic activity. While continental reconstruction and solar-powered, hydrologically mediated erosion do roughly balance each other in the long term on Earth, this situation is not assured for Mars, as Venus and Io may remind us. However, the geomorphic evidence is present on Mars that hydrologically mediated erosion is significant; there are serious questions about the energy source for these processes.

For Mars we can make a similar simple consideration of energy available for erosion, first considering the present environment and only solar-powered components of the hydrologic cycle, though it is difficult to assess how efficient this energy is in actually doing geomorphic work. The total surface area of Mars is about the same as the continental area of Earth, but in any case I will evaluate energy on a per-unit-area basis. The energy available to do geomorphic work is simply the gravitational potential energy deposited on the surface without infiltration. On Mars, all precipitation will be solid and will first rest on the surface, though some could infiltrate if snowpack melts. Any precipitation that falls on an area having a net negative annual mass balance will do no work, as it will sublimate before it is able to flow. The only possibility for erosion is in areas that have a net positive annual mass balance.

There is a simple means to estimate the order of magnitude of balance accumulations without getting into detailed climate models. It is based on the rate of deposition of gravitational potential energy in surface precipitates. This is admittedly a naïve approach to estimate erosion rates, as it does not strictly account for whether the precipitation is liquid or solid, whether it percolates slowly into the ground (and then whether it soon emerges at springs or flows slowly as groundwater to the sea), or whether it accumulates over time into flowing quantities that can do any amount of bed abrasion and sediment transport. One could argue that on Mars any surface ice will have a transient existence or that any substantial accumulations will simply sit there and do no work until eventually it sublimates, leaving no trace of its former existence. One can argue, alternatively, that any planet with prodigious amounts of a condensable volatile and a volatile cycle will indeed accumulate surface condensates some place on the surface with a net positive mass balance over some period of time; the accumulation will ultimately thicken until it overcomes forces that resist flow, and will then flow over the surface, pluck and abrade rock, and transport sediment. I will not defend this second assertion, but simply state its plausibility for volatile-rich planets such as Earth and Mars. As such, Mars should have some level of solar-powered, hydrologically mediated erosion; but it cannot be much at present.

On Mars, the vapor pressure of ice is 0.12 Pa at 200 K (roughly the average annual temperature in mid/high subpolar latitudes); this is four orders of magnitude less than the vapor pressure of liquid water at Earth's global mean annual temperature. In terms of moisture densities, the ideal gas law would indicate characteristic Martian atmospheric water vapor mass per unit area of the planet, under saturated conditions, about a factor of 2,000 less than on Earth. If Martian ice at this global mean temperature was exposed at the surface, and atmospheric winds over the ice were comparable to typical winds on Earth, then sublimation rates (mass per unit area) would be roughly 2,000 times slower on Mars than those typical on

Earth. However, Mars exposes condensed H_2O over a fraction of its surface only about one-tenth of what Earth does; global average sublimation rates on Mars would be 1/20,000 that of Earth. Therefore, area-average condensation rates on Mars should be about 20,000 times less than on Earth; the gravitational potential energy deposited with this ice (per unit area) would average about 1/50,000 that of Earth's. Local Martian polar and mid-latitude alpine snowbelts, such as shady sides of mountains, may accumulate just a few tenths of 1 mm of ice (a few millimeters of snow) per Martian year, and so might have well-above-average ice accumulation.

Global average solar-powered, hydrologically mediated erosion rates, estimated with optimizations favoring erosion, may be only 10^{-5} of the Earth's average continental rates. Fundamentally, this is simply because of the high temperature dependence of saturation water vapor pressure and the low temperatures on Mars; a secondary issue of vastly less significance is the lower gravity on Mars. It is, in fact, almost trivial when estimated this way, though if continued for a billion years it might make a mark on the planet. It is actually apt to be even less effective than that, because most of the time in most places the ice is cold based and ineffectual in causing bed erosion. In the areas of most vigorous precipitation and alpine relief, Martian glacial erosion rates for cold-based glaciers accumulating at this rate might be just *a few tenths to a few tens of millimeters per million years.*

Even extended for billions of years, maximum cumulative glacial erosion depths on Mars would not be substantial. Fluvial erosion should be nothing at all, but this is not what the record of the Martian surface shows. Instead, it shows cumulative global average erosion rates perhaps two or three orders of magnitude lower than on Earth, but certainly not five orders less.

CONFLICT RESOLUTION

There seems to be a fundamental contradiction between the observed geomorphic record of Mars and the erosion rates and hydrologic processes that could be supported by a solar energy source and the current Martian climate and hydrologic regime. This basic result is widely accepted in the Mars science community, even if details of geomorphic interpretations differ. There are three possible alternative (or complementary) inferences to be drawn:

1. Current conditions and the present hydrologic regime does not typify all periods of the Martian past. Mars today is climatically much harsher and geomorphically far more sedate than it once (even recently) was.
2. Martian hydrologically driven geomorphology is powered by geothermal energy sources more than by solar power.
3. Martian geomorphology is driven substantially by some substance more volatile and active than water – carbon dioxide being the only likely alternative medium.

These three possibilities constitute the crux of the great Mars debates. It is not my intention to be diplomatic here, but it seems that all three possibilities taken together explain the Martian geologic record much better than any one (or two) without the others. Indeed, all three seem inescapable. The first is patently obvious, especially when outflow channels and other mega-water features, and those requiring sustained flow of water are considered; the debate then is how and why the past has differed from the present. The second and third possibilities require further explanation.

Although the solar energy that goes into driving the Martian hydrologic cycle is orders of magnitude less than that of Earth, the internally sourced heat flow of Mars is only a factor of 2 or 3 less than Earth's. The morphologies of outflow channels and chaotic terrains, of small-valley networks and gullies, and of some glacier-like flows all indicate major derivation of volatiles from the subsurface; hence, there seems to be significant potential for the involvement of geothermal heat in the hydrogeology of Mars. Geologic observations further indicate, at least in some cases, a highly gaseous type of eruption – carbon-dioxide-charged ground water seems highly likely.

MARS, A DIFFERENT SORT OF PLACE

The production and transport of heat in the mantle and through the crust, as well as the transport and atmospheric storage of solar heat and warming of the surface is central to any large planet's geologic and climatic record. However, the manifestation of internal and external heating in features observable on the surfaces of Mars and Earth is quite different. The absence on Mars of plate tectonics means that a whole range of important geologic processes that occur on Earth are absent or comparatively ineffectual on Mars. Consequently, surface-modifying processes on Earth and Mars have different but overlapping sets of major geologic structures to work with. Absent on Mars (or dating from more than 4 billion years ago) are volcanic island arcs, major mountainous fold belts, and extensive granitoid terrains; abundant on Mars, but rare on Earth, are large impact craters; and ubiquitous on Mars, but restricted on Earth, is thick permafrost.

From a geomorphic standpoint, the most crucial difference between the two planets is the energy balance affecting major volatile repositories and the surface-modifying influences of these volatiles. This balance leaves Martian surface volatiles almost entirely in a regime of solid/vapor equilibria, whereas Earth is primarily in the liquid/vapor region. While a large snowstorm on Earth may deposit 50 cm of snow, an equivalent major dump on Mars, due to the low temperatures there, may deposit 0.5 mm of snow (that is being generous). Across some areas of Mars, especially the polar caps, this trace condensation of water vapor would be dispersed in a far larger amount of dry ice snow (a meter or more).

Liquid precipitation is improbable on Mars – not just now, but probably through its history. It is a simple matter that the Martian surface rarely ever rises above the melting point of ice, and when it does those warm events primarily occur in places where ground ice is not present at the surface. If melting of ice ever occurs on Mars – and there is evidence that it has on occasion happened in the past – then there is always excess ice to buffer its temperature at the melting point, and if the temperature of surface liquid ever rose much above the melting point it would boil away rapidly; the loss of heat during the boiling process would cool the remaining liquid to the freezing point. If we accept that liquid water occasionally occurs at the Martian surface, those events are apt to be rare, isolated, and short-lived, and involve small quantities of liquid held at the melting point of ice. If vapor-saturated conditions then exist above these isolated dribbles of water, the condensates would be in solid form, not liquid.

Whatever the details, it is clear that without major change in the climatic/atmospheric regime, no part of Mars could ever experience a weather-induced episode of rainfall – neither a torrential downpour nor even a brief misting rain is possible. It is thus understood why rainfall runoff features are apparently absent – or rare – on Mars. If any rainfall runoff features exist anywhere on Mars, then they are uncommon, but any such landforms would point to some period of a strongly modified climate or some catastrophic, transient event, such as prompt condensation from a hot steam cloud erupted from a huge volcano or emitted by impact.

From a geomorphic standpoint there is of course one additional crucial difference arising from the instability of H_2O in liquid form at the surface. Positive Martian degree-days are few and occur primarily in short contiguous periods of just a few hours and generally in low-latitude areas or on sunny slopes that lack any surface ice. Where they occur over ice, it is for such a brief period that a trace of water could be produced at best. Melting temperatures often occur at elevations where the surface pressure commonly is below the triple point; thus liquid water either does not exist stably on the Martian surface, or exists in such small quantities for such brief periods that surface-driven melting does practically no geomorphic work.

It is thus understood why the wide-ranging and ubiquitous influences of rainwater infiltration and runoff on Earth's continents are muted or absent on Mars. The default for Mars is that several other types of processes – volcanism, wind erosion and deposition, impacts, and internally driven mobilization of volatiles – all of which occur on Earth but compete with rainfall-related modification and plate tectonic influences – dominate the Martian geologic record. While it would require fairly modest and achievable amounts of climatic warming and atmospheric densification to change this situation on Mars, there is another ready source of liquid water, and that is from the geothermal melting of buried ice.

There is no question that some of the most significant geologic events on Mars are related to internal heating of volatiles. The formation of chaotic terrain and outflow channels, among many other features, are directly connected either to anomalous levels of subsurface heating near volcanoes or to big effects from normal background levels of crustal heat flow. No Mars geologists dispute major roles for internal heat in driving much Martian geology. Some geologists, including Vic Baker's group (which includes me) and Jim Head's group, see internal heating as a pivotal mechanism capable of triggering other momentous events, including flooding, ocean formation, climate modification, and glaciation. Some others see internal heating as being less effective or question whether such a link exists between internal events and the Martian climate and morphology of the surface.

Planetary heat flow is closely related to internal heat production. On Mars, there is no significant tidal heating; there are only two substantial sources of heat flow from the deep interior: (1) the delayed release of gravitational potential energy (due to the initial accretion of Mars, later hypervelocity impacts, and core formation and other events related to planetary chemical differentiation), and (2) release of heat produced by natural radioactive decay (including delayed release of heat from short-lived radioisotopes and that from long-lived radioisotopes of potassium, uranium, and thorium). The capacity of Mars to store heat deep inside and release it later is a significant and poorly known matter.

It is known that Earth is now losing heat that was generated hundreds of millions of years ago, or perhaps heat produced during the initial formation and differentiation of the planet. This is certain, because the amount of heat being released today exceeds by about 25% the steady-state production and loss of radiogenic heat, corresponding to a joule-averaged retention period of around half a billion years (Kargel and Lewis, 1993). The steady-state value of radiogenic production and loss probably is, in fact, close to steady state, because the radioactive isotopes are dominantly located in the crust, where it is more difficult to store the heat for long periods of time. The component of "fossil heat" probably then has a characteristic storage time somewhere in the billions of years range. It is not presently possible to say for sure where the excess component of heat is stored, but it is probably stored everywhere in the planet in some amount but dominantly within boundary layers that do not convect readily; a likely place is beneath the 600-km-deep seismic/density discontinuity that separates the upper and lower mantles. There is in fact some evidence that a substantial reservoir of excess heat is in the Fe–Ni core and the so-called D″ boundary layer of the lowermost mantle. Evidence suggests that Earth loses heat from these or other thermal reservoirs in an episodic fashion every few hundred million years, such that the excess component of heat increases dramatically for short periods, with consequences for increased volcanism and mountain-building activity.

We do not know if Mars stores and suddenly releases heat similarly with episodic increases in volcanism and geothermal activity, but one hypothesis suggests that it does. Although it is difficult to precisely quantify the excess component of stored heat, the steady-state geothermal component due to radioactive decay is much easier to evaluate, and in any case it is apt to dominate for a planet as small as Mars. As in Earth, Martian radiogenic heating components are probably mainly distributed in the crust, where time lag between production and loss is not too great. If we assume a steady state between heat production and loss, with no excess heat, then we effectively have a good lower limit on Martian heat flow. We have enough meteorite samples of Mars to know that K, U, and Th are present in Mars at roughly similar concentrations as in Earth, and so calculation of this lower limit to heat flow is simple. The numbers come out to about 30 mW/m^2. This is about three-eighths of Earth's average heat flow, meaning that thermal gradients in comparable materials (having equivalent thermal conductivities) will generally be about three-eighths as steep on Mars compared to Earth. Since radiogenic heating in recent eons is declining on both planets by about a factor of 2 every 1.3 billion years, these numbers suggest that on Mars, some 1.5 billion years ago (probably the early Amazonian), typical thermal gradients were about the same as on Earth today.

Mars would not likely have an average heat flow much below 30 mW/m^2, because this would imply that Mars has an improbable intracrustal mechanism to fight the transfer of radiogenic heat produced mainly within the crust. Regional differences in geologic history could result in contrasting values of heat flow with higher and lower than average values prevailing in any given region; the causes of this contrast are differences in crustal thickness, rock type, and erosion/sedimentation history, as well as lateral variations in and recency of volcanic activity.

The extent of aqueous activity driven by internal heating would have been more probable – and likely to have occurred at shallower levels – early in Martian history, when thermal gradients would have been higher than today, e.g., eight times greater about 4 billion years ago (three times the present global terrestrial heat flow). Despite likely cold temperatures at the surface, the Martian interior just a few kilometers down could have been even warmer than Earth is today at equivalent depths.

Today the picture is somewhat different. Generally surface temperatures and heat flow are too low to enable liquid water to exist widely except for traces of water at very great depths of 8 km (5 miles) or more. However, depending on the mineral composition and thermal conductivity structure of the Martian crust – related both to composition and the porosity structure – even today warm springs could gush water at the surface and spit steam in areas of volcanic activity or in areas of ancient lake beds where abundant salts can dissolve in the water and reduce the melting temperature. Thick glacial ice masses could be melted at their bases, particularly if they contain much carbon dioxide (which reduces thermal conduction). However, there can be no doubting that the internal heat engine of Mars, like that of Earth, is winding down. Mars has about a 1.5-billion-year lead on Earth due to its smaller size – a fact that has been appreciated since the time of Percival Lowell.

MARS: NOT SO FAR FROM HOME? NO FARTHER THAN NOME?

Mars has its un-Earthly qualities, but one of the most active arenas of Mars research today is the study of terrestrial landscapes, landforms, and habitats that serve as possible Mars analogs. Ten to 30 years ago the research fashion was to apply to Mars knowledge gained about basic terrestrial geological processes known also to occur on Mars, such as impacts, eolian processes, and volcanism at the level of single landforms, such as a sand dune, a lava flow, or an impact crater. Studies of glacial landscapes, fluvial erosion surfaces, and other large-scale landscapes, especially those affected by multiple processes, generally were not supported by the fashion of the day. Today a different fashion prevails. There is a new trend to study whole landscape analogs and active geomorphic process systems, especially if there are multiple interlinked operative processes and there is a connection to life's adaptations to special conditions there. The newly fashionable places to study emphasize biological and geological processes under extreme conditions of aridity, temperature, salinity, or other conditions that are not commonly encountered on Earth but are or might be encountered on Mars. Conditions at the edge of life's tolerances are of special interest. I shall offer here one more proposed analog landscape just to help to fill a gap in current Mars literature and scientific considerations of analogs. An example of this landscape type occurs in the Bering Land Bridge National Preserve (BLBNP), western Alaska – a region where permafrost and seasonal liquid water are both important. I describe this landscape below with Mars applications in mind.

Permafrost terrains involving almost an absence of melting are very rare on Earth. Sublimation-dominated permafrost terrains thus offer critical insights into Mars rarely obtained on Earth. The Antarctic Dry Valley region is one such environment, and with its stunning landform similarities to parts of Mars – especially rock glaciers – and its life in liquid saline lakes, it has been, appropriately, one of the most popular Mars analog sites. The Dry Valleys combine high aridity with cryogenic and in places hypersaline conditions, and yet they are geomorphically still very active and insightful with regard to Mars. More widespread occurrences of Earth's permafrost involve key roles of liquid water; these landscapes have not been so popular in the Mars science community, but my sense is that this resistance will be coming down as greater recognition of given to surface melt features on Mars. The possible role of an active layer has been much discussed in the framework of multiple working hypotheses for the development of single landform types, such as small polygons, but it was primarily on that limited basis, with rare exceptions, such as in the work of Costard and Kargel (1995). With new findings about Mars, we now have a strong basis to consider Alaskan-type permafrost terrains as a possible widespread Mars analog. It is important to explore the geomorphic accuracy and theoretical implications of this analog, because, if validated, it will imply summer surface temperatures at middle and high latitudes far more mild than previously considered likely. I do not offer here a test or validation of this type of analog, but I do offer further points to consider.

BLBNP offers a variety of terrains that in my view have many parallels to what we see on Mars in many areas. It is a region of continuous permafrost but where the upper meter (the active layer) of all lowlands are completely saturated in liquid water for a few weeks of the year. The BLBNP has several terrain types in close association. The first, a lowlands terrain, is what I call, for descriptive purposes, the Mosquito Bog Continuous Permafrost (MOBCOP) terrain. This is the most widespread continuous (distinct from "discontinuous," or patchy) permafrost terrain type on Earth and includes nearly all the Alaska Arctic Coastal Plain, the Yukon Delta and Yukon Flats, and other similar terrains in Canada and Siberia. MOBCOP terrain is characterized by a preponderance of low-gradient surfaces (tens of meters per hundred kilometers); low local relief (of order 10 m); summer temperatures conducive to extensive melting within a thin active layer ($\sim$50–100 cm thick); large accumulations of surface meltwater during the summer; seasonal water flow through sluggish surface streams; intense and widespread thermokarst development through thermal interaction of surface water with ice; and enhanced solar absorption in liquid water and, consequently, accelerated thermokarst development along lake shores and river banks. Sediment aggradation occurs in these terrains by stream overbank deposition during high-water periods and channel-bend (point-bar) deposition during periods when water flow is confined to channels. A key difference from non-permafrost fluvially dominated deltaic and coastal plain

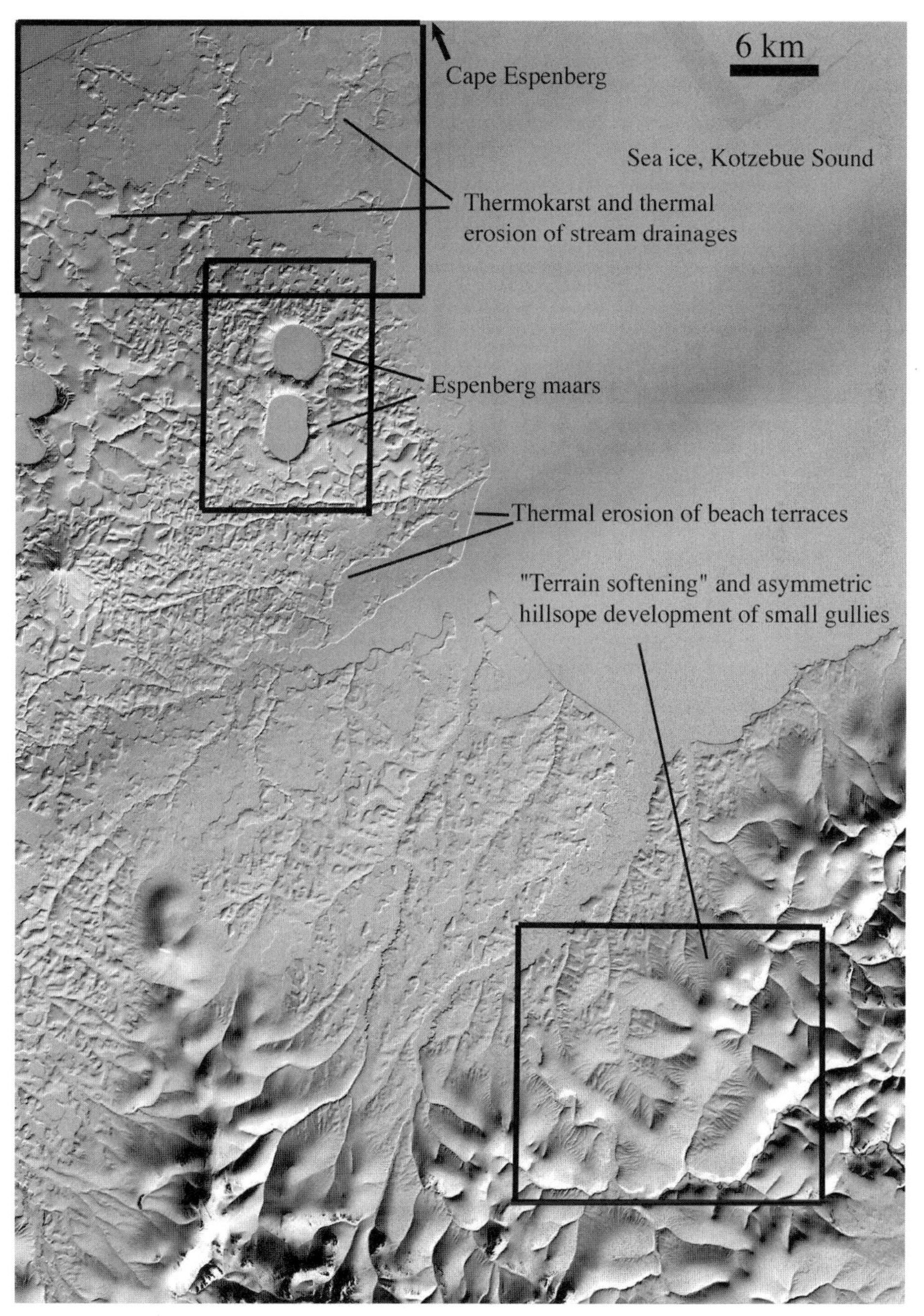

ASTER image, obtained April 9, 2003 under low Sun and uniform snow cover, highlights permafrost terrain features near Cape Espenberg, Alaska. The following pages offer details of box areas. Thanks to Rick Wessels for discovering this spectacular image.

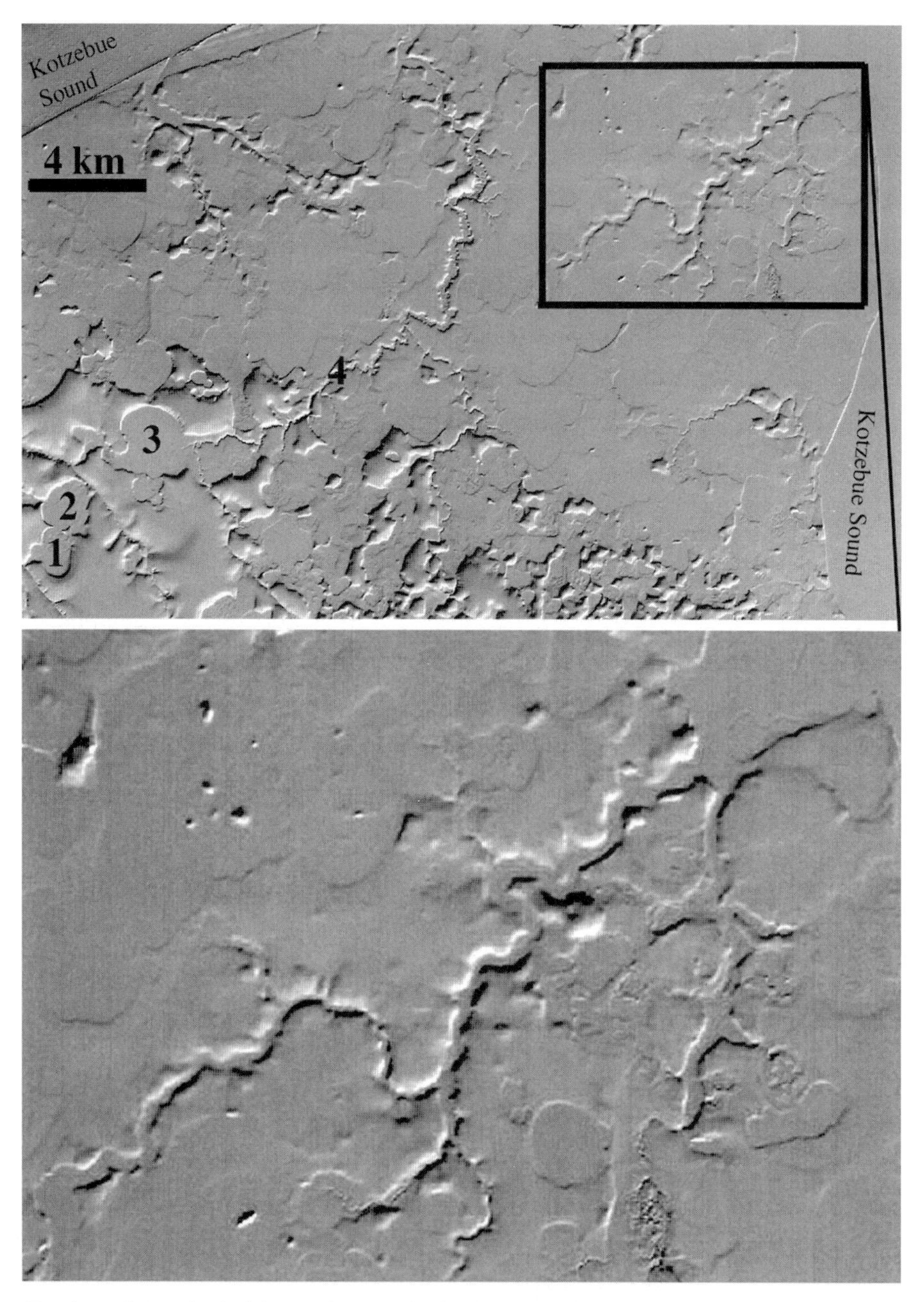

Portion of Level 1A ASTER image obtained over drainage systems in thermokarstic permafrost in Espenberg area of western Alaska. Top panel: Numbers indicate a hydraulically linked drainage system; inset box enlarged in lower panel, which shows a valley network hardly distinguishable from those common on Mars. Destriped by R. Wessels.

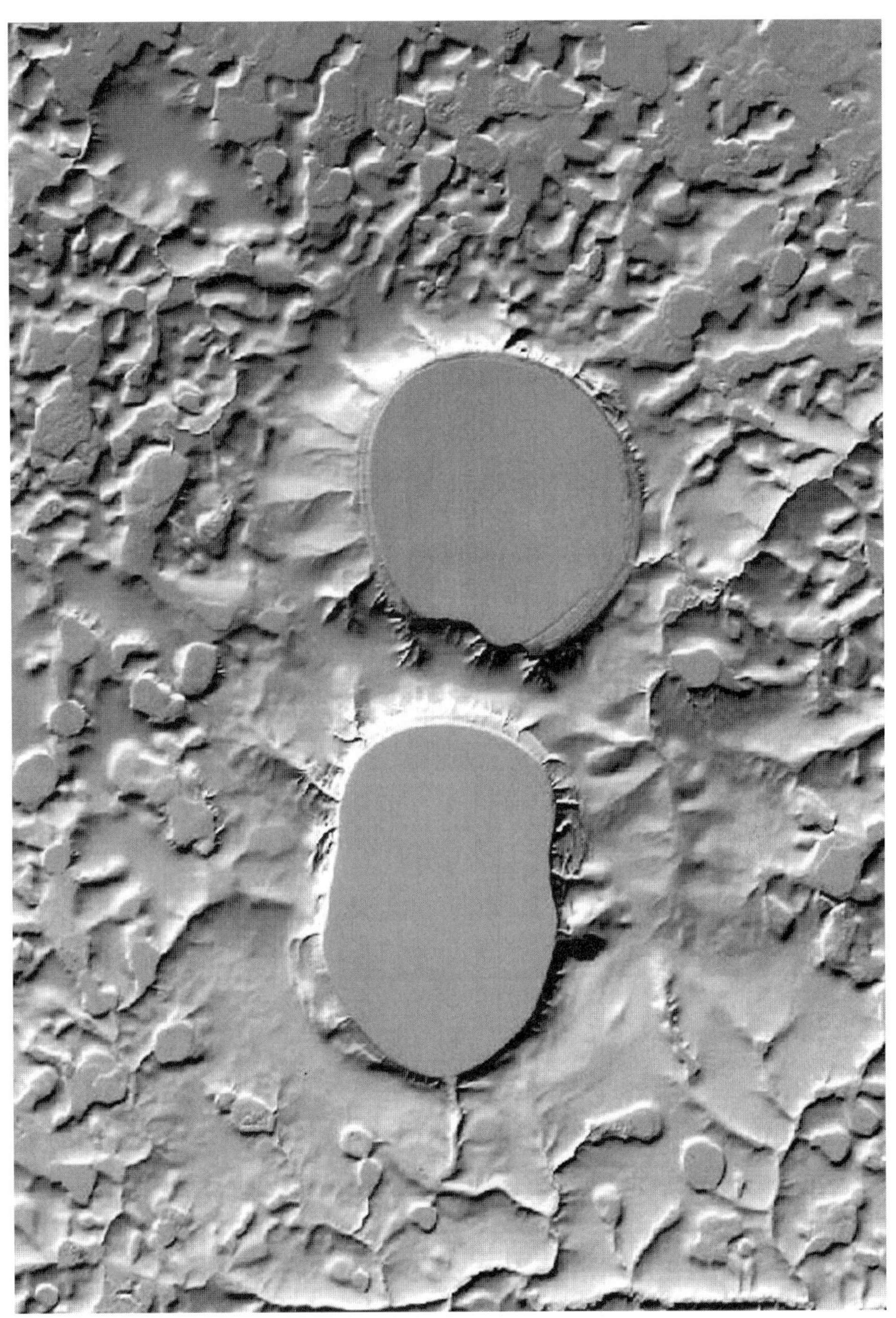

Two of the five largest Espenberg craters, the largest maars on Earth. Lower maar, 5 km long.

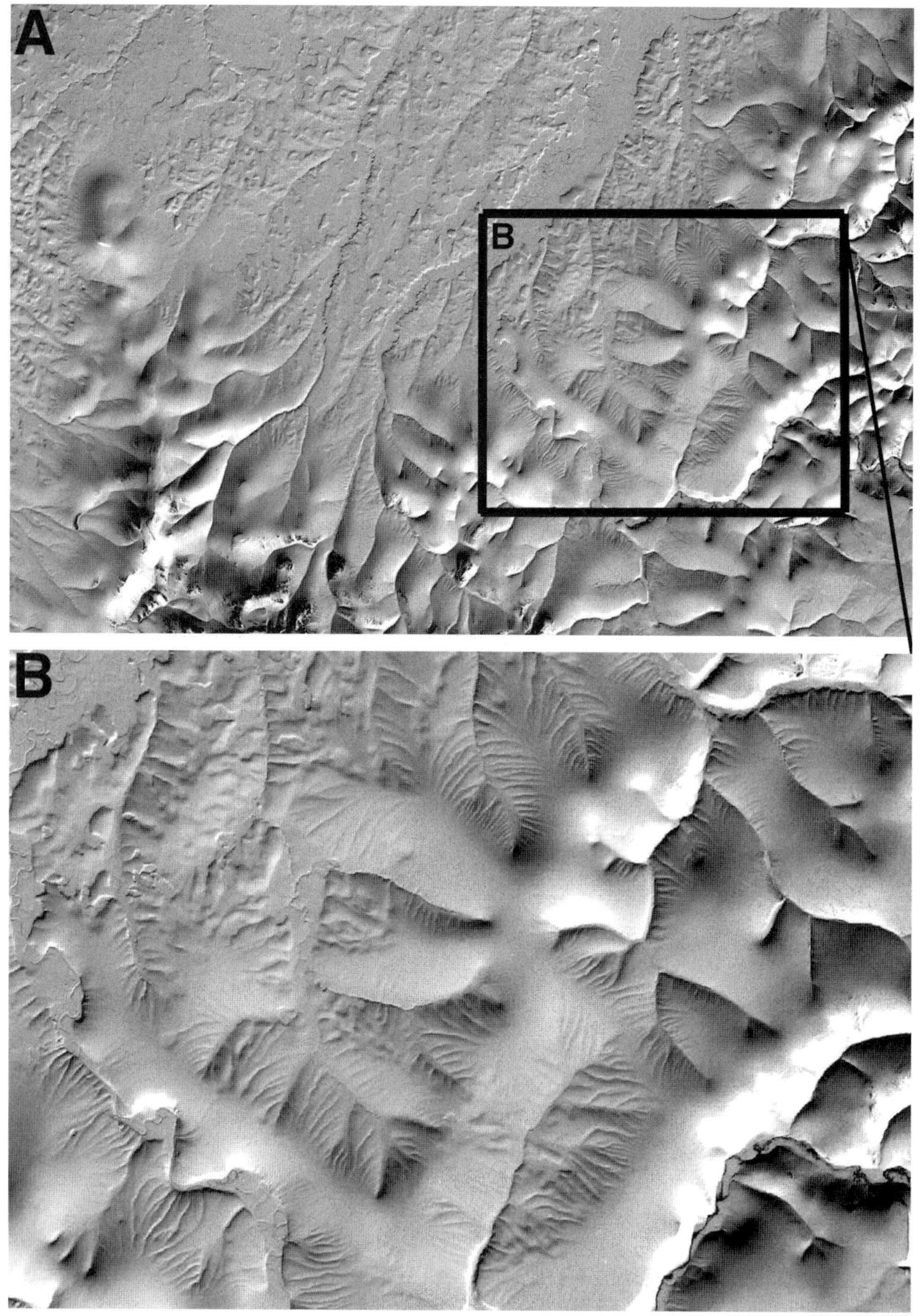

ASTER image, obtained April 9, 2003 under low Sun, shows permafrost "terrain softening" and a north–south hillsope asymmetry in the distribution of small gullies near Cape Espenberg, Alaska.

deposition is that ice is a major component of the deposited material, and ice cementation provides increased material strength and aids the development of steep bluffs. Thermokarst slope development involves thermal erosion, undermining, and collapse, and so steep, low-amplitude bluffs are common. MOBCOP terrains can be riddled with coalesced lakes, dotted with isolated remnant mesas, and dissected by wildly meandering river channels. These terrains can acquire some aspects of the appearance of glacial ice-stagnation topography, but lacking are bouldery moraines and eskers.

A second widespread permafrost unit in BLBNP, Softened Topographic Expression and Relaxation of Continuous Permafrost (SOFTERCOP) terrain, is an uplands terrain with much higher local relief (hundreds of meters) and consequently better drainage and far less accumulated surface water. SOFTERCOP terrain is thus thought to have less thermokarst development related to surface water bodies and a far greater domination of landforms produced by gravity-driven creep due to ice deformation and freeze-thaw. The rocks of MOBCOP terrains may generally tend to be depositional sands, silts, and clays of lakes and streams, whereas those of SOFTERCOP terrains tend to be regolith-mantled bedrock or coarser sediments such as river gravels, old bouldery glacial moraines, and rock glaciers. Although fluvial valleys and gullies are common, regolith creep features, including rock glaciers and other more subtle creep features, dominate SOFTERCOP terrain, producing a softened appearance.

In MOBCOP environments, permafrost ground ice develops by multiple processes, including: burial of frozen lakes, overbank river flood waters, and snow banks; thermal contraction cracking, seasonal meltwater infilling and refreezing to form ice wedges; segregation due to thermodynamically preferred melting of small-grained ice filling small sediment pore spaces and refreezing in larger grains, filling larger pores; brine/ice segregation; and formation of clathrate hydrates by injection of clathrate-forming gases into groundwater or ground ice. Thermokarst development exploits and etches these structures to form some of the major local surface relief and landforms. Ice-cored mounds known as pingoes occur where lakes and saturated sediment freeze from the top down, the bottom up, and from the sides inward, such that high pore water pressures and freezing of the last core of liquid water causes a bulging of the surface. SOFTERCOP terrains may contain less ice than lowland terrains such as MOBCOP, but ground ice and active-layer processes are important in causing hillslope creep, regolith convection, and particle-size sorting, thus yielding stone polygons, stone circles, and stone stripes.

The BLBNP exhibits a variety of features produced by volcanic and geothermal interactions with icy permafrost. Such processes and features have barely been investigated here on Earth, but their relevance to Mars ensures that increasing attention will be paid to this topic. Kargel et al. (2004) have presented an overview of volcanic interactions here, including formation of the largest maar craters known on Earth. These craters were produced by powerful volcanic explosions in permafrost. One possibility is that methane clathrate was involved, with high explosivity caused by either the high vapor pressure of methane or the combustion of methane. If due to combustion, there would be no analog on Mars, but if due to the high vapor pressure of the clathrate guest molecule, carbon dioxide clathrate could take the role of Earth's methane clathrate and result in out-size maar craters. Because of the lower gravity on Mars, maars as large as 10 to 15 km in diameter could be formed.

The biological richness of Alaskan tundra almost surely has no analog on Mars, and whether there is even a microbial community is a big question. However, the possibility of any life at all on Mars brings us to that issue that so intrigued Percival Lowell, as it does our generation ... the biological life of worlds. The final chapters will explore some issues of life on Mars (past, present, and future) and its meaning for life on Earth.

8

Living Mars

"There is almost certainly life on Mars."

Charles Cockell, 2003, in explaining that spores of *Bacillus subtilis* are known to remain viable in Earth orbit for up to six years. Spores of this bacterium almost certainly accompanied the *Mars Polar Lander* from Earth to a low-speed crash landing on Mars in 1999. The spores are there on Mars already. Their viability is questionable.

MAGNIFICENT DESOLATION

Across much of the crazy quilt of human history, two common intertwined presumptive threads wend their way through many cultures: the Earth-like nature of worlds beyond our own and the possibility that life may thrive there. *Apollo* explored a realm utterly barren, unspeakably lifeless, down to the merest molecules of microbes, except for what Humanity's technology has transported there. The lunar reality defies the suppositions of millennia, but only for one dead world. The Moon's "*magnificent desolation*," remarked upon from the lunar surface by Buzz Aldrin, second man on the Moon, is in stark contrast to the bounty of Earth, where every nook and cranny is home to living creatures. Does the lifelessness of our Moon in any way reduce chances that somewhere else out there has Earthly fecundity? Considering the conditions on the Moon, its sterility does not have much bearing on modern concepts of life's possible realm. Where there is no water, there can be no life. So why draw any lessons from any place as parched as the Moon?

Taking a longer view, *Apollo* was a stake in the heart of the ancient speculations that life is ubiquitous in the Cosmos; at least some worlds are unsuited for any life unsustained in artificial habitats. The fact that lunar rocks and *Apollo* astronauts were quarantined, while somewhat silly even then, highlights the serious consideration given until 1969 to the possibility of life on the Moon. (Never mind that the leaky quarantine was a real joke if virulent alien microbes somehow were accidentally returned with lunar samples.) The more pertinent point made now by the astrobiology science community is that another world might possess life if it is more hospitable to aqueous media and complex organic chemistry. A common modern assumption is that if liquid water can exist in any reasonably stable and long-lived planetary environment, life is apt to have originated and evolved there. While there is a scientific basis for this view, it is still a slim basis despite all the work. We should recall a parallel idea that held sway not so long ago: if there exists a solid piece of ground, then 'people' (of sorts) must walk those worlds.

I am somewhat of an agnostic on endemic extraterrestrial life (Martian or otherwise). It has nothing to do with chemical free energies, carbon molecules, impact extinctions, or anything deeply theoretical. It is simply that we have yet to witness the origin of new life solely from abiotic material, either naturally or artificially. In fact, there is no evidence that life has arisen independently more than once in the whole history of Earth. The possibility that the first life on any planet aborts or eats all who try to come later means that the single point of origin of all terrestrial life is not an argument that we are alone in the Universe. But we *could* be alone, a fact also allowed by the lack of documented visitation of alien creatures on Earth (the Raelians and Cydonia Clan notwithstanding). Stephen Webb, in *Where Is Everybody?* (2002, Copernicus Books with Praxis Publishing), considered 50 possible reasons why we might be alone or seem to be alone. Webb considered a true solitary uniqueness of humans in having high intelligence as the probable explanation for the evident absence of alien visits to Earth. About the widely supposed inevitable rise of prokaryotic life and its evolutionary adaptation and advance to eukaryotic life and eventually biological intelligence, Webb wrote,

> *It is a comforting idea, but I know of no definite evidence in its favor, and I believe the silence of the Universe argues against it.*

Despite this uncertainty, questions about extraterrestrial life rise above all other scientific questions that planetary scientists can hope to answer.

Peter D. Ward and Donald Brownlee, in *Rare Earth* (2000, Copernicus Press), also argued that at least advanced multicellular animal life may be a rare phenomenon in the Universe. Their approach was that many conditions that have promoted the evolution of life on Earth, and then of advanced life, are rare in the Universe. Advanced terrestrial life on Earth, according to the authors, depends even on the proximity of our lifeless Moon (to limit Earth's obliquity within moderate bounds and thus limit seasonal climate fluctuations and freeze-out of volatiles at the poles) and the debris clean-up caused by distant Jupiter. The special life-promoting roles of plate tectonics as a volatile and nutrient recycler and continent builder, and the high terrestrial abundance of water compared to that of Mars, are just a few more of many special circumstances, it is argued, that have yielded a very rare and prolific living Earth.

While the perspective of Ward and Brownlee is intriguing, and some of their arguments are certainly compelling, many arguments are not well justified going either way in the discussion. In total the case for and against extraterrestrial life, and especially intelligence, seems weak; but these discussions are always thought provoking. For instance, the stated low abundance of water on Mars is probably not accurate; granted, most of the water is frozen, but Mars may be simply awaiting warmer times from a steadily brightening Sun. Earth's plate tectonic mechanism for recycling volatiles and nutrients and in producing surface relief is without doubt advantageous to life, as cited by the authors; however, plate tectonics is neither likely to be a rare thermal regulatory process for large rocky planets, nor is it the only way to make for effective volatile/nutrient recycling and relief building. Certainly, as our Solar System shows so well, there is no causal link between volcanism and plate tectonics, as Ward and Brownlee almost imply; plate tectonics controls the style and organization of volcanism on planets where plate tectonics occurs, but a shutdown of plate tectonics would not shut off volcanism and all that that means for the production of surface relief and volatile recycling. The role of a large Moon in stabilizing the axial tilt of Earth, while possibly critical for Earth, is not the only common mechanism for having stable obliquity, and the argument about Jovian cleaning of debris can be turned around when it is pointed out that Jupiter is largely responsible for littering the Solar System with millions of fragments of would-be planets that failed to accrete due to Jupiter's influence. The more encompassing argument of Stephen Webb, that we may be alone (not merely "rare") in the vastness of the Universe, is similarly filled with dubious estimations of probability. In fact, the uncertainty of all such estimations of life's frequency is the very point of these authors. Such arguments will not still the debate, but will only to add to it.

NASA's theme, "Follow the Water," is the root of what will naturally evolve as a two-phase strategy to finding life on Mars: First, map the modern and geologically past occurrence of liquid water on the planets, then focus the search for life there (past or present), using our knowledge of water's occurrence as a sort of treasure map. For Mars, we are in the midst of the first phase. At some future, undetermined point we will shift to the second phase. The second phase might begin in Mars exploration around 2015. However, a possible reflight of the *Beagle 2* organic chemistry experiments could speed a transition to Phase 2.

While "Follow the Water" is an efficient precursory approach to finding (or ruling out) life on Mars, it is crucial to recognize that water is not directly correlative with life; some watery realms might not be inhabited or even potentially habitable. Any of several causes could make wet environments unsuited for life: (1) insufficient longevity or stability of wet conditions (e.g., transient flood outbursts are unsuitable for sustenance of life); (2) the composition of the environment may be too toxic, saline, caustic, or acidic; (3) the diffusion of nutrients/wastes may be too slow in low-temperature, high-viscosity superbrines; (4) carbon, phosphorus or other essential elements might not be available; (5) the chemical system may have run down toward thermodynamic equilibrium (= death), and sources of exploitable biochemical energy may be too feeble or unreliable. Even in the case that conditions are ideally suited to life's origin and maintenance, considering that we do not yet know precisely what circumstances are needed for life's origin, or how kinetic factors and time controls the origin, we cannot conclude that life is inevitable given the presence of liquid water.

The modern scientific quest for life on Mars, or anywhere beyond Earth, has some hallmarks of a mature "hard" science. It is highly quantitative and relies on state-of-the-art technology. However, unlike nearly all other sciences, Astrobiology shares a compulsion as narrowly focused and elusive as a life-long quest for the idyllic love. Consider the attention given compared to the rarity of the subject matter. Earth's biomass totals just 2×10^{-10} of Earth's mass and 7×10^{-16} of the Solar System. Earth's biomass (the only biomass known to exist) as a proportion of the Solar System is 200,000 times less abundant than gold. It is easy to see in more ordinary human love some similar devotional aspects shared by the Astrobiologist's yearning to find even a trace of life's precursors in comet dust or its remains in the desert sands of Mars. Sting sings of his "Desert Rose" (1999, A&M Records),

> *I dream of rain, I dream of gardens in the desert sand ...*
> *I dream of love as time runs through my hands ...*
> *I lift my gaze to empty skies above ...*
> *I close my eyes, ... Sweet desert rose... .*

The Astrobiologist's search in the skies above for nothing more than a microbe in otherworldly sands, even a microbe billions of years dead, is not so different from Sting's quest for something rare at best, and known, so far, only to exist in the mind.

THROUGH A GLASS DARKLY

When Galileo started his astronomical observations in 1609, the ideas were already firmly rooted in Western science that the planets are worlds, some like our own and some not, and that the vast real estate in outer space is not entirely barren. Bruno had been burnt alive just years earlier for claiming that other worlds exist and that some may be habitable. Galileo, heedless, turned his telescope to the heavens and reported some amazing things in *Siderius Nuncius* ("Starry Messenger"). Among Galileo's most noted findings was that Jupiter has four satellites, which clearly pay homage to Jupiter. This discovery seemed to place Earth and Humanity somewhere outside the center of Creation. Galileo's heresy that the Sun is blighted by sunspots was the last straw. Surely God would not place Earth anywhere but at the center of everything and bless our world with perfect light. How could this light, which was good enough for God and synonymous with God's wisdom and his very being, not be good enough for Galileo? Or perhaps the concern was that if the reported imperfections are real, then what next would Galileo say about the ruling religious dynasty? Perhaps this new invention clarified flaws in the ruling dynasty as much as it magnified the heavens. Retirement of Galileo and his telescopes out of view and out of contact was the only option.

Galileo's discoveries of worlds unbonded to Earth and Humanity was, however, already public. Bruno's and Galileo's generation was a time of changing intellectual fashions and rising freedom; astronomy was not to be suppressed for long (too long for them, however). The word was out, and anyone could take a spyglass and look heavenward, seeing some new-found element of Truth, thus exposing the ignorance of the established order.

Profound scientific discoveries are always followed by a quest for the greater significance of the findings. In 1638, John Wilkins, an English science writer, based some new ideas largely on Bruno's, Galileo's, and Kepler's publications. Going somewhat beyond their published work but perhaps reflecting discussions he held with professional astronomers, Wilkins wrote, "*The Discovery of a World in the Moone: Or a Discourse Tending to Prove that 'tis Probable that there may be Another Habitable World in that Planet.*" Safe in England from the Catholic Inquisitions, Wilkins speculated on human space travel, believing that it would happen some day, though he recognized that new technology beyond sailing ships would be required.

As knowledge of other planets increased, philosophical attention to the ultimate purpose of these other worlds repeatedly highlighted the probability or inevitability of life beyond Earth. The heavens are universally seen as a source of inspiration for scientists, engineers, theologians, storytellers, and every human who has ever looked skyward and wondered what it would be like to venture there, and then pondered deeper questions, such as *Who am I? Why am I? What is my fate? How are we relevant amidst this Vastness?* The prophet Isaiah wrote, "*Lift up your eyes and look to the Heavens: Who created all these?*" In this twenty-first century AD, scientists and the human population at large still look to the heavens for insights into who we are, and we find mostly humility and the grandeur of unbounded mystery.

In the seventeenth century, these questions became solid points of scientific inquiry, but the spiritual aspect will always remain. Christian Huygens, in *Cosmotheoros* (1698, published posthumously), wrote

> *... all the Furniture and Beauty of the Planets ... seem to have been made in vain, without any Design or End, unless there were some in them that might at the same time enjoy the Fruits, and adore the wise Creator of them.*

This statement well summarizes a widespread belief that life *must* exist somewhere out there. Or else, wouldn't it all seem to be a Cosmic Waste? Huygens' statement may have been partly a pre-emptive first strike against theologians who could have attacked his science, but it also represents the intellectual freedom that had spread almost throughout Europe by the end of the seventeenth century. That freedom was largely won by the sacrifices in blood and civil liberties by astronomers earlier that century, but those sacrifices would have been for naught had their science not been both sound and substantial.

Huygens' sentiment was not unanimous. A more anthropocentric interpretation of the meaning of the Cosmos was offered by Immanuel Kant, "*without man, the whole of creation would be a mere wilderness, a thing in vain.*" The idea that Nature derives value from Human use and appreciation goes back at least as far as the writing of *Genesis*. No matter what is out there, we and our world are minuscule in the vastness of the Universe, a thought that has resounded in philosophical and theological literature around the world. Over a century after Huygens passed, a Scottish mathematician and theologian, Thomas Chalmers, wrote,

> *The Universe at large would suffer as little, in its splendour and variety, by the destruction of our planet, as the verdure and sublime magnitude of a forest would suffer by the fall of a single leaf.*
>
> (*Astronomical Discourses*, 1817)

The end of life on Earth and of Earth itself is about as dreary a thought a human can contemplate, but Chalmers' statement offers unending hope that, even in the worst of all possible calamities, something vastly more magnificent would survive. Chalmers' statement of faith in life and beauty beyond Earth still offers me much solace in our current problematic times. The reality of what exists – that which we admire and, even greater, that which is beyond our observation – is precious beyond the value of all on Earth. It is consoling to believe that the wonders of the living Earth are equaled or exceeded somewhere else. It is heart warming to consider that some consciousness now or in the future will trace our interstellar *Pioneer* and *Voyager* probes back to our civilization's fossilized remains and perhaps honor us, or at least appreciate the vaster Universe, as we do.

Ironically, we love life and the Universe so much that Humanity shall survive vicariously, regardless of our individual fates, through the survival of beings – our brothers in consciousness – whom we assume exist, but from whom we have never heard a whisper of their existence. We call it science, a study of the natural marvelous living Universe ("Astrobiology"). In fact, until we hear that alien whisper, or are greeted by their handshake or roasted in their ovens – our science is no different from religion in this respect of faith.

The thought of this possibly being an inanimate Universe, aside from ourselves and other Earthlings, is disconcerting. There is Huygens' point about a wasted Universe, but even darker would be the thought that, in all the Universe, Humanity – with its ugliness and incompetence to run even one planet – is the closest thing to God, though so clearly falling short of God. Knowing our fallibility, "god-like" is a descriptor few of us would wish to have. Far better it is to believe that there is something greater and wiser.

Alien life would not be the God we seek, but with any luck, maybe there will be greater wisdom nonetheless, perhaps a lower-case god of sorts. And so the concept of alien life pervades much of our modern science-derived thinking. It is scientists' neo-mythology following in the footsteps of the ancient Romans and Greeks and Aztecs. The nobleness of the search, and the ideality of what we seek is well expressed in Carl Sagan's book, *Contact*, and the film that followed. The difference is that in all the past Humanity was unable to ascend to the gods except in the afterlife; now we are poised – possibly – to find these gods either by extending to them an electromagnetic invitation card or by ascending directly to them. What a disappointment, I suspect, we shall find them to be, if we find them. Whatever their faults and frailties, with their evolution by competitive selection of the fittest and most aggressive, they may be no more perfect than ourselves – comparatively ruthless and weak of spirit, but designed to survive. If ruthless, that may be our last conscious thought; if weak in body and technology as well as spirit, God help them fend off Humanity's aggressions. Well, perhaps we shall find our better and stronger selves.

Despite the best efforts of humans to contact or eavesdrop on possible extraterrestrial intelligent beings though projects such as SETI (Search for Extraterrestrial Intelligence), it may well be not human efforts but extraterrestrial efforts that make the contact possible. The thinking is that extraterrestrial intelligence is probably either far less technologically advanced than we – hence unable to hear our electromagnetic calling cards or to make their own probings – or far more advanced.

Speaking about our effort to talk to possible extraterrestrial intelligence greater than human intelligence, *Apollo 11* astronaut Buzz Aldrin summed it, "*If we find out they exist, it will be because of their communication with us*" (NBC *Meet the Press*, February 2, 2003).

The struggle for survival, felt over the bankbook and on the battlefield, is an instinct shared equally by monkeys and microbes. This incessant struggle, however manifested, is a quality of life, a prerequisite for Earth's DNA to propagate successfully against the competition. Competitive aggressiveness is apt to be a universal quality of life, regardless of the molecular basis of life. One can say the same of the inanimate Universe, where Black Holes, without willful desire, gobble up the stars and one another; where forming stars gobble up gas, and then hydrogen atoms eat one another and form helium and heavier elements; where snow crystals in glaciers consume one another in the annealing process, leaving the last survivors larger and more robust; and where, ultimately, thermodynamic death prevails over all animate and inanimate existence. Life just happens to edge out abiotic processes in their fragile domains in terms of aggressive capacity to consume and propagate.

In the face of this gloom, human striving for individual and collective significance is a part of this struggle against irrelevance. The human struggle goes well beyond the universal competition against predators, decay, and one another. The human struggle for significance extends to a quest for the Universe's origin, fate, and meaning. As we search for the significance of the Cosmos and our place within, we extend our own significance across time and space. This behavioral trait is at the root of religion, philosophy, art, and science the world over. On Earth it is arguably a uniquely human quest, our special way to sustain ourselves against that component of our being which is inanimate, unspiritual, and fated to oblivion.

The concept of a God-given soul that can transcend all the ugliness and hopelessness that pervades even the most beautiful things in nature is the ultimate device to immortality. Paul, in his first letter to the Church at Corinth, wrote of his search for God and for his soul in the same thought,

> *Now we see through a glass darkly, then we shall see face to face; now I know in part, then I shall know fully, even as I am fully known.*

Paul was writing in earnest faith that Truth is accessible and will reveal the greater purpose in ourselves, though it may take death and the afterlife to gain this knowledge. This single faith of Paul and Huygens, though expressed to different groups of people in different ways, reveals a timeless need of humans through the millennia to see what is truly relevant and to be relevant. Human beings have struggled within themselves and against one another in attempts to discover this Truth and to disseminate it either by persuasion or force. A key postulate underlying this struggle is that there actually is a Universal Truth to be found, and, once discovered, that it is all important, worth more than mortal life itself. In one form or another scientists also cling to this faith, though scientists usually express it differently than theologians do.

ONE: THE LONELIEST NUMBER

To some, it seems absurd – or just anguishing – to think that Earthlings even could be alone in the Universe. Perhaps so. However, there is nothing learned in modern science that tells us accurately what the probabilities are that life should arise. We only know that life developed at least once in all the vastness of the Universe. No one can begin to evaluate whether other Universes, if they should exist, may harbor life, since nothing is known of the physical laws or existence of those possible Universes.

If there is extraterrestrial life nearby in this Universe, we should be fortunate if it either is benign to terrestrial life or incapable of reaching us aggressively, or if our search for this life is so unsophisticated and our civilization so insignificant or unuseful that we should be ignored. Expressing a slightly different view, Isaac Asimov wrote,

> *There are two possibilities. Maybe we're alone. Maybe we're not. Both are equally frightening.*

It is likely that Earth is the sole planet in this Solar System, and across the expanse to nearby stars, having intelligent, technologically advanced life. The Fermi Paradox refers to alien space farers, their lack of documented arrival on Earth, and the rhetorical question, "*well, then, where are they?*" This answered question ("they seem not to be here") is a powerful argument against extraterrestrial intelligence anywhere nearby but beyond Earth, but it loses some potency when the vaster realms are considered along with the possibility that intelligent life is rare but scattered across the Universe. True, we have the Cydonia clan still claiming that those interesting rock formations on Mars are more than geologically sculpted rocks. And the point of Stuart Clark (*Life on Other Worlds and How to Find It*, Praxis–Springer Publishing) is well taken that no one alive can prove that there is *not* an extraterrestrial intelligent presence within our Solar System, right now, or even here on Earth, either monitoring us, testing us, or ignoring us. While some cultists claim that humanity has extraterrestrial origins, this supposition can make sense only if that origin is pushed back almost 4 billion years to the first life, from which all subsequent Earth life has evolved from common stock. Even so, it is far more likely that, if there is any extraterrestrial life in our Solar System at all, it occurs as microbial life within the Martian subsurface or other warm, wet places, such as Europa's ocean (Kargel et al., 2000).

Whatever Truth is Out There, our search for extraterrestrial life is very much a search for our own identities and the fundamental meaning behind it all, as the Ancients expressed well. Although we have fabulous new knowledge of biochemistry and extraterrestrial environments, the answers are no more evident now than they were millennia ago.

For ecologists, the search for extraterrestrial life is a search for greater beauty and intrinsic value of our planet's own biosphere; the recognition of value in the humblest of our fellow creatures magnifies our own existence by allowing us pleasure and a sense of wonderment, and this recognition promises beauty and truth that will last beyond our own brief period within the grander Universe. For geologists, the search for extraterrestrial life is a search for the beginnings of life on our planet and the documentation of an evolving world here on Earth. For astronomers and cosmologists, the search for chemical clues to life on other planets and of signals from extraterrestrial intelligence is as deep as a quest for God. For all of us, it is a search for ourselves.

While science is about finding answers, scientists revel as much in questions. Even better than antiseptic logical questions and answers are the deeper mysteries, which are shared equally by theists. Albert Einstein, a man of the most profound mathematical logic, wrote in *The World as I See It*:

> *The most beautiful experience we can have is the mysterious. It is the fundamental emotion which stands at the cradle of true art and true science.*

The newly christened academic discipline, Astrobiology, and a new scientific journal by the same name, is dedicated to "*advance our understanding of life's origin, evolution, distribution, and destiny in the universe.*" The technologies and scientific methods of investigation highlight this as a serious field of scientific inquiry, but ultimately it all comes down to the mystery we all seek to solve.

NOTHING BUT PIGEON WHITINGS (OR IS IT THEORETICAL ANTICIPATION?)

This human quest for life's context, shared by billions, also means that one doesn't need a PhD or ThD to ponder vast questions and arrive at immense answers. Witness the uncanny prescient pronouncements by nonscientists throughout the ages regarding the nature of our Universe. These thinkers and their thoughts often transcend religion, philosophy, art, and science. Sometimes the inquirers are impatient with the conservatism of science. Sometimes these people think expansively simply because they wish to participate in discussions that commonly seem to scientists to be beyond a layperson's education and abilities. Commonly these ideas are far off track and lack rigor; their persistent proponents exhibit single-minded dedication to prove a hypothesis to the exclusion of all other possibilities, and they never admit evidence to the contrary. These people become an annoyance to us who are attacked again and again for not giving due consideration. Scientists encounter these wanna-be scientists all the time, and we identify them in seconds by a telltale lack or improper use of jargon, and their simple errors and mischaracterizations of "legitimate" science. On occasion, though, the amateur questions asked – and amateur answers proposed – leapfrog spectacularly beyond the incremental advances offered by the slow grind of science.

Sometimes, for lack of the bias and inhibitions that develop during scientific education, these ideas hit to the heart of significant matters, even if they also may be laden with errors. There was Wilkins' suppositions about future human space travel over three centuries before it happened, based on insights from sailing explorations of the Earth, but written in knowledge that spacecraft would not be propelled by wind or conventional sails. Philosopher Immanuel Kant in 1755 described (in

General History of Nature and Theory of the Heavens) what 41 years later would be developed by Pierre Simon de Laplace (*The System of the World*) as the modern nebula theory of the origin of the Solar System.

Edgar Allen Poe, though not known for his science, in 1848 wrote of the Big Bang in his 150-page prose poem, "Eureka," exactly one century before physicists independently arrived at this concept, and gave the theory that popular title, from science-based theory. Poe had no clue how to prove his idea, but proof came eventually from a survey, by Arno Penzias and Robert Wilson, of the sky's microwave noise. Penzias and Wilson first considered the background radiation to be an artifact related to accumulations of pigeon excrement on the radio telescopes. Ruling out pigeon dung and other potential sources of error, the explosive birth of the Universe was announced, much as Poe had surmised would happen.

Wilkins and Poe offered no means of validation, so their writings cannot be considered science, but rather theirs were works of brilliant work within the gap between science fiction, science, and art, but being not quite any of these. When amateur scientists hit home, such models of the physical Universe and its contents, though commonly based on the slenderest observations, are constructed with gifted insight and pure human imagination trained by everyday experience on planet Earth. Based on "theoretical anticipation," these ideas are what Tom Siegfried (author of *Strange Matters: Undiscovered Ideas at the Frontiers of Space and Time*, Joseph Henry Press) calls "prediscoveries." Eventually, scientific observation catches up and either confirms or – more often – refutes such ideas, which anyway are normally forgotten until they are, if ever, proven correct. And so today we know that Mars is a world both like and unlike our own, and we know there are billions of billions of others, as Epicurus told us 23 centuries ago. We know also that Lowell's canals were purely imaginary, but that Mars is nevertheless a water-world companion of Earth, a fact that Lowell appreciated, even if wrongly.

For several millennia, and more today than ever before, Humanity has been in theoretical anticipation of the discovery of extraterrestrial life. Until late last century there was no realistic hope of proving this conjecture, unless that life were to come to us or loudly announce its presence. In the words of Bruce Jakosky (*The Search for Life on Other Planets*, 1998, Cambridge Press),

> *The exciting aspects about the question of the existence of extraterrestrial life, though, is that it is a question that has an answer, and we have the capability of finding that answer. What would it mean to us as a society to discover that there is extraterrestrial life? Conversely, what would it mean to search for it, and realize that life on other planets may not exist?*

The prospect of a Universe that is lifeless, save for us Earthlings, cannot be validated without a complete search of the Universe, and so the matter of whether there is extraterrestrial life can be solved this century only if the answer is affirmative. Hence, the search for extraterrestrial life to scientists is a bit like the Second Coming of Christ is to Christians, or the first arrival of a savior to Jews and Muslims; many anticipate it, some even actively manipulate events that would somehow force it to happen, but until it happens there is no proving – only faith – that it will happen.

Scientists today announce that we are at the threshold of discovering what has been elusive for millennia. And they just might be correct.

Of course, so far I have avoided even defining what life is. We all know what life is when we see it, at least Earthly life. Nevertheless, there are many descriptions and definitions of life. Most hold that life is capable of responding to the external environment, metabolizing, growing, and reproducing. G.F. Joyce and L.E. Orgel in 1998 defined life more simply as chemical systems that are capable of Darwinian evolution. While this definition catches much of current scientific thought, it may not admit possibilities of life that are evolutionary dead-ends right from the start – simple life that may be capable of thriving under certain stable environments but somehow is coded right from the start in such a way that any chemical perturbation or change in the machinery of the life form destroys its ability to remain alive. Nevertheless, most life probably fits the Joyce–Orgel definition, which probably also adequately rules out chemical systems that would not likely be regarded as living by any other definition. It also renders concepts of life into a distinctly chemical–physical framework, which may not satisfy all theologians but certainly does make the topic prone to laboratory investigation.

GOD'S RECIPE ON THE INTERNET

As I started writing this book it could be said that there was no known successful recipe for life; there is only a set of basic ingredients, genetic codes for a few organisms, and a set of conditions under which we know that life forms on one planet – perhaps 15 million species – are able to survive, grow, and reproduce. But no one could take this information and generate growing, reproducing organisms from the unliving. This may no longer be true, by some definitions of life. In the summer of 2002, Eckard Wimmer (State University of New York at Stony Brook) and colleagues built a polio virus starting from the published code (obtained off the internet), basic chemicals, and rudimentary DNA components available commercially. They constructed the genetic code from the theoretical knowledge of the code, and then the DNA was converted to the RNA of the polio virus. Wimmer provokes, "*We don't need any nature-formed template anymore. We just need the Internet to tell us the sequence of a virus.*" So made, the researchers inserted the polio RNA into a chemical brew mimicking the protoplasm of living host cells. The artificial polio viruses reproduced. The offspring were infectious in laboratory animals, though not quite the same as natural polio (*Science News*).

Viruses are naked genetic codes, constructed of four amino acids, stripped of the nutritive environment, machinery, and protection of cells. They require host cells to perform maintenance services and to provide food. By some definitions, viruses are life forms, and by others they are not fully living, not even parasitic or symbiotic life forms. Whether artificial life was created by Wimmer, using nature's blueprint, is a question of definitions.

Two centuries ago it was believed that organic molecules are produced exclusively by living organisms. In 1828 F. Wohler found that urea, an organic molecule of critical importance to biochemistry, can be synthesized artificially. The abiotic natural chemical synthesis of the first life forms was discussed early in the twentieth century, but understanding life's origins seemed then a remote possibility. The Miller–Urey synthesis of amino acid building blocks of proteins by spark discharge in simple reducing gas mixtures (Miller, S.L., 1953, *Science*, **117**, 528–529) was a seminal early successful experimental study of abiotic organic synthesis relevant to life's origins. Since then there has been extensive consideration of the natural physical environment of the formation of the first life forms on Earth and elsewhere. We have learned that all life on Earth is constructed from a fairly simple set of basic building blocks: about 20 amino acids, 5 nucleotide bases, and a few sugars and other simple compounds bonded and arranged complexly.

Theoretical study and laboratory experimentation has mostly considered various marine environments of life's possible origins, including evaporative lagoons and deep-sea hydrothermal vents. Many diverse natural abiotic synthetic pathways to life have been considered, most of which emphasize some means to concentrate simple organic chemicals and catalyze formation of more complex molecules. The role of phyllosilicate clay–mineral grain surfaces, pyrite surfaces, and concentrated aqueous solutions have been widely considered as the particular points of origin of life or life's chemical precursors.

While ionizing radiation is damaging to highly structured organic systems, and can be deadly to life, the free radicals and quenched species generated by radiolysis can drive the production of simple organics from inorganic materials. The role of ionizing energy sources, such as lightning and interstellar radiation, has been extensively investigated as a source of some amazingly complex organic materials. According to laboratory photolysis of CO_2-bearing ice, abiotic processes can produce a frozen organic soup including hydrocarbons, esters, alcohols, organic acids, and ketones. Radiolysis of simple ices can even produce the amino acids glycine and serine, according to work by Monica Delitsky (JPL) and others. Although their work is nominally applied to icy satellites, comets, and interstellar particles, there should be no doubt that similar radiolytic processes occur on Mars, where cosmic rays, gamma rays from nuclear decay, and solar ultraviolet inevitably must interact with condensed and atmospheric C–H–O–N volatiles. Burial of radiolytic organic chemicals and organics introduced by meteoritic bombardment could result in a situation where significant organic material and oxidants would promote further aqueous organic chemistry and even life, somewhat as was suggested for Europa in 2000 by Christopher Chyba (SETI Institute).

A wide range of studies have shown that biological and abiotic chemosynthesis are intimately linked. (The Gaia theory of Earth, formulated by James Lovelock, goes so far to say that the Earth herself is a complex living object.) Models for the origin and development of life have in common a need for chemical disequilibria and particularly a redox couple to provide a chemical basis for harvesting energy. It is no surprise, then, that large chemical and physical gradients can serve the interests of biogenesis and growth of organisms. Hydrothermal springs and submarine hydrothermal vents have thus figured prominently in chemical-physical model's of life's beginnings. John Baross (University of Washington) and Sarah Hoffman

(Oregon State University), writing in 1985 on the primeval nature of submarine hydrothermal vents in *Origins of Life*, vol. 15, stated that these vents

> *... continue to be a major source of gases and dissolved elements to the modern ocean, as they were to the Archean ocean. Then, as now, they encompassed a multiplicity of physical and chemical gradients as a direct result of interactions between extensive hydrothermal activity in the Earth's crust and the overlying oceanic and atmospheric environments. We have proposed that these gradients provided the necessary multiple pathways for the abiotic synthesis of chemical compounds, origin and evolution of "precells" and "precell" communities and, ultimately, the evolution of free-living organisms.*

Chemical- and particulate-rich black-smoker deep-sea vents indeed have been a big favorite for many evolutionary biochemists, because those vents are known to synthesize a wide variety of organic chemicals, and emit a chemically disequilibrated organic and inorganic chemical stew that sustains complex chemosynthetic ecosystems semi-independent of photosynthesis. The disequlibria is caused by the chemical and thermal contrast between cool seawater and superheated aqueous solutions, where some of the water molecules have been ruptured by magmatic heat and reactions with basalt rock, and where solutes have been produced by high-temperature interactions with rock. The physical and chemical gradients indeed are key. In 1997, M.J. Russell and A.J. Hall (*Journal of the Geological Society of London*, **154**, 377–402) offered a specific variant of this model by proposing that life first emerged from gas-rich iron sulfide bubbles in submarine hydrothermal systems at points of extremely large gradients in pH, temperature, and reduction-oxidation states.

Everett L. Shock (Washington University), a leading biochemical thermodynamicist, has highlighted the synthesis of some fairly complex organic compounds in natural hydrothermal systems; the essential basis of the synthesis is the generation of hydrogen due to high-temperature scission of water and the oxidation of basalt. The freed hydrogen provides the thermodynamic impetus for organic chemical synthesis from CO_2 and bicarbonate, which may exist in the hydrothermal fluid or in seawater. Shock and colleagues find that dissolved carbon compounds can be reduced to a mixture of alcohols, ketones, and carboxylic acids over a wide temperature range characteristic of black smoker fluids.

One theory is that if life arose on the continents, then it did so in hot springs not unlike those at Yellowstone National Park, Wyoming.

Yellowstone National Park, one of Jack Farmer's and Jeffrey Plescia's favorite Astrobiology analog sites for Mars, indeed represents the best in what we might find on Mars. Most important, the geysers and bubbling mud might cough up living microorganisms from depth, and they might even live near the surface in hot springs. However, my guess is that with water the dominant constituent of these eruptions, we should think of Mammoth Hot Springs encased in a volume of ice 10 to 500 times as great as the volume of carbonates, sulfates, chlorides, and other aqueous precipitates. So look for a mound of ice.

Mono Lake, California, an alkaline (pH = 10) soda-rich hypersaline lake in eastern California, has high biological productivity involving very few species. This is typical of hypersaline environments, where very few types of organisms have the ability to survive, but where those that can survive have the benefit of evaporatively concentrated nutrients, as well as of salts, and few competitors. The solutes include abundant dissolved arsenic and selenium compounds, in addition to the major sodium–carbonate–sulfate–chloride solutes. The lake supports chemoautotrophic microorganisms (able to survive on chemicals) fond of respiring arsenic and selenium compounds, instead of free oxygen or sulfur compounds, as most other organisms do. Ron Oremland (US Geological Survey) and I are studying the chemistry and biochemical systematics of this lake. The lake obtained its sodium carbonate-rich composition first by aqueous weathering of sodium-rich volcanic ash and silicic to mafic lava flows, and then evaporation in an arid environment. The near-shore here includes some carbonate tufa cones formed where upwelling mineral-rich water deposited sodium and calcium carbonates; in the background are some volcanic cinder cones. On Mars, aqueous surface weathering may make sodium–calcium–magnesium–carbonate–chloride–sulfate rich brines. The expected important role of sodium carbonate in Martian brines is due in part to the high availability of atmospheric carbon dioxide (about 20 times the partial pressure of CO_2 in Earth's atmosphere) coupled with comparable abundances of sodium in Martian rocks. An expected greater role for sulfates is because of more limited hydrothermal circulation of brines on Mars compared to Earth; this process has resulted in destruction of most of Earth's sulfate (due to reduction of sulfur and formation of iron–nickel sulfides).

This process underlies most of the primary biological production potential of deep-sea hydrothermal vents on Earth. A similar process may sustain chemosynthetic ecosystems on Mars. Although no long-sustained ocean existed on Mars at least since the Noachian, a similar type of chemical and physical regime could exist in permafrost or a subsurface aquifer where volcanism and hydrothermal groundwater circulation occurs. These environments are apt to bring sulfides, including iron sulfide, into contact with sulfates, ferric iron oxides, carbon dioxide, bicarbonate, and carbonate in an environment with high gradients in temperature, salinity, and pH. Jack Farmer (University of Arizona), Jeffrey Plescia (US Geological Survey) and others look to land-based hydrothermal systems, such as the hot springs and geysers of Yellowstone National Park, as analogs of Martian pools where microorganisms may exploit the warmth and disequilibrated sulfurous chemical stew to drive biochemistry. In those systems, sulfur is cycled with biological mediation through sulfide, neutral sulfur, and sulfate forms. Many biochemists favor high-temperature aqueous environments for the origin of life on Earth and other planets, though it is not clear whether life originated there or adapted to such environments from origins elsewhere.

David Dreamer, a UC Santa Cruz biologist, has a very different concept for life's origins. It is based not on the synthesis of the DNA and RNA engine of life, but on synthesis of the physical semipermeable shell or bag which holds the contents of most nonviral life together and helps to maintain the proper internal chemical balance needed to sustain life – the cell membrane. Dreamer thinks that cell membranes, and with them life itself, originated in cool dilute but organically rich aqueous solutions, such as lakes.

Whether viruses are living or not, they cannot survive without cells, which are more complex mechanically than viruses. While some microbes do not possess cell walls, most do. Cell walls are fully or almost fully enclosed semipermeable membranes that contain the contents of the cell and regulate the internal composition and flow of nutrients and wastes between the internal and external environments. Water, oxygen, and many other substances can flow through the membrane, but large organic molecules cannot. Certain organic compounds contained in carbonaceous chondrite meteorites have been found to self organize into water-repelling soap-like bubbles resembling cell walls. The recent experiments by Dreamer and colleagues have formed these micro-bubbles under conditions of cold temperatures, no air, and high radiation. While it was not a Mars analog environment, the experiment, reported at an Astrobiology Conference in 2002, showed a capacity to form enclosed cell-wall-like organic membranes under conditions seemingly more hostile to life than in most laboratory biology experiments. Dreamer and colleagues later described in *Astrobiology* the self-assembly of semipermeable membranes in simple aqueous organic mixtures and the encapsulation of functional organic macromolecules. This is most certainly not life, but these membranes do offer analogs of some of the key machinery of life.

Dreamer and colleagues have shown that self-assembled membranes naturally encapsulate organic molecules, including some as complex as 600-base-pair double-stranded DNA (comparable to some viruses). It might be theoretically possible to fill such membranes with electrolyte solutions and a nucleus of genetic coding, add required organic nutrients to the external medium, add any needed maintenance machinery, and thus form life. This prospect is in many ways even more disturbing than it is exciting, but I will put off those concerns for now. A key aspect of Dreamer's work is that they are not simply simulating one aspect of life's origin in a beaker, but they are using a chemical basis that exists naturally in carbonaceous chondrite meteorites. Amphiphilic organic compounds (composed of a water-loving molecular component and a water-repelling component) are abundant in C2 chondrite Murchison extracts; these same compounds are what form the membranes synthesized by Dreamer and colleagues. Dreamer speculates that polycylic aromatic hydrocarbons (PAHs), also abundant in Murchison, help to stabilize amphiphilic membranes. In relation to Mars, it is noteworthy that PAHs are also abundant in the Martian meteorite ALH84001 – a fact that has been used by some researchers as evidence of biologically relevant chemistry.

The importance of physical containment of an organic-rich broth in some form of semipermeable three-dimensional bag or box or on an open two-dimensional substrate/template is common to several theories of the origin of life. The chemical organization needed for life to arise cannot occur in the absence of solid grain surfaces or vesicles or interfaces between immiscible liquids. William Martin (Universität Düsseldorf, Germany) and Michael J. Russell (Scottish Universities Environmental Research Centre, Glasgow) proposed that vesicles in seafloor hydrothermal iron and nickel sulfide crystalline masses may have formed a "hatchery" of sorts. These structures are thought to have inspired the self organization of self-contained chemical reduction-oxidation cycles of inorganic compounds and eventually the synthesis of complex organic molecules and cell structures (Martin and Russell, 2003, *Philosophical Transactions of the Royal Society, London*, series B, **358**, 59–85). Martin and Russell claim that the propagation of cells formed in these environments, and which evolved subsequently, have conserved the original compartmentalization found in those sulfide minerals. The relevance of vesicle-like sulfide structures in forming some type of cell-wall template seems improbable to me just considering their sizes (the authors showed examples of void structures ranging from 10 microns to 3 mm across, far larger than primitive cells). In effect, the observed vesicles are like two-dimensional surfaces to any pre-cell structure. However, the catalytic effect of sulfide minerals in supporting synthesis of complex organics, and the possible role of intracrystalline and grain-boundary vesicles as a type of semipermeable membrane (with the mineral and grain-boundary structure allowing some molecules to diffuse in and out while containing other synthesized molecules) seems a rather promising concept.

Icy-cold aqueous environments are also considered likely for the existence of life on Mars. Certainly near the surface and upper crustal levels, any liquid water is apt to be in equilibrium with ice. Christopher McKay (NASA Ames Research Center), Peter Doran (University of Illinois at Chicago), Steve Squyres (Cornell University) and many others have pointed to Antarctic saline lakes, many of which are perennially ice covered, as good analogs for Martian life. Whether or not life evolved under such conditions, it seems highly likely that if life ever got started on Mars, it would have evolved many adaptive strategies to cope with cryogenic temperatures and high abundances of dissolved salts.

While astrobiology experiments are still a long way from synthesizing life, there is a science basis for concepts of the natural abiotic synthesis of life here on Earth and elsewhere. At the current state of the art, there is a list of ingredients and certain steps in how they go together, but there is not yet a step-by-step recipe that cooks up life from start to finish. Until a compelling theory and successfully tested recipe for creation of life is devised, thus creating actual living things in the laboratory (a dreadful prospect to many people), or until life is found outside and independent of Earth (an exciting prospect to many; fearful to others), there is but one datum on the frequency of life in the Universe, and there is less-than-predictive theoretical understanding.

As science progresses, many of the key components of life are being synthesized abiotically and naturally. But how does nature put these components together? The famous Drake Equation takes a wild stab at estimating the chances that a suitable environment exists and that all the requisite steps take place. With large uncertainties, the result is that life should be quite common and widespread in the Universe. However, putting together the components of life may be more problematic than making the components, even if it should turn out that careful biomechanical manipulation can make the assembly in the laboratory. When probabilities of poorly understood steps in life's origins and survival characteristics are considered, probabilities for life occurring elsewhere is anyone's guess: it may be as common as three independent origins of life in our solar system (Earth, Mars, and Europa, say) and 50 million types of creatures for every solar-type star, or one chance in 50 million Universes.

PLAYING GOD OR JUST PLAYING WITH FIRE?

Age-old questions of our origins are at the root of many fundamental biology experiments; others are guided by medical research, needed improvements in the efficiency of food production, and building defenses against bioterrorism. These are all legitimate and important reasons that together make a case for pursuit of such experiments. As of this writing, one of my good friends lies on a bed possibly days from death due to a lymphoma. I recognize that everyone wants a cure for cancer; and very few people would wish for genetic cancer research to stop. But beyond hopefully a cure for or treatments for cancer, for Alzheimer's, for Parkinson's, and other comparable achievements, where is all this probing and manipulation of the genetic code leading? I personally believe that experiments dedicated to the formation of a completely artificial biochemical life, or tinkering with the code of natural life and certainly hybridization of humans with other organisms, is like a child playing with fire in a combustible environment. One miscalculation or one rash act and the house burns. Smart people are playing with fire. We are all fascinated, while potential disaster looms. My son, Christopher, says so directly, "*people have become too smart for their own good.*" I'll pick up again later on this and related dark topics, but wish for now to return to Martian wonders.

WATER-WORLD MARS: LIVING WORLD?

We cannot now say that life, past or present, is either probable or improbable within the subterranean watery realms or surface puddles of Mars, or any other place out there. Life either is/was there or it is not/never was. The chance that liquid water currently exists or at one time existed on Mars and in its crust is near 100%, but life is more complex by far than liquid water and simple aqueous chemistry. We have no solid ability to assign a probability to life's origins, try as some may, because life's genesis is not yet understood in its totality or major part by anyone.

Mars is one of perhaps a trillion planets and 10 trillion moon-size (or larger) solid objects in the Milky Way alone. We know that oxygen is among the most abundant products of stellar nucleosynthesis in the Milky Way, and since hydrogen is the most abundant element, hydrogen oxides – overwhelmingly water – are among the most abundant substances if not *the* most abundant condensates. The watery and icy nature of our Solar System, while not shared by all Solar Systems, must be very widespread. In our Solar System we have 12 solid-surfaced worlds (excluding the cores of gas giants) that are roughly Moon-size or larger; five are composed of 50% or more H_2O, and three (Earth, Mars and Europa) possess geologically important amounts of water and ice. We have a minimum of three "water worlds" (Earth, Mars, and Europa: planets or moons with liquid water aquifers, surface or interior oceans, or significant hydrologic cycles). Taking three as a possibly typical number of water worlds per average solar system, there are perhaps 300 billion other water worlds in our galaxy. And how many galaxies? A hundred billion? That makes for a lot of oceans, lakes, rivers, wet-based glaciers, aquifers, and hydrothermal systems even if I overestimate by a factor of 10 or a 100. Will all those places be sterile? I have to admit, that seems improbable and makes life-agnosticism difficult to maintain.

Mars is a world with its own unique evolution and distinctive character shaped partly by the influence of H_2O, but whether Mars harbors life still eludes us. The possible future find of any humble endemic life form on Mars (proven unrelated to Earth's life) would make it likely that there are billions of living worlds in the Milky Way. Contrarily, if we should verify an absence of past or present endemic life on Mars (to a reasonable minimal doubt) and confirm stable aqueous conditions in the past or present, and if this life-negative outcome is further emphasized by a verified absence of life in Europa's ocean, then we would consider more seriously our possible uniqueness. We have not yet reached this point.

Though there is apparently no Martian *ET* who calls home to Mars, might there be at least some scummy little microbial clot that calls Mars "home"? Can any organism cope, perhaps even thrive, with extreme conditions there? If not, did life once exist and then die out on Mars? These old issues have not been resolved, but there are now many new discoveries that bear on them. The Mars exploration program is currently focused around the theme of "follow the water," with discovery of life as the ultimate objective. The Mars program is further structured around three overlapping phases, including investigations of habitable environments, studies of the carbon cycle, and a search for biosignatures or of life itself.

The field of astrobiology burst wide open in the latter half of the 1990s, with NASA now dedicating considerable resources to research into life in extreme terrestrial environments as analogs to possible alien life. The idea that the most Earth-like and potentially life-friendly environments in outer space are "extreme" habitats by Earth standards is well based by what we know of other worlds in this Solar System. Considering Europa's putative ocean, can life survive in concentrated sulfuric acid solution (one chemical model of the ocean), where water is possibly a minor component (Kargel et al., 2000)? Or could it survive in a eutectic multi-component salt solution 10 times saltier than Earth's ocean and −36 °C? If it can, and if there is any place on Earth where similar life thrives, it is apt to be a very extreme, exceptional environment. Certainly life in Earth's extreme environments are places to look for special adaptive mechanism.

Studies of life in extreme environments have greatly broadened our perspective on the habitable zones around other stars and within our Solar System, as was discussed also in the Introduction. However, it may be only rarely the case that life on other worlds just somehow is barely hanging on. In most cases where life occurs, it is probably extremely well adapted and flourishes to the limit allowed by available biochemical energy sources, efficiency of energy utilization, and nutrients. Since those other worlds would have habitats that are not apt to be very like our own planet, any life adapted to those environments would consider Earth's many environments to be either extreme or downright incompatible with life as they know it. There may be worlds where life barely exists down to the last lichen or the final few distressed microbes swimming in the last stable puddle of water. This will hardly ever be the case, because when fateful conditions say that life will fade away, it probably fades rather quickly; and when conditions dictate that life arises, it is not apt to reproduce just barely faster than conditions kill it.

Every living world will have environments that represent the thin outer fringes – the "extreme environments" – of life's existence. The far limits of life's existence on Earth are unlikely to represent the ideal life niches on other worlds. The reason is simple. Think of Earth's vast gene pool from which adaptive attempts can be based. If life can hardly make a living in extreme environments on Earth, even after billions of years of evolutionary adaptation, it may well be incapable of doing well in analogous places on other worlds. Instead, luxuriant life is apt to exist on most worlds where any life at all occurs. The luxuriant growth will take place in niches that are less extreme and less hostile to the existence of complex organic macromolecules (assuming that the life is organic). The assumption here is that life in Earth's most extreme environments has adapted to those conditions based on ancestors who were better suited to more widespread and more benign conditions. One counterpoint often raised is that if a lichen, for example, survives in extreme cold and arid conditions, and there are a lot of the lichens covering rocks extensively, then that is probably a good sign that those are actually the preferred conditions; put that same lichen in the Amazon rain forest or a savanna, and it will likely not survive. This all may be true, but in terms of primary productivity and species and familial diversity, these most-extreme environments for the most part are not superior by any measure; they are tough places to make a living. There are, of course, a few hardy organisms who are able to make a living there and to thrive because there is very little competition.

There is another counterpoint to consider, and this weighs more strongly in my view: some of the most extreme environments on Earth are also among the most unstable environments; they may have a fleeting existence, and biological adaptations to those conditions may be interrupted or terminated repeatedly as the extreme fringes of Earth's environment shift back to more ordinary conditions. For instance, extreme cold, arid climatic conditions (such as in the Antarctic Dry Valleys) cannot last long geologically speaking; a small shift to more humid conditions, and those valleys are overrun by ice; and a shift away from glacial climates leaves these regions with a temperate climate. Each living planet might have its own unique ensemble of habitats where life is especially well suited, and a range of conditions near the fringes of that world's habitability where small numbers of well adapted species can thrive or at least survive.

A prevalent idea remains strong – a holdover from the Warm, Wet, Early Mars geological paradigm, that any Martian life would have existed only very early in Martian history. I take a very different slant on the issue. If there ever was life, there is a strong chance that it still exists in geothermally heated subsurface niches. Such life may have occasionally existed briefly at the surface when flushed out by geysers or springs, but it may be like surface life on Earth existing underground after events such as burial by landslides – it does not last long. I am not inclined to think that surface life has ever had a strong hold on Mars, but if it did, that early period of Martian history was perhaps when it existed. However, the evidence presented in the previous chapter shows that even very recently, and continuing in this present epoch,

Lichens, such as these from Arizona's high desert, were the favorite Earth model of Martian life in the 1940s and 1950s. Lichens, a symbiotic association of fungus and a photosynthesizing algae, can grow under extreme cold and arid as well as hot and arid conditions completely lacking in soil. Aside from the Lowellian conception of an environmentally stressed civilization, lichens represented the first model of Martian life in an extreme environment. These lichens thrive on the area's basalt, but many rock types will do.

Mars has had wet times, transient and isolated though they may be. Whether surface life could rapidly establish itself during such periods is questionable, but it is a possibility. In any event, there is no scientifically compelling reason to confine our view of surface life to early Mars. Indeed, some of the most impressive periods of mega-hydrology on Mars have been in the post-heavy bombardment period, right into the Amazonian. Quite possibly about the time when dense filamentous microbial mats were establishing themselves on Earth's shallow continental shelf marine environments about 2.9 billion years ago (Nora Noffke et al., *Geology*, **31**, 673–676), comparable biological events were occurring on Mars, even though those experiments in biology may have been doomed to brief success. Life adapted to the Martian subsurface occasionally may have made its appearance among flood channels, lakes, and vast glacial ice sheets. I would not rule out the possibility that with each gush and dribble of water onto the Martian surface from subsurface aquifers, there is a fleeting existence of surface life, even today.

THE GREAT BUG SEARCH (*VIKING LANDERS*)

Our first opportunity to look seriously for extant life on Mars was possible with the first successful landers in 1976. The *Viking Landers* were equipped and prepared sparing no expense to detect any life or organic remains of life, if any existed. Five experiments were designed either with life detection explicitly and foremost in mind (Gas Exchange = GEX; Labelled Release = LR; and Pyrolitic Release = PR) or with other objectives in mind but nevertheless with robust capabilities to place constraints on the existence, form, and abundance of possible life (Lander imaging; Gas Chromatograph-Mass Spectrometer = GCMS). The results of these experiments are very powerful in sum and result in the conclusion accepted by most Mars scientists that *Viking* indicated a fairly definitive absence of life at two sites on the Martian surface. Offering an important but perhaps least compelling null finding, the lander cameras indicated no clear evidence of the growth of plant life or the movement of animal life. GEX and LR involved moistening of Martian samples, with the initially provocative results that oxygen was released after samples were warmed and moistened and carbon dioxide produced after samples were warmed and then fertilized with organic compounds, both possible indicators of metabolic activity. Furthermore, PR also showed the generation of carbon compounds by Martian soil. However, PR showed that this same type of activity occurred after samples had been intensely heated above any temperature where living things could possibly survive, thus suggesting that the carbon compound synthesis was abiotic in nature. Most constraining, GCMS showed exceeding low upper limits on the possible amount of organic material (parts per billion and less for many common terrestrial organic compounds) in soil samples. All of the interesting releases of various gases were due, apparently, to abiotic chemistry in highly oxidizing soils. The high oxidation potential is produced apparently by the action of ultraviolet rays on the soil, a result that has since been simulated. For details I refer the reader to an overview by H.P. Klein and colleagues with more technical references (chapter 34 in H.H. Kieffer et al. (eds), 1992, *Mars*, University of Arizona Press).

In the prevalent view of Mars scientists, the summary by Klein et al. is reasonably well on target in its conclusions regarding the *Viking Lander* biology results, i.e., the Martian surface is very nearly or completely lifeless. However, possibly in making sure that their valid crucial point was not lost on readers, Klein et al. made a very biased and unsubstantiated presentation of the broader implications. Their main conclusion was warranted that "*The Viking findings established that there is no life at the two landing sites, Chryse and Utopia.*" They went much further, however, stating, for example, that the processes active on Mars and the Viking results "*virtually guarantee that the Martian surface is lifeless everywhere.*" If I was a betting man, I would bet that Klein et al. are correct, but this conclusion certainly was not substantiated, and it definitely is only one opinion rooted in present conditions at two sites only.

Klein et al. wrote that Antarctic Dry Valleys, while possibly the best Earthly analog of the Martian surface, is nevertheless a poor analog. It may well be that "best" is not good enough. Some Antarctic Dry Valley soil samples, according to Klein et al., are truly sterile, while the majority contain at least traces of organisms, including some endolithic ("inside rocks") microorganisms that are sustained by the moisture of summer snowmelt. However, Klein et al. go on to point out the impossibility of snow and ice melting at the surface of Mars, thus rendering one aspect of the Antarctic Dry Valleys inapplicable to Mars. To a considerable extent, this is a true and accurate statement of most of the Martian surface at most times through geologic history. However, with knowledge gained only recently, this statement is not rigorously true, even currently, for a tiny percentage of the surface; it certainly is not a valid statement for much of the Martian surface for some recent past times.

In fact, the usefulness of the Antarctic Dry Valleys analog is reinforced by a statement of Klein et al. that was intended to do otherwise:

> *[The Antarctic] is a place where a small climatic variation like the one that gives rise to endolithic habitats can make the difference between a livable environment and a nonlivable one.*

A fact knowable only with recent spacecraft data is that many parts of Mars apparently have had melting snow and surface ice in recent times. Thus, while Mars may be utterly uninhabitable at the surface today, the same might not be true of recent times in the past. But more to the point, Klein et al. and the *Viking Landers* did not address the greater possibility of life in the subsurface today and on the surface in the ancient past. In sum, the *Viking Landers* did a great deal to place tight constraints on the existence of present surface life, but large gaps remain in our knowledge such that prolific subsurface ecosystems cannot be excluded.

Not just subsurface life, but surface life is still a possibility, or at least freeze-dried surficial occurrences of life from down below. The *Viking Lander* experiments indeed presented the most conclusive evidence yet regarding surface life on Mars, ruling out the possibility that it exists everywhere; most likely, surface life and organics are either rare or absent, but we do not know which. However, it was recently reported in *Science* that some soils from the Atacama Desert are devoid of measurable organic material and devoid of any culturable microorganisms; this was a first finding of such utterly sterile terrestrial soils. "*In the driest part of the Atacama, we found that, if Viking had landed there instead of on Mars and done exactly the same experiments, we would also have been shut out,*" said Dr Chris McKay (NASA Ames), one of the chief investigators in that study. The biggest problem is that those soils are so severely dry. We can readily conceive of how perhaps 99.9% of Earth's surface has microbial activity or at least dead microbes, and only 0.1% is completely barren; on Mars possibly 99.9% (or much less) is completely barren. Astrobiologists indeed would be interested in finding that 0.1% (or much larger) fraction of life-bearing surface material, if there are any places at all that are life-bearing.

CALLING CARDS FROM MARS: VERIFYING THE SENDER

That rocks falling out of the sky could have originated on extraterrestrial worlds was not accepted by scientists until about 200 years ago. Inventor/Naturalist/President Thomas Jefferson is rumored to have stated, after hearing a lecture by Yale professors Benjamin Silliman and James Kingsley about a meteorite fall in Connecticut, "*I would rather [*or *'It is easier to'] believe that a Yankee professor [*or *'those two Yankee professors'] would lie than that stones could fall from heaven* [or, *'the sky'*]." The quotation has not been traced to source, but it represents a widespread eighteenth- and early nineteenth-century scientific skepticism about the extraterrestrial origin of meteorites. Indeed, the term "meteorite" means literally "weather stone." One hypothesis was that they represent a condensation phenomenon, possibly related to fusion of terrestrial rocks that had been struck by lightning. The "condensation" part of that old hypothesis is correct, for many meteorites; never mind that the condensation was in the extended solar atmosphere called the Solar Nebula!

Comets and asteroids were soon recognized as possible sources of meteorites, although the Moon was considered early in the nineteenth century as a likely origin of most meteorites. A nineteenth-century southern American scientist, J. Lawrence Smith, who is said to have traded slaves for meteorites, and whose name is memorialized in a prestigious award by the National Academy of Sciences, proposed specifically that impacts on the Moon caused rock to be "scraped and propelled" to Earth. Before the end of the nineteenth century, however, it had come to be accepted that the Moon and planets are simply too gravitationally large to lose their rocks. Smith's hypothesis was discarded – until the first of 22 confirmed lunar meteorites was found in 1982. It was 1981 before that long-accumulating evidence finally came together to suggest a Martian origin for a small class of special igneous meteorites, now numbering at least 26.

The first suggestions that a small suite of meteorites, once referred to as the SNC meteorites, are from Mars were made in 1979. The issue came to the forefront of meteoritical research in 1981 with the publication of a provocative paper by Chuck

Wood and his coauthor L.D. Ashwal, "SNC meteorites: igneous rocks from Mars?" Since then, a steady drum beat of independent and corroborative evidence has convinced the Mars science community that these rocks can only be from Mars.

How could scientists become so convinced that particular rocks are from Mars? The short answer, summarized and synthesized from about 700 scholarly papers and abstracts on the topic, is that:

1. Their mineralogy indicates an origin on a large planet that has a high gravitational field where high-pressure minerals such as garnet and amphibole are stable.
2. Mineralogy also indicates that the oxygen fugacity of the home planet's mantle was relatively oxidizing – about like that of Earth's mantle, where coexisting iron oxide phases include magnetite, ilmenite, and other minerals containing doubly- and triply-valent iron, rather than iron metal and doubly-valent iron. Such objects are rare in the Solar System, but iron-oxide-rich red Mars is one.
3. Oxygen isotopes of the rocks indicate that the home planet was not typical of the sources of most meteorites, but rather similar to Earth.
4. The mostly young ages of the Martian meteorites indicate a home planet that has a history of volcanic activity and hypervelocity impacts spanning over 4 billion years, possibly with a spike in volcanism 1.3 billion years ago. Most asteroidal parent bodies have no indications of igneous activity or melting after 4.5 billion years ago, before which short-lived, intense heating processes affected even small asteroids; the asteroids have been cool ever since, except when locally affected by impacts, but larger planets, including Mars, remained volcanically active for billions of years longer.
5. The silicate igneous compositions of these meteorites is similar to those observed by the *Viking Landers* and *Pathfinder*.
6. The abundances of platinum-group metals and other trace metals indicates that the planet of origin has a metallic core.
7. Trapped gases, entrained by impact shock heating, have a molecular composition similar to that observed in the Martian atmosphere by the *Viking Landers*.
8. These trapped gases also have noble gas and nitrogen isotopic compositions similar to that of the Martian atmosphere.
9. Lack of a strong remnant magnetization, except in the oldest Martian meteorite, indicates a home planet with no strong magnetic field, in contrast to other types of meteorites that record an ancient strong field. The oldest meteorites, however, were strongly magnetized by a large planet that presumably had a partly molten core and magnetic dynamo in operation.
10. The home planet has a history of aqueous alteration up to very recent times, including aqueous activity that involved aqueous chemical weathering and open-system aqueous chemical fractionations and salt precipitation from a hydrosphere. Some asteroid parent bodies also exhibited aqueous activity, but only about 4.56 billion years ago when those heating processes were active.

Most of these indicators amount to circumstantial evidence, but the isotopic and chemical gas ratios are about as close to a planetary fingerprint as can be found.

This list does a great injustice to the exquisite quality and variety of information that sings a chorus and refrain that these rocks are from Mars. The evidence accumulated in stepwise fashion, and each time a new type of evidence was obtained or an improved measurement was made, it only reinforced the connection to Mars. In 1981-1983 it was still a classic case of superb detective work by a small number of researchers, and there were good reasons to remain cautious, but since then there has been a steady accumulation of overwhelming pointers to Mars and nowhere but Mars.

At first, geophysicists were adamant that these rocks could not be from Mars for dynamical reasons. If any material was "scraped and propelled" fast enough (5 km/s) to escape Martian gravity, they normally would be pulverized, if not liquefied and vaporized by impact shock. Leading impact dynamicists Ann Vickery and Jay Melosh (University of Arizona) were among the last of the hard-core skeptics in 1983, when they advocated an alternative to an origin on Mars. As the geochemical evidence accumulated, however, the same researchers by 1987 were indicating that a dynamical means existed to eject the Martian meteorites intact so long as the crater was very large (greater than 175 km across) and probably produced by oblique impact. Ever since then, and still with evidence piling up, the Martian origin has not been doubted by scientists.

These meteorites say important things about the global chemical differentiation of Mars into core, mantle, and crust: they tell us when the Mars magnetic field turned off; and they are richly informative about the nature of basaltic and ultramafic magmatism on Mars. Although the Martian origins of these rocks were determined by comparison to properties of Mars known from spacecraft studies, they say all sorts of important things about how Mars operates as a planet that are not indicated by remote-sensing observations of Mars.

There are two things that the Martian meteorites say about the aqueous history of Mars that spacecraft data do not so clearly say. First, the aqueous alteration of these rocks, while significant, is not pervasive; secondly, the Martian water cycle might have supported living microbes. The first of these is compelling; the second is speculative. We will no doubt learn much more as robotic exploration increasingly employs advanced analytical techniques and provides access to more sites and more samples at each site, and as scrutiny of rocks returned from Mars by nature and machines allows us to look deeper into the many processes that have affected these rocks. For now, there is perhaps no Martian rock more important than ALH84001.

ALH84001 AND OTHER ODD ROCKS

In August 1996, a friend and colleague, the late David Roddy, came to my office ashen faced. He was not upset, but busting with excitement, verging on shock, to let me know that David McKay and Everett Gibson (Johnson Space Center) and others were soon to announce that fossilized evidence of microbial life had been found in a Martian meteorite, ALH84001. I had no way to evaluate the claim, but Roddy was wrong about one thing; the announcement would be made by President Bill Clinton.

Indeed, there were newsleaks of the finding, and so NASA and the President sought to reclaim control of the spinnable story. Mars science and space exploration has never been the same since. Mars is now the single-minded focus of NASA's planetary exploration efforts. While I agree with the focus, this story of Martian microbes is the story of perfect spin. ALH84001 is an amazing rock, and through amazing science it whispers an exciting story, as good as any told by an intricate Persian rug, but there are no bugs in this rug. Or perhaps there are.

McKay and colleagues reported 100-nm-size ovoid and tubular objects, some of which occurred in short chains (August 16, 1996, issue of *Science*). The minuscule objects were interpreted as nano-bacterial fossils, associated with heterogeneous, chemically zoned orange-and-black carbonate globules and magnetite chains, which were interpreted as bacterially precipitated minerals similar to those common in terrestrial rocks and soils. Furthermore, these rocks, unlike surface soils analyzed by the Viking Landers, contained organic compounds, including polycyclic aromatic hydrocarbons (PAHs) similar to those found in coal and petroleum on Earth. The researchers undertook significant effort to rule out terrestrial organic contamination. The results were considered in sum to represent microbiological fossils and associated organic compounds and precipitated biominerals. It was in fact a gripping news conference when the details were announced. Instantly, dozens of scientists were preoccupied by the findings – exploring implications, scrutinizing the results, finding fault and recognizing additional supporting evidence. The field of Astrobiology suddenly became a major priority on NASA's agenda, with the field moving from the fringe to the driving force of the planetary program.

Despite the immediate huge impact on planetary science, every aspect of the McKay et al. results has been attacked, with abiotic processes favored by many scientists over any biological activity. The anti-life clan pushes arguments that the nanofossils are too small, that the carbonate globules and magnetite have indications of abiotic origins, that the PAHs are due to contamination or are due to hydrothermal chemistry. Indeed, it is possible to generate PAHs by simple abiogenesis framed around reactions simplified as:

$$6FeCO_3 \text{ (siderite)} + \text{heat} \longleftrightarrow 2Fe_3O_4 \text{ (magnetite)} + C + 5CO_2,$$

where the carbon product can react in aqueous conditions to form PAHs.

The details of the fossil or pseudo-fossil-associated mineralogies and isotopes and chemistries are indeed fascinating and plausibly connected with life. The arguments go round and round without full consensus either way. Much controversy has centered around whether the rock has been affected by high-temperature hydrothermal activity or low-temperature aqueous weathering. There is unanimity, however, that ALH84001 is a fascinating rock, saying much about the early hydrosphere, if not the biosphere, of Mars. The trouble is that much of what the rock says is in geochemical and mineralogical cipher, and we lack all of the codes and sequence of micro- and macro-events to read the record correctly. It is the opinion of most Mars scientists that ALH84001 is telling us nothing directly about a biosphere, though the original proponents and several later converts remain adamant that the evidence comprise an overwhelming indication of microbiological activity.

It is agreed that this rock is more ancient than other known pieces of Mars, having crystallized soon after Mars formed, possibly fewer than 10 million years after the planet accreted. It is an igneous orthopyroxenite, meaning that it formed as gravitationally settled crystals precipitated from a near-surface magma. It contains glassy impact-metamorphosed remains of albitic feldspar, olivine, chromite, pyrite, and apatite, which probably are mostly the crystallized remains of a residual magma that has already crystallized most of its mass. At one time, pyroxene crystals settled to the bottom of what may have been a magma chamber in the throat of a volcano or in a vast volcanic lava lake. The magma continued to crystallize, causing the residue to take on a different composition, and this residual liquid then permeated the mass of pyroxene that had accumulated on the chamber floor. This rock is thus part of the early story of Mars, produced as the original crust of Mars was taking form. Somewhere between about 3.84 and 4.1 billion years ago – close to the time of the great multi-ringed basin-forming impacts on the Moon – a mighty impact shocked Mars in the region where ALH84001 had been formed. The impact probably formed one of the tens of thousands of craters that remain on Mars as battered, weather-beaten, low-rimmed, infilled circular structures, not unlike the 24-km-diameter impact structure of Haughton Dome (Devon Island, Canadian High Arctic).

About 2.5 billion years after this early shock event, fractures in ALH84001 became filled with carbonates and other minerals that apparently were precipitated aqueously. Though the 1.39-billion-year age of the carbonates is young by Solar System standards, it represents the oldest known aqueous alteration on Mars. Of course, we have very sparse data on this aspect. The age of carbonate mineralization in ALH84001 is interesting, as it is nearly the same as the igneous crystallization ages of several other Martian meteorites. It would seem that about that time, an enormous pulse of igneous activity affected Mars by volcanic resurfacing and carbonate and carbon dioxide mobilization. It may well have been a major MEGAOUTFLO event. Radial zoning (variations or oscillations in the content of Ca, Fe, and Mg in the carbonate globules) apparently indicates large variations in the cation content of aqueous solutions that formed these carbonate grains. In general, the cores tend to be Ca- and Fe-rich, and the edges are Mg-rich carbonate. The chemical gradations of composition, visible to the eye and light orange to dark gray colors in photomicrographs of the carbonate globules, are known as chemical "zoning."

Zoned carbonates are especially common among Mg-rich carbonates on Earth. The carbonate assemblage is somewhat similar to zoned carbonates of a polygonally cracked dolomitic limestone (once a desiccated Sun-baked carbonate mud) that I found in the latest Archean aged (2.5 billion years old) Wittenoom Dolomite of Western Australia (J.S. Kargel et al., 1996, *Global and Planetary Change*, **14**, 73–96). While ALH84001 is an igneous rock riddled with carbonate veins and pore fillings rather than a sedimentary rock, the chemical nature of the aqueous fluids may have been similar to the brine that deposited the Wittenoom Dolomite. In the case of the Wittenoom Dolomite, the fluid apparently was seawater affected by hydrothermal

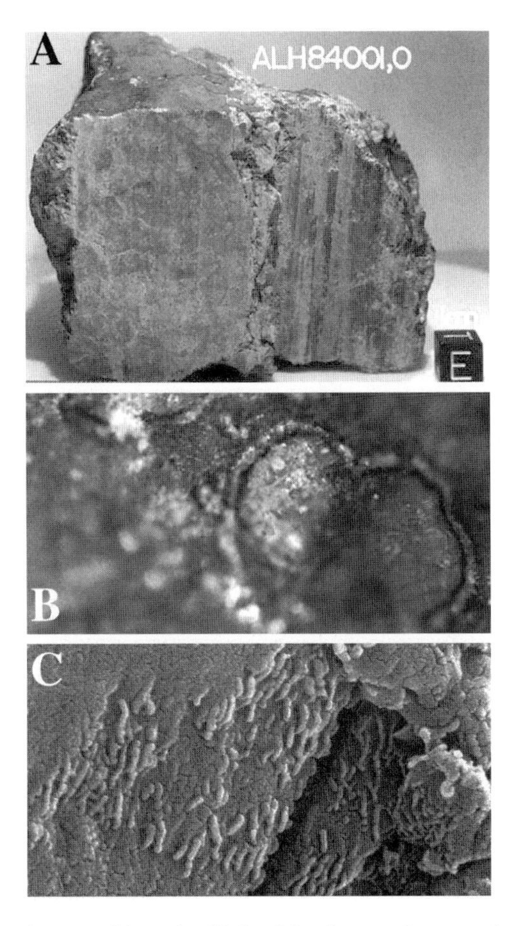

ALH84001 meteorite was blasted off the Martian surface, probably by an oblique impact, 16 million years ago and spent some time as a small asteroid. It landed in Antarctica 13,000 years ago and was promptly buried by snow and ice; it emerged in an ice ablation field, where it was picked up by scientists. Its more interesting history started about 4.5 billion years ago with its formation in the Martian highlands. It suffered impact cratering effects ~4 billion years ago, and then around 1.4 billion years ago it was aqueously altered. Cracks filled with small Ca–Mg–Fe-carbonate globules (B), which formed along with exceedingsly small (100 nm) ovoid objects (panel C), thought by David McKay (Johnson Space Center) and others to be fossilized bacteria, or similar microorganisms from Mars. If these really are fossilized life forms, they are of course of enormous interest. If they are not related to life, they are still of enormous interest, because by every interpretation they are related to early aqueous processes on Mars; questions pertain to how warm the water was and other details. Credits: NASA/JSC.

venting, a chemically reducing atmosphere, and then evaporative conditions. Others have shown that Earth's ocean water at the time in that vicinity was heavily influenced by hydrothermal vent fluids of a deep-sea nature under conditions of low oxygen fugacity. The oxygen depletion is evident because of the high abundances of iron and manganese; modern carbonates on Earth are dominantly fairly pure calcium carbonate, pure calcium-magnesium dolomite, or mixtures of the two, but they normally do not contain much iron or manganese. The similar zoning and Fe-rich carbonates in ALH84001 (even more Fe-rich than those in the mud-cracked Wittenoom Dolomite) argue for even stronger depletion in oxygen on Mars 1.4 billion years ago (at least in the water that contacted ALH84001). Under those conditions, ferrous iron was free to go into solution rather than being immediately consumed by free oxygen to form insoluble iron oxides, as normally happens on Earth today. The water that interacted with ALH84001 could have been formed in a Martian environment similar to that inferred for the Wittenoom Dolomite: volcanic exhalative activity in shallow evaporitic waters. Considering the geologic evidence of standing water during Hesperian and Amazonian times, and the abundant evidence for volcanism throughout Martian history, the analogy seems reasonable.

In the context of life, it is worth noting that the fine laminations present in much of the Wittenoom Dolomite very well could have had biological origins, as such laminae in modern carbonates are almost invariably formed through the mediation of life. Thus, if Mars did have life 1.39 billion years ago, it would not be a far stretch to imprint rocks with evidence of life wherever water interacted extensively with those rocks. The fact remains, however, that abiological processes can form the structures observed in ALH84001.

In addition to carbonates in ALH84001, variable amounts of chloride, phosphate, and sulfate salts, sulfides, and iron oxides were also precipitated along with the carbonates. Thus, briny aqueous fluids of variable composition were flowing through the fractured rock. It is not agreed whether that secondary mineralization (formation of salts and weathering products) occurred from cool brines or high-temperature supercritical aqueous fluid, as there are contrary indications, but the carbon dioxide that went to make these carbonates was most probably derived from the atmosphere, however indirectly.

The abundances of trace lithophile (rock-loving) elements, such as rare earth elements, have proven useful in establishing the origins of carbonates on Earth whose origins may otherwise be difficult to determine. The trace element abundances in the carbonates of ALH84001 are like those of Earth's aqueously deposited sedimentary carbonates and are distinctly unlike those of carbonatites formed from CO_2-rich mantle fluids. While this fact does not bear closely on the issue of whether the nano-ovoids are fossilized bacteria, it does add further strength to the hypothesis that low-temperature hydrospheric alteration occurred, where at least the precursors, if not the observed carbonates, were closely related to surface processes of weathering and aqueous precipitation.

The fractionation of sulfur, hydrogen, and carbon isotopes in ALH84001 and other Martian meteorites indicates that considerable weathering or other low-temperature aqueous processes acted to separate out the heavier and lighter isotopes

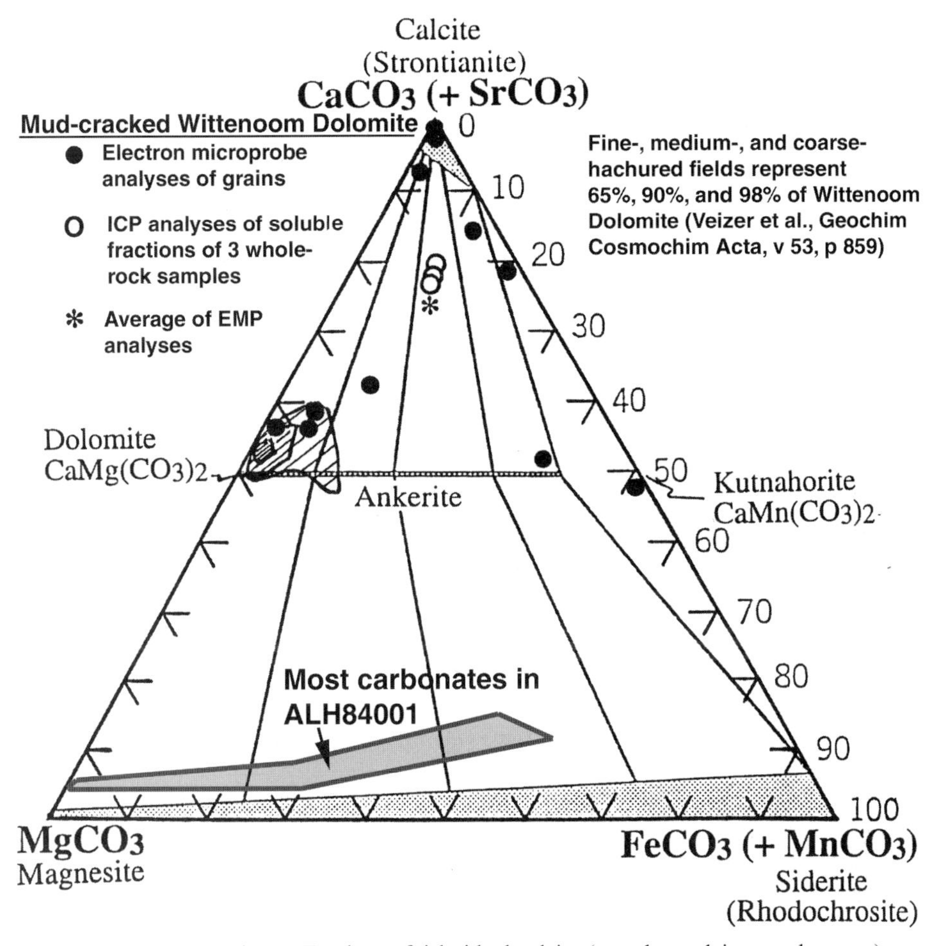

Most carbonate rocks on Earth are fairly ideal calcite (or other calcium carbonates) or dolomite or mixtures of the two, but the late Archean age Wittenoom Dolomite (Western Australia) has chemically zoned mineral grains all over the Ca-rich side of compositional map. Many grains are highly non-ideal, tending to be a messy mixture of carbonates, including large components of iron-carbonate, which may relate to a low-oxygen atmosphere. Although not as magnesium-rich as the carbonates in Martian meteorite ALH84001, the zoning is similar, as is their iron richness. These particular carbonates in the Wittenoom Dolomite form a spectacular, perfectly preserved system of mud desiccation cracks, demonstrating that the rock formed in near-shore saline conditions. The water chemistry could have been formed under hydrothermal conditions, as were the waters that gave rise to the area's banded iron-formations. The non-ideal compositions could be related to rapid precipitation and incomplete chemical equilibration with a brine – a rapidly varying composition. A similar situation may apply to the Martian carbonates. This figure is modified from Kargel et al. (1996, pp. 73–96).

of each element. This is a characteristic of a planet with a full-blown hydrosphere, such as Earth. However, the fairly unweathered condition of ALH84001 and all other Martian meteorites indicates with certainty that these rocks were not pervasively altered, as would occur under warm, wet Earth-like conditions in just thousands to millions of years, depending on temperature and other conditions. The cumulative amount of time that this and other Martian rocks were wet since their origins is probably measured in a few tens to a few hundred years (maybe a few thousand years if the brines were cold and reaction rates slow); we are definitely not talking about a rock that was immersed in an ocean or aquifer for untold eons. Thus, the story so indicated is probably one where liquid water was produced transiently, either due to a short-lived global MEGAOUTFLO climatic excursion or due to a local volcanic or impact heating episode. This does not eliminate possibilities that elsewhere on the planet wet conditions may have prevailed for far longer times. On a cryospheric world prone also to aridity, it may well be that small regional (or depth) differences in temperature and humidity cause large differences in the cumulative duration of wet conditions.

While other Martian meteorites have Ca-sulfate-rich salt assemblages, and the chief occurrence of iron in the altered material is in "iddingsite" (aqueously altered olivine), ALH84001 has abundant iron carbonate and iron sulfate, and iron sulfide components. This indicates a brine that had low oxygen fugacity, which can allow ferrous iron to be a major dissolved cation. This could be due to extensive basalt interactions of the fluid at high temperatures, or it could indicate that the Martian atmosphere and hydrosphere at that time (1.39 billion years ago) was very chemically reducing.

It is possible that a shift to more oxygen-rich conditions occurred on Mars sometime after 1.39 billion years ago (in order to explain the low-Fe sulfates and carbonates of younger Martian meteorites), whereas on Earth a similar atmospheric transition occurred gradually more than a billion years earlier. Alternatively, the different oxygen fugacities of the circulating brines on Mars may simply reflect different local brine histories, such as basalt interactions. In any event, the history of oxygen levels in the Martian atmosphere and hydrosphere, and how it may have deviated from equilibrium with the mantle, is no doubt an interesting one. There could be a connection to life (presumably then at depth), but asteroid-derived meteorites are also replete with indications of variations in oxygen fugacity, so abiotic processes might explain any shifts on Mars, too. For instance, extensive volcanic interactions with ground water and ground ice would generate oxygen from thermally dissociated water; the hydrogen also produced would readily escape into space, leaving the crust more oxidized; this process would be reflected also in the oxygen levels ("oxygen fugacity") of groundwater and salts precipitated from it.

The formation of chemically zoned carbonates in Earth's marine tidal environments and evaporitic lakes is well known. The preservation of large chemical gradients within zoned grains for hundreds of millions and even billions of years is also common. Sedimentary carbonates can have extremely heterogeneous and disequilibrated compositions, as represented by the case of the Wittenoom Dolomite. The fact that ALH84001 has far-ranging carbonate compositions even within a single grain is thus not surprising and is consistent with formation in a low-temperature aqueous environment, contrary to what some researchers have reported with claims that high-temperature processes are required. Sedimentary environments are fully adequate to the task.

Without doubt, ALH84001 has proven to be the biggest attention-getter of the Martian meteorites in the past eight years. However, many other Martian meteorites offer crucial insights into Martian aqueous processes. Low-temperature brines are generally implicated, having caused small degrees of aqueous alteration and precipitation of oxidized salt assemblages. The chief differences of the alteration of the other Martian meteorites are that it occurred later in Martian history and it involved more oxidized brines. None of the Martian meteorites, however, shows signs of having stewed around in liquid water for eons. The transient existence of liquid water in contact with these rocks is consistent with a world that has, at least in those localities from which the rocks were derived, tended toward hyperaridity and/or severe permafrost conditions for almost all of geologic time. Transient wet episodes can be explained by brief geothermal excursions related to nearby impacts or volcanism. Transient local flood episodes or brief global MEGAOUTFLO climate excursions are also compatible with the geochemical and mineralogical evidence. All this is consistent with the theoretical expectation of a sustained presence (even today) of liquid water at greater depths or nearer to active volcanoes.

In sum, while the evidence for fossilized life in the Martian meteorites is ambiguous, the evidence contained in these rocks for transient and repeated brief episodes of aqueous chemical alteration at low temperatures relatively late in Martian history is compelling. These rocks, therefore, add rich new detail to the story of water and ice on Mars and the developing picture of Martian aqueous environments that would perhaps have been suitable for the existence of life on Mars. The mineralogical and other evidence obtained from the Martian meteorites is thus completely consistent with what imaging and other orbital observations have indicated.

MARTIAN ISLAND JUNGLES?

If life occurs beyond Earth it may form a fairly simple and homogeneous community if the physical environment of that world is simple, homogeneous, and stable; or it may be exceedingly diverse on complex, dynamic, and heterogeneous planets, such as Earth. While biologists are probably too focused on DNA and RNA as a basis of life, DNA and RNA do offer a great means of understanding how evolution can occur simply. A team of British and Swedish researchers, for instance, discovered that swapping a single chemical base pair (out of millions) in the genetic code of Salmonella changes the virulence from one where the infected host becomes severely ill with typhoid fever or, like Typhoid Mary, merely carries (and transmits) the infectious agent. A human, in fact, has only about 300 genes that a mouse does not also have (out of about 42,000 genes in a human). Genetically, a human is not really any more complicated than a mouse. A fruit fly may seem to be orders of magnitude less complicated, but in fact its genome contains about one-third as many genes as a human, and many are identical genes. These findings ignore comparisons of the much larger amount of non-coding DNA, whose purpose is not well understood – it may be "junk" DNA, functioning only as molecular scaffolding; it may be designed to make the coding DNA more resilient to mutations due to chemical or biological attack, or it may have some other purpose. But the well-known point is that changes of a few genes can cause huge changes in morphology and functionality. Herein is the power of mutation and natural selection (evolution) to create new varieties of organisms, even entirely new species and genera, from old forms.

If extraterrestrial life is DNA based, it will exhibit the same propensity to evolve as Earth life, if the host environments are as diverse and as unstable as those on Earth. If extraterrestrial life somehow has another biochemical basis, it will no doubt still be responsive to physical environment, and diversity and instability will still be the engines of evolution. "What does not kill us makes us stronger" should be the Universal motto of all life. The genetic malleability, or inherent capacity to evolve, however, cannot be gauged without knowing and understanding the biochemical system. Extraterrestrial life generally will make use of a relatively small and simple collection of basic building blocks. These blocks will be arranged in patterns that are diagnostic, in some sense, of life. Christopher McKay (NASA Ames) has suggested that remote sensing and chemical analysis might discern bio-molecules based on such patterns. The trick, of course, is in decoding the molecular arrangement of these building blocks and showing that they are not reproducible by ordinary chemical abiogenesis.

The initiation of life on Earth is thought by many biochemists to involve the abiotic synthesis first of RNA before DNA (G.F. Joyce, 1989, *Nature*, **338**, 217–224). RNA could have used clays or other minerals as a template and substrate, or it may have formed by exploitation of surface chemistry in lipid vesicles involving pre-RNA organic molecules. Important, plausible steps toward abiotic synthesis of RNA have been reported, but as yet there has been no combination of these steps to show the likelihood of this crucial event in life's origins.

Once formed, many factors may control a planet's biodiversity, an inference that can be extended very simply from Charles Darwin's 1859 classic *On the Origin of Species by Means of Natural Selection*. The long geologic history of Mars has been one of first-order upheavals and periodic increases in physical environmental diversity. Thus, if life has existed there and survived from one upheaval to the next, it may have evolved a considerable biological diversity in response to the pressures caused by geologic and climatic dynamics. Other factors may suggest low diversity, such as a possible restriction of life to subsurface hydrothermal realms, and possible reliance on chemical systems lacking diurnal and annual fluctuations of conditions and lacking latitudinal gradient of conditions.

Physical conditions on Mars have only played around at the fringes of habitability, so that any life may have been selected to survive and thrive in only a small subset of physical environments, perhaps all in particular niches of the subsurface. Possibly the evolutionary leap from subsurface to surface, when climatic conditions were right, was so great and the time so limited, that surface life never existed. Ecosystems might be based in the subsurface around one or a few chemical disequilibria related to igneous activity. Biodiversity might be low.

As volcanism and geothermal activity waxed and waned, and as climate underwent its revolutions, the habitable zone of subterranean life may have simply migrated up and down through the upper crust and expanded and contracted in latitude and oscillated in radius around hotspots, but the physical environments may have changed little. Biodiversity on Mars again could be very low.

The physical dimensions of the inhabited niche are also significant; comparing ecosystems of similar physical environmental attributes, larger ecosystems generally have greater biodiversity than smaller ones – a point emphasized by Darwin and expounded upon by later generations of evolutionary biologists. If suitable habitats are restricted, for instances, to small domains where ground ice and volcanism coexist, ecosystems could be equivalent to low-diversity rocky islets on Earth. A global geothermally heated aquifer, such as considered for early Mars history by Steve Clifford, on the other hand, could produce relatively minor physical diversity on the whole, but minor special niches would exist in various places around the globe, helping to drive the engine of biodiversity.

If Martian life is restricted to rare and physically well-isolated environmental niches, then a given isolated ecosystem may be homogeneous and simply structured, but there may be great variance between ecosystems if natural selection has had eons to operate on independent systems. Terrestrial island ecosystems should then be the guiding insight. If, however, the environments are isolated, but global dust-borne transport of viable, freeze-fried microorganisms causes continual contact between ecosystems, then they could be homogeneous systems. Homogeneity of isolated ecosystems could also apply if there was global commingling of species but isolation occurred recently. Charles Darwin pointed out this in connection with isolated but homogenous arctic-type communities on the isolated peaks of the Pyrenees and Alps and in the North American Sierra Nevada and Labrador. Darwin adapted Louis Agassiz' concept of glaciation, which was a contemporary revolution in Earth science, to explain the stranding of similar arctic communities on isolated peaks. He extended the concept to the global distribution of communities that were pressed to migrate to different elevations as Ice Age climate changed globally but nonuniformly with latitude. He drew the graphic analogy,

> *As the tide leaves its drift in horizontal lines, though rising higher on the shores where the tide rises highest, so have the living waters left their living drift on our mountain summits, in a line rising gently from the arctic lowlands to a great height under the equator.*

If Mars has a subterranean ecosystem, a similar concept will be applicable there, substituting depth for altitude, where the living high tide during warm climate epochs or global heat flow excursions may have left a fossilized drift line high in the crust, shallowing from poles to equator and from geothermal "background" regions toward volcanic hotspots. A key objective of robotic Mars surface exploration now is to search for places where an ancient biodrift line reached accessible places near the surface, or even which now may be depositing biodetritus at rare hot springs.

If Mars is still or ever has been a living world, the first *in-situ* signs of abundant and stable life in the rocks of Mars might be found at great depths. The geothermal gradient on Mars is a factor of about 3 less than that on Earth, but the regolith of Mars would have a great volume of rock where porous and wet conditions might exist, while the low surface temperatures and low thermal gradients of Mars in typical regions would displace any large wet domains of crust to great depths. However, with surface gravity about 40% that on Earth, closure of porosity, where water and microbes may exist, would occur at depths 2.5 times greater than on Earth. Thus, the idea of a wet, porous, and biologically active crust 10—12 km down is not improbable. The pressure at a depth of 12 km is just about 120 MPa (1,200 bars), not much greater than that in Earth's oceanic trenches, which provide viable niches for life. Indeed, experimental work by Anurag Sharma (Carnegie Institution of Washington) and colleagues (reported in 2002 in *Science*, vol. 295) have shown that some microbial life can survive and metabolize at pressures more than an order of magnitude greater, where the high-pressure polymorph, ice VI, is stable.

We can reasonably identify the sources of biochemical energy and the major physical and chemical challenges to life. Hence, we can identify broad adaptive strategies likely to be employed by Martian life. This approach suggests that Martian life is apt to share some similarities to life on Earth. But in detail, the physiology of Martian microbes, and especially any complex multicellular life cannot be predicted. Though few scientists anticipate any organisms as complex as, say, fish and dinosaurs, the rich variety of major biological innovations of terrestrial life forms offers a glimpse as to the possible complexities and divergent evolution of possible extraterrestrial life, including that of Mars. Few serious people outside Hollywood have envisioned, for instance, hominids of any sort on Mars or anywhere but Earth. A potential exists for highly innovative and un-Earthly adaptations. Richard O. Prum and Alan H. Brush (in *Scientific American*, March 2003), writing on the topic of the evolution of the feather, framed the still unsolved problem of major biological innovation, which applies to microbial innovation as well as the evolution of feathers on dinosaurs and birds:

> *Although evolutionary theory provides a robust explanation for the appearance of minor variations in the size and shape of creatures and their component parts, it does not yet give as much guidance for understanding the emergence of entirely new structures, including digits, limbs, eyes, and feathers.*

It is similarly true that the evolution of wholly new biological strategies, such as exploitation of new energy sources by microbes, poses an unmet challenge to evolutionary theory. However, the variety of possible actual solutions that life has somehow adopted is practically unlimited and may go well beyond what people can imagine. Some biochemical fundamentals are apt to be shared by life on Mars and Earth, but, beyond that, there is every expectation that Martian life, whether or not related to terrestrial life by primordial panspermia, will physically be very unlike terrestrial life. To the extent that environmental conditions and stresses are in some cases shared on the two worlds, there may be striking instances of convergent or parallel evolution, but to the extent that Mars is unlike Earth, adaptive solutions and consequent physiology will differ markedly.

It is entirely possible that Mars has had such severe conditions throughout geologic time, with exceptions of short excursions too brief to be exploited by life, that its life has at best been incipient, struggling, and perhaps not even fully meeting common definitions of life. Perhaps Mars is a world of chemical precursors of life or of what John Baross (University of Washington) terms "pre-cell communities." Such quasi-biological activity may have arisen repeatedly during the heavy impact period of Earth's and Mars' history.

The search for life associated with Earth's impact structures – especially life or chemical biosignatures unique to those environments and directly related to the physical and chemical environment established by the impact – is an area of active research. Places such as the Haughton Dome crater (Devon Island) offer rich potential as analogs of possible Martian impact-crater ecosystems. C.S. Cockell and P. Lee (2002, *Biological Review*, **77**, 279–310) have described three phases of biological activity associated with large impacts on Earth following the immediate phase of impact destruction. Basing their assessment on studies of craters such as the Haughton Dome crater, they have defined:

1. Phase of thermal biology, which is expected to be similar to that at hydrothermal sites.
2. Phase of impact succession and climax, where a complex succession of events would involve life taking advantage of and adapting to the physical/sedimentologic/climatic niches occurring at an impact crater following cooling of the crater; this would include any crater lakes that may form, the rim mountains, and the sedimentary, impact melt, and breccia deposits, within and around the crater, including also any impact-related mineralization.
3. Phase of ecological assimiliation, where finally the significance of the crater itself is lost as the basin is eroded or buried and becomes irrelevant to biology.

Mars very well may have exhibited ecological successions repeated hundreds of thousands of times through geologic history. If only life's precursors were involved, there would be some type of chemical succession comparable to that defined by Cockell and Lee for full-up biological ecosystems. If actual life was involved, but the genetic coding for it was inflexible and lacked opportunities for rapid mutation and evolution, or if the ecosystem of Mars are interconnected, then these hundreds of thousands of impacts may have repeated almost the same biological stories of succession again and again. More likely, however (if life has occurred on Mars at all), differences in rock composition, differences in impact parameters, differences in geologic and climatic settings, and isolation of viable environmental niches for life would have produced a different result every time. On Mars, Cockell and Lee's third phase, and maybe the second as well, may involve extinction of all life associated with the impact crater, fossilization of the last surviving life, or kinetic freezing-in of any quenchable biological precurors generated during the phase of post-impact "thermal biology."

Charles Cockell has further indicated that impacts, such as that on Devon Island, create new rock types that serve as microbial habitats. He has shown that impact shock of granite gneiss causes the rock to become extremely porous and translucent. The Devon Island shocked gneiss has been colonized by complex cryptoendolithic microbial communities, including cyanobacteria and other microbes. These rocks thaw for only about five weeks per year, but during that thaw period temperatures rise far above prevailing air temperatures and actually reach benign levels around 22° C. The porosity of these rocks can accommodate and retain snowmelt, while the translucency admits enough visible light to sustain photosynthesis to a depth of 8 mm in the rock, but most damaging UV radiation is blocked at a depth of just 3 mm. Cockell thus suggests that the zone between 3 and 8 mm deep in such shocked rocks could serve as a microbial habitat on Mars.

There is another possibility that may make moot all these considerations of biodiversity and life's niches on Mars. This final possibility, of course, is that Mars is and always has been a nonliving world, at least until the arrival of *Mars Polar Lander* in 1999.

Now I venture my guess about Mars life. It may be a flip of a coin whether Mars has (or had) endemic life or never had it, but I continue on the premise that there is life in the subsurface. I shall start with statements about the igneous life of Mars, because that is where the Martian subsurface has hope of offering sustained warm conditions, and that is where the chemical disequilibria occurs that can be exploited by life. The major volcanic provinces of Mars are large and volcanically isolated systems; even if there is physical global continuity of a deep aquifer that connects these systems, the major igneous provinces will have their own unique geochemical signatures, each of which would probably have evolved over time. The composition of volcanoes depends on the depth of the magmatic roots and fractional degree of partial melting in the mantle, the volatile content of the mantle, the amount of time and extent of cooling and dissolution of crustal rocks in shallow magma chambers, the amount and type of volatiles in the rocks through which the magma traverses, and other aspects that are bound to be as highly variable on Mars as they are on Earth. Thus, one volcano will not be compositionally the same as another.

We already know from lander chemical analyses and remote-sensing observations that a minimum of two distinct common rock types exists – one is basaltic and the other andesitic. However, Martian meteorites indicate further petrologic diversity of igneous rocks, including ultramafic rocks. Any groundwater flowing from one igneous province to another would change composition by precipitating some mineral phases and losing certain gases while dissolving others, and by mixing with waters of separate equilibration. However, maybe such mixing and large-scale brine transport has not occurred in a long time. The volcanic provinces may now, and for hundreds of millions of years, have been hydrologically isolated by frozen intervening expanses, except for surface ice condensations and sublimations.

If there is any life on Mars, the biochemical environment will of necessity vary along with volcanic chemical gradients and discontinuities. Some volcanoes will be acidic and others basic, some less or more oxidized than others, and brine composition will follow the rocks. Some brines will be hot, others cool, and some chilled by ice below the normal freezing point of pure water. Some volcanoes will churn sulfate-rich brines, some will be sulfidic, and others chloride rich; some will be seleniferous or arseniferous, rich in dissolved silica, or in ferrous iron, some in calcium carbonate, calcium sulfate, or magnesium sulfate. Some volcanoes will belch carbon dioxide, others hydrogen sulfide. The high thermal and chemical gradients would ensure that any ecosystems would require wide-ranging strategies to cope. Since many life forms will no doubt be hyperthemophilic – sustained at temperatures where metabolic rates are very high and life spans short – the possibility exists that evolution may be very rapid near volcanoes, though commensurately slow in the cold outskirts.

It will be a jungle among the warm roots of geothermal systems, if there is any life at all. The vast expanses of terrain with background levels of geothermal heat flow are apt to be dominated at great depth by psychrophilic (cold-loving) obligate halophilic (salt-requiring) microorganisms in a globally uniform community. There may exist a system of sulfate-reducing and sulfate-oxidizing microorganisms capable of thriving in eutectic brines at the depressed melting point of ice. Biological communities existing at the limits of habitability where the cold top of the crustal biosphere merges with the cold outer fringes of geothermally influenced volumes of crust are apt to exploit chemical disequilibria of trace solutes that are concentrated by freezing, including arsenic, selenium, and tellurium compounds. Such specialized arsenic- and selenium-compound-respiring microbes (though not psychrophilic) are known in the region of Mono Lake, a volcanic crater lake in California.

The surface, however, is apt to have been quite a different matter from the subsurface, if in fact there ever has been surface life. While I should say nothing of the first half billion years, which in any case is a little known period, the last 4 billion years have been times of dramatic but short-lived environmental upheavals. Seeing how long it took terrestrial life to learn to survive on the land surface, one may reasonably doubt that life would have taken hold at the Martian surface. If I should be wrong about this, the successful surface colonists are likely to be only slightly changed from organisms that perhaps today make their living amid the salt-saturated subterranean ice waters.

JUNGLE LAW

How might Martian organisms, if they exist, compare to those on Earth? Since they probably are mainly or entirely subterranean dwellers, they would live amid empty spaces underground, mostly microscopically small pores, mineral grain boundaries, and thin fractures. Multicellular forms, if any exist, are probably small. Most if not all are probably unicellular. Most would live in a hypersaline liquid aquifer beneath the cryosphere, and in the semi-frozen, semi-liquid slushy- groundwater medium near the base of the cryosphere. Capped by hard-frozen, ice-cemented ground, magmatic gases would be trapped, and so dissolved gases would be abundant.

Hyperthermophiles could be biochemically similar to terrestrial ones. They will more probably be as uniquely exotic as the planet they live on. Psychrophiles (super-cold lovers) may be different from anything known on Earth. At the very low temperatures typical in the Martian crust (away from active volcanoes), brines, if not totally frozen, will be saturated in many salt species. Probably more than cold temperatures, water eight times saltier than Earth's sea and high heavy-metal abundances will be the most stressful environmental parameter. Part of the osmotic challenge will be to keep the internal levels of heavy metals low, while retaining enough solutes to keep from freezing or salting up one's insides.

The cold conditions will mean that metabolism will be extremely sluggish; however, no matter, because a hungry critter will not have to swim rapidly; the high viscosity of the cold salt water, 10 to 100 times more viscous than tap water, means that their prey is moving slowly, too, if it has motility. Motility will require that hydrodynamics be well coded genetically. However, the energy cost of moving and beating viscosity will be high; it may be better for a microbe to creep amoeba-like or to anchor itself and let the sluggish current bring nutrients to it. The currents will be less than typical on Earth, also because of the high viscosity; if thermohaline/density driven, the currents will also be lower because of low gravity. However, that will be no problem, because the high salinities will support high concentrations of nutrients.

So how to anchor? The answer would likely be in precipitation of ice or salts. However, a microbe might not wish to be where the brine is undergoing cooling, because that could mean rapid encasement in a salt-ice eutectic crystalline mixture, if cellular fluids don't freeze first. Of course, anchoring has its price – if somebody else anchors on top, the bug won't get a decent meal, unless it is excrement. So there will be a war of the threads – as with many bacteria and algae on Earth, long slender thread-like growth patterns may allow a microbe to wriggle between the spaces and seize nutrients from above a rude neighbor who sits squarely on top. This competitive struggle is rather like that of the forest, where the competition for photons has driven the evolution of trees.

An effective competitor might have evolved the means to control the location of crystalline precipitates. Salt precipitation nearby (but not inside) would cause the local environment to become enriched in water, which might alleviate some of the osmotic stress. This would work if the temperature is somewhat above the ice-salt eutectic. However, if the environment is at a eutectic, water enrichment would promote ice precipitation, but even that could be used to skillful advantage. A slender thread-like appendage could grab a salt crystallite, causing growth of the salt grain, thus triggering also the growth of a needle-like ice crystal. Orient the ice crystal just right, and the microbe carries a dagger that could pierce a competitor-neighbor's cell wall, thus spilling its contents. Particularly effective would be a critter that does this, eats the spillage, and reproduces in the dead cell structure of its victim. Best of all would be one who has evolved the means to look like a barren piece of rock – virgin real estate. Surprise!

But not to be outdone, mimickry can work both ways, and if a chemical or temperature sensor is added to discern whether the temperature is right at the eutectic or slightly above, a microbe could have the ultimate key to knowing where it is safe from these ice daggers carried by one's enemy, and where one might fool the predator, thus turning the tables on him.

A very potent adaptive device could be employed by bugs who have learned to control aqueous phase equilibria in the most powerful way. Carbon gases – both methane and carbon dioxide, among others – are highly involved in the stability of ice and the ice-like solid phase of water molecules called clathrate hydrate. Carbon gases are the materials most desired, from which organisms build their cell walls and genetic material and metabolic energy factories. Consumption of these gases, and conversion into macromolecular forms that cannot fit inside clathrate cage structures, thereby shifts phase equilibria into a domain less favorable to the existence of solid water. This mechanism may be so well utilized, that the very nature of microbial activity may differ fundamentally in regions where clathrates are stable versus where they are not. Not only can this approach to life help to maintain a liquid environment, but the destruction of clathrates may free many trapped nutrients. Hence, the concept of the clathrate-chewing "ice worm." Sound too fantastic to be possible? Well, there are organisms on Earth – on the Gulf of Mexico sea floor – that do basically this. Roger Sassen and colleagues (University of Texas) have been there to explore a realm where cold bubbling streams of methane and other gases and bulbous lobes of clathrate are being eaten by microbes and worms.

Of course any life will require a source of chemical energy. Earth's biological systems are notoriously inefficient despite billions of years of natural selection favoring organisms that are better able to extract and use energy from the environment. In a classic study of a photosynthetically based ecosystem – a simple vegetation-mouse-weasel chain of an abandoned agricultural field – F.B. Golley (1960, *Ecological Monographs*, **30**, 187–206) found that only 0.62% of incident solar radiative flux was used in photosynthesis. Of that amount, mice consumed 0.43%; of that amount, the mice actually used 2.1% of consumed plant energy to build mouse tissue. At the predatory stage, the weasels proved to be more efficient and, in fact, consumed slightly more energy than the mice provided – presumably from birds, mice that arrived from outside the study plot, or perhaps insects. In any case, the weasels consumed an amount of energy equal to 0.000062% of incident solar flux, and only 4% of that went into making new weasel tissue and weasel offspring (most of the rest having gone into respiration to support life activities and maintenance). Only the energy of two average photons in a hundred million generated new weasel stuff.

The inefficiencies of biological systems are built into the thermodynamics of biochemistry and the hierarchical designs of ecosystems. From the highest predators' vantage, it is somewhat more efficient, because they have all these slave laborers gathering energy on behalf of the predators. The inefficiencies along the food web, however, mean that improvements are possible. Improvements happen all the time in natural evolution. One of the most revolutionary advances in Earth's life's efficiency came with the evolution, probably just a few million years ago, of the C4 photosynthetic pathway in grasses such as corn and sugarcane. Most plants use the C3 photosynthetic pathway. The C3 process begins with the absorption of light energy by chlorophyll, which provides the energy to synthesize ATP (adenosine triphosphate) and NADPH (a reduced form of nicotinamide adenine dinucleotide phosphate), with generation of waste oxygen. In the second part of the C3 process a phosphorylated 5-carbon sugar (ribulose diphosphate) catalyzes a reaction with one carbon dioxide and one water molecule to produce two 3-carbon molecules of 3-phosphoglycerate. The complete C3 pathway is catalytic, because the original 5-carbon sugar is regenerated after further processing into starches and usable sugars; this process requires use of energy from ATP and NADPH generated in the first part of the process. The C4 process differs fundamentally because the end product is fixation of a 4-carbon molecule in two specialized types of cells. The C4 process involves energy costs that are prohibitive in most environments, but the net efficiency of the C4 pathway is said to be greater in habitats that are hot and have high light abundance, especially hot-climate plains having a wet–dry seasonality (e.g., some grasslands).

Some bizarre metabolic pathways may have arisen on Mars due to unique conditions there and to differing elemental abundances in the upper crust due to the specific processes of Martian core formation and differing igneous and hydrothermal fractionations related to an absence of plate tectonics. Some Martian microbes may, for instance, have an alternative lifestyle that favors places that are toxic to nearly everyone else. Chalcophile (sulfur-loving) elements indeed could be somewhat more abundant in the Martian upper crust, and dissolved salts in groundwater may be highly fractionated by cold conditions. Selenium and arsenic, for instance, are dominantly sequestered in the cores of Earth and Mars. Some small fraction (a fraction of a percent of the total planetary endowments of these elements) survived core formation and was retained by the mantle and crust. On Earth plate tectonics and associated igneous processes have offered a means by which a steady rain of sulfides (which selenium and arsenic love) into the core has progressively extracted most of these elements from the mantle/crust system. With plate tectonics lacking on Mars, the same downward transfer of chalcophile elements has not happened on Mars so readily. It would not be surprising if these elements are locally highly enriched in Martian hydrothermal sulfide deposits near volcanoes and in impact melt deposits.

Imagine microorganisms that thrive by respiring selenium and arsenic compounds. Some volcanic brines on Earth are enriched in poisons, such as arsenic and selenium. While wanna-be hot-spring bathers (the human type) are cautioned in Nevada with skull-and-crossbones signs warning of death in the hot springs, some microbes love such places. In fact, they derive nearly as much of their metabolic energy from the reduction-oxidation cycles of As and Se as they do from the more usual sulfur cycle. These environments are rare on Earth, and so are these microbes, but they may be far more common on Mars. If so, a premium may be placed on a microbe's ability not just to *tolerate* high As and Se, but to *utilize* it. This is exactly the line of thought being developed by myself and its chief advocate, Ron Oremland (USGS/Menlo Park, California), where Oremland has considerable experience in studying these microbes and their respiration in Mono Lake, California, and toxic Nevada waters.

The evolution of such adaptations under conditions where metabolism is so low and life spans so long might take longer than the Solar System has provided. Instead, the really powerful innovations may take place among the hyperthermophiles, who then, on occasion, adopt psychrophilic ways. On the other hand, it may be that life's origins incorporate such adaptations right from the start.

CONTACT

The prospect of mucusoid life on Mars is not an outrageous possibility, though I do not know a single reputable scientist who is seriously allowing that Mars may yet harbor intelligent life. That possibility, which I would not lend any serious support, faded from slim credence nearly four decades ago. Microbial life – past and dead life or viable, culturable life – is much more plausible. Life's evidence could abound in the Martian rocks and soils, even right up to the red color of Mars. We just need to find it and show that this evidence is pointing to life, or show that this evidence is lacking or can be produced abiologically.

Even environments that are not now suitable for growing, reproducing life may contain viable microorganisms. Chris McKay (NASA Ames), for instance, has taken part on drilling operations in permafrost in Siberia, the high Arctic, and Antarctica. Those samples, including 3.5-million-year-old Siberian permafrost, have produced microbiological cultures. McKay's group in fact has found that significant metabolism occurs at temperatures as low as –10° C, where nutrients are available in thermodynamically stable, "unfrozen water." McKay is a big believer that ice is the perfect refrigerated storehouse for life on Mars, at least life's corpses. McKay explains his suspicion that "*Mars was alive a long time ago; it has been a tough place to make a living ever since.*" McKay adds that with the climatic conditions on Mars and the long-term stability of permafrost on Mars, *"The preservation of life in ice is much more relevant on Mars than on Earth.*"

Brent Christner and John Priscu have shown that metabolism of certain microbes occurs easily at –15° C, and they recently reported evidence that microbes entrained deep within the Antarctic Vostok ice core, where they were subjected to lower than – 50° C, have nevertheless metabolized. Such a result seems improbable, yet the evidence points in that direction. It is Christner's opinion that possibly even DNA repair of radiation damage could occur at such low temperatures. However, it seems almost inconceivable that certain life maintenance and propagation activities could occur without a free liquid water phase.

McKay states that the death of frozen microorganisms occurs by three processes: starvation, thermal decay (including amino acid racemization and degradation of organic molecules), and radiation damage due to natural low-level radioactivities. Low temperatures actually reduce rates of thermal decay, and since metabolism also declines with temperature, starvation becomes less of an issue at low temperatures. The ultimate limits to frozen life's viability is usually imposed by radiation damage to cells that cannot repair themselves when frozen. In permafrost, the typical survival time against radiation damage is 10 million years. McKay considers that Martian life in permafrost billions of years old will surely be dead. However, McKay's group appears not to have considered that radiation damage will accumulate very slowly in relatively clean polar ice and massive permafrost ice segregations, where K, U, and Th radioactivities are reduced and the ice provides gamma-ray shielding. Thus, there could be northern plains or highlands permafrost ice deposits on Mars that might potentially retain viable deep-frozen microorganisms dating from as far back as the Noachian or Hesperian. The down side of Christner and Priscu's results from the

Vostok ice core is that metabolizing microorganisms are consuming their limited food, and when completely entombed in ice without even brine pockets present, there is no ready means for fresh food to arrive or for the microbe to move to a food source.

The possibility that life exists on Mars or that even habitable environments exists there raises important issues about planetary protection of life on Mars (the danger being forward contamination by Earth's life) and life on Earth (the danger being back contamination when we return samples). There are worries about damage to scientific studies, since, for instance, we would not wish to go to Mars, look for life, find it, and then realize that it is the life we took with us. The more fearsome, though far more remote, possibility would be that some sort of "Andromeda strain" was brought back to Earth, or that we similarly irreparably damage Martian ecosystems. These issues are being dealt with in NASA and ESA by a fairly rigorous set of guidelines for spacecraft going to Mars, although as Charles Cockell pointed out, we have probably already contaminated Mars with Earth's life.

What if endemic life should be discovered on Mars, say some crystallized mucus containing viable, freeze-fried microorganisms in a young lake bed or old icy permafrost? It could happen in the next 10 or 15 years. Would Mankind rally in collective, cooperative exploration? Would biotech companies lead the charge, in order to earn the first rights to exploit the wealth of genetic information and coping strategies encoded in any still-living Martian microbes? Are they and we the product of independent origins, or does Earthly and Martian life share some ancestral root? Is there any useful or dangerous compatibility of life developed and left isolated for 4 billion years on two worlds? Those are questions I take seriously. Less serious, but fun to contemplate: what if extraterrestrial intelligence unexpectedly announces their presence from the bowels of Mars? Would we rally in collective defense or seek alliances to gain advantages over our fellow humans? Will those alien civilizations defend against Earthly intruders? Would we seek interplanetary trade relations? How will historians read the Martian chronicles of the millennia and perhaps eons, and how will we tell our neighbors our story? Would they and we share the same inquisitiveness, or will we see a wisdom to leave one another alone? Have those Martians known of us all along?

What if life, contrarily, should be exceedingly rare in the Universe? Asked at a recent meeting about what it would take to convince him that Mars has no biological past, Chris McKay responded,

> *If we drill a core in deep, old permafrost, and the geophysicists prove that the ice is old, and we don't find evidence of [past] life, then I'll go back to being a motorcycle mechanic. I still have the tools.*

In fact, McKay was not joking. He points out that as a physics student he wondered, as the *Vikings* explored for life and found none, why Mars seemed to be lifeless, or where life might be hiding, since the planet harbors all the elements needed for life. McKay explained that he really would leave planetary science if we could not find life or evidence of past life in this Solar System. "*A rock here and a volcano there is interesting, but only as it pertains to life,*" was McKay's view.

Life on Mars or evidence of past life could be found within a few years, even as early as 2004. It could be within a decade that we discover life – or no life – in a drill core through Martian permafrost. Hence, McKay's career may be culminating this decade. The different prospect of intelligent life occurring in this Galaxy, or of being extremely rare in our corner of the Universe, could be proven this century. Rare is one thing, but what if we on Earth are *alone*? The philosophical implications would be vast. Being alone may seem impossible to some scientists, but I would not dare to be too confident one way or another about it. If we find ourselves to be alone, or reasonably alone, then will the thought of willful burning of a rain forest, of accidental spilling of petroleum on a coral reef or seabird rookery, or of slaughtering other humans by the millions take on a seriousness like never before? Will we suddenly recognize that this is as good as it gets, and realize that deleterious human impacts on the biosphere and planetary life-support system must be halted or minimized? Anyway, will it be too late? Could our economic means of finding a new star and new world already be expired with the wasting of once-plentiful resources on Earth?

We may anticipate such eventualities and ask these hypothetical questions, but until life is found on Mars (or elsewhere) or is shown to be absent in our cosmic neighborhood, these questions are not likely to sway human behavior any more than *Apollo*'s photos of Earth already have. We are, unfortunately, apt to go ahead with our increasingly ruinous impact on Earth, hopefully just to a limit. However, Mars beckons on our near horizon, and with it the prospect that it will be a fresh New World. We just might smarten up.

9

The Future Living Mars

> *This cause of exploration and discovery is not an option we choose. It is a desire written in the human heart. We are that part of creation which seeks to understand all creation. We find the best among us, send them forth into unmapped darkness, and pray they will return.*
>
> President George W. Bush, February 4, 2003, speaking at a memorial to the fallen astronauts of the *Columbia.*

The President's eloquence, special that day, certainly appeals to the human spirit of exploration, which might now be fulfilled by Mars exploration more than any other possible human activity. This hopeful prospect of course begs the question of why, for three decades, we have contented ourselves with Earth orbital human spaceflight operations, a long step back from *Apollo*, and two giant leaps back from where we would be today had we only pressed on from the Moon. The *Columbia* disaster highlights the fact that presently the United States is spending huge sums and taking large risks to fly the same hazardous territory America has flown 144 times. It is always a hazardous journey into low Earth orbit and back to the Earth's surface. Possibly the trip to Mars would carry triple the cost and triple the risk of Earth orbital operations, but the reward would be immensely, disproportionately larger. Alternatively, one-third of the shuttle/space station funding level dedicated to robotic planetary operations would offer a dramatic increase in the pace of true space exploration. However, as things stand, the US is investing heavily in the International Space Station, but Americans, at least, are hardly aware of its mission. Questions have been raised among NASA advocates and space enthusiasts and professionals why we pursue such an expensive engineering feat when the risks are so high and the rewards and public awareness are so low. According to this view, the USA and the world yearn for a more dramatic goal, which Mars can provide. A debate is occurring within Congress whether NASA should be funded merely to maintain the shuttle and build a follow-on crew-ferry craft for more Earth orbital trips, or whether there should be an "*overall vision for the human spaceflight program*," according to one Congressman.

One outcome of the *Columbia*'s disintegration is a current review of American and international space exploration objectives and priorities, especially with regard to human spaceflight. It is too early to say whether this outcome will be a pullback, a continuation of the status quo, or a bold thrust into true uncharted wilderness. We must decide either to go to Mars or to exploit near-Earth space resources, or both; with cost-effective technology, we might build Earth-orbiting solar power plants, fuel Earthbound nuclear fusion reactors with lunar He-3, and extract asteroidal precious metals. Perhaps all things considered, the collective decision will be that we need to go primarily robotically, for a time, in an accelerated Mars exploration program. My sense of the state of the world is that America and other nations will welcome some form of bold and cooperative thrust to throw a light onto the unmapped darkness (that of the troubled human heart as much as anything in outer space) and seek aggressively the benefits of space for the home planet.

The early twentieth-century Supreme Court Justice Oliver Wendell Holmes, Jr, summed up the obligations of an advancing spirit:

> *I find the great thing in this world is not so much where we stand, as in what direction we are moving: To reach the port of heaven, we must sail sometimes with the wind, and sometimes against it, but we must sail, and not drift, nor lie at anchor.*

The day may be fast approaching when we will choose from the best of the cream of volunteers to go forward to Mars. There is a new nexus of sociopolitical forces, both domestic and international, in the USA and around the world, that pushes for the cohesion that would be provided by accepting the technological challenge of human Mars exploration. There is, contrarily, a new quest by some Americans for clear pre-eminence, as expounded by the Project for the New American Century vision for dominance in all spheres. Beyond any narrow political views, Mars exploration may well be forced, in the eyes of some competitive thinkers, by a new Chinese thrust into space; the first Chinese Earth orbital human spaceflight now finally backs up long Chinese claims that their's is a nation poised and destined for the Moon and Mars. While there may be a new space race, Western Mars exploration by humans may instead be assisted by the Chinese space program and considerable remaining Russian expertise. Not only that, but the world may be assisted by India, Brazil, and other nations long on the sidelines of space but which now are pushing forward from their entry into the Space Age. The Indian Space Research Organization has announced a lunar orbiting geochemical mapping mission, due for launch in 2008, and plans robotic missions to other planets. Thus, even India, which currently boasts one of the best Earth resources satellites in the world, is poised to become a major player in planetary exploration.

The reasons for going to Mars include but extend far beyond science and technology; they go straight to the human spirit. Humanity, however, is not a monolithic biological entity, as shown by unending millennia of human competition, conquest, and different approaches in just about everything we do. While fundamental human spiritual and emotional needs have doubtless remained constant since the origin of humanity, the expression of the human spirit – both collective and individual – is amoeboid and heterogeneous, as art and political history, religion and science show well. These dynamics go to the core of social revolution and human conflict as well as to issues of unity and perceptions of "we." It means that when we go to Mars (whoever "we" may be at the time) there will not be any particular pre-ordained ranking of reasons for going there, though we are perhaps close enough in time that we can make some good guesses as to the most important motivators.

Which parts of our human spirit will we take with us to Mars, when finally human beings go there? Of course the crew will carry their own unique experiences, personal reasons, and personal spirituality that make whole people. The crew may include all the common virtues, hopes, dreams, ideals, needs, fears, instincts, and biases of humanity as a whole, but there will be choices and decisions made as to which of these will guide the new Mars society. The day will come when the paramount guiding ethic and goals will be based on the crew's character and collective decision making. Until that day, the public rationale of Mars exploration and the governing force will be whatever basis the chief Earthly space-faring governments establish, as dictated by the Earthly goals and political outlook of the governing bodies. Whatever parts of our spirit and psychological basis are carried onward to dominate the beginnings of a new Mars society, they will be the parts we build and direct here at home. Certainly the basis for conducting Mars exploration will be what we on Earth decide is important and motivating. If the Mars crews are selected and nurtured with a nationally competitive and fighting spirit, that is what will form the basis of the new world, at least at its inception; if they go with an internationally unified one-world spirit, that will be established and may propagate.

Not since the colonization of the Americas has there existed an opportunity to build a new culture and economy and means of governance from the most basic elements of human nature and driving motivations. The exploratory motive, international competitiveness, a colonial and economic domineering attitude, and a can-do pioneering spirit pushed the initial exploration and exploitation of the Americas. These elements of European Psyche had a legacy that shaped, for better and worse, the New World societies, economies, and history during the following centuries, right up through the twenty-first century. This should offer a sobering reminder that our thoughts and decisions today regarding Mars exploration and settlement may have far-reaching ramifications on Earth and Mars for centuries to come. But if any nation should decide that they will imprint themselves on the New World of Mars, and found a colony that will therefore reinforce the home country, that nation will be as sorely disappointed as were the European colonists when they lost their global empires a bit and a region at a time. The new Martians will, in fact, become Martians – no longer Americans or Chinese or Europeans or Earthlings – as soon as they build a self-sustaining economy. It remains to be determined whether they will view themselves as allied with Earthlings, or whether Earthlings will become irrelevant to them, or competitive. Benjamin Franklin, in a 1782 letter to English scientist Joseph Priestly, observed

> *Men, I find to be a sort of beings very badly constructed, as they are generally more easily provoked than reconciled, more disposed to do mischief to one another than to make reparation, much more easily deceived than undeceived, ...*

Whether the character of human beings and the behavioral pattern of the future Mars society will materially diverge constructively from Earth's history may depend on choices made in selecting the first permanent Mars crews. This will be essentially Earthlings' last big choice that will direct the New Martians' future. After that, it will be pretty much up to the New Martians to make of it what they will. It will be interesting to see how this crew selection process unfolds and how it may affect the political and sociological foundations of the new Mars society. Will we seek a broad representation of humanity to venture forth to Mars? Will it be a diverse crew or an ethnically, religiously, politically, and culturally monolithic crew? Will democracy prevail or will a centralized authority attempt to impose a common good? What tolerances will there be for individual behavior and allowances for personal achievement and reward versus communal work for the public good? Will the concept of nationalistic pride become extinct? How will suicidal, murderous, and terroristic tendencies be dealt with, or will these behaviors be successfully selected against right from the start? Will class distinctions be eliminated right from the start, or will it be incorporated into the economic structure of a society based on free enterprise and meritorious reward? Will bigotry be defeated on Mars and left behind on Earth?

Will we – humanity – select and reconstruct a new Martian master race to supersede humanity? Indeed, what human traits are universally shared and coded – even if latently – in human genetics? What improvements on humanity are possible, what "improvements" would be so evil or deleterious in other ways that they must be avoided; what hoped-for improvements are simply wishful thinking? Might some potentially helpful selection criteria exceed political feasibility?

My sense is that the New Mars Society will be far more disciplined than that dominating Earth today. It will be nothing at all like the free-living, almost anarchical society portrayed in *Total Recall*, where a futuristic version of the old Wild West mining culture dominated. My reasoning is simple. It concerns the early stages of human habitation. Lack of discipline, unconstrained greed, and aggressive power plays are apt to result in loss of habitat integrity or other causes of mission failure and loss of the crew.

It is also unlikely that the New Mars Society will carry with it the same class structure as old society on Earth has. As Clint Eastwood, speaking from a battlefield graveyard, simplified it in *The Good, The Bad, and the Ugly*,

> *You see, in this world there's two kinds of people, my friend: those with loaded guns, and those who dig.*

There will obviously have to be leadership, a hierarchy, and a skills structure, and there will have to be a means to enforce discipline. But the loaded gun is not well suited to be that means; bullets and spaceships do not go well together, no matter what Hollywood says. Rather, it will be the common need for survival and collective improvement that best instills discipline. The need for this is evident, but precisely how it will be accomplished successfully is not so clear. The requirement conjures ideal Marxist society, which, as the twentieth century showed, absolutely does not work. Marxism is equally unlikely to work on Mars, regardless of how the crew is selected, because innate human characteristics which doomed Marxism on Earth in the twentieth century would doom Marxism on Mars in the twenty-first century. Nevertheless, a more communal societal spirit than we have on our planet today obviously will be implemented on Mars, as it was during the early colonizing periods in the Americas. Unless and until the planet is terraformed, there will be very limited or no allowance for the disenfranchised mountain man to move off to a new corner of the crater, and no allowance for the leader to disenfranchise anyone, unless that person is shipped back to Earth.

As for organized warfare, that is difficult to imagine being consistent with the survival of any New Martians. General Douglas MacArthur, in his farewell speech at West Point in 1962, quoted Plato (although six Plato experts I contacted have been unable to discover the source of the quotation):

> *Only the dead have seen the end of war.*

It is a key question whether lethal conflict is somehow innate and inalterable in some fraction of humans, or in all of us in particular circumstances, or whether a way will be found to circumvent this widespread tendency or to select against it when we go forward to Mars. To think that somehow a peaceable, prosperous New Mars might develop is possibly just another hopeless Utopian dream. However, if Utopia can be accomplished anywhere, it may be Mars.

MARCHING TO MARS

Crew selection – knowing the immensity of the challenge – is not going to be easy; but neither is the technological endeavor of going to Mars and building the means to stay there. The first step, however, is building the will to go in the first place. John F. Kennedy knew that the goal of human lunar exploration is so difficult that its successful accomplishment would identify the greatest nation in the eyes of the world. Without doubt, this same motivator, with additional motives at play, will be pushing the political decision to go onward to Mars. Whichever nation(s) are involved in the first human flights to Mars, certainly pride and prestige will be motivating factors, along with a profound sense of adventure. Planetary scientist James W. Head, III (Brown University), counseling his students the day after the atrocities of September 11, 2001, considered these emotional factors in that completely different context as he urged a nondefeatist, constructive personal and national response. In the tenth of his recommendations, Head stated:

> *Pride is how we view ourselves. Contribute to a new level of introspection and, through your actions and thoughts, a rededication to the principles that make this country very special, while at the same time working on those things that need*

> *fixing. Prestige is how others view us; perform individually and collectively so that we are worthy of a high level of prestige in the eyes of the world in the aftermath of these events, and in our framing of a response to them.*

Planetary exploration is inextricably tied to pride and prestige at the national level. There is no reason, aside from global politics as usual, for an effort so vast as sending humans to Mars, that every single nation on Earth could not contribute something of material benefit to the effort. Even so, nationalist pride could ride high in a collective global endeavor, just as corporate pride of multiple contributors rode the crest of the *Apollo* lunar exploration wave in the USA. Will there be an internationalist, unified approach? Or multiple nationalism-based efforts? The choices are stark, and the ramifications for the long-range future of humans on Mars are vast.

Global or not, some single nation is going to lead the charge to Mars. It could be one powerful nation or a consortium of nations – or the whole world – who comprises the royal "We" who venture to Mars. We can consider the USA, Europe, China, Japan, and Russia as possible leaders; major contributors are apt to include the other industrialized nations, such as Canada and Australia, and major developing nations, such as Brazil and India.

If the European Union were to lead the human charge to Mars any time in the next two decades, there would be little doubt that a broadly multinational approach would be adopted, simply because there is no other option for them: the burdens would have to be shared and the necessary expertise would be drawn from around the globe, or at least from around Europe, to make the effort feasible. Similarly, Russia and Japan are not apt to go it alone, Russia due to insufficient financial resources and Japan for insufficient technical experience in human spaceflight. The definition of "we" would be very broad, not necessarily (but possibly) for virtuous reasons.

Obviously Russia is well positioned technologically to contribute to a humans-to-Mars mission; if their economy continues to improve, they may well be able to lead a Mars mission or possibly conduct it by themselves; close partnering of Russia with Europe would seem to be the likeliest possibility, considering the international political climate lately. The Japanese could seek any number of partners, but due to lack of experience in human spaceflight, Russia and/or the USA would be the most viable partners.

If China leads the charge, it is difficult to project how inclusive the effort would be, as the shroud of secrecy surrounding the Chinese space program obscures the underlying motivations, which in any case are probably (but not assuredly) similar to the motivations that drive the European and American space programs. Possibly a greater-than-average nationalist sentiment in China would prompt them to try a China-only effort. Their nascent human space experience, however, suggests that China is simply positioning itself to be a major partner in a Mars project to be led by somebody else. I would not be too surprised if I am wrong, however, and China gets to Mars first, even without outside assistance, while the rest of the world sleeps or fights.

If America leads the charge to Mars – which, I think, has a slight edge of being the likeliest possibility – there will be competing domestic political forces pushing for alternative approaches. The arguments will again rest on differing perceptions of need and value. One option will be based on national pride and a particular notion of expediency: America can go it alone, do it America's way, and rely exclusively on American intellect, command of resources, and dedication to high achievement. This unilateralist approach will be advocated by some politicians because it would offer a means of accomplishing the job with only engineering and science-based compromises, without political compromises with foreign powers – and it would show the world who is the top dog. Surely America can do it alone (the only nation where I can say so unequivocally), and I would have to agree that going alone would be simplest in a political sense. But simple choices can have vast consequences. There would be repercussive effects here on Earth and on Mars. A nationalist fervor-based drive to Mars, whether based on Chinese or American or any other nationalism, would immediately be imprinted on the New Mars society by whichever country gets there first or establishes subsequent colonial footholds there. Another approach would be multinational and would appeal to economic sensibilities as well as a fuller perspective of whom constitutes the prideful "we" and which elements of prestige are most valued.

If it is America who secures a permanent base on Mars first, will America reserve the glory exclusively for Americans, or will America offer to share the glory, share the responsibility, and share the burdens with other nations?

Apollo was completed on a basis of pride, and still America reaped considerable dividends in prestige around the world. However, that was when there were only two major space-faring nations who were arch competitors, and each had a flock of allies. America could go to Mars alone, as probably no other nation can, but this option would be perceived around the world as the flagship of Yankee arrogance. America could navigate that route, excluding others; but the consequences would be felt here on Earth as well as on Mars.

Imagine that America or Europe or China launches a flotilla of nations to Mars in a communion of global spirit, making political compromises when needed. The world then will expect that the leading nation leads the way, and world will be happy to make value-added and vital contributions. Personally, I would prefer either a middle road between a fully American and a fully, dependently international approach, or establishment of an international consortium with a compelling power to enforce agreed-upon dedications of effort and timely contributions of hardware from each partner. A poor approach would be one of blind trust that multiple partners would deliver the needed hardware on time, as one nation's failure could jeopardize the whole effort.

These are two fundamental options from which America can choose when finally she makes the decision to go to Mars, if that is the decision. The accepted option will be a measure of America's national self-confidence of her place in the world. The chosen option will provide the basis upon which the world will view America – as a globally conscious, gifted, and gracious superpower worthy of positive respect and honor, or as an insecure or perhaps arrogant self-interested superpower. The issues of prestige and definition of "we" go far beyond Mars exploration; they penetrate to the fundamentals of the historical and future greatness and weakness of America. With America's chosen course to Mars will ride the world's perspective on American ideals, America's economic system, and her trustworthiness as a guardian of freedom and nations' rights to self-respect, life, liberty, and their pursuit of happiness. Of course I write a bit presumptuously here: at the moment, it seems that only America will be capable of sailing to Mars alone; however, the long haul of history shows the rapid ascent and decline of nations, and so it is difficult even to project capability even two decades hence. China, of course, is the new Wild card just brought into play with the launch and successful recovery from Earth orbit of a Chinese astronaut. China may well feel a need to assert her ability to go it alone, but the choices described above for the USA then apply equally to China.

It would be a positive development if China would open its space program for the world to see. Not only would that opening demonstrate a maturity and confidence, but it would allow China to contribute its expertise and discoveries to the world, reciprocating what the world has already given to China. The Chinese could well be closer than 15 years from Mars, as they may have a dedication and a strategic nationalist motive that exceeds that of America and Europe. Whether China forges ahead to Mars cooperatively or in competition with other nations, I personally welcome their effort and wish us all success vicariously through them. The issue of excessive secrecy, however, is troubling. It can only mean one of two things: that there is some hidden motive which the Chinese would not like the world to know about; or, more likely, that they lack self-confidence and fear a public failure. While cultural differences between East and West no doubt play into the Chinese secrecy, the matter of national pride is cross-cultural. Surely it is apparent that national heroes are borne of successes and failures in efforts as bold as human space exploration; success breeds admiration, and failure breeds global empathy as well as admiration for the valiant attempt. Secrecy in this endeavor would forfeit the positive attention that the world would give on the way to Mars, and would hide the glorious effort under a veil of insecurity.

ONE-WAY TICKET TO MARS

When could we go to Mars? When might be tragically too soon? Ever since I can remember, humans have been 10 to 15 years away from going to Mars. We are still 15 years away, in the most optimistic case, especially if it is the US who leads the way. This is based on the simple calculation that it will take, optimistically, 5 years to undo the current huge American budget deficit, and then 10 years to build the plans and capability from the word "go."

I am as impatient as anyone. My impulse says, "Just do it!" But then I think more carefully. We have much to learn about Mars before we go, especially the possibility of endemic life and how the life of two worlds may interact. Many additional things we would like to know could be worked around without knowing the details in advance, but prior characterization of the biological environment is critical. Some desired advance work, fortunately, already is being done robotically. This preparatory work by NASA and ESA is being done openly, for global benefit, a fact that China ought to weigh in considering whether exploratory space programs should be state secrets.

Unfortunately, ever since the Viking Lander results, biological characterization has lagged. Is there any endemic Martian life? How would humans and our fellow Earthlings (microbes especially) impact it? How would endemic "Martians" (microbes especially) treat human bodies on Mars and the biosphere on Earth? I am afraid that in our impatience to visit Mars, we may settle for half answers. In the foreseeable 15–20 years of progress, we will have half answers at best.

We owe it to the astronauts, fellow Earthlings, and possible Martian life to characterize the biological environment there before venturing forth. This is not the Moon we are talking about. My best guess is that Mars either is not a living world, or any endemic life and terrestrial life will have a benign interaction due to incompatible genetic natures or different metabolic schemes; but pity the person who makes that judgment call too soon, and erroneously. We cannot gamble on guesses. NASA has been at the forefront of planetary protection efforts; these efforts must continue to be updated by NASA and the global science community and then applied without shortcuts. At some point a judgment call will be necessary, but it should be after our robots have scratched, drilled, burrowed, analyzed, and examined multiple sites on the surface and in the subsurface of Mars. Only then, without rushing the agenda, should we send robotic sample return craft to Mars. After that is completed and samples are analyzed and tested, and it is safe to send humans, we should send them under conditions of informed consent of the space farers, and allow them to return (if that is the mission design) under the informed consent of Earth's educated global population. For any nation to do otherwise would be, in my opinion, unethical and scientifically unwise. All this is not to imply that any nation is contemplating short-circuiting the needed measures, but I just caution the world's space-faring powers against a politically mandated rush to Mars that could forestall the needed process. The details of an effective planetary protection scheme will require careful deliberation of many experts, and then diligent application by the space agencies.

One argument against going very far in this type of concern is that samples of Mars – the Martian meteorites – have been delivered naturally to Earth, and yet Earth has not been attacked by killer Martian microbes, or if it has been, those effects have long since been accommodated. It is an important point, but not the only point. For one thing, the natural transit to Earth took hundreds of thousands of years under conditions that were not optimal for the survival of microbes. Robotic spacecraft travel for months under conditions where microbial life can remain viable as spores. A human venture would take months of transit under conditions designed to keep things alive and growing.

I do not expect to make many friends by offering recommendations, because each person has his own opinion on when we should go to Mars, how we should do it, and how seriously we should take planetary protection. However, I am compelled to offer my best assessment. We should accelerate the world's robotic Mars programs. NASA's "Follow the Water" and "Astrobiology" themes are right to emphasize if this robotic work is the precursor to human Mars missions. Robotics must be designed, and redesigned if necessary, to meet updated planetary protection measures.

Unless we can get full answers on the life issues from these robotic missions, no nation should consider a quick human jaunt to Mars and back to Earth. After a partial biological characterization of Mars, we could consider sending astronauts, but only on a one-way mission with the designed capability to subsist indefinitely and to grow their presence there. Permanent human habitation would begin from the first human arrivals. The astronauts may be offered a return ticket (which they may or may not use) after they deliver to Earth the knowledge that Earthlings will be safe upon the astronauts' return. This is the selfish view. Better would be to do a sufficiently complete biological characterization for the sake of any endemic Martians, and then go full throttle to Mars when we know that we will not harm them, nor them us. Precursory robotic sample returns require, prudently, much the same caution as with human missions.

If, contrary to my recommendations, some nation sends humans to Mars on a short-term jaunt before we understand the biological environment there, the interplanetary cruise back to Earth might be long enough that the astronauts may learn about the possible existence of endemic Martian microbial life; they may even learn about hazardous interactions with Earth life, even themselves. However, the hazard might exist, but they might not learn in time. The long incubation period for AIDS and the long time required for prion diseases to do their devastating work should signal caution. Furthermore, months and years of space travel will produce ill effects in the astronauts, and isolating the causes will be a challenge.

What if the astronauts should fall ill on the return journey? How would we quarantine them? On a space station? Space stations eventually return to Earth; until they crash, they must be resupplied. Quarantine at a moon base? That would be expensive and requires resupply. Would it be a lifetime quarantine? How could anybody deal with that? The most effective and practical lifetime quarantine would be on Mars, where the astronauts, by design, could raise families and build an infrastructure. The astronauts would venture forth from Earth much as Europeans and Polynesians of the last two millennia ventured across the seas, knowing that return was unlikely. Hopefully and most likely, fears of a Martian Andromeda Strain will not be realized.

Who in their right mind would go to Mars with a one-way (non-suicidal) ticket? Even most astronauts would not go. I would not, either, but so long as it is not a suicide mission, there would be thousands of bright, educated, emotionally stable people who would not give it a further thought, because they have thought it already. They would go knowing that they have a high likelihood of living the rest of their lives on Mars. From these thousands of candidates, we will choose the best. They will be the founders of a new world.

BREATHE EASY

Regardless of the chosen course, making it to Mars will be a tremendous human stride. Getting there, and staying there, will be the toughest technological challenge humanity has ever faced. Fortunately, and crucially, Mars is a world rich with natural resources that will assist our staying power. In August 1989, as President George Bush the First began defending a losing political battle to send humans to

Mars, Vice President Dan Quayle, also the President's chairman of the National Space Council, was quoted,

> *Mars is essentially in the same orbit – Mars is somewhat the same distance from the sun, which is very important. We have seen pictures where there are canals, we believe, and water. If there is water, there is oxygen. If oxygen, that means we can breathe.*

Though lacking eloquence and complete accuracy, and often cited in a long list of famous Quaylisms, this profoundly simple paragraph actually demonstrates a grasp of some fundamentally important discoveries about Mars and its resources for possible future human habitats. Water presents a potentially exploitable resource for its oxygen content as well as the water substance; its resource value will be a primary shaper of any future human exploratory effort. Indeed, "follow the water" is NASA's theme guiding the architecture of the current robotic exploratory program. Water is important not just for future human exploration but for the geological and possible biological history of the planet. The general public and our political leaders understand this.

EXPLORING MARS, NEAR TERM

Humans surely will go to Mars, but clearly a wise decision has been made to reconnoiter the planet with robotic systems first. Since 1960 there have been 34 rocket launches intended to deliver unpiloted spacecraft to Mars. Of these, 10 have been successful missions, and several (the Japanese *Nozumi* mission, the American *Mars Exploration Rovers*, and European *Mars Express* with the mini-lander *Beagle 2*) are still en route to Mars as of this writing. (Some missions, such as the two *Viking* lander/orbiter pairs, involve more than one spacecraft.) Useful but very limited data were returned by a few others (notably the Soviet *Phobos*), but without a science return that can be regarded as mission success. It is evident that Mars is not an easy place to explore robotically, though beyond the Moon it is the easiest planet-size object to deal with. Since the first successful mission in 1965, *Mariner 4*, a running average of the fraction of missions that are successful has not notably changed. So many things must go right; success is a difficult achievement. What has increased dramatically is the observational sophistication and complexity of attempted missions. With that increase, the science return from successful missions has expanded exponentially.

There is a sustained human interest in exploration of the most Earth-like extraterrestrial surface in the Solar System. In today's dollars, the 16 Mars missions launched by the USA, plus Mars scientific research and data analysis, have cost over 5 billion 2003 dollars. While exchange rates make the USSR/Russia's contributions to Mars exploration appear much smaller, in terms of level of effort, the USSR/Russia has spent a comparable sum, and Europe and Japan some more. Mars exploration thus represents a substantial public investment, and it seems destined to continue or increase into the fiscally foreseeable future.

Mars exploration in the 1960s had ill-defined scientific objectives, because the real objective was just getting there, sneaking a peek, and proving the visit to the world. The Soviets had been first in almost everything in space until the first human landing on the Moon scored an upset victory for the USA; and America could not let the Soviets be first to Mars with flybys, landers, orbiters, and eventually humans. Issues of pride and prestige were paramount. It was not just Soviet and American feelings that were at stake; other nations would secure trade and military alliances based in part on who was perceived to be the likeliest winner of the Cold War competition. Scientists educated in Mars studies have now clearly taken command of the mission planning, as some rather impressive science experiment packages have been flown to Mars and will soon be flying. The politicians remain in the background, but the military is barely a hazy silhouette on the horizon, unless it is China's military that looms large in its space exploration program.

A notable recent development is NASA recent adoption of the overarching mission theme, "*Follow the Water!*" This theme hits to the core of why most people think Mars is interesting. Yes, it is based in part on sensationalism (just listen to the Cydonia Clan), but it is also true that water is key to the history of climate, surface geology, and possible life on Mars. Whether there is now or ever has been life, the life-giving substance of water will make Mars habitable by humans and our fellow Earthlings in the future. Though many scientists appropriately object to a tunnel-vision approach to studying Mars by emphasizing life, there is no issue that planetary science can address that is more important to more people than extraterrestrial life. Water in all its physical forms is a key to life: past, present, and future.

The specific agenda of future missions is continually changing, as to be expected in a time of rapidly increasing knowledge of Mars, advancing technology, iterations or revolutions of fiscal policies by major space-faring nations related to political or economic upheavals, interruptions of the exploration strategy by mission failures, and announcements of new plans by other space agencies. A dynamic approach to future mission planning will have to be accepted as a fact of exploring, as it must have been during the classical age of discovery in the Americas. NASA's latest exploration plans, changing at least every other year, are available on the internet, as are those of the European Space Agency. Recent missions and those of the near future are designed largely to search for water and to investigate the geological effects of water and ice. Few missions have been designed with any capacity to search for life, contrary to media reporting (the *Viking Landers* were the chief exceptions that *were* designed for a life quest). However, it is evident that life – Martian life and future human occupation of Mars – is at the root of recent mission design and funding approval. It all goes back to water.

The sounding radar aboard the *Mars Express* orbiter (launch in June 2003, arrival December 2003) is a particularly important experiment that promises to provide a clearer picture of the subsurface distribution of ice and any liquid water or brine that may exist close to the surface. It will provide a means to ground truth some of the remote sensing findings of the orbiter and of previous missions. Of the prior missions, thirteen have been successful. Besides the crucial radar sounder, the *Mars*

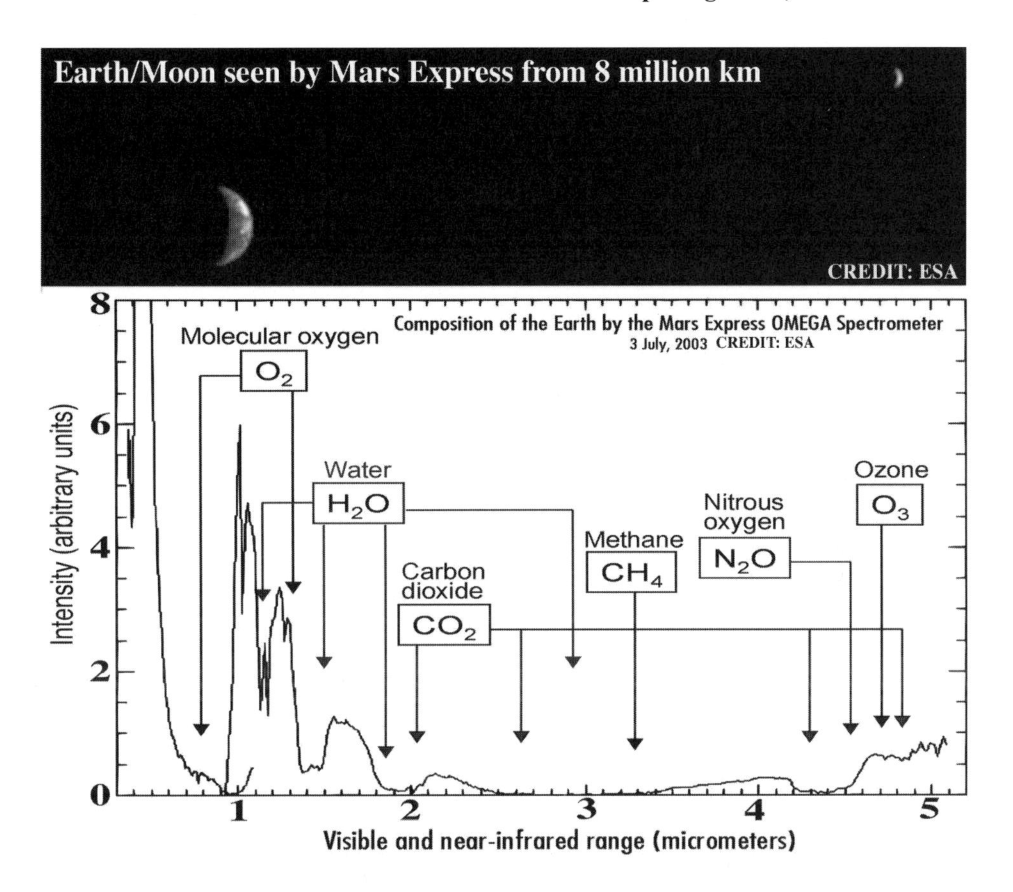

(*Top*) Mars Express took a long view of Earth and Moon while *en route* to Mars. (*Bottom*) A similar view of Earth was obtained by *Mars Global Surveyor* from Mars orbit. This is roughly the view that would be afforded to astronauts from the Martian surface using a small telescope. A large telescope would reveal the shape of Florida, the cyclonic structure of hurricanes, the seasonal greening and browning of Arizona, and the seasonal advance and recession of snow cover at middle and high latitudes. (*Middle*) Mars Express obtained a sample test spectrum of planet Earth, which is sufficient to reveal several key volatiles pertaining to water and life on Earth; increasingly, the distantly measurable chemical species indicate human activities.

Express orbiter is equipped to the gills with a huge chemical analyzing capability to explore the atmosphere's composition and the elemental, molecular, and mineralogical composition of the surface. This mission largely represents to European Mars exploration efforts what *Mars Global Surveyor* and *Mars Odyssey* represent to American efforts – a chance to recover the science that was lost in catastrophic mission failures of the 1990s. In Europe's case the failure was loss of the Russian-led *Mars '96* mission, which carried many European experiments. *Mars Express* and NASA's Mars-bound rovers capitalize on the fortuitous extended life of *Mars Global Surveyor* and *Mars Odyssey*, which will provide data relay. In effect, they are the first comsats in Mars orbit.

Surface landers since the time of the *Viking Landers* have always been targeted for terrains or deposits that are thought have been water lain or water/ice influenced, and this will generally continue. The *Mars Exploration Rovers* (*MER*, launch mid-2003, arrival January 2004), will involve two highly traversable, well-instrumented rovers to be targeted for two different terrains. The decision to build and launch two rovers in one launch window marks a return to the old successful NASA approach of incorporating redundancy and multiple levels. The landers will settle onto the surface *Pathfinder* style with about a dozen bounces on inflatable cushions similar to oversize automotive airbags. It may bounce and roll for a mile before coming to rest. However, unlike the *Sojourner* rover of *Pathfinder*, once each *MER* rover rolls off its landing craft, it will be independent of it. *MER* rovers will have 100 m/day traversability and an advanced capability for *in-situ* chemical and mineralogical analysis. Unlike *Pathfinder/Sojourner*, *MER* rovers will be able to abrade rock surfaces to get at fresh or less weathered material.

The leading landing site candidates for *MER* were winnowed from a very long list of potential sites, all of which were mandated by engineering constraints to reside near the equator (which is energetically much easier to access than higher latitudes). The latest rumors originating in NASA are that the leading *MER* landing site candidates are the paleolake deposits of Gusev Crater and the hematite terrain of Terra Meridiani. Former lead contenders, such as the Athabasca Vallis in the Elysium region and Isidis Basin, are sure to be raised again as potential landing sites for follow-on missions, as will many others as engineering constraints on latitude are relaxed in order to enable the science that really needs to be done.

Gusev Crater is exciting, because there are adequate indications that sedimentary rocks were deposited in standing bodies of water. Not only do we see layered rocks on the floor of the crater, but hydraulically eroded inlet and outlet channels connect the crater to the external environment. There was, in essence, a hydrologic cycle operating at Gusev Crater. The particular sedimentologic depositional mechanisms will be difficult to decipher, however, because the channel morphologies and deposits are so old. Gusev Crater, while clearly affected by dramatic hydrogeological events of the distant past, is so weather-beaten that I wonder how much we will learn of those events so long ago. We are more apt to learn about more recent weathering and erosional processes that have little relation to the dramatic hydrology of the distant past.

The hematite deposits of Terra Meridiani are exceptionally interesting and rather mysterious; are they comparable to the water-lain banded-iron formations of Earth? Are they unrelated to water altogether? Terra Meridiani is perhaps one of the more enigmatic sites on the planet; it may represent a sedimentary deposit somewhat like the great Archaean banded iron-formations (BIFs) of Earth, important iron ores that are also reknowned among geologists and geochemists for their significance relating to chemical deposition in low-oxygen waters prior to the advent of an O_2-rich atmosphere (or during the transition to it). BIFs also contain evidence of some of the oldest life forms on Earth. It would be exciting, to say the least, to discover comparable evidence on Mars, which *MER* just might do if one of the rovers is targeted there. While the most conclusive evidence of life in BIFs is microscopic and geochemical in nature and probably will not be revealed by *MER*, the macroscopic forms of stromatolites would be visible, though inconclusive. *MER* will include a microscopic imager, which will prove to be a useful investigation in many regards, including scrutiny of possible fossils. Of course, we simply do not know what to expect Martian fossils to look like. Stromatolites are as good and as a poor a model as any other. Even if life was never a part of Terra Meridiani's story, the abiotic chemical deposition or alteration processes that caused the ubiquitous hematite mineralization will be extremely interesting; MER is well equipped to study these deposits.

Of all Martian terrains, Terra Meridiani is probably most apt to fulfill the areological expectations raised in the 1958 sci-fi short-story about Mars in *The Outward Urge* (John Wyndham and Lucas Parkes, Ballantine Books): a bleak, flat landscape of hematite-caked sand. In that short story, the areologist among the exploring crew instructed the others, "*Mars is practically all oxides of one kind or another. [Hematite] will be the commonest.*" "Various iron oxides" was the authors' intended meaning, a fact confirmed by the *Mars Pathfinder* magnet experiment, where the magnet pulled magnetized iron-oxide-laden dust from thin air; further confirmation has come from mineralogical mapping by TES (*Mars Global Surveyor*). Wyndham and Parkes based their prediction simply on the distinctive redness of Mars and basic familiarity with *Geology and Mineralogy*, 101. Sometimes basic knowledge is as good as the best.

The details of the formation of Terra Meridiani and its hematite mineralization are bound to be interesting and decipherable by *MER*, regardless of the monotony of large swaths of the terrain. (Though landing within sight and roving distance of distinctive sedimentary layers would be as exciting as a landing site gets.) If this terrain was mineralized by volcanic-hydrothermal processes, as some have suggested, then it will be especially interesting to discover what other potential ore minerals and vital metallic trace constituents are present as a potential natural resource. If it was water lain (with or without a hydrothermal component), then in addition to the possible future resource value, a treasure-trove of information about ancient wet surface environments will be obtained from the chemistry, mineralogy, and sediment stratigraphy of these deposits. Of course fossils of former living organisms would be the jackpot; nobody is expecting this, but I am sure it will be one of the first things *MER* looks for.

The *Deep Space 2* lander probes (lost in 1999) sent from the *Mars Polar Lander* (also lost in 1999) represent a rare shift from engineering mandates to focus on low latitudes. This engineering constraint has never been tighter than with *MER*. This constraint is perhaps a larger compromise on science than the engineers realize; ultimately it will have to be removed or relaxed for subsequent missions in order to access the terrains that are most pertinent to "follow the water." This pressing need has been adequately communicated to the design engineers for future missions, starting next with *Phoenix* (launch 2007), which will land on thinly mantled ice in the Martian high arctic. Nevertheless, we have with *MER* two excellent opportunities to explore Mars in better fashion than ever before possible.

In 2005 we look forward to an orbiter that will image Mars in multiple wavelengths with a resolution as high as 25 cm – our first spy satellite sent to another world. The engineering technology of missions and instrumentation keeps increasing by leaps and bounds even as budgets remain tight.

NASA's and ESA's Mars exploration programs have necessarily undergone almost continual revision; beyond the 2007 *Phoenix* lander, the Mars program is unclear, but it will probably accelerate. Seismic and weather station networks, robotic aircraft, drilling and boring probes, and other interesting missions may be flying after *Phoenix*. A robotic sample-return is in NASA's baseline in 2013. Early sample-returns raise questions about whether we might not yet know enough about the Martian biological environment to risk a sample-return's effects on Earth's biosphere. A viable, endemic Martian microbe might be improbable. However, if it exists, survives transit to and arrival on Earth, and escapes quarantine, it could be highly malicious; we have no way of knowing without a biological characterization of Mars, but I would prefer the necessary biological analysis to be conducted on the Martian surface. Sample-returns eventually will happen and may involve *in-situ* propellant production for ascent. In any event, ambitious sample-return missions have been off in the future about a decade ever since I was a child. Interplanetary robotic sample-returns are difficult and expensive. The Soviets returned lunar samples robotically to salve a failed bid to beat Americans to the Moon with humans, so a technical precedent is now three decades old.

My personal favorite landing site

Gusev Crater and Terra Meridiani offer probably the best possibilities among many potential candidate *MER* landing sites to uncover surface-based physical sedimentologic evidence of liquid water and chemical and mineralogic indications of aqueous chemical activity. These are, of course, important landing sites, and I am especially hopeful for Terra Meridiani, which is a wild card sort of site, seeing that it is such a unique place. Both sites indicate very ancient aqueous activity. There are no indications of any water having been present at those sites for billions of years. To go where water may have been present very recently, we need to go where ice is still present. These sites are at the middle and high latitudes. The middle latitudes are especially interesting for the wide range of ice and water features and indications of recent hydrologic activity. I look to these areas – the lineated valley fill and lobate debris aprons in the fretted terrain in the northern hemisphere and the lobate debris aprons in the south – as the best places to find the ice that will sustain future human exploration and should be explored this decade by robotic systems.

My preferred optimum site for the first establishment of humanity on Mars, and for the next post-*MER* robotic landers, would be southern Argyre Planitia, not least because of the scenic environment and dramatic geology. The geologic history includes pervasive effects of cold-climate processes, including, it seems, glacial and periglacial activity. Argyre Planitia was also the site of an ancient lake, probably one associated with melting glaciers.

The complex geology of Argyre is, of course, fundamentally interesting to me as a geologist, but there is a more general practical reason: complex geology usually means rock diversity, and that will imply a broader suite of ores to sustain an industrial base. The giant Argyre impact must have torn up rocks from tens of kilometers deep – basically a full crustal cross section has been uprooted and exposed. These rocks, though greatly disturbed by the impact, must still be in a decipherable stratigraphic context in the rim mountains. The impact, furthermore, generated enormous heat, which would have caused extensive hydrothermal alteration – always a winner for ore diversity and richness. The rim rocks ultimately may be exploited in huge quantities by an expanding human presence. Of immediate attraction as a possible source of the full range of ore types would be the lake plains and eskers of Argyre Planitia, which are composed of material eroded from the mountains by flowing ice and water. Seeing that the eskers include giant boulders over 10 m across, a single boulder from each major formation in the rim mountains could provide for quite a nice resource base for a small base intent on "living off the land."

We would need to know that buried ice is accessible, and to explore the compositional variation of Argyre's rim mountains and esker boulders. This requires landing a rover near those ridges and being bold enough to traverse them. We may imagine that rovers sent to the eskers will discover an amazing variety of rocks – mantle peridotites, lower crustal igneous cumulates (formed by sinking and accumulation of crystals in slowly cooling magmas), impact breccias from the ancient pre-Argyre highlands, ancient bits of sedimentary rocks from pre-Argyre crater basin-lakes and river beds long since gone, and volcanic rocks. While these might be ordered in some way along the length of the eskers, more than likely they will be a jumbled mix. The dried lake plains surrounding the eskers will be made of fine-grained dust, silt, and sand, perhaps with interbeds of glacial moraines or iceberg-rafted boulders. Wind, water, and ice were the principal agencies of deposition, through the surface layer has seen millions of years of periglacial churning. The sediment source would include the rim mountains (for water- and ice-transported debris) and the global eolian dust system.

THE NEXT GIANT LEAP

Humanity's efforts to explore Mars robotically this decade and last are unquestionably aimed toward a greater future goal – that of human exploration. These missions are helping us to learn about the Martian environment, and they are characterizing potential landing sites and building a capability that will be needed for later piloted missions to Mars. Unmanned rover exploration and radar sounding will be needed to confirm the prediction that southern Argyre Planitia – or any other candidate landing site – is one where ice and water and other resources will be accessible. NASA's and ESA's special interest in ice relates, of course, to the geological and biological roles of water and ice. More than the geological past it is the near future – the next giant leap for Humanity – that is at heart and in mind.

The intense exploration for ice and water is the clearest indication that the major space agencies intend to "live off the land" as much as possible. This will not be *Apollo* for the Red Planet. It will be, or should be, far more. It will be Jamestown and Plymouth.

Humans could land almost anywhere on Mars, but low latitudes would be easiest, requiring least energy and greatest payload mass for each rocket launch from Earth, if it was merely an interplanetary vacation that was sought. The vision for human exploration of Mars presented by Robert Zubrin and his collaborators in *The Case for Mars* could allow inflatable habitat structures placed just about anywhere on the surface. However, staying on Mars and growing a base, using Zubrin's approach or any other strategy, calls for a deliberate selection of landing sites. An immediate environmental hazard, that due to cosmic rays, could be mediated simply by shielding the habitat with regolith, which is available almost everywhere on Mars. With the import of a little bit of hydrogen (or production from Martian water), methanol or methane can be produced for rocket fuel for a return voyage to Earth; it can be manufactured from atmospheric CO_2 available everywhere on the planet. Many basic raw materials for subsistence living and growth of a permanent base would be available everywhere on the surface, but some key materials are not randomly available. Staying on Mars and "living off the land" will put a premium on finding places such as southern Argyre Planitia, which provide a richer-than-average resource base of materials that are most easily differentiated into their constituent elements and oxides. More than any single substance, it is H_2O, particularly ice, that is critical. At Argyre, ice is probably buried, but this must first be proven. The distribution of accessible, near-surface ice is the limiting factor that will restrict our choice of landing site for the first human activities on Mars.

Even before our astronauts need to shield themselves from cosmic rays, even before they land, three vital human resources will be needed. These will be the human qualities of ingenuity, bravery, and steadfastness in the face of adversity. Together, these qualities comprise a boldness of spirit that largely defines what it is to be human. Ingenuity and bravery are self-evident in such a complex endeavor, but every bit as necessary and as difficult to summon and sustain will be single-minded steadfastness. Keeping our eye on the ball in the face of setbacks, even when distracted by competing priorities, was a key for the success of *Apollo*, and it will be equally key in venturing forth to Mars.

Ingenuity, bravery, and steadfastness are not Space-Age qualities of humans, but have been around as long as there have been humans. These qualities have been reinforced by millions of years of natural selection, and are summoned by the best among us when the challenge is greatest. Certainly in any exploration as boldly daring as a human trip to Mars, and especially in going there to stay, there will no doubt be disasters and near-tragedies comparable with the *Apollo 1* fire, *Apollo 13* and *Challenger* explosions, and depressurization catastrophe of *Soyuz 11* and break-up of *Columbia* during entry.

The *Columbia* tragedy on February 1, 2003, underscores an illogical semi-commitment of the United States to space exploration. Compared to a full-up human Mars flight effort, the US Earth-orbital human spaceflight program is carrying perhaps a third of the risk and a third of the financial burden but deriving a tenth of the benefit that a full commitment would entail. Humanity needs to cast our thoughts and set our course for the far shores and uncharted territories, as America once did through the boldness of Lewis and Clark and *Apollo*, as the Europeans did during the classical era of maritime exploration, and as the Polynesians and Arabs did before them. It has possibly taken the *Columbia* tragedy to force a re-examination of our national priorities and direction, but as the re-examination proceeds, I am confident that the overwhelming outcome will be a recommitment of the USA, and with it the world, to achievement of things that are at the hard end of what is barely possible.

As our world leaders make the decision to go to Mars, and as the chosen astronauts take part in training and then embark on their daring trip, all will do so recognizing that danger and tragic setbacks are surely as much a part of the exploratory landscape as Olympus Mons. Our astronauts will be trained in the tasks needed for success, and they will train and prepare for adversity. Even now some of the first future Mars astronauts are school children or college students. Our young people – the hope of the future – are training this very day, as I write, the day after the break-up of *Columbia*; they are learning a hard lesson, as we all are, but one needed nonetheless–that bravery and ingenuity alone are not enough, but must be supplemented with steely steadfastness to persevere. Our nation's leaders, our astronauts, and our children who will become the next generation's leaders and astronauts know that, in today's words of US Senator and Space Shuttle flier Robert Nelson, "*We are in a cause far bigger than any one of us*" (NBC *Meet the Press*, February 2, 2003). This is what presses us onward and outward. This cause transcends politics and national borders. Indeed, it transcends planetary bounds.

The eternal human quest for something better is what led *Homo sapiens* out of Africa, across the Old World, to move on to fill every livable corner of the globe, across Oceania and through the New World. It caused our species to build Babylon, the Niña, the Pinta, and the Santa Maria, and then the *Apollo* moonships. My Grandfather's family in the 1880s sought a better place in life than the one they had in native Norway. They finally settled amid the wolves and wild winters in the wilderness of the rocky, glacier-sculpted north shore of Lake Superior – a land that made more than a passing resemblance to the fjordlands of Norway; with a struggle common to pioneers, certainly they did not find greater security, but it was a land where they knew how to eek a living from the cold waters and rocky soils. The growth of civilizations has largely been spurred and patterned by this search for something new and better; we build on what generations before have handed us, and hopefully find hunting grounds richer, a river wider, or a world somewhere beyond that has just a touch of home. As we now look to our next steps beyond Earth, the same pioneering spirit will be carried forward as we extract sustenance and wealth from Mars matter.

WATERING MARS

In addition to the Right Stuff of our space explorers, the first human presence on Mars will possibly be by courtesy of the world's taxpayers. The political reality is that the high investment needed for Mars operations will not be sustainable for decade after decade. Unless we are to go to Mars on an *Apollo*-style fling, which no one I talk to is in favor of doing, Mars habitats must be developed and expanded with smart use of indigenous raw materials as basic as air, salts, ice, and dust. It would be neither economically nor politically practical for Earth to provide Mars with unending equipment and supplies and it would be especially unrealistic to imagine that resupplies would occur on an increasing magnitude. Century-long British, American and international commitment to sustenance of Antarctic exploration and bases provides an unrealistic model for Mars, because the relative costs of logistical support are disproportionate. Centuries-long European involvement in the settling of the Americas is also an inappropriate model, because two-way trade was a vital aspect of the success of that model. These models do provide a basis for the initial exploration and initial establishment of a human presence, which plausibly can be almost entirely dependent on terrestrial resupply. This cannot last for long unless it is clearly a diminishing level of support or if it is somehow (unexpectedly) balanced by products or commodities produced on Mars and exported to Earth. While I would imagine that smart people on Mars will learn to produce something of value to Earth, it is extremely improbable, given the difficulty and cost of interplanetary travel, that Mars will somehow sustain itself principally by trade with Earth. Scientific knowledge of Mars, and its service as an ultra-long-baseline astronomical site will be valuable to Earthlings. However, self-reliance will have to be an early goal if humans are to establish a permanent foothold on Mars. Earth will maintain an interest in Mars for a time, but expectations on Earth will be that the new Martians will increasingly have to help themselves.

Oxygen, water, rocket propellant, building materials, soil, and raw materials for solar power plants, roadways, and other infrastructure all must necessarily be manufactured and built with minimum reliance on terrestrial imports, though much of the manufacturing technology will initially be built on Earth. Above all other materials, water will be required for drinking and for electrolytic production of hydrogen and oxygen, and regolith will be needed to build shelters against cosmic radiation.

An efficient base might require 4 kg of water per person per day for drinking, washing, and cleaning. This water, however, would be recyclable with simple technology transportable from Earth. It would be helpful to have a two-year supply (29 metric tons for 10 people) on hand to hedge against catastrophic failure of recycling systems; this water can serve double duty as a cosmic radiation shield, and if recycling is efficient it would enable a more liberal use of fresh water. In addition, as considered below, establishment of agricultural self sufficiency will require a lot of water; though a closed agricultural ecosystem will recycle the water, the complete system may require a 730-ton endowment of water. This much is not apt to be supplied from Earth, and anyway is readily available on Mars and can be supplied 100 kg per day in 20 years.

Some industrial chemical and physical processing will consume water. An average, physically active adult human being requires around 900 grams of oxygen per day, an amount chemically bound in and extractable from about 1 kg of water (1.1 liters of ice) or 1.3 kg of Martian air (about 80 m^3 of ambient air at mean surface pressure). In a fully self-sustaining artificial Mars ecosystem, these oxygen requirements would be supplied by plants, which by photosynthesis would recycle the carbon dioxide generated by aerobic respiration. The partly failed experiment at the commercial Biosphere project in Arizona, however, points out some of the challenges of establishing a truly self-contained, self-sustaining ecosystem. In that experiment it was not recognized initially that the mineral and animal component of the soil itself would generate an enormous quantity of carbon dioxide and consume oxygen. Such things will have to be accounted for and accommodated on Mars. A closed ecosystem certainly will not be accomplished in the first few years on Mars, though it must be achieved within the first human generation on Mars.

Until agriculture is fully established, oxygen will have to be provided by a combination of delivery from Earth and indigenous production. Oxygen can be obtained from splitting either or both of two abundant substances: carbon dioxide and water. Industrial schemes have been developed for utilization of both materials. Below I consider the case of water electrolysis and more briefly CO_2 decomposition.

Water will be accessible everywhere on the surface in the form of crystallographically bound water in hydrated salts and hydroxyl in phyllosilicates, which make up a large fraction of ubiquitous Martian dust. At middle and high latitudes, water is additionally available as disseminated ground ice in permafrost and as massive ice of debris-covered glaciers and rock glaciers, and in the polar caps. With deep drilling involving heated drill stems and pipes, water could be obtained as we obtain petroleum on Earth. The obtained water will be extremely saline, so distillation will be necessary.

Without question, deep drilling in geothermally heated permafrost areas, if reasonably thin permafrost can be found, would be the best way to go. If a hot spring could be found, possibly fairly minor engineering would be needed to provide a continuous supply of liquid water. Lacking the good fortune for such a spring, lower levels of geothermal heating might make liquid water accessible within a few hundred meters of the surface. Normal background levels of Martian heat flow, however, would place highly saline water at a depth of about 3 to 6 km – not easy to access. Drilling that deep would require tens of tons of drilling rig and drill fluid mass. A conventional drill rig on Mars will have to be two and half times as massive as an equivalent one on Earth due to the lower gravity there, though rocks and regolith could provide much of the mass. Innovative approaches to water-well drilling will be required because of low Martian crustal temperatures. Although a water well is much desired, more traditional mining methods might be needed to get ice.

Depending on where the base is sited, solid masses of ice or ice-rich permafrost might not be accessible by open-pit or strip mining, groundwater might not be present at accessible depths, and abundant, strongly hydrated dust might not be available at the base site. However, such sites are not apt to be chosen for the first human presence, because H_2O is so vital as a resource.

A water-well drilling process might involve injection of saline ethanol into the drill hole, which would remain liquid and dissolve ice at upper crustal temperatures. The cryogenic solution would serve as the drill fluid, which will increase in density as drilling adds particulates, thereby maintaining drill-hole pressure between the local hydraulic pressure and the lithostatic pressure, as required. After striking liquid water, the drill mud eventually would be extracted from the drill hole and reprocessed for possible later use. Technology developed for drilling in gas-hydrate zones in places such as the Alaska Coastal Plain will be needed to prevent blow-outs of what is expected to be gas-saturated brine. The lack of free oxygen and dominant carbon dioxide clathrate (instead of methane clathrate) will, at least, prevent catastrophic combustion, which is a serious danger when drilling methane clathrate environments on Earth.

More likely, the base would be sited where ground ice is available very near the surface. The latest results from the Mars Odyssey GRS instrument, according to Bill Feldman, suggest that high concentrations of ground ice are very widely distributed above $\pm 50^\circ$ latitude at depths of just 5–10 cm. Permafrost or glacier ice is there for the taking and probably will be the first source of Martian water. The energy advantages of utilizing ice versus mineral-bound water are significant, although as a last resort mineral-bound water in dust deposits could be used at low latitudes.

Even in a water-efficient Mars Habitat, far more water will be needed daily for drinking, food preparation, washing, cleaning, and industrial uses than for human respiration; however, waste water can be recycled, such that oxygen production may constitute one of the major requirements for "new" water. Ten human beings would require 9 kg/day of O_2, derived by electrolysis of about 10 kg/day of water. This amount assumes that CO_2 scrubbers purify the habitat air; otherwise air containing a mixture of CO_2 and O_2 and probably N_2 would be vented and replaced with freshly produced air, thereby increasing the amount of O_2 that must be generated. Some quantity of O_2 will be lost from the habitat, especially during ingress and egress by occupants, and through inevitable leaks. A baseline human habitat of 2,000 m^3 would contain about 570 kg of air oxygen, and with a supporting greenhouse of 30,000 m^3 with a similar artificial atmosphere, the complete habitat would contain about 9 metric tons of oxygen. Hence, the need for almost airtight engineering should be clear. If the full volume of air is lost over the course of each Earth year, 25 kg of replacement oxygen would be required each day, plus any other artificial atmospheric gases. Additional electrolytic O_2 and H_2 would be consumed for surface transportation (using fuel cells), for rocket fuel production, and many other industrial purposes, as well as to keep animals alive. A 10-person base might therefore require electrolysis of 100 kg/day of H_2O for wide-ranging purposes, plus that needed for production of rocket fuel for eventual return to Earth (if that is their desire). This quantity would increase with a rising standard of living and increasing population.

Another 100 kg/day of water will be required to build an agricultural production to the point of self-sufficiency in 20 Earth years (200 kg per Earth day of ice extraction sustained for 20 Earth years represents 1,600 m^3 of ice). An open pit mine in a rock glacier consisting of one-third by volume of ice would grow to about 2 m deep and 55 m in diameter over 20 years. A single typical Martian rock glacier has a volume of order 10 km^3, and at this rate of exploitation it would last 40 million years; there is no risk of running out of ice anytime soon! A fairly primitive, one-person (or robotic), part-time mining operation could yield the required ice, and more, for all such needs. If ice is not so readily available, but salty dust is, a single person or robotic operation still could readily disaggregate and transport the necessary quantity of dust to a distillery and remove the dehydrated waste.

Besides rock glaciers, good base sitings could include icy, dusty, salty dry lake beds. Massive segregation of ice would be helpful, as would massive evaporite deposits. The polygon-cracked plains of southern Argyre Planitia or Utopia Planitia could be fine locations to obtain the raw materials needed for self-sustaining industrial growth if buried ice exists there. The cavernous polygons of Utopia, so long as the ice is not yet totally gone, would provide all these materials plus natural caverns in which much development could be situated safe from ionizing cosmic radiation and proximal to mining operations. However, the "salty-ice pegmatite" model that I presented in Chapter 5, if accurate, would suggest that ices and salts desired for industrial development will be very widely distributed on Mars. The thermodynamic energy required to extract Martian water and dissociate it to H_2 and O_2 is presented in Table 9.1 (Baker, Gulick and Kargel, 1993).

There is a strong argument favoring use of carbon dioxide as an oxygen source: it is without doubt available everywhere on the surface of Mars, since the atmosphere is 95% CO_2. I have emphasized use of ice as a source, because it, too, is present in copious quantities almost everywhere at latitudes greater than 40°, and it can be

Table 9.1 Energy requirements[a] to extract water on Mars

Material	Water extraction energy (kJ/kg)	Extraction + dissociation energy (kJ/kg)	Mass of ore (kg) to yield 1 kg water
Warm spring water, 323 K	–120[b]	12,900	1.0
Pure ice	450	13,500	1.0
Ice, disseminated, 30% by volume	850	13,900	7.3
Ice, disseminated, 10% by volume	1,990	15,000	25
Brucite, $Mg(OH)_2$, pure	5,960	19,000	3.1
90 mass% basalt + 10% brucite	14,600	27,600	31
Kaolinite, pure	9,490	23,000	7.2
90 mass% basalt + 10% kaolinite	36,500	50,000	72
Illite, pure	18,200	31,700	21.8
90 mass% basalt + 10% illite	126,000	139,000	218

[a] Includes any needed energy to heat the ore to the melting point and to melt the ice.
[b] Water extraction energy for spring water can be negative if it is warm, because it is assumed that the excess heat of water above 293 K can be used for heating or other energy applications.

broken by electrolysis with very simple hardware. There is, however, another reason to support use of CO_2 as a raw material – not for oxygen so much as for elemental carbon and CO, as discussed later.

HEATING MARS

Energy will be required for everything done, built, and consumed on Mars. Initial energy sources may be either from nuclear fission sources or solar energy; in either case the first power plants are apt to be built on Earth. However, a sustained and growing human presence will require the new Martians to be increasingly responsible for producing their own energy.

It is worth considering the magnitude of energy required for a self-sustaining base of perhaps ten people. High demands for energy will be placed by electrolysis of water, heating of the inhabited and agricultural spaces, lighting, operation of computers and industrial machines, surface transportation and mining, chemical manufacturing, construction materials manufacturing, and other activities. The energy required just for melting and electrolysis of 100 kg of ice per day would be 1.35×10^9 J/day, or 47 kW if generated 8 hours each day.

Heating the habitat is another large energy sink. When food production begins in earnest, heating of agricultural habitat will place the major demand for heat. A low- or zero-meat diet would require about 1,500 m^2 and 3,000 m^3 of agricultural space per person, all of which needs to be artificially heated. This compares to a generous living/working/indoor-recreation and communal space of 80 m^2 and 200 m^3 per person (2,000 m^3 for 10 people). An initial base might involve one-fifth of this volume, or about 400 m^3 of living space for 10 people, which would grow to the full 2,000 m^3 over the first two decades. During the same period, the agricultural base could be established, along with industrial and storage spaces. Within 20 years of setting up camp on Mars, a nearly self-sufficient base will have been built, covering a total of about 2 hectares (5 acres) of one-story construction for humans, animals, and plants. From that point, the human population could expand, or standard of living can rise. True self sufficiency will no doubt require an increase in population, since available expertise (needed for survival, maintenance, technological redundancy, and economic growth) may increase almost with the square-root of population.

For a mid-latitude region of Mars, the number of heating degree-days will be almost three times that at Barrow, Alaska, averaged over many years. Proper insulation takes new meaning for such an extreme environment. There are high-tech silica aerogel super-insulations with thermal conductivities with an R-factor better than 17.5 for a layer of insulation 2.5 cm thick (0.005 W/m K) for an ambient pressure of 6 millibars; if evacuated, such aerogels can exceed five times that insulating efficiency. An exterior wall and floor filled with a 25-cm layer of insulation would provide at least R175 of insulating power. A square structure with a floor area of 2 hectares (141 m on a side) and interior wall height averaging 3 m would have a total wall, floor, and roof area of about 41,700 m^2. (Building floor-plan shape does not strongly affect this area, but the number of stories obviously does.) The insulation has a very low density, but for this large habitat would have a mass exceeding 400 metric tons. Thus, insulation and most other building materials for the expanded habitat will have to be primarily manufactured on Mars, though insulation for the initial small habitat would be only 5 tons and would be built into the Earth-built habitat. If the average internal temperature is 293 K (68 °F) and annual mean outside temperature is 193 K (–112 °F, for a medium- to high-latitude site), heat losses total around 83 kW for the 2-hectare habitat (annual average, decreasing slowly as the ground beneath the habitat warms up).

Mammalian metabolic heating of the habitat is minor but not totally inconsequential. Each human operates on about 100 W of energy on average, so 10 people would yield 1 kW; add more if there are nonhuman mammals. Of course this small sum of energy comes from food, which is ultimately obtained via photosynthesis; thus, natural sunlight or electrical energy for artificial lighting (from sunlight or fission energy) is the ultimate energy source.

Room with a view

Given the high heating bill, would it pay to have windows? The psychological importance of windows is unquestioned. As it turns out, with such excellent insulation, intelligent passive solar heating using well-designed "smart windows" could provide all the heat required by the habitat; if one is not careful, it could overheat! Any windows, of course, would have to be sturdy, as they would have to confine several tenths of a bar of internal artificial atmosphere (or 0.8 bars for normal air composition), and accidental breakage cannot be tolerated. Triple-pane doubly-coated low-infrared-emissivity and UV-blocking windows have been designed for terrestrial residential use with a U value of 0.23, corresponding to a heat loss of 1.32 J/s m^2 K, and with 70% transmittance of solar radiation (normal incidence). South-facing windows (at 40 °N latitude) of this design, without any further modifications, would have a sharply negative energy balance due to solar gain averaging about +7 MJ/m^2 per Martian day and losses almost twice that magnitude. The mean net loss would be –6 MJ/m^2 (more losses in winter, less in summer), equivalent to –67 W/m^2. A habitat with 30 windows of 0.5 m^2 each would lose on average only 1 kW of heat energy through these windows. This is not a prohibitive amount of energy. Even limited numbers of north-side windows (in northern base siting) would be acceptable, since their daily mean energy balance would be about –10 MJ/m^2 per day (equivalent to –118 W/m^2). Effective greenhouses sufficient to feed 10 people, however, will require far more window area, in the order of 10,000 m^2, which would lose an unacceptable 670 kW of heat energy if the windows are of this fully passive design.

A "smart window" could incorporate external Sun-tracking or stable mirrors (beam-bending technology) and night-time insulating panels. The beam-bending technology could make it so that effectively there would be an equivalent of normal incidence of a full solar flux on the window from Sunrise to Sunset (537 W/m^2 after UV is blocked for 12.3 hours per Martian day). Such windows would not offer good outside views, but for a greenhouse views are not needed. For 70% transmittance of this energy, and allowing for the occasional hazy or partly cloudy days, the windows will admit an average of 320 W/m^2 for 12.3 hours per day, or 14.1 MJ/m^2 of energy per Martian day. Energy losses would be ~3.7 MJ/m^2 by day and 6.6 MJ/m^2 by night. The total daily mean energy balance is almost +4/MJ m^2. But it gets better than this if those night-time losses are reduced by emplacing insulated shades at night. A theoretical balance of over +10 MJ/m^2 is possible. This is a daily mean solar heating rate over the window area equivalent to +113 W/m^2.

At this rate of solar heating, the full heating bill for 83 kW of heating for the 2-hectare complex could be paid for with fewer than 750 m^2 of smart windows on the roof and south-facing side of the complex (for northern hemisphere base siting). Since the greenhouses will require more light than these windows would admit, and we would not want to bake our inhabitants, some not-so-smart windows would also be incorporated. Perhaps smart use of night-time sky views of the starry night and planet Earth would help. Heat balance could be achieved by regulating incoming day-time and outgoing night-time radiation through smart windows, and through effective air circulation. Passive solar heating will diminish greatly during the winter, and since the possibility of dust storms must be accommodated, the base might have over 2,500 smart windows of 0.5 m^2 area each (or fewer larger ones).

Of course this high rate of solar heating requires two mechanical devices per smart window, each set with timers or photodetectors, and it requires these devices to operate tens of thousands of times with little maintenance. Such simple systems have been the bane of planetary exploration for the past few decades. Failed unfurlings of solar panels and radio antennas and jammed filter wheels, instrument platforms, and tape recorders have been the cause of more robotic planetary mission failures and interruptions of science data return than rocket explosions and all other causes of failure since the major bugs in rocket technology were worked out by the mid-1960s. Astronauts have proven adept in fixing or countering such failures, for instance the Skylab space station's solar arrays and Sun shield, but still we would not want the Mars crew maintaining windows for a large part of their time. Nevertheless, +113 W/m^2 is an impressive amount of energy and may make this approach attractive if the mechanical design can be made reliable.

Solar storm shelters

The incorporation of windows is justified from an energy balance perspective, but they would of course never be opened to the outdoor environment, except for one environmental parameter. Windows will always be wide open to cosmic rays. Parts of the habitat that are not windowed could be shielded by polyethylene-regolith bricks, according to some engineering work supported by NASA, or with hydrated salt-bonded bricks. Cosmic rays entering through windows will no doubt damage crops, but not very seriously in their short lives, and the food should be perfectly fine, especially if diets are kept rich in anti-oxidants. Seed stock plants, however, should be fully shielded and illuminated artificially to keep intact their genetic content. Thus, direct solar illumination of food plants is possibly not a severe problem, as long as human farm workers' efficiency can be increased and their time in those spaces can be reduced by use of robotic seeders, weeders, and harvesters.

A lava tube near Flagstaff, Arizona. Squatting 11-year-old Dianna Kargel (recently D. Del Giorgio) for scale. Many lava tubes are much larger, and many are structurally very sound. This lava tube has about 200 m of braided passageways, many with high ceilings, making it as voluminous as a mansion. Here the ceiling is about 2.4 m high, and it ranges to over twice that in this cave. A nearby lava tube is over 1 km long, and some in Hawaii are tens of kilometers long. No sunlight penetrates past ~50 m or so from the entry. Mars has the types of volcanism where lava tubes are to be expected, though they are not as common, apparently, as on the Moon. For service on Mars of this type of natural structure as a habitat shell, natural darkness obviously would require abundant artificial illumination and would prevent passive solar heating and naturally illuminated agriculture. For human psychological reasons, it would be necessary to have some spaces that use natural or engineered skylights, which could also provide emergency egress. A composite approach could use lava tubes for sleeping quarters and supply storage and emergency solar storm shelters, but surface habitats for day-time work and recreation. Heat retention eventually would be fairly good with thick insulation, especially after the massive rock walls were warmed by conduction from the heated habitat.

The background cosmic-ray dose on Mars is not as bad as some researchers have stated. It is reduced significantly at the lower elevations by a larger atmospheric air mass; the annual dose rate (unshielded) is one-fourth to one-half that of space station crews, who have already spent year-long periods in orbit with acceptable health impacts. Even at moderate elevations chronic exposures of agricultural specialists for an hour a day would not be too hazardous (except during solar storms), especially since the agricultural spaces will be about 50% shielded. More serious will be the exposures of outdoor workers, such as field geologists and construction specialists, whose cumulative time outside the habitat will have to be limited to perhaps 1,000 hours per Earth year. Ill effects to people with life-long careers spent largely outdoors are likely, but may be tolerated as an occupational hazard.

In the main (nonagricultural) human habitat and most indoor work spaces, however, cosmic rays will have to be more limited. While one is working and especially sleeping, direct outdoor views through windows are not necessary. Direct window views of the external environment could be limited to hallways and airlocks and other spaces that are used briefly or rarely. Elsewhere, direct line-of-sight through windows could be shielded by water containers (cosmic rays are very well shielded by hydrogen contained in water), while the exterior view could be made available in mirrors, which would be silver-coated water containers.

Solar storm shelters would protect against transient, orders-of-magnitude rises in cosmic ray dose that occur during major solar coronal mass ejections. The storm shelters would be small spaces stocked with food and water and communications links to Earth, and will be heavily shielded with polyethylene-regolith bricks or water-storage containers. During intensified flare activity, the crew would seek shelter for a few days until the storm passes. Better yet, living and working spaces or at least solar storm shelters could be built into lava tubes or caverns produced by sublimation of massive ground ice or dissolution of massive salt accumulations. There are strong indications that these types of structures exist on Mars in many areas.

ELECTRIFYING MARS

Passive solar heating will get the base only so far toward paying the energy bill. Another major power consumer will be electrolysis ($\sim$160 kW daily average). By contrast, if solar radiation is used to support most agricultural lighting needs, only about 2 kW is needed for the 2-hectare 10-person habitat for smart lighting, e.g., systems with lighting tripped by infrared motion/presence sensors; a few kilowatts for stoves, ovens, and water heaters; a few kilowatts for all computer systems; a few kilowatts for deep-space radio communications. Electrical needs for scientific experiments and light industrial manufacturing and artificial lighting of seed plants may total close to 100 kW, though this will eventually increase to much higher levels. Thus, if smart passive solar heating is used, total electrical usage averaged through the year will be of the order of 200 kW, with a diurnal demand cycle ranging from about 150 kW (night) to 250 kW (day). (Electrical needs for the modest initial base will be lower.) Night-time energy will have to be produced during the day and stored in batteries, fuel cells, or fly wheels. If supplied by solar photovoltaics or active solar-thermal systems, an allowance will have to be made for low solar energy production in the winter (if the habitat is sited at a temperate or polar latitude) and during major dust storms. Allowing also for additional energy to go toward rocket fuel production (possibly the activity placing the greatest energy demand), day-time peak capacity (sun at zenith) of about 2 MW will be needed. This requirement will be built gradually over two decades as the base expands toward self-sufficiency.

For 25% visible/near infrared solar energy conversion efficiency of a photovoltaic solar farm, and with glazing to prescreen damaging UV radiation, the light-harvesting area to supply 2 MW peak power will be about 15,000 m^2 (1.5 hectares) of

This solar photovoltaic power station in the Mojave Desert, California, is a model for what a photovoltaic grid might look like on Mars. The desert cities of Barstow, Palm Springs, and Bakersfield derive most of their electrical energy from renewable solar and wind sources.

array area. The full array will cover more than 2 hectares considering also nonphotovoltaic surfaces within the arrays. Mechanical solar tracking would allow a smaller array area, but still it will be over a hectare.

While most commercial residential solar panels these days have conversion efficiencies around 12%, efficiencies as high as 30% are attained in a wide range of photovoltaics, so the 25% assumed above is conservative and allows for dust accumulation and radiation degradation. Humans or robots would have to periodically dust the solar arrays, though the Mars *Pathfinder* spacecraft showed that solar panel performance decreased only slightly (percent level) even after dust storm activity. Some new, high-efficiency photovoltaics are of the thin-film design, including InP/InGaAs cells. Experimental production of thin-film semiconductors by epitaxial growth has been accomplished in the space shuttle wake-shield facility, so this manufacturing effort does not require a large industrial infrastructure. Raw materials for the solar farm would be supplied partly from Mars and partly from Earth; the super-pure semiconductors and high-vacuum manufacturing facilities would likely be supplied from Earth, at least for a while. A hectare-size solar farm could be built gradually on Mars with a fairly small effort over 20 years, and then could continue to expand. Active solar-thermal systems involving solar reflection/concentration arrays (fancy mirror arrays) could heat and melt salt mixtures, which then could flow into water boilers, which might power steam turbine generators.

I would be remiss if I did not mention wind power. I doubt that this will ever be feasible on Mars, but it's worth considering; the number one reason is the low atmospheric density on Mars and the low energy density of Martian winds. Using the ideal gas law and considering the higher mean molecular mass and lower mean temperature of Martian air compared to that on Earth, and taking 6 mbars as the mean surface pressure on Mars, the same speed wind on Mars will carry about 1.2% as much energy as on Earth. Another way to view it is that a wind blowing 20 m/s (45 mph) on Mars would carry as much energy as a slight breeze blowing 2.2 m/s (5 mph) at sea level on Earth. However, there are some slight compensating factors. The lower gravity will mean that lower-mass wind turbines and support structures would still be robust, and their lower mass and lower gravity would mean that frictional rotational losses would be much less than on Earth. The lack of vegetation on Mars would reduce boundary-layer effects that retard wind speeds relevant to wind turbines. Also, there are large regions where katabatic (slope-related) and sublimation winds may consistently (at least seasonally) blow much harder than is common on Earth. Finally, there are low-elevation regions, such as in Hellas (where katabatic winds are also high), where the atmosphere is significantly denser. There remains a problem in some regions (including Hellas) involving CO_2 winter-time condensation on the turbine; this would not be a problem in certain areas where slope winds are expected to be high, such as at the base of the Olympus Mons scarp and in Valles Marineris. In parts of Valles Marineris, mean wind energy densities may very well reach 10% of those common in the Mojave Desert, where wind turbine generators produce a few percent of California's electrical needs.

Just during the course of writing this book, the political feasibility of introducing nuclear power generation, perhaps by radioisotopic thermal generators (RTGs), has been reintroduced, at least for the US political scene. If ever there was a place where nuclear power seems a reasonable option, it is in deep space and some extraterrestrial planetary environments. The political change making this possible was the politics of the Bush Administration. Use of nuclear energy on the Moon, for instance, would not be troubling. (For one thing, it has already been used there during *Apollo*.) However, meltdown issues for scaled-up systems remain a serious environmental hazard on Mars, where a hydrosphere exists with the potential for steam explosions and spreading of radioactive waste, and where life may very well exist in the subsurface or may be propagated in terraforming. Nuclear energy offers a technical alternative to solar, and in many regards could be expedient in the short run, solving the problem of night-time energy storage and planning for winter-time and dust-storm low-power output of solar energy. Solar power, however, would offer energy with less potential for single-point catastrophic failure. (Short of any environmental calamity, loss of the nuclear generator's capacity could jeopardize a mission, if we were dependent on it.) If nuclear energy were to be used on Mars, it would not be a bad idea to develop and utilize solar energy in tandem. All in all, it would seem that solar photovoltaics offer the simplest and least problematic near-term and probably long-term solution.

FARMING MARS

Fertilizers and soils must be manufactured and greenhouses fabricated to initiate significant agricultural production. I foresee a gradual build-up of agricultural capability over the first 20 years, with most food initially supplied from Earth; much of that food resource will be landed on Mars near the base even before the astronauts arrive. Dried and freeze-dried food and packaging amounting to 1 kg per person for 10 people and 10 years, for example, will amount to 40 tons, a reasonable quantity to deliver from Earth in one or a few resupply missions. This 40 tons, however, may fulfill a decreasing portion of imported food requirements over a period of 20 years, while indigenous Martian agricultural production increasingly comes online. A much smaller quantity of consumable water and oxygen will be provided, because the expectation is that human Mars operations lasting more than a few months on the surface will require indigenous production of these consumables starting soon after landing (or even before by robotic systems). Even after the 20-year build-up period, there will likely always be a food reserve for several years' requirements in the event of a massive crop failure – enough food to allow the new Martians to try, try again, and then to seek a return to Earth if absolutely necessary. Return to Earth would not be the goal, however; within 20 years a successful Mars base may never again require a shipment of food from Earth.

The basic elements for soil are available on Mars, but Martian regolith is not pre-made for agriculture. The process will involve sieving to remove boulders, cobbles, pebbles, and coarse sand-size material; remixing the fine fractions in the right ratio to produce an ideal silty loam; mild heat treatment to remove undesirable adsorbed constituents, such as radon gas and peroxides, which will have accumulated to dangerous levels; and addition of water and fertilizers. One and a half hectares of soil averaging 30 cm deep will have a total dry mass of about 9,000 metric tons, thus requiring daily soil production of about 1,250 kg to meet the target of food self-sufficiency in 20 Earth years. Water (from ice) will have to be supplied at a rate of over 100 kg per day. Such massive soil manufacturing is not completely necessary, since hydroponic agriculture offers an option. However, manufacturing of the fertilizers in any case will be necessary, and the physical processing of regolith is among the least challenging tasks the Mars crew will face, since the basic materials are right there and pre-crushed anyway.

A typical composition of dry plant material (young corn plants for example) is about 43% C, 44% O, 6% H, 1.5% N, 1.2% Si, 0.9% K, 0.2% S, 0.2% P, 0.2% Ca, 0.2% Mg, 0.1% Cl, 0.1% Fe, 0.04% Mn, plus essential traces of B, Al, Ni, Cu, Zn, and Mb. Photosynthesis adds C, H, and O to plant material, and legumes add soluble nitrogen by fixation of atmospheric nitrogen. The Martian atmospheric partial pressure of CO_2 is over an order of magnitude greater than that on Earth, so that is not a limiting factor for photosynthesis. It is not clear that nitrogen fixation could occur with Martian air composition, especially if the agricultural facility is not pressurized. It is entirely clear, however, that sulfate, phosphate, nitrate, and potash fertilizers will either have to be added to Martian soils, or modified from the naturally existing abundances. Thus, soil manufacturing will be a major crew activity. Some chemical aspects of fertilizer production from Martian materials is described below. Once the fertilizers are produced, an efficient mulch recycling system will mean that the nutrients are partly preserved, though some amount is not recyclable without chemical reprocessing.

FROM DUST AND THIN AIR, CITIES WILL RISE

The classical Mesopotamian civilization in what now is Iraq was built largely of clay, and its history was written in clay. Though the technology will be different, it is not a stretch to say that future cities of Mars will rise from little more than dust. The need for radiation shielding may favor sites in lava tubes or in caverns left by sublimation of ice wedges or by dissolution of salt beds. The first human Martians thus may be cave men and women. Although these will be high-tech people, they will share some crude but vital technologies with our Neolithic ancestors, including also commonalities in some building materials.

The first human base will use both rigid and inflatable structures, surface transport vehicles, solar (or nuclear) power stations, and materials-processing plants transported from Earth. The growth of the base will involve structures built with Martian materials and built with every intention of being the start of a permanent presence. The first humble, crucial step will be attainment of self-sufficiency of an initial small crew of perhaps 10 people. A very high degree of self-sufficiency will be an engineered evolution requiring perhaps 20 years, though complete independence in every regard may take several more decades as industries are built and the population of New Martians increases. As the Mars base is built, engineers will have to account for unique conditions on the planet, which include some positives as well as challenges. The lower gravity can enable lighter construction, and the low-density atmosphere can enable structures to withstand enormous Martian winds with little danger of blowing over. The abrasiveness of ubiquitous Martian dust will have to be

Anasazi home construction at Mesa Verde and throughout the "Four Corners" area of the south-western United States in roughly the twelfth century marked the ascent of a nascent civilization. This civilization was short-circuited across the region soon after these homes were built, and it was never restarted. The reasons for failure of that civilization are not known, but are thought to be due to drought or possibly disease. The construction is quite solid and consists of locally derived stone slabs with a simple mortar of sand, mud, ground corn cobs, and other local material. On Mars, building construction will similarly involve use of local materials and a combination of very low-tech and high-tech construction technologies.

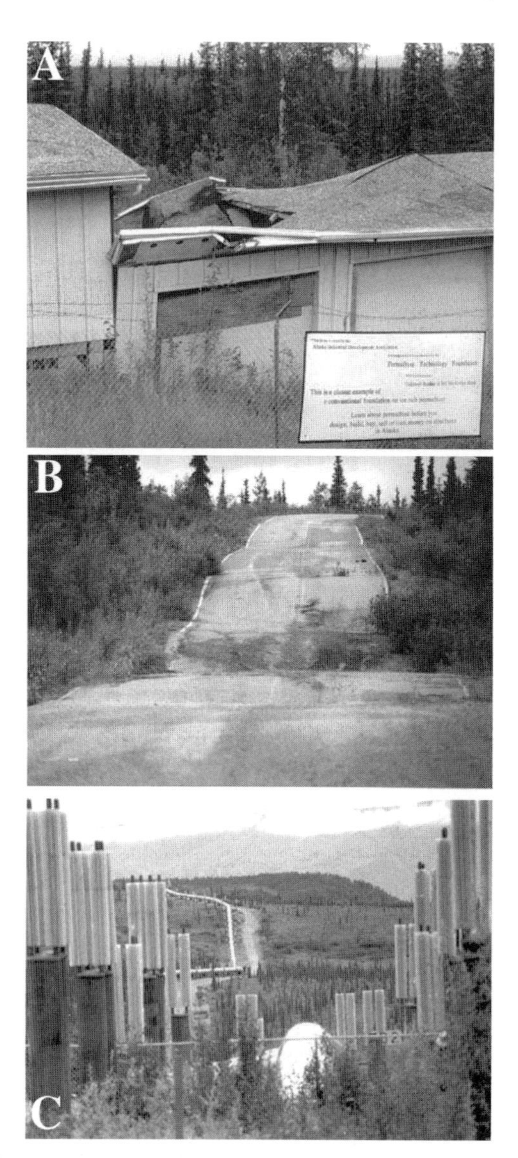

These examples of construction on Alaskan permafrost hold some possible lessons for settlement and construction on Mars. Panels A and B show construction-gone-awry due to thawing of permafrost. In the case of the Alyeska Pipeline, ingenious use of novel technologies have permitted hot oil to be moved across permafrost without melting it. Elevated pipeline sections deal with the problem in some places, while other sections are buried, with the permafrost kept frozen with use of passive liquid-ammonia cooling. On Mars, extremely low temperatures may prevent thawing related to construction, but sublimation could be an issue to face.

considered for anything that has moving parts (including human bodies as well as spacesuits and machines) or requires optical-quality surfaces. The brittle behavior and extreme thermal expansion–contraction of materials at severe Martian temperatures will be a challenge for engineers to deal with effectively. The widespread occurrence of ice-rich permafrost and hydratable salts will mean that volume reductions due to sublimation/dehydration will have to be either avoided or accounted for. The damaging role of permafrost thawing in Alaskan and Siberian construction should be well considered by Mars construction engineers.

The development of a permanent Mars habitation will require use of natural resources well beyond that needed to obtain water and oxygen. In addition to building stone, metals, plastics, ceramics, composites, photovoltaics, cement and concrete, glass, fertilizers, insulation, fabrics, and explosives will be required; a basic chemical industry to produce pure oxide extracts, solvents, lubricants, acids, oxidants, and reductants will be required. If the human presence is to be sustained and grow, all these materials and food will be needed in quantities larger than the amount that is apt to be transported from Earth.

Isolation of pure materials and manufacturing composite ones on Mars will start with basic mineralogical and lithic materials as well as air and ice. While information technology, surface vehicles, and many other complex equipments are likely to be supplied from Earth for many years, a basic mining and chemical industry will be required from an early stage. Mining and ore transportation will require an infrastructure of roads and railways. Plants for physical processing – crushing, grinding, polishing, and separation by particle size, density, hardness, and magnetic properties – and facilities for chemical refinement and manufacturing and energy production will have to be built largely using Martian materials, as will finished products for use in maintaining and expanding the habitat.

Early preparations for these activities have already begun with the general geological and mineralogical exploration of Mars. While we would wish for riches of gold and platinum (not for vanity but for industrial utility of these materials), we will settle happily just for the reliable, abundant reserves of ice and salty dust that we already know are so widely distributed. Ice provides not just drinking and irrigation water and oxygen for respiration, but oxygen and hydrogen for chemical redox processing of ores and chemicals. Air-fall and water-lain dust deposits – accumulations of micron-size silicate and salt materials – are available in many places at all latitudes on Mars, though in some places the deposits are thicker than elsewhere. Additional surface-based geological exploration will be required before we can pin the hopes of human exploration on these materials at a particular site. Ice-laden dust is the ideal ore for almost everything needed to sustain a human presence. From icy dust we will make steel, plastics, glass, and chemicals; from this dust we will irrigate our crops, feed our animals, build our roads, generate our energy, and make our fuel for surface transport and rocket propellants to maintain contact with Mother Earth. The dust of Mars is made of four crucial components: ice, salts, silicates, and oxides. All, along with Martian air, are vital raw materials.

The salt content of Martian dust and the crust is considerable, and global atmospheric circulation has produced a somewhat homogeneous and widely distributed ore, i.e., one that we can depend on being present in some quantity at all sites where dust is not continuously blown away. (Recent rover results suggest that the dust is somewhat heterogeneous.) There are, unfortunately, only indirect estimations of the mineralogical and chemical content of the salts. *Viking Lander* and *Pathfinder Sojourner* soil analyses indicate that the soil contains roughly 0.5% chlorine, 3% sulfur, 0.4% phosphorus, and a trace of bromine plus elements that make up silicate and oxide minerals and volcanic glass. A thermodynamic model by Sidorov and Zolotov predicts that weathered rocks on Mars would yield a soil containing 13.5% anhydrite ($CaSO_4$), 0.2% kieserite ($MgSO_4{\cdot}H_2O$), 0.3% halite (NaCl), 0.5% sylvite (KCl), 1.1% bischofite, and 0.4% Ca phosphates, including apatite ($Ca_5(PO_4)_3(F, Cl, OH)$). At least at high and middle latitudes, the salt assemblage would be fairly hygroscopic; some phases would include larger amounts of hydrate water than indicated in that model. The 84% non-salt mass fraction of Martian soil, if in thermodynamic equilibrium with the atmosphere, would be composed of clay minerals, such as montmorillonite and talc, hematite, quartz, rutile, and pyrolusite. The ideal thermodynamic predictions of salts compare to the salts observed in Martian meteorites: calcium–magnesium–iron carbonates, gypsum ($CaSO_4{\cdot}2H_2O$), and halite. (Considerable new evidence points to the existence of abundant hydrated sulfate salts of Mg, Ca, and Fe both globally and at the Mars rover sites.)

A synopsis of the evidence is that Martian dust is overwhelmingly rich in sulfates and secondarily in chlorides, carbonates, and phosphates. The dust is already pulverized to particle sizes of one to a few microns. The small grain size lends the soil a very high specific area that optimizes chemical reaction rates and dissolution in appropriate solvents. Fresh crystalline volcanic bedrock also probably contributes to Martian dust and regolith. This component is composed of typical common terrestrial minerals in basaltic and andesitic rocks, including olivine, pyroxene, feldspar, amphibole and rarer quartz and other minerals. Volcanic glass and semiglassy microcrystalline volcanic rock, as expected for instance in volcanic ash, is very abundant on Mars and would provide an excellent plant growth medium.

Salts and clay minerals will be generally available on Mars wherever dust has accumulated. However, aqueous processes have operated episodically through Martian history, and these would have left evaporitic deposits and veins of nearly pure salts. A dry lake once watered by flowing streams or groundwater would be the ideal source of these minerals. Not only would the valuable sediment be abundant, but in dry lakes the sediment has been naturally sorted by both grain size (for clastic components) and solubility. Fractional crystallization driven by evaporation or freezing is one of the finest means by which valuable minerals are enriched and sorted into economic deposits. Dry lake beds of California and many other areas have provided part of the foundation of Earth's global industrial economy, particularly a chief basis of the chemical industry. One playa alone, Searles Dry Lake, has produced about forty billion dollars' (2003 $US) worth of chemicals, especially trona ($Na_2CO_3{\cdot}NaHCO_3{\cdot}2H_2O$), and secondarily borax ($Na_2B_4O_7{\cdot}10H_2O$), potash (various potassium salts), and sodium sulfate. IMC Global (which operates IMC Chemicals, Inc.) extract valuable soluble minerals from Searles

Searles (Dry) Lake, southern California. (A) Soda and boron minerals go out by rail from Trona; coal comes in. Tufa towers in the background mark former lake-bottom spires, where warm mineral-rich springs produced mineral precipitation. (B) The town of Trona, named after the mineral primarily responsible for the town's presence on one of the most forbidding parts of the Mojave Desert. (C) A system of brine pipelines delivers water to injection sites and transports brine to filtering and fractional crystallization plants, where various salts are isolated. Although the town obviously struggles, 40 billion dollars worth of mineral have been extracted from Searles Lake since 1874.

evaporite beds by forcible hydraulic injection, which drives groundwater flow through the salt beds. Extracted brine is then refined mainly by filtration and fractional crystallization.

One can envision use of this method if hot water or steam could be injected at rates sufficient to overcome the tendency to freeze and eliminate permeability. On Mars at low latitudes it may be possible to use calcium–magnesium chloride brine or sulfuric acid eutectic solution injection to extract other soluble components, but at those low temperatures most sulfates and carbonates have very low solubility. At middle and high latitudes, temperatures are too cold for such brine solutioning extraction techniques. Terraforming is probably going to be the thing that introduces the feasibility of solution mining. A more practical method on Mars under present conditions might be physical excavation of raw solid ore and transport in bulk to a refinery. This is how it was done in the 1870s during the first exploitation of Searles Lake by John W. Searles, a frustrated Californian gold miner, who made his riches by literally raking borax from the lake bed's surface. It is not difficult to imagine that the first salt deposits on Mars will be mined by an astronaut with a shovel and wheelbarrow or simple earth-moving machinery. Brine solutioning would be used, however, in the refinery to extract the valuable salts and separate out the insoluble constituents.

Salt bricks, ceramics, and Portland cement

Martian adobe-like salt-mud bricks will almost surely be a major construction material for habitats, storage buildings, radiation shielding, and roadways. Saline aqueous solution can be evaporated to a hypersaline brine, then mixed with fine-grained regolith, dust, or sand; moulded into bricks; then frozen. The brine could be the byproduct of solution-extraction of chemicals from soda-ash plants. The brine composition could be controlled such that the net volume change during freezing is zero or a small value, so that the bricks will not fall apart when frozen. The hydration state of salts in the interstitial precipitates can also be regulated by adjusting the salinity of the brine and the solute composition, such that the bricks will remain stably hydrated under ambient Mars surface conditions. A mortar can be similarly prepared. Hydratable mortars and adobe-like bricks can have a sealant applied to prevent crumbling or fracturing that could be caused by changes in hydration state. True Portland cement and concrete can also be prepared from Martian salts and regolith, as can true ceramic bricks, but these will require more extensive chemical processing.

Ice-cemented dust at Martian temperatures may be almost as hard as granite, except where it undergoes freeze–thaw (and sublimation-driven) disaggregation. The exploited material likely will be an ice-cemented lake bed or deposit of ice-impregnated air-fall dust. Mining and processing can be envisioned thus. The frozen dust will be disaggregated with explosives a safe distance from the habitat. The crumbled, blocky masses will then be transported to a distillery adjacent to the habitat. The raw ore will be further crushed to pebble-size material in an industrial jaw crusher and placed into a low-temperature, slightly pressurized melting vessel. The ore will be heated to perhaps 50 °C using electrical resistance heating coils or focused solar radiation. The resulting saline water will be drained and circulated through removable radiators. The cooling solution will progressively precipitate a chemical sequence of salts in the radiators. The salt-encrusted radiators will be periodically removed and replaced with cleaned radiators. The encrustations will be chipped from the radiators and flushed with hot acidic water in a salt distillery; the radiators will then be reused. Salts will be reprecipitated and heated until sulfur dioxide gas is driven off from sulfate minerals, leaving ceramic oxide residues, such as MgO and CaO. The gas will be used in wide-ranging chemical production. The ceramic oxides will be used in glass and porcelain manufacturing, cement and concrete manufacturing, in metal alloy production, for production magnesium-based batteries, and in the chemical industry.

Portland cement can be manufactured from lime with careful further additions of silica, alumina, and other oxides, plus addition of gypsum salt or anhydrite. These materials are dehydrated and powdered. The extremely hygroscopic nature of the resulting mix then makes for the cement qualities. Portland cement can be mixed with sand, pebbles, and water to make concrete. In outdoor applications at middle and high latitude areas of Mars, where brine melting temperatures are not expected to be attained (for instance, in basements or the bases of roadways), a modified recipe for Portland cement and concrete might include addition of any type of hydratable salt, i.e., almost all salts on Mars.

Sulfuric acid production

Perhaps the most ubiquitous, abundant, diverse, and versatile class of materials on Mars are salts. Produced by hydrothermal and low-temperature aqueous chemical activity, and redistributed globally by wind, salts are apparently abundant on Mars and may be accessible almost everywhere in dust. Salts are chemically diverse and exceptionally important as raw materials. Also with Martian air and hydrothermal minerals, a robust chemical industry can produce many of the basic ingredients and products needed for sustenance and expansion of a Mars habitat. A few pertinent examples of chemical processes will make this point.

Sulfuric acid is the largest product (in mass) of the chemical industry in the United States. About 150 kg is produced per person each year. Sulfuric acid is used in fertilizer and gypsum manufacturing, in petroleum refining, as a solvent, an etchant, and many other purposes. The agricultural, semiconductor, construction, and many other industries are dependent on sulfuric acid; it will be similarly important on Mars. Sulfur for the acid is obtained ultimately from a sulfate, a sulfide, or elemental sulfur. Each of these may be found near volcanoes or in ancient volcanic ore deposits. Sulfides are probably contained in Martian dust, and the sulfates surely are major components of dust. The first step in manufacture of the acid is production of sulfur dioxide, which on Earth is usually made by burning of elemental sulfur or roasting of sulfides, the reason being that oxygen is plentiful in the atmosphere. On Mars, there is no *a priori* reason why reduction of sulfates should not instead or also be used to produce SO_2.

Representative reactions are:

$$4FeS_2(s) + 11O_2(g) \rightarrow 2Fe_2O_3(s) + 8SO_2(g) \quad (9.1)$$

$$CaSO_4(s) \rightarrow CaO(s) + SO_2(g) + \tfrac{1}{2}O_2(g) \quad (9.2)$$

$$MgSO_4(s) \rightarrow MgO(s) + SO_2(g) + \tfrac{1}{2}O_2(g) \quad (9.3)$$

$$S(s) + O_2(g) \rightarrow SO_2(g) \quad (9.4)$$

$CaSO_4$ or $MgSO_4$ are thought to be widespread and abundant components of soil salts and might occur in some lake beds on Mars. Their usual water of hydration also would be available for extraction and use, as would the oxygen generated by the decomposition of sulfate salts or electrolysis of water. Decomposition can be achieved by heating. Solid sulfur might occur in massive volcanic-hydrothermal beds, but any abundant deposit of native sulfur will be a lucky find, and probably will not be located near the Mars base; sulfur requires a narrow range of redox and pH conditions, and much of Earth's native sulfur is produced through particular biologically mediated reactions. Sulfide minerals, represented above by pyrite, are probably widespread trace minerals in Martian soil and sands and may occur in massive veins at volcanoes and ancient volcanic and impact deposits.

Once SO_2 gas is produced, it is oxidized at high temperature with a metallic catalyst, such vanadium or platinum–palladium sponge, to form SO_3 gas:

$$SO_2(g) + \tfrac{1}{2}O_2(g) \rightarrow SO_3(g) \quad (9.5)$$

SO_3 then reacts spontaneously with water to form the desired acid:

$$SO_3(g) + H_2O(l) \rightarrow H_2SO_4(l) \quad (9.6)$$

Metal ore smelting

Iron oxides are abundant on Mars and are likely to include also nickel in some cases. They exist mainly as ferric (+3) oxides on the surface, but the Martian meteorites also show abundant and even dominant ferrous iron oxides (especially in silicates). Ferrous iron carbonates or sulfates would be excellent iron ore minerals. (Recent rover analyses have confirmed abundant ferric sulfate.) Siderite would be an excellent iron ore. On Earth, iron oxide ores (along with other metals they contain) are smelted by addition of carbon, usually in the form of coal or wood, which produces carbon oxides and reduces the iron to its metallic state. On Mars, where iron oxides are so much in evidence across the planet, reduction of metal oxide ore minerals to produce free metals can be accomplished by smelting with the addition of carbon produced by thermal decomposition of carbon dioxide, by the addition of graphite (if we are lucky enough to find deposits), or by fluorination. Fluorination would never be considered on Earth because reduction with carbon is so much simpler and cheaper. But on Mars, free carbon may not be available, and organic carbon is a precious material to be kept in organic form if possible.

Production of free fluorine gas, used to extract metals and for other processes, will be an important base of a Martian chemical industry if a fluorine source can be found. The production would proceed through the following reactions working on an original ore of the widespread, common hydrothermal mineral, fluorite (CaF_2), which is probably as abundant on Mars as it is on Earth in areas affected by volcanic-hydrothermal activity:

$$CaF_2(s) + H_2SO_4(aq) \rightarrow 2HF(aq) + CaSO_4(aq) \quad (9.7)$$

F additionally can be obtained with much lower yield from fluorapatite, which probably is abundant in Martian rocks and regolith. HF acid solutions attack potassium silicates (such as feldspar) to produce salts, including KF. A molten salt mixture of KF and HF then can be subjected to electrolysis in a steel vessel with a carbon anode to yield $H_2(g)$ and $F_2(g)$.

According to a study by B. Agosto (Lunar Industries) of lunar metals production, metals could be extracted using a feedstock of ilmenite (separated magnetically or by density from soils), with this reaction taking place efficiently at about 725 K – a process that could be conducted also on Mars:

$$\tfrac{5}{2} F_2(g) + FeTiO_3(s) \rightarrow 2FeF_2(s) + TiF(g) + \tfrac{3}{2} O_2(g) \quad (9.8)$$

When the product gas is cooled to 560 K, the TiF gas condenses and thus can be separated from the oxygen gas and the refractory solid FeF_2. Once isolated, the metal fluorides can be reacted with alkali metals, for instance:

$$FeF_2 + 2Na(s) \rightarrow Fe(s) + 2NaF \quad (9.9)$$

The alkali fluoride can be electrolyzed again in a molten eutectic salt mixture, such that the alkali metal is conserved and the fluorine recycled to reaction (9.8). If elemental carbon is available as a natural graphite deposit, that would be better for smelting of metals, but massive graphite deposits are apt to be rare and may be discovered only after considerable geological exploration. Carbon can be generated also by dissociation of CO_2 gas for the purpose of smelting metal ores.

Fertilizer manufacturing

Ammonia will be another basis of a Mars chemical industry. Ammonia is made by reacting hydrogen gas with nitrogen gas at high temperature with a metallic catalyst:

$$3H_2(g) + N_2(g) \rightarrow 2NH_3(g) \quad (9.10)$$

Nitrogen is contained in the Martian atmosphere at a level of around 5%. However, carbon dioxide and oxygen in the atmosphere will have to be removed before the reaction. This can be done easily by condensing dry ice at low temperature or by reacting it with CaO or MgO at high temperature. Oxygen can be removed by oxidizing metals or carbon.

Soluble ammonium salts will be needed to manufacture fertilizers:

$$2NH_3(aq) + H_2SO_4(aq) \rightarrow (NH_4)_2SO_4(aq) \quad (9.11)$$

Alternatively, ammonium sulfate can be manufactured by a two-step reaction:

$$2NH_3(aq) + CO_2(aq) + H_2O(l) \rightarrow (NH_4)_2CO_3(aq) \quad (9.12)$$

$$(NH_4)_2CO_3(aq) + CaSO_4(aq) \rightarrow (NH_4)_2SO_4(aq) + CaCO_3(s) \quad (9.13)$$

The ammonium sulfate and calcium carbonate products can be separated using their different aqueous solubilities.

Soluble phosphate fertilizers will be manufactured with natural fluorapatite feedstocks if it can be found in pegmatite rocks or silicic hydrothermal veins. Soluble calcium dihydrogen phosphate can be manufactured by:

$$2Ca_5(PO_4)_3F(s) + 7H_2SO_4(aq) \rightarrow 3Ca(H_2PO_4)_2(aq) + 7CaSO_4(aq) + 2HF(g) \quad (9.14)$$

The calcium sulfate can be further used in reaction (9.13) to manufacture ammonium sulfate, and the hydrofluoric acid can be used elsewhere in the chemical industry.

Urea can be used both as an animal feed supplement and as a fertilizer and can be readily produced by:

$$2NH_3(g) + CO_2(g) \rightarrow (NH_2)_2CO(aq) + H_2O(l) \quad (9.15)$$

Explosives and rocket propellants can be produced from salts and water. Chlorates and perchlorates of sodium, potassium, and ammonium can be prepared by electrolysis of water and the common alkali and ammonium halide salts.

Synthetic petrochemical industry

On Earth, petroleum and natural gas serve this basis, but it is improbable that any rich deposits of these exist on Mars, which probably lacks the biological history (or magnitude of biology) that Earth has, and it lacks the type of sedimentary basins where petroleum and natural gas originate by metamorphism of biogenic organic material. If I should be mistaken, then by all means such fossil fuels would serve the Martian petrochemical industry. Plants and animals also contain a rich variety of organic compounds, which can be processed in such a way that alkanes, paraffins, alcohols, complex sugars, and a rich variety of polymers could be synthesized. Controlled high-temperature decomposition of dried plant material in a low-oxygen atmosphere, for example, may be useful. An alternative option would be to start with H_2 (from water electrolysis) and C and/or CO, which can be manufactured in many ways. CO is naturally present as an atmospheric trace constituent and is potentially extractable. Thermal decomposition of carbon dioxide is another means to yield CO. At 1,800 K and low pressures, for example, about 1% of CO_2 decomposes to CO and free oxygen; CO yield increases rapidly with rising temperature. Since the properties of CO_2 and CO are so different, they can be easily separated, but the initial dissociation of CO_2 is very energy intensive.

The Fischer–Tropsch process is heavily utilized in the terrestrial chemical industry. It involves the catalytic hydrogenation of carbon monoxide in the presence of a heated catalyst of Fe, Co, or other metals. Alcohols, paraffins, and straight-chain alkene compounds (olefins) are commonly produced. Typical reactions, with products and reactants in the gas phase, include:

Alkane production:

$$(n+1)H_2 + 2nCO \rightarrow C_nH_{2n+2} + nCO_2 \text{ (iron catalyzed)} \quad (9.16)$$

$$(2n+1)H_2 + nCO \rightarrow C_nH_{2n+2} + nH_2O \text{ (cobalt catalyzed)} \quad (9.17)$$

Alkene production:

$$2nH_2 + nCO \rightarrow C_nH_{2n} + nH_2O \text{ (cobalt catalyzed)} \quad (9.18)$$

Alcohol production:

$$2nH_2 + nCO \rightarrow C_nH_{2n+1}OH + (n\text{–}1)H_2O \text{ (cobalt catalyzed)} \quad (9.19)$$

For example, methane can be produced by reaction (9.17) as such:

Methane synthesis:

$$3H_2 + CO \rightarrow CH_4 + H_2O \text{ (cobalt catalyzed)} \quad (9.20)$$

Alkanes with up to 20 carbon atoms are routinely produced. Mixtures are normally generated, so distillation is required to yield pure products. Many different catalysts can be used with different reaction product yields generated under different conditions. Carbon nitride catalysts, for instance, permit enhanced production of alcohols. In general, cobalt catalysts usually yield water as a final product, and iron catalysts produce abundant CO_2, while the hydrocarbons and organics can be yielded by either catalyst. Possibly cobalt would be preferred for Mars, where the water can be recovered for other uses, and where the products include more use of H_2 and less CO (H_2 being more readily obtained than CO, unless the atmosphere can serve as the source for CO). Cobalt also involves lower reaction temperatures, but reaction rates are slower than with iron. Fischer–Tropsch reaction temperatures are quite low, in the range of 473 to 593 K, the pressures required are modest (less than 1 to 200 atm) and the reactions are always exothermic, so that they are energy-efficient processes. In fact, the reactions occur at much lower temperatures and in low-grade vacuums, but rates can be too slow to be of human use.

While not all the hydrocarbons and organic compounds are straight-chain structures, most Fischer–Tropsche products are, so further processing is needed to yield large quantities of complex organics, plastics, and many other desired polymers. High-mass branched-chain alkanes, for instance, can be manufactured by co-condensation of low-mass straight-chain alkanes with alkenes (a process called alkylation).

Reaction of alkanes with halogen gases (yielded from halide salts) produces halogenated alkanes, such as methyl chloride:

$$CH_4 + Cl_2 \rightarrow CH_3Cl + HCl \quad (9.21)$$

Similar reactions can produce important industrial chemicals (and pollutants), such as trichloroethane and carbon tetrachloride, which can be further used to generate CFCs and related compounds for refrigeration and as possible future artificial greenhouse gases for Mars.

The oxidation of alkanes is, of course, important for the functioning of civilization here on Earth, where we heat our homes with natural gas, drive gasoline-powered cars, etc. The lack of abundant free oxygen on Mars and the high value of organic carbon and hydrocarbons means that alkanes will not be used in that manner on Mars. However, carefully controlled oxidation of alkanes can yield other important industrial oxidation products, including alcohols, ketones, aldehydes, and acids, including for instance formaldehyde, ethanol, acetone, acetic acid.

High-weight acids can be reacted with alkali to make soap. A simple plastic is polymerized formaldehyde (polyoxymethylene, $OH{\cdot}(HCHO)_n{\cdot}H$), which was once used, for example, in automobile door handles, until it was recognized that combustion yields toxic substances.

Teflon is manufactured by fluorination of chloroform, $CHCl_3$, which is produced by fluorination of methane in a reaction similar to (9.21). This process results in formation of $(CF_2)_2$ (tetrachloroethylene), which then polymerizes to form teflon (C_2F_4).

Precious metals

Gold, silver, copper, platinum, palladium, rhodium, ruthenium, and rhenium, besides being used in art and jewelry, have become essential industrial commodities for their properties of high electrical conductivity (copper and silver), resistance to oxidation (platinum and gold), chemical inertness (platinum), hardness and durability (platinum and palladium), ductility (gold, copper, and silver), chemical catalysis (platinum, palladium, ruthenium, rhodium, and rhenium), and lack of spectral aborptivity and good optical polish (silver). Other precious metals include gallium and germanium, which are essential in the manufacture of the most efficient photovoltaics and semiconductors, among other uses. The industrial uses of these metals on Earth will eventually find parallel uses on Mars. Eventually Mars will provide what the New Martians will require of these metals, in addition to more common essential metals, such as manganese, vanadium, titanium, iron, nickel, and cobalt.

In 1994 I showed that Earthlings extract, use, and price metals in close correlation with their abundance in the Earth's upper continental crust, the accessible reservoir of raw materials over the past several centuries and up to this point in our development (J.S. Kargel, 1994, "Metalliferous asteroids as potential sources of precious metals", *Journal of Geophysical Research*, **99**, 21,129–121,142). That only makes intuitive sense, but it is a fact that is supported by hard data, and it is a fact that yields the simple and profound inference that humans will find and develop uses for metallic substances (and all other natural resources) to whatever degree nature makes them available. This will also be true on Mars, with uses eventually expected to adapt to the particular element and mineral abundance patterns present in the Martian crust. Initially, the accessible reservoir of raw materials will be the upper few meters of crust at the local level, and initially industrial uses of materials will be inherited from the technologies that are adapted to the availability of Earth's raw materials. To the extent that Mars is very Earth-like in element abundances, this will not be a bad initial economic model for Mars, too. But Mars is not Earth; Mars has undergone different types and degrees of geologic processing, and we can only expect that some significant differences in mineral resources and ore deposits will be discovered the more we explore. It will take time for the industrial and economic maturity of Mars to bring about a unique Martian economic valuation of resources based on global-scale geochemistry and petrology, but this relationship will probably emerge on Mars rather rapidly.

Just to be sure, I would not expect Mars ever to be a chief supplier of precious metals to Earth; transport costs are simply far too great and will remain so. Martian metals will be for the Martian economy. The same does not apply to asteroids, which have bright prospects as sources of precious metals in Earth's future – a point I have written about considerably.

The metallic cores of Earth and Mars have sequestered comparable fractions of many metals, leaving comparable small traces in the mantle. The abundances of precious metals available in the upper Martian crust, however, may differ in significant ways from those in Earth, and so the economics and industrial use patterns will differ to some extent from that on Earth. On Earth, three processes have acted separately or together to concentrate metals, such as platinum and gold, in deposits that are rich enough to mine at a profit:

1. In igneous systems, especially massive subsurface magma chambers and deep lava lakes of basaltic and ultramafic magmas, crystallization and gravitational settling of sulfide minerals naturally extract most of the magma's precious metals in so-called stratiform ores of layered igneous complexes.
2. In hydrothermal systems, hydrous fluids leach rocks and extract their loads of precious metals, then concentrate them further in sulfide mineral deposits.
3. In sedimentary systems, nuggets and flakes of gold and platinum are deposited and concentrated in sands and gravels due to the high densities of these grains.

Mars has had all these processes available to it to generate rich metal ore deposits. However, the lack of plate tectonics means that (a) subduction-related hydrothermal processes will not have occurred (except perhaps early in crustal development), and (b) recycling of crustal sulfides to the mantle will not have occurred. These two facts will have opposing significance with regard to metal ore deposits. My guess is that the ice richness of the Martian crust will substantially compensate for the lack of subduction-related fluids, and that eons of water-mediated mineralization will locally result in profoundly rich ores, which may be more widespread than on Earth. The continuous stability of ice for billions of years in large volumes of the Martian crust may have a dramatic effect on how precious metals and many other raw materials are distributed in the Martian crust. On Earth, groundwater flow and stream flow continuously wash soluble minerals back to the sea; even in permafrost regions, one needs to step back only a few brief million years when ground ice was not stable. Thus, Mars has a cold-trapping mechanism for mineral storage that Earth lacks. The significance of this mechanism will have to be determined by field studies before we can bank on it.

There is ample evidence that Mars has had extensive volcanic interactions with ground ice and water, and so Mars is expected to have its own style of intense hydrothermal activity. The lack of plate recycling to the mantle and lack of intensive crustal erosion means that hydrothermal ores have probably been concentrated into the upper crust more than on Earth. Less extensive sedimentation on Mars probably means that alluvial gold deposits, such as those of California, are not so important on Mars, although if there are really rich crustal ores, just one cycle of erosion, sediment transport, and deposition could provide for a true Martian Mother Lode. The massive shield volcanoes of Mars and other vast lava fields suggest that huge magma chambers have probably produced unprecedented huge stratiform deposits.

Those metal deposits may have been remobilized and further enriched by hydrothermal activity. Alluvial and eolian resorting of eroded material from those deposits may have locally produced some rich ores.

A peculiarly Martian process might have further yielded huge enrichments of certain precious metals in some Martian ores, and resulted in a strong temporal evolution in the precious metal content of stratiform ores. The Martian core apparently has almost completely solidified, unlike Earth's core, which is 95% liquid. There are recent measurements suggesting that possibly a thin layer of the outer core remains liquid. We know that the Martian core once was mostly liquid, like Earth's, because ancient Martian crust is magnetized from a dipole magnetic field generated by a molten core. The gradual solidification of the Martian core would have extracted most of the core's gold and platinum into the solid core, as it has on Earth, but the most "chalcophile" (sulfur-loving) elements, such as palladium, would have been concentrated into the last sulfur-enriched dregs of the outer core. We suspect, from geochemical and theoretical studies of the Earth, that outer core material can become entrained in lower mantle rocks, which can then yield magmas that reach the surface, thereby carrying precious metals into the crust. If this happened on Mars, those lavas would initially have been gold- and platinum-enriched and later would have evolved palladium-rich compositions. Further natural processing by the mechanisms listed above could then yield rich ores of those metals. Any volcanic rocks derived from near the core/mantle boundary thus may be extraordinarily enriched in palladium, which may be as plentiful in the Martian crust as silver is on Earth.

Finding the precious metals on Mars will take systematic geochemical and mineralogical exploration, which has already begun. Except for iron, which will generally be far more abundant and widespread on Mars than on Earth, economically exploitable sources of metals will be distributed according to a well-known function, where the richest ores will occur in the smallest and most dispersed deposits, and more mundane ore deposits will be larger in volume and more widely distributed. In getting established, water will be the most critical resource, and small-scale metal mining will have to be conducted in the local environment by making due with whatever mineral deposits geology has provided. As Mars civilization grows, as transportation infrastructure is put in place, and as industrial demands for metals increase, the richest ore deposits will be exploited through familiar strip mining, underground mining, and open-pit mining methods of ore extraction.

There will be some key differences in mining and ore processing techniques. I already mentioned smelting issues. Use of aqueous cyanide leach solutions (a common processing method on Earth) would not be effective on Mars without preheating of the ores, because those solutions would freeze solid under Martian crustal conditions. For small-scale production, particularly rich ores might be leached by ethanol- or methanol-based leach solutions, and concentrated sulfuric acid-eutectic solution would be perfectly stable as a liquid at lower latitudes. However, ordinary leach methods could be used in an enclosed, heated building. Hydraulic mining of placer deposits of course would not work, again because liquid

Open-pit mine in Arizona shows many familiar features, including a terraced pit; cork-screw road to provide access to mining equipment and personnel; haulers and excavators; exterior plastic-lined leach piles and leachate-processing facilities; explosives storage sites; administrative and crew maintenance facilities, such as cafeterias, sleeping quarters, medical, and sanitation facilities; environmental monitors; crew rescue equipment; a roadway or rail to the outside; and a fueling depot. Martian open-pit mining would require similar facilities with suitable technological adaptations. Oblique air view.

water is not available in needed quantities. In any case, a new ore-processing chemical technology will need to be developed to deal specifically with Mars.

SPACESHIP MARS?

The successful exploration, habitation, and eventual economic self-sufficiency of humans on Mars is a glorious and hopeful prospect. Most people in the developed world probably recognize Mars as part of human destiny; this is particularly true in the USA, where a pioneering spirit was a necessity of survival for several centuries. Exploration of Mars is widely regarded as a noble and pure expression of the best in humanity. We look forward to the detailed exploration for Martian life; to the elucidation of the causes and processes of global climate change and derivation of lessons for Earth; to other types of science that could be done on Mars, for instance astronomy done with ultra-long-baseline radio telescope arrays. We look forward to the technical challenge and the communion of human spirit that will develop as we go on to Mars and establish a sustainable, growable presence there.

Humanity at Mars is not likely to end with exploration any more than exploration of the Americas was the end of European interest in the New World. According to one popular view of humans on Mars, we shall not go there to live in a shell in perpetuity. The ultimate dream of many space dreamers is to create on Mars an Earth-like environment. Terrestrial gravity on Mars cannot be reproduced, but the global scale engineering of a more tolerable climate and atmosphere is, in theory, possible. Kim Stanley Robinson described *terraforming* in a fictional trilogy, and before that Arthur C. Clarke wrote about the terraforming of several planets, starting his series with *2001: A Space Odyssey*. It may be possible to trigger an artificial MEGAOUTFLO episode, unleashing latent geologic forces of Mars and again releasing torrents of volatiles that repeatedly have established comparatively clement and hydrologically active conditions. If terraforming other worlds should ever be both possible and necessary, there is no place in our Solar System better technically suited than Mars. Whether we should do it is a separate question.

This sci-fi concept, though clearly beyond current technology, is not totally lacking engineering and geological sense. A sensational approach, just slightly possible, might be to use hydrogen fusion explosions (H-bombs) or an asteroid diversion to rupture a confined, carbon dioxide-saturated aquifer. Once initiated, a violent, autocatalytic venting process lasting several months to a year may bring on a hundred thousand years of clement conditions beneath a densified carbon dioxide atmosphere. It is not clear that this process would work, because there are certainly many impacts on Mars vastly more energetic than any conceivable nuclear explosion, and yet most impacts do not trigger MEGAOUTFLOW cycles. There is, first, the unknown magnitude of an event required to produce a critical failure of the permafrost cap; and, second, the possibility that once the volatiles are unleashed, carbon dioxide may tend to snow-out at the poles and cause a rapid collapse of the incipient greenhouse. It is possible that a critical threshold of degassing rate is needed to prevent snow-out.

Possibly an easier and less controversial way to initiate an enhanced greenhouse atmosphere would be through the use of highly efficient artificial greenhouse gases. Traces of laughing gas (nitrous oxide, N_2O) is no laughing matter when it comes to the greenhouse warming of Earth. Industrial activity has increased the Earth's N_2O abundance by about 17%, from 270 to 316 ppb, contributing about 10% as much warming as increases in CO_2 has, though N_2O concentrations are three orders of magnitude lower than CO_2. It may be possible to manufacture chlorofluorocarbons and a slight trace of methane and vent them into the atmosphere. A modest greenhouse-warmed surface, while still perhaps as cold as the Antarctic Dry Valleys (or optimistically as warm as Barrow, Alaska), might destabilize CO_2-clathrate permafrost and CO_2 polar caps and cause it to belch forth carbon dioxide. Atmospheric pressure may build to 320 millibars just from sublimation of half a kilometer of dry ice (or equivalent clathrate hydrate) from each polar cap over an area 500 km in radius. This pressure is five times that needed to prevent the boiling of blood and saliva in an unpressurized human body. This amount of CO_2 is also enough to aid the greenhouse effect of chlorofluorocarbons and to increase water vapor and its greenhouse influence in warming the surface. Thus, oxygen, but no pressure suits, would be required to sustain humans walking free on the surface.

Effective terraforming would require control of the atmosphere and climate beyond that attainable by simply unleashing carbon dioxide. Most crucially, the artificial chlorofluorocarbons and methane might be enough to alter conditions in the upper atmosphere in order to prevent global CO_2 cloud formation and snow-out of carbon dioxide at the poles. This would be needed to sustain a dense CO_2 atmosphere. These compounds would have to be replenished on a century timescale, because they would gradually be destroyed by solar radiation-driven photolysis. The source of chlorofluorocarbons and methane would be a basic chemical industry set up with nothing but ubiquitous, plentiful soil salts, ice, and the natural atmosphere as the raw materials, and a solar or nuclear energy source to drive the processing. The same chemical industry using the same raw materials would generate fertilizers to sustain plant growth.

Over a few centuries, the carbon dioxide polar caps and near-surface clathrate hydrates would sublimate, and a more habitable surface would progressively evolve as one process and its positive feedbacks cascades into another, driving the densification and warming of the atmosphere. Unless an artificially triggered MEGAOUTFLOW event could alter the surface in the space of a single year, the slower chemically driven process would take perhaps one century before humans could walk around on the surface with nothing but ordinary cold-weather clothing, sunglasses, sunscreen, an oxygen canister, an oxygen mask, and plentiful drinking water. There would remain problems with peroxides and other carcinogenic and irritating substances that have built up in the soil, but with simple life support systems far short of space suits, and perhaps a diet rich in anti-oxidants, it seems likely that Mars would not be intolerable to human life.

No quantitative global circulation climate models have been developed for the unnatural case of terraforming described here, but it is fun to speculate on what might be the result for artificial plant-based ecosystems. A 300-mbar CO_2 atmosphere (warmed also by artificial chlorofluorocarbons and methane and natural water vapor), might take a few centuries to build. The plentiful CO_2 would counter deleterious effects on plant growth due to low sunlight intensity, pervasive cloudiness, and cool temperatures. The net result might be cold-climate plant growth at characteristic terrestrial rates. In wet climates, such as alpine areas and mid/high latitudes, dominant conifer forests might be sustainable if they are engineered to tolerate high soil and ground water salinities. Food crops may be grown preferentially at about $\pm 40°$ latitude, where water from abundant melting ground ice is available, but reasonably warm temperatures prevail and sunlight is plentiful all the year. An interesting riparian desert ecosystem, or irrigated croplands, could be established in areas between ± 30 and $40°$, where snow and rain are apt to be nil but slowly melting deep ground ice would produce abundant springs. The equatorial belt would be an exceedingly dry desert, generally with no accessible ground ice or ground water for irrigation; this is where Mars would remain natural Mars.

It might even be possible, with no small trouble and time, to transform the atmosphere to one rich with oxygen, containing also a trace of UV-protective ozone. With time, plants nourished in the dense CO_2 atmosphere would generate oxygen, which might be sufficient to sustain certain types of oxygen-respiring animal life

Gray-scale rendition of a painting of future human activities on Mars, by Michael Carroll.

(though not humans for a long time), thus allowing the introduction of some microorganisms needed in a more complete ecosystem. With 600 mbars of CO_2 (about the same density as Earth's atmosphere at low elevations) and a circumpolar arboreal ecosystem, the oxygen content might increase at about 10 ppm per year, so that after one century the oxygen levels would reach 0.1%, enough for some aerobic microorganisms. Unlike highly reduced lunar rocks and soils, which would rapidly oxidize and remove free oxygen, the exposed parts of the Martian crust are already oxidized, so oxygen produced is oxygen added to the atmosphere, until an animal respires or fresh rock surfaces are generated and weathered. At this rate, it would be several millennia before humans would find enough oxygen to breathe the air freely, though CO_2 toxicity would remain a problem.

10

New World

(Why Mars? The Case for Earth)

SMART AND WISE?

To some readers any terraforming concept will seem fanciful and futuristic beyond consideration. However, there are many smart engineers and scientists who feel that terraforming is possible, especially on Mars, and are serious about their topic. If terraforming is possible, is it smart and apt to succeed in generating the type of planet humans would find materially useful? Rarely are machines ever built that do the right thing, first time, without serious problems. It would be doubtful that Mars (a global machine) or any other planet could be engineered to desired characteristics at the first attempt, though an approximation to desired characteristics may be more probable to obtain. Practice makes perfect, and it might be that one or two goofed-up planets would train us in the right engineering skills.

Imagining that the technical difficulties can be mastered and desired planetary responses predicted without real-world testing, unintended consequences could be dramatic. Possibly terraformers would have to contend with, in the most optimistic scenario, centuries of melting permafrost and boggy conditions that could make surface construction unstable across many parts of the planet. Climate could be so dynamic that designing to particular conditions may be futile. While these could be serious engineering and ecological issues, I do not consider them to be arguments against terraforming, if Mars is presently a place without life. If humanity can inadvertently engineer Earth's climate, I suppose that humanity could deliberately engineer the Martian climate by copying the processes that nature itself has used to change Martian climate. The key would be to induce a small change that then precipitates a geoclimatic chain reaction of bigger processes. Boggy thawing permafrost would be inevitable, but if engineers can build a petroleum pipeline through Alaskan permafrost, my guess is that engineers will also manage to deal with the Martian mud bogs.

If the myriad technical difficulties can be overcome and terraforming Mars is possible, there is – far beyond *smart* – the question of whether terraforming would be *wise*, as Carl Sagan pointed out the distinction. Would it be successful, good, and right, all in one? Would it have repercussions that we would marvel at and be proud of, or would it bring on unanticipated bleakness for humanity's future and for Mars itself? There would be one answer if Mars is today actually a living world; it would not likely be possible to terraform Mars without causing immense impacts on native ecosystems. With Mars possibly hosting the only other ecosystem in the Solar System, would it be ethically acceptable to damage – possibly destroy – that ecosystem and displace it with human beings and our food and pets and pests? I cannot imagine an affirmative answer, but ethics has had comparatively little moderating influence on what humans have done in the past to propagate ourselves. More hopefully, the terrestrial organisms that would be introduced and those native to Mars would occupy such different niches that nondestructive coexistence will be possible. Perhaps a biochemical deal could be reached: "You live down there in the volcanic plumbing, and we'll live up here on the surface." It may be that, after concerted searching, we discover that Mars is not and never has been a living world; any ethical dilemma would be moot. Even with a dead Mars, there remains Buzz Aldrin's otherworldly "magnificent desolation" factor. However, total preservation of raw, untamed Martian magnificence is not apt to restrain humans for long. If we go to Mars at all, and I think we will as long as Earth's civilization holds together, some degree of terraforming is likely, as even certain small changes may prove advantageous. However worth while these issues are to discuss today, they do not now bear on our shoulders, though related issues do.

TERRORFORMING?

The concept of terraforming another planet – even if it should prove impossible – raises an entirely different issue that deserves serious discussion. Why do we wish to terraform in the first place (if we so wish)? There may be many reasons, some legitimate, some potentially so powerful that one day terraforming may become compelling above all environmental objections and cost considerations. I actually have no great interest in pursuing these particular reasons in any but a superficial or fictional context, because to me mega-terraforming seems an issue that might have to be confronted head-on only several generations from now. Pretending to weigh in on a political battle that might take place generations from now (or never) is not one of my pressing concerns. I would not shirk my responsibilities to the future; it's just that I can't predict or control it. Rather, we are faced with getting our children through college, beyond the threats to Humanity caused by hatreds among humans, circumventing global bioterrorism, and mitigating industrial threats to a habitable environment on a world that still is wondrously Earth-like and accessible – Earth itself. Proper planetary stewardship now is the pivotal issue for me: this generation must deal with Earth this life/this century in such a way that we and our distant descendants will have something wondrous still here on Earth; the long-range fate of

Mars can be dealt with by succeeding generations. However, the question of future terraforming of Mars bears on how we consider our role here on Earth today.

The most compelling possible reason for terraforming, and the most dire possibility, is that Mars will eventually serve as the sole principal refuge of the inhabitants of Mother Earth. The dark nature of this prospect, of course, is that it implies that Earth somehow may become uninhabitable, either due to natural causes (such as mega-impact) or human activity. Richard S. Williams, Jr (2000) described what he called "a modern Earth Narrative" based on the twin concepts of geological and biological evolution. Ominously, Williams forewarned: "*The failure of humans to comprehend and understand the Earth Narrative, especially the place of humans in it, presages dire consequences for the Earth's biosphere.*"

No matter how much engineering our descendants may undertake to create artificial habitats on Mars, or in eventually terraforming the planet, the fact remains, as Edward O. Wilson wrote in *The Diversity of Life* (1992, W.W. Norton & Company):

> *Humanity coevolved with the rest of life on this particular planet; other worlds are not in our genes.*

This has been true almost to this point in time, but it will not pertain henceforth in an age when the human genome has already been recorded and is being decoded, and when insertions of genes from other natural or artificial gene pools are already possible. It might be possible, for instance, to produce plants with greater tolerance to cold or ultraviolet rays or salty soil. This could be essential if environmental engineering does not completely alter Martian conditions to ideal terrestrial conditions. Humans could be manufactured in one clever generation to have smaller bones, since the lower gravity would demand less of the skeletal system. The heart could be smaller, as it would be pumping blood of lower weight. Muscles could be smaller. Most organs could be proportionately smaller to service a smaller overall body. The brain, of course, would best be kept large. And so we can envision a human body recalling old episodes of *Star Trek*. These changes could compress into one generation the magnitude of genetic changes that natural selection and normal mutations succeed at doing in half a million generations. These and many other changes to the genetic code of terrestrial organisms would make sense for Mars, just as bacterial genetic insertions into strawberries making them more frost-resistant make sense to strawberry farmers. But this prospect raises seriously disturbing issues, and the more so when human engineering and global biospheric scale engineering is considered.

"*We are at the dawn of a new era!*" goes the cliché. As the Sun was coming up this morning I reached page 17 of *The New York Times*, which reported that an American and Canadian team of biologists is debating the formation of a human–mouse hybrid. While advanced genetic engineering might make us and our fellow creatures more able to cope with unique Martian conditions, I doubt that the artificial adaptations could be as snugly fitting to Martian conditions as life presently fits Earth's many environments. I may underestimate the genius of humans and the marvels of genetic engineering to create designer food animals and plants and quasi-human creatures crafted for Mars. It is also important to recognize many sound reasons for conducting genetic manipulations of organisms on Earth and this would be true also for an extension of Earth's life on Mars. Who can criticize the Indian government's hopes to clone a couple of the last 50 Asiatic cheetahs surviving in Iran to help to save a species from extinction and reintroduce this magnificent cat in India? Who can have a friend with terminal Parkinson's and call for a halt to stem-cell research? Not so important, but much desired, would be a cup of low-caffeine transgenic coffee that actually tastes like coffee. Some might desire the new glow-in-the-dark zebra fish, which incorporates a coral gene that gives it a natural fluorescence. The dilemma is that at some point we must acknowledge the unwise nature, possible unintended consequences, and inevitable evil abuses of genetic manipulations, including creation of ecologically devastating recombinant organisms, or transgenic species developed as weapons. At what point do risks exceed benefits? Far over this line, any successful effort at biosphere-scale genetic engineering here on Earth would result in a biosphere and gene pool that is nothing like what we still have. Few if any people would allow that human beings have the right to remake Creation, but the gene-ie is out of the bottle already. The longer we go with nearly a complete lack of regulation of these world-altering technologies, the more we risk irreversible damage of incalculable proportions.

Whether looking for a cure for terroristic smallpox or terraforming Mars or other grandiose applications, the original intention of genetic reshaping of humans and our fellow creatures might be fine, but where is nature and individual free choice through it all? With the end of natural selection and the advent of programmed selection by central authorities (governmental or corporate or private concerns) we may see on our near horizon the end of humanity as we know us. That prospect is much closer than Mars, much closer than we may think. The ethics questions raised by new genetic engineering possibilities extend quite a bit beyond cheating on tests or simple questions like, "Do we drop The Bomb or not?" When human–mouse hybrids are possible and being considered, whether for medicine or terraforming or any good purpose, the sky is the limit for dreadful consequences. Given human history, how can we even imagine an absence of accidental horrors and deliberate abuses or tragically misguided weaponization of these new technologies? Might important components of the natural biosphere, even Humanity, be decimated or rendered extinct? Regardless of the noble goals, does anybody really believe that somehow these technologies will remain under control and within ethical realms, or that any accident would necessarily have a minor scope?

Henry C. Kelly, president of the Federation of American Scientists, in a forewarning to biologists that society and government would control their activities if they do not regulate themselves, wrote for *The New York Times* (July 2, 2003),

> *Within a few years it may be possible for an inexperienced graduate student with a few thousand dollars of equipment to download the gene structure of smallpox, insert sequences known to increase infectiousness or lethality, and produce enough material to threaten millions of people.*

Kelly noted that all of this could be done with resources dedicated for the purpose of curing disease. He was wrong about one thing, and that was not to write *b*illions of people. Threats to the viability of civilization and even the possible extinction of *Homo sapiens* are at our doorstep, thanks to modern technology and our inability or unwillingness to regulate it. Apparently, with sales of the glowing fish at stake, as though coral reefs and tropical lakes have been lacking in their creativity, there is simply too much money to be made and too many feelings to be hurt to bother with such regulation.

To understate it, there are serious ethical questions surrounding many experiments in life creation and genetic engineering and many other fields of science. Nevertheless, the experiments go on seemingly heedless of the ethics discussions, which seem to be applied more as oil to annoint the technology and salve the public's worries than for any debate that could potentially halt investigations or applications. Extrapolating a bit into the future, and without even considering the potential horrific outcomes of, for example, a transgenic human–animal hybridization-gone-awry, one can foresee a dangerous near future. Hybridization or any serious re-engineering of human genes would mean that humanity would be fundamentally changed in our notion of who we are. Peter Ward, in *Future Evolution* (Time Books), even considers the directed evolution of a super-hominid species derived, but separate, from our own. That is not too-far-out science fiction, since it could be initiated early this century. It seems doubtful that such a species, if successful and allowed to propagate its numbers, would tolerate us mere humans to continue, unless super-humanity is embued with an enlarged moral sense, or unless a small sample of mere humans are to take a place beside the gorillas of Rwanda or in future zoos. A scenario modified from *Planet of the Apes* is not so preposterous. I am happy to see that Peter Ward is capable of unabashed optimism when considering such potential. I am not really with him in that optimism, much as I would like to be.

Dismaying though some of these prospects may seem for terraforming Mars, the arguments against intensive bioengineering, even of humans, are less compelling because the reasons for it are stronger and there is perhaps no natural ecosystem to be impacted. For example, as Percival Lowell pointed out, coping with the lower gravity places a lower demand for body strength, as long as the New Martians do not intend to revisit Earth. Fewer demands for metabolic energy would mean that each "human" would need to consume less food and water, and respiration could be effective at lower oxygen contents than is contained in Earth's atmosphere; perhaps 10% O_2 in a 600-mbar atmosphere would be sufficient. Taking advantage of lower gravity would require human anatomical and biochemical adaptations. Any successful terraforming and human inhabitation of Mars would doubtless result in natural adaptations over the course of a few tens of thousands to millions of years. These changes might be speeded through genetic engineering, with only one generation needed to accommodate many beneficial adaptive changes and another generation to fine tune the New Martians. (Never worry whether the modified humans could still see one another as being attractive enough to mate with; that could all be done in a laboratory; human sexuality and genitalia just dangle in the gears of progress and could be genetically eliminated or neutralized.)

Getting back to the ethics and political practicality of it – whether talking about bioengineering humans or microbes on Earth or Mars – to keep any modicum of control, the experiments, applications, and products must be controlled by either a responsible designer-gene corporation (a novel concept) or a responsible central government. The alternative is to have bioengineering anarchy. The fact is, humans, with all our fallibilities, are already making the initial inalterable changes to the biosphere, with not much of a chance for nature to resume her time-tested regulatory authority. My urgent sense is that this is an issue for the present. The technologies for redesigning the world are advancing and spreading like wildfire as we sit. An absolute power is out of control. Recalling that absolute power corrupts, and that corporate entities are not accountable to the people as a democratic government is, it would seem that we – the people – have a serious decision to take upon ourselves collectively or to have that power forever seized from us and used with inalterable consequences that will affect all Earth inhabitants. These are serious issues that we face in this generation for the first time, yet in the seats of power there is no semblance of a decision-making process regarding these issues.

The powers wielded by the bioengineering entitities are overwhelming and becoming more so the longer we sit and read with fascination about glow-in-the-dark fish. Whatever the noble purpose and guiding ethic now for most genetic engineering, it will soon be purely money and the control of people and wealth, or the power to doom one's enemies. This, of course, has been true of all major technological innovations from the beginning of technology, but the potential for ill consequences of these genetic engineering technologies certainly raises the risks. The power now is in the hands of individual people to direct the future course of evolution, to erase or create species and whole orders and families of organisms, and even to threaten Humanity. Will democracy and freedom, such as we enjoy, survive in this kind of world? Maybe I underestimate the benefits and practicality of ultimate power over life in the hands of a few self-annointed all-wise leaders. The gene-ie clearly is out of the bottle and might not be put back in, but we at least need a serious discussion on how we might deal with these new technologies and their social, environmental, and security implications.

Perhaps I am too pessimistic. For a New World on Mars or this Old Terran World, maybe I just need to *Imagine* the possibilities, as John Lennon did for Earth. The hurdles to wholesale redesign of Earth are of course – fortunately – very large, though one by one they are being leaped at an Olympic pace. If we need to reconsider what we are doing to Earth, the time is not too soon to think about Mars, and how we may or may not approach it differently. We could create a new living world on Mars, at first in our bubble habitats, but one day maybe across that globe. With a new artificial environment, the New Mars society could create a war-free system dedicated to the production of goods and to the reproduction of revamped forms of life, including revamped big-brained, small-bodied humanoids. Mars can be a living global machine functioning with New Man, created in Man's Image of Mc.E. Mouse, perhaps, functioning like perfect cogs on an oiled wheel.

While it seems that the possibilities for cheap cynicism are unbounded, so does it seem are the real possibilities for something incomprehensibly dreadful if we are not careful. I lack a solution, but I believe that people can create a solution if they see the need to act. Besides having an intangible gloom, I also have an irrepressible optimism that somehow we shall evade these problems through wise stewardship of technology and life on Earth. While my pessimism may be too dire, and my optimism may be a case of "irrational exuberance" (to use a modern economic term), my thinking is that by thinking, and by seeing the new dangers that lie before us, we can avoid those dangers and paint an optimistic portrait of our future and make that come true. This is how it has always been; only the present dangers are different. This essential difference is why old politics and old adaptive strategies might not be sufficient to the modern world if we wish to survive, to retain the best of what remains viable from past traditions and progress, and to move forward to a better future, both on Earth and on Mars.

THE SIXTH EXTINCTION

Much of that still is (barely) in the science fiction realm. Terraforming has another immediate relevance. The main argument for terraforming Mars would be that Earth somehow has lost its habitability. Any forced abandonment of Earth for Mars or the stars would imply a catastrophe of the first order on this planet, loss of millions of species, and loss of everything human that we cherish. This dismal portrait of life's future in the Solar System is tough even to contemplate with anything short of Chalmers' acceptance of fate, however dire.

I don't lose sleep over an inevitable Sun-Gone-Red-Giant 7.5 billion years hence, or of the rock band Sound Garden's "Black Hole Sun" careening through the Solar System at some unpredictable future time. I don't even worry about a cosmic *Armageddon*, with comets and asteroids smashing and vaporizing us. A comet as large as Hale-Bopp careening into the Solar System and striking the Earth is an improbable event any time in the next 10 million or so years. While impacts are probably responsible for the extinction of more biological families of organisms than any and possibly all other causes, with just 30 years' notice, humans probably could make Mars a celestial ark for many terrestrial life forms, and soon transform the ark into a self-sustaining base on which planetary and genetic engineering could sustain and further develop life. That's the hopeful side of Mars terraforming. I admit, I am perhaps not the biggest advocate of remaking planets, so there may be a tinge of negativism. It is, however, a possibility tailored for Hollywood to reconsider at least once a decade.

I do worry about the next 25 to 50 years (something we might affect). My real concern is the possibility that through our own preventable actions, this generation might make terraforming another planet a future necessity; potentially worse, through thoughtless or greedy waste of our resource base and loss of a sense of collective humanity and good stewardship of Mothership Earth, we might make terraforming impossible. This is the rub. We are not being good stewards. Humans are the cause of an ongoing mass extinction. Amazingly, many politicians and writers – for lack of education in the matter or for motives only they know for sure – deny that such an extinction is occurring or deny its significance to Earth and Humanity. Edward O. Wilson, in *The Future of Life* (2002, Vintage Books), summed up the ignorance, *"... a few authors – none of them biologists – still doubt that extinction is occurring on a large scale."*

The immediate relevance of Mars as a refuge for Earthlings comes through our attitudes regarding the planet we live on at this moment. A likely reason that could force abandonment of Earth and adaptation of life to Mars would be human activity of some detrimental type – we throw a climatic switch we never knew existed, or progressively destroy the ecological connectedness of the biosphere. If these disturbances occur gradually enough, we might make the jump to Mars, especially with near-Earth asteroids and the Moon as a resource base to build the interplanetary transportation infrastructure needed to build the Martian ark. However, there is no assurance that human tinkering with Earth's climate system, with the genetic code, with biodiversity, or with weapons of mass destruction might not cause a catastrophic loss of habitability on such a short timescale that the economy and biosphere could not sustain itself long enough to build that ark on Mars or ships to the stars. Worse than nuclear weapons and poison gases would be microbes engineered to consume or poison all else that lives; it would, of course, be an engineering antimarvel fated to oblivion itself.

Far short of these potential future calamities, already Humanity is doing quite enough damage. Edward Wilson wrote (in *The Diversity of Life*),

> *The Sixth Great Extinction spasm of geological time is upon us, grace of Mankind. Earth has at last acquired a force that can break the crucible of biodiversity.*

And what a crucible it has been! Careful consideration of the fossil record and preservation potentials shows that biodiversity has been on a slow exponentially increasing trend for at least 600 million years, with downward blips due to giant impacts and other events causing mass extinctions. Civilization inherited more diversity in the biosphere than ever existed before in Earth history, and amazingly has managed a sharp reversal.

The downward blips in Earth's gene pool are followed by orgies of diversification and recovery of diversity after about 10 million years. It is a sobering thought that human impacts on biodiversity are comparable to giant cometary and asteroidal impacts approaching the scale of the Chicxulub impact that doomed dinosaurs and most other life on Earth. However, the inevitable recovery of diversity after each calamity in the Earth's rock record means that humans probably will not be the doom of life on Earth, though we may provide our own doom. Besides human impacts on the gene pool and the effects this has on our ability to survive and respond to calamitous changes, nonbiological resource depletion is a further issue that could strip civilization of the economic and technological capability to make any eventual transfer of life to Mars or anywhere else in the Solar System and beyond. If this should occur, then the Terran gene pool will be stranded on Earth for better or worse.

The irony is that while many of us dream wondrously of terraforming other planets, we are, bit by accelerating bit, "lunaforming" Earth. We are taking what is so marvelously Terran and, through ravenous consumption and enslavement to human purposes, transforming it to concrete, fenced and walled compounds, and moonscapes. We are making our atmosphere resemble the Venusian atmosphere by small degrees. If we are so foolish as to strand ourselves on a seriously damaged Earth, or if we are forced to terraform other worlds to replace a lunaformed Earth, we shall only have ourselves to blame. The "tribe of scientists" (as Carl Sagan called us), and particularly the vast-thinking dreamers among us, will share in this blame, as scientists are creating the technologies that destroy species and may even ruin worlds, even as we are responsible for observing the progress of Earth's global degradation. The impact of humans on this world's life was summed by E.O. Wilson in *The Future of Life*:

> *We,* Homo sapiens, *have arrived and marked our territory well. Winners of the Darwinian lottery, bulge-headed paragons of organic evolution, industrial bipedal apes with opposable thumbs, we are chipping away the ivorybills and other miracles around us.*

Believe it or not, there is some optimism here. Humans indeed are chipping away at the viability of hundreds of thousands of species and whole ecosystems; as importantly, we have at last recognized this fact and have begun efforts to preserve whole ecosystems. One must see that there is hope. Given what seems to me an inevitable return of some semblance of natural order, whether by the wise recognition of the limits of Earth, or forced by our exceeding these limits, there is every likelihood that natural life eventually will reassert and reprosper – with ups and downs – for hundreds of millions of years longer. One can certainly take heart in the view that the deleterious human influence on the Universe is really quite insignificant, and our puny effects are certainly to be washed away through the hand of God or gravity and thermodynamics. However, it clearly remains that human beings have some significant responsibility for this planet. The sooner we recognize that we are tinkering with Earth's global system, which we barely understand, and the sooner we act to alter human actions that are most detrimental to the system, the more likely humans will remain a part of the system and go on to prosper throughout our little corner of the Milky Way.

Whether Humanity expands through the Milky Way may or may not depend on how successful we are at Mars. By one perspective, Mars is the key stepping-stone either within one step's reach, or it is further than one step. Dr Louis Friedman, cofounder (and now executive director) of the Planetary Society with the late Carl Sagan, recently told *The New York Times*, "*Mars becomes the experiment. If we can't make it on Mars, then Earth is our limit, and we are going to have to re-examine our relationship to our home planet.*"

The possibility that humans may be Earth-bound forever, apart from our brief forays to the Moon and Earth orbit, is a vast thought the equal of any concept of Humanity as a Galactic species. The possibility is that humans will re-adapt to living with small impacts on a world where the natural realm reprospers; that could be either a delightful or a dreadful prospect for Humanity, depending on how the re-adaptation occurs, exactly what forces it, and how rapidly the change from present trends occurs. Present trends and detrending are the issues. Peter Ward, in *Future*

Evolution (Times Books) wrote about the possible future paths in evolution under situations that are either devoid of humans or directed by humans. In either case, Humanity, since the dawn of civilization, has been acting increasingly as a global antibiotic, and effects to date will affect the future path of evolution. Ward wrote, "*As a result of our antibiotic influence, and the current pulse of extinction it has engendered, many currently living species will die. Some, however, will survive and thrive, becoming the rootstock of a new biota.*"

So now, on with the optimism. ...

A POUND OF LIGHT

Benjamin Franklin, writing to Joseph Priestly, his friend in British science, at the height of the Revolutionary War said in a letter dated February 18, 1780:

> *The rapid progress of true science now... occasions my regretting sometimes that I was born so soon. It is impossible to imagine the height to which may be carried, in a thousand years, the power of man over matter.*

I dare say Franklin would be mostly pleased, as I am, and somewhat frightened, as I am, over the progress of science in the quarter millennium since he wrote. The wonders of science are plain to see. Much of the past quarter century's progress is traceable directly to Franklin's lightning-chasing kite experiments. The ravaging disregard for nature's ways and human dignity – some of the ways in which our power over matter is utilized – is the disappointment. The joys are in scientific discovery and in the improvement in the quality of human life, and, more rarely, the correcting of old wrongs done to the environment.

An automobile bumper sticker popular in the western United States says "*Earth First! Mine the Asteroids Later!*" It is, of course, poking a little fun at the expense of environmentalists, being a take-off on the militant-extremist environmentalist group *Earth First!* I would adopt this slogan as my own, but for the opposite meaning. The brightest possible future for Humanity is apt to be one where we finally accept the limits of Earth, strive successfully to protect the biological diversity of this planet, and gradually learn to exploit outer space for the benefit of Earthlings without destroying everything we touch. I am, on the whole, optimistic that we shall make the right choices after our present time of tough lessons. Proper stewardship of any planet involves what for some could be a religious realization, and for others could be just common sense. Carl Sagan, projecting into a religious viewpoint (not necessarily his own), stated this necessity well in *Billions and Billions*:

> *The natural world is a creation of God, put here for the purposes separate from the glorification of 'Man' and deserving, therefore, of respect and care in its own right, and not just because of its utility for us.*

Thoughtful extraction and use of extraterrestrial resources is likely to be one of the human behavioral and economic adjustments that Humanity will make to enable protection of the Earth with simultaneous increase in quality of life, including material standard of living. I concede that my outlook is highly biocentric, though I probably have plenty of company. John S. Lewis, in *Mining the Sky*, asked, "*What is worth more – a pound of gold or a pound of light?*" He answered his question, "*A pound of light, hands down.*" Of course Lewis was thinking like Albert Einstein merged with the CEO of an electrical utility. In fact, light is worth about 60,000 times its weight in gold, according to Albert Einstein's special theory ($E = mc^2$) and my monthly electric bill. Lewis's point was then made: Earth-orbiting solar power plants could deliver otherwise implausible amounts of energy to Earth with relatively slight environmental impact. No more oil spills; no more global warming; no more smog-choked cities. This concept was first proposed by Arthur C. Clarke in 1945 and first developed quantitatively by Peter Glaser. According to a NASA review based on the premise of Earth-based prefabrication of solar energy production units, that concept may not be economical anytime soon. However, Lewis has pursued a different concept that makes strategic use of extraterrestrial resources so as to minimize the exhaustive fight against gravity.

The most economical source of raw materials for Earth-orbiting power stations would not be the Moon, certainly not Mars, and most definitely not Earth, but rather near-Earth asteroids, says Lewis. It turns out that Earth's surface is so deep in a huge gravity well that expedient use of asteroidal water (for propellant) and regolith beats Earth launches of raw materials and rocket fuel by a huge margin. The technology development needed to support asteroid resource utilization is next to nothing compared to the cost that would be needed to move millions of tons of materials into Earth orbit. According to my studies (published in 1994 in the *Journal of Geophysical Research* and other venues), well-chosen asteroids could supply semiconducting elements needed to make the enormous orbiting photovoltaic arrays and carbon to form a graphite skeleton. Asteroids also could supply enough precious metals to Earth that our industrial economy would never look the same, with presently expensive and critically essential metals such as palladium dropping in price by up to two orders of magnitude and platinum becoming cheap and plentiful enough to use industrially on a large scale. New uses of certain platinum metals, ludicrous to envisage today because of high prices, would be conceived and implemented as freely as we use silver and copper. These metals would become vital to many segments of our global economy.

Imagine: one kilometer-size asteroid containing $6 trillion worth of precious metals, or merely $300 billion worth at collapsed market prices! It may well take quite a lot of imagination to foresee this happening, and far more than imagination to make it happen, but the trends of resource depletion on Earth, rising demand, and technological innovation would all seem to conspire to make asteroid mining likely this century. This is possibly a vital step to take this century if we are to retain high economic productivity and a spacefaring capability and to have the stars in our future.

Near-earth asteroids could also provide the propellant and raw materials that would be needed if we are to extend human presence to Mars in a major way. The vision laid out by Lewis for steam-propelled asteroid mining vessels traversing the inner Solar System is bold; certainly it is challenging. His vision is plausible, perhaps implementable this century, with a rudimentary start within a decade if we should simply decide to get on with it. Martian applications of asteroid-based deep-space transportation technologies could be ready for use by the time we are ready to go to Mars in earnest. The same thin-film photovoltaic technologies being considered for Earth-orbiting solar power stations would also be useful in Mars development.

Unlike asteroids, which I view as stepping stones to the stars, Mars exploration and development will probably be primarily for its own intrinsic value, though Mars may provide the psychology and political base for our next momentous push that finally compels us to our destiny. Mars has a large enough gravity well, and a sufficiently hostile environment (short of major terraforming), that it will be difficult to serve as a vital economic trading post or as base camp to the stars. While some Mars enthusiasts will no doubt disagree, I do not see a strong role of Mars in serving Earth's economy, although with brainy New Martians, Mars-grown technology could prove a worthy trade in exchange for certain raw materials and an industrial infrastructure that Mars will need from Earth. I rather suspect that the future of Mars for humans may be as a great wilderness to be explored but preserved.

FATEFUL, FAST-FORWARD LOOK AT TWO WORLDS

Living, dying planets

Already I have taken a view of Mars looking backward in time by billions of years and forward by possibly a century or two. Now I would now like to take a very long-range forward view of life and geology on Earth and Mars, assuming for now that humans either do not terraform Mars, or that terraforming happens but eventually Mars reverts back to the natural order. It is necessarily an effort in scientific conjecture, but I have the comfort that others have been through this exercise before. I add just incrementally to their predictions.

There are good reasons why planetary and Earth scientists commonly look back through the geologic eras but rarely look far to the future of planets. Future predictions require a more accurate understanding of the system than does simply an explanation or model of what happened in the past, as written in the geologic record. Any future models, furthermore, may require eons to validate, and few scientists are willing to wait that long. But some general trends are readily evident. Many of these atmospheric, climatic, and biological trends are well summarized in scientific form in many papers by James Kasting (NASA Ames Research Center), and also in more popular form by Peter D. Ward and Donald Brownlee in *The Life and Death of Planet Earth* (Times Books). Ward and Brownlee outline an intriguing quasi-symmetrical evolution of the diversity and complexity of Earth's life, where the climax of biological productivity took place a few hundred million years ago, and where the diversity and complexity of life is also due to start declining in the next few hundred million years. I would like to add a glaciologist's perspective, extend the projections to Mars, and then look to the distant future of relic fossils of life on both planets. I would not object if the reader were to regard this section as a science fantasy. Nonetheless events something like those indicated are likely to take place; we will have to content ourselves that, unlike traditional science, the predictions are not falsifiable, as no one has the ability to wait 7.5 billion years to see it happen. I justify this indulgence, first, because it is fun to consider; but, second and more important, because these future portraits help us to understand the stages of evolution of Earth-like planets. This thought experiment may help us to understand what to predict when the first star trekkers or robotic star ships finally explore new realms of our Galaxy. We may even discover this century that an incipient Venusian biosphere may have experienced – 4 billion years ago – what lies in store for Earth some 2.5 billion years in the future, long before a Red Giant Sun finally melts away the fossil remnants of life.

The basic concept has two components: one (which Percival Lowell hit on) is that the terrestrial planets are regarded as having geological life spans. They progressively mature, age, and slowly die, but they do so at different rates depending on size. Mars effectively is an older, dying planet, but it is not as geologically old as the still smaller, colder Moon. Second, the Sun, like others stars, also undergoes a life cycle; it is undergoing an inexorable brightening that will have inevitable huge impacts on the habitability of Earth and Mars; the brightening will even affect the survival of fossils of life. While planets and stars are complex systems, not unlike the degree of complexity of living organisms, we can be sure of basic trends in aging of all of these systems. We can draw some fairly definite conclusions that elevate the subject of future life and death of worlds beyond mere conjecture, even if details are impossible to know. Some of the aging mechanisms are actually quite simple and amenable to quantitative understanding, even if the complete systems are difficult to understand in detail.

Some systems are so comparatively simple that accurate projections of future behavior can be made on either an empirical or a theoretical basis with little uncertainty and controversy. Radioactive decay is one such simple system where the basic nuclear physics is understood and the empirical details known. The inexorable decline in numbers of atoms of radioactive elements is known with certainty (statistically), and with their declining numbers is a decline in heat production and geologic activity by this source, which involves transformation of some mass into energy according to Einstein's $E = mc^2$. While the coupling of radiogenic heat to geologic systems is complex and uncertain, we know that the internal heat production available to drive the geology of Earth, Moon, and Mars currently is declining by a factor of 2 every 1.3 billion years. This energy half-life is not constant, but is slowly increasing since it is the sum of several different isotopic half-lives, and the longer-lived isotopes are coming to dominate while the shorter-lived ones become increasingly irrelevant. Regardless of the messy details of how each planet uses its heat to drive geologic activity, this activity is destined to decline in vigor, and with it, geologic activity must decline over time for any world, like Earth and Mars, that is heated primarily by radioactive decay.

Heat production scales with planetary mass (the product of volume and density), and heat loss scales with surface area, so per-area heat flow at planetary surfaces scales roughly with planetary radius times density. This assumes two things: (1) that radioactive decay is the main source of energy – true in our Solar System except for Io, Europa, and a few other outer planet satellites, and the gas giants themselves, which involve mainly tidal or accretionary heating; and (2) the abundances of heat-producing elements is roughly the same in rocky worlds – not strictly true, but this assumption works roughly. This is the explanation for the absence of volcanic activity on the Moon for the last few billion years, the low but probably continuing volcanism of Mars, and the prodigious volcanism on Earth. Since other types of geologic activity, such as faulting, mountain uplift, weathering related to exposure of fresh rock at planetary surfaces, and sedimentary infilling of basins also relates to global heat flow, we can expect that all of these types of internally driven processes will decline into the distant future. That is definite. There is no escape, though bumps and valleys in geologic activity will attend the slowly dying plate tectonic cycle.

Earth may be about where Mars is now, in terms of intensity of internally driven processes, in another 1.5 billion years, and Mars ought to be going dormant some time toward the end of this same period. All other things equal, hot spring activity, mega-flood outbursts, and many other types of activity will decline. However, all other things are not equal, as there is another major source of energy that drives geology. It is the Sun.

The Sun is a simpler system than planets. It obviously has a profound influence on the habitability of planets regardless of the complex details of planetary geological behavior. The Sun is a typical Main Sequence star in its middle age. Such stars undergo a steady brightening as they get older and their fusion cores gradually convert hydrogen to helium, thus making the fusion core become denser, and the star's core more compact and subject to rising pressure and temperature; fusion of remaining hydrogen occurs at a steadily rising pace. With increased heat production, the outer envelope of a Main Sequence star must expand in order to accommodate the "gas law" and to increase the efficiency of energy radiation to space. The Sun, like every Main Sequence star, shines more brightly as it ages; the Sun's rate is presently about 0.8% brightening per hundred million years. Planetary surfaces, bathed in solar radiation, should become warmer over the eons, so long as their atmospheres do not lose their greenhouse gases in compensation. Earth, for example, should have exhibited a 30 K temperature increase since formation, and 25 K or so since the start of the rock record.

Until now, Earth has not exhibited a clear long-term climate warming despite solar brightening. This is because of the regulating influence of water-mediated chemical weathering and carbon dioxide uptake/release by rocks when carbon dioxide becomes too abundant. Earth's global temperature has not been as strongly coupled to solar luminosity as it would be for a simple emitting sphere (such as a spherical asteroid), where the global average surface temperature, T, is given by an emissivity-modified Stefan–Boltzmann equation for a Sun-illuminated rotating sphere, $T^4 = Q/4\varepsilon\sigma$, where Q is the solar energy flux for a subsolar point on Earth's surface, ε is the emissivity (about 0.8), which is related to albedo, and σ is the radiation Boltzmann constant.

The global system, however, encompassing more than temperature, is closely coupled to solar flux, and CO_2 is a key component of the system. It's hard to say whether CO_2 is the culprit in crazy climate or the savior of the planet. We can be thankful to carbon dioxide for the fact that the Earth's climate is warmed far above

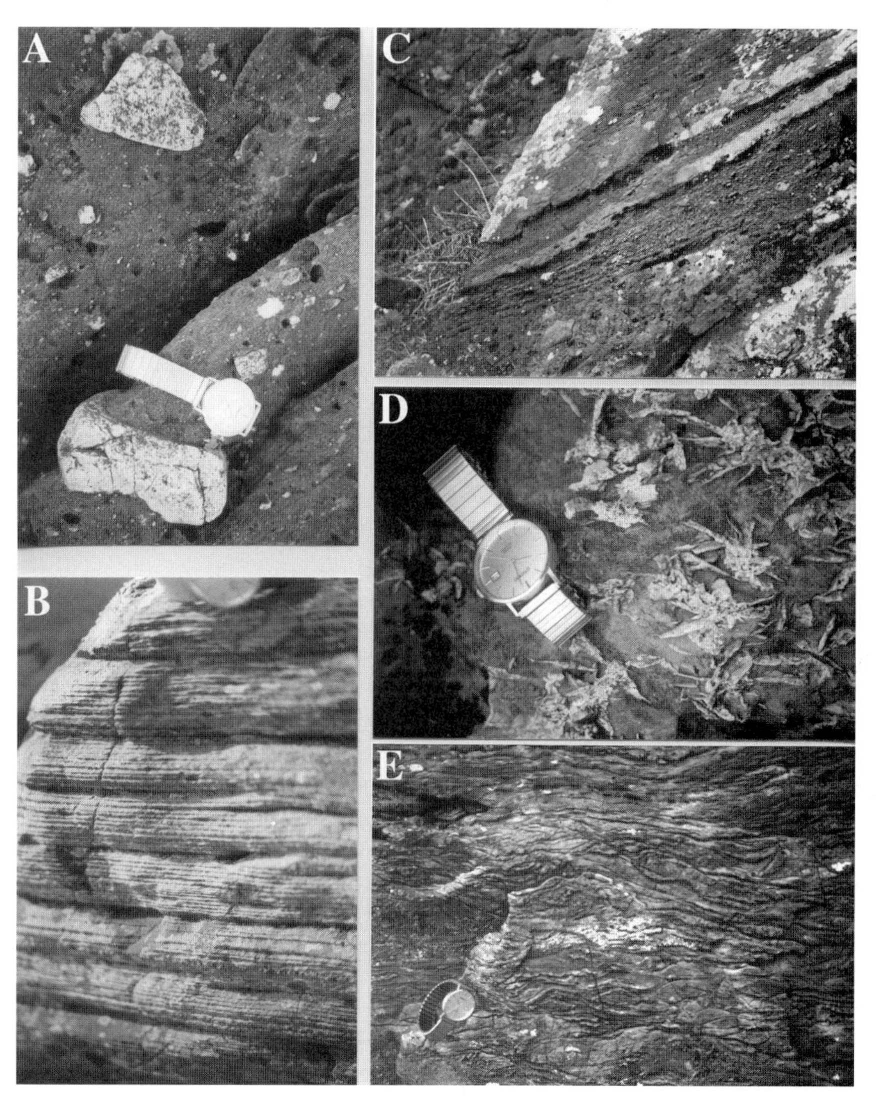

A thick sequence of "Snowball Earth" glacial deposits (~700 Ma) and overlying carbonates believed due to a "hot-house" rebound from snowball conditions, occurs in almost pristine, unmetamorphosed condition on Garbh Eileach, watern Scotland, in the Port Askaig Formation. Unpublished work by myself with Charles Sonnett, Joseph Schreiber, Jr, and Don Tarling has emphasized aqueous deposition of much of this sequence. (A) Striated, faceted boulders and cobbles, believed to be iceberg rafted dropstones, require wetbased glaciers during much of the snowball period. (B) Varve-like deposits (interpreted by my group as tidally modulated near-shore marine deposits). (C) Size-sorted, water-deposited sands and gravels, with cross-bedding. (D) The overlying Islay Limestone, which here is dolomitic, contains what have been interpreted by others as silicified gypsum crystals (pictured here, interpreted as evaporites), ice-wedge or sand-wedge polygons, and (E) algal structures, including stromatolites.

what it would be without it, and to the slow uptake of atmospheric CO_2 by the solid Earth for not being roasted by now. The Earth has responded to solar brightening by reacting carbon dioxide, in aqueous solution, with dissolved ions leached from rocks, thereby sequestering CO_2 as calcium-magnesium carbonate (iron-magnesium carbonate, too, early in Earth history), which has precipitated to form limestone and dolomite rocks. The reduced load of atmospheric CO_2 has almost compensated for solar brightening. This mechanism tends to stabilize climate also by opposing other processes which tend to unsettle the climate. For instance, when Earth has twice suffered nearly global glacial conditions (2,500 and 700 million years ago), the reduced availability of liquid water resulted in reduced uptake of CO_2; as volcanoes gradually added CO_2 to the atmosphere, as they always do, greenhouse warming finally ended the glacial epochs with a sudden shift to a warm climate, which ultimately settled down as the excess CO_2 was withdrawn from the atmosphere. While those era-making climatic acrobatics are partly related to CO_2, without it, Earth never would have escaped from global glaciation, because the brightness of ice and snow was then reflecting too much solar radiation back into space. Earth today would be as cold as Mars, but almost completely ice covered.

This mechanism has the capacity to continue to have a role in the regulation of climate into the indefinite future, but as the master of climate this mechanism has declined, because CO_2 has been drawn down to such a low level already. As the Earth warms and CO_2 is drawn down further, another greenhouse gas, water vapor, will increase in abundance, even as the Sun continues to brighten. Unlike CO_2, there is no buffering response by the solid Earth due to warming caused by H_2O vapor. Called a "positive feedback," the more the planet warms by the Sun, the more water vapor goes into the atmosphere, and that vapor further warms the planet, which drives more water vapor into the atmosphere; it is not without its limits, due partly to cloud formation, but this is a profound feedback loop. As CO_2 plays a decreasing greenhouse role, and H_2O an increasing one, the Earth will warm faster than predicted by simple consideration of incident solar radiation as the eons wear on. This will impact life in a serious way, but even more important will be the continuing decline of atmospheric CO_2, which will reach crisis levels for photosynthesis over the next few hundred million years.

Without attendant changes to the atmosphere of Earth and Mars, solar brightening at the projected rate would cause a temperature rise of roughly 0.6 K (1.1 °F) per 100 million years on Earth and 0.4 K per 100 Ma on Mars. That may seem a minor forcing, but continued over a few hundred million years it will remake both planets, particularly as water vapor will increase and CO_2 and snow/ice cover will decrease in response.

Earth's final ice age?

Ward and Brownlee have suggested that a new ice age is likely to occur soon after Earth recovers from anthropogenic global warming. This is almost surely true, since the present warm interglacial is possibly within one to several millennia of ending, according to the geologic record of glacials and interglacials, and according to the known climate forcings due to Earth's orbital obliquity variations. Five centuries is a long estimate for the sum of the future duration of fossil fuels burning and marine/ biospheric uptake of excess atmospheric CO_2. There is a not insignificant possibility that a crash from anthropogenic CO_2-hothouse into a natural new ice age may occur that fast, just a few centuries. If so, the rate and magnitude of global cooling, by going from a hothouse sustained by unnaturally rapid emissions of CO_2 to a full glacial, would be as fast or faster than Earth has ever seen in its geologic history.

The cycle of glacial and interglacials, as Earth has experienced for 2.5 Ma, is likely to continue, but only for another few million years. The long-term geologic conditions (particularly continental tectonic plate locations) are shifting away from situations conducive to glaciation, and solar brightening is occurring to such a degree that atmospheric CO_2 drawdown from already-low levels will no longer be able to fully compensate. Minor ice-sheet activity may continue in East Antarctica for a few tens of millions of years and small-scale cirque glaciation may continue in the highest mountains outside of Antarctica. Alaskan mountain ranges are apt to be sustained and may even grow, especially when what is now southern California and Baja California slams into Alaska in another 30 to 40 million years, and glaciers will remain a part of the Alaskan story at the highest elevations. However, Earth will generally become much less icy. Permafrost will disappear outside Greenland, Antarctica, Ellesmere Island, small parts of Siberia and Alaska, and a few high mountains.

The grip of ice on all these areas is apt to lift completely within the next 100 million years. Antarctica will migrate away from the pole, making even the Transantarctic Mountains and any other mountains there unsuited for the least bit of glacier ice, and albedo feedbacks and the brightening Sun might conspire to make Earth again resemble the Cretaceous Period; the warming will only continue.

The last possibility for a minor glacial epoch, much less significant than today's Antarctic glaciation, may occur in about 250 million years. That will be when Antarctica, after a brief retreat from the south pole, is projected to return there; full cover by an ice sheet or even semi-continental ice cover of Antarctica is not likely then. Global average temperatures will have increased by at least 1.5 K (2.7 °F) due to solar brightening, plus a greater amount due to decreased global albedo relating to loss of snow and ice cover. Global temperatures more likely will be at least 5 K (9 °F) warmer than now, and Antarctica will be much warmer still. Lacking a low albedo covering of snow and ice, nucleation of a continental ice sheet is improbable. After termination of the present era of glacials and interglacials, after the Antarctic and Greenland ice sheets disappear, nothing comparable is likely ever to return anywhere on Earth. In a quarter of a billion years relatively minor mountain ice cap glaciation in Antarctica may recur. Other continents will have either a tropical or searing desert climate, perhaps with cirque glaciers on the highest peaks formed at the mountain sutures of a reassembled supercontinent. This may be the last time ice glaciers ever show their face on Earth.

Or a new snowball Earth?

The picture above of declining potential for future glaciation would be true if the geologic and climatic history of Earth during the last quarter of a billion years can be adapted to the next quarter billion. Probably this is valid reasoning: the most probable sequence. However, going just a bit further back in time, around 700 million years, we come to the anomalous Snowball Earth condition, where ice gripped every continent at all latitudes (W.B. Harland, 1964, *Geol. Rundsch.*, **54**, 45–61; P.F. Hoffmann et al., 1998, *Science*, **281**, 1342–1346). A similar period of all-out mega-ice-age affected the planet about 2,500 million years ago. The climatic calamity causing these bizarre periods, in which virtually the entire planet was shrouded in ice, is hard to imagine. Surely it involved a precipitous drop in Earth's atmospheric greenhouse gas contents. Earth still harbors a latent capacity to fall into a snowball condition, but that possibility is fading under the brightening Sun. The global warming effect is comparatively small, but it drives CO_2 into the solid Earth, effectively reducing the sometimes destabilizing effects of a fluctuating CO_2 content. What will increase over the next few hundred million years is the propensity of Earth to enter short-lived episodes of CO_2-driven hothouse conditions, which may occur when volcanism and tectonic activity cause buried carbonates to degas. The reason for the lack of snowball potential is that, between rare hothouse episodes, Earth's atmosphere restabilizes at far reduced CO_2 content, while remaining warm through the brighter Sun and the greenhouse effects of water vapor and other gases. Unlike atmospheric CO_2, H_2O abundance is positively correlated with atmospheric temperature. A small increase in CO_2 can then have a huge effect.

The snowball Earth is often regarded as a state from which, once started, it is difficult to escape because of the reduced albedo. Both the vast extent of ice and difficulty of escape from these conditions pose a quandary for life, obviously. While snowball Earth clearly did not kill all life on Earth, it must have been as close to an annihilation as Earth could have sustained. For example, Steve Warren and colleagues at the University of Washington have modeled the possibilities for photosynthesis in the oceans under ice-covered conditions (2002, *Journal of Geophysical Research*, **107**, 31.1 to 31.18). The possibilities seem quite limited, even in the tropics, though patchy and thin sea ice might be possible if the ice surface was dusty (possible if there were unglaciated and windy continental deserts) and if ice shelves were not effective in spreading across the seas. However, their conclusion was sufficiently dismal that they called on special geothermally heated refugia to save life on Earth. Speculatively, it could also involve desert oases of a severe permafrost sort. The snowball-escape mechanism could also involve dust, but it is more probably due to the gradual build-up of atmospheric CO_2 and other greenhouse gases by volcanic and other emissions. The build-up is possible under snowball conditions because of reduced weathering and reduced uptake of CO_2 by rocks.

According to my observations and work done with Charles Sonett and Joseph Schreiber (University of Arizona, both emeriti), the Eocambrian snowball period was not quite as universally and unrelentlingly hostile to liquid water as is sometimes considered. The Port Askaig Formation (Scotland) is a glacially deposited tillite dating from that snowball period. It includes glacial moraine-type sediments containing sands and gravels that were obviously sorted by flowing water, dropstones dumped into an ocean or lake by icebergs, and laminated sediment that probably represent annual varves or tidal rhythmites laid down in shallow water. The laminated sediments contain features that appear to require sand sorting in braided outwash streams. Such deposits are interbedded with morainal material which contain multi-faceted boulders, which clearly were scraped along the beds of wet-based ice sheets. These multiple indicators underscore a climate that allowed liquid water in many glacial environments, even in the midst of this period. Nevertheless, the Port Askaig Formation represents deposits of a continental ice sheet, far from any mountain range, at a low latitude, flowing into the sea. Not to minimize it, that was an extreme situation, from which I am happy that Earth, like a planetary Houdini, could escape.

That snowball period culminated in deposition of a massive global carbonate deposit formed when the Earth was suddenly released from the grip of the snowball. The extreme oscillations of climate during and following the Snowball Earth, and less extreme oscillations common through much of Earth's history, underscore the potential of planetary climates to behave in erratic ways with highly nonlinear responses and feedbacks. Part of that erratic behavior is due to the episodic and sudden transfer of greenhouse gases from the solid Earth to the atmosphere, then back to condensed form. As Mars quite clearly has comparable mechanisms, even if different in detail, it should not be surprising that Mars can undergo equally drastic climatic gymnastics.

THE SLOW DEMISE OF LIFE ON EARTH

Regardless of the ups and downs of ice, the times of ice on Earth are coming to a close. As the Sun continues to brighten, dramatic changes will occur to continental life on Earth, and then also in the oceans. Ward and Brownlee, summarizing work by climate modelers such as James Kasting (Pennsylvainia State University) and others, and adding their own conjectures and informed insights, paint a vivid portrait of the future of Earth's life and explain how a sequence of extinctions of various life forms will occur. While the next hundred million years may not see dramatic changes in biodiversity beyond those in response to human-caused extinctions and others that normally mark evolution, the next 500–800 million years certainly will see revolutionary changes to the biosphere.

Surface-dwelling animal life on Earth, as we know it, will have a bright future perhaps only for another hundred million years. During this period, but after human beings have left the scene (if we do), the competitive struggle to adapt to changing conditions will probably be the same engine of biodiversity as it is at present, though total biological productivity may have already climaxed and will be declining from here on out. Ward and Brownlee project that one ecosystem after another will become extinct over the next few hundred million years, with the most dramatic extinctions due to CO_2 starvation of plants and the animals that rely on them; warming itself will play a secondary role at first. I shall interject a few more details into the portrait developed by Ward and Brownlee.

Due to the interlinked components of the biosphere, the extinctions are apt to occur in a "stick-crash-flash" fashion. By that, I mean that there will be a repeating cycle of (1) a relatively stable period, (2) a climatic or geologic event then setting off a crash of biodiversity on one or another continent or sea (or globally); then (3) a brief biological flash, or adaptive re-radiation, to partly fill vacated niches; followed again by a stable period and repetition of the cycle. Though extinction will be unsteady, it will be, over the long haul, relentless. By around 800 million years from now (the timing uncertain by a few hundred million years) Earth will become, once again, a world of microbial life and lichen-like life (only). Ward and Brownlee offer the shocking view that Earth has already seen the best, and that life here is already in an inexorable decline of productivity. The decline may have started from a peak in the Carboniferous (Mississippian/Pennsylvanian) period of tropical coal swamps and the spread of diverse forms of plant and animal life on land, with Humanity arising partly through the challenges of living on an increasingly difficult planet.

Though biological *productivity* may already be declining after a peak a few hundred million years ago, the trend in *biodiversity*, especially diversity of highly specialized forms, appears to be lagging and is still on the rise. The Earth at the advent of civilization 10,000 years ago had greater biodiversity at the familial level than ever before in geologic history. Humans have set biodiversity back a few tens of millions of years. The travails of giant impact, volcanic orgies, global ice ages, and other catastrophes each had their devastating impacts on life, but life's familial diversity rebounded after each catastrophe within 10 million years going back to the Cambrian period. To read Ward and Brownlee is to see that this exponential rise in life's richness may have reached an end, with biodiversity now to follow productivity on a downward trend. It is not just because of human impacts on the biosphere, but far more stunningly and inevitably it is due to nuclear physics, helium generation in the Sun's fusion core, and consequences for solar luminosity.

Among the first ecosystems to go will be tundra and other psychrophilic ecosystems. Icy climates, a feature of Earth episodically throughout geologic time, are on their way out for sure, if one looks beyond the next quarter billion years. *Homo sapiens*, a species that is partly a product of Pleistocene climatic upheavals, including glacial climates, is now living virtually at the end of the cold. High alpine and permafrost ecosystems will be increasingly rare, smaller, and less diverse than those of today. Their plant life will be stressed more by CO_2 starvation than by cold, though frost resilience will still be necessary on the highest peaks for the next several hundred million years, after which no place on Earth will ever again see the frost point or a single tundra flower blossom. The final demise of the last alpine tundra ecosystem will be when the last alpine refugium is eroded down. Frost-tolerant organisms may occasionally re-evolve in different forms in newly formed alpine systems using brand new genetic innovations or latent genes surviving in warm-adapted descendants of tundra species. However, over time, these genes will become increasingly useless or detrimental; their mutational degradation will eventually cause all organisms to lose the ability to adapt to conditions where cellular water expands and is chemically fractionated upon freezing. Areas with an alpine/permafrost climate, and even a temperate climate, will become rarer and smaller and shorter-lived than ever before, so any frost-adapted organisms will increasingly be a novelty, until there is a last frosty night somewhere on planet Earth.

Each ecosystem will be successively driven toward reduced numbers, reduced diversity, and reduced overall productivity. Ecosystems will probably lose large predators first. We already see this in many human-stressed ecosystems around the world, and nature will have a similar continuing effect. Alpine systems will be the last refugia of hairy mammal-like creatures and temperate-climate conifers. But these gene pools will not last as long. As each individual alpine system is eroded down, its ecosystem will disappear with it because its animals and plants will be unable to make the geographic leap to the next isolated alpine system.

CO_2, more effectively sequestered in carbonate form due to higher temperatures, will become a more precious atmospheric trace gas; forests will decline. Grasslands (or further adaptations of them utilizing recent innovations in photosynthesis that make more efficient use of CO_2) will spread to replace the forests. Animal life will find it increasingly difficult to cope with heat over more of Earth's surface no matter what adaptations they evolve, and those who manage to survive will find decreased plant food at the base of the food web. Oxygen also will be taken up more effectively by rock weathering, and will not be as quickly replaced by the fewer plants; hence, animals will find less free oxygen to sustain respiration. The atmosphere will tend increasingly toward one wholly dominated by nitrogen plus argon and a rising amount of water vapor.

A magnificent mountain range, like today's Himalayas, but with glaciers restricted to elevations above 6,000 m, may exist a quarter of a billion years from now where today's American mid-Atlantic states and Africa collide during the eventual re-closure of the Atlantic Ocean. Cool-climate alpine ecosystems may be sustained between the glaciers, but at productivity levels far below those of today due to atmospheric CO_2 partial pressure one-sixth of what it is today at half that elevation. A reassembled supercontinent, called Pangaea Ultima by Christopher Scotese (University of Texas, Arlington), may be partly fringed by subduction zones and moist, green-wrapped, glacier-capped stratovolcanoes, such as those of the Cascades of the US Northwest and the Andes of South America, but cool conditions allowing ice caps will be far less common than they are today.

Those are logical, general predictions. Taking the absolute numbers and other details with a grain of salt, events like these are qualitatively practically inevitable. May I take a quarter-billion-year leap of fantasy? New York City's eroded, redeposited, metamorphosed remnants may support the last stand of hairy descendants of mammals, who may leap from rocky ledge to ledge, on all eight legs, amidst rocks composed of what once were post-hominid glacial sediments of the Earth's final mega-ice age. These animals might not be oblivious to the rocks' origins, as there may be geologists among them. They, or alien space visitors (perhaps descendants of humans returning from Mars), may discern among the metamorphosed sediments the eroded and oxidized fragments of rusted skyscraper steel and conglomeratic bits of what were granite gravestones and concrete interstate highways once ploughed by the last continental ice sheets into the long-gone Atlantic continental shelf. Those fossils and trace fossils of ancient life will by then have been squeezed and uplifted between converging continents in mighty, quasi-vertically inclined stacks of layered rocks, like books on an endless warped shelf. The only glaciers, the last ice glaciers Earth will ever have, may be small debris-choked cirque glaciers hanging precariously to the highest peaks, rather like the disappearing "snows of Kilimanjaro" today. In the vast continental interior of the mountain-fringed Pangaea Ultima, a hot, hyper-arid desert stands sandy and practically lifeless where once there were corn fields of Indiana, lilly-dotted lakes of Minnesota, and arctic tundra of the Yukon. When the plate collision terminates, this last high alpine refugium of hairy creatures subsequently will be worn down to rugged hills after a further 50 million years; the last great extinction of large animals will hit with a hard finality. Environmental stresses, especially reduced primary productivity, will no longer support re-emergence of large ground-dwelling animals of any type.

Looking to half a billion years from now, the heat and hunger crunch by then will be turned up. Some parts of the Earth's land surface will have lost all multicellular animal life, even arthropods, especially in the hyper-arid continental interior of Pangaea Ultima. In cooler and wetter climates and lower altitudes on the continental windward periphery, a mammal-derived assemblage of animals may survive, having genetically radiated from today's opportunistic mammals capable of surviving human onslaught – descendants of coyotes, rodents, skunks, and macaques. However, high metabolic rates will not be a good adaptive strategy, due to sparse primary vegetational production and the difficulty of heat rejection under high prevailing temperatures. Mammals will disappear on one continent after another. These extinctions will be followed later by those of all other land animals. Reptiles, then probably insects and other arthropods, will go last of all. While heat would tend to drive animals to higher elevations, the food is more plentiful in the lowlands, where the highest CO_2 partial pressures can better sustain plants. Anyway, Earth, now lacking the energy she once used with abandon, will make mountains more sparingly.

Over the half billion years from our time, survivors of the relentless rising heat waves may take increasingly to the night, with larger eyes, infrared vision, and ecolocation increasingly well adapted for night vision above ground; many more animals than at present will take to living their daylight hours underground. Some will sleep underground by day, and hunt or forage on the surface by night; others will sleep underground by night and hunt underground by day as their prey returns to their shelters; some will emerge on the surface by day only briefly in desperate escapes from burrowing teeth. Most creatures spending an appreciable time above ground during the day will be sluggish and adapted to radiative or evaporative heat regulation; large animals will shed at night their metabolic and conducted daytime heat; profuse sweating will be a good strategy in environments where liquid water is always readily available. Sweating hairless sloths with elephantine ears may be one image to consider.

While the arid continental interiors will also prevail, there will be many watery places, because as global temperatures rise, so will ocean evaporation rates and atmospheric fluxes of water vapor onto the continents. Erosion rates relative to geological constructional rates will increase. As Earth's heat engine slowly winds down, rates of volcanism and tectonic activity will slowly decrease, so that Earth's continental surfaces will be more leveled; mountains will be fewer, and low-gradient coastal plains wider and more common. Coastal plains will have larger boggy areas and meandering, oxbow-straddled rivers. Karstic lake-dotted landscapes will be made from today's carbonate platform deposits and rocks formed by half a billion years of seasalt extraction by evaporation.

Metabolic heat will be a liability for creatures whether living above or below the ground. Few creatures, except those with large radiative or evaporative cooling devices, will retain the type of perpetual hyperactivity of today's primates, except possibly during night-time temperate and polar winters, which in any case will not be much cooler than summer. Large, fast-charging predators dependent on large quantities of meat, such as today's big felines, long-distance carnivorous runners, such as canines, and herbivores designed to outrun the large and fast predators, are dying families. They will change through evolution or will have no place in that far-future world. Amphibians and freshwater fish will be increasingly stressed and unable to cope. The abundance of needed dissolved oxygen will decrease with rising water temperature. With more sluggish water runoff on worn-down continents and decreasing oxygen levels due to reduced plant activity, dissolved oxygen will be in short supply.

While green hills will continue to grace Pangaea Ultima's edge, land-based plant life will dwindle, and with it animal life, and then marine animals and phytoplankton. Half a billion years from now hyperthermophilic archaea – such as that in hot springs and in submarine "black smoker" vents – may be sustained in the oceans. Submarine chemosynthetically based communities, including animals, will be less affected by the environmental changes on the continents. The global productivity of these submarine communities will be reduced by slightly diminished geothermal fluxes emanating from a slowly cooling Earth and reduced chemical fluxes due to lower rates of oceanic basalt production and lowered water:rock ratios; the reduction from today's chemosynthetic energy availability, however, will be only a few tens of percent, an amount that might be offset by evolutionary increases in the efficiency of use of this energy. Today's seafloor chemosynthetic communities, however, include a component of organisms who live from organic detritus originating as photic-zone plankton and phytoplankton; these components of benthic fauna will be far reduced from what exists today.

Ward and Brownlee write of a second era of microbial life, starting perhaps 800 million years from now, when all complex land life has finally ended with barely a hungry whimper. This period of a biological Earth will likely last until the Sun has brightened a further 20%, about 2.2 billion years in the future. While the rising solar insolation will carry just a fraction of the blame, exponentially rising water vapor – a greenhouse gas – will aggravate the Hell that Earth is slowly becoming.

Perhaps 2.5 billion years from now, the oceans themselves will boil away, and the last hardy hyperthermophilic holdouts of life will become extinct on Earth, leaving only fossil remnants under a torrid H_2O-dominated atmosphere. The efficient greenhouse warming of the Earth due to the radiative properties of water vapor will induce a runaway warming; carbonate rocks will decompose and release their CO_2 back into the atmosphere, furthering the greenhouse spiral of destruction. Over time, perhaps another billion years, the Earth's inflated atmosphere will lose its water vapor due to photodissociation, but the Sun will just keep brightening and the CO_2 in the atmosphere will further prevent a return to cool conditions. By that time, the plate tectonic mechanism of thermal regulation of the Earth's mantle temperature will probably have ceased functioning, and the planet will settle into a spasmodic "Venusian" mode of intense volcanism separated by hundreds of millions of years of virtually nothing at all happening to the surface. Sometime between 4 and 6 billion years from now, Earth will have become indistinguishable from the Venus of today.

EARTH'S END GAME

Around 7.5 billion years from now, the mean global temperature of Earth will be over 500 K. Metamorphic rocks will have lost much of their water. Earth's life will then be a long-dead phenomenon of the distant past. However, highly altered fossil evidence of past life will still abound in the metamorphosed carbonate rocks and kerogen remnants. But that is about when things start getting really bad. Mars, by this time, will be no refuge; being just slightly cooler than Earth, Mars will be destined to follow Earth through most of the horrific future events I describe next, some of which are well known to solar physicists, and others that are described here for the first time.

As the Sun's core starts to deplete its last reserves of hydrogen, a solar crisis will occur. Over a 70-million-year period, the Sun will balloon to a Red Giant object with a radius exceeding Earth's present semi-major axis. We know this from astronomical stellar studies, which show the Sun, according to the Hertzsprung–Russell diagram of star luminosity versus temperature (color), to be on the so-called "main sequence" road to an ultimate red giant. This evolution is understood from models of hydrogen fusion in the sun's fusion core and effects on the outer envelope of the Sun. There is no doubt: a solar red giant phase is inevitable, and it will ravage Earth, and Mars too. Icy moons as far out as Jupiter and Saturn will become toasted clinkers, while even Neptune's Triton will lose its ices and become something like a broiled carbonaceous CI chondrite. Even Kuiper Belt objects, though having current Earth-like temperatures during the brief red giant culmination, will come to resemble Sun-dried mud bricks, since they are too small to retain atmospheres and liquid water. Mercury and Venus, alas, will become gigantic blazing meteors for several centuries as they are swallowed by the Sun even before the maximum bloating of our dying star.

Modeling by Klaus-Peter Schröder (University of Sussex, Brighton, England) and colleagues has shown that Earth probably will not be consumed by the Sun; the bare survival of Earth, despite a Sun that grows to a radius larger than Earth's present orbit, is due to solar mass loss related to the outflow of an accelerated solar wind from the distended outer envelope of the Sun, and then an immense series of explosions during the red giant stage of evolution. Earth, to maintain angular momentum about the Sun, will move to a higher orbit in accordance with Newton's and Kepler's Laws, so chances are good that it will escape being swallowed; it will merely be burnt to a crisp, then melted, and partly vaporized under a Sun that will expand until it finally spans up to 124 degrees of the sky. (This assumes that tidal forces do not cause Earth to spiral into the expanding Sun.)

According to work reported in the 2003 meeting of the Division for Planetary

Science (American Astronomical Society), Bruce Fegley (Washington University), Laura Schaefer and I showed that the red giant sun will melt the Earth's crust and upper mantle (and Mars', too). A huge magma ocean will extend from the surface and will probably grow to hundreds of kilometers deep. If the Earth remains in asynchronous rotation, an equatorial magma ocean will extend around the Earth's circumference to the middle latitudes, while solid continents – hotter than the current Venusian surface and almost at the melting point of basalt – will occur near the poles.

Earth's rotation might become tidally locked to the Sun, depending on the unknown tidal dissipation function of the largely molten Earth. If so, there will be a hot side (with magma ocean) and a cold side of Earth, though the huge size of the Sun will mean that the Sun's limb actually shines partway around the antisolar hemisphere. Even at red-giant maximum, any residual super-volatiles outgassed from Earth's mantle or from metamorphosed sedimentary rocks would condense on the cold antisolar spot, a North America size region, which might become colder than any place in the Solar System's history. (As much as one ocean's worth of water, plus other volatiles, might remain in the Earth's mantle today and might be outgassed during the red giant phase.) In that tidal-locking scenario, nitrogen, argon, sulfur dioxide, carbon dioxide, and water will condense in the strangest, coldest antisolar ice cap, fringed perhaps by a liquid-water ocean. It may have been over seven and a half billion years since Earth sported an ice glacier calving into the sea, but for a brief time it may happen again.

During the final few million years of the Red Giant Sun something amazing will happen, according to our work, with only details differing depending on whether Earth becomes tidally locked or not. As the magma ocean heats up, there may be a new "ice age" of a very different sort; the "ice" will consist of condensed oxides and possibly even metals of sodium, potassium, silicon, iron, and magnesium evaporated from the magma ocean. In the asynchronous rotation model, the more volatile metals (Na and K) will condense close to the poles probably as oxides; if Earth becomes tidally rotationally locked to the sun, they will condense all along the terminator around the perpetual dark spot. A bit sunward of the alkali metal-snow zone, silicon and silicon dioxide, iron oxide, and magnesium metal or oxide will rain or snow from the sky. Some of the denser mineral accumulations will founder and sink, whereas many substances will float in the mantle and form land masses. Solid accumulations of the softer materials may flow glacier-like into the magma ocean, where they will dissolve in the ocean and complete a volatile cycle. The partial evaporation of the magma ocean will leave residues enriched in refractory calcium, aluminum, and titanium oxides, some of which will form buoyant masses which will float like icebergs or wash up on the hot continental shores of the magma ocean. Some of the evaporitic material will sink to the bottom of the mantle or into the metallic core, releasing further violent torrents of internal heat long stored as gravitational potential energy, while some may accumulate on the magma ocean seafloor like mud.

It is just slightly possible that the final aqueous phase of Earth may arise during the red giant's end game. A true hypersaline water ocean and water-ice glaciers may exist in a narrow band along the dusky zone just beyond where the limb of the

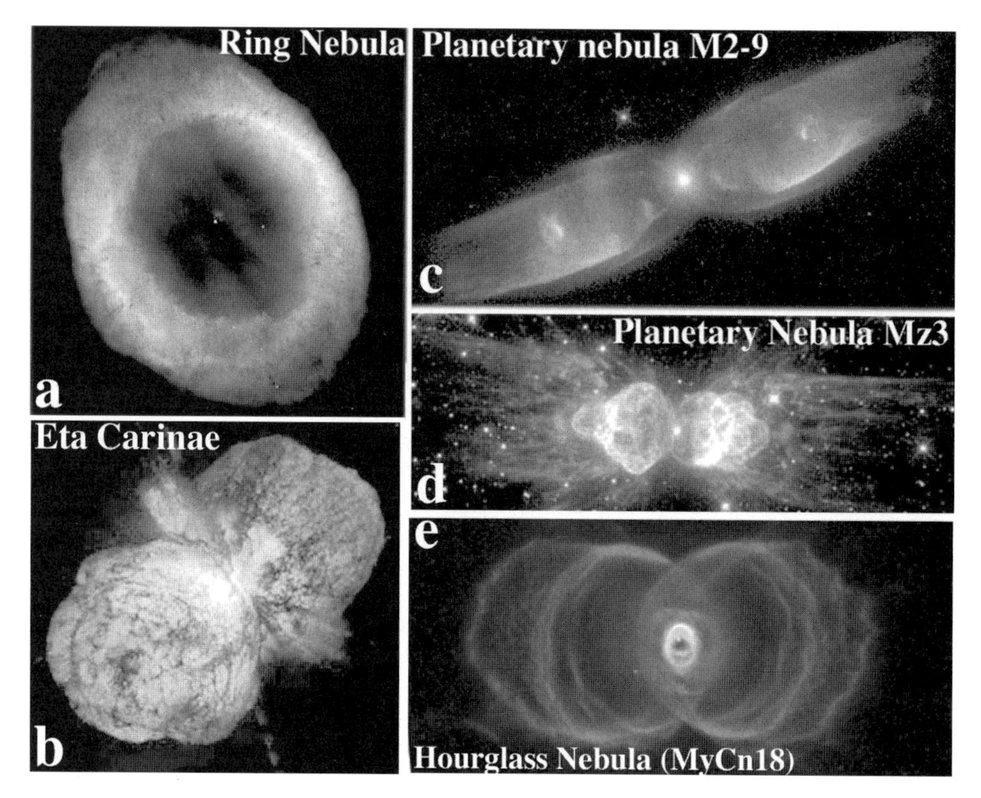

Dying stars near the termination of the Red Giant or Red Supergiant phases of their evolution culminate their brilliant careers with unimaginable explosions, which populate interstellar space with hydrogen and heavy elements, from which new stars and planets can form. The Sun will experience explosions comparable to those shown in panels a, c, d, and e, each of which were once Sun-like stars. Jetting generally occurs along the polar axis, with outflow inhibited along the equatorial midplane due to constriction by a leftover disk of circumstellar gas and dust. Hubble WFPC2 images show: (a) M2–9, 100 light-years away, started with an explosion about 1200 years ago and is now seen as the glow of supersonic ionized oxygen and nitrogen, presumably with hydrogen and other elements, jetting at speeds up to 320 km per second. (b) Eta Carinae, billowing clouds of gas and dust expanding at up to 670 km/second, resulting from an explosion 158 years ago of a star 100 times as massive as the Sun. This star briefly became one of the brightest objects visible in the southern skies. (c) M57, the "Ring Nebula," is actually a cylindrical barrel, one light-year across, of jetting helium, oxygen, and nitrogen (with other elements). M57, now one light-year across, was produced by an explosion that occurred thousands of years ago, with the axis of jetting occurring along a dying star's poles (now a white dwarf, at nebula's center) and in the direction of view. Credits to NASA and: (a) B. Balick (University of Washington, V. Icke (Leiden University), and G. Mellema (Stockholm University); (b) J. Morese (University of Colorado); (c) Hubble Heritage Team (AURA/STSci/NASA); (d) R. Sahai (JPL) and B. Balick (University of Washington); (e) R. Sahai and J. Trauger (JPL). Color versions of these and other Hubble images are available on the internet, e.g., at http://heritage.stcsi.edu/.

While Hubble images of stellar explosions are as dramatic as one may imagine anything can be, images of possible planetary systems caught in the act of forming are among the most stunning confirmations of nebula theories of the Solar System's origin going back to LaPlace and before. The structure seen above in Ab-Aurigae, viewed looking down more or less along the polar axis of the system, is probably gravitationally induced clumping of gas and dust related to formation of gas-giant type planets; it is a Solar System in formation. The disk of Beta Pictoris, viewed almost in the plane of the disk, shows just the type of radial and "z-axis" structure predicted by Solar Nebula models of star and disk formation. Nevertheless, there is some clumping perhaps related to planet formation. These images may portray what our Solar System probably looked like at its birth; and it gives a hint of regeneration possible in the aftermath of the death of our Solar System; the Sun will explode, and much of its matter, including Earth, will be dispersed, then reincorporated into new stars and new planets. Credits: NASA/Hubble Space Telescope project.

bloated Sun shines directly. This prospect – possible only if Earth's rotation becomes tidally locked and only if the Earth's atmosphere does not efficiently advect heat to the dark antisolar spot – recalls the end of the Earth described in 1895 by H.G. Wells in *The Time Machine*. Some of Wells' details are probably not far off if tidal locking can occur. However, it is doubtful that any lichenous and crabby forms of life, as suggested by Wells, would exist here at this time, since life on Earth would have become extinct billions of years before the red giant. It is just remotely possible that the final aqueous oceanic phase of Earth could be exploited by future space travelers. I more suspect that this ocean would be as nasty as can be, with concentrated sulfur dioxide, arsenic, and selenium and other heavy metals and poison gases. This ocean would be utterly frozen on its dark glacial side and boiling on the more sunward side; ripped and seared by solar coronal mega-outbursts; the shores tormented by tsunamis due to Earth's melting and quakings; and the seafloor unsettled by massive magmatic eruptions and the sea surface fogged in and ashen by condensations of liquids and solids of many types. Inquisitive but sane space travelers, if there should be some paying a visit, would be tempted to send only robotic probes from their exploration bases perhaps situated safely among the Kuiper Belt comets or on pleasant Sun-dried Pluto.

Whether rock-encrusted evidence of former life on Earth (and Mars) survives the red giant phase depends on whether and when tidal locking of rotation occurs. If it does occur and takes place before the complete metamorphic destruction of fossils, then fossils might survive in the cold spot beneath the expected ice cover. Eight-billion-year-old carbonate rocks and slightly younger hominid-bearing sandstones might survive more or less intact beneath the antisolar ice cap. Elsewhere, the great Paleozoic carbonate platforms will melt and thermally decompose on the Earth's hot side, as will massive salt beds, deposited 6 billion years before the red giant on the seafloor of the drying ocean. These chemical sediments will yield a potent brew of gases, which will contribute to the antisolar ice cap if Earth is tidally locked. Outside the dark spot's protective ice cap, any lingering fossilized and metamorphosed trace of ancient life will soon be melted and vaporized.

This will be truly an era of "Fire and Ice," satisfying Robert Frost's dual poetic perishings. Any alien visitors examining Earth and Mars might never suspect what once was, unless they excavate beneath the ice cap. While the details of this model are entirely uncertain, the broad-brush outlines of it are invevitable. As Ashley Davies (JPL) refers to this model, it is "*Earth and Mars in a bad neighborhood*," or, as Karl Hibbitts says, "*Earth and Mars in the red-light district*." Donald Brownlee, in an interview for *New Scientist* magazine, summed what we might see if we could venture there: "*We might see a pink planet with a melted surface and a hundred atmospheres of oxygen. Most people would say that's a bizarre planet, but it's not – it's just our own future.*"

The harsh drama of this ultimate period in Earth's and Mars' evolution, with bizarre ice glaciers, silicate- and metal-oxide glaciers, and an evaporating magma ocean, will be brief: just a few million years. Then things start to become downright dreary to an extreme. The Sun will finally run completely out of hydrogen fuel, the core will collapse further, and a final series of enormous explosions will blast the outer envelope of the Sun away, ripping at Earth and Mars. The Sun's outer envelope will finally explode massively, its core collapsing to a white dwarf star, which will fade over a million-fold almost overnight. If Earth and Mars survive this final solar blast at planetary dignity, they will suddenly be plunged into a global cold comparable to that now felt by Pluto. From there it will just get colder as the white dwarf Sun slowly fades. The magma ocean will start to crust over in a matter of months, sending lava crusts to the bottom of the magma ocean until soon the Earth is mostly rock-solid. The final cooling of the interior will be slow, and so volcanism may persist occasionally for a few billion years. After that, taking hundreds of billions of years to cool internally completely, basically nothing further will happen of interest. The surface and then the interior will slowly cool to cosmic background temperature, which already will be only 2 K (–456 °F) by the time the red giant phase ends. It is the inevitable cosmic thermodynamic death, which may be alleviated only by the ultimate collapse and total annihilation to non-existence of the Universe (if the Universe is what cosmologists describe as "closed"). For Earth and Mars, the best possible fate may be that the collapse of the Red Giant Sun and the attendant blast may disintegrate our worlds, with the debris cast into interstellar space and eventually reincorporated into a new star and new solar system and, maybe, new life.

No matter how I put it, and regardless of the uncertainties, the Earth's fate is sealed and dismal, whether it becomes engulfed in the dying Sun, is melted and then plunged into unimaginable cold, or is blown to smithereens. This dismal future recalls a more cosmically comprehensive statement by particle physicist and Nobel laureate Steven Weinberg (University of Texas), "*The more the universe seems comprehensible, the more it also seems pointless*" (*The First Three Minutes*, Basic Books, 1977). While I can appreciate his perspective, it is one I choose not to share. Although I cannot lay claim to any knowledge about the point of the Universe or the "why?" of our existence within it, I must believe, exuberantly so, that there is a point and a value to our being. Looking around especially at planet Earth today and for its last 4 billion years as a living world, it is not a difficult belief to maintain if one looks beyond thermodynamics, particle physics, and planetary geology; theologians and philosophers across the ages and around the world have arrived at the same conviction. It is a matter of faith; without that faith the significance of the Universe indeed would collapse to Weinberg's pointlessness. Though some might prefer different terminology, it all boils down to a bare essence, as quoted in Chapter 1, of the Universe telling of the glory of God. While a strict theistic mysticism is not required for proper appreciation, most people, even the most ardent atheists, would probably agree that the Universe speaks of the glory of something grand beyond comprehension. And so, in the molten red giant phase, future Earth has its place in the cosmic and post-cosmic grandeur.

MARS BEFORE THE END

Returning to my subject of Mars, and assuming for now that the natural course of planetary evolution prevails, a sequence of events will unfold on Mars that parallels but sharply trails that of Earth. In general, as Earth becomes less habitable, Mars will become more habitable. But it is not that simple. Some 1.5 billion years from now a 15% rise of solar flux will drive an increase in global Martian temperature. The brighter Sun will drive into the atmosphere more of the planet's stored CO_2 and a little more H_2O; the greenhouse effect will take hold a bit more than it does today. Climate will be well on its way to sustained Earth-like Ice Age conditions, but just too little and almost a little too late for surface life. Solar luminosity at Mars will not reach that which prevailed on Earth 4.5 billion years ago until 4.5 billion years into our future, and it will not reach today's Earthly incident flux until another 2 billion years after that. At that point, Mars will have only another billion years or so before the Sun becomes a red giant and experiences a fate very similar to the end of the Earth, as described above.

Before Mars finally reaches current Earth-like surface temperatures – and it certainly will attain that threshold – it will have long previously become volcanically and tectonically extinct. That will not be an interesting period for life. During the run-up to Earthly conditions, glacial and fluvial erosion under relatively mild permafrost conditions will have infilled every crater and leveled every mountain, with insufficient internal energy to drive production of new relief. When stable year-round wet conditions finally prevail on Mars, the surface will already have been worn down to a flat global plain. There could be a shallow global ocean or a subsurface aquifer, but it will be fairly inactive and offer life little in the way of dissolved nutrients. Any exposed permafrost and shorelines will thaw, but on this volcanically and tectonically inactive Mars there is likely to be little or no life to take advantage of the wet conditions. The lack of volcanic recycling of gases will mean that carbon dioxide and water itself will be increasingly bonded into minerals, and the lack of efficient erosional and nutrient cycling might doom any effort for this mostly rock-bonded hydrosphere to make Mars a living world once again if it has become lifeless (or always was lifeless). The only hope may be for errant asteroids or comets to impact, creating surface relief, generating thermal gradients and hydrothermal circulation, and releasing nutrients and disequilibrated reducing gases from impact-melt glass for a few million years, which might be just enough time for life to get started again.

These conditions will not suddenly arise. There will be a slow transition from today's icy Mars to this future moist-but-boring Mars. Possibly the most interesting time for surface life will be about 1.0 to 1.5 billion years from now, before the final volcanic and tectonic extinction of the planet, but after substantial solar brightening and global warming has occurred due to the brighter Sun, the sublimation of dry-ice polar caps, and a more humid atmosphere. The types of conditions where today's small gullies form during seasonal warmth will become extremely widespread for a time, and glaciation will be common. Life may well get a foothold on the surface, though conditions at most locations might still be as severe as those of today's Antarctic Dry Valleys. That would indeed pose a revolution in Martian evolution. Short of that possibility, however, more subtle changes to any existing biosphere will take place but nothing nearly as dramatic as the disastrous events on Earth.

An ironic and somewhat frustrating aspect of this next eon is that the rate of decline of geothermal heat will exceed the rate of surface warming under the brightening Sun. Thus, the crust will become more uniform in temperature – increasingly frozen throughout, even though the surface will see lengthening seasons of summer thaw. The geothermally heated subterranean Martian biosphere, if it has one today, will become progressively more dissected by solid-frozen regions of low heat flow. Any Martian life will, over the next 1.5 billion years, become isolated into separate refugia located deep in the crust near a dwindling number of hotspots, forced to exploit dwindling fluxes of life-giving gases. Biological productivity will decline, and within each refugium biological diversity will decline in accordance with ecological laws that scale number of species to size of the ecosystem. Ecologically, it would be fascinating to watch, if we could. Though the diversity of species and abundance of individuals within each refugium will decline, evolution in these isolated ecosystems will proceed divergently. Mars might become the ultimate in "island" type isolated ecosystems, with all sorts of novel evolutionary experiments taking root and radiating into more than the mind can imagine, a veritable montage of microbial Hawaiis and Galápagoses until their final slow freezing extinction some

2 billion years before the crust finally, gradually thaws throughout from the top down due to the brightening Sun.

The distant future of Mars during the red giant phase of the Sun will be not much different from events on Earth, lagging just slightly in time and not reaching quite as extreme an extent. Temperatures will be cooler, but still sufficient to generate a magma ocean. Tidal damping and locking of rotation to the Sun could result in a Martian cold spot and cold trapping of ices and recondensed metals and metal oxides; the lower temperatures over Martian history will mean that a larger proportion of these volatiles will escape the photodissociation and global loss that will remove most of Earth's volatiles. However, tidal locking is not likely. Mars will be just cool enough that even lacking tidal locking, the poles may remain cool enough to retain some fossil evidence of former life perhaps beneath kilometers of crust. Nevertheless, any alien visitors during or following the red giant phase might discern very little difference in the histories of Earth and Mars until they drill into the cold-spot crusts of the two worlds and discover geological and paleontological relics of former existence.

If there is any good news for life in the red-giant traumas of our Solar System it is the news that the exploding Sun holds for future solar systems. The Sun will produce large amounts of carbon by the triple-alpha fusion process, and if the final explosion of the Sun is sufficiently violent, some of its carbon and oxygen (and potential for formation of organic molecules and life) will be blasted into interstellar space, where new galactic clouds, new stars, new planets, and maybe new life will form partly from the dispersed ashes of our star.

STILL WAITING FOR THE GOOD NEWS ...

So where is that promised optimism? So far it sounds like an even mixture of all-out doom and bare survival of crystallized relics of pond scum. The good news is far more immediate than eons henceforth, when admittedly there is no prospect for good news in this Solar System. The excellent news is that the Earth – what Carl Sagan called the *"Pale Blue Dot"* – is still a wonderful, biodiverse planet, a good home to tens of millions of species. Earth is fully adequate to human needs. Quite possibly Earth is the most amazing living world in this corner of the galaxy. As far as we can clearly see for millions of years, there is no high-risk natural threat that would be apt to render Earth uninhabitable. We worry about global warming, mega-earthquakes, perfect storms, kilometer-size impactors, and the waxing and waning of ice sheets. Natural disasters are always with us, and global change is the only constant for planets; massive change is tough to deal with, but foreseeable changes over the next half billion years are not going to snuff out life in its entirety. Foreseeable changes over the next half million years are not apt to destroy civilization unless they unleash the folly and full wickedness of Humanity. We can worry that if humans do not satisfactorily meet the challenging issues of today – especially rapacious and inequitable overconsumption – we will suffer an ascending spiral of conflict, inexorable drastic economic decline, the rise of abusive governments, and then perhaps extinction. But that very statement embodies an optimism that we *can* overcome our problems and avoid a dismal fate, at least up to the limit that cosmology allows. It really is not such a tall order to solve our problems; we simply have to recognize the problems, acknowledge our blessings, and then let the rational and far-sighted side of our beings control our destiny. If we can be guided politically by those requirements, then we can argue and squabble over details, but the future will be secured.

Running away to Mars is no solution. By comparison to our lovely world, Mars is a planet already ravaged by extremely hostile conditions judged from any human point of comparison. It is a place to explore, and ultimately one day possibly to exploit – a place upon which we may build new nations or one nation; or perhaps it is simply an amazing world to observe in awe of its deadly silence and majesty, or a world to enlighten us for its possible subterranean life. Mars is not a place to bank the fate of Humanity, because in so banking, we give up on the ideal planet for terrestrial life – planet Earth.

The chief high-risk threats to Earth's habitability pertain to human activities. We, as human beings are smart enough to consider the effects we have on the world and to avoid the pitfalls if we wish to. The outstanding question is whether our choices will be the smart equal of our raw intelligence. As Carl Sagan wrote (*Pale Blue Dot*, 1994, Random House),

> *Many of the dangers we face indeed arise from science and technology – but, more fundamentally, because we have become powerful without becoming commensurately wise. The world-altering powers that technology has delivered into our hands now require a degree of consideration and foresight that has never before been asked of us.*

So where is that elusive optimism? It is that most people, faced with making difficult choices for survival of themselves and their children, and given the needed information, will make the right choices. Furthermore, democracy, literacy, and the technology needed for information transfer has spread across much of the globe into the seats of power and to individual households. My hunch is that consideration and foresight will prevail before it is too late, and that enlightened democracies will decide that human activities that take the planet sharply out of the environmental realm we evolved with are not in our children's interests.

Left to nature, solar and geologic influences on Earth's climate and biosphere will be very gradual, and indeed have been going on for billions of years. The magnitude of environmental change likely to occur by nature's hand – such as the Younger Dryas climatic excursions – are apt to be survivable by a smart and wise species. Life has survived the "dance of the continents," mega-impacts, fundamental changes in the composition of the atmosphere, alternations and zeroing out of the geomagnetic field, rapid alternations of ice ages and interglacials, CO_2 hothouse conditions, and even global ice cover of "Snowball Earth." Life takes hits but overall is resilient, as it has survived for 4 billion years and has radiated across the Earth. Up to basic limits imposed by long-term solar evolution, and absent major human interventions and hindrances, major biological adaptations to large-scale climate change in the next hundred million years will come even more readily in the future than they have in the past, simply because the familial-level gene pool of the Late Cenozoic is larger than during any previous era.

Individual humans are no longer under the degree of constant threat once experienced from lions, mosquitoes, droughts, floods, and ice ages. There are problems, to be sure, that ravage whole countries, but the most serious threats we face are the effects of our own making. Human creativity means that for every human-caused threat there is a human solution, even if it is as simple as not engaging in the activity that threatens us. The fact is, we have choices that our ancestors never had: choices to do the planet and ourselves good or ill. A hopeful outcome is possible, but only if Humanity acknowledges the severity of human impacts on the physical environment and biosphere of our planet and if we decide to address these problems in effective ways.

For those of us who wish to venture on to Mars and beyond, as I certainly do, we would do well to heed the advice of Edward O. Wilson,

> *We did not arrive on this planet as aliens. Humanity is a part of nature, a species that evolved among other species. The closer we identify ourselves with the rest of life, the more quickly we will be able to discover the sources of human sensibility and acquire the knowledge on which an enduring ethic, a sense of preferred direction, can be built.*

Human activity has started to engineer our planetary environment, partly by design and partly by accident. Let us take stock of the fact that some global changes are taking our planet away from stable, baseline conditions faster and further than experienced anytime since the dawn of civilization. These changes, so far, are not apt to reduce Earth to an alien planet, but the fact remains that we do not fully understand all the global knobs and switches we fiddle with. If the global changes happening now are not enough already, the potential exists that we could trigger calamitous, rapid changes that our economy would be ill-equipped to deal with. We know that there are several physical thresholding phenomena that act as climatic triggers or tripwires for global change. Methane releases from marine clathrates, chaotic or thresholded changes in marine circulation (e.g., that responsible for the Younger Dryas climatic oscillations and, on a smaller scale, El Niño), and other processes can initiate global climatic shifts in periods as short as a few months to a few years. We know from the geologic record that these abrupt changes can cause the onset or demise of glaciations and mass extinctions that exceed the catastrophe of the dinosaur-exterminating Chicxulub impact. In my field of cryospheric sciences, there is already recognition that in 1988 there was an abrupt increase in the rates of recession of many of Earth's glaciers. No climate mechanism has been found to account for that abrupt change, yet it seems to be real; it may have analogs in the glacio-sedimentological and ice core records dating from the previous full interglacial. The 1988 change required a climatic trigger. What was this trigger? Did human activities trip it, and what other tripwires are there still waiting to be tripped?

Near-Earth space, including asteroids, the Moon, Mars, and radiant energy from the Sun, has much to offer Humanity. Continuing improvements in human welfare may be possible, especially if we can successfully make use of extraterrestrial resources and ultimately take the great leaps to Mars and then to other star systems. However, to lean on this prospect like a crutch, as though it is our only hope, is to gamble the best thing going – Mother Earth. If we are good planetary stewards we can have the best of everything, including Mars. Some traditional attitudes of Native Americans are positive models in striving toward a unity of the human realm with the natural Universe. Native American animism is not so much a religion as it is a life-pervading philosophy that sets human beings within nature, and sees our fellow creatures as having real, actual lives, instead of being industrial chemical cytoplasm and nuclear material within fur, feathers, scales and cellulose. Instead of seeing our Manifest Destiny as the master of a concrete-and-steel planet and Cosmos, we should see ourselves as the joyous stewards of a still-lovely planet and star-trekking visitors to the Universe beyond. If we are poor stewards of Earth, we could lose all, including the concrete and steel and the means to venture beyond Earth.

US Senator Bob Nelson, a champion of space exploration who flew aboard the Space Shuttle *Columbia* in 1985, even in the tragic light of the loss of *Columbia* during atmospheric entry in 2003, is pressing America's space program and the world on to Mars as soon as possible, pointing out that, "*The nation needs a vision, and the space program is the way we can fulfill that vision*" (NBC *Meet the Press*, February 2, 2003). The ultimate challenge, according to the Senator, will be to determine if there ever was life on Mars and then to determine its fate. Did climate change spell the demise of life there, even the end of ancient civilizations? Does Mars bear any lessons that we would do well to heed here on Earth?

While I could not disprove the possibility that there were civilizations on Mars, I have already indicated that there is no compelling evidence to suggest that they ever existed. However, there certainly have been climatic upheavals on Mars that were of a magnitude that would challenge any but the most adaptable, wise, and technologically advanced civilization, if it existed. With Humanity already pressing our planet's global climate system to an extent never experienced in the whole duration of civilization, we would be wise to gain from any fundamental insights that bear in any way on the enormous economic and scientific issues swirling around the issue of global climate change on Earth. Regardless of whether there were civilizations on Mars, that planet offers a dramatic lesson in what can happen. Venus offers yet a different example at another extreme. We know from Earth's geologic history that comparable climatic shifts have occurred in the distant past. In fact, the past 10,000 years have been a time of unusual climatic stability – a stability that humans are rapidly undoing.

Some of our political leaders would have Humanity reconcile ourselves to global change of unprecedented and accelerating magnitude and reconcile ourselves to the economic upheavals and extinctions that will attend epoch-making climate change. These leaders do not understand or admit that these upheavals may involve the demise of the nations most responsible for the upheavals, or that the extinctions possibly could include our own if the changes enter a chaotic realm. Third-quarter 2004, pre-election US politics and corporate bottom-lines are just too important to be concerned about the fate of civilization. To a considerable extent, reconcilation to some change is necessary and practical, because dramatic further climate change on Earth is now inevitable, and in any case change is not in and of itself always a bad thing. How far and how fast it goes and how unsettling global change will be, however, are matters of collective human choice as well as of the whim of a chaotic climate system not yet understood. We could seek a way into the future that extends (for as long as is still possible) this rare period of climatic stability that we have enjoyed for the ten thousand years of civilization's ascent. Understanding the causes of the dramatic climate shifts in Earth's geologic past and those on Mars and Venus is perhaps the best way to understand the knobs and buttons of our planet's climate system, and to devise effective climate-change mediation schemes. Fidgeting with the knobs and buttons, and turning up the heat dial of a climate system that no one fully understands is probably not the way to regain stability or to sustain conditions that have given rise to civilization.

Foremost, planetary exploration is uniquely capable of driving home the fragility of planets and the unique attributes of Earth that must be protected. Mercury and Shuttle astronaut and former US Senator John Glenn emphasizes that "*the reason we're up there is ... to do things of value for people here on Earth*" (NBC *Meet the Press*, February 2, 2003). Extraction of insights that may help to save our own planet from ourselves is certainly a qualifying deed. Providing constructive goals for the young also qualifies.

Mars is a fascinating world. It is the other place in this Solar System most apt to be inhabited someday (soon) by Earthlings, whether for just a time or permanently. Mars exploration should be pursued aggressively, in my view, as it is something within our means to do, can teach us much, and is inherently a peaceful outlet for human aggressive creativity. I believe that humans are starbound, if we are able to avoid the pitfalls that may doom our species. Mars will be visited along the way, but it is not the end or purpose or even the primary stepping stone in our space faring. If we go to Mars with humans, we should go there with a plan to stay or to gradually build a permanent presence, if it is first verified that Mars is not a living world already. All my concerns about genetic engineering and bioterror weapons and global climate change are going to be issues that we face here on Earth and which the New Martians will also face in parallel. My concerns are not so much about solving the problems of a New World; the New Martians will solve those problems. My concern is foremost that we take responsibility for Earth and, as we reach out to Mars and beyond to the stars, we do so in the best spirit that humanity can offer. We can "reach beyond ourselves," individually and collectively, with the compassion of Mother Teresa, with the can-do attitude of *Apollo*, in the exploratory spirit of Lewis and Clark, out of the love for nature here on Earth that guided Ansel Adams, and with Carl Sagan in his quest of the Cosmos. Only then can we know that we have done well.

11

The Late News

One of the greatest pleasures and a chief challenge I have faced in writing this book has been the rapidity with which discoveries have been made and hardware has flown in the global Mars exploration programs. Just as the finishing touches were being made, the *Spirit* and *Opportunity* landings took place, and the European *Mars Express* arrived in orbit. The *Beagle 2* landing unfortunately failed, as did the *Nozumi* mission. Nevertheless, we have an international flotilla of probes – five craft operating simultaneously from the surface of Mars or low orbit. On the political front, no sooner had *Spirit* landed that President Bush gave a surprise announcement providing a long-term vision for the American space program. This brief chapter is about these developments, where it is not just year by year but literally day by day that key findings are announced.

SPIRITUAL EXPLORATION AND AN OPPORTUNITY FOR ALL

After a half-year journey, on January 3, 2004, the *Spirit* rover bounced onto the Martian surface on the sediment-infilled floor of the ancient crater called Gusev. This vehicle, its complement of scientific instruments, and its Earth-bound guides makes *Spirit* the first robust robotic field geologist ever flown to another world. Although its mission on the surface has barely begun as of this writing, it has already produced some stunning observations and unexpected discoveries. *Spirit* has produced images of rocks and soil far more detailed than any prior images taken of any planetary surface by a robotic craft. The detailed panoramic images taken by that craft and its sister, *Opportunity* (operating on the other side of the planet since January 24, 2004), are paralleled in detail only by the *Apollo* lunar panoramas shot by astronauts. In spectral range and radiance calibration they provide a quantitative analytical potential for Mars vastly beyond any lunar images produced during *Apollo*. The degree of interactive scientific sensory and analytical capability afforded by the instrumentation of these roving robotic geologists and their Earthbound guides was something *Apollo* astronauts never had while on the surface.

The operation of *Spirit* and *Opportunity* certainly remains cumbersome; and although these rovers' eyes can see at wavelengths well beyond the range of human eyes, their imaging observations in many other regards (particularly data volume and processing speed) are far less than offered by the unaided imaging and image-processing capabilities of human beings. By comparison to sampling capabilities and surface investigations by *Apollo* astronauts, these rovers are highly inefficient, taking hours to days to do what astronauts in the field and scientists back in Earth-based labs can do in seconds to minutes.

Of course we all are most excited by the search for signs of life. The search for life is not actually the purpose or design of these missions or those of the three craft orbiting overhead. Certain kinds of fossil life could exist on Mars, and if they left macroscopic remains the rovers might discover them, but far short of that unexpected discovery it is accurate to say that these rovers are exploring for signs of ancient climates and ancient roles of liquid water – expected findings that will bear on questions of possible past life. Nor, of course, can these rovers return rocks to Earth for scrutiny by more powerful Earth-based laboratories. Nevertheless, the unprecedented robotic mobility and measurement capabilities of these Mars Exploration Rovers herald a new age in planetary exploration.

What is also historic is the unprecedented availability of raw and processed data. NASA has broken new ground in providing the people's data to the people who paid the bill and to the world as a whole. Never before has so much of the scientific data been available so rapidly and so completely and in such a totally public way. Indeed, every "wired" Fifth Grade child in the world can explore right along with the scientists, making discoveries, offering conjectures, and sharing in the pain of a rover seemingly lost and then the joy of seeing it recovered. Without doubt, this is NASA's finest hour since *Apollo* in public responsibility, and it is a model in data dissemination that many scientists and the general public would like to see repeated in future missions.

SPIRIT EXPLORES GUSEV CRATER

The imagery thus far has revealed details of the chemical, mineralogical, and physical clastic products of Martian weathering processes. The data and definitely the data analysis thus far are insufficient to provide an overview of the Martian weathering regime, but the observations have motivated some new ideas and clarified old ones. The eroded voids and pock marks common in rocks in Gusev Crater are similar to those seen at other landing sites in much less detail. The new detail now available makes it clear that if these are vesicles in volcanic rocks they are highly modified and enlarged by weathering processes. Fractures that slice across rocks are also erosionally modified by weathering, and the cracks themselves appear consistent with shattering related to volume changes – possibly due to eons of thermal expansion and contraction of the silicate minerals comprising the rock, or due to growth in the fractures of salt or ice crystals, or due to expansion of salts as the result of hydration. However, more compelling has been evidence of a lack of pervasive or deep aqueous alteration of rocks at this landing site; the minerals include olivine and pyroxene, which would be altered to clay minerals and other materials in just a few hundred thousand years of weathering, even under the typically arid climates found on Earth. These early findings do not rule out episodes of wet conditions, nor do they yet tell us anything about the conditions and processes during the origin of the thick deposits that partly fill Gusev Crater, but it is a result that dissuades us from thinking of Devon Island (high Arctic), the Mojave Desert, or the Alaskan tundra as analogs of this locality on Mars for any geologically significant period at any time in the billions of years since Gusev Crater was breached and infilled with sediment. However, no analysis by *Spirit* has yet offered any clear clues about that initial landscape-setting event, because the rocks and soils so far analyzed were emplaced after that event.

The first soil analysis reinforces this view of sharply restricted aqueous alteration (or none whatsoever). It shows a composition very much like the dust clods and soils analyzed by prior spacecraft, with the added new discovery that the mineral olivine is a component of the soil. Such fine-grained olivine would be rapidly altered if it was in any but the very driest and coldest deserts on Earth. These soils look very much like basaltic or andesitic volcanic rocks, but physically they have been disaggregated. The disaggregation of rocks and formation of soils is itself very interesting, and we may yet discover a link to aqueous processes for a component of the soil. As has been known since the *Viking* days over a quarter century ago, salt-forming elements – and presumably salts – are abundant in the soil. Chlorine (forming chlorides), sulfur (forming sulfates), and even bromine are known to be important constituents, and now we have further evidence of carbonate in the soil. Carbonates could be due to weathering and attack of the finest rock powders by a slightly humid and cold atmosphere, but most likely the salts are dust components derived from aqueous evaporitic deposits formed somewhere on the planet. That source may be far away, but there could also be a local source in Gusev Crater.

Other possible evidence of ancient water activity exists in *Spirit*'s observation in Gusev Crater, but the clues are indistinct and ambiguous; it will take a detailed study to be more certain of such interpretations. Possible evidence of lake beach terraces is seen on the flanks of distant hills. These hills are not too far distant and, with a little luck, *Spirit* might just make it there to study any beach gravels and search for evidence that those gravels might have been rounded by wave action, piled up in characteristic beach landforms, sorted according to grain size by wave and current action, and aqueously altered by contact with liquid water.

The advanced analytical capabilities of the *Spirit*'s APXS (Alpha Particle X-Ray Spectrometer) and MBS (Mössbauer Spectrometer) have led to some fascinating early results. Ni and Cr are significant minor elements in the Martian soil, as they are in mantle-derived volcanic rocks on Earth. This is a definite indication of derivation of this soil (or a large fraction of it) from rocks that were previously directly formed by melting of the mantle (i.e., no remelting of crustal rocks) and even actual pieces of the Martian mantle. The inference that salts are contained in this assemblage, as indicated by high S and Cl abundances, is consistent with the existence of a weak cemented "crust" inferred from the airbag retraction mark. It is also consistent with

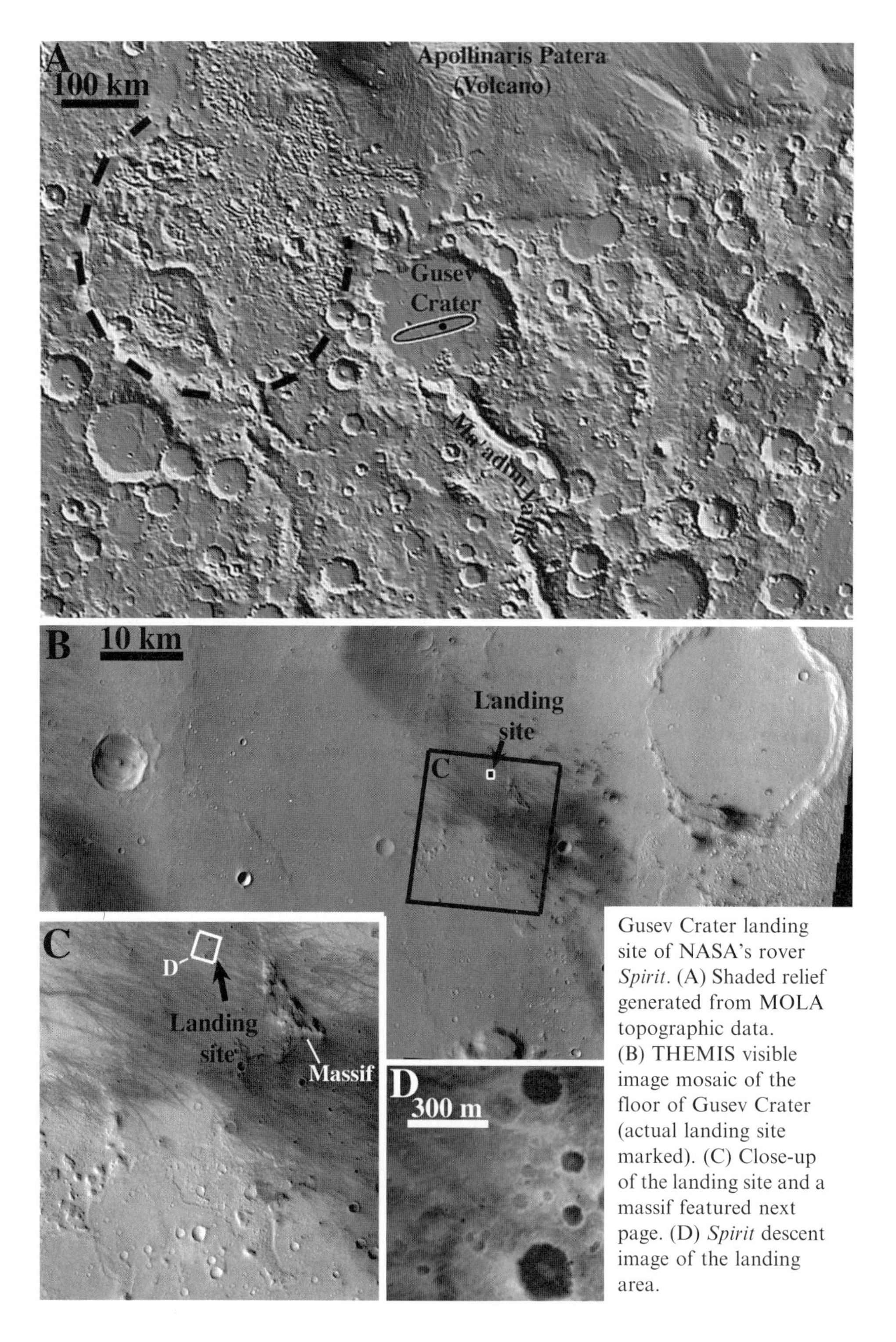

Gusev Crater landing site of NASA's rover *Spirit*. (A) Shaded relief generated from MOLA topographic data. (B) THEMIS visible image mosaic of the floor of Gusev Crater (actual landing site marked). (C) Close-up of the landing site and a massif featured next page. (D) *Spirit* descent image of the landing area.

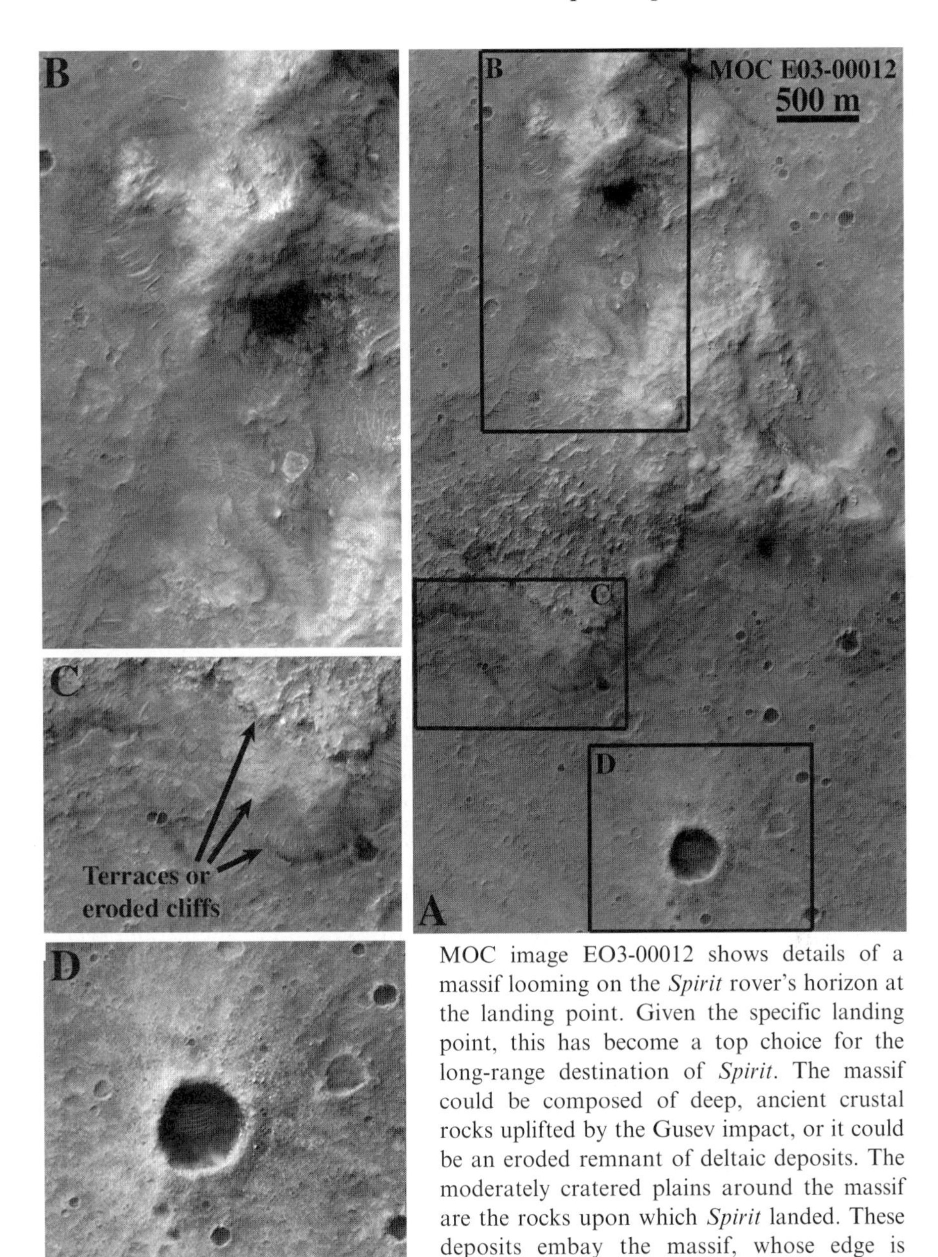

MOC image EO3-00012 shows details of a massif looming on the *Spirit* rover's horizon at the landing point. Given the specific landing point, this has become a top choice for the long-range destination of *Spirit*. The massif could be composed of deep, ancient crustal rocks uplifted by the Gusev impact, or it could be an eroded remnant of deltaic deposits. The moderately cratered plains around the massif are the rocks upon which *Spirit* landed. These deposits embay the massif, whose edge is marked by eroded cliffs or terraces. These features may bear morphological, sedimentological, mineralogical, and chemical evidence that a lake once existed here. Did waves wash these hypothesized island shores? Did the lake immediately freeze? Did it precipitate salts as it dried or frozen solid? Was it a mud lake instead? Spirit is equipped to address questions such as these. To get at a deep record of sedimentary deposition, *Spirit* will no doubt look at ejecta thrown up by small impacts. The crater enlarged at lower left excavated boulders up to 10 m across from up to 50 meters deep.

the microscopic imager results, which show what appears to be a weakly cemented assemblage of fine sand-size particles. The first MBS analysis of this same soil revealed minerals including both divalent (ferrous) iron and trivalent (ferric) iron. Ferrous iron would be contained dominantly in silicates, but some is also possible in ferrous sulfate and ferrous chloride salts and in some common iron oxide minerals such as magnetite.

Most exciting, a Mini-TES analysis of the spectrum of soil and rocks (a composite of both) showed strong thermal emission features caused by carbonate minerals and a signal due also to water of hydration. These are all signs that this is a region of salt concentrations. It remains to be determined whether these salt abundances are consistent with background abundances in global dust-derived soil and the low local abundance of rocks, or whether the salts are anomalously abundant at this particular site. If there proves to be an anomaly, we should be able to verify this by analyzing additional soils and rocks. It would be not surprising – but certainly highly informative and exciting – to discover a possible outcrop source of salty dust, such as impact-excavated strata of evaporitic salt beds in Gusev Crater.

The first rock analysis shows that one of the larger nearby rocks (named Adirondack) is olivine-rich and also contains pyroxene and magnetite. This combination is consistent with the official MER team's interpretation of the rock as an olivine basalt, but I will go out on a limb and suggest that the ratios might be more consistent with an olivine-dominated ultramafic rock known as a peridotite or picrite. Quantitative analysis of the APXS results – not reported as of this writing – will determine the rock type. If it is a peridotite, it would seem likely that this rock is a piece of impact crater ejecta derived from a distant impact that was big enough to excavate through the crust and into the mantle. Such a crater would likely be about 180–400 km in diameter. If I am wrong, and this actually is an olivine basalt, then its point of origin is most likely the nearby shield volcano Apollinaris Patera or lava flows from this volcano that might underlie the sedimentary fill in Gusev; in the latter case, it could be ejecta from a small nearby crater. Nevertheless, the odds are that this rock has been near its present location for hundreds of millions if not billions of years. It would be improbable that it just arrived a few thousand or a few million years ago, although there is no way presently to know for sure.

A positive identification by MBS of ferrous-iron-bearing olivine in the soil was a surprise. It indicates that this soil either has not been subjected to extensive aqueous alteration or includes a wind-deposited component of ground-up volcanic rocks or small chips of volcanic rocks (or mantle peridotites) that have not been subjected to extensive aqueous alteration. An interpretive summation of these results is error prone at this early stage, but I would imagine that these findings are another indication that this small piece of Martian soil has not experienced nearly as complex and convoluted a history as most soils on Earth. It is much more comparable to immature soils in geologically fairly simple sites, such as young or active volcanoes in Hawaii, where basaltic volcanic rocks are formed by partial melting of the mantle, then physical disaggregation occurs with incomplete aqueous chemical alteration of the fragments. However, the probable ancient age of Gusev's landscape means that, despite an immensity of time and its effects on this soil, the cumulative effects are as though it has not been heavily touched by water.

Imaging evidence of physical disaggregation of rocks near the lander includes accumulations of coarse sand-size particles around Adirondack and other rocks. This is a direct indication that at least a component of these soils is locally derived (even if those rocks are not themselves local in origin). However, it is inevitable – and supported by evidence – that a large fraction of most Martian soils, including that at Gusev Crater, is contributed by airfall dust swept around the planet in global dust storms. Hence, the soils include a component of material that represents most of the Martian surface, most of its geologic history, and most of its rock types – and another local component. Therefore, the soils are not just crumbled up local rocks, although they clearly contain a fraction of local crumbles. Their high volume of minerals, such as olivine, that are prone to chemical weathering under aqueous conditions means either (1) that these soils were not washed in Martian lakes or reacted with Martian rainfall for any great length of time, or (2) that such a lake has not been present for a very long time. Of course, geomorphology and the crater-pocked surface of the sediment fill of Gusev Crater tells us that there has not been a lake there for a long time.

This emerging picture is entirely consistent with global geology, where locally there have been extensive interactions of the surface with liquid water at various times in the past, but on average it has been a very dry and very cold planet for the vast majority of geologic time. No doubt this is an oversimplified picture. As the rover mission continues, details will fill in, and perhaps this summary view of Mars will be overturned. Many new details will doubtless emerge if *Spirit* is able to make its way across the ejecta blanket of a nearby crater and sample many rocks representing the stratigraphic layers at this site.

OPPORTUNITY ROCKS

On January 25, 2004, Mars Exploration Rover *Opportunity* bounced into a small crater, about 22 m across, in Meridiani Planum's hematite region. We knew already from orbital observations by the Mars Global Surveyor TES instrument that hematite (a crystalline form of ordinary iron oxide rust) was an important mineral in this unique area, and that the rocks are layered. The common association on Earth of massive amounts of hematite with aqueous sedimentation is by itself suggestive of aqueous deposition. This hint is consistent with the topographic confines of the hematite region to a small highland basin where water could have pooled. There is also the fact that the layered plains are smooth and appear to blanket the cratered terrain and embay craters at the edge of the terrain. It is as if a lake once occupied this region, and that the deposits formed through freezing or evaporation-driven precipitation of minerals (with perhaps hydrothermal or volcanic influences on chemistry), including hematite. This prospect, along with the fact that it lies within the equatorial band allowed by engineers for these missions, is the reason the site was chosen. On a very small-scale basis, the existence of hematite and rock layering was confirmed as some of the first significant results of the mission.

The soil studied by *Opportunity* on the floor of the small crater is also unique in our experience on the Martian surface to date. The soil is definitely at least partly of local/regional origin, as indicated by a high abundance of hematite. The soil also has less carbonate and less water of hydration than the surface in Gusev Crater. Thus, we have landed on, and are exploring, two completely different mineralogical terrains. The soils also look completely different even to untrained eyes. It will be a matter of analytical details to puzzle out the reasons for the differences and to ascertain (or refute) a key role of liquid water and the climatic conditions and specific processes by which these deposits were formed. One fact that was a bit puzzling to me, and contrary to my expectations based on Earthly analogs, is that carbonate is so deficient at *Opportunity*'s site. I had expected something resembling the carbonate–hematite deposits of Earth's banded iron-formations, an ancient type of rock formed mostly between 1.8 and 2.6 billion years ago when Earth's atmosphere was making a transition to an oxygen-rich composition. Banded iron-formations most commonly include abundant carbonate. This has not yet been the case in *Opportunity*'s exploration.

The most exciting find so far in *Opportunity's* little crater is a thin sequence of layered rocks on the rim. As of this writing, the rover is traversing this sequence of rocks. A thin, almost rhythmic sequence of fine-grained layers suggests four possible modes of deposition: by wind, by water, by volcanic ash clouds, or by impact ejecta clouds. A chief clue is the presence of a coarse component of spherical granules in this deposit. These very much resemble impactite spherules, such as seen around Meteor Crater, but at this moment it remains a possibility that these are small carbonate or salt spherules grown layer by layer on a wave-washed beach of a saline lake (called oolites); that they are volcanic lapilli formed in ash eruption clouds; or that they are secondary aqueous alteration products such as concretions. There remains another highly improbable but extremely important possibility: they are fossils. My bet is that, before this text is printed, the origin of the layers and their spherules will be known. Meanwhile, the scientific world awaits resolution of this suite of possibilities.

EXPRESSLY EUROPEAN

On Christmas Day, 2003, *Mars Express* went into orbit. It released a small lander named *Beagle 2* after Charles Darwin's ship. The lander was designed to search for organic chemical traces of past life or abiotic organic building blocks that never made it into life. Indeed, *Beagle 2* was the first spacecraft since the *Viking Landers* designed specifically to look for signs of life. While the lander disappeared without a trace, joining a long list of such disappearances, *Mars Express* made it safely into orbit and began transmitting stunning results from its high-resolution stereo camera and its radar. The representative selection of key results has been utterly phenomenal. For the first time we have a high-resolution imaging system in orbit

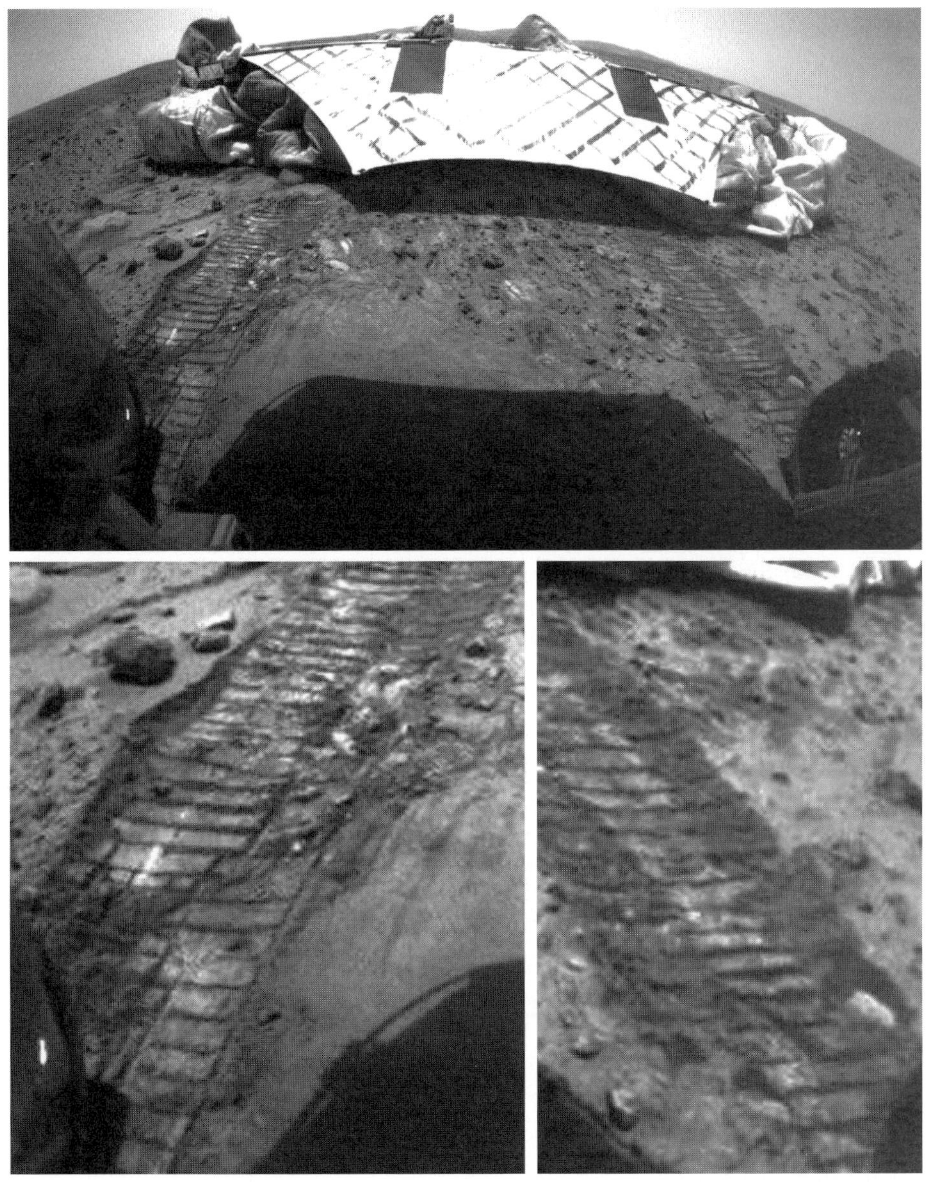

Spirit rover's first post-deployment image taken looking back toward the landing craft and its first tracks in the Martian soil. The cohesiveness of the soil is intriguing. Is it due to electrostatic charging of fine powdery dust (like lunar soil, which took bootprints very well), or is the soil actually damp? Dissolved chloride salts could stabilize liquid water, but the air is still too dry to maintain it as a liquid – some of the water would readily evaporate, leaving a crust of hydrated salts. Truly stable, concentrated liquid sulfuric acid brine could explain these and many aspects of Martian geochemistry, but would pose an additional formidable environmental challenge to *Spirit* and to future astronauts if such conditions are widespread.

that is specifically designed for stereo/topographic imaging. The craft's camera has a spatial resolution similar to that of THEMIS, while focusing its efforts on different wavelengths and stereo capabilities. Thus, THEMIS, MOC and HRSC offer complementary capabilities in Mars imaging.

With little doubt the next revolution in our understanding of Mars will come over this next year from two completely different investigative angles: the MER rovers and the *Mars Express* ground penetrating radar. Following the expected upheaval in Mars science due to analysis of these new data sets, a deeper understanding will emerge from data fusion from all the instruments that have recently flown – a process that will take many years and indeed decades to complete.

Mars Express also represents something else of fundamental value, which has little to do specifically with science: Europe has joined America in the exploration of another world. Indeed, with recent missions both successful and attempted, Mars exploration is truly global in nature. One planet is exploring another.

This sense of a global community is slightly reduced by the slow and highly controlled release of data from *Mars Express* that hinders a sense of public participation, unlike that engendered by the immediacy and completeness of data releases and the drama of the rover missions. From a pure public relations perspective, *Mars Express* has so far failed the global public, although I am certain that the results to be forthcoming will rival anything produced previously in Mars exploration.

AN AMERICAN PRESIDENTIAL DIRECTIVE

On January 14, 2004, President George W. Bush made a stunning announcement:

> *Inspired by all that has come before, and guided by clear objectives, today we set a new course for America's space program. We will give NASA a new focus and vision for future exploration. We will build new ships to carry man forward into the universe, to gain a new foothold on the moon, and to prepare for new journeys to worlds beyond our own. ... With the experience and knowledge gained on the moon, we will then be ready to take the next steps of space exploration: human missions to Mars and to worlds beyond.... We do not know where this journey will end, yet we know this: human beings are headed into the cosmos.*

The new initiative involves development of a new Crew Exploration Vehicle capable of Earth orbital, lunar, and interplanetary flights. It will fly with human crews as early as 2014, with the first extended lunar missions in the 2015–2018 time frame. The approach is to refine indigenous lunar resources into rocket fuel for travel to further realms and for use in construction. For example, oxygen could be produced by extracting it either from south polar ice or from titanium-rich glass and ilmenite (TiO_2), a mineral abundant in certain rocks and soils of some lunar maria (including the orange soil discovered by Jack Schmitt on *Apollo 17*). An extended return to the moon would be used to sustain and expand our presence in outer space, including Mars. Robotic systems as exploratory precursors were specifically mentioned – a point that planetary scientists universally agree would benefit both scientific understanding and prospects for success of human exploration and settlement.

This new directive is the closest parallel to Kennedy's announcement in 1961 that America would go to the Moon within the decade. Bush's announcement is the first believable long-term, open-ended vision ever espoused by an American president. The directive, of course, is permeated by politics, as was the 1961 directive. In the case of President Bush, the venture, if fully realized, will constitute a substantial positive legacy that goes well beyond anything that has gone before. However, for all of us, the Moon/Mars venture represents an immensely constructive endeavor, if it can be accomplished within the existing and foreseeable fiscal environment.

The President touts international cooperation, although comments made after the President's speech by NASA Administrator Sean O'Keefe makes clear that this will be fundamentally an American-led endeavor. The Moon/Mars initiative has excited Russia. Various proposals have been made in Russia either to join the Americans or to beat them to Mars. The more sober commentators are advising a unified effort. However, Chinese financing and Russian technological capabilities would provide a means whereby those nations could mount a credible competitive alternative. Already, Russian experts and corporate leaders are lining up on both sides of the issue – compete with America, or cooperate. Of course there is a middle ground: cooperate in robotic precursors, for instance in coordinating missions and sharing data, and engage in Olympics-like peaceful competition in the human realm.

For all the seriousness the world has given this new Presidential directive, it has met with a luke-warm reception (at best) in the US Congress. Its members, both Republican and Democratic, are largely concerned about the initiative's deleterious impacts on the US federal budget deficit and on other spending priorities. There is a drastically different fiscal environment now compared to when JFK made his famous announcement. In the early 1960s, the US federal budget was operating at a deficit or surplus oscillating between roughly –1% and +1% of total Gross Domestic Product. Today, the federal budget – even before the inclusion of the Moon/Mars initiative – is operating at a deficit of –5% of GDP, and reasonable extrapolations have it worsening into the future. This is viewed by most economists and the World Monetary Fund as unsustainable and a danger to the US and global economy. Thus, political pressure is being brought to bear that would reign in the deficit. The Moon/Mars initiative would be counter to that effort. In the Bush plan, the early stages of the initiative will be financed mainly by reallocating funds from the already-underfunded space shuttle and space station programs. In a flash of realism, the President has proposed that the go-nowhere shuttle program will be terminated by 2010, and that the international space station also will be terminated a few years later. Although these programmatic terminations must be made if we are to achieve truly great things and go where no one has gone before, they come at a dear cost. The *Hubble Space Telescope*, for instance, will be an early casualty, since a *Hubble* servicing mission will be forbidden.

In discussions with other planetary scientists, I have noticed a palpable and widespread skepticism based in part on the feeble funding to be allocated ($1 billion spread over five years) and reallocated ($11 billion over five years), when in fact the total project costs will no doubt exceed a quarter trillion dollars over the next 25 years, and I would expect it to be closer to half a trillion (at least $10–20 billion per year average). Thus, the direct new funding over the next five years will be exceeded at least 50-fold by the actual per-year average costs over the next few decades. Including the reallocated funds, the actual annual average costs will exceed those planned by a factor of 4 to 9. Hence, the realism of what has been proposed, during this era of half-trillion-dollar annual budget deficits, has been questioned by one colleague after another. It is without doubt a case of grand plans with deferred payments.

Whether or not this project would actually constitute a back-breaking budgetary initiative, there is a widespread public perception (beyond that of scientists) that it would be. While few scientists bemoan the uprooting of the shuttle and space station programs, there is a general view that President Bush's Space Exploration Initiative will be no more successful than his father's, when a comparable dismal fiscal environment prevailed.

Nevertheless, in the planetary science community there is a general view that this directive has the potential of reshaping the American space program to something that is much more aggressive and future oriented than anything we have seen since Jack Schmitt and Gene Cernan lifted their boots from the lunar surface. One thing is certain, unless Congress intervenes this year: the shuttle and station programs will soon be terminated. This is a pragmatic decision. Aside from honoring international commitments, one can question why we would continue to build the international space station, only to end it almost as soon as it is built. One reason to maintain and build the station would be to use it for long-term human physiology experiments, including 3-year-long stays by astronaut volunteers to determine some of the physiological impacts of interplanetary space travel during a Mars mission. However, this chief goal could be completed within the time frame for station phase-out indicated by the President.

If the Moon/Mars initiative is aborted for fiscal or any other reasons, it will be clear that the leading initiative in human spaceflight will have been ceded to China and Russia, who would probably be pressed by economic and technological realities to join forces. This fact alone, considering international political environment, is enough to make me think that the USA very well may be forced, despite the federal deficit, to follow through on the President's directive. My own slant, as an American and as a world citizen, is that I hope we can follow through on the President's words, although I seriously doubt its feasibility at this time.

Notwithstanding the dubious financing of the project, there are some major positives about the President's announcement. An essential highlight of this initiative is its organic construction. It is not, as the President says, like the 1960s Cold War race to the Moon and back. If successful, this will be an open-ended expansion by humans and our machines into the Cosmos. The President specifically stated that extraterrestrial resources will be used along the way. This is the only way to make for a permanent and expanding human presence in space. It is the only way we can build a Solar System economy and move outward to the stars. It is the first time this approach has ever been proposed by a President of the USA.

As long as an internationalist and especially a civilian agenda dominates, the President's initiative offers great hope. It will give the children of our nation and the world something positive and forward-looking, and will offer a rational basis for optimism about the future and a reason to become high achievers. It will help to renew the world's admiration and respect for America. It will offer the world a rallying point to unify their goals and aspirations. The Russian *Energiya* rocket is already designed and would be capable of the heavy lift needed for the Moon/Mars initiative. European and Russian expertise in robotic planetary exploration can be brought to bear, as can the growing capabilities of Japan, China, Canada, India, Brazil and other nations.

If we allow it, through this program we will become a better people, both in terms of our technical abilities and our global perspective. The potential benefits are far reaching. A very serious risk is that through neglect of fiscal realities and domestic needs, and failure to develop an economy that can sustain the much greater investments needed for success than so far identified by the President, both the economic machine and the political mandate to sustain the long-term investment in this program will falter. If the cost is so prohibitive that we fail to honor our domestic and international commitments here on Earth, and the project fails for the predictable response that would ensue, it would be a clear sign of an America in decline. On the other hand, if the project can achieve a solid grip in a solid future-minded budget, the world will also seek to work with America and to share the glory and the burden; the world will draw broad inferences from the degree to which America will either share opportunities to collaborate or designs the project purely around her own abilities. With luck my doubts will not be warranted by future progress. A cynical perspective, which I do not share, is that the President's initiative is really a back-door means to cancel all American human spaceflight activity. Also, cynical views have been expressed also that this initiative veils a military expansion into space. From a purely scientific vantage, I would have to agree that the shuttle and station are clearly not delivering their promise of great and valuable science that is unachievable by cheaper robotic means, and those programs are not doing America or the world any service worth their cost. However, other nations are clearly on the ascendancy in human spaceflight. Although the political environment is not the Cold War of the 1960s, the stakes are similar; perceptions of the ascent and decline of powers will be defined through this century by progress in space. Certainly fear of potential failure is not what defines a great power, but failure to deliver on boasts of greatness would certainly define a power in decline.

Optimism is where I prefer to be, but caution is what I must express. If the Moon/Mars initiative is successful, the implications for the long-term future of humanity are vast. For the near term, the magnitude of America's contribution to the world through this enterprise will be measured by the degree to which we give to the world and invite the world to contribute, rather than the degree to which America tries to be great for her own sake and isolated fortunes. We must now ask ourselves what we want the twenty-second century to look like, because it is being forged now. Can we avoid foolish and greedy use of our time and Earth's natural resources and avoid being stranded on Earth in a state of constrained and ever-tightening individual and global aspirations? Do we want to avoid an environment whereby the overly competitive struggles of the twentieth century propagate into the Cosmos through this and the next century? Do we want one-world, one-Solar System governance, and if so, who will that entity be and what values will be carried forward into the Cosmos? I do not pretend to have answers, I have just an abundance of caution as well as optimism that we can find the right path forward. I hope, more than all else, that we do not go to Mars with an attitude of throw-away worlds.

WET INK

Motivated by the deluge of new spacecraft data, our concepts of Mars are racing ahead almost faster than I can type and the ink can dry. Here I offer a brief summary of some highlights from the 2004 Lunar and Planetary Science Conference (LPSC), including presentation of the first in-situ direct evidence of abundant water on Mars. Abstracts of these presentations can be obtained from the Lunar and Planetary Institute on CD-ROM or on the LPI website, http://www.lpi.usra.edu/meetings/.

In the theoretical realm, dramatic advances have been made in the Mars community's computational and conceptual understanding of recent Martian climate changes. Robert Haberle (NASA Ames) suggested that orbital cycles of eccentricity and the longitude of perihelion drives oscillating pole-to-pole (and middle latitude) transfers of ice between hemispheres, while cycles of obliquity drive oscillating bipole-to-tropics-to-bipole transfers of ice. Currently, Mars is in a phase of fairly low obliquity and dominantly polar concentrations of ice; perihelion currently is in the southern hemisphere summer, and there is a net annual flux of water vapor and ice from the southern polar cap to the northern one. However, 25,000 years ago, perihelion occurred during the northern summer, and the net flux was the reverse of what it is today. According to Mars geomorphologists this transfer is evident at southern middle latitudes in a widespread, fine-scale pitted terrain, believed due to partial sublimation of a surface ice mantle.

The idea of cyclic pole-to-equator-to-pole transfers of ice is not a new idea – just newly confirmed – and was first suggested by Bruce Jakosky and Michael Carr in 1985. The most up-to-date computations of solar insolation and applications to permafrost active-layer processes was given at LPSC by Mikhail Kreslavsky (Brown University) and colleagues, who have found that the most recent large climatic anomaly (relative to current conditions) was 9.45 million years ago in the northern hemisphere and 5.0 million years ago in the southern hemisphere. This precision is now possible thanks to new theoretical computational findings by Jack Laskar and colleagues concerning Martian obliquity and eccentricity evolution.

Laskar et al. (2004) found that the average long-term obliquity of the Martian rotational axis relative to its orbital plane is about 37.6°, compared to 24° today. Thus, Martian climatic seasonality on average is far greater as a long-term average than it is today. Across the expanse of hundreds of millions of years, obliquity may

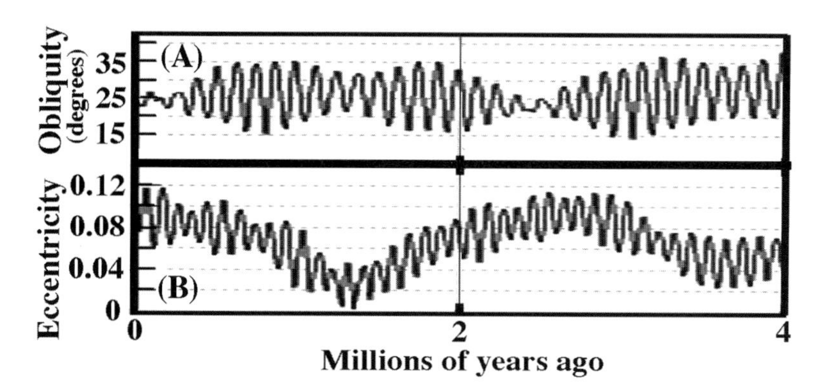

Recent history of Martian orbital obliquity and eccentricity according to calculations by J. Laskar, B. Levrard, and J. Mustard, "Orbital forcing of the martian layered deposits", *Nature*, v. 419, pp. 375–377, 2002.

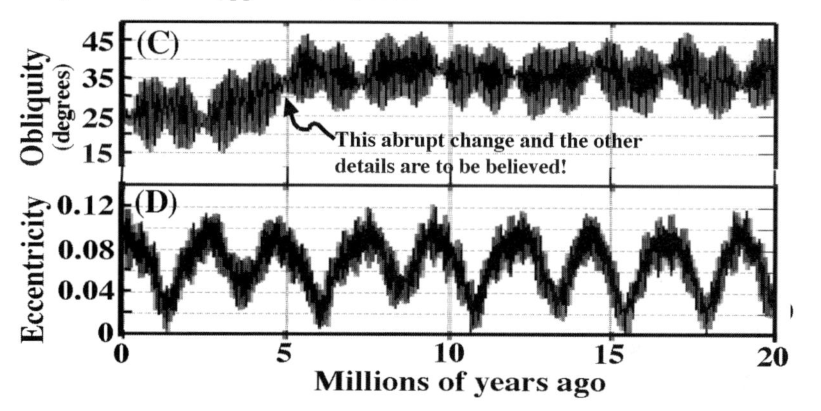

A new set of very precise computer simulations of Martian orbital obliquity (A), eccentricity (B), longitude of perihelion, and solar insolation was performed by Jacques Laskar and presented at the 35th Lunar and Planetary Science Conference. Panels C and D courtesy of Jacques Laskar et al.; for further details see their abstract in the 35th LPSC proceedings. Such results have motivated interpretations of surface geology in terms of cyclic global transfers of volatiles (panels E, F, G).

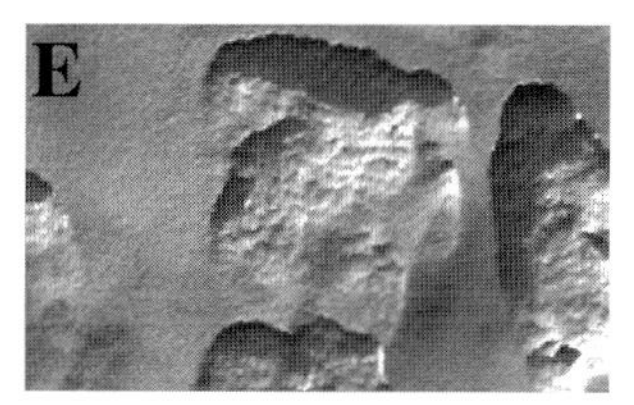

Pitted mantle south of Hellas: MOC M2101239. Width 2 km. Sun from top/right in all images.

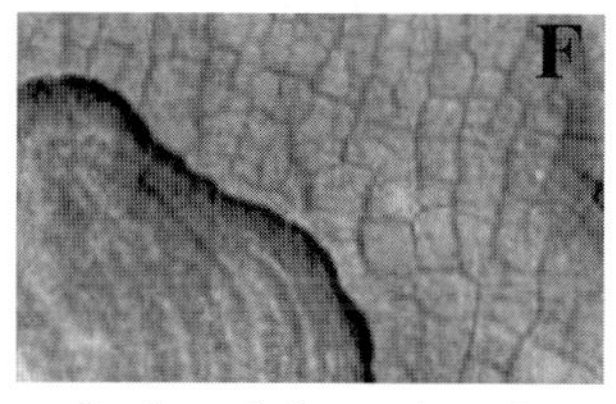

Scalloped thermokarstic pits and polygons, Utopia

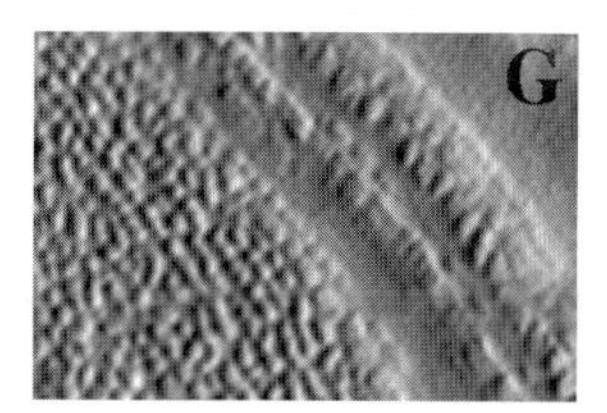

Planitia: MOC M0401631. Width 1 km.

vary chaotically to as much as 82°, which would cause a complete loss of polar caps and formation of equatorial ice caps. Ice (and with it, the loci of glacial and periglacial and meltwater processes) has been in nearly constant flux, a dynamic that has been modeled by Michael Mischna; obliquity cycles are marked by drastic shifts of ice abundance around the globe. Laskar's calculations also confirm an earlier result that about 5 million years ago there was an abrupt decrease in orbital obliquity from around 37° (with large 10^5-year periodic oscillations around that value) to 24° (with large 10^5-year periodic oscillations). A key point made by Jim Head (Brown University) and others that much of the periglacial record of the modern Martian middle latitudes can be traced to a climate shift caused by that obliquity shift as well as more recent and more modest changes due to the 10^5-year obliquity cycle. Laskar's simulation data can be obtained through his LPSC abstract, and a complete report (Laskar, J., Gastineau, M., Joutel, F., Robutel, P., Levrard, B. and Correia, A., 2003, Long term evolution and chaotic diffusion of the insolation quantities of Mars) can be downloaded from the internet at: http://hal.ccsd.cnrs.fr/view/ccsd-00000860/.

At least a hundred presentations concerned Martian deltas, alluvial fans, debris flows, ice-wedge polygons, and other features believed to have been formed with the involvement of liquid water. Rebecca Williams gave a compelling presentation of Martian alluvial fans – very nearly identical in form and maturity and almost of the same scale as a variety of fans in the Mojave Desert (southwestern USA). Those particular examples occur in an equatorial crater informally named Mojave. Jeffrey Moore (NASA Ames) followed that paper with a survey of similar well-developed fans in at least 50 impact craters mostly at middle southern latitudes. While Williams characterized "her" fans as being geologically very young (indicated by very few superposed impact craters), Moore considered most fans to be late Noachian or Early Hesperian – over 3 billion years old. It is unclear whether there is any chronological conflict of interpretation, but certainly issues of chronology will have to be sorted out. The bigger point, however, is that a role of liquid water has become the accepted interpretation of these features. These findings have reinforced ideas developed by hundreds of scientists over the past three decades; they have left no rational base for ideas of "anything but water." Even Nick Hoffman (the "White Mars" advocate), swayed by rover results, has accepted a much wetter perspective, according to his private comments.

The most stirring results of the 2004 LPSC meeting involved numerous reports, using multiple orbital and rover-based instruments, of sulfate salts. Hydrated (or hydroxyl-bearing) sulfates of ferric and ferrous iron, magnesium, and calcium are now recognized as major components of the surface layers of many regions on Mars. William Feldman and coworkers, including myself, reported that magnesium sulfate hydrates have a stability with respect to Martian temperature and humidity consistent with the hydrogen abundances of the upper meter at low latitudes known from the neutron and gamma ray data produced by *Mars Odyssey*. Many other reports of sulfate hydrate minerals were presented based on more direct analyses of orbital data. For example, evidence supporting a widespread distribution of abundant magnesium sulfate was reported by researchers analyzing data from the OMEGA experiment aboard *Mars Express*.

The most exciting new results, judged from the overflow crowd of more than 700 who packed every seat and every square meter of standing room in the conference room and into the hallway, were given by the Mars Exploration Rover teams. Starting with an overview presentation by Steve Squyres (Cornell), the PI of the instrument package aboard the rover, the audience listened to one paper after another showing multiple independent proofs or added evidence indicating a major role of liquid water. Squyres, in a subsequent news briefing, stated, "*We think Opportunity is parked on what was once the shoreline of a salty sea.*"

Without doubt, the conference highlight, presented for the first time in detail, was the discovery of the hydroxylated ferric iron sulfate mineral known as jarosite ($KFe_3(SO_4)2(OH)_6$. That mineral was reported by the German-led team for the Mössbauer spectrometer as a major constituent of the layered outcrop at the *Opportunity* landing site. Ordinarily, such a paper would be greeted as touting just one more mineral on Mars. But jarosite contains water (in the form of hydroxyl anion, OH^-); more importantly, it can only form with the involvement of many times its mass of liquid water, and specifically acidic water. Jarosite's widespread presence on Mars, along with a lack of carbonates, has long been predicted according to a narrow geochemical model, which by no small chance apparently is correct, at least for this site on Mars (Burns, 1987; Remolar et al., 2002; Benison and Laclair, 2003). One by one, the other teams presented results that pointed toward or were consistent with the presence of salts in *Opportunity*'s little layered outcrop in Meridiani Planum. More than 40% of the mass of one bed consists of hydroxylated salts, such as jarosite, and hydrated salts, believed to include gypsum ($CaSO_4{\cdot}2H_2O$) and either kieserite ($MgSO_4{\cdot}1H_2O$) or hexahydrite ($MgSO_4{\cdot}6H_2O$). There are only three possibilities: it is an evaporite bed produced by direct aqueous-acid chemical precipitation; a sedimentary bed soaked and pervasively altered by acid-evaporitic brine; or an eolian dune deposit made of minerals that had been deposited under aqueous-acid evaporitic conditions.

The scientists running *Opportunity*'s instruments also reported that the spherules ("blueberries") found to be eroding from the laminated salt-rich rocks actually consist largely of the ferric iron oxide hematite. This mineral, too, normally forms in the presence of liquid water. The general belief at this point is that much of the abundant hematitite in Meridiani Planum is in the form of these "blueberries," and that they formed as ferric iron concretions due to reactions in water-soaked rocks. Almost gone were earlier preferences for explanations involving impact melt or volcanic lava droplets. A strong case for an actual salty sea was presented by Phil Christiansen (ASU), who reported that orbital observations by TES, THEMIS, and MOLA show the hematite (now known to be associated mainly with "blueberries") of Meridiani Planum is confined below a topographic level that is best explained as a lake/sea level. Christiansen favors a model where goethite is a precursor mineral, which then altered to hematite.

Melissa Lane, working with the local mini-TES dataset of *Spirit* lander in Gusev Crater on the other side of Mars, reported results pointing toward hydrous ferrous iron sulfates, possibly instead of carbonates, as a component of the soil crusts found there. Overall, however, the evidence for a role of water at Gusev Crater was neither

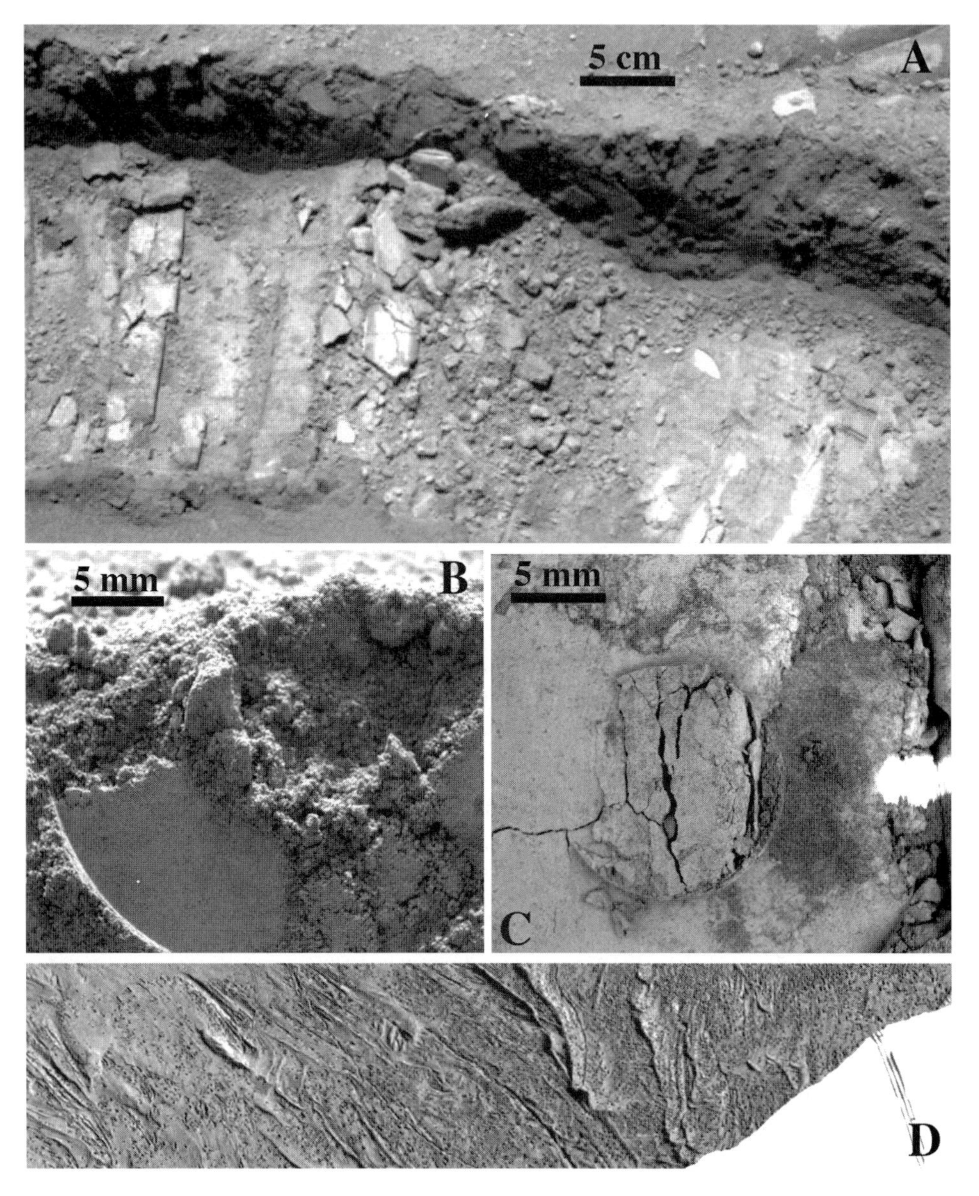

Physical evidence of highly cohesive soil at both rover landing sites. A and B are soils at the Gusev site and C and D are soils at the Meridiani site. Two types of cohesion are indicated at each site. A shows soils clods produced during trenching experiments with a spinning wheel. B is a microscopic imager scene of a Mössbauer spectrometer imprint on what seems like fine-grained and perhaps slightly damp tacky sand. C is a similar imprint on soil at the Meridiani site; soil clods and compaction are evident.

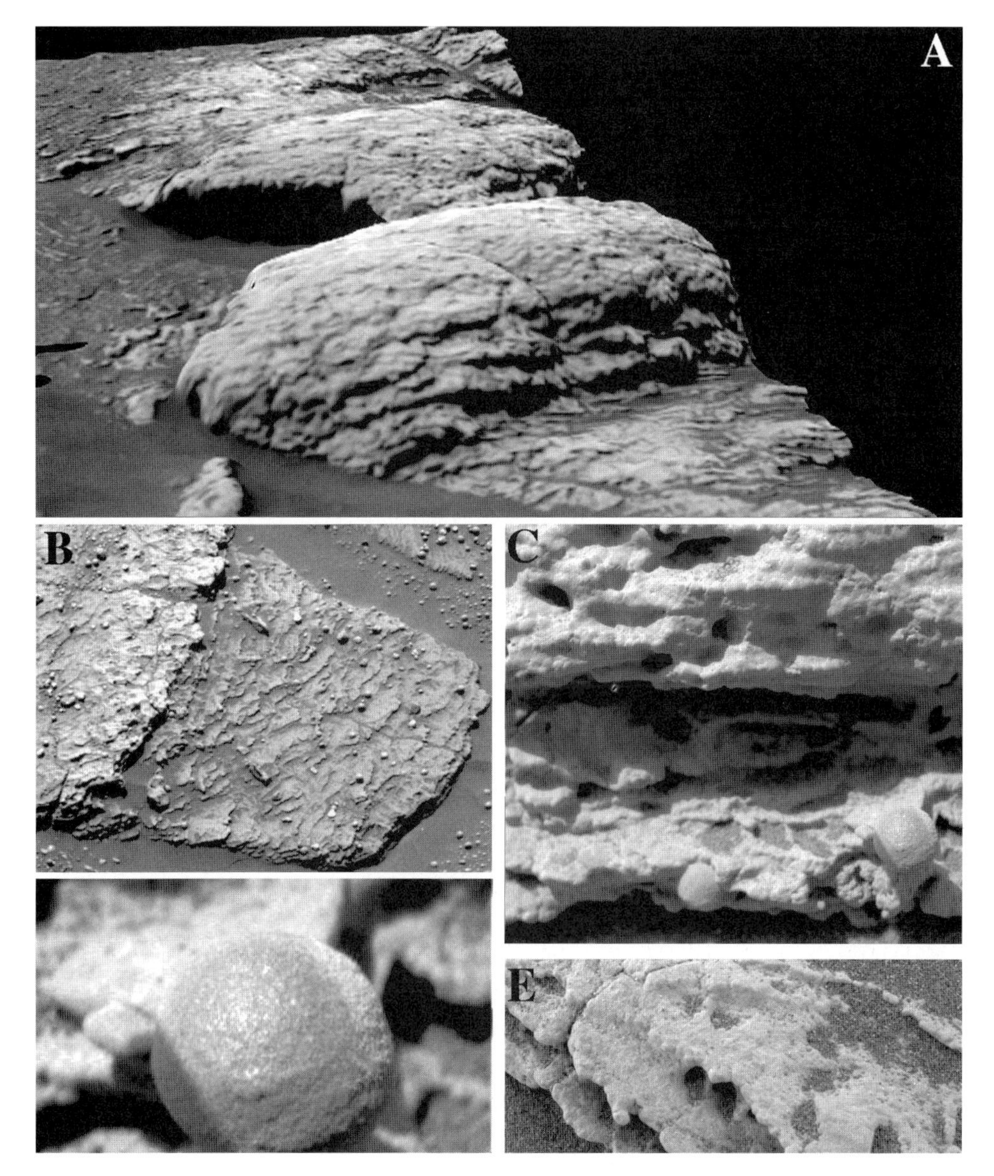

Rocks in the vicinity of Meridiani Planum are layered, according to observations by cameras aboard *Mars Global Surveyor* and *Mars Odyssey*; those layers are on the order of 10 m or more thick. Here we see *Opportunity* images of layers just 2 mm thick. (A) Perspective view afforded by stereo imaging by the pan camera; (B) looking down on the layers; (C) a microscopic imager scene spanning 3 cm left to right; note the microkarstic "caverns" eroded into the layers and the spherules eroding from them. (D) Close-up of a 4-mm spherule. (E) Microscopic imager scene of another set of layers. Data courtesy of NASA/JPL/Cornell/USGS and the MER imaging and microscopic imager teams.

so abundant nor as varied as the evidence for a watery past at Meridiani Planum. Although these rover-based results concern only two minuscule areas of the planet, they provide essential ground-truth supporting the results and calculations of people working with orbital data. The findings all go together to support a new consensus that Mars was and remains transiently wet, generally icy, and very, very salty near its surface. In fact, if one could take a pinch of Martian surface dust or sedimentary rocks, chances are it would have a distinct flavor, which might vary from region to region. More common than the shocking salt taste of halite would be the bitter-salt taste of epsomite, the acrid-metallic taste of ferrous sulfate, acid-sour taste of jarosite, and the faint sweet taste of gypsum.

I and my coworker, Giles Marion (Desert Research Inst., Reno, Nevada), presented at LPSC a suggestion that acid-sulfate chemistry can explain the Meridiani mineralogy data and can even explain the occurrence of highly cohesive and possibly damp, tacky soils at both rover sites, of what appear to be partly corroded rocks at the Gusev site. Sulfuric acid-rich brines may form ponds and marshes on today's Mars, with liquid brine forming seasonally and daily. The polygonated form of rocks at the Meridiani outcrop has the appearance of the solution-etched cracks and veins common in carbonate rocks, although, as mentioned, any original carbonate in these rocks probably has been destroyed or replaced by sulfates. The polygonal crack system could be due either to solution etching or to cycles of expansion and contraction related to salt hydration and dehydration. The polygon network looks like the veins and solution polygons that characteristically affect both dolomites and gypsum rocks.

Without exception, jarosite requires highly acidic water in its formation; and, once formed, any neutral water that comes in contact with jarosite rapidly becomes acidic. Jarosite is common, for instance, in acid-mine drainage sites, an analog for Mars surficial geochemistry that was offered in my LPSC presentations. Aqueous chemical processes in such sites cause an approach toward a thermodynamic balance involving reactions (with cations in solution or available from other solids) and products such as:

Ferric iron sulfate salts

- Jarosite (confirmed by *Opportunity* team):
  $3Fe^{3+} + K^{+} + 2SO_4^{2-} + 6H_2O = KFe_3(SO_4)_2(OH)_6 + 6H^{+}$
- Schwertmannite (expected/predicted)
  $8Fe^{3+} + SO_4^{\ 2-} + 14H_2O = Fe_8O_8(OH)_6SO_4 + 22H^{+}$
- Copiapite (likely):
  $Fe^{2+} + 4Fe^{3+} + 6SO_4^{2-} + 21H_2O = Fe^{2+}Fe_4^{\ 3+}(SO_4)_6(OH)_2.20H_2O$

Ferric iron oxides (consistent with Fe^{3+} sulfate-acid equilibria)

- Ferrihydrite:
  $10Fe^{3+} + 60H_2O = 5Fe_2O_3.9H_2O + 30H^{+}$
- Goethite:
  $Fe^{3+} + 2H_2O = FeO(OH) + 3H^{+}$
  $KFe_3(SO_4)_2(OH)_6 = K^{+} + 2SO_4^{2-} + 3FeO(OH) + 3H^{+}$

- Hematite:
  $2Fe^{3+} + 3H_2O \rightarrow Fe_2O_3 + 6H^+$
  Goethite = hematite reaction

It is generally believed that hematite is not a direct aqueous precipitate but is a diagenetic alteration product, with goethite and intermediate oxy-hydroxides occurring before transformation to hematite. With jarosite and hematite already reported by the MER team at the Meridiani site, other reactions expected to have taken place include these (with hydrogen ion and sulfate anion contributed by sulfuric acid):

- Olivine weathering: $(Mg, Fe)_2SiO_4 + 4H^+ = 2(Mg, Fe)^{2+} + H_4SiO_4$
- Calcium carbonate dissolution: $CaCO_3 + 2H^+ + SO_4{}^{2+} = H_2CO_3 + CaSO_4$
- Magnesium carbonate weathering: $MgCO_3 + 2H^+ + SO_4{}^{2+} = H_2CO_3 + MgSO_4$
- Siderite or iron carbonate weathering: $FeCO_3 + 2H^+ + SO_4{}^{2+} = H_2CO_3 + FeSO_4$

In addition, the hydrolysis reactions working on silicate minerals produce silicic acid, which precipitates as silica gel, which then opalizes or turns to chert – another predicted mineral constituent of Martian evaporites. Carbonate minerals are not expected in strongly acidic environments, unless acid-sulfate equilibrium has not been achieved. Since carbonates rapidly alter to sulfate minerals in such systems, any residual carbonate or a documented complete lack of it would be extremely important as an indicator of the extent of aqueous chemical processes. Magnesium silicates also are destroyed during acid-sulfate weathering, but they take much more time to react than the carbonates. If the results at the Meridiani site are extended globally, acid-sulfate processes could explain the hitherto unexpected absence or rarity of carbonate minerals. The formation of hematite and other iron oxides as products of this type of chemistry can even explain the ochre colors of Mars. Such acid chemical processes were proposed for Mars at least as far back as 1987 by the late Roger G. Burns. Quantitative adaptation and extension of this type of geochemical model to low temperatures on Mars is a major ongoing area of my work with Giles Marion (Marion et al., 2003; Kargel and Marion, 2004).

Acidic conditions on Mars could arise locally from oxidation of sulfide minerals or from volcanic emissions of SO_2 (which reacts under oxidative conditions to sulfuric acid). Other sulfate-producing processes important on Earth have at their root hydrogen sulfide gas from coal, gas, and petroleum deposits. Sulfate formation on Earth occurs with usual mediation by biological activity. It is not clear how any of these processes would proceed in the absence of biochemistry, but certainly sulfates are produced in both chondritic and Martian meteorites and so must also occur on Mars. Global acidity could arise also from the continuous destruction of sulfate salts by ultraviolet photolysis of sulfate salts in the eolian active layer of windblown dust and sand. Acid hydrolysis of silicate minerals then would be part of a continuous cycle involving (1) acid formation with sulfate salt destruction and (2) acid neutralization with sulfate salt formation.

With volcanic and hydrothermal processes important on Mars, and plate-tectonic destruction of ore bodies not taking place, I would expect Mars to be an otherworldly, unbelievably rich motherlode of a planet, with rich sulfide ore deposits and fractionally crystallized brines rich in rare minerals. However, that presupposes that Mars will become a place of interest to human life. The other side of the acid-sulfate coin is that Mars could become something like the Solar System's largest "Superfund" toxic-waste clean-up site, a rather hopeless one at that. If the acid-mine drainage analog is correct, Mars potentially may offer a mix of metal sulfides, sulfates, acids, and neutral salts – a rather delicious brew if one were an acidophilic, psychrophilic, chemoautotrophic microorganism. Sulfate-reducing and sulfide-oxidizing microbes could live in one big, happy global community.

If there is a liquid niche remaining near the surface of Mars, or especially at the MER landing sites, the possibility of transplanetary biological contamination must be considered. As Andrew Schuerger (University of Florida) told me, "*It is likely that a small number of hardy terrestrial microbes have landed on Mars as hitchhikers on the landed spacecraft.*" Schuerger has been leading a study of the survivability of terrestrial microbiota in the nooks and crannies of spacecraft under prevalent conditions of UV radiation, pressure, temperature, and humidity on Mars (Schuerger et al., 2003, 2004). While many species must have survived the trip aboard Mars landers, all the species recovered so far from lab tests would not survive Martian conditions; but the data so far are incomplete; questions remain, and Schuerger's study continues.

There is only the obvious definite conclusion from all this recent observation and analysis: Mars was definitely warm and wet (by comparison to today's Mars). Beyond that, our ideas of tomorrow will no doubt be different from today's. However, as a final provocation, I would like to offer a synopsis of the Meridiani Planum results from *Opportunity* and the orbital missions, and how we may have seen the tip of a story that is as dramatic as any in Earth history. I draw on an analog from the Permian to recent history of New Mexico (the great Permian Basin and associated lake basins). That was a time when a deep inland salt sea, a barrier reef, a system of large saline lakes, and huge river deltas preceded the age of dinosaurs and the Rocky Mountains. The Capitan barrier reef (named for El Capitan, a peak of the Guadalupe Mountains, and the namesake of a prominent Martian rock outcropping analyzed by *Opportunity*) produced thick sequences of carbonate and evaporite rocks. It is in these rocks that aqueous carbonic and sulfuric acids have more recently dissolved caverns, such as Carlsbad Caverns. Gypsum beds in these rock formations either formed directly by precipitation from the salty lakes, or formed by episodic acidification of the sea, which could have then altered the carbonate sediments to sulfate. The seas and lakes at that time swarmed with life, as indicated by fossils foraminifers, corals, brachiopods, mollusks, crinoids, and algae. I would not make too much of the Martian hematite "blueberries" as possible fossils of Martian life, although we are nowhere near ruling out this unsubstantiated but remote possibility. I should conclude with ideas that are more substantiated; the Permian of New Mexico and acid-mine drainage situations can be kept in mind.

I found the particular minerals reported at Meridiani Planum especially compelling, because of: (1) long-standing thermodynamic calculations that these minerals should be stable due to equilibrium between the atmosphere and silicate

Opportunity provided pancam mosaics of an intensely polygonally fractured outcrop. The fractures could be due to freeze–thaw of ice, but ice is not even nearly stable here. It could be due to impact, but the evidence for hydratable sulfate hydrate salts is strong at this locality, and expansion/contraction due to hydration/dehydration is a likely cause. The fracture network form and scale is similar to dissolution-etched polygons in terrestrial dolomites, limestones, gypsum, and other slightly water-soluble salts.

The Early Permian set the stage for the mid-late Permian Basin carbonate reef and evaporite sequences. A paleogeographic map (courtesy of Ron Blakey, NAU), A, shows the locations of three sets of field photos this page and the next. B–D are deep-water sediments of the Delaware sub-basin in an area that included deep-water carbonates and landslide/turbidite deposits (B) shed from the nearby Capitan Reef. El Capitan looms in the background of C. A portion of a submarine landslide deposit is shown in D. Paleogeographic map (A) adapted slightly from work of Ronald Blakey (http://jan.ucc.nau.edu/~rcb7/) and rendered in black and white with permission from that author. Is this an analog of Meridiani Planum?

Gypsum is abundant in the Permian Basin (New Mexico), and it takes many morphological forms. White Sands (A) are gypsum dunes made of highly active (B) windblown playa precipitates. Interdune areas (C) reveal cross lamination (D) and effluorescent gypsum crusts (E). The ultimate source of White Sands' $CaSO_4$ is the Permian Basin's evaporite sequence. The beds, such as in F (San Andres Formation), are broken in polygons: mud-crack-like structures are due to dehydration/hydration and dissolution (G–I).

rocks represented by *Viking* soil analyses and Mars meteorites; (2) long-standing knowledge that traces of gypsum occur in some Martian meteorites; (3) theoretical equilibrium of hydrated magnesium sulfates under Martian atmospheric humidity and temperature conditions and consistency with neutron spectroscopic measurements of hydrogen abundances in the Martian surface materials at low latitudes (reported at LPSC by Bill Feldman and coworkers, including myself); (4) spectroscopic detection of globally widespread kieserite and epsomite ($MgSO_4 \cdot 7H_2O$) in *Mars Express* OMEGA data, presented by J.-P. Bibring and the OMEGA team; and (5) the morphological appearance of *Opportunity*'s little outcrop, including its polygonated, mud-crack-like nature.

Very few rock types develop the particular 3-dimensional polygonated form as seen in *Opportunity*'s outcrop, but gypsum rock and dolomite (Ca–Mg carbonate) and other slightly water-soluble rocks characteristically do. My first impression was that this fracturing was due to impact shattering, but the MER team's LPSC presentations on mineral and chemical analyses, and the appearance of solution weathering have convinced me that we are dealing with classic gypsum-type rock textures in Meridiani Planum's rocks. Gypsum and other salts on Earth commonly form laminated structures either by direct aqueous deposition (most common) and by eolian reworking of gypsum (and other salt) deposits. A spectacular case of cross-bedded, dune-forming gypsum sands occurs in the dune field of White Sands National Monument. In that instance, saline Lake Lucero is actively precipitating gypsum. Gypsum for the dunes is provided by the seasonally dry bed of that lake (and also the lakebed of the larger ancient Lake Otero, of which Lucero is a remnant). The calcium sulfate originates from the gypsiferous rocks of the Permian age (250 Ma) San Andres Formation, an ancient shelf-sea deposit that was roughly coeval with the inland sea of the Permian Basin (whose rocks now are a major source of petroleum and natural gas). Thus, the gypsum dunes of White Sands has a long trace back to a classic series of evaporite basins.

Although we have seen up close just one small rock outcrop of an ancient sulfate-salt-depositing system on Mars, it is my view that the *Opportunity* rover and the MER team have discovered at Meridiani Planum another classic evaporite basin. Many details are different from the evaporitic Permian basins of the southwestern US, including a dominant role on Mars of chemical equilibria involving strong acids and Fe–Ca–Mg sulfates (versus comparatively dilute acids and Ca–Mg carbonates and some Ca sulfates in the Permian of New Mexico); a critical role of life in the Permian of New Mexico (its uncertain role on Mars); and a plate tectonic background to the New Mexico story (lacking on Mars). However, the concept of big, continental saline lakes and complex aqueous-acid chemistry is shared on both planets. Many rich details remain to be discerned and integrated into a full story, but so far *Opportunity* seems to be leading us beyond our wildest imaginations.

Bibliography

NON-TECHNICAL ... TO GET YOU STARTED

Hartmann, W.K. (2003) *A Traveler's Guide to Mars*. Workman.

Kargel, J.S. and R.G. Strom (1992a, December) Ice ages of Mars. *Astronomy*, 40–45.

Kargel, J.S. and R.G. Strom (1996, November) Global climatic change on Mars. *Scientific American*, 180–189.

Lunine, J.I. (1999) *Earth: Evolution of a Habitable World*. Cambridge University Press, 319 pp.

Morton, O. (2004) Mars: Planet Ice. *National Geographic* (January).

Sawyer, K. (2001) A Mars never dreamed of. *National Geographic* (February), 30–51.

Sheehan, W. (1996) *The Planet Mars, A History of Observation and Discovery*. University of Arizona Press, 271 pp.

TECHNICAL ... TO GET YOU ACQUAINTED

Allen, C.C. (1979) Volcano-ice interactions on Mars. *J. Geophys. Res.*, **84**, 8048–8059.

Baker, V.R. (1982) *The Channels of Mars*. University of Texas Press, Austin.

Baker, V.R. (2001) Water and the Martian landscape. *Nature*, **412**, 228–236.

Baker, V.R. and D.J. Milton (1974) Erosion by catastrophic flood on Mars and Earth, *Icarus*, **23**, 27–41.

Baker, V.R., R.G. Strom, V.C. Gulick, J.S. Kargel, G. Komatsu and V.S. Kale (1991) Ancient oceans, ice sheets and the hydrological cycle on Mars. *Nature*, **352**, 589–594.

Baker, V.R., V.C. Gulick and J.S. Kargel (1993) Water resources and the hydrogeology of Mars. In J.S. Lewis (ed.), *Resources of Near-Earth Space*. University of Arizona Press, pp. 765–798.

Baker, V., R.G. Strom, J.M. Dohm, V.C. Gulick, J.S. Kargel, G. Komatsu, G.G. Ori and J.W. Rice (2000) Mars, Oceanus Borealis, ancient glaciers and the MEGA-OUTFLO hypothesis. *Lunar Planet. Sci.*, XXXI, Abstract 1863 (CD-ROM).

Battistini, R. (1987a) Indices morphologiques d'anciens glaciers dans les basses latitudes de Mars: Leur relation avec l'hydrolithosphère. *Inter-Nord*, no. 18, publication of CNRS, pp. 399–411.

Battistini, R. (1987b) La notion d'hydrolithosphère sur Mars: son rôle dans la morphologenese des basses latitudes. *Bull. Soc. Géol. France*, **8** (1), 49–57.

Benison, K.C. and D.A. Laclair (2003) Modern and extremely acid saline deposits: Terrestrial analogs for Martian Environments? *Astrobiology*, **3**, 609–618.

Boynton, W.V., W.C. Feldman and 23 others (2002) Distribution of hydrogen in the near surface of Mars: Evidence for subsurface ice deposits. *Science*, **297**, 81–85.

Burns, R.G. (1987) Gossans on Mars: Spectral features attributed to jarosite. *Lunar Planet. Sci.*, **XVIII**, 141–142.

Carr, M.H. (1986a) The Martian drainage system and origin of valley networks and fretted channels. *J. Geophys. Res.*, **100**, 7429–7507.

Carr, M.H. (1986b) Mars: A water-rich planet? *Icarus*, **56**, 187–216.

Carr, M.H. (1996). *Water on Mars*. New York, Oxford University Press.

Carr, M.H. and J.W. Head, III (2003) Oceans on Mars: An assessment of the observational evidence and possible fate. *J. Geophys. Res.*, **108** (E5), 5042; doi:10.1029/2002JE001963.

Christensen, P.R. (2003) Formation of recent Maryian gullies through melting of extensive water-rich snow deposits. *Nature*, **297**, 81–85.

Christensen, P.R., Bandfield, J.L., Clark, R.N., Edgett, K.S., Hamilton, V.E., Hoefen, T., Kieffer, H.H., Kuzmin, R.O., Lane, M.D., Malin, M.C., Morris, R.V., Pearl, J.C., Pearson, R., Roush, T.L., Ruff, S.W. and Smith, M.D. (2000) Detection of crystalline hematite mineralization on Mars by the Thermal Emission Spectrometer: Evidence for near-surface water. *J. Geophys. Res.*, **105** (E4), 9623–9642.

Christensen, P.R., Morris, R.V., Lane, M.D., Bandfield, J.L. and Malin, M.C. (2001) Global mapping of Martian hematite mineral deposits: Remnants of water-driven processes on early Mars. *J. Geophys. Res.*, **106** (E10), 23,873–23,885.

Clark, B.R. and R.P. Mullin (1976) Martian glaciation and the flow of solid CO_2. *Icarus*, **27**, 215–228.

Clifford, S.M. (1993) A model for the hydrologic and climatic behavior of water on Mars. *J. Geophys. Res.*, **98**, 10,973–11,016.

Clifford, S.M. and T.J. Parker (2001) The evolution of the Martian hydrosphere: Implications for the fate of a primordial ocean and the current state of the northern plains. *Icarus*, **154**, 40–79.

Colprete, A. and B.M. Jakosky (1998) Ice flow and rock glaciers on Mars. *J. Geophys. Res.*, **103** (E3), 5897–5909.

Costard, F.M. and J.S. Kargel (1995) Outwash plains and thermokarst on Mars. *Icarus*, **114**, 93–112.

Craddock, R.A. and A.D. Howard (2002) The case for rainfall on a warm, wet early Mars. *J. Geophys. Res.*, 10.1029/2001JE001505.

Dobrovolskis, A. and A.P. Ingersoll (1975) Carbon-dioxide-water clathrate as a reservoir of CO_2 on Mars. *Icarus*, **26**, 353–357.

Fairen, A.G., J.M. Dohm, V.R. Baker, M.A. de Pablo, J. Ruiz, J.C. Ferris and R.C. Anderson (2003) Episodic flood inundations of the northern plains of Mars. *Icarus*, **165**, 53–67.

Farmer, J.D. (1998) Thermophiles, early biosphere evolution and the origin of life on Earth: Implications for the exobiological exploration of Mars. *J. Geophys. Res.* **103**, 28,457–28,461.

Feldman, W.C., M.T. Mellon, S. Maurice, T.H. Prettyman, J.W. Carey, D.T. Vaniman, D.L. Bish, C.I. Fialips, J.S. Kargel, R.C. Elphic, H.O. Funsten and D.J. Lawrence (2004) Contributions from hydrated states of $MgSO_4$ to the reservoir of hydrogen at equatorial latitudes on Mars. *Geophys. Res. Lett.* (submitted).

Fisher, D.A. (1993) If Martian ice caps flow: Ablation mechanisms and appearance. *Icarus*, **105**, 501–511.

Forget, F. and R.T. Pierrehumbert (1977) Warming early Mars with carbon dioxide clouds that scattered infrared radiation. *Science*, **278**, 1273–1276.

Glen, J.W. (1952) Experiments on the deformation of ice. *J. Glaciol.*, **2**, 111–114.

Goldspiel, J.M. and S.W. Squyres (1991) Aqueous sedimentation on Mars. *Icarus*, **89**, 392–410.

Grant, J.A. (2000) Valley formation in Margaritifer Sinus, Mars, by precipitation-recharged ground-water sapping. *Geology*, **28**, 223–226.

Greeley, R., N. Lancaster, S. Lee and P. Thomas (1992) Martian eolian processes, sediments, and features. In H. Kieffer, B. Jakosky, C. Snyder and M. Matthews (eds), *Mars*. University of Arizona Press, Tucson, AZ, pp. 730–766.

Gulick, V.C. (1998) Magmatic intrusions and a hydrothermal origin for fluvial valleys on Mars. *J. Geophys. Res.*, **103**, 19,365–19,387.

Haberle, R.M. et al. (2000) Meteorological control on the formation of Martian paleolakes. *Lunar Planet. Sci.*, **XXXI**, Abstract #1509.

Harbor, J.M. (1992) On the mathematical description of glaciated valley cross sections. *Earth Surface Processes and Landforms*, **17**, 477–485.

Hartmann, W.K. et al. (1999) Evidence for recent volcanism on Mars from crater counts. *Nature*, **397**, 586–589.

Head, J.W. (2001) Mars: Evidence for geologically recent advance of the south polar cap. *J. Geophys. Res.*, **106**, 10,075–10,085.

Head, J.W. and S. Pratt (2001) Extensive Hesperian-aged south polar cap on Mars: Evidence for passive melting and retreat and lateral flow and ponding of water. *J. Geophys. Res.*, **106**, 12,275.

Head, J.W. III and D.R. Marchant (2003) Cold-based mountain glaciers on Mars: Western Arsia Mons. *Geology*, **31**, 641–644.

Head, J.W. III, H. Hiesinger, M.A. Ivanov, M.A. Kreslavsky, S. Pratt and B.J. Thomson (1999) Possible oceans in Mars: Evidence from Mars Orbiter Laser Altimeter data. *Science*, **286**, 2134–2137.

Head, J.W. III, M. Kreslavsky, H. Hiesinger, M. Ivanov, D.E. Smith and M.T. Zuber (1998) Oceans in the past history of Mars: Tests for their presence using Mars Orbiter Laser Altimeter (MOLA) data. *Geophys. Res. Lett.*, **24**, 4401–4404.

Hiesinger, H. and J.W. Head (2000) Characteristics and origin of polygonal terrain in southern Utopia, Mars: Preliminary results from the Mars Orbiter laser Altimeter

(MOLA) and Mars Orbiter Camera (MOC) data. *J. Geophys. Res.*, **105**, 11,999–12,022.

Hoffman, N. (2000) White Mars: A new model for Mars' surface and atmosphere based on CO_2. *Icarus*, **146**, 326–342.

Howard, A.D. (1981) Etched plains and braided ridges of the south polar region of Mars: Features produced by basal melting of ground ice? *NASA Tech. Memo.* 84211, pp. 286–288.

Howard, A.D. and J.M. Moore (2004) Scarp-bounded benches in Gorgonum Chaos, Mars: Formed beneath an ice-covered lake? *Geophys. Res. Lett.*, **31**, L01702; doi:10.1029/2003GL018925.

Hu, F.S., H.E. Wright Jr, E. Ito and K. Lease (1997) Climatic effects of Glacial Lake Agassiz in the midwestern United States during the last deglaciation. *Geology*, **25**, 207–210.

Hynek,B.M. and R.J. Phillips (2001) Evidence for extensive denudation of the Martian highlands. *Geology*, **29**, 407–410.

Hynek, B.M., Arvidson, R.E. and Phillips, R.J. (2002) Geologic setting and origin of Terra Meridiani hematite deposit on Mars. *J. Geophy. Res.*, **107** (E10), 5088; doi: 1029/2002001891.

Jakosky, B.M. and M.H. Carr (1985) Possible precipitation of ice at low latitudes of Mars during periods of high obliquity. *Nature*, **315**, 559–561.

Jakosky, B.M. and R.J. Phillips (2001) Mars' volatile and climate history. *Nature*, **412**, 237–244.

Jakosky, B.M., Henderson, B.G. and Mellon, M.T. (1995) Chaotic obliquity and the nature of the Martian climate. *J. Geophys. Res.*, **100**, 1579–1584.

Jöns, H.-P. (1986) Arcuate ground undulations, gelifluxion-like features and "front tori" in the northern lowlands on Mars – what do they indicate? (abstract), *Lunar Planet. Sci. Conf.*, **XVII**, 404–405.

Joyce, G. F. (1989) RNA evolution and the origins of life. *Nature*, **338**, 217–224.

Joyce, G.F. and L.E. Orgel (1998) The origins of life: A status report. *Am.Biol. Teacher*, **60**, 10–12.

Kargel, J.S. (2003) Earthly and otherworldly glaciers on Mars: Expressed subsurface subpolar ice and "plate tectonic" south polar ices. *Fall Meeting of the Amer. Geophys. Union*, San Francisco.

Kargel, J.S. and J.S. Lewis (1993) The composition and early evolution of Earth. *Icarus,* **105**, 1–25.

Kargel, J.S. and J.I. Lunine (1998) Clathrate hydrates on Earth and in the solar system. In C. de Bergh, M. Festou and B. Schmitt (eds), *Solar System Ices.* Kluwer Academic Publishers, pp. 97–117.

Kargel, J.S. and G.M. Marion (2004) Mars as a Salt-, Acid-, and Gas-hydrate world, *Lunar Planet. Sci.,* **XXXV**, Abstract 1965 (CD-ROM).

Kargel, J.S. and R.G. Strom (1990) *Lunar Planet. Sci. Conf.*, **XXI**, 597–598. Lunar and Planetary Institute, Houston.

Kargel, J.S. and R.G. Strom (1991) Terrestrial glacial eskers: Analogs for Martian sinuous ridges. *Lunar Planet. Sci.*, **XXII**, 683–684.

Kargel, J.S. and R.G. Strom (1992b) Ancient glaciation on Mars. *Geology*, **20**, 3–7.

Kargel, J.S., V.R. Baker, J.E. Beget, J. Lockwood, T.L. Pewe, J.S. Shaw and R.G. Strom (1995) Evidence of continental glaciation in the Martian northern plains. *J. Geophys. Res.*, **100**, 5351–5368.

Kargel, J.S., Kaye, J., Head, J.W., III, Marion, G., Sassen, R., Crowley,J., Prieto, O., Grant, S. and Hogenboom, D.L. (2000a) Europa's crust and ocean: Origin, composition, and the prospects for life. *Icarus*, **148**, 226–265.

Kargel, J.S., K.L. Tanaka, V.R. Baker, G. Komatsu and D.R. MacAyeal (2000b) Formation and dissociation of clathrate hydrates on Mars: Polar caps, northern plains and highlands. In *Proc. Lunar Planet. Sci. Conf.*, **31**, Abstract 1891.

Kargel, J.S., R. Wessels, J.E. Beget, T. Eddy, S. Lloyd, D. Macaulay, M. Proch, J. Skinner and K.L. Tanaka (2004) Alaska permafrost analogs of Martian small valley networks, thermokarst, terrain softening, terraces, and volcanic craters. *Lunar Planet. Sci.,* **XXXV**, Abstract (CD-ROM).

Kasting, J.F. (1991) CO_2 condensation and the climate of early Mars. *Icarus*, **94**, 1–13.

Kieffer, H.H. and A.P. Zent (1992) Quasi-periodic climate change on Mars. In H.H. Kieffer et al. (eds), *Mars*. University of Arizona Press, Tucson, pp. 1180–1233.

Komatsu, G., J.S. Kargel, V.R. Baker, R.G. Strom, G.G. Ori, C. Mosangini and K.L. Tanaka (2000) A chaotic terrain formation hypothesis: Explosive outgas and outflow by dissociation of clathrate on Mars? *Lunar Planet. Sci.,* **XXXI**, Abstract 1434 (CD-ROM).

Kreslavsky, M.A. and J.W. Head (2000) Kilometer-scale roughness of Mars: Results from MOLA data analysis. *J. Geophys. Res.*, **101**, 26,695–26,711.

Kreslavsky, M.A. and J.W. Head (2002a) Mars: Nature and evolution of young latitude-dependent water-ice-rich mantle. *Geophys. Res. Lett.*, **29**, 10.1029/2002GL015392,.

Kreslavsky, M.A. and J.W. Head (2002b) The fate of outflow channel effluents in the northern lowlands of Mars: The Vastitas Borealis formation as a sublimation residue from frozen ponded bodies of water. *J. Geophys. Res.*, **107** (E12), 5121; doi:10.1029/2001JE001831.

Kreslavsky, M.A. and J.W. Head, III (2004) Periods of active layer formation in the recent geological history of Mars. *Lunar Planet. Sci.*, **XXXV**, Abstract #1201 (CD-ROM).

Lachenbruch, A.H. (1962) Mechanics of thermal contraction cracks and ice-wedge polygons in permafrost. *Geol. Soc. America*, Special Paper 70, 69pp.

Laskar, J., M. Gastineau, F. Joutel, B. Levrard and P. Robutel (2004) A new astronomical solution for the long-term evolution of the insolation quantities of Mars. *Lunar Planet. Sci.*, **XXXV**, Abstract #1600 (CD-ROM).

Lewis, J.S. (1996) *Mining the Sky*. Addison-Wesley.

Lewis, J., M.S. Matthews and M.L. Guerrieri (eds) (1993) *Resources of Near-Earth Space*. University of Arizona Press.

Lucchitta, B.K. (1981) Mars and Earth: Comparison of cold-climate features. *Icarus*, **45**, 264–303.

Lucchitta, B.K. (1982) Ice sculpture in the Martian outflow channels. *J. Geophys. Res.*, **87**, 9951–9973.

Lucchitta, B.K. (1984) Ice and debris in the fretted terrain, Mars. *J. Geophys. Res.*, **89.** (Lunar Planetary Science Conf., 14th Proceedings, Pt. 2, Supplement, B409–B418.

Lucchitta, B.K. (2001) Antarctic ice streams and outflow cannels on Mars. *Geophys. Res. Lett.*, **28** (3), 403; 2000GL011924.

Lucchitta, B.K., H.M. Ferguson and C. Summers (1986a) Sedimentary deposits in the northern lowland plains: Mars. *J. Geophys. Res.*, **91** (suppl.), E166-E174.

Lucchitta, B.K., H.M. Ferguson and C.A. Summers (1986b) An ancient ocean on Mars? In *Reports of Planetary Geology and Geophysics Program – 1985*. NASA TM-88383, pp. 450–453.

MacGregor, K.C., Anderson, R.S., Anderson, S.P. and Waddington, E.D. (2000) Numerical simulations of glacial-valley longitudinal profile evolution. *Geology*, **28**, 1031–1034.

McGee, W.J. (1894) *J. Geology*, **2**, 350–364.

Malin, M.C. and M.H. Carr (1999) Groundwater formation of Martian valleys. *Nature*, **397**, 589–591.

Malin, M.C. and K.S.D. Edgett (2000) Evidence for recent groundwater seepage and surface runoff on Mars. *Science*, **288**, 2330–2335.

Malin, M.C. and K.S. Edgett (2003) Evidence for persistent flow and aqueous sedimentation on early Mars. *Science*, **302**, 1931–1934.

Mangold, N. and P. Allemand (2001) Topographic analysis of features related to ice on Mars. *Geophys. Res. Lett.*, **28** (3), 407; 2000GL008491.

Marchant, D. et al. (2002) Formation of patterned ground and sublimation till over Miocene glacier ice in Beacon Valley, southern Victoria Land, Antarctica. *Geol. Soc. Amer. Bull.*, **114**, 718–730.

Marion, G.M., D.C. Catling and J.S. Kargel (2003) Modeling aqueous ferrous iron chemistry at low temperatures with application to Mars. *Geochim. Cosmochim. Acta*, **67**, 4251–4266.

Martin, H. and W. Whalley (1987) Rock glaciers: Part I, Rock glacier morphology, classification and distribution. *Progress in Phys. Geog.*, **2**, 260–282.

Mellon, M.T. (1995) The distribution and behavior of Martian ground ice during past and present epochs. *J. Geophys. Res.*, **100**, 11,781–11,799.

Mellon, M.T. and R.J. Phillips (2002) Recent gullies on Mars and the source of liquid water. *J. Geophys. Res.*, **106**, 23,165–23,179.

Mellon, M.T., Jakosky, B.M. and Postawko, S.E. (1997) The persistence of equatorial ground ice on Mars. *J. Geophys. Res.*, **102**, 19,357–19,369.

Milkovich, S.M., J.W. Head III and S. Pratt (2002) Meltback of Hesperian-aged ice-rich deposits near the south pole of Mars: Evidence for drainage channels and lakes. *J. Geophys. Res.*, **107**; 10.1029/2001JE001802,.

Milton, D.J. (1974) Carbon dioxide hydrate and floods on Mars. *Science*, **183**, 654–656.

Mischna, M., M.I. Richardson, R.J. Wilson and D.J. McCleese (2003) On the orbital forcing of Martian water and CO_2 cycles: A general circulation model study with simplified volatile schemes. *Jour. Geophys. Res.*, **108** (E6), 5062, doi:10.1029/2003JE002051.

Mustard, J.F., C.D. Cooper and M.K. Rifkin (2001) Evidence for recent climate change on Mars from the identification of youthful near-surface ground ice. *Nature*, **412**, 411–414.

O'Connor, J.E. and J.E. Costa (2004) The world's largest floods, past and present: their causes and magnitudes. *USGS Circular* 1254, pp. 1–19.

Paige, D.A. (1992) The thermal stability of near-surface ground ice on Mars. *Nature*, **356**, 43–45.

Parker, T.J., R.S. Saunders and D.M. Schneeberger (1989) Transitional morphology in west Deuteronilus Mensae, Mars: Implications for modification of the lowland/upland boundary. *Icarus*, **82**, 111–145.

Parker, T.J., D.S. Gorsline, R.S. Saunders, D.C. Pieri and D.M. Schneeberger (1993) Coastal geomorphology of the Martian northern plains. *J. Geophys. Res.*, **98**, 11,061–11,078.

Pieri, D. (1980) Martian valleys: Morphology, distribution, age and origin. *Science*, **210**, 895–897.

Remolar, D.F., R. Amils, R.V. Morris and A.H. Knoll (2002) The Tinto River Basin: An analog for Meridiani hematite formation on Mars? *Lunar Planet. Sci. Conf.*, **XXXIII** (CD-ROM).

Ross, R.G. and Kargel, J.S. (1998) Thermal conductivity of solar system ices, with special reference to Martian polar caps. In C. de Bergh, M. Festou and B. Schmitt (eds), *Solar System Ices*. Kluwer Academic Publishers. pp. 33–62.

Sagan, C. (1997) *Billions and Billions*. Random House.

Sagan, C., O.B. Toon and P.J. Gierasch (1973) Climatic change on Mars. *Science*, **181**, 1045.

Scott, D.H., M.G. Chapman, J.W. Rice Jr and J.M. Dohm (1992) New evidence of lacustrine basins on Mars: Amazonis and Utopia Planitae. *Proc. Lunar Planet. Sci.*, **22**, 53–62.

Seibert, N.M. and J.S. Kargel (2001) Small-scale Martian polygonal terrain: Implications for liquid surface water. *Geophys. Res. Lett.*, **28**, 899–902.

Sharma, A. et al. (2002) Microbial activity at gigapascal pressures. *Science*, **295**, 1514.

Smoluchowski, R. (1968) Mars: Retention of ice. *Science*, **159**, 1348–1350.

Squyres, S. and M. Carr (1986) Geomorphic evidence for the distribution of ground ice on Mars. *Science*, **231**, 249–252.

Squyres, S.W. and J.F. Kasting (1994) Early Mars: How warm and how wet? *Science*, **265**, 744–749.

Strom, R.G., S.K. Croft and N.G. Barlow (1992) The Martian impact cratering record. In H.H. Kieffer et al. (eds), *Mars*, pp. 383–423. University of Arizona Press, Tucson.

Sugden, D.E. and B.S. John (1976) *Glaciers and Landscape*. London: Edward Arnold.

Tanaka, K.L. (1986) The stratigraphy of Mars. *Proc. 17th Lunar Planet. Sci. Conf.* (Part 1), *J. Geophys. Res.*, **91** (suppl.), 249–252.

Tanaka, K.L., W.B. Banerdt, J.S. Kargel and N. Hoffman (2001) Huge, CO_2-charged debris flow deposit and tectonic sagging in the northern plains of Mars. *Geology*, **29**, 427–430.

Tanaka, K.L., J.S. Kargel, D.J. MacKinnon, T.M. Hare and N. Hoffman (2002) Catastrophic erosion of Hellas Basin rim on Mars induced by magmatic intrusion in volatile-rich rocks. *Geophys. Res. Lett.*, **10**, 1029; 2001GL13885.

Tricart, J.L.F. (1998) Environmental change of planet Mars demonstrated by landforms. *Z. Geomorph. N.F.*, **42** (4), 385–407.

Turcotte, D.L. and G. Schubert (2002). *Geodynamics*. Cambridge University Press.

Ward, W.R. (1992) Long-term orbital and spin dynamics of Mars. In H.H. Kieffer et al. (eds), *Mars*. University of Arizona Press, Tucson, pp. 767–795.

Whalley, W. and H. Martin (1992) Rock glaciers, II. Rock Glaciers: Models and Mechanisms. *Prog. Phys. Geog.*, **16**, 127–186.

Williams, R.M.E., K. Edgett and M. Malin (2004) Young fans in an equatorial crater in Xanthe Terra, Mars. *Lunar Planet. Sci.*, **XXXV**, Abstract #1415 (CD-ROM).

Williams, R.S. Jr (2000) A modern Earth narrative: What will be the fate of the biosphere? *Technology in Society*, **22**, 303–339.

Wilson, E.O. (1992) *The Diversity of Life*. W.W. Norton & Company.

Zuber, M.T (2001) The crust and mantle of Mars. *Nature*, **412**, 220–227.

Zuber, M.T., D.E. Smith, S.C. Solomon, J.B. Abshire, R.S. Afzal, O. Aharonson, K. Fishbaugh, P.G. Ford, H.V. Frey, J.B. Garvin, J.W. Head, A.B. Ivanov, C.L. Johnson, D.O. Muhleman, G.A. Neumann, G.H. Pettengill, R.J. Phillips, X. Sun, H.J. Zwally, W.B. Banerdt and T.C. Duxbury (1998) Observations of the north polar region of Mars from the Mars Orbiter Laser Altimeter. *Science*, **282**, 2053–2060.

Zubrin, with R. Wagner (1996) *The Case for Mars: The Plan to Settle the Red Planet and Why we Must*. The Free Press.

Index